Evolutionary
Biology

Evolutionary Biology

ELI C. MINKOFF Bates College

ADDISON-WESLEY PUBLISHING COMPANY
Reading, Massachusetts • Menlo Park, California
London • Amsterdam • Don Mills, Ontario • Sydney

Library of Congress Cataloging in Publication Data

Minkoff, Eli C.
  Evolutionary biology.

  Bibliography: p.
  Includes index.
  1. Evolution. I. Title.
QH366.2.M56   1983   575   82-11317
ISBN 0-201-15890-6

Reprinted with corrections, March 1984

ISBN 0-201-15890-6
EFGHIJ-HA-89876

To Dr. Hasye Cooperman,
a true scholar

*One can only understand the essence
of things when one knows their origin
and development.*

HERACLEITOS

*Nothing in biology makes sense,
except in the light of evolution.*

DOBZHANSKY

Preface

This book deals with evolution, the process of change in biological systems. Yet, in a different sense, it deals with all of biology, because the changes in biological systems affect all aspects and components of these systems—there is nothing in biology that does not evolve. If one asks how biological systems work, a physiologist or biochemist might respond by describing various parts of the system and the actions of each part. But if one asks how these systems came to work this way, and not some other way, one is asking an evolutionary question. This is why Dobzhansky has written, "nothing in biology makes sense, except in the light of evolution." Evolution is the context into which all other biological subjects fit. All other biological subjects may be viewed as filling in the details of a grand scheme, and evolution is this all-encompassing scheme.

The writing of this book grew out of my teaching a course in evolution at Bates College. In searching for a textbook for the course, I found that there were several good books treating evolution from the standpoint of genetics, or paleontology, or some other component discipline, but no one book covered all the necessary subjects together. I thus found myself using a series of three or more texts to cover different aspects of the course, and some subjects, like the history of evolutionary biology, or the origin of life, were not to be found in these readings. Much that I had learned and was eager to teach was not covered in any textbook at all! And, to compound the difficulty, many of the books available at the time were not addressed to the needs of students.

Many students, and even many teachers, fail to realize the interdisciplinary character of evolutionary biology. Hardly a field of human endeavor does not somewhere impinge upon evolutionary studies. Evolutionists must first learn genetics if they are to understand the changes in the hereditary make-up of populations, and they must learn ecology if they are to understand the interactions of these populations with one another and with their environments. They must also learn paleontology (and, therefore, also some geology) if they are to appreciate the magnitude of changes that have happened in the past. These are the obvious disciplines from which evolution draws and to which it contributes in return. But there is far more. Modern evolutionists seeking to understand the origin of life and the evolution of biological molecules must learn astronomy, geology, physics, analytical chemistry, and biochemistry. To understand various forms of adaptation at all levels and in various organisms, one must know zoology, botany, microbiology, physiology, embryology, cell biology, and more biochemistry. To analyze evolutionary trends, gene frequencies, developmental changes, taxonomic nearness, or macromolecular sequences, a certain knowledge of mathematical and statistical techniques is needed. Many concepts in evolution, such as character displacement, or even natural selection itself, require mathematical formulation for their thorough understanding. In order to study and interpret the history of our own species, we need to know geology, anthropology (both physical and cultural), sociobiology, psychology, archaeology, and history proper. Many issues in both evolution and in taxonomy (the theory of biological classification) have philosophical implications, and discussions of these issues often invoke philosophical principles associated with such names as Popper, Kuhn, Hempel, and others. And the history of science itself has much to learn from evolutionary biology—a science with a rich and complex history, tied in many cases to other scientific developments, and to cultural and intellectual movements outside the sciences. The modern evolutionist cannot fail to be versed in the history of evolution, for various older theories, including Lamarckism, catastrophism, and especially progressionism continue to be resurrected from time to time. I am still amazed that so many learned men and women equate Darwinism with our descent from apes, or Lamarckism with the inheritance of acquired characteristics. It is now 40 years since Julian Huxley proclaimed a ''Modern Synthesis''; yet, it is amazing how slow many textbooks have been in incorporating all of its lessons, perpetuating many pre-1940 errors and attitudes. Just look at the several high school biology texts, and even a few textbooks of evolution, that offer the Linnaean hierarchy as proof that evolution has taken place; yet Linnaeus and his contemporaries used the hierarchy to argue exactly the opposite.

The writing of this book comes at an exciting time, when many new developments are being made in evolutionary biology. Modern biochemical techniques, such as the study of enzyme polymorphisms by electrophoresis, have revealed a far greater variation in natural populations than had previously been expected. Unmanned flights to other planets are providing new sources of information about the history of our solar system, and the origin of life within it. Many new studies are currently being made of genetic variation within populations, mechanisms of species formation, biogeography, bacterial evolution, human evolution, and other fast-developing topics. Even taxonomic theory and the

intellectual history of evolution are the subject of numerous recent papers, books, and entire journals. Many of these newer, rapidly developing topics have yet to make their impact on textbooks devoted to evolution in general. I have attempted to be comprehensive by including them all. Unfortunately, a thorough treatment of every topic would demand a book much longer than the present one.

This book is intended for a course that deals with all aspects of evolution. The book would be of greatest benefit to students who have had an introductory course in biology, but it may also prove useful to more advanced students and to those in other disciplines. The book could also be read by beginners with no training in biology, for it presumes no prior knowledge and defines all new terms when they are first encountered. Since it is written for teachers and their students, I would appreciate receiving comments and suggestions from those who have used it. Evolutionary biology is a rapidly evolving subject, and I intend that future editions of this book shall coevolve with it.

ACKNOWLEDGMENTS

Evolution is such a vast field that this book could not have been written without the helpful advice of many others. In alphabetical order, I would like to thank the following persons for their helpful comments or for reading portions of the manuscript: D. I. Axelrod, A. E. Balber, J. S. Bhorjee, B. J. Bourque, R. M. Chute, H. Cooperman, J. W. Creasy, J. D'Alfonso, P. H. Gingerich, J. R. Jungck, J. S. Law, H. Levy, E. Mayr, N. Minkoff, G. Nelson, T. S. Parsons, R. J. Peters, L. F. Pitelka, D. M. Raup, J. M. Savage, B. J. Stahl, W. C. Steere, D. W. Tinkle, J. E. Wahlert, J. E. Welch.

I would also like to express my thanks to the helpful staff at Addison-Wesley, especially to James Funston, Arthur Ciccone, and Laura Skinger Lane.

Contents

1

Introduction

This first chapter introduces several important concepts used in later chapters. New terms will be printed in **bold-face** when they first appear in the text. An important aid to study is the glossary at the end of this book, which lists most of these terms.

A. THE MEANING OF EVOLUTION

Biology is the study of life (the Greek word *bios* means ''life''). **Evolution** is the process of change in living populations. The study of evolution is one of the fundamental branches of biology, known as **evolutionary biology.** It is the study of biology as a historical science, or the study of living systems as they change through time.

The nonhistorical study of isolated living systems as they now exist represents the nonevolutionary aspect of biology—the search for a mechanism (an ''efficient cause'') that would answer one fundamental question: ''How does it work?'' Any further question—like ''How did it come to work that way and not some other way?''—takes us into the realm of evolutionary biology, in which living systems are seen as changing through time, reaching their present state through a series of historical changes.

All living systems are products of evolutionary history; even their present mechanical operations are the details of an ongoing evolutionary process. Evolution is thus the context into which all biological sciences fit. It is the most general theory, of which everything else is but a special case. As Theodosius Dobzhansky (1973) once remarked, ''nothing in biology makes sense, except in the light of evolution.''

Modern evolutionary biology is a synthesis of many theories; foremost among these are Darwinian natural selection and Mendelian genetics. The historical development of modern evolutionary theory is explained in unit I of this book. Unit II emphasizes **microevolution,** that is, the manner in which evolutionary change takes place at and below the species level. It is this aspect of evolution that stands closest to genetics and to ecology. Unit III emphasizes **macroevolution,** or **phylogenetics.** This is evolution on the grand scale of geological time, whose results are studied by paleontologists or comparative morphologists or biochemists. Where units II and III emphasize *how* changes *can* occur, unit IV studies those changes

1

that really *did* occur. It thus applies the principles of the first three units to the study of life's history on Earth, including the study of human evolution.

B. LIVING SYSTEMS

Properties of Life

We have said that biology is the study of life, and that evolution consists of changes in populations of living things, but we have not yet defined "life" or "living." We all know that people, clams, trees, and bacteria are alive, but rocks, radios, and cars are not. Wool, sea shells, and lumber, though not alive, owe much of their organization to the life that produced them. How do we distinguish bacteria from rocks and machines, or living bodies from their lifeless products?

Most people would answer these questions by describing processes carried out by living things, but not by lifeless ones. The living person breathes; the corpse does not. The living person has a heartbeat and a pulse; the corpse has neither. But certain nonliving things may mimic life in fascinating ways: the car's engine has something rather similar to a pulse; it even "breathes" air into the carburetor and "exhales" various exhaust gases. The characteristics of life that distinguish life from both corpses and complex machines may be listed as follows:

1. **Metabolism** Living things extract certain chemicals from the environment and release others in return. The environment is thus selectively depleted of certain substances and enriched in others. Energy is extracted from the environment in this process: the substances used are generally more energy-rich than those released as wastes. Much of this energy remains stored in the living cells until death and decomposition effect its return. Combustion engines differ in that most of their energy is typically put to immediate use. They can also be turned off and then on again, but the total cessation of metabolism in a living thing is irrevocable.

2. **Motion** Living things often convert some of their energy into motion. External motion provides the most immediate evidence of life to the ordinary observer. Most cells (those called eucaryotic) also exhibit an internal turnover of materials called **cytoplasmic streaming,** or **cyclosis.**

3. **Selective response (irritability)** All living things can distinguish certain external stimuli from others and respond selectively. Most animals can be prodded into moving away or striking back; plant roots will also turn away from noxious stimuli. Even bacteria can distinguish "food" from nonfood objects and can respond by producing certain chemicals in greater quantities and others in lesser quantities.

4. **Homeostasis** Not only do organisms respond selectively to stimuli, but their responses follow a pattern: potentially harmful conditions are removed or ameliorated and more favorable conditions are restored to within habitable limits. Toxic chemicals are metabolized into harmless ones; threats to life are generally undone. Some self-regulating machines may mimic this ability somewhat, but rarely do they approach the complexity of homeostasis in even the simplest of organisms. Some scientists (vitalists) see this as "purposeful" or as indicating the existence of "vital forces," as explained later in this chapter.

5. **Growth and biosynthesis** All living things pass through phases in which they increase their own matter quantitatively at the expense of their environment—they incorporate some of their surroundings into themselves. Even nongrowing organisms use energy to synthesize chemicals unavailable from their environments. Few machines can mimic either property, and chemical systems that mimic "growth" (e.g., crystals forming from solutions) are all relatively simple.

6. **Reproduction** Perhaps the most unique property of life is reproduction, by which living systems tend to make copies of themselves. Not only do frogs produce more frogs and trees produce seeds that grow into other trees, but even bacteria divide so as to produce more bacteria. No machine makes others like itself, though it would defy no rule of logic to imagine one that did. Reproduction is a process exclusive to living systems, though not exhibited by every one of them: castrated individuals and neuter ants (workers) are still considered alive until their metabolism ceases, and postreproductive senility is not the same as death.

7. **Population structure** All organisms belong to populations of similar organisms. Among sexual organ-

isms, these populations are interbreeding communities, but even asexual clones may be considered as populations whose membership is defined by common descent.

8. **Hereditary information** All organisms differ from nonliving systems in possessing hereditary information, derived from antecedent organisms in the same population. All organisms—or very nearly all—are also capable of either a sexual or a parasexual process in which hereditary information is transmitted to other organisms in the same populations. In all known organisms, the hereditary material is a nucleic acid that is capable of exhibiting novelty in the form of mutations (Chapter 6).

Using this list, we can define life as follows: a thing is living if, at some time during its existence, it exhibits metabolism, irritability, homeostasis, growth, and usually also some form of motion, and if it belongs to a population of similar organisms, of which at least some must be capable of reproduction.

Any definition of life may be strained by the consideration of viruses. In some ways, viruses behave as if alive, while in other ways they behave as mere fragments of other organisms. They exhibit very little that can be called metabolism, irritability, homeostasis, or any but the most rudimentary form of motion. Their claim to being alive rests largely in their ability to grow and reproduce; however, they can do so only if a host organism is present. The virus then subverts the host's reproductive machinery to replicate the viral nucleic acid and manufacture more viruses. Considerable evidence suggests that viruses originated as fragments of the hereditary material of other organisms.

Cellular Organization

When Robert Hooke (1635–1703) examined a thin slice of cork under his microscope, he noticed that it was divided into compartments, which he called "cells." These compartments were really empty spaces formerly occupied by the living cells, but soon other workers were able to examine the cells themselves. Then, in the 1830s, some German microscopists proposed that all animals and plants were composed of cells. Today's evolutionists study many features of cells that have remained more constant. Several cellular details, especially those involving chromosome structure and gene sequences, provide

important information on the nature of evolutionary processes.

We now know that there are many types of cells, but they all have certain features in common. All cells, for example, contain a **cell membrane,** also known as the **plasma membrane,** consisting of a central bilayer of **lipids,** or fat-soluble molecules, accompanied by globular proteins. All cells contain **ribosomes,** which function in protein synthesis; these ribosomes are rich in **RNA (RiboNucleic Acid).** DNA, RNA, proteins, and lipids are among the types of molecules present in all cells.

Most familiar organisms are made of cells that we call **eucaryotic** (sometimes spelled *eukaryotic*). Each eucaryotic cell has a distinctive **nucleus,** surrounded by two layers of membrane that constitute the **nuclear envelope.** Within the nucleus, the hereditary material, or **DNA (DeoxyriboNucleic Acid),** is arranged in lengthy coiled sequences called **chromosomes.** The chromosomes, of which there are usually two or more, contain protein as well as DNA. The cellular material outside the nucleus is called the **cytoplasm;** it contains many structures, or **organelles,** not found in bacterial cells: **vacuoles, golgi complexes, endoplasmic reticulum, mitochondria,** and in many cases **plastids;** each of these is further explained in Item 1.1 on page 10. Hairlike **cilia** or whiplike **flagella** may also be present; these always have a complex internal structure involving a "9 + 2" arrangement of **microtubules.** Cell division, or mitosis (Chapter 2), is aided by **centrioles,** which have a similar ninefold structure.

A generalized animal cell is shown in Fig. 1.1. Plant cells differ from animal cells in possessing cells walls and plastids, in usually lacking flagella or cilia, and in possessing a single large vacuole rather than many small vacuoles. The organelles of animal and plant cells are listed in Item 1.1.

Bacteria and blue-green algae have much simpler (procaryotic) cells lacking nuclear envelopes, mitochondria, plastids, and certain other organelles. Other differences between procaryotic and eucaryotic cells are explained in Chapter 26.

Higher Levels of Organization

Some organisms contain only one cell, but most organisms are composed of many different kinds of cells, arranged in several *levels of organization.* In these organisms, groups of similar cells are organized into *tissues.*

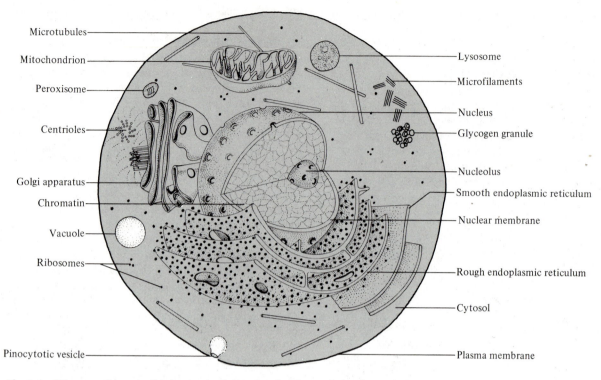

Microtubules

Mitochondrion

Peroxisome

Centrioles

Golgi apparatus

Chromatin

Vacuole

Ribosomes

Pinocytotic vesicle

Lysosome

Microfilaments

Nucleus

Glycogen granule

Nucleolus

Smooth endoplasmic reticulum

Nuclear membrane

Rough endoplasmic reticulum

Cytosol

Plasma membrane

Fig. 1.1 Diagram of a generalized animal cell. Reprinted with permission from John W. Kimball, *Biology*, 4e, © 1978. Reading, Massachusetts: Addison-Wesley (Fig. 5.10).

The cells of different tissues behave differently, even though they generally carry the same hereditary information. Several tissues may together form a complex *organ,* such as a stomach or a leaf. The organs are grouped together to form *organ systems,* such as the vascular or circulatory system. All the organ systems together make up the *organism.* Each tissue or organ may have its own evolutionary history, for all levels of organization are subject to evolution.

Living organisms generally occur in *populations.* Many of the changes studied by evolutionary biologists occur at the population level or at still higher levels. The populations are grouped into *species.* The kinds of organisms do not grade into one another continuously, because only certain discrete species exist. (We will explain why in Chapter 14.) Above this level, related species are grouped into *genera,* related genera into *families,* related

families into *orders,* related orders into *classes,* related classes into either *phyla* or *divisions,* and related phyla or divisions into *kingdoms.* The animal kingdom is divided into phyla, the plant kingdom into divisions. From kingdom to species, the hierarchy of terms is as follows:

Kingdom
  Phylum or Division
    Class
      Order
        Family
          Genus (plural, *genera*)
            Species (same spelling in singular and plural)

One of these groups, at any level, may be called a **taxon** (plural, *taxa*), and the study of these taxa constitutes the science of **taxonomy.** The theoretical basis of taxonomy will be discussed in Chapter 22. Each taxon is given a name, usually composed of Latin or Greek roots. Certain conventions of **nomenclature** (the naming of taxa) may be summarized as follows:

1. Names of families, orders, classes, phyla or divisions, and kingdoms are single words, which are capitalized and treated as plural. Family names of animals always end in -*idae*. Among plants, fungi, and bacteria, the names of families end in -*aceae* and the names of orders in -*ales*. There are no standard endings for the names of kingdoms, phyla, divisions, classes, or animal orders.

2. Names of genera are single words, which are always capitalized, *italicized* (or underlined in typing), and treated as singular. There are no standard endings.

3. Names of species are always two words long and are always *italicized*. The first word, always capitalized, is the name of the genus. The second word is nearly always lower-cased: *Homo sapiens*.

4. Finer or intermediate ranks are sometimes added to the hierarchy. Most commonly, the prefix *sub-* is used to indicate an intermediate rank just below another: a subfamily is just below a family, and a subclass is just below a class. In the same way, *super-* indicates a rank just above another: a superfamily is just above a family.

5. English words derived from the names of taxa are neither capitalized nor italicized: insects belong to the class Insecta, mammals to the class Mammalia, canids to the family Canidae, felines to the subfamily Felinae, and the hippopotamus to the genus *Hippopotamus*. Note how this allows us to distinguish the English word from the corresponding formal name.

Students often encounter taxonomic terms (names of taxa) in the form of a list to be memorized. Having committed one set of terms to memory, they then find it shocking to realize that other books or other teachers use a somewhat different set of names. Sometimes these differences are over terminology only: a particular animal phylum may be called Bryozoa by some people and Ectoprocta by others (and Polyzoa by still a third group), but all zoologists now recognize such a taxon at the phylum level, regardless of the name they apply to it. Sometimes

there is agreement that a taxon exists but disagreement over its rank: some zoologists consider the Hemichordata to constitute a phylum in their own right, while others have listed them as only a subphylum within the phylum Chordata. The tendency to make finer distinctions or to make the same distinctions at a higher taxonomic rank is called **splitting**; the opposite tendency, to recognize fewer distinctions or to recognize the same distinctions at a lower taxonomic level, is called **lumping**.

Some taxonomic disagreements are more deep-seated: they involve true disagreements over the membership or even the existence of taxa. One such disagreement involves kingdoms, the highest taxa of all. The traditional distinction between animals and plants was for centuries expressed by a two-kingdom classification. Ernst Haeckel (1834–1919), who studied certain one-celled organisms known as Radiolaria, added a third kingdom (Protista) for all one-celled creatures. His suggestion fell largely on deaf ears, and for nearly another century most biologists continued to classify all organisms as either animals or plants.

The much more fundamental distinction between organisms composed of eucaryotic cells with well-formed, membrane-limited nuclei (most organisms) and those composed of procaryotic cells lacking such nuclei (bacteria and blue-green algae) was noted by Stanier and Van Niel (1942) and repeatedly by Stanier *et al.* (e.g., 1976). Copeland (1938, 1956) was the first to express this distinction in a formal classification. His work inspired other attempts at kingdom-level classification (Item 1.2 at the end of this chapter), culminating in the five-kingdom classification of Whittaker (1959, 1969). This classification, the most widely adopted system in use today, is followed in this book in modified form, with the plant kingdom redefined to include all organisms with plastids. A somewhat different five-kingdom classification by Margulis (1974) differs principally in being a horizontal classification (Chapter 22), in which transitional forms are placed together with their ancestors, rather than a vertical classification in which transitional forms are placed together with their descendants (Fig. 1.2).

Five kingdoms are recognized in this book:

Procaryotae—organisms with procaryotic cells.

Protista—eucaryotic unicellular organisms lacking cell

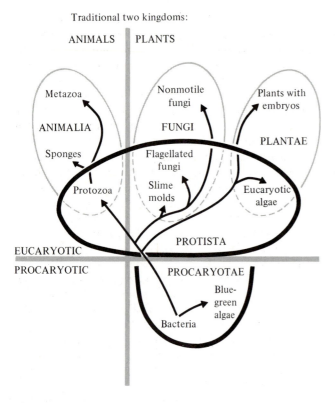

Traditional two kingdoms:

Fig. 1.2 A phylogeny of organisms for comparison with the classifications shown in Item 1.2.

walls or plastids, not forming syncytia, usually capturing food by phagocytosis, and usually motile with the aid of contractile filaments.

Fungi—eucaryotic organisms without plastids, usually forming multinucleated masses (syncytia) from the aggregation of many cells, usually with cell walls containing chitin (exceptionally cellulose), typically with absorptive nutrition and reproducing by means of spores.

Plantae—eucaryotic organisms containing plastids.

Animalia—eucaryotic organisms developing from multicellular blastula stages.

Further descriptions of these kingdoms are given in Chapters 25–26.

C. THE STRUCTURE OF MODERN SCIENCE

The Nature of Science

Science is seen by many people as a body of knowledge. Certain objects and certain data concerning these objects are thought to be "scientific." Scientists, however, prefer to think of science as a process or method of investigation that can be applied to any object. The moon can be the subject of science or of poetry. Likewise, the poet or the poem can be studied by scientific methods.

Crucial to the **scientific method** is a type of inference known as **hypothetico-deductive reasoning.** The first step in this method of reasoning is the statement of a **hypothesis,** which is simply an educated guess or an idea to be tested. A **deduction** is then made from this hypothesis. The logical consequence of the deduction is then subjected to a test, and it is either confirmed (**verified**) or disconfirmed (**falsified**). Falsification of the predicted consequence thereby falsifies the hypothesis, but verification of the consequence—or even of many such consequences—does not in any way verify the hypothesis. In fact, there is no way of verifying hypotheses, only of falsifying them. Much-tested hypotheses are dignified by calling them **theories,** but even a single disconfirmation can falsify a widely held theory. In response to such disconfirmation, a much-tested theory will often be restated in a modified form instead of being discarded entirely. In this way, normal science proceeds in small steps.

Scientific progress can also occur in large jumps, similar to the "macromutational" changes postulated by certain evolutionists (Chapter 6). Such **scientific revolutions** (Kuhn, 1962, 1970) come about when a complex and coherent set of theories, called a **paradigm,** is replaced by another. A paradigm, as conceived by Kuhn, includes not only theories but also goals and research methods held to be important, as well as problems held to be interesting. By implication, a paradigm also excludes certain other problems as being unimportant, uninteresting, or insoluble. The redirection of research to investigate these problems requires the establishment of a new paradigm, which often means rejection of the old one.

Not every statement can be a scientific hypothesis. Matters of taste ("spinach is good") or aesthetic judgment ("the moon is beautiful tonight") are excluded; so are moral or ethical statements ("thou shalt not kill"). These are excluded from the domain of science because no testable consequence can be deduced from any of them—they are, in Popper's (1959) terminology, not **falsifiable.** Other statements that cannot possibly be falsified are also considered "unscientific." Among these are "two plus two equals four," which is always true, or "the soul is immortal," which is untestable. Neither statement could possibly be disproven—no conceivable observation would necessitate rejecting either one as false. For a statement to be a scientific hypothesis, there must be some conceivable observation capable of proving it false. Thus "humans evolved from apes" is falsifiable, since a complete fossil record showing humans always to have existed in their present form would falsify it.

Theoretical or unseen entities, like atoms and genes, belong to science only to the extent that we can make falsifiable statements about them. From hypotheses about genes, we can deduce consequences to be expected of progeny; failure to observe these consequences can falsify our hypotheses. Statements about gods, demons, or unseen forces are part of science only to the extent that their postulated activities have falsifiable consequences. But the existence of demons or of God can never be falsified, and both are thus outside the realm of science.

Different branches of science are recognized by the means used to test hypotheses. A fundamental distinction between **naturalistic science** and **experimental science** was first made by Francis Bacon, who called the former "natural history" and the latter "natural philosophy." Experimental scientists test hypotheses by manipulating conditions in frequently unnatural ways. They pride themselves on being able to control as many conditions as precisely as they can—hence the popular caricature of the "mad scientist." Naturalists do not manipulate conditions but only wait for suitable conditions to occur naturally. Unlike experimentalists, who generate their own data, naturalists are only able to gather the data that nature provides. They can study dinosaurs consigned to the irretrievable past, long-term and large-scale phenomena like cliff erosion and continental drift, or stars too large and too remote to permit any experimental manipulation. Naturalists may also study phenomena like "behavior under natural conditions" that cannot be realized in the laboratory or manipulated without violating the stipulation of "natural conditions." Social scientists are naturalists because human societies resist experimental control and because manipulation distorts their behavior. Experimentalists often chide naturalists as being less than thoroughly scientific, since they have not controlled all the variables. Naturalists tend to look upon experimentalists as being narrow, since behavior in the laboratory is often very unnatural, thus frequently unsuitable as a model for behavior generally. Models are used in many fields of science, but they are always vulnerable to the criticism that they do not accurately reflect conditions in nature. Astronomy, geology, evolutionary biology, ecology, and the social sciences are largely naturalistic; physics, chemistry, and physiology are largely experimental. Modern biology is an amalgamation of both types of science, and modern evolutionists and ecologists conduct experiments whenever the problem before them lends itself to experimentation.

A parallel distinction has been made by Windelband (1895) between nomothetic and idiographic sciences. **Nomothetic science** deals with repeatable phenomena and has as its goal the formulation of "laws" about them. (The word "nomothetic" translates from the Greek as "law-making.") Repeatability implies interchangeability: any beaker of nitric acid, under specified conditions, is expected to behave just like any other; so, too, with electron beams or mouse livers. Laws like "acids and bases neutralize each other" are expected to hold, within certain boundary conditions, at any place and at any time. One might say that these laws are nonhistorical, first because they pertain (whenever the specified conditions are met) throughout all time, and second because the individual acids and bases carry few if any idiosyncrasies revealing of their history.

Idiographic or **historical science** deals with unique phenomena that are not interchangeable. The planet Mercury, the Gobi Desert, the genus *Drosophila,* and the French Revolution are all unique. Their unique features bear telltale idiosyncrasies indicative of their past history. The Russian Revolution differed from the French because Russian history up to 1917 differed from French history up to 1789. The planet Mercury, even if removed farther from the Sun, would still bear telltale markings showing that it had once been closer. Laws about planets, deserts, insect genera, or revolutions often specify "all other things being equal'' or similar phrases, like ''tend to'' or ''usually.'' For example, ''violent revolutions *tend to* be followed by dictatorships,'' but mitigating circumstances could greatly alter the outcome. To the extent that laws can be formulated in the idiographic sciences, the distinction between idiographic and nomothetic science is not always clear-cut. The idiographic sciences are largely naturalistic: neither planets, deserts, insect genera, nor revolutions are easily manipulated, and ''laws'' in the idiographic sciences are formulated by comparative study rather than by experimentation.

Overall Philosophies about Living Things

Scientists and philosophers have often pondered the nature of life. Are living things fundamentally different from nonliving ones, or are they only a bit more complex but basically similar? The former view is held by **vitalists**, the latter by **mechanists**; their contrasting viewpoints are often reflected in their evolutionary theories.

The basic premise of **mechanism** is that life is nothing more than a special sort of chemical process, a process that can be studied by physical and chemical means. The characterization of that process in purely physical and chemical terms is the greatest unsolved problem facing all mechanists.

The opposing theory of **vitalism** stresses the uniqueness of life. Life, to a vitalist, is more than just a complex chemical process; something else must also be present. This ''something else'' has received many names, including *vital force, vital principle,* and *élan vitale.* Particular vitalists have identified the mind or the soul as vital principles; others have given to their vital principles such strange names as ''entelechy.'' Vitalists view their vital principles as not subject to chemical analysis but obeying a higher or at least a different order of law. The description of the ''vital force'' in scientific terms is the one overriding problem facing all vitalists.

Mechanists and vitalists recur throughout history; no age is free from the debate between them. Influential vitalists have included Aristotle (384–322 B.C.), Louis Pasteur (1822–1895), Hans Driesch (1867–1941), Erwin Schrödinger (1887–1961), and in some ways Pierre Teilhard de Chardin (1881–1955). Influential mechanists have included Democritos (fourth century B.C.), René Descartes (1596–1650), Eduard Büchner (1860–1917), Ivan Pavlov (1849–1936), and Jacques Loeb (1859–1924).

The debate between mechanists and vitalists shows no signs of resolution. The existence of a ''vital principle'' may indeed be untestable if we define it as something characteristic of life yet not susceptible to chemical analysis. The vitalists have never demonstrated that they had found one. But if they ever did, how would they analyze it other than in chemical terms? On the other hand, mechanists have not succeeded in explaining all of life's processes in purely chemical terms. Meanwhile, the question ''does a vital principle exist?'' seems untestable. In Kuhn's (1970) terminology, the conflict between mechanism and vitalism is a conflict between different paradigms and is thus inherently incapable of rational solution. Hein (1972) concludes that mechanism and vitalism are two attitudes, both basic to human nature, and are thus likely to persist indefinitely.

At present, mechanism seems to enjoy a better reputation among scientists. Why? There is no evidence against the existence of a vital principle, except the negative evidence that none has yet been discovered. Then why do the mechanists care? Why not leave the vitalists to the harmless folly of believing in undiscovered vital principles? Mechanism's present appeal is that it is currently more fruitful or productive of research. Education in the mechanist paradigm includes lessons in the folly of vitalism: vital principles are (by definition) not subject to chemical analysis, and the vitalists have proposed no other form of analysis as a substitute. Therefore a belief in undiscovered vital principles is an impediment to research, at least within the mechanist paradigm. Yet the vitalists continue to believe in undiscovered vital principles. The inability of mechanists to account for the difference between living and nonliving systems convinces vitalists that a vital principle may yet exist. The ''purposiveness'' that vitalists see in homeostasis (page 2) seems, according to vitalists, to defy any mechanistic explanation. Clearly, the vitalists see in the mechanist paradigm precisely the sort of anomaly that is often the prelude to a paradigm shift by ''scientific revolution.''

We have no definitive answer to the metaphysical question of whether a vital principle exists. Maybe the question is unanswerable. At present, the philosophy of mechanism seems to offer a better working premise. So let us conduct research *as if* no vital principle existed, analyzing life in physical and chemical terms wherever possible. In this way, we may someday find the physical and chemical basis of life in purely mechanistic terms, or we may on the contrary find data that cannot be explained in these terms, data that would provide a means of investigating the vital principle scientifically.

An important assumption of all mechanistic theories is the notion that many biological processes are describable in purely physical and chemical terms —that they are **reducible** to chemical and physical processes. Some mechanists go still further, holding that *all* biological processes are reducible to physical and chemical processes, a philosophical outlook known as **reductionism.**

Reductionism includes several distinguishable though related beliefs. First, there is the belief that more complex objects (living organisms) are made of simpler ones (molecules or atoms) and nothing more. Even informed vitalists would today admit that organisms contain molecules; it is the denial of other possible constituents that characterizes these beliefs as reductionist. A second type of reductionism holds that complex processes are describable in terms of simpler ones, that the behavior of a living organism can be described wholly in terms of the behavior of its constituent molecules. A third type of reductionism, which deals with laws and theories, suggests that biological laws and theories can be reformulated in strictly chemical terms, meaning that they can be *reduced to* chemical laws. A fourth type of reductionism holds that biological laws or organic behavior can be predicted or *deduced from* strictly chemical laws. To a reductionist, history and economics may be reduced to (i.e., explained in terms of) social behavior, social behavior to individual behavior, individual behavior to organ behavior, organ behavior to cellular behavior, cellular behavior to molecular behavior, and molecular behavior to atomic behavior. Thus the social sciences can all be reduced to sociology, sociology to psychology, psychology to animal physiology, animal physiology to cell physiology, cell physiology to chemistry, and chemistry to physics.

Reductionism is best described as a belief or an attitude. It might also be described as a paradigm insofar as attempts to reduce one theory to another have been fruitful of research. One example most often mentioned is the reduction of classical thermodynamics to statistical mechanics in early–twentieth-century physics.

Vitalists have always been against reductionism, since it deprives them of such nonreducible phenomena as vital forces. Reductionists have for this reason always been mechanists, but not all mechanists are reductionists. There are several schools of antireductionist mechanism; dialectical materialism (see below) is one of them.

Antireductionism is also called **holism, compositionism,** or **organicism.** Crucial to an antireductionist theory is the notion of **emergence.** Emergence may be defined as the belief that, as one ascends a scale of increasing complexity (say, from atoms to cells to organisms to societies), phenomena not present at the simpler levels will gradually appear, or *emerge.* Water is wet, liquid, and capable of both dissolving ionic compounds and quenching thirst, yet the hydrogen and oxygen of which it is composed have none of these properties. Molecules are lifeless, but the molecular aggregations that we call cells are sufficiently complex to carry on the processes that we recognize as characteristic of life. New properties such as these emerge at the more complex levels because of the way in which constituent parts are put together. Supporters of emergence are fond of saying that the whole is more than just the sum of its parts because of these newly emergent properties. "Society," says Emile Durkheim (1895: 103), "is not a mere sum of individuals," since the behavior of society differs greatly from, and is in no way predictable from, the behaviors of its individual constituents in isolation.

What we have characterized thus far may be termed **logical emergence,** since it assumes emergent properties to be only the logical consequences of higher levels of complexity. **Historical** or **chronological emergence** goes a step further in assuming also that simpler phenomena came first, and that increasing levels of complexity followed one another in historical sequence. Emergence is then viewed as a historical, evolutionary process in which life arose as an emergent property of cellular organization. Vitalists who believe in emergence view the vital principle as an emergent property of cellular organization. This viewpoint seems common to all those vitalists who consider their vital principle to have a natural as opposed to a supernatural origin.

Belief in emergence is also quite compatible with mechanism; a large number of working scientists, including many evolutionists, appear to hold both viewpoints, rejecting reductionism on the one hand and vitalism on the other. Reductionism does, however, thrive in such fields as biochemistry, where the attempt to explain biological processes in purely chemical terms is probably the greatest single stimulus to research.

Because it plays such a prominent role in the thought of Soviet evolutionists, some mention should be made of **dialectical materialism.** This is really several theories in one, including a thoroughgoing materialism that denies the existence of God or of other immaterial forces in the Universe. More important to the present discussion, dialectical materialism rejects vitalism but views the properties of life and society as historically emergent. Conflict, says the theory, is inherent in all levels of organization; resolution of any such conflict results in the emergence of a new and higher level of organization, with new laws and new conflicts. Beyond this, there is a strongly Marxist view of history as resulting from class conflict, with the establishment of a classless, communistic society seen as the ultimate outcome.

CHAPTER SUMMARY

The chapter summary, appearing at the end of each chapter, contains the salient highlights of that chapter in concise form. Careful reading of the summary is recommended as an important means of review.

Evolutionary biology is the study of life as it changes through time.

The properties of life include metabolism, biosynthesis, irritability, homeostasis, growth, reproduction, hereditary information, and mutability. The last four of these are not easily mimicked by nonliving systems. Living things are composed of cells. Most cells are eucaryotic; they have a nucleus surrounded by a well-defined membrane. Bacterial cells are procaryotic.

Living things are arranged into kingdoms. Biologists have recognized as few as two kingdoms or as many as five. This book recognizes five: Procaryotae (bacteria and blue-green algae), Fungi, Plantae (plants), Protista, and Animalia (animals). These kingdoms are subdivided into phyla (or divisions), the phyla into classes, the classes into orders, the orders into families, the families into genera, and the genera into species. Names of species contain two words; the first word is the name of the genus.

Science is a method of inquiry using hypothetico-deductive reasoning to test falsifiable hypotheses; no statement is scientific unless it is falsifiable. Sciences may be either naturalistic or experimental or both.

Life may be seen either as a mere physical-chemical process of high complexity (mechanism), or as a phenomenon dependent on some ''vital force'' not subject to physical or chemical analysis (vitalism). Either view is defensible, but in research even vitalists must adopt mechanistic assumptions in order to make their statements falsifiable.

Mechanism is a philosophical ally of reductionism, the explanation of wholes in terms of their parts, or of large-scale processes in terms of small-scale ones. Reductionists attempt to explain biological processes in terms of chemical and physical laws only. The opposite philosophy of compositionism holds that biological or other complex processes obey certain laws related to the ways in which smaller components are put together to form larger ones, and that these laws are not predictable from those of chemistry or physics alone.

FOR FURTHER KNOWLEDGE ▪

ITEM 1.1   *The Organelles of Animal and Plant Cells*

Name	Description	Typical structure
Cell wall	Not strictly an organelle, the cell wall is an extracellular structure secreted by plant (not animal) cells. Plant cell walls are made largely of cellulose (a polysaccharide), together with other chemical substances, such as lignins and pectins.	

Name	Description	Typical structure
Cell membrane	Also called plasma membrane, this organelle determines what substances may enter the cell and what may leave. The "unit membrane" structure contains two opposing layers of lipid molecules (a bilayer), together with various protein molecules of varying sizes.	
Endoplasmic reticulum	This unit membrane structure has a composition similar to that of the cell membrane, but folded into an extensive network of vesicles. A large number of ribosomes may adhere to one surface, giving a "rough" appearance.	
Mitochondrion	All mitochondria contain two layers of unit membrane. The inner layer is folded into a series of *cristae,* which are thought to contain many of the enzymes for the production of energy-rich ATP via the Krebs citric acid cycle.	
Chloroplast	This chlorophyll-rich organelle is the site of photosynthesis in plant cells only. Plants also contain a number of other plastids of somewhat similar but less complex structure, containing other pigments or, in some cases, starch.	
Nucleus	The hereditary information of the cell is contained in this organelle in the form of strands that become visible as *chromosomes* during cell division. There is also an RNA-rich *nucleolus,* believed to be the site of ribosome production, and a *nuclear envelope* consisting of two layers of unit membrane.	
Ribosome	This RNA-rich organelle is the site of protein synthesis (see Chapter 2).	
Golgi complex	This series of stacked vesicles assembles newly formed protein into larger "packages" bounded by membranes.	
Vacuole	This unit membrane surrounds a droplet of water, food, or a secretory product. Animal cells usually have many vacuoles, but in plant cells these often coalesce to form one large, central vacuole.	
Lysosome	This unit membrane surrounds a solution of protein-digesting enzymes.	
Peroxisome	Similar in appearance to a lysosome, the peroxisome contains enzymes that reduce peroxides to water.	

Name	Description	Typical structure
Microtubules	These tubular protein structures form a "skeleton," which contributes to the structural stability of the cell. During cell division, they also function to draw the chromosomes apart.	
Centrioles	Formed of microtubules arranged in a ninefold pattern, this pair of organelles is also important during cell division (Fig. 2.3).	
Flagella (or cilia)	These organelles of motion contain a "9 + 2" arrangement of microtubules. The microtubules slide along one another to produce the motion. The cells of multicellular plants generally lack these organelles.	
Contractile filaments	These filaments, often scattered in bundles, contain such proteins as actin and myosin. They are believed responsible for cytoplasmic streaming (page 2).	

ITEM **1.2**   *Kingdom-Level Classifications*

A.   TRADITIONAL TWO KINGDOMS
(Linnaeus, 1758)
KINGDOM PLANTAE
  bacteria
  algae
  fungi
  plants with embryos
KINGDOM ANIMALIA
  protozoans
  sponges and metazoans

B.   THREE KINGDOMS
(Haeckel, 1866)
KINGDOM PROTISTA
  bacteria
  protozoans
  sponges
KINGDOM PLANTAE
  algae
  fungi
  plants with embryos
KINGDOM ANIMALIA
  metazoans

C.   FOUR KINGDOMS
(Copeland, 1956)
KINGDOM MONERA
  bacteria and blue-green algae
KINGDOM PLANTAE
  plants with embryos
KINGDOM PROTOCTISTA
  eucaryotic algae
  fungi
  protozoans
KINGDOM ANIMALIA
  sponges and metazoans

D.   THREE KINGDOMS
(Dodson, 1971)
KINGDOM MONERA
  bacteria and blue-green algae
KINGDOM PLANTAE
  eucaryotic algae
  fungi
  plants with embryos
KINGDOM ANIMALIA
  protozoans
  sponges and metazoans

E.   FIVE KINGDOMS (VERTICAL)
(Whittaker, 1969)
KINGDOM MONERA

  bacteria and blue-green algae
KINGDOM PLANTAE
  multicellular algae
  plants with embryos
KINGDOM FUNGI
  fungi
KINGDOM PROTISTA
  protozoans and one-celled algae
KINGDOM ANIMALIA
  sponges and metazoans

F.   FIVE KINGDOMS (HORIZONTAL)
(Margulis, 1974)
KINGDOM MONERA
  bacteria and blue-green algae
KINGDOM PLANTAE
  plants with embryos
KINGDOM FUNGI
  nonmotile fungi only
KINGDOM PROTISTA
  eucaryotic algae
  flagellated fungi; slime molds
  protozoans
KINGDOM ANIMALIA
  sponges and metazoans

2

Basic Principles of Genetics

A good understanding of modern evolutionary theory is impossible without a thorough understanding of genetics. No biological change can have any evolutionary significance unless it is inherited. Evolution is thus constrained by the ability of genetically inherited information to undergo change and to perpetuate the change once it occurs.

In this chapter, we review some of the basic principles of modern genetics. The structure, function, and transmission of genetic material is emphasized, and certain other topics (mutations, population genetics) are left to be covered in later chapters. Those who have had a thorough course in genetics may bypass this chapter.

A. STRUCTURE OF THE GENETIC MATERIAL

In all known organisms, the hereditary characteristics are determined by genetic material in the form of **nucleic acids**. The determination of the chemical structure of this genetic material was one of the major achievements of twentieth-century genetics.

Basic Constituents

The molecular structure of the genetic material is rather complex, but it is composed of three types of basic constituents: phosphate groups, sugars, and nitrogen bases (Fig. 2.1). The sugars in question are five-carbon (pentose) sugars. Two kinds of pentose sugars occur in nucleic acids: **ribose** and **deoxyribose**. A given nucleic acid contains either one or the other of these two sugars. Nucleic acid containing ribose sugar is called **ribonucleic acid** or **RNA**. Nucleic acid containing deoxyribose is called **deoxyribonucleic acid** or **DNA**. In either case, the structural framework or "backbone" of the molecule is formed by an alternating sequence of sugars and phosphate groups.

Protruding from the sugar-phosphate backbone are a series of so-called **nitrogen bases**, attached to the sugars. There are five common nitrogen bases, and several more uncommon ones. The five common bases are **adenine, guanine, cytosine, thymine**, and **uracil**, commonly abbreviated as **A, G, C, T**, and **U**. All five of these bases contain **heterocyclic rings**, meaning that nitrogen and carbon atoms both contribute to the ringlike structure. Adenine and guanine, which contain two such rings, are called **purines**; cytosine, thymine, and uracil,

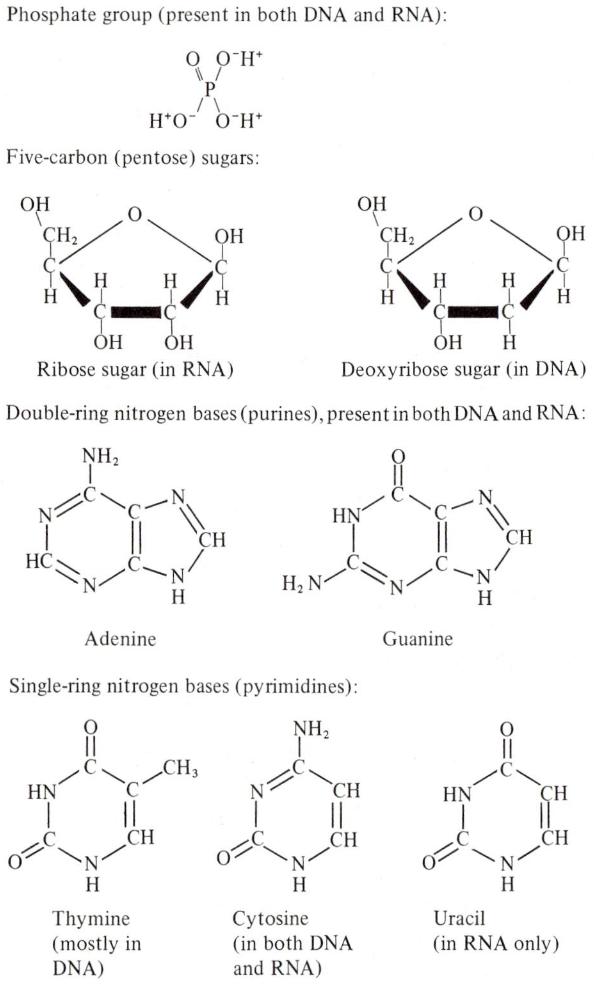

Phosphate group (present in both DNA and RNA):

Five-carbon (pentose) sugars:

Ribose sugar (in RNA)   Deoxyribose sugar (in DNA)

Double-ring nitrogen bases (purines), present in both DNA and RNA:

Adenine   Guanine

Single-ring nitrogen bases (pyrimidines):

Thymine (mostly in DNA)   Cytosine (in both DNA and RNA)   Uracil (in RNA only)

Fig. 2.1 The building blocks of DNA and RNA.

which contain only one heterocyclic ring, are called **pyrimidines.** Thymine occurs principally in DNA; uracil occurs only in RNA. The other three common bases occur in both DNA and RNA.

Deoxyribonucleic Acid (DNA)

The chromosomes of all organisms, except for a few viruses, contain deoxyribonucleic acid (DNA). This DNA exists in a double-stranded form, with each strand held together by an alternating backbone of phosphate groups and deoxyribose sugar. From each strand, a series of nitrogen bases protrude. The two strands are held together by weak intermolecular forces, called **hydrogen bonds,** between the nitrogen bases of one strand and the nitrogen bases of the other.

The distance between the two backbones is rather constant and is just big enough to accommodate one purine and one pyrimidine. A purine on one strand must therefore be matched with a pyrimidine on the other strand, and vice versa. Adenine, furthermore, cannot pair with cytosine, because the parts of each molecule would not fit together to form hydrogen bonds; the same is true between guanine and thymine. Thus the only possible pairings are between adenine and thymine, or between guanine and cytosine. In these cases, the opposing nitrogen bases are of suitable size and configuration for hydrogen bonding to take place.

The two strands of DNA are like the rails of a ladder, with the rungs represented by the hydrogen bonds connecting the nitrogen bases of the opposing strands. But the ladder is twisted in space to form a **double helix** (Box 2–A) and a person climbing the ladder would make one complete revolution (360°) for each ten rungs climbed.

The complementary pairing of bases (adenine with thymine, or guanine with cytosine) makes possible the **replication** of the DNA molecule, because each strand contains all the information needed to synthesize the opposite or complementary strand. The two strands separate, and each preexisting strand directs the synthesis of a new complementary strand. Wherever adenine occurs on the preexisting strand, thymine is inserted on the new complementary strand. Where guanine occurs, cytosine is inserted; where cytosine occurs, guanine is inserted; and where thymine occurs, adenine is inserted. The new complementary strand thus comes to have the same base sequence as the former *partner* of the preexisting strand that directed its synthesis. If we represent the two complementary base sequences as + and − , each + strand directs the synthesis of a − strand, and vice versa.

Ribonucleic Acid (RNA)

Four important differences distinguish RNA from DNA. Ribonucleic acid contains ribose sugar instead of deoxyribose. Ribonucleic acid contains uracil in place of

BOX 2–A *The Structure of DNA*

(A) The structure of DNA, drawn with the molecule untwisted into a flat, ladderlike configuration. The two "backbones" of the molecule, formed by alternating phosphates and sugars, represent the uprights of the ladder. The rungs of the ladder are formed by the hydrogen bonds between complementary nitrogen bases, drawn as dotted lines. The sequence of nitrogen bases in any single strand is not predictable in any uni-

form way; it varies from one gene to another. The pairing of complementary bases, on the other hand, is very regular: adenine pairs only with thymine, and vice versa; guanine pairs only with cytosine, and vice versa.

(B) In reality, the ladderlike structure shown in (A) is twisted into the form of a double helix, as shown here. The molecule undergoes a complete 360° twist every ten base pairs. It was this ladderlike arrangement that Watson and Crick first described in 1953.

Symbols Used:

(P) Phosphate group

⬠ Deoxyribose sugar
A   Adenine
G   Guanine
C   Cytosine
T   Thymine

One nucleotide

(A)

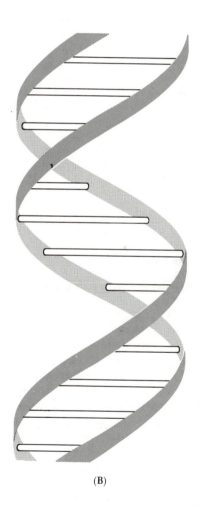

(B)

(continued)

BOX 2–A   *(continued)*

(C)

(C) Detail of a sequence of four nucleotides. The particular sequence shown here is only one of 256 possibilities for a sequence of this length.

thymine. Ribonucleic acid is usually single-stranded instead of double-stranded. And ribonucleic acid exists in three common forms instead of the one form in which DNA typically occurs.

Messenger RNA, or **mRNA,** contains one long helical strand. The linear sequence of bases is determined by the complementary strand of DNA from which the mRNA was synthesized.

A second form of RNA is called **transfer RNA,** or **tRNA.** There are several dozen forms of tRNA known, all of them characterized by a fairly large number of unusual nitrogen bases, other than the four (A, G, C, U) that normally occur in RNA. All forms of tRNA can be represented in a "cloverleaf" configuration, shown in Fig.

2.2; the actual three-dimensional shape of each tRNA is both more intricate and more compact than this.

The third type of RNA is **ribosomal RNA,** or **rRNA.** This type of RNA is synthesized in the nucleolus and is most characteristic of ribosomes.

The role of mRNA, tRNA, and ribosomes in protein synthesis is described in the next section.

B. GENE FUNCTION AND PROTEIN SYNTHESIS

Proteins are among the most important of all biological molecules. Not only are proteins important as cellular

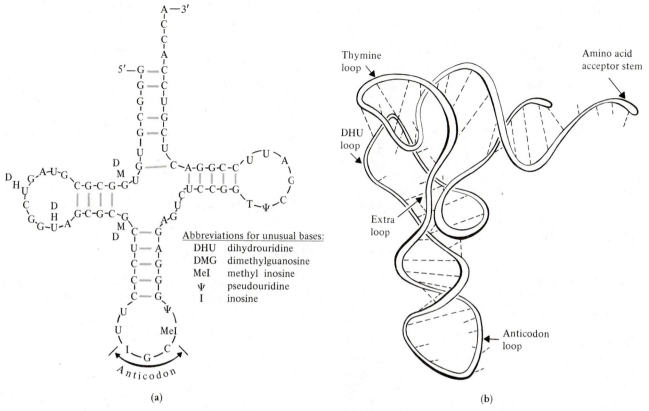

Abbreviations for unusual bases:

DHU	dihydrouridine
DMG	dimethylguanosine
MeI	methyl inosine
Ψ	pseudouridine
I	inosine

(a)   (b)

Fig. 2.2 The structure of transfer RNA (tRNA) molecules from yeast cells. (a) Cloverleaf structure of the tRNA coding for alanine. (b) Proposed three-dimensional structure of the tRNA coding for phenylalanine. Part (a) is redrawn in part after Donald I. Patt and Gail R. Patt, *An Introduction to Modern Genetics,* © 1975. Reading, Massachusetts: Addison-Wesley (Fig. 7.18).

and extracellular constituents, but they may also function as enzymes in the synthesis or utilization of other biological materials. (An **enzyme** is an organic substance, usually a protein, that speeds up a chemical reaction without itself being altered or "used up" in that reaction. Chemicals that function in this manner are generally called **catalysts.** An enzyme is simply an *organic catalyst.*) The great intricacies of living systems are all the results of enzyme-controlled activities, and it has often been said that proteins, including enzymes, are therefore the chemical basis of life. Since all biological molecules are synthesized with the aid of enzymes, it is important for us to examine both the structure of proteins and the manner in which protein molecules are synthesized.

Protein Structure

The building blocks of proteins are small molecules known as **amino acids.** Each amino acid contains an **amino group** ($-NH_2$) and an **organic acid group** ($-COOH$), plus a distinctive side chain represented symbolically by the letter R. Different amino acids have different side chains. There are some twenty amino acids commonly found in proteins (See Item 2.1 on page 33).

The amino acids are capable of being linked to one another in long chains known as **polypeptides.** In the formation of a **peptide bond** between two successive amino acids, a molecule of water is split out. One amino acid, cystine, contains an amino acid portion at either end, linked together by two sulfur atoms in between (a so-called **disulfide bridge**). A **protein** consists of one or more polypeptides. If two or more polypeptides are present, they are held together by these disulfide bridges. Further details of protein structure are shown in Item 2.2.

The portion of a nucleic acid molecule that directs the synthesis of a single polypeptide is called a structural gene, or **cistron**. The sequence of amino acids that make up the polypeptide is determined by the sequence of nitrogen bases contained in the cistron. These sequences are important, because the proper function of any enzyme depends on its molecular shape, which in turn depends on the exact sequence of its amino acids.

In addition to cistrons, there are certain other sequences in each DNA molecule that do not appear to determine the amino acid sequence of any polypeptide. Some of these may function as "spacers," and others are believed to function as **regulatory genes**, which control the transcription of other genes.

Transcription

The synthesis of messenger RNA is called **transcription**. Transcription occurs with the aid of certain enzymes by the addition, one at a time, of units known as **nucleotides**. Each nucleotide consists of a nitrogen base attached to a sugar and one or more phosphate groups. The sequence of nitrogen bases in the DNA molecule determines the sequence in which complementary nucleotides are added to make up the mRNA molecule. For example, the presence of guanine in the DNA sequence ensures that the next nucleotide added to the growing mRNA molecule will contain cytosine. G is thus transcribed as C, C as G, T as A, and A (adenine) as U (uracil).

Transcription of genes takes place in the nucleus. Not every gene is transcribed at the same time. Some are transcribed almost continuously, others only at certain times; a few control the transcription of other genes without themselves being transcribed. Also, in multicellular organisms, a very large number of genes are transcribed in certain types of cells only. Unless you have diabetes, each of your cells possesses a gene for insulin production. Yet only in the islets of your pancreas is this gene transcribed and insulin produced! One of the central problems of the field known as **developmental biology** is the discovery of the processes that control the transcription and subsequent expression of genes in the various types of body cells.

Translation

Following transcription in the nucleus, the messenger RNA thus formed migrates from the nucleus into the cytoplasm through one of the pores in the nuclear envelope. In the cytoplasm, the mRNA is **translated** into a sequence of amino acids that form a polypeptide chain, i.e., part of a protein. Messenger RNA molecules may contain several copies of the same "message." Each copy is translated over and over again to produce many molecules of protein with the same specified amino acid sequence.

Each amino acid is determined by a sequence of exactly three nucleotides in the mRNA "message"; this sequence of three nucleotides is known as a **codon**. The transfer RNA molecule contains a matching set of three complementary nucleotides known as the **anticodon**; among all possible tRNA molecules, only those having the complementary anticodon sequence can pair successfully with the codon. Transfer RNA molecules having the same anticodon sequence always bind to the same type of amino acid molecule. The sequence of amino acids is thus determined by the sequence of tRNA molecules, which in turn is determined by the sequence of codons in the messenger RNA molecule that was transcribed from a DNA sequence.

Translation of mRNA into protein occurs only in the presence of ribosomes. Translation is initiated by the combining of a messenger RNA molecule with a transfer RNA molecule on the surface of a ribosome. Each successive codon in the messenger RNA is then translated in its turn. The codon pairs up with a matching anticodon, carried by a tRNA molecule to which an amino acid is already attached. A **peptide bond** now forms between the new amino acid and the existing amino acid sequence. The ribosome shifts along the mRNA molecule, somewhat like the slide of a zipper, carrying with it the elongating end of the polypeptide, to which new amino acids continue to be added. Three specific codons are capable of terminating the "message"; the appearance of any chain-terminating codon causes the polypeptide to be released and the other molecules to separate from one another.

The Genetic Code

The exact sequence of nucleotides that determines each amino acid is known. A "coding dictionary," giving correspondences between codons and amino acids, is shown in Item 2.3.

Study of this "coding dictionary" reveals certain regularities. Of the many possible amino acids, only twenty can be specified at the time of translation; other amino acids, when they do occur, must be produced by modification of these twenty. Each of the twenty amino acids is specified by at least one codon; most are specified by two or more. This last phenomenon, known as **degeneracy,** is largely due to the "wobble" of the third nucleotide in each codon, which as a consequence is much less specific than the other two nucleotides. The code as a whole is thought to be universal, for it is identical in all organisms for which it has been determined. Also universal is the fact that each codon "word" contains three nucleotide "letters," and that the code is "commaless," meaning that the starting point of each codon is determined only by the fact that it follows the final letter of the previous codon.

C. CELL DIVISION

The hereditary material in the nucleus of the cell is normally not visible, even under the microscope. During cell division, however, the hereditary material becomes more easily visible, and is seen to consist of a series of discrete **chromosomes.** In animals and plants, the majority of cells are **diploid,** meaning that their chromosomes occur in pairs. The **gametes,** or egg and sperm cells, are **haploid**—their chromosomes occur singly and are not paired.

Mitosis

The usual form of cell division in diploid cells is called **mitosis.** In mitosis, the duplication of hereditary material is followed by a single division, which preserves the diploid chromosome number in both daughter cells. This orderly process (Figs. 2.3 and 2.4) also ensures that all the hereditary information carried in the chromosomes of the parent cell is passed to each daughter cell.

The first and longest phase of mitosis is called **prophase.** During prophase, the nuclear envelope and nucleolus appear to dissolve, and the boundary between nucleus and cytoplasm becomes indistinct. Thin microtubules begin to radiate out from each pole, forming between the two poles a vaguely football-shaped structure known as the **spindle.** The centrioles are usually located at opposite ends of the spindle. The spindle enlarges during prophase as the centrioles migrate to opposite sides of the cell. Meanwhile, the chromosomes become shorter, thicker, and increasingly distinct as their threadlike material becomes more tightly coiled. Soon it is evident that each chromosome is two-stranded, consisting of two identical strands known as **chromatids.** The sister chromatids of each chromosome are held together by a structure called the **centromere.** By convention, the number of chromosomes is determined by counting the centromeres, so at this stage, the normal diploid chromosome number is still present, though the number of strands (chromatids) is already twice this number.

The second and shortest phase of mitosis is known as **metaphase.** It is during this phase that the chromosomes all line up halfway between the two poles along a **metaphase plate.**

The third phase of mitosis is known as **anaphase.** Anaphase begins with the splitting of each centromere in half, or the separation of previously formed halves, an event that doubles the number of chromosomes present. The sister chromatids, which now form distinct but identical chromosomes, begin to separate, as the centromeres are pulled to opposite poles by the spindle fibers (microtubules).

As the single-stranded chromosomes reach opposite poles of the cell, mitosis enters its final phase, called **telophase.** The chromosome strands uncoil and straighten, so that the chromosome appears to lengthen and grow thinner, until it becomes so thin that it can no longer be seen easily. A new nuclear envelope is reconstituted around each new diploid set of chromosomes, and a nucleolus reappears.

Division of the nucleus is complete; division of the cytoplasm (**cytokinesis**) now begins. In most animal cells and in most other cells that lack a cell wall, the cytoplasm simply constricts near its middle to form a dumbbell shape, after which it pinches in two. In most plant cells, the cell wall is too stiff to permit this dumbbell-shaped stage, and cytokinesis thus occurs differently: a series of tiny vesicles (they look like bubbles) form across the cell between its two daughter nuclei; these bubbles are the

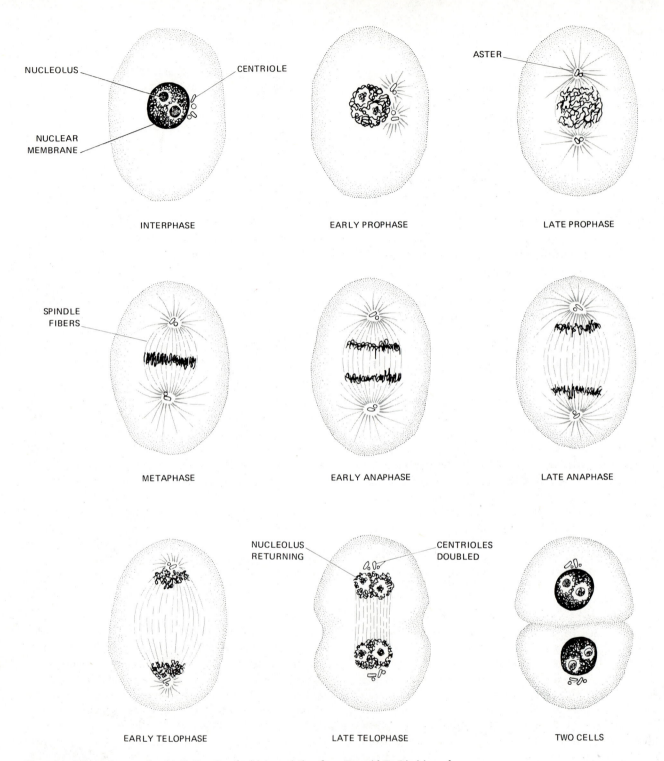

Fig. 2.3 Mitosis in an animal cell. Reprinted with permission from Donald D. Ritchie and Robert Carola, *Biology,* © 1979. Reading, Massachusetts: Addison-Wesley (Fig. 8.2).

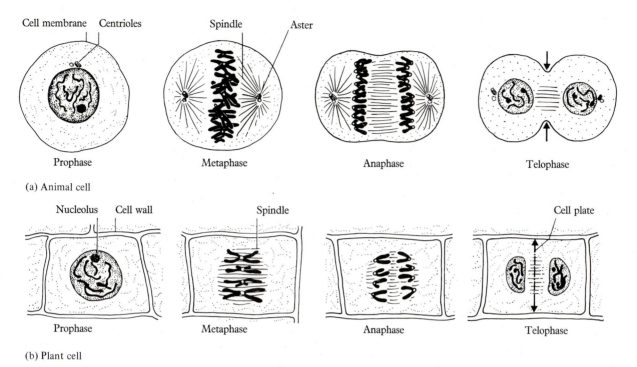

Fig. 2.4 Comparison of mitosis in animal and plant cells. Reprinted with permission from Donald I. Patt and Gail R. Patt, *An Introduction to Modern Genetics,* © 1975. Reading, Massachusetts: Addison-Wesley (Fig. 2.7).

beginning of the new cell wall that will separate the daughter cells.

Following cytokinesis, the cell enters **interphase,** the long interval between mitotic divisions. In slowly dividing tissues, interphase may last weeks or even longer, whereas all of mitosis may take only an hour or two. Interphase is much shorter in rapidly dividing tissue and may last less than a day. Interphase is a cell's most active period metabolically, the time of greatest absorption, contraction, secretion, or biosynthesis. It is also during interphase, before the next mitosis, that DNA replication takes place; it is for this reason that the chromosomes are already double-stranded when they become visible during the next succeeding prophase. Interphase is divided into a G_1 phase, which precedes DNA replication, a synthesis phase (S phase), in which DNA replicates, and a G_2 phase, following replication.

Meiosis

Meiosis may be defined as cell division in which the chromosome number is halved. It contrasts with mitosis, in which the chromosome number is maintained without any change. In brief, meiosis consists of a single duplication of chromosomes, followed by the partitioning of these chromosomes among four daughter cells, with each receiving half the original chromosome number, or one member of each chromosome pair. The events of meiosis are shown in Figs. 2.5 and 2.6.

The **first prophase** of meiosis, or **prophase I,** is similar in many ways to the prophase of mitosis: microtubules emanate from the centrioles and form a spindle, which enlarges; the nuclear envelope and nucleolus gradually dissipate; the chromosomes undergo coiling in such a way that they appear to both thicken and shorten. The

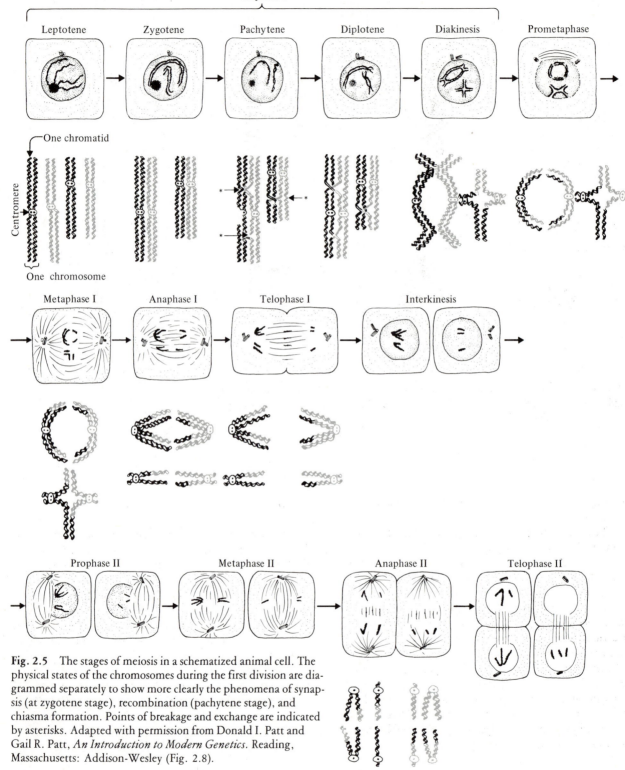

Fig. 2.5 The stages of meiosis in a schematized animal cell. The physical states of the chromosomes during the first division are diagrammed separately to show more clearly the phenomena of synapsis (at zygotene stage), recombination (pachytene stage), and chiasma formation. Points of breakage and exchange are indicated by asterisks. Adapted with permission from Donald I. Patt and Gail R. Patt, *An Introduction to Modern Genetics*. Reading, Massachusetts: Addison-Wesley (Fig. 2.8).

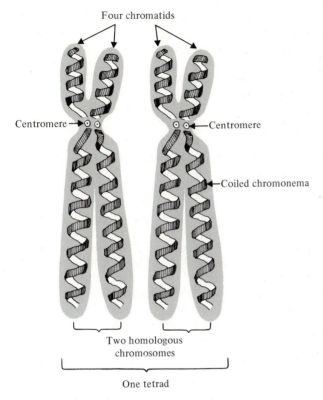

Four chromatids

Centromere →

← Centromere

← Coiled chromonema

Two homologous
chromosomes

One tetrad

Fig. 2.6 A tetrad drawn at the zygotene stage of prophase I of meiosis. Such a tetrad contains four chromatids (strands) or a pair of homologous chromosomes.

major difference concerns the chromosomes themselves: each double-stranded chromosome associates with another chromosome of similar morphology, called its **homologous partner,** to form what is called either a **bivalent** or a **tetrad.** Each bivalent contains two chromosomes belonging to a **homologous pair,** and since each chromosome is double-stranded, the bivalent contains a total of four chromatids.

Metaphase I is marked by the lining up of bivalents along the metaphase plate. Next, in **anaphase I,** the homologous chromosomes separate without any division of their centromeres. **Telophase I** consists of the grouping of the chromosomes near opposite poles of the original spindle. Often this passes directly into **prophase II,** marked by the appearance of new spindles, usually at right angles to the old one. In some cases, a brief **interkinesis** stage may intervene between telophase I and prophase II.

Metaphase II is marked by the lining up of double-stranded chromosomes along metaphase plates. **Anaphase II,** which resembles the anaphase of mitosis, is marked by the splitting of the centromeres and the migration of the single-stranded chromosomes thus formed to opposite poles. **Telophase II** is very much like the telophase of mitosis, in that chromosomes become less distinct while nuclear envelopes and nucleoli reappear into view.

Cytokinesis during meiosis is rather variable. In some cases, the dividing cell remains whole throughout meiosis and subsequently undergoes two successive events of cytokinesis, dividing the original cell into four daughter cells. In other cases, a cytokinesis following telophase I divides the diploid cell into two haploid cells, and a second event of cytokinesis takes place following telophase II, dividing each of these haploid cells once again to produce four daughter cells from the original diploid cell. The four daughter cells are usually about equal in size, but in the formation of eggs (ova) in many animal species, the daughter cells may be decidedly unequal, consisting of a large egg and three other cells, called **polar bodies.**

In most eucaryotic life cycles, the fusion of gametes (egg and sperm cells) restores the diploid chromosome number that was halved during meiosis. Through the alternation of meiosis and fertilization in each generation, the genes of the population are constantly being reshuffled into new combinations. This is one important source of variation (Chapter 7), without which no evolution could take place.

D. GENETIC TRANSMISSION IN DIPLOIDS

Basic Terminology

Before discussing the manner in which hereditary material is transmitted from one generation to the next, we must first introduce certain standard terminology. The initial generation in any breeding experiment is called the **parental generation** and is designated **P.** The successive **filial generations,** or generations of offspring, are designated F_1, F_2, and so forth. The visible characters of an organism are its **phenotype.** The hereditary characters of an organism are its **genotype.** The genotype is not always evident phenotypically, but it can often be determined through breeding experiments.

The portion of hereditary material that codes for a particular polypeptide is called a **gene,** and its position on the chromosome is called its **locus.** The alternative states of a gene are called **alleles.** A genotype containing two identical alleles (*AA* or *aa*) is said to be **homozygous** for that trait, and the individual is called a **homozygote.** A genotype containing two unlike alleles for a given trait (*Aa*) is said to be **heterozygous** for that trait. Any trait that is displayed phenotypically in heterozygous as well as homozygous condition is called **dominant;** a trait that is expressed phenotypically only when homozygous is called **recessive.**

Single-Gene Inheritance

The basic principles of genetic transmission in diploids were discovered by Gregor Mendel in 1865 (Chapter 6). In his first series of seven experiments using pea plants, Mendel studied the inheritance of one character at a time. One of his experiments is shown in Box 2–B. The dominant allele in this case is symbolized by *T* (''big *T*'') and the recessive allele by *t* (''small *t*'' or ''little *t*''). Both *TT* and *Tt* plants are phenotypically tall, but *tt* plants are short.

The homozygous parents in this type of cross can each produce only one type of gamete: all gametes from the tall parent contain *T*, and all gametes from the short parent contain *t*. The F_1 plants are therefore uniformly *Tt* (heterozygous), and they are phenotypically tall.

The F_1 plants produce two types of ovules (female gametes), containing either *T* or *t*. Two types of pollen, *T* and *t*, are also produced. Four combinations of gametes are thus possible and should all be expected in approximately equal numbers: *T* pollen with *T* ovules, *T* pollen with *t* ovules, *t* pollen with *T* ovules, and *t* pollen with *t* ovules. The first three of these combinations contain at least one *T* and should thus produce tall plants. Only the last combination, *tt*, should produce short plants. The F_2 plants should therefore be both tall and short in about a 3:1 ratio. This 3:1 ratio is indeed what Mendel found, not only in this case but in six other similar cases as well. Other geneticists have since found that all other sexually reproducing diploid organisms show similar cases of single-gene inheritance.

The principle of **particulate inheritance** states that genes keep their separate identities even when they occur together. For example, the recessive allele in the cross above remains unmodified while hidden in the F_1, and when the F_1 produces gametes, the recessive allele separates from its dominant partner and can express itself in a predictable fraction of F_2 phenotypes. This instance of particulate inheritance is sometimes called **Mendel's first law** or **the law of segregation.**

Two-Gene Inheritance

From single-gene inheritance, Mendel proceeded to the investigation of the inheritance of two characters at a time. One of his crosses using two genes is shown in Box 2–C. In this cross, a plant with yellow, round peas is mated to one with green, wrinkled peas. The reciprocal cross, yellow-wrinkled × green-round, would yield the same results. The F_1 plants were heterozygous for both traits (*YyRr*). They were all round and yellow, confirming Mendel's earlier finding that both round and yellow were dominant traits.

Mendel allowed the F_1 plants to self-fertilize and produce an F_2 generation. Both parental phenotypes, yellow-round and green-wrinkled, were represented, as were two new combinations: green-round and yellow-wrinkled. The proportions of the four phenotypes were 9 yellow-round: 3 green-round: 3 yellow-wrinkled: 1 green-wrinkled. Thus, with color ignored, the ratio of round to wrinkled was 12:4 = 3:1, and there was similarly a 3:1 ratio of yellow to green. Furthermore, the 3:1 ratio of round to wrinkled seeds held equally among the green seeds and among the yellow ones, indicating that the two pairs of traits had assorted independently. This is called the principle of **independent assortment,** sometimes known as **Mendel's second law.** Formally, this principle states that a *diploid organism heterozygous for two traits produces gametes in which all four combinations of these two traits are equally represented.* (An important proviso, added many years later, was that the two genes had to be located on different chromosomes in order for this principle to hold.)

The cross above enabled Mendel to predict that several different genotypes would be represented for several of the F_2 phenotypes. For example, the yellow-round phenotype can be produced by any of four genotypes (*YYRR, YyRR, YYRr,* and *YyRr*). By self-fertilizing each of the F_2 plants and producing an F_3, Mendel was able to confirm the genotype of each F_2 plant; his results were entirely according to expectation.

BOX 2–B. *A Single-Gene Mendelian Cross (Monohybrid Cross)
in Pea Plants*

Genotypes: *TT, Tt, tt*

    T = tallness (dominant allele)
    t = shortness (recessive allele)

Phenotypes: Tall, Short

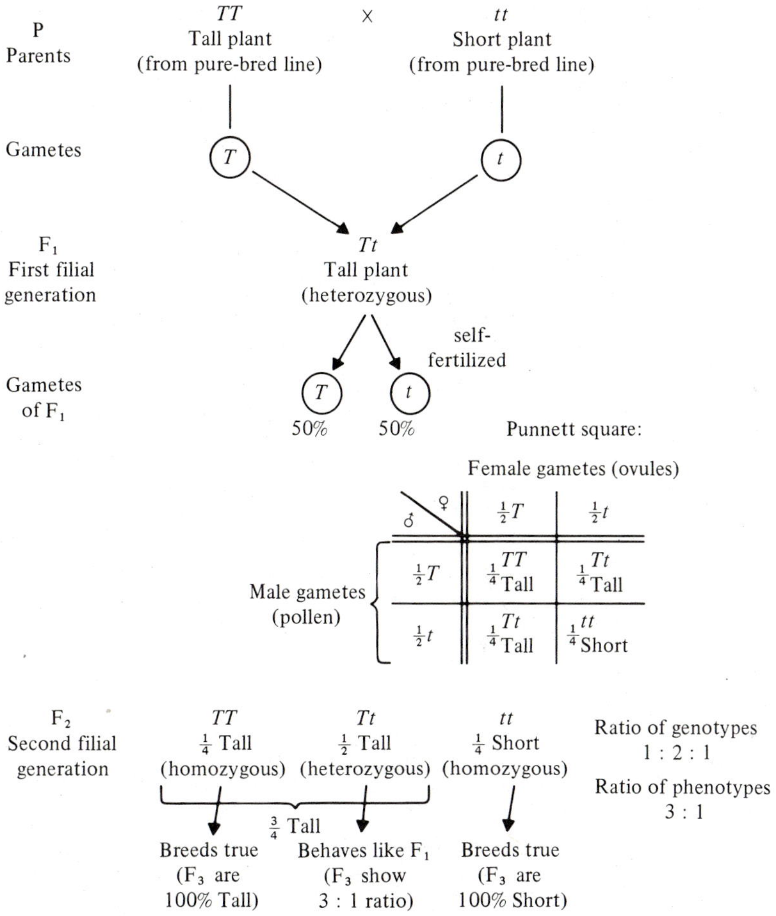

P
Parents

TT × tt
Tall plant   Short plant
(from pure-bred line)   (from pure-bred line)

Gametes

T t

F_1
First filial
generation

Tt
Tall plant
(heterozygous)

self-
fertilized

Gametes
of F_1

T t
50%   50%

Punnett square:

Female gametes (ovules)

♂ \ ♀	$\frac{1}{2}T$	$\frac{1}{2}t$
$\frac{1}{2}T$	$\frac{1}{4}\,^{TT}$Tall	$\frac{1}{4}\,^{Tt}$Tall
$\frac{1}{2}t$	$\frac{1}{4}\,^{Tt}$Tall	$\frac{1}{4}\,^{tt}$Short

Male gametes
(pollen)

F_2
Second filial
generation

TT Tt tt
$\frac{1}{4}$ Tall $\frac{1}{2}$ Tall $\frac{1}{4}$ Short
(homozygous) (heterozygous) (homozygous)

Ratio of genotypes
1 : 2 : 1

Ratio of phenotypes
3 : 1

$\frac{3}{4}$ Tall

Breeds true   Behaves like F_1 Breeds true
(F$_3$ are   (F$_3$ show   (F$_3$ are
100% Tall)   3 : 1 ratio)   100% Short)

BOX 2–C *A Two-Gene Mendelian Cross (Dihybrid Cross) in Pea Plants*

R = round seeds (dominant allele)
r = wrinkled seeds (recessive allele)

Y = yellow seeds (dominant allele)
y = green seeds (recessive allele)

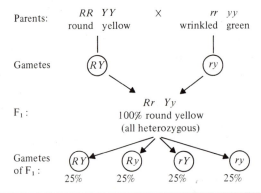

Parents: $RR\ YY$ × $rr\ yy$
round yellow   wrinkled green

Gametes: RY ry

F_1: $Rr\ Yy$
100% round yellow
(all heterozygous)

Gametes of F_1: RY Ry rY ry
25% 25% 25% 25%

Punnett Square

Male gametes (pollen)	Female gametes (ovules)			
	¼ RY	¼ Ry	¼ rY	¼ ry
¼ RY	$RRYY$ round yellow	$RRYy$ round yellow	$RrYY$ round yellow	$RrYy$ round yellow
¼ Ry	$RRYy$ round yellow	$RRyy$ round green	$RrYy$ round yellow	$Rryy$ round green
¼ rY	$RrYY$ round yellow	$RrYy$ round yellow	$rrYY$ wrinkled yellow	$rrYy$ wrinkled yellow
¼ ry	$RrYy$ round yellow	$Rryy$ round green	$rrYy$ wrinkled yellow	$rryy$ wrinkled green

Each cell in Punnett square contains 1/16 of the offspring, on the average

F_2: Genotypes   Phenotypes

1/16 ($RRYY$) + 2/16($RrYY$) + 2/16($RRYy$) + 4/16($RrYy$) = 9/16 round-yellow
1/16($RRyy$) + 2/16($Rryy$) = 3/16 round-green
1/16($rrYY$) + 2/16($rrYy$) = 3/16 wrinkled-yellow
1/16($rryy$) = 1/16 wrinkled-green

Polygenic Inheritance

The last few examples have dealt with all-or-none characteristics, also known as discrete characteristics. The peas, for example, were short or tall but not middle-sized. There are many other characteristics, however, which are continuous: human height, weight, and strength are examples. In these characteristics, intermediate values can always exist between any two measurements. Genetic control of these characteristics is believed to require many genes of small but additive effect. This type of inheritance is called **polygenic inheritance.**

A simplified example of polygenic inheritance is shown in Item 2.4 at the end of this chapter. Each allele in this simplified example contributes a small additive effect; the amount added (above some minimum threshold) is proportional to the number of capital letters in the genotype. There is no dominance in this example; the presence of dominance would change the exact values but not the overall distributions.

The example in Item 2.4 uses only three genes, but in a more realistic example, using many more genes, the distribution would be considerably "smoothed out," resembling the dashed line shown in the lower diagram. The presence of an environmental effect would smooth out the distribution even further. Such an environmental effect exists, for example, in the case of human standing height, which depends not only on many genes but also on the quality and quantity of protein and other nutrients consumed during the growing period.

Multiple Alleles

We have thus far assumed that each gene exists in two allelic states, and that one of these is always dominant and the other recessive. Although this is the most frequent situation among diploid organisms, other types of inheritance are also known. Two alleles, for example, may be **co-dominant,** meaning that the heterozygote has the phenotype of both alleles present together at the same time. Another possibility is **incomplete dominance,** in which heterozygotes are somewhat intermediate phenotypically between the homozygous conditions.

The human A,B,O blood groups (Item 2.5) are an example of a system of **multiple alleles.** In this particular system, the allele A results in the production of a group A antigen, and allele B produces a group B antigen. These two alleles are co-dominant, so that the heterozygote AB produces both A and B antigens. Either of these alleles is dominant to a third allele o, which is recessive and produces no antigens.

Linkage and Mapping

The cytological basis for Mendel's law of independent assortment is that different pairs of chromosomes assort independently during meiosis (Item 2.6). Genes that are located on different pairs of chromosomes will therefore assort independently, too. The similar behavior of genes and chromosomes was cited by Sutton and Morgan as evidence for their belief that genes were located on chromosomes (Chapter 6).

When genes do occur on the same chromosome, they are said to be **linked.** If linkage were complete and unbreakable, then the cross shown in Fig. 2.7 would have only the two parental types represented among the F_2. The fact that the new combinations, or recombinant genotypes, *do* occur can be attributed to the phenomenon known as **crossing over.** In a certain percentage of cases, the homologous chromosomes break and recombine during meiosis, as shown in Fig. 2.8.

The frequency of crossing over between two genes is greater if the genes are farther apart. The relationship is not strictly linear, because certain points along a chromosome are more apt to break and form crossovers, and the presence of a chromosomal inversion (Chapter 7) or of the centromere inhibits crossing over. Within these limitations, the frequency of crossing over may be used to determine the approximate "distance" between genes, and to arrange them in a linear sequence; this process is known as "mapping," or more properly as "linkage mapping." The result of such mapping may be represented as a linkage map, such as that shown in Fig. 2.9.

Sex-Linked Inheritance

The two most frequent types of sex determination in diploid organisms are the so-called XX-XY and XX-XO types. In both these types of sex determination, the normal females have two X chromosomes, and the male has only one. In males of the XO type, the X chromosome is simply unpaired (the "O" was originally a zero). In nor-

Fig. 2.7 A cross involving two linked genes in corn (maize, *Zea mays*). From *Principles of Genetics*, 5e, by Sinnott, Dunn, and Dobzhansky. Copyright © 1958, McGraw-Hill Book Co. Used with permission of McGraw-Hill Book Company.

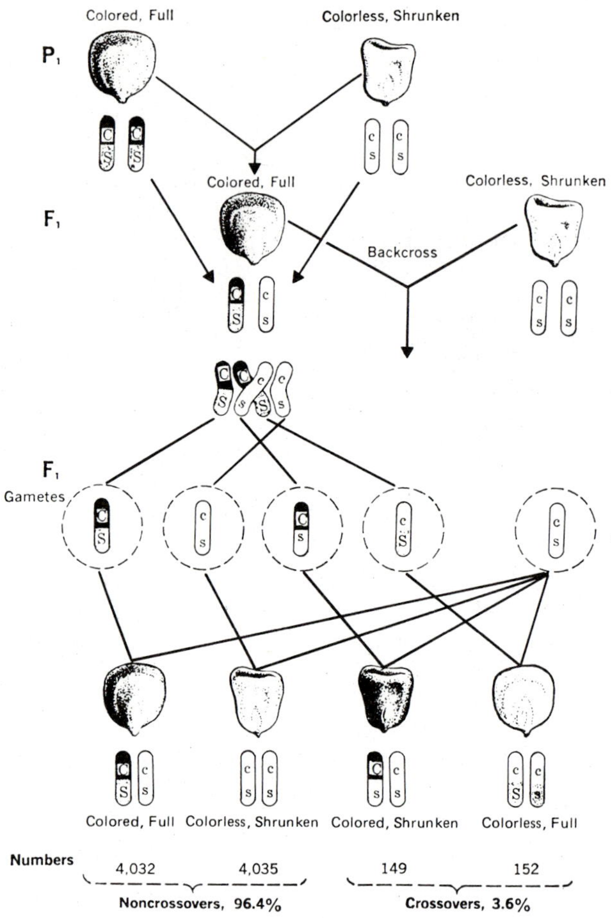

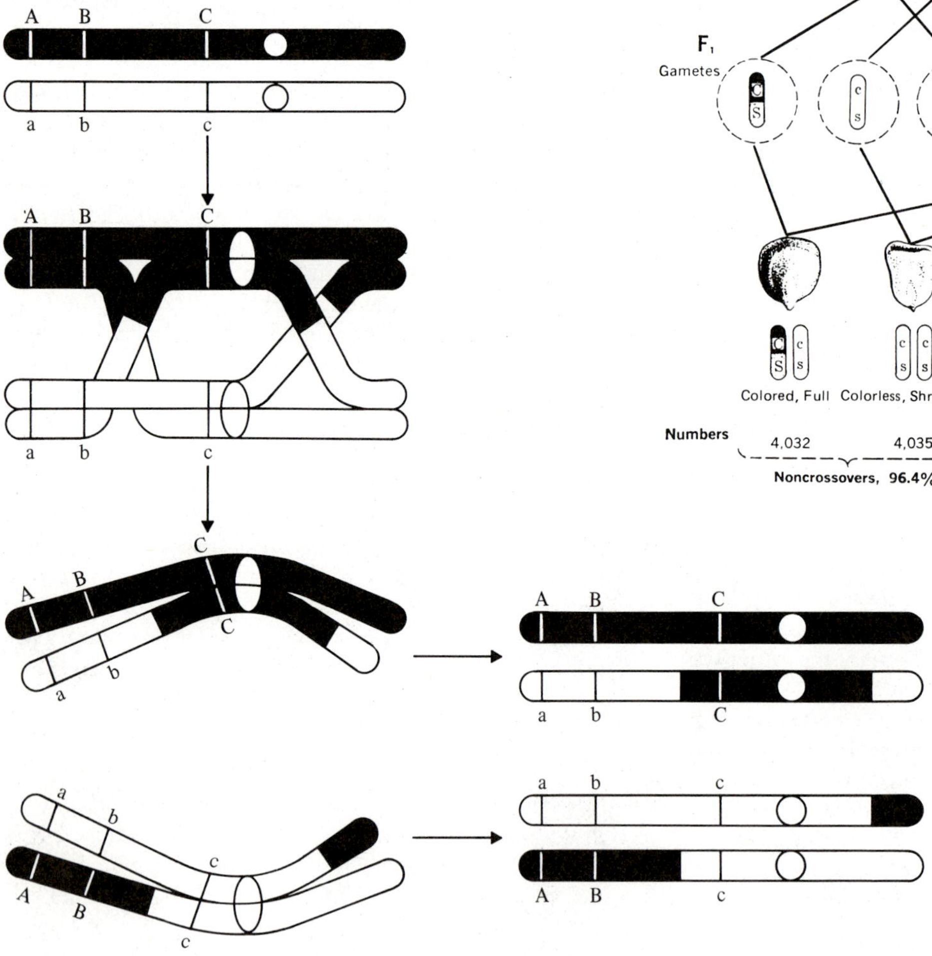

Fig. 2.8 Relation between genetic recombination and crossing over during meiosis, shown diagrammatically. Recombination between B and C results from a crossover between them, but no such recombination occurs between A and B.

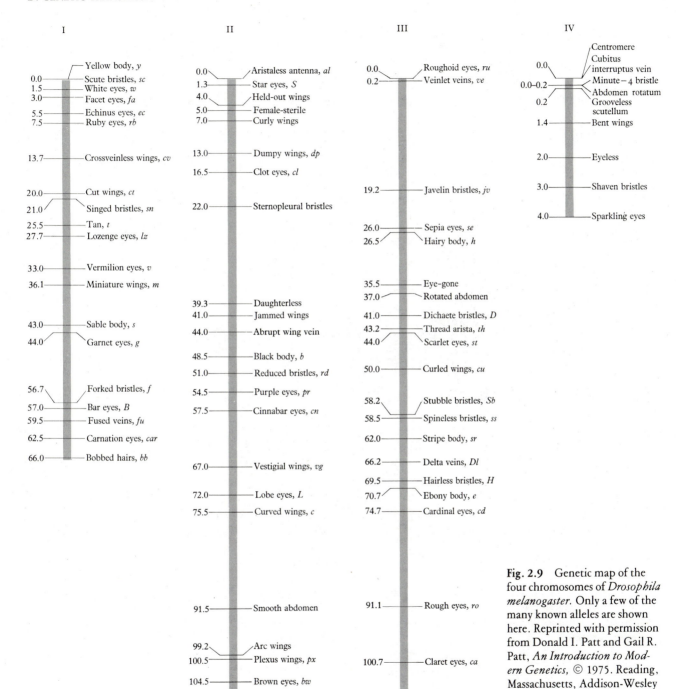

Fig. 2.9 Genetic map of the four chromosomes of *Drosophila melanogaster*. Only a few of the many known alleles are shown here. Reprinted with permission from Donald I. Patt and Gail R. Patt, *An Introduction to Modern Genetics,* © 1975. Reading, Massachusetts, Addison-Wesley (Fig. 4.3).

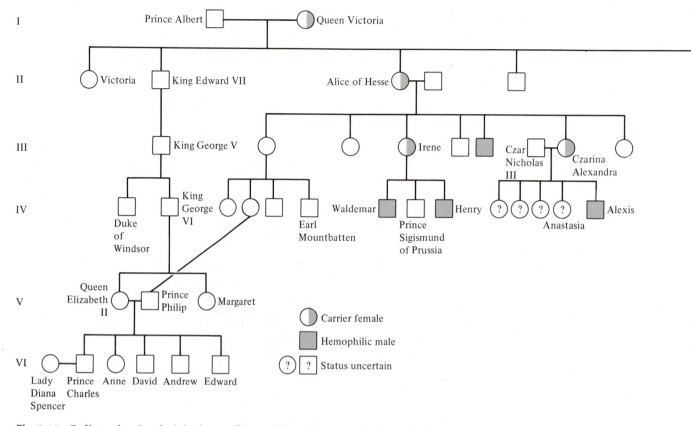

Fig. 2.10 Pedigree showing the inheritance of hemophilia, a human sex-linked trait, among the descendants of Queen Victoria. Many of the affected males died before reaching reproductive age.

mal males of the XY type, the X chromosome pairs with a chromosome known as the Y chromosome; this Y chromosome has very few genes.

Most of the genes located on the X chromosome follow the sex-linked pattern of inheritance. The Y chromosome is largely "blank" and may be represented in crosses by the letter Y. The X and Y chromosomes are known as sex chromosomes; all the other chromosomes are called **autosomes.**

Males are functionally haploid, or **hemizygous,** for genes located on the X chromosome; these genes are phenotypically expressed in males whether they are dominant or not. Females possess two X chromosomes and can thus express recessive traits phenotypically only when they are homozygous. In the case of sex-linked in-

heritance shown in Item 2.7, males are white-eyed if they possess a single w gene, but females are white-eyed only if they are homozygous, ww.

Among human traits, both colorblindness and hemophilia ("bleeder's disease," resulting from a lack of clotting ability) are inherited as sex-linked recessive traits (Fig. 2.10). Males can exhibit these traits by possessing a single gene in each case. These genes (and the entire X chromosome as well) are acquired by a male from his *mother* and are passed on to his *daughter*. Heterozygous females are called **carriers** of these traits, because they can pass them on to their sons or grandsons, yet the females are not themselves affected. A female can exhibit a trait such as colorblindness only in the rare event that her mother had been a carrier and her father colorblind.

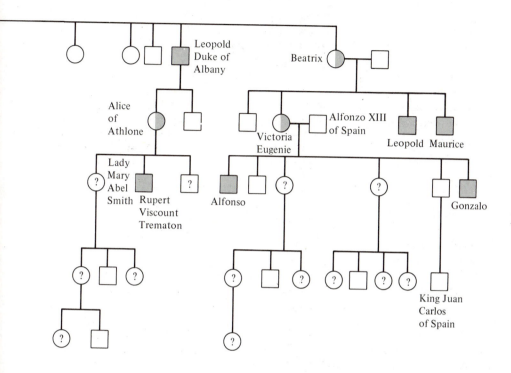

Adapted with permission from Donald I. Patt and Gail R. Patt, *An Introduction to Modern Genetics*. Reading, Massachusetts: Addison-Wesley (Fig. 5.13).

CHAPTER SUMMARY

Cells containing paired sets of chromosomes are called diploid; those containing only a single set of unpaired chromosomes are called haploid. In mitosis, the normal form of cell division, the chromosomes are each duplicated and then distributed into two daughter cells in such a way that the original chromosome number is preserved. In meiosis, the chromosomes are first duplicated and then distributed among four daughter cells in such a way that each daughter cell receives a 50% sample of all the genes while the chromosome number is halved.

Deoxyribonucleic acid (DNA) contains two helically twisted strands of alternating sugars and phosphates in the shape of a double helix. Attached to these ''back-

bones'' are the nitrogen bases adenine, guanine, cytosine, and thymine. Since adenine can pair only with thymine, and guanine only with cytosine, the sequence of bases on one DNA strand can be used to synthesize a new, complementary strand during the process known as replication. Ribonucleic acid (RNA) is usually single-stranded and contains uracil in place of thymine. Messenger RNA molecules are linear; their synthesis, directed by DNA, is known as transcription. Transfer RNA molecules serve as adaptors, inserting the proper amino acid in response to a codon, a sequence of three nucleotides in a messenger RNA molecule. The selection of a particular amino acid in response to a codon is known as translation. A linear sequence of codons is translated as a linear sequence of amino acids, known as a polypeptide. Proteins contain one or more such polypeptides, and many of these proteins function as enzymes (organic catalysts).

Mendel's first law, the law of segregation, simply states that inheritance is particulate and that genes do not blend; a recessive allele hidden in a heterozygote can separate out during gamete formation and become expressed in heterozygous condition during succeeding generations. Mendel's second law, the law of independent assortment, states that traits carried on different chromosomes will assort independently, and that a diploid organism heterozygous for both traits produces gametes with all four combinations of these two traits in equal numbers.

The inheritance of continuously variable traits can often be explained on the assumption of polygenic inheritance, in which several to many interacting genes each have small but additive effects.

Linked genes may be detected by the fact that they fail to assort independently. The frequency of crossing over, or recombination between two genes, may be used as a measure of the distance between those genes on a genetic map of the chromosome.

Sex-linked inheritance involves in most cases males that are functionally haploid and females that are functionally diploid for such traits. Sex-linked, recessive traits are in these cases fully expressed in the male sex because there are no dominant alleles present to mask them.

Students unfamiliar with the material of this chapter are particularly urged to reread the text and boxes several times, and to try to work out as many of the problems as they can.

PROBLEMS

Genetics as a subject is best learned by working out problems such as these. Students are strongly advised to attempt these problems on their own, and to consult with their instructors if they cannot solve them.

1. In cattle, the "polled," or hornless, condition is dominant over the horned. A certain prize bull mated to three cows produces the following offspring:
   a) with cow A (horned): a polled calf.
   b) with cow B (horned): a horned calf.
   c) with cow C (polled): a horned calf.
   What are the genotypes of all four parents? What further

offspring might be expected from each of the three matings shown?

2. In the fruit fly, *Drosophila,* vestigial wings (v) are recessive to the normal or "wild type" (V). What type of offspring would you expect from each of the following crosses?
   a) $Vv \times vv$
   b) $VV \times Vv$
   c) $Vv \times Vv$

3. In snapdragons, normal flower shape is dominant over the peloric, or "snap," shape. Flower color is determined by a pair of alleles on a different chromosome: the heterozygotes are pink, and the homozygotes are either red or white. Work out a cross between a white, peloric flower and a red flower homozygous for normal shape. Show genotypic and phenotypic ratios for both F_1 and F_2.

4. In the snapdragons of the previous problem, an F_1 plant is back-crossed with each of its parents. Show what types of offspring would be expected from a back-cross with each of the following:
   a) the white, peloric parent
   b) the red, normal parent

5. In the fruit fly, *Drosophila,* white eyes are caused by a sex-linked recessive allele, and vestigial wings are caused by an autosomal recessive. A white-eyed, vestigial-winged male is crossed with a female homozygous for both normal wings and red eyes.
   a) Show both genotypic and phenotypic ratios for F_1 and F_2.
   b) Why was it unnecessary to state above that the two genes are on different chromosomes?
   c) Suppose the cross had been between a white-eyed male homozygous for normal wings and a vestigial-winged female homozygous for red eyes. Would the results be the same?
   d) Suppose the cross had been between a white-eyed, vestigial-winged female and a male homozygous for both dominant traits. Would the results be the same? Explain.

6. In humans, an Rh-incompatibility exists if an Rh-negative mother has an Rh-positive child. Assume that this result is controlled by a single gene (actually, several are known), and that the Rh-positive type is dominant.
   a) Show genotypes and phenotypes of mother, father, and child in a case of Rh-incompatibility.
   b) If the child of the cross above marries someone who is Rh-negative, should they expect any Rh-incompatability problems? In what percentage of offspring? Does it matter which sex has which genotype?

7. In some breeds of cattle, incomplete hairlessness is caused by a recessive allele (h), and the normal condition is dominant (H). Farmer Smith buys a bull from Farmer Jones and mates this bull to several cows, all of them offspring of his prize blue-ribbon bull. Of the 46 calves produced by these matings, 7 are incompletely hairless, and Smith accuses Jones of having sold him the bull that was responsible. Jones claims that his bull was only partly responsible, but Smith replies that in all his 40 years as a cattle breeder he has never seen the incompletely hairless condition in any of his animals.
   a) Who is right?
   b) What crosses (matings) would you propose to test your hypothesis? What results would you expect?
   c) Of the 39 calves that are normal in appearance, how many are heterozygous?

8. A woman who has just given birth accuses a certain man of being the father of her baby, and she sues for child support.
   a) If the woman has blood type AB, the baby has blood type A, and the accused man has blood type B, what would you say? *Could* he be the father? *Must* he be the father?
   b) The woman's ex-husband has blood type O. Might he be the father?

9. Two babies were born in the same hospital during a bomb scare. In the confusion, it was feared that they might have been interchanged. Mr. and Mrs. Green had blood types AB and B, respectively, and Mr. and Mrs. Gray both had blood type A. The first baby tested had blood type A, and the second baby had type O. Who belongs to whom?

10. In mice, one type of yellow coloration seems to be dominant over black, but in a curious fashion: the black mice breed true, as expected, and crosses of yellow mice with black mice yield a 1:1 ratio of colors, as if yellow were heterozygous. No pure-breeding yellow mice are ever produced, and a cross between two yellow mice always produces a 2:1 ratio of yellow to black, with a slightly reduced litter size. Discuss the possible reasons for this.

11. Work out a cross between a hemophilic man of blood group O and his wife, a carrier for hemophilia, belonging to blood group AB. Their daughter Sally does not have hemophilia and belongs to blood group B. Sally marries a man who has blood type A and whose father has blood type O. The man's family has no history of hemophilia. What offspring might they expect to have?

12. In fruit flies, the gene for wing shape (V, normal; v, vestigial) and the gene for body color (b, black; B, gray) are located on the same chromosome, 18 crossover units apart (i.e., the frequency of crossovers is 18%). A fly of genotype $BBvv$ is mated to $bbVV$, and the F_1 are crossed to a doubly recessive fly (test cross). What proportions would you expect in the F_2? Explain.

13. In *Drosophila*, the following genes are all located on chromosome III:
   hairy body (h), at crossover position 26.5
   scarlet eyes (st), at position 44.0
   ebony body (e), at position 70.7.
   Assume that wild-type flies are homozygous for all three dominant alleles, and that a triple-mutant strain is available that is homozygous for all three recessive mutants. Work out a cross to verify the crossover position of these three loci. What F_2 proportions do you expect?

14. In maize (corn), a colored kernel (C) is dominant to a colorless kernel (c), and a fully inflated kernel (S) is dominant to a shrunken one (s). The two loci are on the same chromosome, with a 4% frequency of recombination between them. Show in full a cross between a plant homozygous for both dominant alleles and a plant homozygous for both recessive alleles. Perform a test cross by back-crossing the F_1 plants to their doubly recessive parent. Show all genotypes, phenotypes, and expected frequencies in the F_2 generation.

FOR FURTHER KNOWLEDGE

ITEM 2.1 *The Amino Acids of Proteins*

The characteristic amino group (NH_2) and carboxyl group ($-COOH$) are shown in their ionized form.

Reprinted with permission from Donald I. Patt and Gail R. Patt, *An Introduction to Modern Genetics*, © 1975. Reading, Massachusetts: Addison-Wesley (Fig. 8.5).

(continued)

ITEM 2.1   *(continued)*

Glycine (gly)	L-Alanine (ala)	L-Valine (val)	L-Isoleucine (ile)

L-Leucine (leu)	L-Serine (ser)	L-Threonine (thr)	L-Proline (pro)

L-Aspartic acid (asp)	L-Glutamic acid (glu)	L-Lysine (lys)	L-Arginine (arg)

L-Asparagine (asn)	L-Glutamine (gln)	L-Cysteine (cys)	L-Methionine (met)

L-Tryptophan (try)	L-Phenylalanine (phe)	L-Tyrosine (tyr)	L-Histidine (his)

ITEM 2.2   *The Structure of Protein Molecules*

Primary structure The primary structure of protein molecules is simply the sequence of amino acids. These may be represented as a linear chain with a zig-zag backbone (A). This repre-sents the actual sequence of the first six amino acid units in the α-chain of human hemoglobin. The same sequence can also be represented this way:

$$Val-Leu-Ser-Pro-Ala-Asp-$$

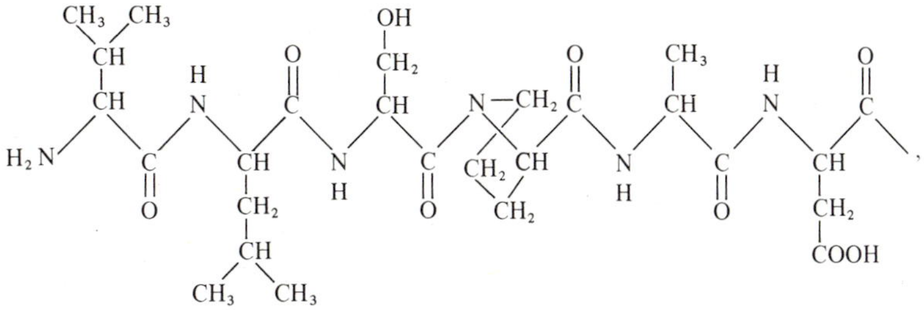

(A)

Secondary structure Except for silk fibroin, few proteins have the linear structure shown in (A). Most fibrous proteins, like keratin, have their sequence twisted into the form of a so-called alpha helix, held together by weak molecular forces (hydrogen bonds) between amino acid units spaced exactly one turn away from each other along the helix. One turn of the alpha helix corresponds to slightly less than three amino acid units. Only the backbone of the alpha-helical spiral is shown in (B), together with a few of the hydrogen bonds (dotted vertical lines) that hold the molecule in this helical configuration.

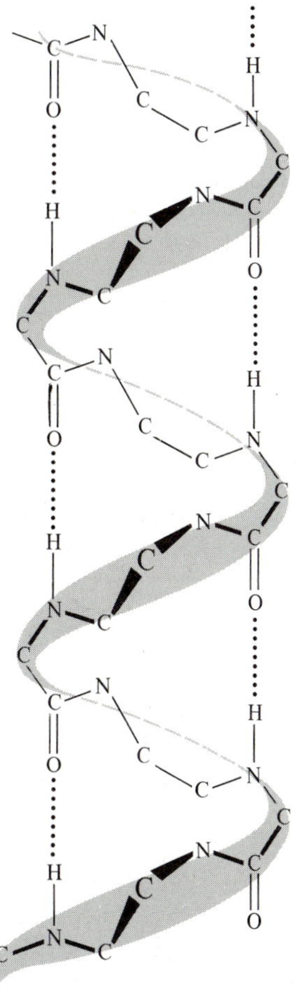

(B)

(continued)

ITEM 2.2 *(continued)*

Tertiary structure Although fibrous proteins have very long alpha helices, many proteins are further twisted into more compact, **globular** shapes. There are many factors that determine the exact shape of protein molecules, their *tertiary structure* in space. Perhaps the most important factor is the combining of cysteine units to one another (C), forming the amino acid

cystine (note the change in spelling), and holding nonadjacent portions of the molecule close together by disulfide linkages (S—S).

Two or more different polypeptides may be united by these disulfide linkages into a single protein, or a single polypeptide may be bent back on itself and held in shape by these linkages. The three-dimensional (tertiary) structure of the myoglobin molecule is shown in (D).

Sulfhydryl group

Two cysteine residues
(in polypeptide chains)

(C)

Loss of two hydrogens
(dehydrogenation)

Disulfide bridge

Residue of cystine, a
double-headed amino acid.
The two polypeptide chains
are now connected by a
disulfide bridge.

Quaternary structure Some proteins have yet a fourth level of structure, in which two or more molecules must combine with one another in order to function properly. This fourth level of structure is known as quaternary structure.

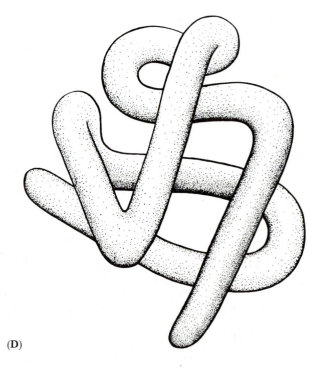

(D)

ITEM 2.3 *The Genetic Code for Translation of mRNA Codons into Amino Acid Sequences*

The three chain-terminating codons are indicated by ''term.''
See Item 2.1 for amino acid abbreviations.

First base	Second base								Third base
	U		C		A		G		
U	U-U-U	Phe	U-C-U	Ser	U-A-U	Tyr	U-G-U	Cys	U
	U-U-C	Phe	U-C-C	Ser	U-A-C	Tyr	U-G-C	Cys	C
	U-U-A	Leu	U-C-A	Ser	U-A-A	term.	U-G-A	term.	A
	U-U-G	Leu	U-C-G	Ser	U-A-G	term.	U-G-G	Trp	G
C	C-U-U	Leu	C-C-U	Pro	C-A-U	His	C-G-U	Arg	U
	C-U-C	Leu	C-C-C	Pro	C-A-C	His	C-G-C	Arg	C
	C-U-A	Leu	C-C-A	Pro	C-A-A	Gln	C-G-A	Arg	A
	C-U-G	Leu	C-C-G	Pro	C-A-G	Gln	C-G-G	Arg	G
A	A-U-U	Ile	A-C-U	Thr	A-A-U	Asn	A-G-U	Ser	U
	A-U-C	Ile	A-C-C	Thr	A-A-C	Asn	A-G-C	Ser	C
	A-U-A	Ile	A-C-A	Thr	A-A-A	Lys	A-G-A	Arg	A
	A-U-G	Met	A-C-G	Thr	A-A-G	Lys	A-G-G	Arg	G
G	G-U-U	Val	G-C-U	Ala	G-A-U	Asp	G-G-U	Gly	U
	G-U-C	Val	G-C-C	Ala	G-A-C	Asp	G-G-C	Gly	C
	G-U-A	Val	G-C-A	Ala	G-A-A	Glu	G-G-A	Gly	A
	G-U-G	Val	G-C-G	Ala	G-A-G	Glu	G-G-G	Gly	G

Modified from Donald I. Patt and Gail R. Patt, *An Introduction to Modern Genetics,* © 1975. Reading, Massachusetts: Addison-Wesley (Fig. 8.3).

ITEM 2.4 *A Simple Case of Polygenic Inheritance*

The length of an ear of corn is undoubtedly controlled by numerous genes, as well as by a number of environmental (including nutritional) factors. For the sake of simplicity, let us ignore environmental factors, and let us assume that ear length is controlled by three genes. Let us further assume that genotype *aabbcc* has some minimum length, say 18 cm. For each capital letter in the genotype, we add another 2 cm, so that

aaBbcc is 20 cm long. These effects are strictly additive, so that *AaBbcc* is 22 cm long, and *AABBCC* is 30 cm long. There is no dominance, meaning that *Aa* has a phenotypic effect (+2 cm) exactly intermediate between that of *aa* (+0 cm) and *AA* (+4 cm).

Note that in the cross (A), an approximately normal distribution (actually a binomial distribution) is achieved in only two generations. More intricate models, allowing for dominance, etc., also produce distributions that are approximately normal.

ITEM 2.4   *(continued)*

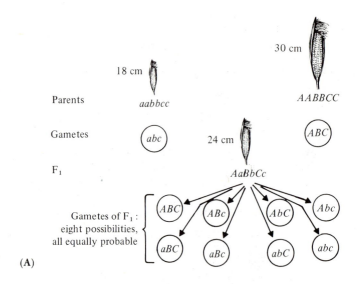

(A)

	ABC	*ABc*	*AbC*	*aBC*	*Abc*	*aBc*	*abC*	*abc*
ABC	*AABBCC* 30 cm	*AABBCc* 28 cm	*AABbCC* 28 cm	*AaBBCC* 28 cm	*AABbCc* 26 cm	*AaBBCc* 26 cm	*AaBbCC* 26 cm	*AaBbCc* 24 cm
ABc	*AABBCc* 28 cm	*AABBcc* 26 cm	*AABbCc* 26 cm	*AaBBCc* 26 cm	*AABbcc* 24 cm	*AaBBcc* 24 cm	*AaBbCc* 24 cm	*AaBbcc* 22 cm
AbC	*AABbCC* 28 cm	*AABbCC* 26 cm	*AAbbCC* 26 cm	*AaBbCC* 26 cm	*AAbbCc* 24 cm	*AaBbCc* 24 cm	*AabbCC* 24 cm	*AabbCc* 22 cm
aBC	*AaBBCC* 28 cm	*AaBBCc* 26 cm	*AaBbCC* 26 cm	*aaBBCC* 26 cm	*AaBbCc* 24 cm	*aaBBCc* 24 cm	*aaBbCC* 24 cm	*aaBbCc* 22 cm
Abc	*AABbCc* 26 cm	*AABbcc* 24 cm	*AAbbCc* 24 cm	*AaBbCc* 24 cm	*AAbbcc* 22 cm	*AaBbcc* 22 cm	*AabbCc* 22 cm	*Aabbcc* 20 cm
aBc	*AaBBCc* 26 cm	*AaBBcc* 24 cm	*AaBbCc* 24 cm	*aaBBCc* 24 cm	*AaBbcc* 22 cm	*aaBBcc* 22 cm	*aaBbCc* 22 cm	*aaBbcc* 20 cm
abC	*AaBbCC* 26 cm	*AaBbCc* 24 cm	*AabbCC* 24 cm	*aaBbCC* 24 cm	*AabbCc* 22 cm	*aaBbCc* 22 cm	*aabbCC* 22 cm	*aabbCc* 20 cm
abc	*AaBbCc* 24 cm	*AaBbcc* 22 cm	*AabbCc* 22 cm	*aaBbCc* 22 cm	*Aabbcc* 20 cm	*aaBbcc* 20 cm	*aabbCc* 20 cm	*aabbcc* 18 cm

ITEM 2.4   *(continued)*

If there is environmental variance (caused by nutrition, soil type, etc.), it will tend to "smooth out" the corners of the histogram, so that it more closely approximates the "normal curve" shown in (B) as a dotted line. An increasing number of genes (more than three) also has the effect of producing a histogram that more closely resembles the normal distribution.

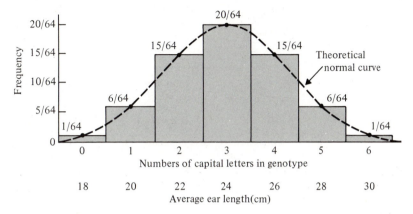

(B) F_2 represented as a histogram.

ITEM 2.5   *The A,B,O Blood Group System in Humans*

The letters A and B represent two protein substances (*antigens*) found in blood cells; O represents the absence of either antigen. Individuals can produce antibodies to protein substances that their body tissues recognize as foreign; thus a person normally has such antibodies to all antigens *except* those produced by his or her own body.

Blood Type	A	B	AB	O
Corpuscular antigens	A	B	A and B	neither
Serum antibodies	anti-B	anti-A	neither	anti-A and anti-B

(continued)

ITEM 2.5   *(continued)*

Blood Type	A	B	AB	O
Reaction to anti-A serum	agglutination	no reaction	agglutination	no reaction
Reaction to anti-B serum	no reaction	agglutination	agglutination	no reaction
Can *give* blood transfusion to	A, AB	B, AB	AB	A, B, AB, O ("universal donor")
Can *receive* blood transfusion from	A, O	B, O	A, B, AB, O ("universal recipient")	O
Possible genotypes	AA, Ao	BB, Bo	AB	oo

ITEM 2.6 *The Cytological Basis for Mendel's Laws*

These diagrams are schematic only, but they are intended to
represent meiosis in pea plants, with a diploid chromosome
number of $2N = 14$.

CYTOLOGICAL BASIS FOR SEGREGATION

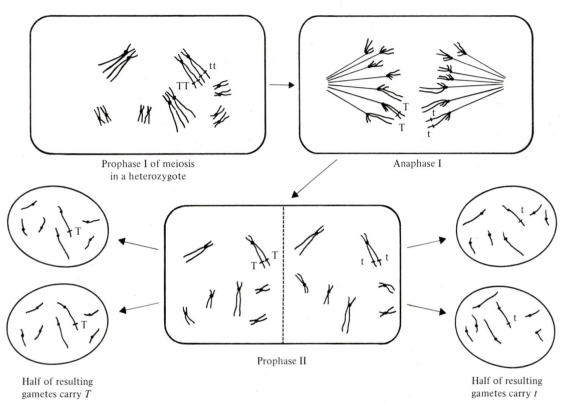

Prophase I of meiosis
in a heterozygote

Anaphase I

Prophase II

Half of resulting
gametes carry *T*

Half of resulting
gametes carry *t*

ITEM 2.6 *(continued)*

CYTOLOGICAL BASIS FOR INDEPENDENT ASSORTMENT

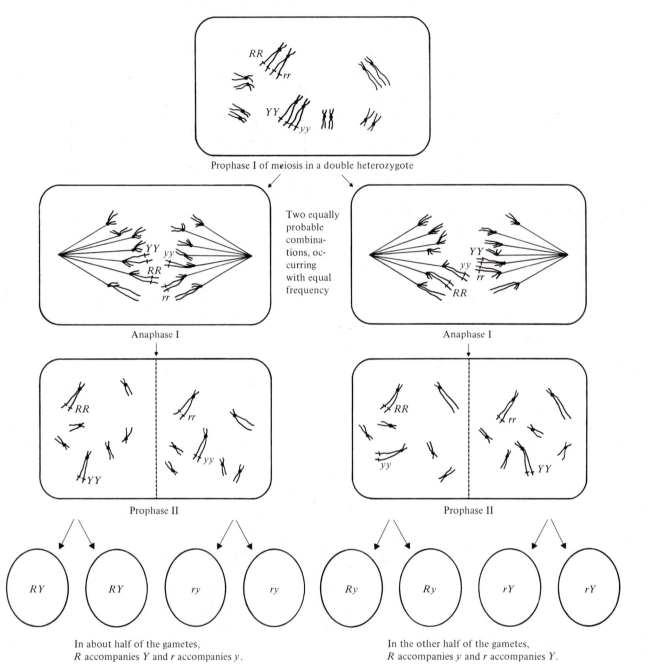

Prophase I of meiosis in a double heterozygote

Two equally probable combinations, occurring with equal frequency

Anaphase I   Anaphase I

Prophase II   Prophase II

In about half of the gametes,
R accompanies *Y* and *r* accompanies *y*.

In the other half of the gametes,
R accompanies *y* and *r* accompanies *Y*.

ITEM 2.7 *Sex-Linked Inheritance in Fruit Flies*

The allele for red eyes (*W*) and the allele for white eyes (*w*) are carried on the X chromosome. The hook-shaped Y chromosome carries neither allele, allowing complete expression of whatever genes are present on the X chromosome. Note that the cross of a white-eyed male with a red-eyed female produces results very different from that of the reciprocal cross (red-eyed male × white-eyed female).

W = gene for red eyes (wild type), on X chromosome
w = gene for white eyes, on X chromosome
Y = Y chromosome (hook-shaped)

Cross of white-eyed male × red-eyed female:

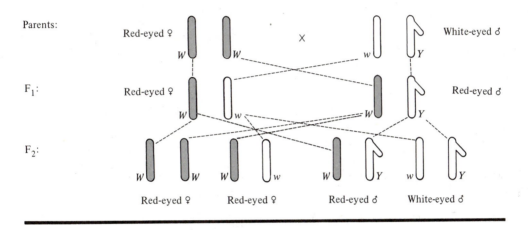

Reciprocal cross (white-eyed female × red-eyed male):

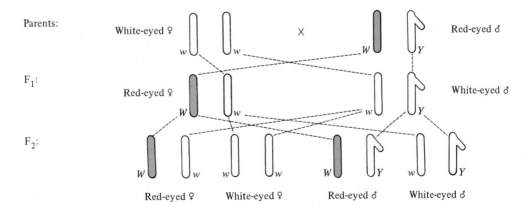

3

Biology Before Darwin

"In science, as in all other departments of inquiry, no thorough grasp of a subject can be gained, unless the history of its development is clearly appreciated." Archibald Geike's remark, with which he begins his own book *The Founders of Geology,* pertains at least as much to our subject as to his. Yet it is amazing that the history of evolutionary biology has so often been neglected or distorted. The contributions of Darwin's predecessors have been slighted by some writers and uncritically magnified by others. Darwin himself has been frequently misunderstood; the mere belief in evolution is often incorrectly cited as his greatest achievement, a distinction that should properly go to his theory of natural selection. All evolutionary theories, from Darwin's to the present day, are permeated with discussions of numerous earlier theories. Evolutionary biology has a much longer and richer history than most people realize.

A. THE ANCIENT MATERIALISTS

The first theories of evolution came from a materialist tradition that began in the Greek colonies of Asia Minor over 2000 years before Darwin's time. This tradition originated with Thales, the reputed founder of Greek philosophy. His disciple, Anaximander, was called the earliest evolutionist of all time by Osborn (1894). Anaximander taught that life arose in the water, that simpler forms of life preceded more complex forms, and that humans originated from fish that forsook the seas and came up to live on dry land. The theory that simpler forms of life preceded the more complex forms was repeated by Empedocles, Democritos, and Lucretius. To this theory, Empedocles added the theory that all matter was composed of the four elements earth, air, fire, and water. Two great forces, love (or attraction) and hate (or repulsion), acted upon these four elements; when love predominated, the elements would come together to form material things. Animals were first formed as unattached arms, legs, eyes, "heads without bodies, and limbless trunks," brought together in random combinations under the force of attraction. Many became grotesque monsters with extra heads or none, too few limbs, no eyes, or parts in the wrong places, incapable of proper functioning. Only by chance and after many repeated attempts did the more harmonious combinations arise.

The monstrous or disharmonious combinations perished from their own inability to carry out life functions. Democritos repeated the teachings of Empedocles, to which he added the theory of atoms. He taught that all matter was composed of rapidly moving atoms, which could be neither created nor destroyed. The creation and destruction of material things were nothing more than the coming together and disassembly of their constituent atoms.

The fourth century B.C. saw the eclipse of the materialist tradition, but Epicurus revived the tradition, particularly the teachings of Democritos. The Roman poet Lucretius finally codified and wrote down all the aforementioned theories in the form of a long poem, *De Rerum Natura* (''on the nature of things''). Here are contained the ideas of Empedocles, Democritos, and Epicurus in their most fully developed form. The atomic theory is fully explained, along with the theory that nothing is ever created out of nothing. Plants appeared on the Earth before animals, and humans appeared last of all. From bodies formed by the random combination of parts, only the harmonious or well-adapted forms survived (Box 3–A). Centaurs, if they ever existed, would long since have died out, for the human teeth could never chew enough grass to nourish the horselike body and limbs. Lucretius thus had grasped the essential concepts of adaptation and natural selection during the first century B.C.

B. THE CLASSICAL TRADITION

The classical tradition, which culminated in the philosophical writings of Plato and Aristotle and in the subsequent biological writings of Linnaeus, was the most productive paradigm for biological thought for more than 2000 years. As a series of beliefs to which many authors contributed, it formed a complex tradition rather than just a single theory. The classical tradition emphasized

---

BOX 3-A *A Very Early Theory of Natural Selection (circa 55 B.C.)*

''In those days the earth attempted also to produce a host of monsters, grotesque in build and aspect—hermaphrodites, halfway between the sexes yet cut off from either, creatures bereft of feet or dispossessed of hands, dumb, mouthless brutes, or eyeless and blind, or disabled by the adhesion of their limbs to the trunk, so that they could neither do anything nor go anywhere nor keep out of harm's way nor take what they needed. These and other such monstrous and misshapen births were created. But all in vain. Nature debarred them from increase. They could not gain the coveted flower of maturity nor procure food nor be coupled by the arts of Venus. *For it is evident that many contributory factors are essential to the reproduction of a species. First, it must have a food supply. Then it must have some channel by which the procreative seeds can travel outward through the body when the limbs are relaxed. Then, in order that male and female may couple, they must have some means of interchanging their mutual delight.*

''In those days, again, many species must have died out altogether and failed to reproduce their kind. Every species that you now see drawing the breath of life has been protected and preserved from the beginning of the world either by cunning or by prowess or by speed. In addition, there are many that survive under human protection because their usefulness has commended them to our care. The surly breed of lions, for instance, in their native ferocity have been preserved by prowess, the fox by cunning and the stag by flight. The dog, whose loyal heart is alert even in sleep, all beasts of burden of whatever breed, fleecy sheep and horned cattle, over all these . . . man has established his protectorate. They have gladly escaped from predatory beasts and sought peace and the lavish meals, procured by no effort of theirs, with which we recompensate their service. *But those that were gifted with none of these natural assets, unable either to live on their own resources or to make any contribution to human welfare, in return for which we might let their race feed in safety under our guardianship—all these, trapped in the toils of their own destiny, were fair game and an easy prey for others, till nature brought their race to extinction.*''

From Lucretius, *De Rerum Natura* (''On the nature of things'') Book V, lines 837–880 (emphasis supplied). Translation copyright © R. E. Latham, 1951. Reprinted by permission of Penguin Books, Ltd.

abstract forms, whereas the earlier tradition had emphasized material things. The classical tradition emphasized human affairs, whereas the materialist tradition had emphasized nature. And although the materialist tradition had flourished largely in Italy and Asia Minor, on the colonial fringes of the Greek world, the classical tradition was for its first few hundred years characteristically Athenian.

The classical tradition was begun by certain Greek intellectuals who called themselves sophists, from the Greek word *sophos,* meaning "wise." These men cared more for persuasive argument than for truth. Protagoras, their leader, often argued first one side of a question and then the other, showing that either side could be made to appear just. "Man is the measure of all things," he declared. Many Athenians were horrified by this ethical relativism, but they nevertheless sought the services of the sophists for use in the law courts.

Socrates and Plato

Socrates (470?–399 B.C.) and his student Plato (427?–347 B.C.) were impressed by the persuasive abilities of the sophists but appalled by their lack of an ethical standard. Their attempt to develop standards of ethics and of correct logic led to the development of ideas that have become the cornerstones of Western philosophy. Their main concerns were logical, ethical, and political, but many of their ideas have had a profound and lasting influence on biology for well over 2000 years.

Plato's greatest influence was exerted through his **Theory of Forms,** which has also been called the Theory of Ideas. To Plato, the imperfect and changeable world of everyday experience was only a poor imitation of a more real, permanent, and perfect world: the world of Forms. The beauty of a woman or a flower may fade with age, but beauty itself is eternal, since it belongs to the world of Forms. Today my arguments may be just, whereas tomorrow they may be unjust, but justice itself belongs to the world of Forms and is permanent. The triangle I draw has sides that are not perfectly straight or perfectly thin, but the ideal triangle has three sides that are perfectly straight lines of zero thickness.

In its treatment of living or nonliving objects, the theory of Forms dismisses idiosyncratic variations as unimportant or illusory. Your chair may differ from mine in color, shape, material, presence or absence of arms, etc.,

because each is but an imperfect replica of an eternal Form, or **eidos,** the ideal or heavenly chair, which exists in the world of Forms. This *eidos* of a chair embodies all of the features an object must possess in order to be a chair, and all material chairs partake of this *eidos* to a greater or lesser degree. The permanence of the *eidos* might be illustrated as follows: if all the chairs in the material world were destroyed, the *eidos* would yet remain, and it would be possible to build a new chair by copying this heavenly form.

Aristotle

Aristotle (384–322 B.C.) was Plato's most prominent student and also the foremost biologist of classical antiquity. His physiological ideas, though imperfect, were not supplanted by any more modern ideas until Harvey's work in the seventeenth century. Aristotle's descriptions of animals, especially of marine species, were far better than any others until the eighteenth or nineteenth century. But Aristotle also made some mistakes: he thought, for example, that the brain was an organ for cooling the blood, though Democritos had earlier advanced the more correct notion of the brain as the seat of thought.

Aristotle also described the embryonic development of a number of species, pointing out that the embryonic forms were very different from the mature organisms that followed them: the acorn, for instance, was nothing like a miniature oak tree, for it contained no leaves, roots, or bark. This kind of embryological development, from early stages bearing little resemblance to the adult, was later called the theory of **epigenesis.**

Aside from these descriptive studies, Aristotle's influences on biology were threefold: he applied Plato's theory of Forms to the living world; he originated the concepts of genus and species; and he introduced the *scala naturae* into biology, as described later in this chapter.

The earliest concepts of genus and species, which form the basis for all classification, are Aristotle's. However, the two concepts were not always applied to the same categories. Aristotle would have said that Lassie was a species of the genus *collie,* that collies were a species of the genus *dog,* and that dogs were a species of the genus *animal.* The concept of a *species* comes directly from Plato's theory of Forms; the word "species" is often used to translate the Greek word *eidos,* or form.

Plato's theory of Forms, applied by Aristotle to the living world, became decidedly antievolutionary, and it is sometimes known as the theory of Types or of essentialism. My cat has four legs, two eyes, and a tail, and he purrs and says "meow." These are the features by which we recognize that he is a cat; they are part of his "essence"; they are essential features that reflect his species, or *eidos*. My cat may also be gray and white, overweight, long-haired, and lazy. These are inessential features: they reflect no *eidos*. They neither detract from nor contribute to cattiness, for other cats with equal claim to the title may be black, slender, short-haired, and frisky. The essential features are important; they conform to an *eidos,* or type; they are indicative of the species. The inessential or "accidental" features are to the Aristotelian unimportant; to the Platonist they are even an illusion or an imperfection.

Both Plato and Aristotle held the vitalistic belief that living organisms differed fundamentally from non-living things in possessing a *psyche,* or soul (without the theological connotations that the latter term has since acquired). Aristotle went beyond this, however, to postulate the existence of three kinds of *psyche.* The lowest and most universal kind is the "vegetative soul," common to humans, animals, and plants; this carries out such functions as growth, nutrition, and reproduction. Plants possess this soul alone, but animals and humans have additionally an "animate soul," which controls bodily movements. Humans alone are distinguished by the further possession of a "rational soul," which controls rational behavior and thought.

The three kinds of *psyche,* as well as the creatures possessing each, are arranged in a hierarchy or scale. Plants, having only a vegetative soul, are the lowest. Animals, having both a vegetative and an animate soul, are next on the scale. Humans, having both these together with the highest type of soul, the rational soul, are of course highest on the scale. Among animals, those with red blood (vertebrates) are higher than those without (invertebrates), and among them the quadrupeds that give birth to live young are closest to humans and therefore highest. This scale of being, which came to be an important unifying concept in biology, was known under many names: chain of being, scale of being, ladder of perfection, échelle des êtres, or *scala naturae* ("scale of nature"). The concept still exerts influence today: whenever we speak of "higher" or "lower" forms of life, we are implicitly referring to this scale. Although some have tried to interpret it with hindsight as an evolutionary concept, the *scala naturae* was to virtually all of its supporters a static hierarchy of varying degrees of "perfection," a decidedly nonevolutionary concept.

The Great Chain of Being

For a long time, organic species were thought to be arranged along a single linear scale, the *scala naturae;* the history of this concept has been recounted by Lovejoy (1936). As a biological concept, it originated with Aristotle, but this biological concept became wedded in early Christian times to an older, purely metaphysical concept: the perfection and ultimate goodness of Creation. The basic argument was that any object capable of existence is more perfect when it exists than when it does not. Thus the ideal or most perfect world contains not only perfection but all degrees of imperfection as well. From this purely metaphysical argument, the early Christians took the chain of being as living proof of God's omnipotence and supreme goodness: any object that has the property of existence is more perfect than an otherwise identical object that lacks existence; therefore God created all degrees of imperfect creatures, down to the lowest, because the omission of any one of them would have detracted from the Creation.

The choice of classical perfection over materialistic flux was largely a matter of theological preference in early Christian times. Platonism first, then Aristotelianism later, proved to be valuable allies of Christian theology. Those components of Aristotelian philosophy that could be reconciled with Christian theology were incorporated into Christian teachings, but the writings of other Greek philosophers were rejected. Philosophy became the handmaiden of theology, and science, which was incubating within the womb of philosophy, was largely neglected.

From medieval times through the eighteenth century, the *scala naturae* continued to receive the support of many great intellectuals, including theologians from Thomas Aquinas to Edmund Law, poets such as John Milton and Alexander Pope, philosophers like John Locke, Gottfried Leibniz, and Immanuel Kant. Biologists who supported the concept, especially in the eighteenth century, included Carl Linnaeus, Georges Buffon, and Charles Bonnett.

As knowledge of animals and plants increased, so did the details of the *scala naturae* become somewhat clearer: humans were followed by other mammals, then birds, then cold-blooded vertebrates, invertebrates, and plants, customarily listed in this order, from highest to lowest. The discoveries of the Renaissance seemed to fill in gaps in the scale, reinforcing the belief that any remaining gaps were only apparent, not real, and that other forms would soon be found to fill them.

The Renaissance and the Revival of Botany

Renaissance humanism brought with it renewed interest in the study of the human form and in the study of nature. The new interest in the human form showed itself in art as well as medicine, and several Renaissance artists taught themselves anatomy by dissecting cadavers illegally. Andreas Vesalius (1514–1564), the greatest of Renaissance anatomists, was also an accomplished artist. The landmark discovery of the circulation of the blood by William Harvey (1578–1658) ushered in a new era in the study of physiology. The mechanistic and overtly reductionist philosophy of Descartes was in part an outgrowth of the new astronomical theories of Copernicus and Galileo and the physics of Newton. This philosophy inspired a new, reductionist tradition in physiology, with emphasis on experimentation and analysis in purely mechanistic terms. This continues to be the dominant philosophy of research in physiology, molecular biology, and biochemistry to this day.

The revival of botany contributed even more to the history of evolutionary biology. This revival grew directly out of the revival of medicine, for botany was considered the handmaiden to pharmacology. Most medicines at the time were plant products, obtained by grinding the leaves or other plant parts with a mortar and pestle. The early botanical texts, known as herbals, had therefore a strong bias: they were devoted primarily to the correct identification of plants having medicinal value.

The bias toward the study of medicinally important plants raised one important problem, whose solution led eventually to the development of **taxonomy**, the science of classification. Related species of plants often differ significantly in their medicinal properties; often a plant of pharmacological importance is very similar in appearance to a closely related plant of no medicinal value

whatever. This single problem at once gave paramount importance to the matter of correct species identification, requiring attention to such details as differences in leaf margins, stem size and shape, manner of branching, or smooth versus hairy surfaces. Details such as these proved very difficult to describe accurately in words alone, especially when matters of sickness and health were at stake. Herbals therefore became illustrated, and the illustrations paid careful attention to details indicative of species differences. Many artists of the period show the same careful attention to detail in their painting and woodcuts (Fig. 3.1).

Gutenberg's introduction of printing by movable type allowed the spread of many new ideas. The new botany and the new Protestant religion both proclaimed that Germany no longer had to follow the traditional examples of Mediterranean culture. Two of the German "fathers of botany," Otto Brunfels (1488?–1534) and Leonard Fuchs (1501–1566), converted to Protestantism. For the next 200 years, progress in biology would be dominated largely by the Protestant countries of Northern Europe.

The sixteenth and seventeenth centuries witnessed much progress in the naming and describing of an increasing number of plants, no longer restricted to those of medicinal importance only. The study of animals made slower progress, meanwhile, and thus lagged a good bit behind. In England, Nehemiah Grew (1641–1712) was the first to recognize that stamens were male flower structures, that pistils were female structures, and that most flowers contained both: they were hermaphroditic and reproduced "like snails." But the greatest classifier of animals and plants of this period was another Englishman, John Ray (1627–1705). Ray was both a naturalist and a clergyman, the first of many Englishmen who combined both pursuits. His numerous perceptive observations added greatly to our understanding of animals, plants, and geology. He was the first to distinguish dicotyledonous and monocotyledonous plants, having two leaflike structures in the seed versus only one. He was also the first to distinguish between those vertebrates whose hearts possessed only a single ventricle (fish, amphibians, reptiles) and those that possessed two ventricles (birds, mammals). His distinctions were based as often on internal anatomy or embryological development as they were on the external characters of adults. One of his

Fig. 3.1 "Large piece of turf," watercolor (1503) by Albrecht Dürer. This kind of attention to biological detail was very important in the growth of both botanical illustration and botanical taxonomy. Reprinted with permission of the Albertina Museum, Vienna, from the Photo Archive of the Austrian National Library.

most progressive groupings brought together all animals possessing hair, breathing with lungs, having a heart with two ventricles, and giving birth to living young—a group that Linnaeus later named Mammalia but that Ray had been the first to recognize. This grouping was especially progressive in including bats and whales along with their four-footed brethren.

Ray anticipated a number of later naturalists in his theoretical discussions as well. His belief in the fundamental reality of the species anticipated Linnaeus. He declared the number of species to be finite, since God had rested on the seventh day of creation, and the species themselves to be immutable. Yet he recognized the possibility of variation within species, especially the "degeneration" of domestic breeds. His definition of species sounds surprisingly modern: plants belong to the same species as long as they arise from the seed of plants like themselves, just as cows and bulls arise from the seed of other cows and bulls and are therefore the same species. He also gave support to the belief in the *scala naturae* by his declaration that distinct species were connected by intermediates since "Natura non facit saltus" (nature does not proceed by jumps). Ray's geological contributions are described in Chapter 4.

Linnaeus: The Height of Classicism

The Swedish botanist **Carl Linnaeus** (1707–1778) is considered to be the father of modern taxonomy (see Item 3.1 on page 59). His major work, the *Systema Naturae,* described all the animals and plants known in his day. Though the "natural system" he proclaimed was innovative in its results, Linnaeus may truly be said to represent the height of the classical tradition, and his major achievement lay in the codification of that tradition. Linnaeus used the Aristotelian terms *genus* and *species,* but he now applied these at definite levels in his hierarchy. He also recognized three **kingdoms** (Animalia, Plantae, Mineralia), and he divided the animal kingdom into six **classes** (Mammalia, Aves, Amphibia, Pisces, Insecta, and Vermes). Each class was divided into **orders,** and each order contained one or more genera. (The *phylum* was added later by Cuvier, the *family* by Michel Adanson and by Buffon.) Linnaeus also established **binomial nomenclature,** the designation of each species by a two-word name, of which the first word was always the name of the genus.

The theoretical justifications that Linnaeus gave to his system were often Aristotelian. Each species was considered immutable because it had been created by God and because it reflected an unchanging, heavenly type. Ray's earlier views on species were largely repeated, including the statement that "Natura non facit saltus."

Linnaeus frequently quoted from the Bible and from Aristotle, as well as from the works of earlier botanists.

Yet Linnaeus's system was also very innovative. The plants, for instance, were no longer classed as trees, shrubs, and herbs but were grouped into orders on the basis of a new sexual system: plants having the same number and arrangement of stamens, pistils, petals, and other flower parts were now grouped together. The various members of the rose family, for example, were now grouped together for the first time, for despite the fact that these plants may grow as small herbs or large trees, they always bear flowers with the same recognizable structure. It was this new sexual system, conditioned by his earlier studies on plant sexuality, that first established Linnaeus's reputation as a keen identifier and perceptive classifier of plants. His new sexual system allowed Linnaeus to incorporate newly discovered plants into his system very easily. It was this sexual system that was later attacked by Buffon.

C. THE BREAKDOWN OF THE CLASSICAL TRADITION

The classical tradition was a magnificent edifice indeed, and it had stood for more than 2000 years. When it finally came to an end, there was no Samson to bring it crashing down all at once, but it fell rather into a gradual decline, and slowly it became a ruin.

Of the many factors that contributed to this decline, some were of long standing. The geographical explorations of the Renaissance and the discovery of "new worlds" stimulated both an interest in nature and a turning away from the writings of classical antiquity. The strange animals and plants brought back by explorers fascinated people in all walks of life, and since the ancients had not written of such creatures, there was no recourse but to study them directly. The interest in nature spread to the world of art: painters of the sixteenth and later centuries depicted more and more background, until in some cases the background became the dominant element (many of Brueghel's paintings, for example).

Francis Bacon (1561–1626) and the new philosophy of empiricism did much to undermine the authority of Aristotle by insisting on the direct observation of nature itself. The Protestant Reformation was quick to attack both Aristotelian and papal authority, preaching instead that every man should interpret the Scriptures for himself and follow the dictates of his own conscience. Toward the end of the classical era, evidence began to accumulate that would eventually prove the Earth to be much older than the Bible had allowed. The mere raising of such a question is evidence of a willingness to challenge authority.

Buffon and His Era

Even the authority of Linnaeus was now challenged. Among his severest critics was the French naturalist Georges Buffon (1707–1788). A wealthy aristocrat and keeper of the Jardin du Roi under Louis XV, Buffon stimulated an immense interest in natural history throughout France and indeed all across Europe. Unlike most previous naturalists, who wrote in Latin, Buffon wrote in French and for a general audience. A household could scarcely be considered educated without owning a copy of his 44-volume encyclopedic work, *Histoire Naturelle*. In its style, this work was clear, comprehensible, and very popular.

Buffon's criticisms of Linnaeus were several. He objected especially to any arbitrary subdivision of nature, for he believed that the *scala naturae* was a continuum. This belief caused him to embrace for some time the theory of **nominalism,** according to which collective groupings such as families and genera are arbitrary mental constructs only. Buffon also protested that the sexual system was arbitrary in preferring flowers to leaves or stems. Flowers bloom for only a short period, he observed, and are then often lost; of what importance could they then be? Some of Linnaeus's classes were also ridiculed as being too heterogeneous: how could the crayfish be included among the Insecta, or the snail under Vermes?

Buffon's writings were hardly confined to zoology and botany. He translated Newton's *Fluxions* into French and speculated that biology would some day be reduced to a branch of physics. Buffon also wrote on geological subjects, like John Ray before him, but we shall defer these ideas to the next chapter.

Many of Buffon's contemporaries took the *scala naturae* as a signal to search for new species to fill in the remaining gaps in the system, the "missing links" in the chain. The duck-billed platypus (*Ornithorhynchus*) was, upon its discovery, mistakenly hailed as a link between the birds and the mammals. But the gap most keenly noticed was that between apes and humans. Early studies of the orang-utan (*Pongo*) and other great apes did much to stimulate speculation about the nature of the missing forms, and reports, often inaccurate, of naked Hottentots or of still more primitive tribes devoid of spoken language added much further interest. These were the beginnings of the science of anthropology at the hands of Lord Monboddo (1714–1799) and Charles Bonnet (1720–1793) and in the early writings of Buffon. Charles Bonnet was further responsible for an embryological theory of **preformation,** according to which each adult is present as a miniature form, encapsulated within the embryo. Development of the adult from the embryo was to Bonnet a process of growth and of the realizing of potentials inherent in the embryonic form. For this unfolding of potential during development, Bonnet coined the term *evolution* (*e-*, out, + *volutio,* turning or folding), meaning literally an unfolding.

It was in this era that the *scala naturae* was transformed from a fixed scale of perfection into a new moving scale of progress—an escalator rather than a ladder or stairway. The French mathematician and naturalist Pierre Louis Moreau de Maupertius (1698–1759) was apparently the first to suggest this change, but the idea was later reiterated by Charles Bonnet, by Lord Monboddo, possibly by Buffon, and most especially by Lamarck. Voltaire, on the other hand, attacked both the *scala naturae* and its supporters. This type of programmed evolution, or unilineal ascent of nature's scale, is quite different from the branching type of evolution later favored by Darwin.

The French Revolution, and the Formation of National Traditions

The French Revolution contributed greatly to the breakdown of the classical tradition. The "Age of Reason" had sought *order*—a static or final perfection which was "Heav'n's first Law." Naturalists like Linnaeus had sought to discover order in God's creation; theologians and philosophers of the age had done likewise. Social and political theorists looked for orderly, well-governed systems, thinking that these would remain stable and unchanged, once established. Haydn and his contemporaries wrote orderly music. But now order was giving way to a new goal: progress. The French Revolution represented progress. The Industrial Revolution represented progress. The era of romanticism in both literature and music was instilled with the idea of progressive change, as opposed to the older more static perfection. Governments and social systems were no longer expected to remain static—they, too, were supposed to undergo continual change in the direction of progress. The belief in progress, brought about largely by the French Revolution, is now part of our general cultural heritage, and we may therefore tend to forget what a revolutionary concept it once was. Nothing, it was said, could hold back progress or stand in its way.

One apostle of "progress" was Jean Baptiste Lamarck (1774–1829). A strong believer in the *scala naturae,* Lamarck turned this fixed scale of perfection into a new moving scale of progress: an escalator rather than a ladder or stairway. The *scala naturae* became a dynamic series, through which each species ascended to the pinnacle of organic creation. New species were continually being created at the lower end of the scale by spontaneous generation, so that no gaps appeared as each species ascended the scale ahead of the one beneath it. Lamarck became the first to list the animals in an ascending sequence, from one-celled creatures to man, rather than the other way around. He also introduced the terms "vertebrate" and "invertebrate," which have been used ever since, and he suggested the recently coined word "biology" (*biologie*) as the name for the new science that he saw emerging from the fusion of the classical subjects of zoology and botany.

Following the French Revolution, the classical tradition was replaced by at least three distinct national traditions, each in a different country. The political disruptions of this era undoubtedly contributed to the decrease in communication between scientists in different countries, and this in turn fostered the development of separate national traditions: an environmentalist tradition, which was strongest in France; a theist tradition of "Natural Theology," principally in England; and a "romantic" tradition of *Naturphilosophie,* centered in Germany. These and other national contributions to evolutionary theory may be listed as follows:

England:
  Population thinking, statistics (Malthus, etc.)
  Adaptation as perfection, design (Paley, etc.)
  Natural selection (Darwin, Wallace)

Statistical population genetics (Fisher, Haldane, etc.)

France:

Temporalizing of *scala naturae* (eighteenth century)

Influence of the environment (Voltaire, Buffon, Lamarck, Geoffroy)

Germany:

Unity of life, common structural plans (*Bauplan*) reflecting Platonic archetypes (Goethe, Oken, etc.)

Geographic variation (Humboldt, von Buch, etc.)

Appropriation of evolution into leftist political thought (Haeckel, Engels, Marx, etc.)

Scotland:

Uniformitarianism (Hutton, Lyell)

Russia:

Genetics of wild, natural populations (begun by Chetverikov, continued principally in the United States by Dobzhansky, etc.)

United States:

Chromosomal theory of inheritance (Sutton, Morgan, Sturtevant, Bridges)

Molecular genetics, discovery of role of DNA, etc. (with British help, e.g., Watson and Crick)

Public interest in vertebrate paleontology (since last half of nineteenth century; Marsh, Cope, Osborn, Matthew, Simpson, etc.)

One scientist of this era, Georges Cuvier, does not fit into any of these three traditions; he might best be considered as a final extension of the classical tradition, or else primarily as a paleontologist.

Georges Cuvier

A contemporary of Lamarck, Georges Cuvier (1769–1832) was a Swiss-born French scientist who also lived through the turmoils of the French Revolution. As if striving to retain the old order, Cuvier reasserted his belief in the fixity of species, and his geological theories, to be discussed in the next chapter, embodied both the notion of cataclysmic upheaval and the concept of progress. Acknowledged as the foremost comparative anatomist of his era, Cuvier, together with Felix Vicq d'Azyr (1748–1790), advanced the "principle of correlations." According to this principle, all bodily parts had to be designed to function harmoniously with all the others. Using this principle, Cuvier made the bold claim that from

a single bone he could reconstruct all the others, one by one, and from these the muscles and internal organs, and finally the entire animal. His studies of fossils made extensive use of this principle, aided, of course, by the similarity of those fossils to the bones of closely related living forms.

Cuvier, an admirer of Aristotle and of the classical tradition, was ironically responsible for one of the most dramatic blows to this classical tradition. His anatomical studies led Cuvier to reject the traditional *scala naturae* as a unifying concept. Rather, he said, there are four distinct patterns on which animals are constructed, constituting four distinct branches (*embranchements*) of the animal kingdom. These four **phyla** (singular, *phylum*), as they came to be called, were given the names Vertebrata, Articulata, Mollusca, and Radiata (Box 3–B). A scale of higher and lower forms might exist within each of these phyla, but between the phyla there were distinct gaps that no form of life could cross. One could thus speculate about "higher" and "lower" Mollusca or Articulata, but in no way could a mollusc and a vertebrate be compared, nor could either one be declared "higher" or "lower" than the other. This concept arose, it seems, from Cuvier's comparative anatomy, in which he (and others) were beginning to discover that vertebrates were all built on a common plan, arthropods and segmented worms ("Articulata") on a very different plan, molluscs on yet a third plan, and "Radiata" on a fourth. This last phylum is the only one of Cuvier's four that would today be regarded as unnatural: modern zoologists would now divide most of these forms between the two distinctly unrelated phyla Coelenterata and Echinodermata.

Cuvier's decisive dismemberment of the chain of being was at best a *coup de grâce*. The classical tradition had already suffered mortal wounds from a far stronger challenger, the theory of environmental determinism, which we shall now consider.

D. LAMARCK AND ENVIRONMENTALISM

The influence of the environment was by far the most important force that finally brought the classical tradition to an end. Since the geographical explorations of the

BOX 3-B *Cuvier's Four Phyla ("Embranchements")*

The four phyla of Cuvier are listed below with examples of each.

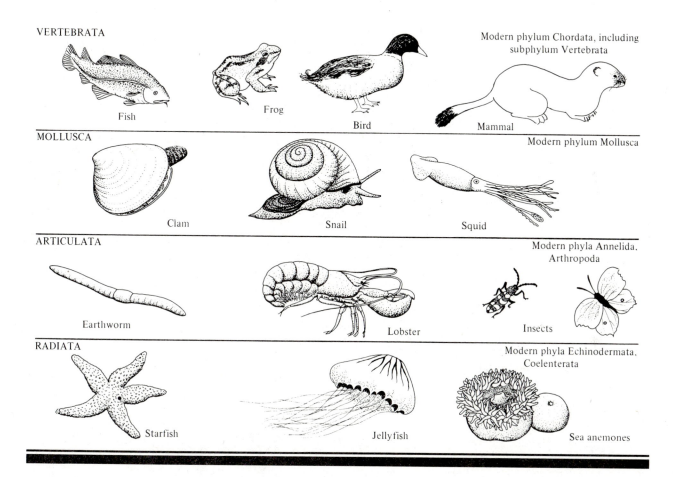

VERTEBRATA
Modern phylum Chordata, including subphylum Vertebrata

Fish Frog Bird Mammal

MOLLUSCA
Modern phylum Mollusca

Clam Snail Squid

ARTICULATA
Modern phyla Annelida, Arthropoda

Earthworm Lobster Insects

RADIATA
Modern phyla Echinodermata, Coelenterata

Starfish Jellyfish Sea anemones

Renaissance, Europe had become accustomed to the idea that each new and different country had its own strange inhabitants, plant and animal as well as human.

The British philosopher John Locke (1632–1704) theorized that the human mind begins as a *tabula rasa,* a blank slate on which experience writes. James Mill (1773–1836) and Jeremy Bentham (1748–1832) were among other English followers of this tradition, which emphasized also the importance of education. It was in France, however, that environmentalism reached its highest development, largely at the hands of Voltaire (1694–1778). Helevetius, Condorcet, and especially Rousseau (1712–1778) extended the implications of environmentalism into the theory of education. Man came to be viewed as the product of his upbringing, an idea that, largely through Rousseau's influence, came to pervade such nineteenth-century literature as the novels of Dickens, Dostoevski, and Zola.

It was probably Voltaire who transmitted the environmentalist outlook to the growing number of French naturalists, especially to Buffon. Tropical animals were now viewed as products of their warmer climate. The influence of climate was soon extended to animals in general. Buffon noted that animals were suited to their respective climates, and he added the further observation that climatic change would cause migration of the old inhabitants and their replacement by immigrants from elsewhere.

The environmentalist outlook was strengthened further by the success of the French Revolution and the growing influence of Rousseau. Botany came under strong environmentalist influence at the hands of Augustin Pyrame de Candolle (1778–1841) and his son Alphonse de Candolle (1806–1893). The use of plants in the study of geography was also strongly advocated by the great geographical explorer and climatologist Alexander von Humboldt (1769–1859), who wrote an entire book on the subject, as did the younger Candolle.

The concept of **adaptation,** the suitability of organisms to their environmental circumstances, came to form an important part of the growing environmentalism. The explanation of adaptation became one of the most fundamental problems in nineteenth-century biology. Strict environmental determinism provided one explanation, but the mechanism by which the environment produced these adaptations was not yet known. Environmentalists thus hastened to propose various mechanisms.

The simplest mechanism of all, that of direct environmental induction, was proposed by the French zoologist Étienne Geoffroy Saint-Hilaire (1772–1844). He believed that temperature alone could directly bring about physiological changes that resulted in adaptations suitable to the temperature, a fact borne out to some extent by modern physiological experimentation. Geoffroy went further, however, to postulate that all adaptation was environmentally induced and was furthermore passed on to the offspring; hardly any biologist would support these ideas today. In marked contrast to the fixed, unchanging species of Aristotle or Linnaeus, Geoffroy viewed the species as something quite plastic, molded almost without limitation by environmental influences. For this view he was sharply attacked by nearly all his contemporaries. An even greater controversy was stirred by Geoffroy's insistence on a single unity of structure throughout the animal kingdom, which was vehemently opposed by Cuvier, who insisted on four distinct types. This bitter quarrel divided much of France during the 1830s.

Isidore Geoffroy Saint-Hilaire (1805–1861), the son of the aforementioned, pursued his father's idea that species were rather plastic. The younger Geoffroy took to the study of abnormal and malformed embryos (teratology), on which he wrote a treatise that had later influence on Charles Darwin.

Erasmus Darwin (1731–1802), a grandfather of Charles Darwin, proposed a different mechanism to account for adaptation. An organism habituated to certain circumstances would through its own volition seek to develop those structures most suitable to those circumstances. The importance of volition as a source of change bears great similarity to the views of Lamarck, though it now seems that both men came upon this idea independently.

Jean Baptiste Pierre Antoine de Monet, chevalier de Lamarck (1744–1829, Fig. 3.2), is perhaps the most famous—although ironically one of the most misunderstood or poorly understood—predecessors of Darwin.

Fig. 3.2 Jean Baptiste Lamarck (1744–1829).

Originally a botanical assistant at the Jardin du Roi, Lamarck was later influential in changing the name of the institution to Jardin des Plantes following the French Revolution. An ambitious man, Lamarck sought to advance his career, but the botanical chair was already held by Antoine Laurent de Jussieu, who was in perfect health, whereas the chair in vertebrate zoology was held by Cuvier. Lamarck therefore assumed charge of the invertebrates almost by default. Fortunately, he excelled in the study of these animals, on which he soon became an expert. Lamarck greatly advanced our knowledge of the invertebrates, describing their variety and their internal structure in far more detail than had ever been done before.

Lamarck is now remembered largely for his *Philosophie Zoologique* (1809), in which he discusses his theory of evolution. He begins by describing the *scala naturae* in some detail, emphasizing the continuity of forms by which the major groups of animals are connected (see Item 3.2). He describes the progressive "degradation" of form, particularly of the sense organs, that we find on descending the *scala naturae* from the highest forms (mammals) to the lowest ("infusoria"). But this view, says Lamarck, is wrong: the succession of living forms should be viewed from the lowest forms to the highest, with each species ascending the scale up to the highest rank. The *scala naturae* was now a constant flux reminiscent of the Greek philosopher Heracleitos, with species continuously ascending in what Lamarck called "La marche de la Nature" (the parade, or flow, of nature). The continuity of the scale is maintained by spontaneous generation, which continually produces new forms at the bottom of the scale, thus replenishing it. The overall pattern that Lamarck saw was thus a gradual ascent of the linear scale of nature's parade. The major feature of this continuous parade was the progressive acquisition of ever more perfect form, especially more perfect sense organs. Lamarck, in an addendum, did suggest the possibility of branching evolution, but for the most part he emphasized the unilineal progression up the scale—La marche de la Nature.

The gaps between the major classes of animals were as annoying to Lamarck as they had been to previous supporters of the *scala naturae*. Lamarck was quite inventive in his arguments for the continuity of the scale, placing the barnacles as transitional forms between the Arthropoda and Mollusca (on account of their shells), and placing the squid and cuttlefish as transitional from the molluscs to the Vertebrata on account of their highly developed senses and their possession of an internal shell, which Lamarck incorrectly assumed to be the precursor of the backbone.

Lamarck also advanced another argument to account for the apparent discontinuities: the various species do not *exactly* form a graded series, because each is adaptively modified with respect to its way of life. These adaptations come about, said Lamarck, by the use of an organ through exercise. This important concept goes far back in history (Zirkle, 1946); popular belief in the accumulated heredity of such acquired characteristics was widespread in Lamarck's own time.

According to Lamarck, a blacksmith hammering at his anvil would strengthen his right arm and would pass on to his son a somewhat stronger arm than he himself had at birth. If his son also became a blacksmith, he would strengthen the right arm still further, and so on, eventually producing a race of blacksmiths with well-developed right arms. Turning to a zoological example, Lamarck said that in a similar way, the giraffe's long neck can be attributed to a similar progress. An ancestral giraffelike animal with a short neck could easily browse leaves off the easy-to-reach lower branches of trees. But in a time of scarcity, having already browsed the lower leaves, the animal had need (*besoin*) of a means to reach the higher leaves. Voluntarily the animal stretched its neck to reach the leaves, and its neck became somewhat longer. Its offspring was born with a slightly longer neck, and the process was repeated until the long-necked giraffe had evolved. Note that this process was assumed to be gradual, adaptive, and dependent on the animal's exercise of volition. Note also the important role played by unfavorable conditions, also an important element in Darwin's later thought.

Opinions as to the origins of Lamarck's evolutionary beliefs have varied widely. Spontaneous generation has been regarded as the basis for these beliefs or as a mere corollary to them. Lamarck's disbelief in extinction has likewise been hailed as the source of his evolutionism or again cited as a mere corollary. The importance of volition in Lamarck's thought has been denied by Ernst Mayr (1972) and attributed to Lyell's misreading of the word *besoin* as "want." August Weismann's views, equating

Lamarckism with the inheritance of acquired characteristics have exerted an undue influence for nearly a century.

The theory of environmental determinism represented most squarely the rejection of the classical tradition. The focus of this theory was the explanation of adaptation, a phenomenon with which the classical tradition had never come to grips except teleologically. The environmental determinists attributed adaptation to the effects of environment, especially of climate. The ultimate failure of environmentalism lay in its failure to provide an adequate mechanism by which climate could exert its influence—or rather, it provided too many different mechanisms, none of which proved satisfactory under scrutiny. Creationism, as we shall see, provided a much simpler explanation in the benevolence of an omniscient and omnipotent Creator.

E. NATURAL THEOLOGY

The roots of this creationist theory go back to biblical times, and to the assumption of a benevolent and omnipotent Creator. The purposefulness of organic structure was also considered important by Aristotle, who emphasized **teleology,** the determination of structure by functional end results.

The **argument from design,** as it came to be called, was first formalized by Thomas Aquinas. The argument uses the harmony or design in nature as evidence for the existence, omnipotence, and benevolence of God. The German philosopher Leibniz noted that Newton's laws of motion perpetuated an existing order, but that God was needed as the Creator of this order and in particular as the initial source of heavenly motions. The philosophy known as **theism** used this argument extensively as proof not only of God's existence but also of his benevolence. A favorite theist analogy was that the Universe functioned as a well-tuned clockwork, designed and set into motion by God. The clergyman-naturalist John Ray, a supporter of theism, wrote a book entitled *The Wisdom of the Creator, as Manifested in the Works of His Creation.* This book anticipated many of the biological arguments that William Paley would later use in support of theism.

The leading theist arguments were criticized by the Scottish philosopher David Hume (1711–1776) and by Immanuel Kant (1724–1804). The Universe is more com-

plex than a ship, said Hume, but it is not infinitely more complex. Thus, if a human architect can design a ship, then a finite Divine Architect, powerful but not all-powerful, wise but not infinitely wise, could have designed the Universe. With arguments such as these, Hume ridiculed the major theist arguments of his day.

But theism flourished, despite Hume's criticisms, and was actively supported, especially in England, by a large body of clergy. Foremost among these was the Reverend William Paley (1743–1805), who in 1802 published a book entitled *Natural Theology, or Evidences of the Existence and Attributes of the Deity.* Paley's books were widely read and discussed, especially at Cambridge, where several of them became a required part of the curriculum. It was at Cambridge that Darwin would later read these works with avid interest and become strongly influenced.

Paley's originality and his importance to the history of evolution lay not only in his theist arguments but more especially in the many examples of design that he used to bolster those arguments. Many of his examples were instances of remarkable adaptation among living organisms. The perfection of the eye's mechanism, no less the hand's, were but two of Paley's examples. The bills of woodpeckers, the fangs of snakes, the feathers of birds (and the specially constructed down feathers in young birds)—all these and more were mentioned and described in detail as showing purposeful design. Some designs even anticipated their future utility, as in the migration of birds before the onset of winter or the separation of the two atria of the unborn fetus's heart by a valve designed to close at birth. These anticipatory adaptations later proved particularly important in that none of the various environmentalist theories could account for them. Paley and the other proponents of Natural Theology thus unwittingly provided Darwin with a long series of well-documented examples of remarkable adaptation in nature. But to Paley, these examples and others showed only that the Universe was a magnificent contrivance, a clockwork of far greater intricacy and perfection than any ever devised by man. "The marks of *design* are too strong to be gotten over. Design must have a designer. That designer must have been a person. That person is God."

Natural Theology grew in influence during the early nineteenth century. The eighth Earl of Bridgewater endowed a series of treatises by the best scientific minds of the age, designed to show "the power, wisdom, and goodness of God as manifested in the Creation." Eight of these *Bridgewater Treatises* were published, including one by the Scottish anatomist Sir Charles Bell entitled *The Hand: Its Mechanism and Vital Endowments as Evincing Design.* These Bridgewater Treatises represented the culmination of scientific support for Natural Theology, though belief in this antievolutionary theory would last for quite some time. In fact, much of Darwin's early opposition came from the defenders of Natural Theology.

F. ROMANTICISM AND *NATURPHILOSOPHIE*

The romantic tradition meant different things to different people and at different times. Originally a literary movement begun by Goethe and Rousseau, romanticism came also to influence scientific thought in the nineteenth century. In the 1830s, this influence was largely conservative and persistently Aristotelian in its outlook. Later in the century (Chapter 5), it came to be merged with progressivism and with leftist politics. In both cases, its influence on science was far-reaching.

As a literary tradition, romanticism was widespread across Europe: Goethe and Schiller in Germany, Rousseau in France, and Coleridge, Shelley, and Byron in England all belonged to this tradition. As a scientific tradition, however, romanticism reached its highest development in Germany at roughly the same time that environmentalism was flourishing in France and Natural Theology in Britain.

The pre-Darwinian romantic tradition emphasized the "unity of all Creation." Thus, in contrast to both the environmentalist and Natural Theological approaches, there was little emphasis placed on the adaptations unique to each species. Instead, there was a great emphasis on the study, especially the anatomical study, of those features that were common to the most diverse types of animals or plants—features that tended to unite them and provide evidence for their unity. For despite their differences, related organisms were always built on the same structural plan (**Bauplan**), reflecting the same **archetype.** Each archetype was interpreted as a variation

on a theme: the various animal phyla, for instance, represented so many themes, on which the archetype or *Bauplan* for each class was a variation. These variations were in turn used as new themes, subject to further variation. The justification usually given for this system was that the Creator did not simply create endlessly varying forms at random but, rather, started with a *Bauplan* or "theme," which He then varied within limits to produce subsidiary themes, and so on. Evidence of structural similarity was sought and found: mammals, birds, reptiles, amphibians, and fishes, for example, had skeletons built on the same plans, with many of the same bones, but with variations in proportion and position and occasionally in the addition or loss of individual elements. It was evidence of this sort that paved the way for the later acceptance of Darwinism, even though this school of thought was, if anything, creationistic and certainly not evolutionary!

Johann Wolfgang von Goethe (1749–1832), the great German poet and playwright, was undoubtedly the founder of this scientific school of *Naturphilosophie,* as it came to be called. In addition to his theory of colors (*Farbenlehre*), three of Goethe's scientific achievements seem to stand out: the metamorphosis of plants, the vertebral theory of the skull, and the existence of a human "intermaxillary" bone. All three of these reflect Goethe's belief in the unity of Creation. His theory of plant metamorphosis can be summarized in three words: "Alles ist Blatt" (all is leaf). By this, Goethe simply meant that all other plant parts (stems, flowers, fruits) can be interpreted as modified leaves—an idea that gave rise to the concept of evolutionary homology nearly a century later. Goethe also interpreted the various skull bones as parts of highly modified vertebrae; the skull as a whole was but four or five such modified vertebrae. In these two theories, we have the search for structural unity—a common *Bauplan* (structural plan) or Platonic archetype: plant parts are all built on the *Bauplan* of the leaf, and the skull is built on the *Bauplan* of the vertebrae. Goethe's work on the "intermaxillary" bone—now called the *premaxilla* or incisive bone by most anatomists—arose from the claim that humans were fundamentally different from other mammals in having only a single bone on each side of the upper jaw instead of two, a claim that would have contradicted the unity of plan between humans and other mammals. Goethe studied a number of human fetuses and showed that they indeed have two separate bones, as do all other mammals, but that these fuse into a single bone shortly before birth. Thus a single bone corresponds to two others and results from their fusion. Fusions of this sort, or the modification of vertebrae to form skull bones (now, incidentally, a discredited theory), were easily explained by the theory of common archetypes or *Baupläne* but were not so easily explained as resulting either from environmental adaptation or from the "Wisdom of the Creator." A strong supporter of the unity of nature, Goethe strongly supported Étienne Geoffroy Saint-Hilaire in his dispute with Cuvier (Section D of this chapter).

The chief scientific disciple of Goethe was Lorenz Oken (1779–1851), a founder of the school of "idealistic morphology." Oken was somewhat of a mystic, who sought to unify all of natural science by showing that all of life was built on a common *Bauplan*. His search for a universal *Bauplan* led him to postulate that all life was but a modified form of a primeval slime, or "*Urschleim.*" This German word was soon translated into Greek, where it became **protoplasm**, the general name for cellular material.

German scientists of the 1830s continued to search for evidences of a common *Bauplan* wherever they could, usually in the anatomy of the skeleton. The most intensive search was for a universal *Bauplan*—the structural plan that would unify all of organic Creation into a single whole. Nor was this search fruitless, for soon the so-called **cell theory** emerged as just such a universal *Bauplan*. The botanist Matthias Jakob Schleiden (1804–1881) showed that all plants are composed of cells, and the zoologist and anatomist Theodor Schwann (1810–1882) showed that the same was true of animals. Robert Brown (1773–1858) discovered that all plant and animal cells were characterized by the presence of a nucleus, and Rudolf Virchow (1821–1902) proposed his theory of the cellular origins of many diseases.

The one great Englishman that belongs to this largely German tradition was the anatomist Sir Richard Owen (1804–1892), first a friend and later a bitter opponent of Charles Darwin. Owen was perhaps the foremost anatomist of his time, and his contributions to descriptive anatomy and paleontology are quite numerous. Also significant is his coining of two new terms arising out of the *Bauplan* theory: homology and analogy. **Homology**, as Owen first used the term, meant essential resemblance, "essential" in the Aristotelian sense that it reflected an archetype or a Platonic *eidos*. Homology was thus resemblance indicative of a common structural plan (*Bauplan*). Any other kind of resemblance was called **analogy**. Analogous resemblance was "inessential"; it reflected no archetype, *eidos,* or *Bauplan,* but it often showed instead the results of similar and convergent adaptation (Chapter 18). Owen and other British anatomists now described numerous resemblances among different animals, carefully distinguishing homology from analogy. It was Charles Darwin who later gave to these two words their modern, evolutionary meanings.

These were the major paradigms of biological thought during the time that immediately preceded Darwin's *Origin of Species*. The classical tradition was effectively dead and required no refutation. The German school of *Naturphilosophie* received very little comment from Darwin, either friendly or unfriendly, though Darwin did explain Owen's concepts of analogy and homology in evolutionary terms. Darwin's criticisms were thus directed largely at the environmental determinists on the one hand and Natural Theology on the other. But some of the strongest influences on Darwin were geological, not strictly biological, and it is to these that we now turn.

CHAPTER SUMMARY

From Anaximander to Lucretius, the materialist tradition of ancient Greece and Rome produced a number of evolutionary theories. The atomists took from Empedocles a belief in natural selection among random combinations of body parts, with only the harmonious combinations surviving.

The classical tradition produced in ancient times Plato's theory of Forms, which Aristotle adapted to the living world. Each species was seen to have an unchanging ideal Form (*eidos*); each individual was assumed to represent an imperfect copy of this *eidos*. Another classical theory was the linear arrangement of species along a continuous hierarchy or scale known as the *scala naturae* or "chain of being."

The revival of botany in the Renaissance began with plants of medicinal importance. From the need to distinguish closely related species, the science of classification arose. Geographical exploration resulted in the discovery of new animals and plants, which had to be classified. The greatest classifier was Linnaeus, whose *Systema Naturae* listed all the animals and plants known at that time. Linnaeus used a sexual system, based on flower parts, for the correct identification of plant species. He also codified the classical tradition into a theory of classification that proclaimed the fixity of species as a consequence of their objective reality.

The breakdown of the classical tradition grew slowly out of geographical explorations, out of the direct study of nature, and out of Buffon's attacks on Linnaeus. The breakdown was greatly accelerated at the time of the French Revolution; the static perfection assumed in earlier times was replaced by a search for "progress." Lamarck and others changed the *scala naturae* into a dynamic stream of progress. Cuvier's four separate phyla dealt the final blow to the *scala naturae* as a unifying concept.

With the breakdown of the classical tradition, three distinct national traditions emerged, each with a different contribution to evolutionary thought. The strongest of these was the environmentalist tradition in France, imported from England by Voltaire. French scientists emphasized the effects of climate and other environmental factors on the characteristics of organisms. Buffon explained that organisms had simply migrated to the climates to which they were best suited. The elder Geoffroy attributed adaptation to the direct induction of characters by outside environmental forces. Lamarck rejected this type of induction; instead, he favored the adaptation of organisms to their environment through the inherited effects of either exercise or disuse.

The English tradition of Natural Theology arose from a theist argument, the so-called argument from design. Proof of God's existence was sought in the intricacy and purposiveness of nature; the emphasis was thus on adaptation. The principal exponent of this school was William Paley, who brought much of British science, especially anatomy, under theist influence. Long catalogues of adaptations in the living world were described, each in great detail, as illustrating design beyond the capacity of human craft and therefore indicative of God's superior design. Charles Darwin was later able to use these examples of adaptation to support his own theory.

The romantic tradition of *Naturphilosophie* flourished largely in Germany, though it had adherents elsewhere. It was founded on the theory that all forms of life were united by a common plan. The anatomists and idealists of this school looked for evidence of common structural plans (homologies). Goethe correctly suggested that all plant parts were modified leaves, and that the skull was made of modified vertebrae. From this school developed the concept of homology, or essential resemblance indicative of a common Bauplan or Platonic archetype. Darwin later explained that homology resulted from common ancestry.

FOR FURTHER KNOWLEDGE

ITEM 3.1 *Carl Linnaeus (1707–1778)*

"The father of modern taxonomy"

Carl Linnaeus was born in 1707 in rural Sweden, the son of a country pastor. He studied first at Lund, then at Upsala, where he became apprenticed to Olaf Celsius, a botanist whose nephew later invented the centigrade thermometer. He was fascinated with the study of plants and spent many long afternoons on walks through the woods and fields, where he could study them firsthand. He was impressed with the writings of Tournefort and Rivinus and also with Grew's theories on plant sexuality. So interested was he in this last subject that he wrote a paper on it and was soon elected to the Swedish Academy of Sciences on its account. In 1732, the Academy selected him to go on an expedition to survey the plants of Lapland. He sought to further his studies, but the opportunities in Sweden were limited, and so he went to Holland, taking his M.D. degree at the University of Harderwijk in 1735.

(*continued*)

ITEM 3.1 (*continued*)

While in Holland, Linnaeus arranged to have published the first edition of his most important work, the *Systema Naturae* ("system of nature" or "natural system"), a work describing all the animals and plants known at the time. This work, which he had largely completed before he left Sweden, brought Linnaeus great fame throughout Europe, and to its continual revision Linnaeus devoted much of his subsequent lifetime. The *Systema Naturae* underwent twelve editions, meanwhile growing from a small pamphlet to a major two-volume book. Linnaeus's other works included *Flora Lapponica, Fauna Suecica, Flora Suecica, Fundamenta Botanica, Classes Plantarum, Genera Plantarum, Philosophica Botanica, Species Plantarum,* and others.

In 1738 Linnaeus set up medical practice in Stockholm, which he abandoned in 1741 to assume a professorship in medicine at the University of Upsala. The next year he attained his life's ambition, the professorship in botany at Upsala. He devoted the remainder of his very productive life to the amassing of a great botanical and zoological collection, and to the continual revision of the *Systema Naturae.*

As Linnaeus's reputation grew, more and more students came to Sweden to study under him. More and more naturalists from all over the world, including his former students, sent specimens to Linnaeus to be described. His collections grew and grew until they became the biggest in all of Europe. So great and unchallenged was Linnaeus's reputation as a naturalist that others continued to use his names; scientists later agreed that the works of Linnaeus should form the starting point in the selection of zoological and botanical names, a practice that still continues.

Upon Linnaeus's death in 1778, his intellectual heritage passed on to Johan Christian Fabricius (1743–1808), a Danish zoologist who described many insects and developed a new classification based on their mouthparts. Under the terms of a prior agreement, Linnaeus's professorship at Upsala passed to his son, Carl the Younger, who was unfortunately not worthy of the task. Linnaeus's immense collections fell into disuse, and upon the death of Carl the Younger, they were put up for sale. A group of Englishmen promptly formed the Linnean Society, raised the necessary sum, purchased all of Linnaeus's collections along with his library and memorabilia, and arranged to have them shipped to London, an act for which Sweden has never forgiven the British.

ITEM 3.2 *Lamarck's Linear Arrangement of Animal Classes*

Lamarck's classes, arranged in "nature's order"		Modern phyla or classes to which they correspond	
INVERTEBRATE ANIMALS	1. Infusorians	Phylum Protozoa	
	2. Polyps	colonial Coelenterata (plus Rotifera, Porifera)	
	3. Radiarians	solitary Coelenterata (plus Echinodermata)	
	4. Worms	Phyla Platyhelminthes, Nematoda, etc.	
	5. Insects	PHYLUM ARTHROPODA	Class Insecta
	6. Arachnids (spiders)		Class Arachnida (plus Diplopoda, Chilopoda)
	7. Crustaceans		Class Crustacea
	8. Annelids	Phylum Annelida	
	9. Cirrhipedes (barnacles)	(included in class Crustacea)	
	10. Molluscs	Phylum Mollusca (plus Brachiopoda, Tunicata)	
VERTEBRATE ANIMALS	11. Fishes	PHYLUM CHORDATA	Classes Agnatha, Placodermi, Chondrichthyes, Osteichthyes
	12. Reptiles		Classes Amphibia, Reptilia
	13. Birds		Class Aves
	14. Mammals		Class Mammalia

4

Geology Before Darwin

A. GEOLOGY BEFORE 1830

Prescientific Speculations

From the earliest times, people have wondered about the markings on the face of the Earth: the hills and valleys, the rivers and mountains, the seas and islands. The origin of these natural features seemed synonymous with the origin of the Earth. It was no wonder, then, that prescientific speculations about the Earth were closely tied to speculations about **cosmology**, the nature and origin of the Universe as a whole.

The northern fringes of the Mediterranean basin are active geologically. The infrequent earthquakes were sufficently frightening to last long in the collective memories of ancient civilizations. Legends concerning such upheavals were so vivid and oft-repeated that the frequency of these cataclysmic events would have been exaggerated in the minds of ancient scholars. It is no wonder that early beliefs were generally of **catastrophism,** the explanation of geological processes as the results of cataclysmic upheavals. Many ancient creation legends are likewise catastrophic for much the same reason.

One important scientific observation made by the ancients was that marine fossils, usually meaning bivalve shells, were sometimes found in the hills at fairly high elevations above the sea. Xenophanes noticed such marine shells in the hills of Asia Minor (modern Turkey) in the sixth century B.C. The Greek historian Herodotus made a similar observation during a visit to Egypt in the fifth century B.C. Xenophanes cited these marine shells as proof that the hills had once been covered with water, the Earth thus having undergone cataclysmic changes since its origin. The earlier views of Thales and Anaximander (Chapter 3) involving the origin of land from water may reflect similar observations, known to them but unrecorded. The biblical account of Noah's flood may likewise reflect the finding of marine shells in the hills of Turkey or Syria.

In the many centuries that followed, two geological concerns remained foremost: the age of the Earth and the nature of fossils. Medieval scholars generally dismissed fossils as the products of inorganic processes or as false signs of antiquity placed in rocks either by Satan in his attempt to lure man away from true faith or by God in order to test that faith. A few Renaissance scholars, Leonardo da Vinci among them, became interested in fossils upon finding marine shells in the hills of Italy. Bernard

Palissy (c. 1510–1589), a French potter, used many artistic designs inspired by fossils. Nicolaus Steno (1638–1687) studied many fossils, mostly marine shells, and demonstrated that they were very similar to, but different in detail from, the shells of related species now living. The true nature of fossils continued to be debated into the very early nineteenth century, when Cuvier and others finally laid the issue to rest.

Throughout this prescientific period, the age of the Earth was assumed, without much question, not to be very great. An Irish bishop, John Ussher (1581–1656), was the first to propose an age in years: by adding up the ages of various persons mentioned in the Bible at the time of birth of their offspring, Ussher calculated that the Earth had been created in the year 4004 B.C. Archaeological studies seemed to bear this out: the oldest Egyptian civilizations known seemed only a few thousand years old. Most convincing was the evidence recovered from Egyptian tombs, which showed that the flowers and other plants buried with Egyptian mummies were indistinguishable from modern species.

But the slowness of most everyday geological processes was gradually becoming apparent to many careful observers. Among the most perceptive of these was the English clergyman-naturalist John Ray (Chapter 3), who in 1663 wrote the following words (emphasis added): "*Either* the world is a great deal older than is imagined . . . , *or*, in the primitive times, the creation of the earth suffered far more concussions and mutations in its superficial part than afterwards." Here the great debates of eighteenth- and early–nineteenth-century geology were anticipated, and the lines were clearly drawn. The few thousand years that Bishop Ussher had allowed for Earth's history were inadequate to allow for the postulated changes in the Earth's crust at currently observed rates. Catastrophists rejected the currently observed rates and assumed that the rates of upheaval had once been much greater. Their opponents, called **uniformitarians,** maintained a belief in the constancy of many observed rates and were thus driven into assuming a much older age for the Earth. But uniformitarianism was not yet a theory to be proclaimed openly. Cautious thinkers realized that it would be seen as an affront to biblical chronology and cosmology. For this reason, those who first expressed uniformitarian ideas often hastened to condemn their own carefully presented theories, disavow

them, or put them into the mouths of heathen characters like de Maillet's fictional sage, Telliamed.

Georges Buffon's *Théorie de la Terre,* published in 1749, contains a curious mixture of catastrophic origins with largely uniformitarian processes acting since the Creation. Naturalistic explanations are invoked throughout. Environmental determinism is used as an argument against the possibility of extinction and thus in support of the *scala naturae:* if climate changed, it would change gradually, and the inhabitants would survive by migration to a more favorable region. Furthermore, Buffon anticipated many later authors. For example, he explained the tendency for each species to multiply without limit, anticipating Malthus, and noted also that this tendency is balanced by a number of destructive forces that hold the species in check. He anticipated Darwin's observations on domestic animal breeds, noting that they were in each case derived from wild progenitors. He also anticipated Darwin's theory of Pangenesis (Chapter 6). He estimated the age of the Earth at some 72,000 years, far below any modern figure but well above all previous estimates. He divided the Earth's history into seven definite periods (Item 4.1 on page 69). Like de Maillet, he believed in the reality of fossils as indicators of past life. At least three separate issues can here be perceived: the antiquity of the Earth, the reality of fossils, and the possibility of extinction, the last of which Buffon denied. It was not until Lyell's great book of 1830 that any one scientist would hold our modern position on all three of these issues at once.

Werner's Neptunism

Before belief in the antiquity of the Earth could become widespread, a geologic time scale had to be established. The early attempts at this failed, however, because they confused **lithology** (rock type) with stratigraphic age. Several of these early attempts are compared in Item 4.2.

The most influential geologist of this period was Abraham Gottlob von Werner (1750–1817), a professor of geology and mineralogy at Leipzig. Werner's theory of Earth history is called **neptunism**. It postulates that most, if not all, of the Earth's rocks formed as chemical precipitates out of a "universal ocean." His stratigraphy (Item 4.2) confuses lithology with age; in modern terminology, Werner believed all igneous rocks to be Paleozoic and all sedimentary rocks to be Tertiary. Werner also generalized his system as pertaining to the entire world, not just to Saxony. It was perhaps Werner's greatest failing that he spent nearly all his life in Saxony and never traveled; geology as a modern science was founded largely by men like Lyell, who traveled extensively.

Hutton's Uniformitarianism

The real founder of the theory of uniformitarianism was a Scottish gentleman farmer-turned-scientist, James Hutton (1726–1797). According to story, he was standing by a stream in his native Scotland when he saw some soil break off a bank and begin washing downstream toward the sea. Looking back to the stream's source, he suddenly realized that the hills were slowly being eroded by the very process he had witnessed. Thus was he convinced of the immense slowness of geological processes, especially of erosion, a process whose effects were emphasized by all subsequent uniformitarians. The slowness of these processes at their present, uniformitarian rates required immense spans of time, but Hutton went beyond this to postulate *infinite* time in both past and future, with "no vestige of a beginning, no prospect of an end."

Hutton read his ideas before the Royal Society of Edinburgh in 1785. Three years later they were published in the society's *Transactions* under the title *Theory of the Earth,* a direct translation of Buffon's earlier title, *Théorie de la Terre.* In 1795, these ideas were expanded and published as a two-volume work, *Theory of the Earth, with Proofs and Illustrations.* Hutton's writing was unfortunately dry, far less exciting than Buffon's, and so the work was neglected by many. Furthermore, unlike Werner, Hutton was not a teacher and could therefore not rely on any students to advocate his ideas. Fortunately, a British mathematician and friend of Hutton's, John Playfair (1748–1819), restated Hutton's various theories in a much more readable volume entitled *Illustrations of the Huttonian Theory of the Earth* (1802).

Some scientists have wondered about Hutton's views on biological issues. Fortunately, he did leave a posthumous work, *Principles of Agriculture,* published in 1799. This little-read work, which deals largely with farming, contains a rather clear though brief statement of the theory of natural selection!

Fig. 4.1   Georges Cuvier (1769–1832).

Cuvier's Catastrophism

Despite Hutton, the majority of practicing geologists up to 1830 continued to support catastrophism. Perhaps the most forceful exponent of catastrophism was the French naturalist Georges Cuvier (1769–1832, Fig. 4.1), whose zoological theories were described in Chapter 3. Cuvier's studies on the fossils of the Paris basin convinced him that fossils were real, a fact that nobody has since doubted. He also realized that species could become extinct, and thus he came to differ with Lamarck on this important issue of the time. Cuvier's catastrophic theories are an attempt to account for these facts, as well as for the organization of rocks into strata and the discontinuities in the fossil record. In the Paris basin, Cuvier also discovered that certain fossils characterized particular strata and could thus be used for correlation. But unlike William Smith, who discovered the same principle independently and earlier (but published it later), Cuvier put this idea to no practical use.

Cuvier proposed to explain the layering of rocks into strata by claiming that each stratum was laid down rapidly during a flood or similar catastrophe. Since there were many strata, there must have been many such catastrophes; Cuvier viewed Earth history as a series of cataclysmic upheavals. The fossils in each layer represented animals that perished in the catastrophe which laid down that stratum, explaining why certain fossils characterize particular strata. The last such catastrophe was the flood of Noah mentioned in the Bible. Recall that Cuvier's four phyla of animals emphasized the discontinuities between them; here his geological theories emphasized the discontinuities between successive strata. We now realize, incidentally, that most strata are deposited very slowly, over long periods of time, and that the burial of fossils in them is anything but catastrophic.

Cuvier went so far as to unite his zoological, geological, and theological beliefs into a single coherent theory. God, said Cuvier, created life according to a Divine Plan, which included four major types of animals (Radiata,

Mollusca, Articulata, Vertebrata). Dissatisfied with his first Creation, God destroyed it by a great catastrophe, such as an earthquake or a flood, laying down a new geologic stratum and burying all the life of that period. God then proceeded with a somewhat better Creation, creating now better radiates, better molluscs, better articulates, and better vertebrates, but still following the same plan. Thus each of the four phyla would be represented in all the various strata (Box 4–A), and the Mollusca of successive strata would be progressively improved over those of earlier strata; the same would also be true of the other three phyla. This belief in successively better creations came to be known as **progressionism.**

BOX 4-A   *Cuvier's Stratigraphy*

A *highly diagrammatic* portrayal of Cuvier's conceptions. Sketches are illustrative, and do *not* represent actual species found by Cuvier.

Note that Cuvier's four phyla (Box 3–B) are represented in each stratum and are symbolized by A, B, C, and D. According to Cuvier, the animals in a given stratum (A_1, B_1, C_1, D_1) perished in the catastrophic destruction that caused the stratum to be laid down, and another, more perfect (i.e., more modern) creation succeeded them, only to perish again.

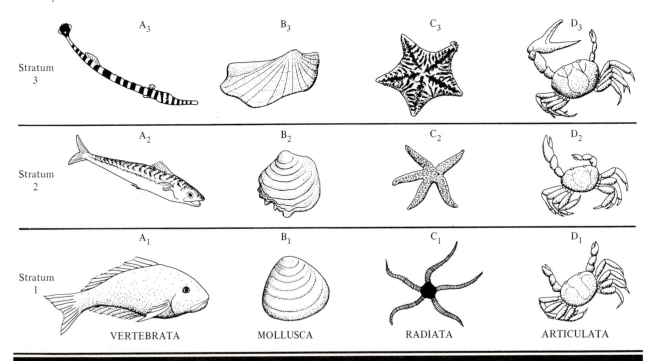

William Smith and Stratigraphy

Stratigraphy is the study of stratified (layered) rocks, and of the time periods they represent. Our modern theory of stratigraphy begins with William Smith (1769–1839) and his rejection of Werner's neptunism. Smith would not have considered himself a geologist at first: he was a surveyor and civil engineer in the early years of the British Industrial Revolution. A practical man, Smith cared little for theories such as Werner's because he could not put them to practical use.

One practical problem Smith faced was that of distinguishing between strata, a problem made more difficult by the fact that two or more distinct strata could be very similar in lithology. Smith discovered the principle of correlation by fossils, which states (1) that fossil assemblages can be used to characterize and identify particular strata, and (2) that rocks of the same age will have similar fossil assemblages. Smith also discovered that the strata always appeared in a definite sequence, with the oldest layers on the bottom and successively younger layers closer to the top. This principle, the **law of superposition**, had been discovered earlier by Nicolaus Steno and by Cuvier. But it was Smith who first put these ideas to practical use in the identification of similar-looking strata, the finding of coal-rich strata, and the planning of routes for barge canals. He also drew some of the earliest geological cross sections, which resemble the view of exposed rocks along the walls of such a canal.

Smith was interested in putting his ideas to use, not in publishing them. In 1799 he circulated among some friends a manuscript outlining his ideas. They urged him to publish this material, and he finally did in 1817, under the title *Stratigraphical System of Organized Fossils*. Smith's real major work, a geological map of England, was published in 1815; it was the first accurate geological map of any entire country.

Smith is often regarded as the founder of invertebrate paleontology, because the fossils he studied were nearly all invertebrate. Invertebrate paleontology and stratigraphy have been taught as closely related subjects ever since.

B. CHARLES LYELL AND THE UNIFORMITARIAN PARADIGM

Up to 1830, the majority of practicing geologists were catastrophists. But beginning in that year, the weight of

Fig. 4.2 Charles Lyell (1797–1875).

opinion swung to the uniformitarian position, and the modern science of geology was born, with the principle of uniformitarianism as its cornerstone. The scientist responsible for this change was Charles Lyell. Of all scientists, it was Lyell who seems to have had the most influence on Darwin.

Charles Lyell (1797–1875, Fig. 4.2) was born in the same year that Hutton died. Like Hutton, he was Scottish; his well-to-do family lived on a large estate in the Scottish countryside. The fact that his family wealth enabled Lyell to travel freely and extensively undoubtedly enriched his understanding of geology.

Lyell was the first to establish geology as an empirical science, free from speculations as to the Earth's origin or interior composition, matters that Lyell correctly regarded as insoluble in his own day. The divorce of geology from cosmology was clearly the decisive step, but without a theory as to the Earth's origins, how could one proceed to interpret any rock or formation? Here Lyell invoked the principle of uniformitarianism, saying that we must interpret ancient structures as formed by processes now operating. If we see the effects of contemporary processes, then similar effects in the past can be attributed to similar causes. Also, one could estimate the time of geological processes by extrapolating from present rates, for Lyell assumed (wrongly in this instance) that such rates would never vary.

The theory of uniformitarianism teaches that the forces that have shaped the face of the Earth in the past are the very same forces that can be seen at work today; thus the science of geology permits no explanation based on either miracles or cataclysms. There are three distinct senses in which the term *uniformitarianism* is used:

1. The principle that natural laws are invariable with the passage of time;

2. The use of this principle in studying past geologic history—"the present is a key to the past," to use Archibald Geikie's famous aphorism; and

3. The belief, now no longer held, that *rates* of geological processes are also invariable with time.

Lyell was a strict uniformitarian in all these respects, more so even than Hutton. But slow processes, it was argued, could hardly have a noticeable effect on mountains. To counter these objections, Lyell showed the benefit of his many travels in pointing out numerous instances of coastlines changed within historic time by the effects of wave erosion, habitations buried by volcanic ash and rubble, or coastlines that had noticeably subsided and sunk beneath the sea. One striking example of the latter was used by Lyell as his frontispiece illustration (Fig. 4.3): the Temple of Jupiter Serapis at Puzzuoli, near Naples, Italy, of which three columns still stand vertically. The base of each column shows that a certain amount of weathering has taken place by the abrasive action of sand and water; above this level, a portion of each column has been more severely weathered by the burrowing activities of a bivalve mollusc, *Lithodomus*. The temple, only 2–3 thousand years old at most, was presumably built above the water level but was subsequently submerged and has since been uplifted to its present elevation above the sea. The most heavily eroded portion of each column was apparently in contact with the sea water, but the base of each column was evidently buried in the bottom sediment and so escaped the ravages of the burrowing *Lithodomus*.

Lyell went still further, criticizing Cuvier's catastrophism and also his progressionism. He coined the term **nonprogressionism** to represent his own belief that there could be no progress, nor could there be any overall change, under uniformitarianism. The progress of life, either by successive creation (as in Cuvier's theory) or by transformation of one species into another (as in Lamarck's theory) could not be made consistent with Lyell's uniformitarian principles. Thus Lyell predicted that all

Fig. 4.3 The Temple of Jupiter Serapis at Puzzuoli, near Naples, Italy. This illustration is taken from the frontispiece illustration of Lyell's *Principles of Geology* (1830 edition). The temple, built by the Romans, was partly buried beneath the sea during historic times and has since been uplifted once again. A portion of each column shows the effects of the rock-boring clam, *Lithodomus,* whereas the bottom few meters was buried beneath the sea bottom and thus escaped the action of these clams. Lyell used this picture to demonstrate that geologic processes at work in the modern world could have significant effects over time. Reprinted with permission from Leonard G. Wilson, *Charles Lyell, The Years to 1841: The Revolution in Geology,* © 1972. New Haven and London: Yale University Press.

forms of life, including mammals, should be preserved in rocks of all ages, a prediction partially confirmed by the discovery of a few Jurassic mammals. Lyell's greatest hostility was directed at Lamarck, whom he criticized for denying the possibility of extinction as well as for his belief in evolution. Cuvier and other paleontologists, after all, had described extinct species, and by this time Buffon's belief—that these species had survived in some remote and unexplored part of the world—had become increasingly untenable. Lyell further argued against the possibility of evolution as follows: de Candolle has written that ''Nature is constantly at war with herself,'' and thus there will always be individuals who perish by disease or by the action of predators. In a variable species, however, it would not be the typical individuals who succumbed most often to premature death, but rather the deviant ones—either too great or too small, too thin-legged or too thick-legged—that would most often be the victims of predation or accident. Here we have, in Volume 2 of Lyell's *Principles of Geology,* a theory of natural selection, confined to the case of stabilizing or normalizing selection (Chapter 11) and used as an argument *against* the possibility of evolution!

Lyell's great work, the *Principles of Geology,* was originally published in two volumes, one of them in 1830 and the other in 1832. The work was revised throughout Lyell's lifetime, and it underwent more than a dozen editions. A staunch believer in nonprogressionsim, Lyell was at first critical of Darwin's theory when it was published in 1859, but subsequently he accepted Darwinism and even incorporated the theory into the last few editions of his own *Principles of Geology.*

Lyell did for geology what Newton had done for physics. In explaining phenomena using only known and measurable forces, he single-handedly established geology as a modern science, an effort for which he was later knighted by Queen Victoria. Geology was thus the first of the historical sciences to be placed on modern footing.

C. PROGRESSIONISM AND GEOLOGY, 1830–1860

The generation that intervened between the first appearance of Lyell's *Principles* (1830) and the publication of Darwin's *Origin of Species* (1859) witnessed great progress in geology. The principle of uniformitarianism, championed by Lyell, became firmly established, though it did for a time have a strong opposition in such men as William Buckland, Adam Sedgwick, and Roderick Murchison.

Many naturalists of this period also wrote on biological themes. Those who wrote books for popular consumption often combined biological and geological theories together, and books of this sort sold very well. With the coming of the Industrial Revolution, ''progress'' had become a watchword of the age, and the theories of Frenchmen like Cuvier, Buffon, and even Lamarck came to be widely read. Thus, while the professional geologists were slowly being won over to Lyell's theories, popular works of the period tended to emphasize progressionism.

Louis Agassiz and Nonevolutionary Progressionism

Jean Louis Rodolphe Agassiz (1807–1873) was a charismatic figure of the mid-nineteenth century. He was a student of Cuvier's, and like Cuvier, he was Swiss-born. As a zoologist, he was especially interested in fishes, and he undertook a great expedition to Brazil, during which he amassed an enormous collection. Throughout his life, he remained true to Cuvier's progressionism, including the belief in successively better creations. He also supported Cuvier's scheme of classification, in which all animals were grouped into four distinct phyla.

Agassiz's greatest contribution to geology was as a champion of the theory that a Great Ice Age had once covered all of Europe. Parallel scratches in rocks, similar to those formed by existing glaciers, could be seen over the face of Europe, even in places where no glaciers now exist. Most impressive were the large boulders, called ''erratics,'' which differed in lithology from any nearby formations. If its lithology was sufficiently distinctive, the source of such an erratic boulder could often be determined. In these cases, the direction of the boulder's transport was always the same as the scratches gouged in nearby rock formations through glacial action. Glacial ice was furthermore the only agency known that could transport such large boulders so far from their sources, sometimes hundreds of kilometers away.

Agassiz, a gifted speaker, now lectured to European audiences, thrilling them with his scientific tales of a Great Ice Age. He also toured America, and in 1846 he moved to Cambridge, Massachusetts, where he founded Harvard's great Museum of Comparative Zoology.

Agassiz clung throughout his life to many of Cuvier's theories, including his four phyla. He continued to uphold Cuvier's belief in progression through successive creations. Though he lived 14 years beyond the publication of Darwin's *Origin of Species,* Louis Agassiz never accepted the theory of evolution.

The "Minor Evolutionists"

During the first half of the nineteenth century, a number of so-called "minor evolutionists" anticipated Darwin's later theories in various respects. Often they had only one or two elements of Darwin's theory. Many of these "minor evolutionists" believed in progressionism; none of them appreciated the importance of Lyell's *Principles* or its "great gift of time." Among these "minor evolutionists" were the British naturalists Patrick Matthew and Edward Blyth and two American-born physicians, William Wells and J. Stanley Grimes. All four had important elements of Darwin's theory, though none had more than a few. None shared Darwin's insight into the importance of reconciling evolution with uniformitarianism. And none of their writings captured the public's imagination.

Vestiges of the Natural History of Creation

In 1844, an anonymous book was published that would soon capture the public's imagination. The *Vestiges of the Natural History of Creation,* as it was called, presented what appeared to be a well-thought-out scientific theory on the creation and subsequent evolution of the Earth and its inhabitants. At last the public thought it had found a definitive or at least a very entertaining work on all the major scientific issues of the day. The book was written in a very clear and appealing style, and it immediately became a commercial success. And if the book's contents did not provide sufficient reason for the public to read it, the mystery of its authorship did. It had obviously been written by a person well versed in science and fully in command of the English language. Yet no member of the scientific establishment was suspected of holding such Earth-shaking ideas. One of the most scandalous suggestions was that the book might have been written by Queen Victoria's consort, Prince Albert, who was known to have an interest in scientific subjects. Gos-

sip concerning the identity of the author was widespread, only helping to further publicize the book and ensure its commercial success. The book's publisher, Robert Chambers (1802–1871), refused to reveal the author's identity. Yet in his will, Chambers finally revealed what some had already come to suspect: he himself had authored the work.

Robert Chambers was a Scottish printer and publisher. Two aims seemed to guide most of his publishing efforts: a patriotic and sentimental devotion to his native Scotland and a desire to further the education of the masses. In accordance with the former, he published books praising the scenic beauty of Scotland, and he also reprinted much of Robert Burns's poetry. In accordance with the latter, he often included articles explaining some scientific subjects in everyday terms within the pages of his weekly newspaper. He also published many inexpensive pamphlets on educational subjects, again directed at the masses of working people. His acquaintance with the reading tastes of the masses certainly contributed to the commercial success of the *Vestiges.*

The ideas contained in the *Vestiges* included the nebular hypothesis of Kant and Laplace (Chapter 25), a progressionist geological theory reminiscent of Buffon, a theory of progressive evolution, and a number of other very controversial ideas. Here, too, were contained an account of an experiment in which life had supposedly been created using electricity (!), and a chapter on what would now be called the occult sciences. There was much in the book to attract the public's attention, and there was also more than enough to ensure the book's condemnation by all serious scientists, by the Church—indeed, by all the respected intellects of the day. One of the book's critics was an anatomist named Thomas Henry Huxley, who would later become famous as a defender of Darwin and who would later also apologize for having judged this prophetic book so harshly.

The lasting effects of the *Vestiges* were these: more than any other work, it had stimulated the public's thinking on the subject of evolution and laid the groundwork for the theory's later acceptance at the hands of Darwin. The public, as a result, had become accustomed to thinking about evolution; the idea no longer seemed shocking. The *Vestiges* also served as a lightning rod for criticism that would later be vented against Darwin, and in drawing off the most vehement criticism, it may have blunted some of the future anti-Darwinian attack. But seeing these criticisms launched against the *Vestiges,* Charles Darwin was driven to withhold his theory more than a decade longer, and to express his results in terms as mild and carefully reasoned as possible when he finally did publish. It is for this reason primarily that the *Origin of Species* seems in so many places to bend over backwards in an effort not to offend.

CHAPTER SUMMARY

Before 1830, most geological theories drew their ideas from speculations on the origin of the Earth. Most of these theories supported a belief in catastrophism, the explanation of Earth history in terms of cataclysmic upheavals and other events unlike the everyday occurrences of modern times. In particular, the finding of marine fossils well above sea level gave rise to a variety of speculations, including the origin of the land from the sea, the disappearance of a sea that was once more extensive, and the precipitation of rocks from a "universal ocean," this last being Werner's theory of neptunism. The reality of fossils was frequently questioned, and the possibility of extinction was denied. The age of the Earth was estimated to be rather small, only a few thousand years at most.

The use of fossils to identify and correlate stratified rocks was proposed independently by Georges Cuvier and by William Smith. Smith put his ideas to practical use, and he also discovered the law of superposition, which states that the older layers are always laid down on the bottom and the younger ones on top. Cuvier thought that each stratum represented the destruction of all organic life in a great catastrophe, followed by a new and increasingly perfect creation, which would repeat the process until the stratigraphic record became one of successively more perfect creations; this theory was progressionism.

The "great gift of time" was given to the historical sciences by the principle of uniformitarianism, which states that natural laws are invariant with time, and that the present is therefore a key to the past. The true founder of uniformitarianism was James Hutton, but the theory did not become widely accepted until championed in its modern form by Charles Lyell, beginning in 1830. Lyell also divorced geology as a science from the speculations of cosmology. He rejected Cuvier's progressionism and instead believed in the constancy of all life forms, a theory he called nonprogressionism. It was Lyell who first made geology into a modern science.

Some geologists who published after Lyell still clung to progressionism. Louis Agassiz supported most of Cuvier's theories, including a nonevolutionary progressionism. He also supported the Ice Age theory. A number of "minor evolutionists" correctly perceived a few of the key elements of Darwin's later theories, but none had them all, and none were reconciled to uniformitarian geology. Another pre-Darwinian theory of evolution was expressed in the book *Vestiges of the Natural History of Creation,* which weathered much of the criticism that would otherwise have been vented even more forcefully on Darwin. This book acquainted the public with evolution and paved the way for the acceptance of Darwin's later theory.

FOR FURTHER KNOWLEDGE ▰▰▰▰▰▰

ITEM 4.1 *Buffon's Seven "Epoques" of Earth History (1778)*

1. The formation of the Earth and planets in a molten state, from matter ejected from the sun during its collision with a comet.

2. The age of mountain-building, when the Earth was solid but too hot to touch.

3. A stage when the Earth was covered by a "Universal Sea."

4. A stage of volcanic upheavals and withdrawal of the Universal Sea.

5. A period in which elephants and other tropical beasts roamed over all the continents, which were all connected by land bridges.

6. The period in which the present continents became separate from one another and climatic zones were formed.

7. The present period, "when the power of man has seconded that of nature."

ITEM **4.2** *Some Early Stratigraphical Divisions*

(*Reminder:* All stratigraphic columns should be read from the *bottom up.*)

Lehmann	Arduino	Werner	Modern names	
Unconsolidated and loosely consolidated sedimentary material, formed since the time of Noah's flood	Quaternary or Alluvium: earthy and rocky materials that rest on all the preceding	Alluvium (*Aufgeschwemmtes-gebirge*): "Mountains formed since the Flood by local, special events (earthquakes, volcanoes, flooding by rivers and the sea), and consisting of unconsolidated materials."	Cenozoic	Quaternary
Stratified, fossiliferous rocks, formed during Noah's flood	Tertiary: low mountains and hills, generally composed of clays, sand, gravel, and volcanic materials	*Flötzgebirge:* "Mountains formed at the time of Noah's flood, and consisting of non-crystalline rocks with regular layers; mineral deposits as beds (sedimentary rocks, coal)."		Tertiary
	Secondary: mountains built of marble and limestone	"Transitional series"	Mesozoic	
Primary crystalline rocks, containing no fossils, formed at the beginning of the Earth	Primary: mountains containing metallic ores and crystalline rocks	*Urgebirge* or *Erzgebirge:* "Mountains formed at the formation of the Earth and consisting of crystalline rocks with steep and unsystematic layering; mineral deposits as veins."	Paleozoic	

5

Darwinism

A. DARWIN'S LIFE

Of those thinkers mentioned in the past two chapters, none compares in importance to Charles Darwin (1809–1882) as an evolutionary theorist. Darwin was in many respects a self-taught naturalist, though he read avidly and was thus influenced by many earlier naturalists through their writings. Perhaps the strongest early influences on Darwin were those of Rev. William Paley (Chapter 3) and of Charles Lyell (Chapter 4). Paley's writings drew Darwin's attention to many marvelous and precise adaptations. Lyell's *Principles of Geology* stimulated Darwin's thinking as a geologist. The extrapolation of modern processes, not just into the past but over the long course of past geologic time, was an important mental habit Darwin learned from Lyell's geology.

Early Life and Education

Charles Robert Darwin was born on February 12, 1809, the very same day as Abraham Lincoln. His father was Robert Darwin, a well-to-do country physician in Shrewsbury. Both of his grandfathers had been distinguished: Erasmus Darwin (1731–1802), Robert's father, had also been a doctor by profession but was better known in those days for his verse, which was usually judged to be entertaining; it was only after the *Origin of Species* that anyone presumed to inspect the scientific contents of Erasmus Darwin's writings. Darwin's other grandfather, Josiah Wedgwood (1730–1795), was a well-known potter and designer of fine porcelain ware, one of the leading figures in the Industrial Revolution in England. This Josiah was the grandfather not only of Charles Darwin but also of Emma Wedgwood, whom Darwin would later marry.

Young Charles was an undistinguished student in his home town of Shrewsbury. He cared little for his studies, which were largely in the Greek and Roman classics. He loved the outdoors, especially hunting and riding. His preference for sports over studies, which remained with him through college, once provoked his father to exclaim, "You care for nothing but shooting, dogs, and rat-catching, and you will be a disgrace to yourself and all your family."

What little aptitude Charles Darwin showed was for science, and so his father sent him to Edinburgh to follow a family tradition in the study of medicine. Darwin found the lectures dull and the clinical aspects far too

71

"ghastly" in those days before anesthesia; a class visit to a morgue proved especially repulsive to him. Out of dutiful respect for his father's wishes, he stayed at Edinburgh for two years but meanwhile neglected his studies. Instead, he kept company with some local naturalists, including a geology professor who espoused Werner's neptunism and a botanist, Robert Grant, who told him of Lamarck's theories, which he later recalled listening to in "silent astonishment."

When it had become abundantly clear that Charles Darwin would not become a doctor and had never shown any interest in law, his father decided that Charles should study for the ministry. To this Charles consented, provided it was a *country* parsonage for which he was destined, for this would certainly give him time to enjoy the great outdoors and to carry on his nature-loving hobbies. So Charles was transferred to Cambridge, where he spent three years studying for the ministry. Religion at Cambridge was nearly synonymous with the Natural Theology of the Rev. William Paley (Chapter 3), and Darwin was so thrilled with two of Paley's books that he read a third even though it wasn't required. He also hobnobbed with naturalists, including Adam Sedgwick (1785–1873), the clergyman-turned-geologist, William Whewell (1794–1866), the great philosopher of science, and John Stevens Henslow (1796–1861), a botanist who became one of Darwin's lifelong friends. During his free time, Darwin went hunting in the local woods and collected beetles. He took far more interest in his life as an outdoor naturalist than he did in his studies.

The Voyage of the *Beagle*

In 1831, after receiving his Bachelor's degree from Cambridge, Darwin received a letter. A ship was fitting out for a two-year voyage (it would later take five years) and needed a naturalist. In those days it was customary for any ship on a long voyage to have on board a naturalist who would collect specimens wherever the ship went and study the local faunas, floras, and geology. One of Darwin's professors, J. S. Henslow, had been offered the position himself, but in declining it, he had nominated Darwin instead. He knew that Darwin would like to go on such a voyage, because Darwin had expressed just such a desire after reading one of the travelogues of Alexander Humboldt. But Darwin's father refused: the voyage was a worthless diversion from which young Charles

could not at all profit, and it would distract him from his chosen profession. Robert Darwin even dared his son to find "any man of common sense" who would favor the voyage, promising his consent if Charles could. Disappointed, Charles wrote Henslow declining the offer and instead took a trip to the house of his Wedgwood cousins some 30 kilometers away.

Darwin's uncle, Josiah Wedgwood, was the son of the famous porcelain maker Josiah and also the father of Emma Wedgwood, Charles Darwin's future wife. Charles told his uncle of the voyage that his father had made him turn down, but to his delightful surprise, Uncle Josiah was strongly in favor of the voyage. Here was a "man of common sense" who could help Darwin obtain his father's consent! Uncle Josiah wrote a letter answering each one of Dr. Robert Darwin's objections and stating in particular that "the pursuit of Natural History . . . is very suitable to a clergyman," a statement reflecting the mood of the times and the influence of Paley's Natural Theology. On the day that the letter was sent, however, Uncle Josiah decided to go to Shrewsbury personally and plead young Darwin's cause; Josiah was so persuasive that Darwin's father consented that very day.

His Majesty's Ship *Beagle*, under the command of Captain Hugh FitzRoy, was to undertake a geographical survey of the South American coastline, mapping out for the Admiralty the correct longitude and latitude of the various coastline features. The ship, which then would circumnavigate the globe, was also to return to the island of Tierra del Fuego three natives that FitzRoy had brought with him to England on a previous voyage. The voyage (Fig. 5.1) would last five years (1831–1836) and would shape young Darwin's entire career.

To keep himself occupied during the long intervals at sea, Darwin took along two books. One was by Paley; the other was the first volume of Lyell's *Principles of Geology*. Henslow had urged Darwin to read Lyell's book, but not to believe a word of it. Darwin had gone on a geological field trip with Adam Sedgwick during the previous summer, and upon arriving at the ship's first destination, St. Jago in the Cape Verde Islands, Darwin was immediately impressed with its volcanic geology. Geology continued to be Darwin's major preoccupation along the entire voyage, and contrary to Henslow's advice, he became firmly convinced of the truth of all that was in Lyell's *Principles*.

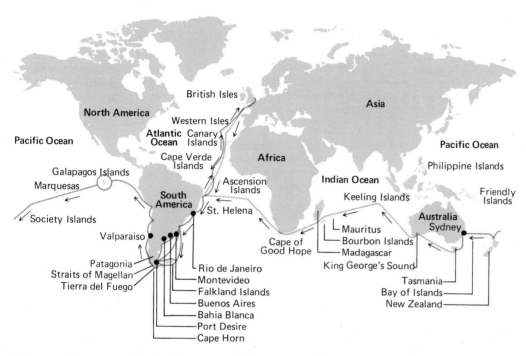

Fig. 5.1 The voyage of the *Beagle*. Reprinted with permission from William C. Schefler, *Biology: Principles and Issues,* © 1976. Reading, Massachusetts: Addison-Wesley (Fig. 16.10).

The ship spent much of its time surveying and resurveying the South American coastline between Brazil and Tierra del Fuego. During this time, Darwin made several trips inland to do collecting. Here he witnessed a luxuriant and unique fauna and flora, very different from that of other tropical regions. He studied its latitudinal variation from north to south, but he was more impressed with its uniqueness. In Montevideo, he purchased a copy of the second volume of Lyell's *Principles,* published since the *Beagle* had left England. In Argentina, Darwin hunted gaucho-style, using a bolo. He also discovered fossils of animals similar to the living South American forms; these he sent to Sir Richard Owen, Britain's leading anatomist, for identification.

Early in the voyage, Darwin showed himself a strong believer in the Anglican creed; he was visibly shocked when a fellow crew member expressed disbelief in the literal truth of Noah's flood. At Cambridge, Darwin had earlier written that he had not the "least doubt" about "the strict and literal truth of every word in the Bible."

It was probably his geological observations and the reading of Lyell's *Principles* that convinced him otherwise, at least regarding Noah's flood.

In the summer of 1834, the *Beagle* sailed through the Straits of Magellan; it would spend the next year along the coast of Chile. Here the Andes Moutains, part of the great Cordilleran system, come within easy reach of the coast, and Darwin made several inland trips to examine their geology. These mountains are still relatively young and persistently active, and on February 20, 1835, Darwin witnessed a rather severe earthquake that devastated a nearby city. The seeing of such geological activity in the present convinced Darwin that similar events in the past had led to other episodes of mountain-building, perhaps to the uplift of the Andes as a whole. Darwin would later witness several more earthquakes and study the effects both of geologically recent uplifting and of subsidence, along the coast and on several offshore islands. His theory that coral atolls develop atop undersea moutains as they subside was thought out on the coast of

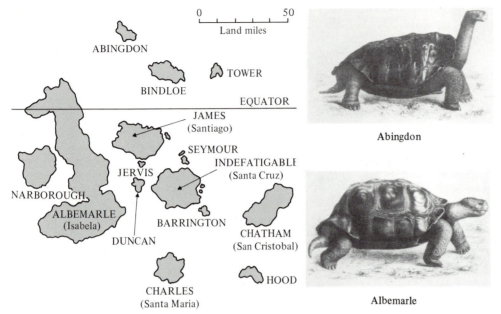

Fig. 5.2   The Galapagos Islands and some of the giant tortoises characteristic of each island. The map is adapted with permission from John W. Kimball, *Biology,* 4e, © 1978. Reading, Massachusetts: Addison-Wesley (Fig. 32.6). The tortoises are from Stebbins, *Processes of Organic Evolution,* 3e, © 1977, p. 10. Reprinted with permission of Prentice-Hall Inc., Englewood Cliffs, N.J.

South America, before Darwin had ever seen a coral reef. In addition to his geological observations, Darwin studied the local fauna of the region, which differed in particulars from that of the Argentine Pampas and Brazil but still included the same peculiarly South American elements.

The peculiarity of the South American fauna would figure prominently in Darwin's later work. In his *Origin of Species,* Darwin would later argue that the great similarity among South American animals betrays a common heritage whose effects outweigh even those of climate (see quotation, page 85–86): South American animals bear strong resemblance to one another, even when they inhabit different climates, and they meanwhile show no relationship with the faunas of Africa or Australia under closely similar climates. Such matters were undoubtedly already on Darwin's mind as the *Beagle* left the Peruvian coast late in the summer of 1835.

On September 15, 1835, Darwin and the *Beagle* arrived in the Galapagos Islands, an archipelago more

than 800 km (500 mi) from the South American coastline. Darwin was immediately struck by the volcanic nature of the islands; they were geologically very similar to the Cape Verde Islands, which the *Beagle* had visited on its very first stop. Yet the fauna of these islands was strikingly peculiar: there were no land mammals, save for one species of mouse on one island only. No frogs or other amphibians were present. Several quite peculiar reptiles were present, including iguanid lizards, distant cousins of other species found in South America. Most spectacular among the reptiles were the giant tortoises (Fig. 5.2), which not only were peculiar but differed as well from island to island: the local governor of the islands even showed Darwin how to distinguish the various kinds and to determine the island from which each had come.

Perhaps the greatest faunal diversity on these islands lies in their land birds, which are mostly finches of the subfamily Geospizinae, sometimes called "Darwin's finches" (Fig. 5.3). This subfamily occurs nowhere else, but other groups of finches inhabit the mainland of

Fig. 5.3 Darwin's finches (subfamily Geospizinae), an example of adaptive radiation (compare Chapter 18). (a) Males and females of each species. (b) Family tree showing probable lines of descent. (1) large ground finch (*Geospiza magnirostris*); (2) medium ground finch (*G. fortis*); (3) small ground finch (*G. fulginosa*); (4) sharp-beaked ground finch (*G. difficilis*); (5) cactus ground finch (*G. scandens*); (6) large cactus ground finch (*G. conirostris*); (7) vegetarian tree finch (*Camarhynchus crassirostris*); (8) large insectivorous tree finch (*C. psittacula*); (9) Charles Island tree finch (*C. pauper*); (10) small insectivorous tree finch (*C. parvulus*); (11) woodpecker finch (*C. pallidus*); (12) mangrove finch (*C. heliobates*); (13) warbler finch (*Certhidea olivacea*); (14) Cocos Island finch (*Pinaroloxias inornata*). Part (a) is reprinted with permission from David Lack, *Darwin's Finches,* © 1947. New York: Cambridge University Press (Fig. 3). Part (b) is reprinted with permission from Jeffrey J. W. Baker and Garland E. Allen, *The Study of Biology,* 3e, © 1977. Reading, Massachusetts: Addison-Wesley (Fig. 20.20).

(a)

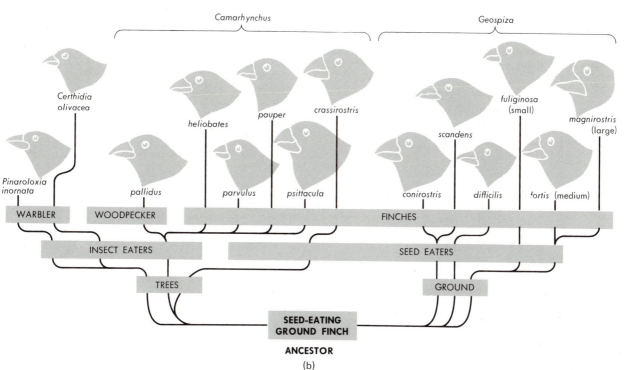

(b)

South America. More than a dozen finches inhabit the Galapagos Islands, and each has its own pattern of distribution among the several islands, many species being confined to a single island or even part of an island. A good deal of variation was evident—for instance, in beak size and shape. The habits and diets of these birds likewise varied: some lived on the ground and some in trees; some ate insects and others seeds. Several species of finch were filling the roles normally occupied by totally unrelated birds on other islands or continents. The hummingbird niche, for example, was occupied by a small finch with a long, recurved beak, suited to nectar-feeding, and very rapid wingbeats that enabled it to hover over a particular flower, just like a true hummingbird. A flycatcher finch had a very erratic flight and a wide gape, which enabled it to catch flying insects on the wing in the fashion of a true flycatcher. A woodpecker finch had a chisellike bill for drilling holes in trees and even a woodpeckerlike tail whose stiffened feathers enabled the bird to prop itself up on a vertical trunk. On one island there was also a curious cactus finch having no exact equivalent among birds, but it was perhaps more similar to the desert reptiles of other places.

Darwin was quite familiar with the theories of Lamarck and other environmentalists who would attribute most differences in adaptation to the effects of climate. But here was a group of islands lying close together, all within the same climatic zone, yet each having distinct species. Why should the birds and tortoises differ from island to island? Why, on the other hand, should they be so similar to one another—why should all the birds be finches, for example? And why should there be so many species with clear affinities to South American groups? If climate, geology, or physical conditions were responsible, then why was there no clear resemblance to the fauna of the geologically similar Cape Verde Islands off Africa? It was probably here in the Galapagos, or shortly thereafter, that Darwin adopted his belief in ''descent with modification,'' but for the moment, at least, he kept these views hidden.

The *Beagle* completed its circumnavigation of the globe, spending some time in Australia and New Zealand, where Darwin observed a fauna even more peculiar than that of South America. The ship also stopped at various islands, including Tahiti, Mauritius, St. Helena, Ascension, and others, so that Darwin became acquaint-

Fig. 5.4 Charles Darwin as a young man and as an old man. Reprinted courtesy of the Burndy Library, Norwalk, Connecticut.

BOX 5-A *Darwin's Theory of Coral Reefs*

Corals are tiny animals (phylum Cnidaria) whose skeletons may form reefs. Darwin knew (from reading) that corals grew only near the surface, yet the reefs formed by their skeletons often reached great depths. He used this knowledge to assert his theory of coral reefs (illustrated here with Darwin's original diagrams). An island with sea level at A–A may acquire an original growth of coral along its fringes, forming what Darwin called a *fringing reef.* If the island slowly sinks, the coral will sink with it but continue to grow upwards near the surface. By the time that sea level reaches A'–A', a *barrier reef* will form a ring around the original island. A lagoon (C—note ship) forms behind the barrier reef in still waters not favorable to the growth of coral. If the island sinks still further until the sea level reaches A"–A", then the original island disappears, leaving only a ringlike *coral atoll* (often a group of small islands forming a ring) surrounding a shallow, quiet-water lagoon.

Darwin explains that he thought out this entire theory while on the coast of South America, before even seeing his first

coral reef. Nevertheless, his theory has stood the test of time, and it still forms the basis of our modern theory. The modern theory also allows the sea level to rise with glacial melting. We also now realize that corals' ability to grow only near the surface is the result of their dependence on symbiotic algae.

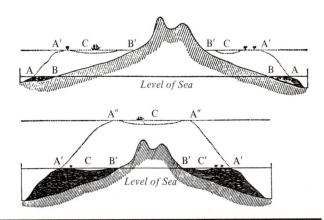

---

ed with many other island faunas. The ship also stopped in Brazil for one final time, in order to check its longitude determinations. "On the 2nd of October [1836] we made the shores of England; and at Falmouth I left the *Beagle,* having lived on board the good little vessel nearly five years."

From the *Beagle* to the *Origin*

The years after the *Beagle* voyage were busy and eventful for Darwin. Aside from his marriage to his cousin, Emma Wedgwood (January 29, 1839), he was kept busy largely with the publication of the results of his *Beagle* voyage. Foremost among these results was a theory for the origin of coral reefs, in which Darwin attributed the circular shape of atolls to the continued growth of coral on the fringes of a subsiding island (Box 5-A).

Darwin was not neglecting what would later prove to be his most notable work, however. In 1837, he opened his first notebook on the subject of species, in which nearly all components of his later theory were already clearly set down. He had already come to believe in what he would later call "descent with modification,"

that is, a branching evolutionary process in which related species share a common ancestor. Somewhat horrified at his own conclusions, Darwin wrote, "I have reached the conclusion that species are after all (It is like confessing a murder!) mutable." Darwin soon inquired of various animal breeders of the methods they had used in producing the desired characteristics of their several domestic breeds (Fig. 5.5). He also joined a London pigeon-fanciers club, again with an interest toward the origin of the various domestic breeds.

One problem apparently plagued Darwin more than any other: he had discovered in the transmutation of species an important phenomenon, yet he knew of no mechanism that would suitably explain it. Others before him had proffered theories of transmutation, or what we would now call evolution, yet each had failed for want of a mechanism. This was Darwin's objection to their theories: the environmental determinists had proposed a mechanism at variance with Darwin's observations from his *Beagle* voyage. Lamarck's theory had been roundly criticized by Lyell; many other theories failed by being too strongly linked to catastrophism. Many of Darwin's

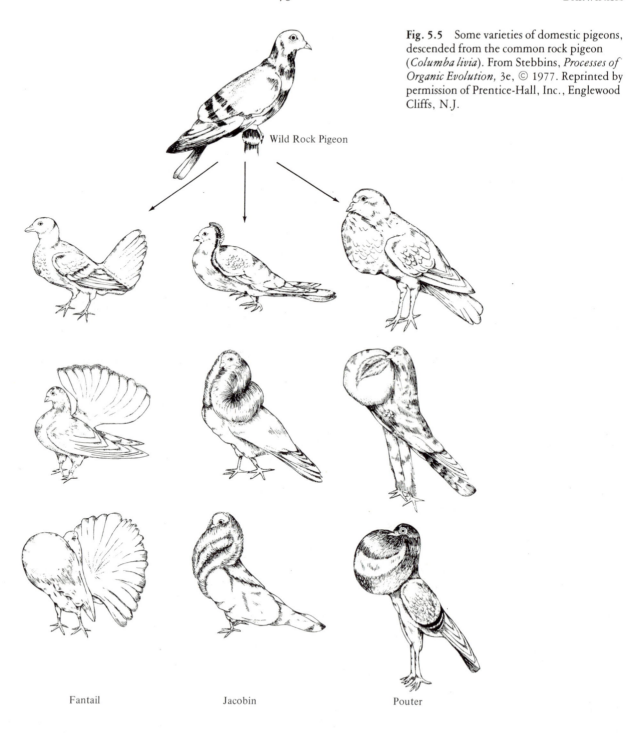

Fig. 5.5 Some varieties of domestic pigeons, descended from the common rock pigeon (*Columba livia*). From Stebbins, *Processes of Organic Evolution*, 3e, © 1977. Reprinted by permission of Prentice-Hall, Inc., Englewood Cliffs, N.J.

Wild Rock Pigeon

Fantail   Jacobin   Pouter

predecessors had proposed no mechanism at all! Darwin's notebooks of 1837 and his discussions with animal breeders beginning that same year show that he had already begun to search for a mechanism.

In 1838, Darwin read *An Essay on Population* (1799) by Thomas R. Malthus (1766–1834). This, Darwin later indicated, was the inspiration for his theory of natural selection, though Eiseley (1958, 1959) doubts this story, saying that Darwin had already thought of all the important ideas himself and had written them down in his notebook before he ever read Malthus. Perhaps Malthus stated very clearly what Darwin was already mulling over in his mind but could not express; perhaps, too, he was already inclined to believe in natural selection before he read Malthus, and the reading of Malthus became especially memorable to him as a revelation of the ideas that he already held unconsciously. Malthus, an economist, wrote that all species had an inherent tendency to increase in geometric proportions, outstripping their available resources and in particular their food supply. Land and other limited resources would thus increase in value, while the population would become increasingly crowded and increasingly miserable unless it could check its own growth. Here Malthus distinguished "preventive checks," from "positive checks"; the former included delayed marriage and "prudential restraint from marriage," both of which he advocated, and birth control, which the Reverend Malthus clearly foresaw but did not advocate. Malthus warned that unless the preventive checks were instituted, population would increase to the point where the positive checks, such as famine, war, and epidemic diseases, would operate. He therefore concluded that any aid to the poor was counterproductive and would tend to compound their misery, and that human affairs could in general not be improved.

Darwin was hardly in sympathy with Malthus's social views; he was rather a liberal humanitarian on most social issues of his day, though he shied away from any controversy, political or otherwise. Darwin did, however, adopt Malthus's argument that a population would tend to increase geometrically and outstrip its resources, and that many individuals would perish as a result. This principle applied to all living species, but its effects were anything but random: the more hardy or better-suited individuals would have an advantage, and the weaker or less-suited would be at considerably greater risk. Darwin saw

an analogy to this in the practice of animal breeders: they modified domestic species at will through **artificial selection,** selecting only those individuals for breeding that have the most desirable characteristics. The analogous process in nature, which Darwin believed to proceed far too slowly to observe, he called **natural selection**.

Here was a mechanism that Darwin could believe in. What was more, it was a uniformitarian mechanism, whose effects in the past could be inferred from its operation in the present. The possibility of change through such selection was amply demonstrated to Darwin by the success of animal and plant breeders over the years.

In 1842, Darwin consolidated all his ideas on species and wrote out a manuscript version of his "Essay on Species," which he expanded and rewrote in 1844. It seemed that he was preparing to write a book on this subject, but suddeny he changed courses and set the work aside. Instead, we find that Darwin began a series of studies on barnacles, sessile crustaceans of the subclass Cirripedia, and he soon became an expert on them, publishing four monographs on both living and fossil barnacles during the years 1851–1854. Even then, he did not attempt to publish his evolutionary ideas on species until he was forced into doing so by Wallace in 1858 (see below), some 22 years after he had returned from the *Beagle*. Why did Darwin set aside his early "Essays," and why was he reluctant to publish? A very obvious reason for both of these can be found: it was Robert Chambers's publication in 1844 of the *Vestiges of the Natural History of Creation* and the subsequent criticism and ridicule that were heaped on this book (see Chapter 4).

Darwin's studies on barnacles convinced him that his theory of "descent with modification" was fully consistent with the fossil record of this group, and furthermore that classification could be understood to be based on genealogical principles. These observations helped Darwin strengthen his argument when he finally did publish the *Origin of Species*.

For 23 years following his return from the *Beagle*, Darwin withheld his views on the transmutation of species from the British public. He might have withheld them even longer, had not his hand been forced by Alfred Russel Wallace (1823–1913). Wallace was a naturalist working in the Malay Archipelago, roughly corresponding to the modern nation of Indonesia. Like Darwin, he arrived independently at the conclusion that species were mutable, and that natural selection was the force principally responsible. Again like Darwin, he credited Thomas Malthus with the inspiration of his theory. By his own account, he was stricken with malarial fever on the island of Ternate, one of the Moluccan Islands south of the Philippines, now governed by Indonesia. As he lay sick and rather near death, he became delirious. In this delirium, it occurred to him that death, whose necessity Malthus had predicted, would fall unequally on the several varieties of a variable species. Thus, the more-favored varieties would be naturally selected, and the less-favored varieties would perish in greater numbers and eventually become extinct. Fortunately for science, Wallace recovered from his fever, wrote down his theory, and dispatched it to London in a letter.

In 1858, Darwin received this letter, which Wallace had addressed to him. Wallace had asked Darwin to transmit his manuscript to Lyell for presentation before the Linnean Society if Darwin found it worthwhile. Darwin read Wallace's manuscript and immediately passed it on to Lyell and to Joseph Hooker (1817–1911), a botanist who had been privileged to read Darwin's "Essay" of 1844. Darwin wanted very much to avoid a fight over priority and offered to destroy all his writings and allow Wallace to take credit, lest it be thought that he had stolen the idea from Wallace. Darwin expressed these feelings to Hooker and to Lyell, in whose trust he placed not only Wallace's manuscript but also his own earlier "Essay" and some letters in which he had mentioned his thoughts on evolution long before reading Wallace's paper. He did point out that he had earlier thought of every point Wallace mentioned, and that furthermore he had supported his ideas with much more evidence than had Wallace. Lyell and Hooker wished to do justice to both Darwin and Wallace, but they also saw it as a duty to science that Darwin's lengthier and more detailed account be published in addition to Wallace's. Accordingly, Wallace's paper was published by the Linnean Society in 1858, together with remarks by Hooker and Lyell explaining the independent discoveries, excerpts from Darwin's "Essay" of 1844, which Hooker had previously seen, and excerpts from a letter that Darwin had more recently written to the American botanist Asa Gray, showing that his ideas had remained unchanged since the 1844 "Essay." Lyell and Hooker also expressed their desire to see the appearance of "Mr. Darwin's complete work."

Darwin's original intention was a much lengthier and more thoroughly documented work, but he reluctantly agreed to prepare for publication "An Abstract of an Essay on the Origin of Species by Natural Means of Selection." Darwin's publisher, John Murray, wisely insisted that Darwin drop the word "Abstract" from the title. Because of the Linnean Society papers of 1858, the *Origin of Species* was widely talked about even before it was published. The publisher was sold out of the entire first printing on the date of its issue, November 24, 1859, and the second printing likewise sold out a month later. All told, the book went through six British editions and various American editions based on these, and it was translated into many other languages.

B. *ON THE ORIGIN OF SPECIES*

The most famous of all evolutionary writings, Darwin's *Origin of Species* is also the most influential book ever written on any biological subject. No other biological book, either before or since, has ever had such a profound effect on all avenues of research or as much influence outside the realm of biology. No other biological book has ever provoked as much comment or controversy or spawned even half the intellectual progeny. Nor has any other book on any biological subject so changed our outlook on nature or on our own humanity.

Darwin's full title reveals what the book is about: *On the Origin of Species by Means of Natural Selection, or the Preservation of the Favoured Races in the Struggle for Life.* The book's major thesis is that species have evolved from preexisting species through a branching process ("descent with modification"), and that natural selection has been the primary means by which evolutionary change has been achieved.

Natural Selection

Darwin's first four chapters are devoted to the mechanism of evolutionary change. Darwin put these first, presumably, because he felt that no theory of evolution could hope to be taken seriously without a credible mechanism—the major fault he had found with all previous evolutionary theories. The book begins on secure footing: Chapter 1 deals with the variation of domestic animals and plants and the manner in which animal and plant breeders have produced these variations by generations of careful selection. Nobody could doubt the existence of domestic varieties, and their derivation from earlier stock, ultimately from wild stock, is in many cases a matter of historical record. Change has thus occurred in these domestic species through the agency of **artificial selection**: breeders have consciously selected the best of their flocks, meaning those closest to the breeders' own goals or ideals, and allowed these to reproduce while preventing the other members of the species from doing so. The effect of methodical selection of this sort is easily documented from the breeders' annals, and it demonstrates very clearly the simple fact that selection can produce change. Casual selection would also have much the same effect in the long run, only more slowly.

Chapter 2 begins with an attempt to show that wild species, like domestic species, are likewise variable, and that much of this variation is inherited. "Any variation which is not inherited is unimportant for us." From here, however, the chapter drifts into a discourse on nominalism, in which Darwin looks "upon the term species as one given arbitrarily, for convenience sake." Here Darwin put too much credence in Linnaeus's assertion that the species were immutable because of their reality, by which Linnaeus meant that each had a fixed and eternal *eidos*. Darwin rejected immutability in favor of evolution, and in adopting Malthus, he had also embraced population thinking and rejected the Platonic *eidos*. It was for these reasons that he felt it necessary to deny the objective reality of species by declaring species divisions to be arbitrary, thus adopting the same nominalist position that Buffon had at one time embraced (Chapter 3). The strong influence of nominalism is seen in Darwin's ambiguous use of the word "variety," pertaining sometimes to individual variations and at other times to subspecies, domestic breeds, and other definable populations.

Chapter 3 shows clearly the influence of Malthus. "There is no exception to the rule that every organic being naturally increases at so high a rate, that if not destroyed, the earth would soon be covered by the progeny of a single pair. . . . Hence, as more individuals are pro-

duced than can possibly survive, there must in every case be a struggle for existence, either one individual with another of the same species, or with the individuals of distinct species, or with the physical conditions of life. It is the doctrine of Malthus applied with manifold force to the whole animal and vegetable kingdoms; for in this case there can be no artificial increase of food, and no prudential restraint from marriage." This argument is carried through the chapter and is documented in two ways: first, by theoretical calculations showing that even the slowest breeders have a prodigious capacity for increase, and second, by reference to cases where new species had been introduced to a region by human agency. These introduced species, kept in check in their own native lands, expanded promiscuously and without restraint in the absence of such checks. This one phenomenon, which Darwin illustrates with several examples, provokes a discussion of the forces that normally keep a population in check. Here Darwin shows that he was an ecologist before the word "ecology" was coined, and that he understood the complex interrelations between species under all conditions.

All species have an inherent tendency for prodigious increase, but they are normally kept in check by such forces as predation, disease, the limited availability of food, and the limited availability of places suitable to the germination of seeds or the laying of eggs. Now, if a species is subject to heritable variation (as most species seem to be), "can we doubt . . . that individuals having any advantage, however slight, over others, would have the best chance of surviving and of procreating their kind? On the other hand, we may feel sure that any variation in the least degree injurious would be rigidly destroyed. This preservation of favourable variations and the rejection of injurious variations, I call Natural Selection." Natural selection is no more than a differential capacity for survival and reproduction. It is Darwin's reliance on natural selection as the major force of evolutionary change that most clearly distinguishes his theory from other evolutionary theories.

Darwin is now ready to deprive Paley's supporters of their most forceful type of argument. He builds his case for the plausibility of natural selection by comparing it with artificial selection: "As man can produce and certainly has produced a great result by his methodical and unconscious means of selection, what may not nature effect? Man can act only on external and visible characters: nature . . . can act on every internal organ, on every shade of constitutional difference, on the whole machinery of life. Man selects only for his own good; Nature only for that of the being which she tends. . . . Under nature, the slightest difference of structure or constitution may well turn the nicely-balanced scale in the struggle for life, and so be preserved. How fleeting are the wishes and efforts of man! how short his time! and consequently how poor will his products be, compared with those accumulated by nature during whole geological periods. Can we wonder, then, that nature's productions should be far 'truer' in character than man's productions; that they should be infinitely better adapted to the most complex conditions of life, and should plainly bear the stamp of far higher workmanship?'' ''Natural Selection . . . is a power incessantly ready for action, and is as immeasurably superior to man's feeble efforts, as the works of Nature are to those of Art.'' Here Darwin is supporting a statistically mechanistic force, natural selection, by the very ''stamp of far higher workmanship'' that the Natural Theologists had used in support of God's existence.

Darwin supplies several illustrative examples to show that natural selection accounts very well for characters that appear very slight in importance, for rudimentary organs, and for the degeneracy of unused organs. Darwin also outlines how flower structure can be altered by selection of the most efficient means of pollination, and how sexual selection, the selection of one sex by the other, can effectively develop ornate structures or weapons peculiar to one sex. Yet, in his chapter on the ''laws of variation,'' he declares: ''Our ignorance of the laws of variation is profound.'' Darwin could document that heritable variations were indeed produced, but the simplest laws of heredity or of variation lay unknown to the world of science until Mendel was rediscovered in the year 1900.

Some of Darwin's interpreters, including both supporters and opponents, misunderstood Darwin's major contribution, the theory of natural selection. Most of these misunderstandings were based and continue even nowadays to be based on a failure to appreciate the nature of population thinking. No variation can ensure 100% survivorship or 100% fertility or 100% survival among offspring. And few variations are 100% fatal under any and all conditions. Most variations fall between these extremes. Just as low-ranked teams may win some games and high-ranked teams lose some, so, too, may the less-favored variations sometimes survive, though in general the more-favored variations will survive in greater proportion.

Much has been made of the expression, ''survival of the fittest,'' which originated not with Darwin but with Herbert Spencer. First there were those, uninitiated into population thinking, who thought Darwin had claimed that fitness ensures survival in every case—this view is clearly based on insufficient understanding of Darwinism. Another objection, made especially by philosophers, is that ''survival of the fittest'' is what logicians would call a **tautology:** a statement that is true for all cases simply because the terms are so defined. (''All bachelors are unmarried'' is such a statement, true because of the way we define ''bachelor.'') According to this objection, the ''fittest'' are recognized only by the fact that they have survived, and so ''survival of the fittest'' becomes ''survival of the survivors,'' which is a tautology. But Darwin never spoke in these terms, arguing only that some variations would confer ''advantage'' and that others would be ''injurious.'' In fact, if we define natural selection as the *differential contribution of heritable variations to the next generation,* all of Darwin's claims on behalf of selection would be just as valid, even without reference to the ''fittest.''

Some writers have supposed that Darwin depicted a brutally mechanistic world in which ''kill or be killed'' was the order of the day—''nature red in tooth and claw,'' to use Tennyson's expression. Nothing could be further from the truth. The expression ''struggle for existence'' was used by Darwin only ''in a large and metaphorical sense, including dependence of one being upon another, and including (which is more important) . . . success in leaving progeny.'' Darwin's examples include a literal struggle between ''two canine animals in a time of dearth,'' and the struggle of a desert plant against the drought. ''A plant which annually produces a thousand seeds, of which on an average only one comes to maturity may be more truly said to struggle with the plants . . . which already clothe the ground. The missletoe [*sic*] . . . is disseminated by birds . . . ; and it may metaphorically be said to struggle with other fruit-bearing plants, in order to tempt birds to devour and thus disseminate its seeds. . . . In these several senses, which pass into each other, I use for convenience sake the general term of struggle for existence.''

Objections Answered by Darwin

Charles Darwin was sufficiently perceptive to foresee objections that would inevitably be raised against his theory of natural selection. To these various objections, Darwin devoted four chapters (Chapters 6–9) of his first edition, and in the sixth edition he added one more in response to other objections his critics had raised. Darwin's answers to these objections were often the best that could have been expected in the 1860s. The least satisfactory explanations were those that required knowledge of twentieth-century genetics.

The first difficulty that Darwin addressed was the general absence of transitional forms between species or between higher taxa, such as classes, for any evolutionary theory seems to require their existence. In the case of higher taxa, Darwin's reasoning appears quite cogent: transitional forms are often short-lived, succumbing to the competition of their more advanced descendants. In branching evolution, furthermore, the common progenitor of two classes ought not to be transitional between the modern representatives of those classes; it may instead differ markedly from either. Darwin's explanation in the case of species is far less satisfactory because of his nominalist claim that species are nothing but ''well-marked varieties.'' He mentions the case of two related species occupying adjacent geographic ranges; as one travels between these geographic areas, one species becomes increasingly rare and the other increasingly common, without any actual intergradation between them. The modern explanation would rely on the reproductive isolation of the two species (Chapter 14), a phenomenon that Darwin's nominalism did not permit him to see. He did suppose that the species ranges might formerly have been discontiguous, and he thus came very close to our modern theory of geographic speciation (Chapter 14), but no sooner did he state this position than he retracted from it.

The second difficulty noted by Darwin was the ability of natural selection to account, on the one hand, for characters of only small selective advantage and, on the other, for highly perfected organs. His answers seemed weak to many of his critics, but subsequent research seems in general to have borne them out. In the

case of characters of ''trifling importance,'' Darwin correctly notes (1) that such characters may well have an importance that we, in our ignorance, have failed to detect, and (2) that even small advantages may turn into big ones under appropriate circumstances. The presence of down on peaches makes them less liable to attack by beetles, says Darwin, and the color of the fleshy part is sometimes correlated with resistance to certain diseases. Infrequent epidemics of these diseases would select strongly for a trait that botanists usually consider unimportant.

The case of highly perfected organs was immediately seized upon by Darwin's opponents: how could natural selection produce such highly perfected organs as an eye or a wing? In particular, how could the early stages evolve if they could not confer any selective advantage until the organ had been perfected? Of what use would a retina be without a lens, or a lens without a retina, or either one without a transparent cornea? No gradual series of steps, ran the objection, could account for such highly perfected organs. The same objection, incidentally, had been used against Lamarck: how could a bird exercise (and thus enlarge) its wing until *after* the wing had already been perfected as a flying structure? Darwin's own admission speaks to the seriousness of this matter: ''If it could be demonstrated that any complex organ existed, which could not possibly have been formed by numerous, successive, slight modifications, my theory would absolutely break down.''

Darwin surmounts this difficulty by pointing out the plausibility of the early transitional stages, including the reasons for selection at each stage. The eye, for instance, exhibits among the invertebrates a series of gradations, from ''an optic nerve merely coated with pigment'' to eyes as highly developed as those of any vertebrate. We shall treat this case more extensively in Chapter 18.

Often in the course of evolution, an organ that has evolved in connection with one function comes at some later time to serve a totally different function. Several cases of highly developed organs are best explained by means of such functional shifts, each frequently destroying evidence of the old function, once the new function has been evolved to any degree. Darwin gives a good example of this phenomenon: one group of barnacles on which he had worked possessed no gills, but they did have ''two minute folds of skin . . . which serve, through the means of a sticky secretion, to retain the eggs until they are hatched within the sac.'' The majority of barnacles, however, have expanded these folds of skin into gills, losing also the adhesive glands and allowing the eggs to lie freely within the sac. Now, if the former group had become extinct, ''and they have already suffered far more extinction . . . , who would ever have imagined'' that the gills ''had originally existed as organs for preventing the ova from being washed out of the sac?''

Darwin's chapter on instinct is remarkably modern in many ways, and yet the evidence Darwin marshals is appropriate to his own age. He documents, again by analogy with the artificial selection of domestic animals, that instincts are inherited, and that animal breeders have successfully selected certain behavior patterns in preference to others. Most of the chapter is devoted to a discussion of social insects, which exhibit among the most intricate of instinctive behavior patterns (Chapter 24). The intricacy of behavior patterns is handled in the same way as the intricacy of structure in organs such as the eye. The most vexing problem for an evolutionist involves the behavior of the sterile castes of workers or of workers and soldiers in some cases. To explain the characteristics of these sterile workers, including their behavior, Darwin invoked both group selection and kin selection (Chapter 24). He then concluded, ''no amount of exercise, or habit, or volition, in the utterly sterile members of a community could possibly have affected the structure or instincts of the fertile members, which alone leave descendants. I am surprised that no one has advanced this demonstrative case of neuter insects, against the well-known doctrine of Lamarck.''

Darwin's chapter on hybridism suffers from several defects: first, there is Darwin's nominalism and his consequent lack of a clear definition of species. Second, the terms ''hybrid'' and ''hybridism'' are used somewhat indiscriminately, as was then customary, to refer sometimes to crosses between distinct individuals from the same population and at other times to crosses between individuals drawn from different populations or species. Darwin cites several examples of reproductive isolating mechanisms but fails to see their importance. He also uses the existence of occasional hybrids between species to strengthen his belief that species are no different from

varieties. From the modern standpoint, this chapter is perhaps the least satisfactory of the entire book.

Darwin returns to the subject of transitional forms in a later chapter, raising on his own the objection that such forms ought to have existed and should have been preserved in the fossil record. ''This, perhaps, is the most obvious and gravest objection which can be urged against my theory,'' he admits. The gaps in the fossil record, as it was then known, were so great that they had given Cuvier reason to believe in catastrophism and successive creations. Darwin exploits these gaps to his own purpose of explaining the absence of known transitional forms. Darwin also warns us not to expect the transitional form between two classes to be intermediate between the present representatives of those classes. The fantail and pouter pigeons, he argues, are descended not from each other but from the common rock pigeon; there never existed a transitional form intermediate between the fantail and the pouter.

Darwin emphasizes the great expanse of geologic time given to him by the uniformitarian geologists. Anyone who can read Lyell's *Principles of Geology* and yet not admit the immensity of geologic time, Darwin asserts, ''may at once close this volume.'' The immense time supported by Lyell is required for so slow a process as natural selection to act. The immense time is also needed to explain ''the imperfection of the geological record,'' that is, the paucity of fossils from any given age. Only the hard parts of animals and plants can be preserved as fossils, argues Darwin, and these only in very favorable conditions, and even if they are fossilized, their subsequent exposure and discovery is itself unlikely. In this, Darwin has largely been borne out by subsequent research: the number of living species is about a million, and the number of species that have ever lived is assuredly much greater, each containing thousands if not millions of individuals in each generation. ''What an infinite number of generations, which the mind cannot grasp, must have succeeded each other in the long roll of years! Now we turn to our richest geological museums, and what a paltry display we behold!'' Darwin's stratigraphy is likewise modern when he tells us that between the successive formations there occurred in each case an amount of time, of unknown duration, during which new rocks were not deposited but may instead have been eroded. Darwin's knowledge of paleontology served him well here, as when he was able to mention the common and nearly ubiquitous Chthalaminae, a subfamily of barnacles occurring in all the world's oceans in untold numbers. Though this group had already been discovered in Cretaceous rocks, it was nevertheless unknown from the entire Tertiary period, of about 65 million years' duration, during which it must certainly have occurred. Darwin mentions now a further difficulty, the fact that whole families or orders first appear in the fossil record with apparent suddenness. This sudden appearance, we are told, is perhaps illusory, for we have good fossil collections only from a small part of the world, and these groups may have evolved over long periods of time elsewhere, only to be discovered on their subsequent migration to Europe. But the greatest difficulty here is facing the simultaneous, sudden appearance of many groups at the base of the Paleozoic. To this, Darwin has no satisfactory answer; a modern hypothesis addressing this problem will be explained in Chapter 18.

Summarizing the "imperfections of the geological record," Darwin states that "following out Lyell's metaphor, I look at the natural geological record, as a history of the world imperfectly kept, and written in a changing dialect; of this history we possess the last volume alone, relating only to two or three countries. Of this volume, only here and there a short chapter has been preserved; and of each page, only here and there a few lines. . . . On this view, the difficulties above discussed are greatly diminished, or even disappear."

Descent with Modification

Darwin's last several chapters are written both with a view toward documenting the fact of evolution and exploring some of its consequences. Here Darwin marshals some of the strongest evidence in favor of what he calls "descent with modification," showing that this theory explains the data best. In many cases, Darwin fights with a two-edged sword, directing one side of his argument against the environmental determinists and the other side against the **theory of special creation**. The latter theory, an outgrowth of Natural Theology, held that each species was the product of a separate and independent act of Divine Creation. "Let us now see whether the

several facts and rules . . . better accord with the common view of the immutability of species, or with that of their slow and gradual modification, through descent and natural selection." Note that, in emphasizing the gradual nature of natural selection, Darwin is also prophetically arguing against the views of such early twentieth-century geneticists as de Vries.

Darwin believes, he assures us, "in no fixed law of development," such as orthogenesis (Chapter 18). He explains that "a species when once lost should never reappear, even if the very same conditions of life . . . should recur," because the necessary ancestor no longer exists. This explanation anticipates Dollo's "law of irreversibility" (Chapter 21). Darwin marvels only slightly at extinction, considering it to be only one small step beyond rarity, and due to the same selective forces. The classification of living and fossil forms together in the same scheme of genera, families, and orders is explained with reference to the theory of descent with modification. The apparently simultaneous change in the faunas of different continents can be explained as resulting from the migration of the most successful forms to all continents, often supplanting older relatives. If this event were repeated, it would appear that the faunas of different continents had undergone the very same sequence of changes independently. "The . . . great leading facts of paleontology seem to me simply to follow on the theory of descent with modification through natural selection," says Darwin. "And they are wholly inexplicable on any other view," for the theory of special creation could not account for any succession of forms, whereas the environmentalist theory would definitely allow, if not require, the reappearance of lost species if the same conditions of life should recur.

Passing from geology to geography, Darwin mentions some of the very arguments that first drove him to believe in evolution, devoting two entire chapters to them. "In considering the distribution of organic beings over the face of the globe, the first great fact which strikes us is, that neither the similarity nor the dissimilarity of the inhabitants of various regions can be accounted for by their climatal and other physical conditions. . . . The case of America alone would almost suffice to prove its truth: . . . if we travel over the vast American continent, from the central parts of the United States to its extreme southern point, we meet with the most diversified conditions; the most humid districts, arid deserts, lofty mountains, grassy plains, forests, marshes, lakes, and great rivers, under almost every temperature. There is hardly a climate or condition in the Old World which cannot be paralleled in the New. . . . Notwithstanding this parallelism in the conditions of the Old and New Worlds, how widely different are their living productions!'' Here Darwin is arguing strongly against environmental determinism, under which similar climates ought to be inhabited by similar faunas and floras. ''In the southern hemisphere, if we compare large tracts of land in Australia, South Africa, and western South America, between latitudes 25° and 35°, we shall find parts extremely similar in all their conditions, yet it would not be possible to point out three faunas and floras more utterly dissimilar. Or again we may compare the productions of South America south of lat. 35° with those north of 25°, which consequently inhabit a considerably different climate, and they will be found incomparably more closely related to each other, than they are to the productions of Australia or Africa under nearly the same climate.''

Another ''great fact'' that Darwin cites ''is the affinity of the productions of the same continent or sea, though the species themselves are distinct at different points and stations.'' Here he can argue against environmental determinism and Natural Theology at the same time. Each continent, for example, has its own distinct type of flightless bird: Africa has the true ostrich, Australia has the cassowary and the emu, and South America has a bird known as the rhea. In different parts of South America, the distinct species we encounter are always rheas, never cassowaries, emus, or true ostriches. Why should the Creator never have created the ostrich in South America? Why would the same environment, as nearly as we can tell, produce an ostrich in Africa and a rhea in South America? Why would different South American environments always produce rheas and never emus or ostriches? Why would the Creator have produced three types of flightless birds, when the one most perfect type surely would have sufficed? The South American rodents, under whatever climate, are always members of the suborder to which the guinea pig and chinchilla belong, a suborder that occurs nowhere else. Why are none of the more usual types of rodents present? ''If we look to the islands off the American shore, however much they may differ in geological structure, the in-

habitants, though they may be all peculiar species, are essentially American. We may look back to past ages . . . , and we find American types then prevalent on the American continent and in the American seas. We see in these facts some deep organic bond, prevailing throughout space and time, over the same areas of land and water, and independent of their physical conditions. The naturalist must feel little curiosity, who is not led to inquire what this bond is.'' Creationism would not have predicted any such bond, and environmental determinism could not account for its independence from physical conditions. ''This bond, on my theory, is simply inheritance, that cause which alone, as far as we positively know, produces organisms quite like, or . . . nearly like each other. . . . On this principle of inheritance with modification, we can understand how it is that sections of genera, whole genera, and even families are confined to the same areas, as is so commonly and notoriously the case.''

Reflecting on related species produced by descent with modification, Darwin shows how the place inhabited by the ancestral species would have the appearance of a ''center of creation.'' Certain other facts of geographic distribution, especially of mammals, are explained with appeal to the climatic fluctuations of the Ice Age. In particular, the generally southward migration of Northern Hemisphere faunas is explained by this agency.

In preparing to deal with oceanic islands, Darwin carefully documents the possibilities, infrequent though they may be, of dispersal across great distances. Beetles caught aboard the *Beagle* while at sea are mentioned, as are floating islands and similar debris. Much is made of dirt, including large seeds or pebbles equally heavy, clinging to the feet of birds, or of snails clinging to the feet of water birds. The ability of seeds to germinate after long periods of immersion in seawater is also discussed.

It is in his chapter on the inhabitants of oceanic islands that Darwin draws some of his best evidence favoring evolution and opposing the theory of special creation. Oceanic islands, he notes, have few species, compared with equal-sized areas of continent, and yet the proportion of **endemic** species, meaning species present nowhere else, is very high. Environmental determinism provides no answer, for the success of innumerable introduced species on oceanic islands proves that conditions had been suitable all along. And the creationist, writes Darwin, ''will have to admit, that a sufficient number of the best adapted plants and animals have not been created on oceanic islands; for man has stocked them . . . far more fully and perfectly than has nature.'' Having driven the creationists to the impious revelation that nature is not perfect, Darwin explains that island faunas and floras are deficient because the islands, however suitable, are simply out of reach to many species. As evidence, he cites the fact that islands are most deficient in precisely those groups that can least tolerate dispersal across salt water. For example, the ''general absence of frogs, toads, and newts on so many oceanic islands cannot be accounted for by their physical conditions; indeed it seems that islands are particularly well fitted for these animals; for frogs have been introduced into Madeira, the Azores, and Mauritius, and have multiplied so as to become a nuisance.'' Having used one edge of his argument against the environmentalists, Darwin directs the other edge against the theory of special creation: ''As these animals and their spawn are known to be immediately killed by sea-water, on my view we can see that there would be great difficulty in their transportal across the sea, and therefore why they do not exist on any oceanic island. But why, on the theory of creation, they should not have been created there, it would be very difficult to explain.''

On different oceanic islands, the percentage of endemic species, present nowhere else, is generally high, but the actual percentage will differ from place to place and from group to group. Marine birds, migratory birds, and bats are frequently present on islands, but they only seldom form endemic species. The highest degree of endemism is found among those groups least able to achieve repeated dispersal, for the first species to reach an island will generally proliferate, in the absence of its usual competitors, and become modified into a group of related species, all endemic to their new island home. Note that this probabilistic argument requires population thinking for its very formulation.

But Darwin's most decisive argument is yet to come, and it comes, typically, from material with which Darwin is personally familiar. It is a perfect example of Darwin's double-edged sword, with one edge directed against environmental determinism and the other against special creation.

> The most striking and important fact for us in regard to the inhabitants of islands, is their affinity to those of the nearest mainland, without being actually the same species. Numerous instances could be given of this fact. I will

give only one, that of the Galapagos Archipelago, situated under the equator, between 500 and 600 miles from the shores of South America. Here almost every product of the land and water bears the unmistakeable stamp of the American continent. There are twenty-six land birds, and twenty-five of these are ranked by Mr. Gould as distinct species, supposed to have been created here; yet the close affinity of most of these birds to American species in every character, in their habits, gestures, and tones of voice, was manifest. So it is with the other animals, and with nearly all the plants, as shown by Dr. Hooker in his admirable memoir on the Flora of this archipelago. The naturalist, looking at the inhabitants of these volcanic islands in the Pacific, distant several hundred miles from the continent, yet feels that he is standing on American land. Why should this be so? Why should the species which are supposed to have been created in the Galapagos Archipelago, and nowhere else, bear so plain a stamp of affinity to those created in America? There is nothing in the conditions of life, in the geological nature of the islands, in their height or climate, or in the proportions in which the several classes are associated together, which resembles closely the conditions of the South American coast: in fact there is a considerable dissimilarity in all these respects. On the other hand, there is a considerable degree of resemblance in the volcanic nature of the soil, in climate, height, and size of the islands, between the Galapagos and Cape de Verde Archipelagos: but what an entire and absolute difference in their inhabitants! The inhabitants of the Cape de Verde Islands are related to those of Africa, like those of the Galapagos to America. I believe this grand fact can receive no sort of explanation on the ordinary view of independent creation; whereas on the view here maintained it is obvious that the Galapagos Islands would be likely to receive colonists, whether by occasional means of transport or by formerly continuous land, from America; and the Cape de Verde Islands from Africa; and that such colonists would be liable to modification;—the principle of inheritance still betraying their original birthplace.

In his penultimate chapter, Darwin turns to several related points of supporting evidence from the fields of taxonomy (classification), embryology, and comparative anatomy. The facts he mentions here had been explained by earlier theories without much difficulty, but Darwin now gives them new, evolutionary meaning. In some cases, phenomena that seemed capricious or confusing under earlier theories make far greater sense when considered in the light of Darwin's evolutionary theory. It is this chapter, dealing with peculiarly biological problems, that seems to have had the greatest impact on biological research for succeeding generations, for evolution had breathed new meaning into some formerly less meaningful subjects.

Darwin discusses classification at some length. Our classifications, he reminds us, are arranged as a series of groups within groups: genera within families, families within orders, and so forth. Why? Are these groups merely arbitrary, arranged for human convenience alone? Darwin states that a higher principle is operating: the members of a genus share a common descent and thus have more characters in common; the members of a family share a more remote ancestor, from which they have inherited fewer characters that all now share. Classifications, insofar as possible, should be genealogical. The best classifications already were, even if their authors had not realized this fact. "Naturalists try to arrange the species, genera, and families in each class, on what is called the Natural System. But what is meant by this system? Some authors look at it merely as a scheme for arranging together those living objects which are most alike, and for separating those which are most unlike. . . . But many naturalists think that something more is meant by the Natural System, . . . that it reveals the plan of the Creator; but unless it be specified . . . what . . . is meant by the plan of the Creator, it seems to me that nothing is thus added to our knowledge. . . . I believe that something more is included; and that propinquity of descent,—the only known cause of the similarity of organic beings,—is the bond, hidden as it is by varying degrees of modification, which is partially revealed to us by our classifications." Darwin's remarks have relevance to our age as to his own, for both phylogenetic (genealogical) and nonphylogenetic schools of classification exist at present; Darwin clearly favored the phylogenetic approach.

Passing to anatomy or, more broadly, **morphology** (the study of form), Darwin noted those forms of resemblance to which Owen had given the name *homology* (Chapter 3). "What can be more curious than that the hand of a man, formed for grasping, that of a mole for digging, the leg of the horse, the paddle of the porpoise, and the wing of the bat, should all be constructed on the same pattern, and should include the same bones, in the same relative positions?" Homologous resemblance had been thought to reflect the existence of a common *eidos,*

archetype, *Bauplan,* or "unity of type." But Darwin's explanation was simply that these resemblances reflect the common inheritance of the same organ, though variously modified, among the descendants of a single form.

Serial homology, meaning homology among repeated parts, is also discussed, with reference to several examples used by the German romantic school of *Naturphilosophie,* or "idealistic morphology," as they often called themselves. Goethe's vertebral theory of the skull is noted with approval, as is his theory that all flower parts are really modified leaves. "How inexplicable are these facts on the ordinary view of creation! . . . Why should similar bones have been created in the formation of the wing and leg of a bat, used as they are for such totally different purposes? Why should one crustacean, which has an extremely complex mouth formed of many parts, consequently always have fewer legs; or conversely, those with many legs have simpler mouths? Why should the sepals, petals, stamens, and pistils in any individual flower, though fitted for such widely different purposes, be all constructed on the same pattern?" Darwin's explanation, of course, is that repeated parts, originally similar, have become variously modified in the course of phylogeny.

Darwin recognizes the importance of embryological characters in classification or in the determination of homologies. Serially homologous structures often resemble one another more closely in the embryo than in the adult. Embryology can often provide clues for classification: "Even the illustrious Cuvier did not perceive that a barnacle was, as it certainly is, a crustacean; but a glance at the larva shows this to be the case in an unmistakeable manner."

Rudimentary or vestigial organs, useless to their possessors, are mentioned, and the creationist theory is brought to task for failing to explain why it should have pleased the Deity to endow organisms with such functionless vestiges.

Darwin concludes his book with the following image-rich passage:

> It is interesting to contemplate an entangled bank, clothed with many plants of many kinds, with birds singing on the bushes, with various insects flitting about, and with worms crawling through the damp earth, and to reflect that these elaborately constructed forms, so different

from each other, and dependent on each other in so complex a manner, have all been produced by laws acting around us. . . . Thus, from the war of nature, from famine and death, the most exalted object which we are capable of conceiving, namely, the production of the higher animals, directly follows. There is grandeur in this view of life, with its several powers, having been originally breathed into a few forms or into one; and that, whilst this planet has gone cycling on according to the fixed law of gravity, from so simple a beginning endless forms most beautiful and most wonderful have been, and are being, evolved.

C. THE RECEPTION OF DARWINISM

Darwin's *Origin of Species* created a storm. It instantly became the number one topic of conversation in England; hardly any person could avoid having an opinion about it. Some favored the theory, often hesitatingly or with reservations. Others denounced the theory, sometimes bitterly. Most theologians opposed the theory, though not all of them understood it. The public was generally divided on the issue, usually for reasons that had little to do with science. Among those people who read the book and understood it, many did accept Darwin's theory.

Scientific opinion was initially divided. Of those scientists whose reputations were already established, many opposed the theory, including especially Britain's most respected anatomist, Sir Richard Owen. Charles Lyell's initial reaction was one of dissension, though he did wish the theory to get a fair hearing and so encouraged its publication. Established scientists who accepted the theory often did so reluctantly: botanists like Hooker and Asa Gray were among these, and Lyell, too, was later won over. Evolution had finally made sense of anatomy, and many anatomists (except Owen) quickly rallied to Darwin's support: Huxley, Haeckel, Gegenbaur. Wallace and other zoogeographers also favored Darwinism, though Wallace, like Lyell and Asa Gray, had reservations when it came to human evolution. But the overwhelming majority of young scientists, those who established their reputations after 1860, were strongly in favor of evolution, and ultimately they decided the theory's fate.

During the late nineteenth century, increasing nationalism in Europe and America's preoccupation with its own affairs (the Civil War, Reconstruction, westward expansion, and industrialization) ensured that Darwinism was somewhat differently received in each country. In England, the theory remained closest to its original form, but in Germany it was swept up into the Romantic movement. We shall thus consider each country separately.

Darwinism in England

When Darwin submitted the *Origin of Species* for publication, he expressed the wish that his theory be well received, not by the public or by the scientific community at large, but by three particular men of science whom he greatly respected: Charles Lyell, Joseph Hooker, and Thomas Henry Huxley. Recall that Lyell and Hooker had arbitrated the situation before the Linnean society in 1858, when Wallace's paper was published together with excerpts from Darwin's unpublished writings.

Thomas Henry Huxley (1825–1895; Fig. 5.6) was at the time a relatively young anatomist with an early training in medicine. Several years' faithful service as ship's surgeon aboard H.M.S. *Rattlesnake* had been rewarded by his election to the Royal Society. His only other claim

Fig. 5.6 One of Darwin's English supporters, Thomas Henry Huxley.

to distinction was his incisive attack on the *Vestiges of the Natural History of Creation,* which probably explains Darwin's desire to convince him especially. Huxley, it turns out, was convinced upon his first reading of the book, at which time he is said to have exclaimed, "How stupid [of me] not to have thought of it earlier!" He immediately wrote to Darwin, pledging his wholehearted support in the struggle he foresaw: "I am sharpening up my claws and beak in readiness," he wrote. True to his word, he became Darwin's most influential advocate —"Darwin's bulldog," as he referred to himself. He also had exactly the qualities that Darwinism needed at the time: a thorough knowledge of anatomy, which was perhaps Darwin's greatest weakness, and a very gifted ability as a public speaker, especially important because Darwin had voluntarily withdrawn from public appearances, owing partly to ill health and partly to a very nonaggressive temperament.

Huxley's abilities as a public speaker first became evident at the 1860 meeting of the British Association for the Advancement of Science, held at Oxford. The bishop of Oxford, Samuel Wilberforce (1805–1873), was ready to crush Darwinism, having been duly coached by Sir Richard Owen (1804–1892), now Darwin's strongest opponent. Wilberforce, who was known as "soapy Sam" because in every skirmish "I always came out clean," attacked Darwin's theory and Darwin himself, both with remarkable unfairness, for more than half an hour before a packed crowd. He concluded his speech by introducing the next speaker, Huxley, and asking, "Was it through his grandfather or his grandmother that he claimed descent from a monkey?" Huxley, seated beside a friend, softly exclaimed, "The Lord hath delivered him into mine hands!" Huxley rose to speak and declared that he had listened to the bishop's remarks but could find no new fact or argument in them, save that concerning his own ancestry. "If . . . the question is put to me," declared Huxley, "would I rather have a miserable ape for a grandfather, or a man highly endowed by nature and possessing great means and influence and yet who employs those facilities and that influence for the mere purpose of introducing ridicule into a grave scientific discussion—I unhesitatingly affirm my preference for the ape." Huxley later reported that "there was unextinguishable laughter among the people, and they listened to the rest of my argument with the greatest attention. . . ."

In his many public appearances, Huxley carried the Darwinian message clearly and convincingly to the British public. He also made a lecture tour of the United States, where he was enthusiastically received not only in Boston and New York but also in Tennessee, where the antievolutionism of the 1920s had not yet surfaced. Huxley's skills as a speaker and writer were also shared by two grandsons, biologist Sir Julian Huxley (1887–1975) and novelist Aldous Huxley (1894–1963).

Although Darwin had carefully chosen to omit human evolution from his *Origin of Species,* it was on this very subject that most of the debate centered. Huxley rose to the occasion and published in 1863 his most influential book, *Evidences as to Man's Place in Nature,* in which he boldly displayed the evidence, largely anatomical, that linked humans and apes closely in two related families of the same order. He also adroitly included an account of the life habits of the several modern species of apes, material that makes the book fascinating and interesting even to those who may disagree with it. His most influential later book, *Manual of Comparative Anatomy of the Vertebrated Animals*, became widely used as a textbook, and a whole generation of biologists were instructed by it.

Joseph Hooker, as Darwin had wished, was also quickly won over. It was largely through his influence and that of Asa Gray that Darwinism was in a short time victorious in the field of botany.

Alfred Russel Wallace (1823–1913), Darwin's codiscoverer, gave many lectures supportive of Darwinism. He balked, however, when it came to human evolution, for which purpose he felt some other explanation, perhaps divine intervention, was still necessary; this last opinion was echoed by Asa Gray and by Charles Lyell. Natural selection, said Wallace, could not account for the brains of savages having intellectual capacities (for civilization, poetry, etc.) far beyond the demands placed upon them by their savage existence. In his later years, Wallace drifted afield and came to support such less scientific movements as Mesmerism. The man who had stimulated Wallace's career by their joint trip to Brazil, Henry Walter Bates (1825–1892), lent support to Darwin's theory through his discovery of the phenomenon of *mimicry* (Chapter 10).

Darwin himself avoided public controversy, secluding himself instead at Down. Since his return from the *Beagle,* he had been periodically afflicted with various illnesses, of which the most serious was probably Chagas' disease, a tropical malady that he presumably acquired from an attack by a *Reduvius* bug in Chile.

Hostility to Darwinism often arose from religious convictions; the reservations expressed even by scientists like Sedgwick and Lyell seem largely to have been so motivated. The public as well perceived Darwin as an affront to their religious beliefs, though some clergymen sought reconciliation of Darwinism with Scripture. Benjamin Disraeli, the Conservative prime minister, addressed Parliament with the question, ''is man an ape or an angel? I, my Lord, am on the side of the angels.'' The wife of the bishop of Worcester, confronted with Huxley's remarks, echoed the feelings of many when she declared: ''Descended from the apes! My dear, let us hope that it is not true, but if it is, let us pray that it will not become generally known.'' Huxley, however, made certain that it did become generally known, and within little more than a decade, the majority of Britons were squarely on Darwin's side.

One of the holdouts was the satirist Samuel Butler (1835–1902), author of *Erewhon* and *The Way of All Flesh.* His major criticism, based on an insufficient appreciation of population thinking, was that Darwin had explained the most miraculous of adaptations as resulting from mere chance—''the law of higgledy-piggledy,'' as he denounced it. Butler, who understood Lamarck sympathetically, was unalterably opposed to Darwin. It was debatably through Butler's influence that a much greater British satirist, George Bernard Shaw (1856–1950), also opposed Darwin's theories.

Scientific criticism of Darwin was rather sparse, but two critics in particular may be mentioned: Fleeming Jenkin (1833–1885) and St. George Mivart (1827–1900). It was Jenkin's criticism of natural selection, based on the then-accepted theory of blending inheritance, that moved Darwin to write his *Variation of Animals and Plants Under Domestication,* which we shall discuss in the following chapter. Mivart, a devout Catholic, was more formidable as an opponent because he accepted evolution, yet denied natural selection. His arguments, often reminiscent of Paley's, pointed to cases of perfection that he argued coud not have evolved gradually by natural selection, but only suddenly, by an act of Divine Providence. Darwin felt these criticisms very keenly, and it was largely in response to them that he rewrote much of the sixth and last edition of his *Origin of Species,* add-

ing an entire chapter devoted to a response to Mivart. Mivart was not alone in accepting evolution but rejecting natural selection; belief in multiple evolutionary mechanisms, including Lamarckian use and disuse, soon became commonplace, causing Darwin to write in 1872: ''I now admit . . . that in the earlier editions of my 'Origin of Species' I perhaps attributed too much to the action of natural selection or the survival of the fittest.''

Although Darwin had omitted human evolution from his *Origin of Species,* the British public was clamoring for his views on the subject. It was largely for this reason that Darwin wrote *The Descent of Man, and Selection in Relation to Sex.* The first part of the *Descent* is less purely Darwinian than the *Origin* and shows strongly the influence of Huxley and other anatomists, as well as of Haeckel and German romanticism. Here Darwin documents the evidence, largely anatomical, for the inclusion of humans in the evolutionary scheme: vestigial structures, such as the projecting point (now called ''Darwin's point'') on the ear, and the striking similarity of human embryos to those of other mammals. The greater part of the book, however, is devoted to an elaboration of the theory of sexual selection, by which Darwin proposes to account not only for **sexual dimorphism** (anatomical differences between the sexes) but also many peculiarly human traits, such as higher intelligence.

Darwin lived to see his theories widely accepted in his own lifetime. He died a quiet death in 1882, and he was buried in Westminster Abbey, not far from the grave of Newton.

Darwinism in America

In the 1860s, Harvard was the leading institution for the teaching of natural history in general or biology in particular. Thanks to Louis Agassiz, who had come to America in 1846, it also was rapidly building the largest natural-history collection in North America; the era of the great municipal museums of New York and Chicago had not yet arrived. Darwin's foremost American supporter, Asa Gray (1810–1888), was director of the university's herbarium, later named in his honor. Ironically, Darwin's foremost American opponent, Louis Agassiz, was also at Harvard, directing the nearby Museum of Comparative Zoology. Agassiz was true to the teachings of his mentor, Georges Cuvier, but many if not most of his students, as well as his son Alexander (who succeeded him as museum director), were won over to the theory of evolution.

As the western United States, especially California, was becoming more settled, the building of a transcontinental railroad became a matter of great importance. The geological field parties that mapped out the best route for the railroad tracks usually had a paleontologist among their number. The founder of vertebrate paleontology in America was Joseph Leidy (1823–1891), professor at the University of Pennsylvania. A student of his, Edward Drinker Cope (1840–1897), was the principal paleontologist on these early surveys of the territory through which the transcontinental railroad tracks were being laid. Cope amassed a great paleontological collection, which later became the nucleus around which the collections of the American Museum of Natural History in New York were built. Cope became an evolutionary theorist, espousing largely Lamarckian ideas. He also founded the first journal, *The American Naturalist*, devoted to the subject of evolution.

Cope's great rival was Othniel Charles Marsh (1831–1899), professor at Yale and founder of the Peabody Museum at that institution. The rivalry between them was bitter, and they often perpetrated mischief on each other's field parties. Their quest for priority of discovery caused many hasty (often inadequate) descriptions to be published, but also encouraged the discovery of many new fossils that would document the evolutionary history of many vertebrate groups.

Another American evolutionist was Alpheus Hyatt (1838–1902), whose views on the evolution of ontogeny are discussed in Chapter 22.

When Darwinism was at its low point in the first decade of the twentieth century, its two major adherents were both American: the ichthyologist David Starr Jordan (1851–1931) and the paleontologist Henry Fairfield Osborn (1851–1935), a disciple of E. D. Cope.

Social Darwinism

A political or sociological theory known as "**social Darwinism**" had its roots in England but flourished in the United States. The originator of this theory was Herbert Spencer (1820–1903), a British philosopher whose belief in evolution predates Darwin's *Origin of Species* by several years. Spencer, an indirect disciple of Auguste Compte's "positivism," anticipated not only evolution but even natural selection. It was Spencer who coined the phrase, "survival of the fittest," which Darwin adopted in the fifth edition of the *Origin*. It was also Spencer who popularized the word "evolution" to refer to Darwin's theory of descent.

Where Darwin had been most careful to point out that natural selection was often a peaceful matter of "success in leaving progeny," Spencer emphasized the imagery of a combative struggle from which only the fittest would survive. It was this theory that Spencer then proceeded to apply to human societies, interpreting much of history as a record of "progress" brought about by competition (economic, military) and by the "survival of the fittest."

William Graham Sumner (1840–1910) was the foremost American disciple of Spencer and the founder of social Darwinism in America. His book, *What Social Classes Owe to Each Other*, advances the theory that success in human affairs is a consequence of competition and "struggle for existence." A fervent opponent of socialism, Sumner and his followers urged the theory of laissez-faire on the grounds that any interference with "free competition" was contrary to nature. Successful businesses, he argued, were successful because of their superiority in competition, and any attempt to restrain them or to lessen competition would only result in mediocrity. This theory was favorable to the growing capitalism of the era—the era of the "robber barons." It is not surprising that many successful capitalists of the era, notably Andrew Carnegie, strongly favored Sumner's theories and hence Darwin's.

Evolution Outside the English-Speaking World

Outside the English-speaking world, the reception of Darwinism or of evolution in general often depended on conditions unique to each country, including the political and intellectual climate generally, the availability of Darwin's works in translation, and the support or opposition of particular individuals.

Darwinian evolution never quite took hold in France. The initial French reaction to Darwin was indignation at the ingratitude to such French evolutionists as Buffon, Geoffroy, and especially Lamarck. It should also be noted that, at the time when evolution was the major topic of scientific discussion in England, Germany, and the United States, France was paying far more attention to Louis Pasteur's discoveries in microbiology, especially to his controversy with Pouchet (see Chapter 25) on spontaneous generation. But even after this episode was over, French scientists tended not to become drawn into the debates over evolution that were raging elsewhere.

Another reason for the poor reception of Darwin in France was the tendency of French evolutionists to propose their own theories, partaking of some Darwinian elements and often Lamarckian ones as well. Lecomte de Noüy, Lucien Cuénot, and Pierre Teilhard de Chardin are but three French authors of twentieth-century evolutionary theories incorporating many non-Darwinian elements.

In the Islamic world, the attack against Darwinism was led by Sayyid Jamal al-Din al-Afghani (1839–1897). He denounced British colonialism, philosophical materialism, and human descent from apes, all in the same breath. Instead he advocated faithfulness to the Koran and pan-Arab unification against Western colonialism.

Ernst Haeckel and German Darwinism

The most distinctive national school of evolution was the German. Here, Darwinism not only fell on fertile soil but also merged with a home-grown tradition, the tradition of romantic *Naturphilosophie*. Many other aspects of German Darwinism resulted from the convictions of Darwin's strongest German advocate, Ernst Haeckel.

Ernst Heinrich Haeckel (1834–1919; Fig. 5.7) was trained as a zoologist. He came from a devout Catholic family, and his student period was one of much political unrest. The students, generally liberal to radical in their political ideas, were opposed to the conservative monarchy. The church, however, allied itself with the monarchy, and it was largely for this reason that Haeckel became alienated from religion altogether, becoming an atheist in the eyes of many of his contemporaries.

Haeckel was completing his doctoral thesis on the Radiolaria, an order of Protozoa, when Darwin's *Origin of Species* was published. He hastily rewrote part of his thesis to accommodate to the new theory and instantly became a strong advocate of evolution. He succeeded in convincing his fellow students and professors, including the great anatomist Carl Gegenbaur (1826–1903).

Fig. 5.7 Darwin's strongest German advocate, Ernst Haeckel.

Where Darwin had been reluctant to speak in public, and where even Huxley tried to behave as a gentleman, Haeckel was bombastic. His strident manner won many converts but many enemies as well. Darwin himself, upon later meeting Haeckel, declared that he could hardly tolerate Haeckel personally. Although Darwin had carefully distinguished his own theory from other evolutionary theories, Haeckel combined these theories freely, even listing Lamarck and Goethe with Darwin as founders of evolutionary thought.

Haeckel was also responsible for another characteristic of German Darwinism: its fondness for embryological study. This came about largely from Haeckel's theory of recapitulation (see also Chapter 22). Haeckel ignored von Baer's principles and went back to the earlier ideas of Meckel. Among the new terms he introduced were **ontogeny**, the "history of the individual," and **phylogeny**, the "history of the race." "Ontogeny recapitulates phylogeny," he declared, "and phylogeny is the mechanical cause of ontogeny." The theory in this form came to be discredited in the years following World War I, but its immediate impact was to stimulate an enormous research interest in descriptive embryology. Hypothetical ancestors could not readily be obtained, but embryos could;

thus, to study an earlier stage in phylogeny, one had only to study embryology at the appropriate time. Large numbers of eager students and professors came to their embryology laboratories, hoping to confirm evolution and to find their own ancestors.

Haeckel had no patience for subtleties. In combining Darwinism with *Naturphilosophie* and with embryology, he had effected a change that gave German Darwinism something it lacked in England. Part of this change involved a political alliance with leftist ideas, quite the opposite of the social Darwinism with which evolution became associated in America.

Largely through Haeckel's influence, the German-speaking world quickly became won over to evolution, more quickly, indeed, than in England. Haeckel himself wrote several books on the subject, including *Natürliche Schöpfungsgeschichte* (translated as "natural history of creation," or perhaps more properly as "the history of natural creation"). Haeckel stressed human evolution as part of organic evolution. He even expressed a belief that some day a new, more general science would encompass all aspects of biology. For this new discipline, Haeckel coined a new name, "oecólogie," whose spelling is now "ecology." Haeckel was also the first person to represent phylogenetic diagrams on an actual tree.

Besides Haeckel, Darwin's other proponents in Germany included explorer-naturalists like Moritz Wagner (1813–1887) and Karl Semper (1832–1893), botanists like Matthias Schleiden (1804–1881) and Wilhelm Hofmeister (1824–1877), the materialist philosopher F. K. C. L. Büchner (1824–1899), and a long list of zoologists: Karl Vogt (1817–1895), Fritz Müller (1821–1897), Carl Gegenbaur (1826–1903), Julius von Sachs (1832–1897), August Weismann (1834–1914), Karl Claus (1835–1899), and Anton Dohrn (1840–1909).

In Tsarist Russia, the reception of Darwinism was mixed. Many Russian scientists of that era wrote in German and were strongly influenced by intellectual trends in Germany. Among Darwin's Russian supporters were the comparative embryologist Alexandr O. Kovalevskyi (1840–1901) and his younger brother, Vladimir Kovalevskyi (1842–1883), a paleontologist who pioneered in the study of horse evolution and translated some of Darwin's works. The zoologist Ilya (or Elie) Mechnikov (1845–1916) followed the evolutionary approach in much of his embryological researches but demurred when it came to natural selection, which he considered at best an untested hypothesis. Opposition to Darwin came from the Estonian-born embryologist Karl Ernst von Baer, who lived in St. Petersburg (now Leningrad) but continued to publish in German.

Darwinism and Marxism

Among the many supporters of Darwin's theories on the radical end of the political spectrum were the Marxists, who forged their own alliance with Darwinism. Both Marx and Engels were living in England by the time the *Origin of Species* was published, but they nevertheless belong to the German Romantic tradition, whose reception of evolution was based more on Haeckel's writings than on Darwin's.

Karl Marx (1818–1883) was primarily an economist. He appears not to have ventured into biology except in his correspondence with Engels. He did come to read and support Darwin's theory (usually in Haeckel's version), but he greatly resented the Malthusian element in natural selection, for he was unalterably opposed to Malthus on nearly every issue of economics.

Friedrich Engels (1830–1895) was more diverse than Marx in his interests and far more scientific. His acquaintance with biology seems to be largely through the writings of Haeckel, particularly the *Generelle Morphologie*.

Marx and Engels collaborated first on their publication of *The Communist Manifesto*. Their later collaboration, *Das Kapital*, contained an indictment of the social order then existing. Marx wanted to dedicate this work to Darwin, but Darwin politely refused, citing as a reason his own inability to read German. Engels's principal scientific work, *Dialektik der Natur* (Dialectics of Nature), published posthumously, has become the cornerstone of evolutionary biology in the Soviet Union.

CHAPTER SUMMARY

Charles Darwin, the son of a country physician, always enjoyed the outdoors. Throughout his education, first at Edinburgh and then at Cambridge, he associated with local naturalists and often neglected his studies. Paley's natural theology and Lyell's geology were among the early influences on him. From 1831 to 1836, the voyage of H.M.S. *Beagle* took Darwin to South America, the Galapagos Islands, and around the world. It was on this voyage that Darwin observed many phenomena, especially those of geographic variation, that convinced him that a branching type of evolution had occurred. Each continent that Darwin visited seemed to have had its own separate history, despite a similar range of climates, and the islands that Darwin visited always seemed to be inhabited by modified animals of a type found on the nearest continent. Neither special creation nor environmental determinism seemed to offer adequate explanation for these observations.

After returning from the *Beagle,* Darwin published many geological and zoological observations, but he kept his evolutionary views largely to himself. From observing the practice of animal breeders and possibly also from reading Malthus, Darwin struck upon natural selection, or differential reproduction, as the key mechanism responsible for most evolutionary change. Still, he did not publish until his hand was forced in 1858 by the simultaneous discovery of both evolution and natural selection at the hands of Alfred Russel Wallace.

On the Origin of Species, Darwin's major work, was published in 1859. In it Darwin set forth the theory of natural selection: heritable variations were always occurring, and if those that contributed to survival and reproductive success were perpetuated in greater proportion, the result, followed over countless generations, would be constantly improving adaptation. The result, Darwin hypothesized, would be a type of "descent with modification," in which relatives shared a more recent common ancestor. "Propinquity of descent" was seen as the basis of shared homologies and therefore of classification, while analogies were seen to be based on similarities of adaptation only. Evolution was thus explained as a gradual, branching process, whose operation in the present could be extrapolated into the geologic past. Darwin could document that the practice of artificial selection by animal and plant breeders had greatly altered their stocks within historic times. He reasoned that the gradual accumulation of heritable variation by natural selection over the course of geologic time could account for far greater change.

Most British and American scientists were soon won over to Darwinism, though in some cases reluctantly or with reservations. Major support came from the British anatomist Thomas Henry Huxley, from the American botanist Asa Gray, from the philosopher Herbert Spencer, from explorer-naturalists like Wallace, Bates, and Wagner, and from the aging Lyell. British opposition came principally from Sir Richard Owen and from the clergy. American opposition was led by Louis Agassiz. Public lectures by Wallace, Lyell, Gray, and especially Huxley carried the tide of Darwinism across both England and America. In America, Herbert Spencer's followers spawned a theory of "social Darwinism" in which unrestrained competition was hailed as Nature's own way of rewarding the stronger at the expense of the weaker.

In Germany and Russia, Darwinism merged with the earlier Romantic tradition of *Naturphilosophie.* Ernst Haeckel was Darwin's chief spokesman in Germany, though he also listed Goethe and Lamarck as evolutionists of equal standing and added his own theory of recapitulation ("ontogeny recapitulates phylogeny"). He combined Darwinism freely with Lamarckism and other theories. Marxist evolutionary thinking owes much to Haeckel's influence on Friedrich Engels. Anatomists like Carl Gegenbaur and most embryologists were strong supporters of Darwinism. In France, the acceptance of evolution was never an acceptance of Darwinism; national pride in the writings of Buffon, Lamarck, and others continues to ensure greater French support for home-grown theories than for Darwin's.

6

Genetics and the Modern Synthesis

Of the many roots of our modern evolutionary theory, two of the strongest are Darwinism and Mendelian genetics. The development of the science of genetics and its eventual fusion with Darwinian natural selection are the subject of the present chapter. The reader of this chapter is assumed to be familiar with the basic principles of genetics presented in Chapter 2.

A. GENETICS BEFORE MENDELISM

Prescientific Beliefs

Genetics was hardly a science before 1900. Except for Mendel's unrecognized work, observations on heredity tended to be casual and unsystematic. This was the prescientific period of genetics, before it received its first paradigm at the hands of Gregor Mendel.

Casual observations on the inheritance of humans and domestic animals had been recorded since ancient times, but there was no accepted method of analyzing heredity before 1900. Animal and plant breeders had been controlling the heredity of their stocks for many generations, but largely without any theory to guide them in their efforts. In nearly all comparisons between parents and their offspring, perhaps the most common fault was the failure to take into account the hereditary background of the parents.

In this prescientific phase of unsystematic observation, the most frequently noticed phenomenon was that, in general terms, offspring usually resembled both of their parents to a greater or lesser extent. This gave rise to the belief that the hereditary traits of both parents had somehow "blended" together in a process similar to the mixing of fluids, or more particularly of "bloods"; this later became known as the **blending theory of inheritance.** Casual observation seemed to confirm this theory, for when two animals were mated, the young shared some of the mother's characteristics with some of the father's. No attempt was made to quantify the results, to simplify them by studying only one character at a time, or to carry out observations for more than a single generation. Expressions like full-blooded and half-blooded, as well as references to the "blood" of a particular nationality or race, are still heard today on occasion; they all date back to the now-discarded theory of blending inheritance.

Occasionally, however, a strange or unusual trait would reveal itself as so striking that it could hardly escape attention. Such traits often defied incorporation into any theory at all, for sometimes they were transmitted to the offspring, but at other times they "skipped" one or several generations and showed up among the grandchildren or even more distant offspring. It was not until Mendel's rediscovery that this type of inheritance was satisfactorily explained.

Darwin's Theory of Pangenesis

Charles Darwin was meanwhile laboring under a great difficulty: he had no theory of heredity to support his evolutionary theory. A British inventor and mathematician, Fleeming Jenkin (1833–1885), criticized Darwin on this very point. His argument, with racist overtones, was roughly as follows: suppose a particular country were inhabited by a dark-skinned race. Even if white-skinned Europeans had a competitive advantage over the native inhabitants, a lone, shipwrecked European would have no hope of establishing his own race through natural selection, for he would have to mate with one of the native inhabitants, and their offspring would be only half-blooded mulattoes of intermediate color. These children, in turn, could marry only other natives, or at best they might marry each other. In each generation, the heritable characters brought by the lone European would dissipate still further and become further mingled with the native conditions, never again to be reconstituted. Natural selection, said Jenkin, would be impotent to overcome this tendency of superior traits to become lost among the vast numbers of inferior ones, just as a drop of ink placed in a lake would have no hope of discoloring the lake, for it would quickly become so diluted that it would effectively disappear.

The argument above, as Darwin correctly recognized, depends on an assumption of blending inheritance. In strict mathematical terms, the variance of a freely interbreeding population would be halved in each generation of blending inheritance.

Darwin's response to this criticism was his book *On the Variation of Animals and Plants under Domestication* (1868). Here he put forth his own incorrect theory of inheritance, which he called the "provisional hypothesis of pangenesis." Similar theories had been proposed in the past by Buffon and by others, but Darwin's treatment of pangenesis was the most thoroughly considered in the light of available evidence.

According to Darwin's theory of pangenesis, every organ of the body produced minute hereditary particles, caled gemmules: the liver produced liver gemmules, the leg produced leg gemmules, and so forth (Fig. 6.1). Using this hypothesis, Darwin could account for the inherited effects of use and disuse by assuming that an enlarged or diminished organ produced either more or fewer gemmules. He could also account for the noninheritance of mutilations such as in polled cattle, whose horns had been removed, by saying that the horns had already produced their horn gemmules before their removal. Darwin thus anticipated one of Weismann's later observations but gave it a different interpretation. Darwin thought that the gemmules were carried through the blood from every organ of the body and were collected together into the gametes during gamete formation.

Darwin's theory of pangenesis was not well received, though Darwin thought that his viewpoint would someday be vindicated. Instead, Darwin's theory of pangenesis met with rejection by August Weismann and by Darwin's own cousin, Francis Galton (1822–1911). Galton, the father of the British Eugenics movement (Chapter 28), transfused blood from a black rabbit into a white one. According to Darwin's theory, this should have transferred some black gemmules, thereby altering the heredity of the white recipient. But upon crossing this white rabbit to another white rabbit, Galton found that their offspring were entirely white, with no trace of blackness, thus refuting Darwin's theory experimentally.

Weismann's Germplasm Theory

August Weismann (1834–1914) was a German biologist, one of the new generation of scientists who accepted Darwinian evolution early in their careers. But even though the scientists of his generation readily accepted evolution, many of them failed to embrace Darwin's proposed mechanism, preferring Lamarck's theory of use and disuse instead. Weismann was keenly aware of the controversy between the supporters of Lamarck and of Darwin, and he sought to resolve the controversy by means of an experiment.

Weismann noticed that Lamarck's theory depended on the inheritance of acquired characteristics. Lamarck had noticed this himself and considered such inheritance to be a very small detail in his much grander theory of

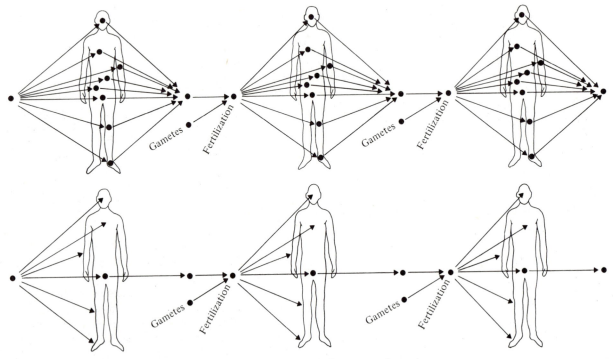

Fig. 6.1 Darwin's theory of pangenesis contrasted with Weismann's later germplasm theory.

upwards or progressive evolution. It was Weismann who coined the phrase *the inheritance of acquired characteristics,* wrongly identifying it as the essence of Lamarck's theory. Our understanding of Lamarck was for several generations greatly hindered by a casual acceptance of Weismann's mistake.

Weismann's experiment was simply to measure and then amputate the tails of 22 successive generations of laboratory mice. The offspring of these mice were all born with tails that grew to the normal length, showing that the experimentally acquired, tailless condition was not inherited. Weismann declared that he had disproved Lamarckism. He became an overzealous supporter of Darwinian natural selection, to the exclusion of nearly all other evolutionary forces. Darwin, meanwhile, had made far less extravagant claims on behalf of natural selection, and even these were toned down in Darwin's later writings (see Chapter 5). But Weismann preached "the om-

nipotence of natural selection," a theory to which he gave the name **neo-Darwinism.**

Weismann's more lasting contribution was his theory of the germplasm, which arose directly from his mouse-tail experiment. Every organism, according to this theory, consisted of two very distinct components: germplasm and somatoplasm. The **germplasm,** or germ cells, had two functions: they produced the germplasm of the next generation, and they also produced the **somatoplasm,** or perishable body, containing the overwhelming bulk of the body's tissues. Heredity, according to Weismann, was simply the *continuity of the germplasm* (Fig. 6.2). The somatoplasm, or perishable body, had no lasting hereditary effect; changes that affected only the somatoplasm, such as amputation of the tail, had no hereditary consequences since the germplasm was unchanged. The function of the somatoplasm was simply that of a vehicle for the germplasm; a British writer,

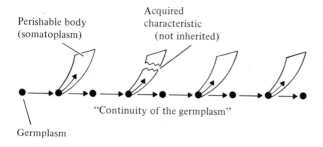

Perishable body (somatoplasm)

Acquired characteristic (not inherited)

"Continuity of the germplasm"

Germplasm

Fig. 6.2 Weismann's germplasm theory in diagrammatic form. Acquired characteristics cannot be inherited because the somatoplasm is merely a perishable body that has no effect on future generations.

Samuel Butler (1835–1902), satirized the germplasm theory by declaring that "a hen is but an egg's way of making another egg."

Weismann's germplasm theory was important also as a refutation of Darwin's theory of pangenesis (Fig. 6.1) and in the laying of the groundwork for the later distinction between genotype and phenotype. The phenotype (Chapter 2) consisted of the characteristics of the somatoplasm, whereas the genotype consisted of the characteristics of the germplasm. The genotype and environment could both influence the phenotype, but neither the phenotype nor the environment could have any direct effect on the genotype. Weismann thought that natural selection was the only mechanism capable of bringing about changes in the germplasm.

B. MENDELIAN GENETICS

Gregor Mendel

Gregor Mendel (1822–1884) is generally acknowledged as the "father of genetics" (see Item 6.1 on page 109). His work, first described in 1865, was generally ignored until 1900, however.

Mendel succeeded where others had failed, partly because of a fortunate but intentional choice of experimental material, and partly because of his systematic form of observation and record-keeping. For his experi-

ments, Mendel chose the garden pea, *Pisum sativum*, a species that would grow easily under cultivation. Mendel knew that he could obtain several varieties of peas in local markets. He knew that the flowers were hermaphroditic, i.e., that the sexes were together in the same flower. One petal was much larger than the rest, and it nearly enclosed the entire flower. By completely enveloping the flower in this enlarged petal, Mendel could ensure self-fertilization; by removing the stamens and dusting with pollen from another flower, he could also achieve cross-fertilization. Another precaution that distinguished Mendel's work from nearly all that had gone before was his use of true-breeding plants of known parentage. He self-fertilized each of his plants for several generations, and in his further studies he used only those plants that produced offspring like themselves for a number of generations.

Mendel's first series of experiments studied the inheritance of one character at a time. Wisely reducing his system to the simplest case, Mendel was thereby able to study the phenomena of heredity that had eluded so many of his predecessors. In this way, he studied seven pairs of opposite traits: round versus wrinkled seeds, green versus yellow seeds, tall versus short stems, axial versus terminal flower position, inflated versus constricted pods, green versus yellow pods, and colored flowers (red-violet, with a gray seed coat) versus white flowers and seed coat. In each of these seven cases, one trait invariably showed itself in the phenotype of all the F_1 plants; Mendel called this trait **dominant** and the hidden trait **recessive**. There was no blending and no intermediacy: the offspring of a tall × short cross were always just as tall as their tall parents, if not taller. In all seven cases, the F_1 plants, when allowed to self-fertilize, produced an F_2 with a 3:1 ratio of dominants to recessives. From this, Mendel was able to explain the pattern of single-gene inheritance shown in Box 2–B, and he was furthermore able to make predictions about the genotypes of his F_2 plants and to test these predictions by raising an F_3 generation. The hereditary material did not blend; inheritance was shown to be **particulate.**

Mendel also performed several crosses between plants differing in two characters at a time; one of these is shown in Box 2–C. In these crosses, he was able to establish the principle of independent assortment (Chapter 2) and the random union of gametes. He also crossed plants differing in three characters at a time but discovered no new principles. His results always accorded with his expectations.

Mendel's brilliant work was completed in 1865. But the time was not yet ripe, and his work lay unrecognized until its rediscovery in 1900.

Mendel's Rediscovery

In the spring of the year 1900, Mendel's work was rediscovered independently by three researchers in three separate countries: Erich von Tschermack (or Czermack), a Czech living near Vienna, Austria; Carl Correns, a Danish citizen studying at the University of Tübingen in Germany; and Hugo De Vries, a professor at the University of Amsterdam in Holland. Curiously, each of the three had conducted his own breeding experiments independently and arrived at essentially the same results, and each had later run across Mendel's paper after his own work was essentially complete.

The next ten years witnessed a flurry of activity in many countries, during which the basic principles of the new science of genetics were quickly established. The same Mendelian principles were soon applied to the heredity of all sexually reproducing diploids, including humans. Important contributors included Punnett, Bateson, and many others in England; Weinberg in Germany; De Vries in Holland; Johansson in Sweden; and Sutton, Morgan, Castle, and East in the United States. Bateson and other British geneticists discovered many instances of Mendelian inheritance in humans, and in 1909 the British physician Garrod initiated the new field of biochemical genetics with his book on human metabolic defects, *Inborn Errors of Metabolism.*

Neither Carl Correns nor Erich von Tschermack made a very large contribution beyond his role in Mendel's rediscovery. But Hugo De Vries did, and his contribution was in the form of a theory of evolution.

De Vries's Theory of Mutations

Mendel had studied the transmission of inherited differences in peas, but he had left one important question unanswered: where had these inherited differences come from, and how had they originated? The same question had occurred to Darwin in a different context. The first fruitful attempt to answer this question resulted in DeVries's theory of mutations. The theory was wide of its mark, and its excesses came to be criticized in later generations, but at the time it represented an important step forward. Despite its distinctly anti-Darwinian outlook, its eventual reconciliation with Darwinism led to our modern synthetic theory of evolution (Section D, below).

Much of DeVries's work was with a flower known as the evening primrose, *Oenothera Lamarckiana.* In a field of white-flowered primroses, a red flower suddenly appeared. Animal breeders had noticed such ''sports'' in the past but had no theory to account for their origin. De Vries, given his background in genetics, was naturally driven to ask one fundamental question: would the newly originated difference be inherited? He crossed his red flower with white flowers and soon found that the red color was inherited as a simple Mendelian dominant. He subsequently discovered other instances of inherited differences that appeared suddenly and without apparent cause. To these new variations DeVries gave the name **spontaneous mutations.** The word ''spontaneous'' originally meant that the new variations were not environmentally induced, and that they occurred without respect to their possible adaptive value.

DeVries expanded these observations into a complete theory of evolution by mutations. Where Darwin, Lamarck, and others had emphasized the adaptive nature of variation, DeVries tended to emphasize the opposite, namely, its randomness. Also, where Darwin and others had focused their attention on small, gradual differences, De Vries focused attention instead on abrupt and often drastic changes, which he thought might be capable of making new species. De Vries was unaware of the unusual behavior of the chromosomes of *Oenothera.* The so-called Renner complexes that these chromosomes form will be described in Chapter 7.

C. THE CHROMOSOMAL THEORY OF INHERITANCE

The great classical period in the history of genetics lasted roughly from 1900 to 1940. During this period, Mendel's principles were confirmed and expanded, and new principles were discovered. The genetics of diploids became perhaps the best-understood area within biology and attracted many of the best minds of the era. During this period, the great unifying theme of genetics, the core of its paradigm, was the so-called **chromosomal theory of inheritance,** which simply stated that the genes were located on chromosomes or were themselves parts of chromosomes.

Walter Sutton's Hypothesis

Walter Sutton (1877–1916) was at the turn of the century a graduate student at Columbia University in New York. His early training had been in cell biology, which in those days emphasized staining methods, cell morphology, and the details of mitosis and meiosis (Chapter 2). One of Sutton's professors, E. B. White, was one of the foremost cytologists of his day; together with several other graduate students and professors, they frequently discussed new advances in the biology of their day. One new advance that greatly interested Sutton was contained in a paper published by Theodor Boveri in 1898. Boveri had studied the fertilization of sea urchin eggs and had shown that the chromosome number, halved during meiosis, was restored to its full diploid number during fertilization by a fusion of the former sperm nucleus with the former egg nucleus.

In 1900, Mendel's paper was rediscovered, and Walter Sutton was one of the persons most excited by this new and promising field. Given his prior training in cell biology, Sutton asked a critical question: *where* in the cell were the "genes" located? Sutton noticed that Mendel's genes occurred in pairs, were inherited discretely, and segregated from one another during meiosis. Both parents, furthermore, made rather equal contributions to heredity, yet their contributions to the fertilized egg (zygote) were decidedly unequal: the egg was in most species many times larger than the sperm, often thousands of times larger. The genes, Sutton reasoned, could not reside in the cytoplasm, because cytoplasm seemed to blend and was not discrete, because cytoplasm divided carelessly and inexactly during cytokinesis, and because the mother's contribution to heredity would be far greater than the father's if the genes had a cytoplasmic location. The nucleus, on the other hand, divided carefully and exactly, and the contributions of both parents to the nucleus of the zygote were equal. Noting Boveri's work, Sutton pointed specifically to the chromosomes: here were discrete bodies within the nucleus that were inherited in pairs, and they segregated during meiosis according to a complex procedure that ensured equal distribution of chromosomes among the daughter cells. Exactly half of the chromosomes were transmitted to each gamete, and the diploid number was later restored during fertilization. In short, the chromosomes behaved exactly in the manner of Mendel's genes. They even assorted independently of one another, just as Mendelian inheritance required.

It was here, however, that Sutton encountered a limitation that his hypothesis required upon Mendel's law of independent assortment: genes located on separate chromosomes should assort independently, Sutton maintained, but genes located on the same chromosome should be **linked.** Sutton thus predicted several important phenomena that had not yet been observed but would later be confirmed: all the genes located on the same chromosome should always be inherited together as a group, and the number of linkage groups should be equal to the haploid chromosome number carried by each gamete. Sutton did not successfully predict the crossing over of linked genes, but in other respects his hypothesis of the chromosomal location of the genes turned out to be correct. What was more remarkable, Sutton published his thoughts on the subject, including all the arguments above, in the year 1903, when very few genetic experiments had yet been carried out.

Morgan and *Drosophila*

Not only had Sutton identified the chromosomes as the probable carriers of the genes, but he was also able to convince his own professors of the truth of this hypothesis, which later became the chromosomal theory of inheritance. One of the professors most influenced by Sutton's theorizing was Thomas Hunt Morgan (1866–1945), a Kentucky-born embryologist and evolutionist. Morgan soon converted one of his rooms into a genetics laboratory, among the first in the country, and became one of the leaders of American genetics.

Most fortunate was Morgan's choice of an experimental organism, suggested to him by William E. Castle: the fruit fly, *Drosophila melanogaster* (Fig. 6.3). Here at last was a small yet phenotypically complex organism that could be raised in the laboratory in large numbers: hundreds could be accommodated in a small milk bottle. Its generation time, moreover, was very short (about two weeks), so that a cross could be begun and an F_2 obtained

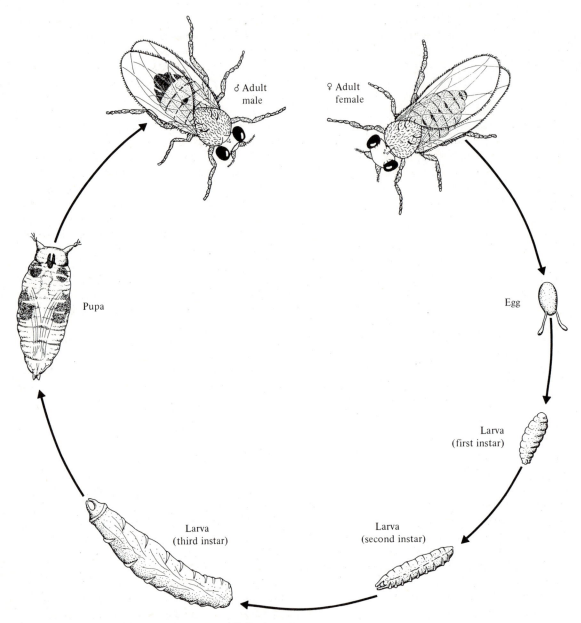

Fig. 6.3 Life cycle of the fruit fly, *Drosophila melanogaster*.

and analyzed well within the limits of an academic semester. Previously, a geneticist had to have the time and patience of a monk in a monastery—Mendel's vocation was no accident. But now long-term experiments in genetics could be carried out more easily, and large numbers of progeny could be sampled. *Drosophila* quickly became the best-understood organism genetically, a status only recently challenged by such microorganisms as the bacterium *Escherichia coli.*

Another fortunate aspect of *Drosophila* was its low chromosome number, 8. With only four linkage groups to be discovered, Sutton's prediction of linkage was soon confirmed, and the additional phenomenon of sex linkage was soon discovered. Linkage, in fact, was reported (but misinterpreted) by Bateson in 1906 (Fig. 2.7), and it was Sutton and Morgan who first suggested the correct interpretation for Bateson's data.

Morgan's greatest personal achievement was the discovery of sex-linked inheritance. In a quest for the "new mutations" that De Vries's mutation theory had predicted, he was looking at hundreds of flies in search of something new. The first fly to draw his attention was a white-eyed male; the normal, or "wild type," eye color in *Drosophila* is red. Morgan mated his white-eyed male to a red-eyed ("wild type") female and obtained red-eyed F_1 flies, showing that red was dominant. Upon breeding his F_1 flies to produce an F_2, Morgan noted that the expected 3:1 ratio was obtained, but all the white-eyed flies were males (see Item 2.7 at the end of Chapter 2). Morgan at first suspected that white eyes might be a trait confined to males, but another cross falsified this hypothesis: he crossed an F_2 white-eyed male to an F_1 (heterozygous) red-eyed female and obtained a 1:1 ratio of eye colors in each sex. *Drosophila* was already known to have the XX-XY type of sex determination, like that in humans, and Morgan noted that he could explain his crosses only if he assumed that the genes for white versus normal eye color were carried along with the X chromosomes. This was the first instance in which a specific gene had been identified with a particular chromosome.

Subsequent research by Morgan and his students, including A. H. Sturtevant, established a linear linkage map for each of the *Drosophila* chromosomes, confirming in part the chromosomal theory of inheritance. Abnormalities of sex determination were studied intensively by Calvin Bridges, another of Morgan's students; each of

these abnormalities gave predictable results from both a genetic and a cytological point of view. Morgan, his students, and their students in turn dominated much of American genetics for half a century—an influence that is still strong. In 1933, Morgan was awarded the Nobel Prize, the first of many geneticists to receive this award.

Confirmation of the Chromosomal Theory

The chromosomal theory of inheritance was increasingly confirmed by the research of Morgan and his students and by other geneticists. Arthur H. Sturtevant first used crossover frequencies as measures of mapping distance and produced the first genetic map of the *Drosophila* sex chromosome. In time, a full genetic map was obtained for each of the chromosomes (Fig. 2.9). The detection of new mutants was greatly aided by the discovery in 1927 of induced mutations by H. J. Muller, who found that exposure to x-rays greatly increased the rate of mutation.

Calvin Bridges discovered an infrequent type of abnormal meiosis in *Drosophila*. By breeding these so-called **nondisjunction** females, Bridges could obtain several chromosomal abnormalities, each of which could be traced genetically if the inheritance of sex-linked traits was studied simultaneously. The behavior of the *genes* always paralleled the behavior of the *chromosomes,* thereby confirming the association between the two, at least in the case of sex-linked traits. By studying other chromosome abnormalities, notably deletions (Chapter 7), Bridges and others were able to confirm other correspondences between the behavior of genes and the behavior of chromosomes.

In the early 1930s, several workers (Creighton & McClintock, also Stern) were finally able to demonstrate crossing over both cytologically and genetically at the same time. In carefully arranged crosses, these workers were able to demonstrate that the recombination of linked genes was always accompanied by the breakage and reunion of microscopically observable chromosomes.

By around 1940, the chromosomal theory of inheritance had been fully confirmed and accepted by geneticists throughout the world, except in the Soviet Union. Trofim D. Lysenko had established a stranglehold over Soviet genetics that banned the chromosomal theory of inheritance until 1963 (Item 6.2).

D. CONTROVERSY AND SYNTHESIS

The Era of Controversy, 1900–1940

For somewhat more than an entire generation, two large groups of twentieth-century scientists held opposing theories—or at least opposing attitudes—summarized in Box 6–A. One group were the Mendelian geneticists who followed DeVries's mutationist theory; the other, less vocal group was composed largely of field naturalists who often resented the success of the new Mendelian geneticists. Each group held half of the truth, and the eventual resolution of their differences led, around 1940, to the modern synthetic theory of evolution.

The early Mendelian geneticists, for their part, emphasized discrete variations that arose spontaneously, without regard to any adaptive significance. As laboratory scientists who seldom did research outdoors, they tended to deny the role of the environment and of geographic variation. They pointed to physiological changes in mammalian pelage (coat color) and argued from this that geographic variation was purely a phenotypic response with no genotypic basis. Another of their arguments against the genotypic basis of geographic variation was simply the fact that it was gradual and continuous, whereas the simple Mendelian traits that they always emphasized were discrete and discontinuous. Those who thought about the origin of species usually followed DeVries's claim of sudden species origin by saltation. As correct as these geneticists were about the physical basis of heredity, they were wrong about geographic variation and the origin of new species.

BOX 6–A   *The Era of Controversy*

Beliefs commonly held by Mendelian geneticists of the period circa 1900–1930 and by contemporaneous naturalists (with certain exceptions). Beliefs printed in CAPITAL LETTERS are still held to be largely correct.

Mendelian geneticists	Naturalists
Typological thinking: natural populations are uniformly ''wild type.''	POPULATION THINKING: NATURAL POPULATIONS ARE EXTREMELY VARIABLE.
VARIATION IS DISCRETE; changes are discrete and sudden; INHERITANCE IS PARTICULATE.	VARIATION IS CONTINUOUS; CHANGES ARE SMALL AND GRADUAL; inheritance is by blending.
VARIATION ARISES AT RANDOM.	VARIATION IS ADAPTIVE, SHOWING INFLUENCE OF ENVIRONMENT. This influence is often by direct induction & inheritance of acquired characteristics.
Geographic variation is a phenotypic response only, and is thus unimportant.	GEOGRAPHIC VARIATION IS GENOTYPIC (AND IMPORTANT).
SPECIES ARE REAL, DISCRETE.	SPECIES ARE VARIABLE (ESPECIALLY GEOGRAPHICALLY); TRANSITIONAL FORMS CAN SOMETIMES BE FOUND BETWEEN NAMED SPECIES*
New species arise suddenly, by saltation.	NEW SPECIES ARISE GRADUALLY.

*In these cases, however, adoption of the polytypic species concept (Chapter 14) has resulted in the lumping together of taxa that had previously been described as separate species.

Naturalists of the same period were a bit more diverse in their opinions, but still a majority of them held views directly opposed to those of the geneticists. They were basically field naturalists to whom the environment and organic adaptations to it were all too real to be denied. Geographic variation was also a phenomenon all too real and all too familiar to them. The naturalists were convinced that geographic variation had a genetic basis, i.e., that geographical differences would be inherited despite their gradual pattern of variation. Many naturalists therefore rejected Mendelian genetics and continued to believe in blending inheritance; others became neo-Lamarckians in an attempt to explain adaptation. Only a few still clung to Darwinian natural selection as the basis for evolution; these included Henry Fairfield Osborn of the American Museum of Natural History in New York and David Starr Jordan and Vernon Kellogg, both of Stanford University in California. The naturalists had no consistent theory of speciation, but they usually insisted that new species originated gradually. Many of these naturalists supported the theory of sympatric speciation (Chapter 14). In an age when the typological species definition became once more popular (no doubt through de Vries's insistence on the discreteness of species differences), these naturalists often delighted in naming ''new species'' on the basis of unbelievably small differences, often with only one or a few of its specimens in hand.

The Coming of the Modern Synthesis

Many issues divided the geneticists from the naturalists in the early decades of the twentieth century. One by one, these issues were resolved by the further accumulation of data, and the reconciliation gradually extended, until the last major issue was resolved around 1942.

In England, the more visible controversy was initially between two schools of geneticists, the Mendelians and the biometricians. The Mendelians, led by Bateson and Punnett, held variation to be discrete, and they were able to establish Mendelian inheritance as the basis for all genetic variation. The biometricians, led by Galton, sided with the naturalists on one important issue: they held variation to be continuous instead of discrete. This particular issue was resolved by Nilsson-Ehle and East, who independently discovered the polygenic type of inheritance shown in Fig. 6.4. The explanation of continuous variation as resulting from the inheritance of discrete

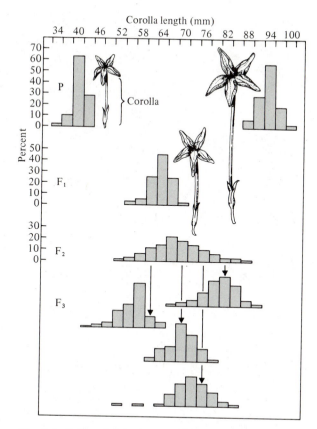

Fig. 6.4 Polygenic inheritance of flower morphology in tobacco plants (*Nicotiana longiflora*). Reprinted by permission from Ayala and Kiger, *Modern Genetics*, Menlo Park, Ca.: The Benjamin/Cummings Publishing Company, Inc., Fig. 15.15, p. 517.

genes did much to reconcile the two British schools of genetics. The influence of the biometricians, however, ensured that British genetics maintained a strong interest in continuous variation, in statistics, and (through Galton's own influence) in human variation.

Meanwhile, in the United States, the geneticists and naturalists were further apart. Genetics was hailed as the ''new'' or ''modern'' biology, the ''wave of the future,'' a distinction which the naturalists often envied or resented. Genetics became a required course for biology majors in most colleges and universities, and geneticists

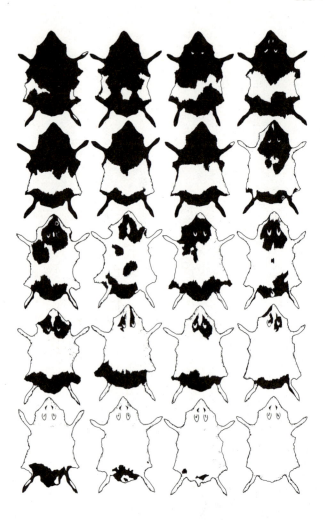

Fig. 6.5 The effects of modifier genes on the coat color of laboratory mice. All the mice depicted on this page are homozygous for the "spotting" trait; the differences among them are the results of selection for various "modifier" genes, producing a nearly continuous range of variations between extremes. Drawn from photographs. Reprinted by permission from Ayala, Francisco J. and Kiger, John A. Jr., *Modern Genetics,* Menlo Park, Ca.: The Benjamin/Cummings Publishing Company, Inc., Fig. 15.3, p. 501.

came to hold important positions on many faculties. Thus the naturalists of succeeding generations were quite aware of genetics. They had been required to take courses in the "new" genetics as part of their training and were thus conditioned to analyzing even their field results in genetic terms.

Studies in mammalian genetics by Dice (1940) and others provided important evidence by showing the polygenic inheritance of continuous coat color variations (Fig. 6.5). This was especially important, because mammalian coat color often varies geographically, and the results suggested that geographic variation might also be explainable in Mendelian terms, even though it, too, was continuous. Important evidence was also being gathered

from studies in which plants from various localities were transplanted and grown side by side under similar environmental conditions (Fig. 6.6). The plants maintained many of their differences, showing that these differences were more than mere phenotypic responses to the environment.

In England, the reconciliation between geneticists and naturalists was begun in 1930 by the geneticist and statistician Ronald A. Fisher. Fisher, a biometrician, disbelieved the randomness of variation, and he was looking for some "antichance factor" that would explain adaptations and other nonrandom events in evolution. The "antichance factor" turned out to be an old friend: Darwinian natural selection. Fisher's book, *The Genetical Theory of Natural Selection* (1930), was a landmark, for it represented the first reconciliation of Mendelian genetics with Darwinian natural selection. Fisher showed the two theories to be quite compatible, provided certain assumptions were made about mutations: they had to be rare, generally small in their effects, most often harmful, and usually recessive. Fisher's arguments were firmly grounded in natural selection and other Darwinian principles. Most mutations, he noted, had probably occurred before; few were wholly novel. Over many generations, natural selection would have ensured that the most favorable genotypes had become most widespread. Thus most mutations would be both rare and harmful, because they had occurred before and had been selected against. Other genes that controlled mutation rates would generally be selected to keep these rates very low. Fisher also assumed that the phenotypic expression of genes in heterozygotes was controlled by "modifier" genes (compare Fig. 6.5), and that dominance was controlled in this

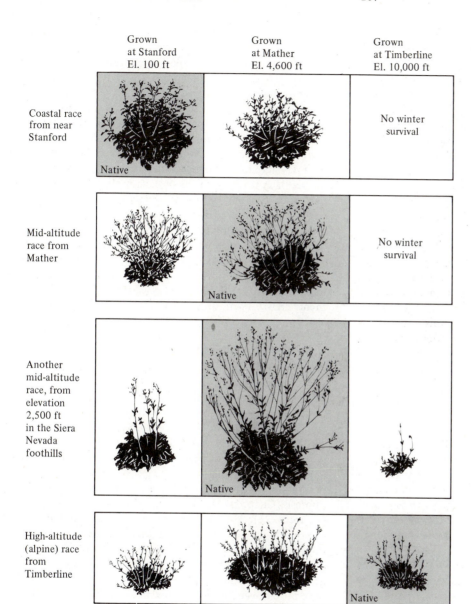

| | Grown at Stanford El. 100 ft | Grown at Mather El. 4,600 ft | Grown at Timberline El. 10,000 ft |

Fig. 6.6 Responses of three altitudinal races of the cinquefoil plant, *Potentilla glandulosa,* to three different environments in California. After Clausen, Keck, and Hiesey, Carnegie Institute of Washington, Publication 520, 1940.

way. Modifier genes would be favored by selection, said Fisher, if they minimized the phenotypic expression of harmful mutations in the homozygous condition, and if they maximized the phenotypic expression of beneficial alleles in heterozygotes. The rareness, recessiveness, and small phenotypic effect of harmful mutations had all been brought about by selection. Fisher also showed that

small mutational changes were likely to be least harmful, and the more drastic changes were likely to be increasingly harmful in proportion to their severity. The integrity of a harmonious genotype, in other words, is like that of a well-tuned mechanical device such as a racing car. Fisher is also responsible for an important theoretical argument against sympatric speciation (Chapter 14) and for many important contributions to statistics. It was Fisher who invented the statistical method known as the analysis of variance, and who first clearly distinguished between population parameters and sample statistics.

The theoretical treatment of population genetics was greatly extended by Sewall Wright and by J. B. S. Haldane, among others. Increasingly, these population geneticists were able to take the phenotypically complex variations of the naturalists and analyze them statistically in terms of discrete Mendelian genetics.

Natural populations were studied genetically by a new generation of scientists, who began to analyze both geographic variation and species differences in genetic terms. This proved to be the most crucial reconciliation of earlier controversies. Pioneering in this study were E. B. Ford and H. B. D. Kettlewell in England, whose work on industrial melanism showed that the actual gene frequencies varied geographically and in an adaptive manner. In the United States, studies on the genetics of natural, geographically varying populations were carried out most vigorously by the Russian-born field-entomologist-turned-geneticist Theodosius Dobzhansky and his students. Dobzhansky and his students studied a number of chromosomal polymorphisms in the field, many of which varied geographically (see Fig. 7.1). Dobzhansky and his students were also much impressed with the genetic variability of natural populations.

Beginning around 1940, Clausen, Keck, and Hiesey studied the geographic variation in several plant species and found that each altitudinal race had its own set of phenotypic responses to the several habitats studied (Fig. 6.6). The genotypic basis of these differences was proved by the fact that the differences were maintained when the plants were raised side by side in an experimental garden.

The explanation of geographic variation in terms of genetics, beginning around 1937, was soon followed by an explanation of the ultimate problem of species origins. The geographical theory of speciation (Chapter 14) was fostered in England by Julian S. Huxley and in the United States by Ernst Mayr. The geneticists and the naturalists were at last reconciled.

The Modern Synthetic Theory, 1940–

It was Julian Huxley, a grandson of Thomas Henry Huxley, who proclaimed first a "New Systematics"

(1940) and subsequently a "Modern Synthesis" (1942). Soon many other classic textbooks (Item 6.3 at the end of this chapter) joined the field. The general return to Darwinism was evident in many of their titles, several of which contained the very words "origin of species."

One by one, each field of evolutionary biology was brought within the orbit of the modern synthetic theory. Dobzhansky (1937) showed that genetics was consistent with this new theory; Mayr (1942) showed that systematics was similarly consistent with it. Simpson (1944, etc.) showed that the major facts of paleontology were consistent with the new synthetic theory, too, and Rensch (1947, 1960) showed that the phenomena of microevolution were adequate to explain macroevolution as well. Stebbins (1950) summarized the entire modern synthetic theory as it pertains to plants, and Mayr (1963) provided a similar compendium of the modern synthetic theory as it pertains to animals. Genetic aspects of the modern theory have been ably summarized by Dobzhansky (1970).

New Advances—And New Questions

The remainder of this book deals with the modern synthetic theory of evolution, including many of the newer advances that have been made since the 1940s. Among these newer advances are the experimental studies on chemical evolution and the origin of life (Chapter 25), the biochemical determination of distances among relatives (Chapters 19, 22), the measurement of natural population variability by such methods as electrophoresis (Chapter 12), the theory of island biogeography (Chapter 23), the punctuated equilibrium model in paleontology (Chapter 21), and a number of other exciting advances. A few newer studies of speciation (Chapter 14) have also brought certain aspects of the theory of geographical speciation into question. New controversies have arisen over the significance of genetic variability in natural populations (Chapter 12), over sociobiology (Chapter 24), and over taxonomic theory (Chapter 22), among various other issues.

The coming into its own of molecular biology, beginning in the 1950s, has added greatly to our understanding of evolution and has also raised some exciting new questions. How is the genetic material organized on a molecular level, and how did it attain this organization through evolution? How did the genetic code evolve?

What is the relation between genotype and phenotype, or in other words, how does the genotype act during development to produce a phenotype? Where does the environment get its chance to act on the phenotype? If the genetic code (Chapter 2) is so redundant, what percentage of mutations change one codon to another coding for the same amino acid? What role, if any, do these "silent mutations" play in evolution? How frequent are "quiet" mutations, which result in an amino acid substitution but without discernible phenotypic consequence? What proportion of the genotype *is* phenotypically expressed, and what role, if any, does the remainder of the genotype serve? Tentative answers to these and other fascinating questions will be offered in succeeding chapters, but the definitive answers still lie in the future.

CHAPTER SUMMARY

Before 1900, the understanding of genetics was hampered by the failure to carry out breeding experiments for more than a single generation, by the failure to interpret results statistically, and by the failure to distinguish genotype and phenotype. Mendel's experiments estab-

lished our modern theory of genetics, just as Weismann laid the groundwork for Mendel's rediscovery and for the distinction between genotype and phenotype. Among Mendel's rediscoverers, De Vries was most original in his own theory of evolution through mutations. De Vries overemphasized the importance of random mutations in evolution, to the extent of neglecting those forces responsible for adaptation.

In the early decades of the twentieth century, research in genetics was dominated by the chromosomal theory of inheritance. Among the major advances made during those decades were the introduction of *Drosophila* as an experimental species, the discovery of sex-linked inheritance, the construction of linkage maps, and the final confirmation of the chromosomal theory.

During the heyday of the chromosomal theory, geneticists were in many cases operating under typological misconceptions about evolution, while naturalists, many of whom rejected the particulate inheritance of continuous characters, labored under equally false misconceptions. The reconciliation of the two schools, around the year 1940, brought about the modern synthetic theory of evolution—the theory to which the rest of this book is devoted.

FOR FURTHER KNOWLEDGE

ITEM 6.1 *Gregor Mendel (1822–1884)*

Born as Johann Mendel in the Czech mountain village of Heinzendorf near the German border, Mendel repeatedly distinguished himself as a student. His forebears were largely peasants, and it was to escape poverty that he entered the priesthood, which he described as "a profession in which [I] would be spared perpetual anxiety about a means of livelihood." In 1843, at the age of 21, he entered the monastery of Saint Thomas in Brno (or Brünn), taking the new first name of Gregor.

The Augustinians at Brno trained many high school teachers for the Moravian countryside. Mendel was ordained a priest in 1847, but he failed his examination for certification as a high school teacher. Because of his inclination to science, he was sent to the University of Vienna, where he enrolled in courses in botany, mathematics, and the physical sciences. Among his teachers in Vienna were the plant physiologist and paleobotanist Franz Unger (1800–1870) and the physicist

(continued)

Christian Johann Doppler (1803–1853), discoverer of the effect of velocity on the wavelength of light or sound received from a moving source.

In 1856, Mendel began his experiments on the crossing of peas, which he grew in the monastery garden. He frequently referred to these plants as "my children" and apparently delighted in the shock value that this little joke had on his visitors. A dedicated scientist throughout his life, Mendel kept a carefully bound notebook of meteorological observations, including data on temperature, humidity, barometric pressure, ozone content (!), depth of local ground water, and a series of careful drawings of sunspots. In 1862, Mendel was one of the founders of the Brünn Society for the Study of Natural Science (Brünn Naturforschende Verein). His famous paper on plant hybridization was read to this society, as were various meteorological reports, including a detailed account of a tornado that struck Brno in 1870. An avid reader of many naturalists, Mendel purchased all of Darwin's major works for the monastery library. The monastery's copy of *On the Origin of Species* is one of many works of natural history that bear marginal notes in Mendel's elegant and easily identifiable handwriting.

In 1863, Mendel concluded his researches on peas, begun

in 1856. Twice in 1865 he spoke before the Brünn Society for the Study of Natural Science on his experiments, and in 1866 his paper, Studies on Plant Hybridization (Versuche über Pflanzenhybriden), was published in the society's Proceedings. In 1866, Mendel began a correspondence with Karl Nägeli (1817–1891), a Swiss botanist whom he wanted very much to please. Unfortunately, Nägeli had little patience for the work of an amateur, even a good amateur, and he did little to encourage Mendel.

Mendel's administrative duties meanwhile grew, even as his botanical studies seem to have run into a snag. In 1868, he was unanimously elected as Prelate, or administrative head, of the monastery. From 1874 onward, Mendel's major attention was definitely occupied with quite another matter: the Austrian government sought to tax monastery lands, a law that Mendel thought very unjust and fought bitterly. Except for his meteorological observations, which he continued until the week before his death, Mendel seems to have abandoned science from about 1870 onward. He died in 1884 at the monastery. After his work was rediscovered, a statue was erected in the monastery garden in 1910, and the square in front of the monastery was renamed Mendelplatz (Mendel Square) in his honor.

ITEM 6.2 *Lysenko and Lysenkoism*

A bleak chapter in the history of science was the suppression of genetics in the Soviet Union from 1936 to 1963 under the leadership of Trofim Denisovich Lysenko. Lysenko supported the inheritance of acquired characteristics and vigorously opposed the chromosomal theory of heredity as "bourgeois capitalist idealism." Lysenko succeeded in convincing an entire nation that genetics was a capitalist science, and that geneticists had to be persecuted.

Aside from an unbelievable amount of name-calling, the scientific content of Lysenko's criticism of genetics was twofold: first, that Engels had supported the inheritance of acquired characteristics (so had Darwin, at least as much as Engels), and second, that winter wheat could be made to grow more rapidly by exposing the barely sprouted seeds to the cold and sowing them in the spring, a process called "vernalization" (Yarovizatsiya). Lysenko first noticed the latter phenomenon in a paper published in 1928 by the Azerbaidjan Experiment Station, though the process had been described much earlier by Klippart (1858) in Ohio:

To convert winter into spring wheat, nothing more is necessary than that the winter wheat should be allowed to germinate slightly in the fall or winter, but kept from vegetation by a low temperature or freezing, until it can be sown in the spring. This is usually done by soaking and sprouting the seed, and freezing it while in this state and keeping it frozen until the season for spring sowing has arrived.

In 1936, at a meeting of the Lenin All-Union Academy of Agricultural Sciences, Lysenko and his supporters denounced not only their opponents but also Mendel, Morgan, Weismann, the entire chromosomal theory of heredity, and the very existence of genes. Using the argument that through vernalization he had succeeded in transforming the Kooperatorka variety of winter wheat into a spring variety, Lysenko concluded that only the inheritance of acquired characteristics had determined the winter habit. From this one argument, he generalized to the nonexistence of genes, chromosomes, or mutations in any organisms, and he went on to denounce the theory that "chromosomes of cells contain a hereditary substance separate from the rest of the organism (genotype)." Geneticists were depicted as evil tools of capitalism (later also of fascism), and many geneticists were arrested and sent off to prison camps, beginning in 1940.

Lysenko's persecution of his opponents was carried out at first in the journal *Yarovizatsiya* (vernalization), which he founded in 1936. Lysenko succeeded in acquiring political support for his own movement, which he presented as the salvation of Soviet agriculture. In this he was strongly supported by Stalin and later by Khrushchev. The 1938 International Congress of Genetics, which was to have been held in Moscow, was postponed until 1939 and was then held in Scotland.

At the 1948 meeting of the Lenin All-Union Academy of Agricultural Sciences, a new and more drastic phase of Lysenkoism was initiated, apparently with Stalin's support. Numerous arrests followed, and all those who failed to support Lysenko and his close associates were dismissed from their posts. Stocks of *Drosophila* were destroyed. Carefully selected varieties of wheat, corn, and other crops were vigorously destroyed, much to the later detriment of Soviet agriculture. Books and periodicals that contained any of the "Mendelist-Morganist" genetics were removed from libraries, and many of those periodicals that were permitted had entire articles cut out of them. Darwin, too, was attacked for his theory that competition takes place within species as well as between species; this belief in particular was labeled "capitalist" and rigorously denounced. Lysenko's address to the Lenin Academy received the personal approval of Stalin, and Lysenko's own slate of candidates for election to this Academy was summarily approved by Stalin, even before the planned elections were held. "Bourgeois genetics" was identified with imperialism and racism. The existence of chromosomes was declared a fiction, and even as late as 1963, the role of DNA in heredity was consistently denied.

It was not until 1963, after Khrushchev had resigned, that genetics could again be taught in the Soviet Union. A long process of reeducation had now to begin, for an entire generation of Soviet scientists had been brought up without any knowledge of genetics.

ITEM 6.3 *The Modern Synthesis: Some Classic Works*

What follows is a list of the more important titles only of the classic works that were instrumental in the formation of the Modern Synthesis of around 1940. They are listed, as far as possible, in chronological order; complete bibliographic references are given in the Bibliography at the end of this book.

Formative Ideas:

Chetverikov (1926, in Russian): On certain aspects of the evolutionary process from the standpoint of genetics (article)

Fisher (1930): *The genetical theory of natural selection*

Wright (1931): Evolution in Mendelian populations (article)

Haldane (1932): *The causes of evolution*

Dobzhansky (1933): Geographical variation in lady-beetles (article)

The Synthesis Arrives:

Dobzhansky (1937): *Genetics and the origin of species* (revised in 1942, 1951)

J. Huxley (1940): *The new systematics*

J. Huxley (1942): *Evolution, the modern synthesis*

Mayr (1942): *Systematics and the origin of species*

Simpson (1944): *Tempo and mode in evolution*

(1949): *The meaning of evolution*

(1953): *The major features of evolution*

Rensch (1947): *Neuere Probleme der Abstammungslehre*

(1960): *Evolution above the species level*

Comprehensive Summaries of a Synthesis Already Established:

Stebbins (1950): *Variation and evolution in plants*

Mayr (1963): *Animal species and evolution* (also the 1970 abridged edition, *Populations, species and evolution*)

Dobzhansky (1970): *Genetics of the evolutionary process*

7

The Sources of Variability

INTRODUCTION TO UNIT II

The next few chapters deal with the core material of the modern evolutionary synthesis: the processes and mechanisms of evolution at and below the species level. This is population biology in a broad sense, including the variability both within and between populations, the behavior of genes within populations, the ecology of populations, and the differentiation of populations into species. Evolution at this level, properly termed **microevolution**, is often defined as the process of change in biological populations. **Populations** (also called local populations or **demes**) may be defined as all the individuals within a species sharing a common gene pool or common descent; among sexual organisms a population may be defined also as all the organisms that freely interbreed with one another in nature.

Modern evolutionists must be familiar with concepts derived from genetics and from ecology. Many of the questions asked by modern evolutionists can hardly be posed without an understanding of these subjects. A comparison of Darwin's evolutionary mechanism with the processes listed by one modern evolutionist (Box 7–A) shows how important genetics has become: three of the five factors listed would have been subsumed by Darwin under the simple heading of ''variation,'' showing a great change in emphasis. The last factor listed, reproductive isolation, is entirely new in the sense that it does not occur at all in Darwin's theory.

A central problem to much of microevolution is the nature and origin of species, the very problem that eluded Darwin's grasp. Chapter 14 is devoted specifically to this problem, but in a broader sense the entire unit is concerned with the processes that produce changes both within and between species. The last factor, reproductive isolation, is of crucial importance to this problem.

A. VARIABILITY AND ITS IMPORTANCE

The study of variation is essential to any understanding of evolution, because variation provides the raw material on which natural selection acts. Unvarying populations, if they ever did exist, simply could not evolve. Ever since Darwin emphasized the importance of inherited varia-

BOX 7–A *Evolutionary Theories of 1859 and 1966: A Comparison*

The table below listing the "major features" of evolutionary theory serves to contrast Darwin's theory (1859) with the modern synthetic theory. The listing on the left is Darwin's theory, shorn down to its barest essentials. "Use and disuse" would not now be considered a Darwinian mechanism, but Darwin did invoke it several times in his *Origin of Species*.

Two major changes are evident. First, genetics plays a far larger role in modern evolutionary theory than it did in Darwin's time; what Darwin called "variation" is now listed as three or more distinct phenomena—more than half of the list. The second major change is the introduction of an entirely new mechanism, reproductive isolation, that had no counterpart in Darwin's theory. (Another factor that might now be added to Stebbins's list is genetic drift.) One obvious similarity between the two theories is that natural selection figures prominently in both of them.

The Major Features of Evolution

Darwin's theory (1859)	Modern synthetic theory (Stebbins, 1966, 1977)*
1. Variation	1. Single-gene mutations
2. Natural selection	2. Genetic recombinations
3. (Use and disuse)	3. Changes in chromosome structure or chromosome number
	4. Natural selection
	5. Reproductive isolation

*From Stebbins, *Processes of Organic Evolution*, 3e, © 1977. Reprinted by permission of Prentice-Hall, Inc., Englewood Cliffs, N.J.

tion for evolution, most evolutionists have insisted on this point.

Yet, despite the acknowledged importance of variation, many of the early twentieth-century geneticists greatly underestimated the inherent variability of natural populations. The very term "wild type" implied that alleles so designated were normal in a state of nature, and other alleles (called "mutants") were little more than laboratory freaks. The uniform homozygosity of natural populations was widely assumed. Selection, it was believed, would ensure the ubiquity of *the* one optimal genotype, and the elimination of all other genotypes. The great variability of natural populations was discovered only through the efforts of geneticists who investigated natural populations or of naturalists who were thoroughly trained in genetics. A pioneer in this field was the Russian geneticist S. S. Chetverikov. Beginning in the 1930s, the new leader in this field was Theodosius Dobzhansky (1900–1975), another Russian-born geneticist strongly influenced by Chetverikov's work. In the United States and elsewhere in the Americas, Dobzhan-

sky and his students took their genetic training—and much of their genetics laboratory—with them into the field, where they performed test crosses, made chromosome preparations, and examined both flies and chromosomes under the microscope. It was through efforts such as these that the immense variability of natural populations first became known. An even greater degree of variability was later revealed by electrophoresis (Chapter 12). Most genetic loci were shown to contain several to many distinct alleles, a phenomenon known as polymorphism (Chapter 12). These alleles were often present in the wild as heterozygous combinations. The chromosomes were likewise found to be highly variable and often heterozygous. Again, natural populations proved to be highly polymorphic, containing several to many distinct chromosome types. Furthermore, the proportions of these various chromosomal types varied geographically and continuously (Fig. 7.1). In the present chapter, we shall consider how this variability originates. We shall defer to Chapter 12 the problem of how such variability manages to persist in the face of selection.

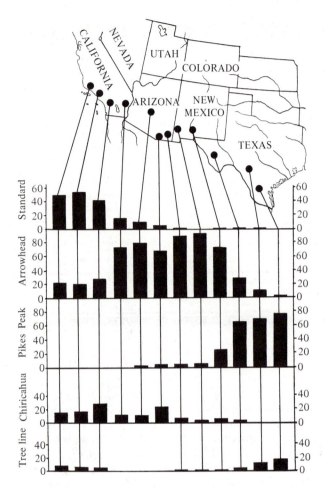

Fig. 7.1 Geographic variation in some chromosome polymorphisms of *Drosophila pseudoobscura*. Reprinted with permission from Th. Dobzhansky, *Chromosomal Races in Drosophila pseudoobscura and Drosophila persimilis*. Carnegie Institute of Washington, Publication 554, 1944.

B. THE EXPRESSION OF GENES AND THE CONTROL OF PHENOTYPE

One of the fundamental tenets of the modern synthetic theory of evolution is that natural selection operates on the phenotype rather than the genotype. No genetic change can be influenced by natural selection unless it first produces some phenotypic change. It is largely for this reason that modern evolutionary biologists must be aware of the manner in which phenotypes are controlled.

With few exceptions, every phenotype is controlled in part by the influence of the environment, and in part by the expression of genes. In some cases, of course, one or the other of these two influences may predominate, but the total absence of either influence is believed by many biologists to be rare.

The immediate role of a gene, or DNA sequence, is in the production of messenger RNA, whose translation in turn produces a polypeptide sequence. These polypeptide sequences, or products of translation, are collectively referred to as the **epigenotype**. With a few exceptions (such as hemoglobins), these polypeptides are important because of their function as enzymes (see Chapter 2). One of the great differences between bacteria and higher organisms is that bacterial phenotypes are hardly if at all removed from the epigenotype, whereas the phenotypes of higher organisms are much more complex and many more steps removed from the epigenotypes that produce them.

Not all of the genotype is transcribed and translated into a portion of the epigenotype, nor are all the transcribable genes ever transcribed at the same time. All eucaryotic genomes thus far studied have highly repetitive portions containing numerous identical copies of the same nucleotide sequences. These repetitive sequences are not themselves transcribed or translated into proteins, but they are believed to play an important role in the developmental regulation and control of genes that are transcribed. Our study of these developmental controls is still in its early stages.

Some genes, directing the most essential life processes (like protein synthesis), are **facultative**, meaning that they are transcribed at all times and are relatively insensitive to external environmental conditions. Other genes are called **conditional** because their enzyme products are synthesized only when the need for them arises as the result of environmental conditions. For example, the gene directing the production of the enzyme β-galactosidase in the bacterium *Escherichia coli* produces this enzyme only under those conditions in which the splitting of lactose is likely to be an efficient source of energy for these bacteria. Those conditions include the presence of lactose (the substrate of this enzyme), the absence of a more efficient energy source, such as glucose, and the absence of excessive quantities of galactose, one of the breakdown products.

In multicellular animals and plants, the necessary conditions for the expression of a particular gene may include its presence in a particular type of differentiated cell. Every one of your cells, for example, has genes controlling eye color; yet their ultimate products (eye pigments) are normally evident only in your eyes. Some genes are far more widespread in their effects. The gene for white eyes in fruit flies, for example, also results in a white stomach and other white internal organs, a smaller than average body size, reduced viability, and a number of other features. Effects such as these are called **pleiotropic; pleiotropism** may be defined as a condition in which a single polypeptide has many different phenotypic effects.

The opposite phenomenon is known as **polygeny**, the interaction of two or more genes to produce a phenotype together, especially one that neither gene could produce by itself. In a sense, polygeny and pleiotropism are opposite sides of the same coin (Fig. 7.2). Some geneticists use the term polygeny to include all cases of gene-gene interaction; others restrict its usage to the specific case of multiple genes with small additive effects (Chapter 2).

An important special case of polygeny is that in which one gene (or its product) modifies the expression of another gene, a phenomenon known as **epistasis**. Some geneticists restrict the term epistasis to cases where one gene completely "masks" the effects of another, usually by modifying the other gene's product in such a way that we can no longer tell what the masked gene had done. Two of the several types of epistasis are shown and explained in Items 7.1 and 7.2, page 134.

Fig. 7.2 The interaction of genes, diagrammatically shown. Each arrow indicates one condition or trait affecting another. The ultimate phenotypic characters, shown on the right, bear anything but a simple one-to-one relationship to the original gene products. After Stebbins, *Processes of Organic Evolution,* 3e, © 1977. Reprinted by permission of Prentice-Hall, Inc., Englewood Cliffs, N.J.

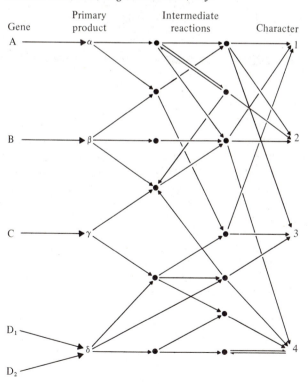

C. MUTATIONS

To Hugo De Vries, who introduced the term to genetics, a **mutation** was *any sudden, heritable change.* The term thus covered a multitude of sins: changes in single genes, changes in parts of chromosomes, changes by the loss or addition of entire chromosomes, and even some changes that were later attributed to such nonmutational causes as hybridization and asexual reproduction. In order to simplify the situation, H. J. Muller suggested that the term "mutation" be restricted to the simplest case, that of single-gene mutations. Usage still varies, but many evolutionists do include the several types of chromosomal aberration and numerical changes listed in Box 7–B. Dobzhansky (1970), for example, defines a mutation as any change in the genotype other than by genetic recombination.

Single-Gene Mutations

By far the most frequent mutations are those that affect single genes, including the so-called **point mutations**, which affect single base pairs only. A mutation in

BOX 7–B   *Types of Mutations*

I. "POINT MUTATIONS" AFFECTING SINGLE GENES
  A. Base pair substitutions
      Transitions: the replacement of one purine by another or one pyrimidine by another, e.g., by tautomeric shifting
      Transversions: all other base pair substitutions.
  B. Frame-shift mutations, in which the reading of the genetic message beyond the mutation is impaired
      Duplications: the addition of one or more base pairs.
      Deletions: the omission of one or more base pairs.
II. CHANGES IN CHROMOSOME STRUCTURE (CHROMOSOMAL ABERRATIONS)
  A. Aberrations in which the total chromosome length is not affected
      Inversions: Turning of a chromosome segment end-to-end, so that ABCDEFG becomes ABEDCFG. Inversions may be either *pericentric*, meaning that the centromere is included in the inverted portion, or *paracentric*, meaning that the centromere is not included.

Translocations: Shifting of a chromosome segment to a different part of the genome. *Homologous translocations,* also called *transpositions,* shift the segment to a new location within the same chromosome. *Heterologous translocations* shift the segment to a different chromosome. Many heterologous translocations are reciprocal, with two chromosomes exchanging portions.
  B. Aberrations in which total chromosome length is altered
      Duplications: Repetition of a chromosome segment. This may take place at either end (*terminal duplication*), so that ABCDEFG becomes either ABABCDEFG or ABCDEFGFG, or the duplication may take place within a chromosome's length (*intercalary duplication*), such as ABCDCDEFG.
      Deletions: Omission of a chromosome segment. Again, this may take place at either end of a chromosome or within its length. A change from ABCDEFG to ABCFG would be considered an *intercalary deletion.* A change to either CDEFG or ABCDE would be considered a *terminal deletion.*
III. CHANGES IN CHROMOSOME NUMBER
      See Box 7–F.

---

this sense may be defined as a change from one allele to another at the same locus.

  Point mutations may be classified either as **transitions** (substitution of one purine for another or one pyrimidine for another) or as **transversions** (replacement of a purine by a pyrimidine or vice versa). Many transitions result from tautomerism, the existence of rare, alternative forms of chemical structure in equilibrium with the normal form. Cytosine, for example, normally pairs with guanine, but it also exists in a rare, tautomeric form that pairs instead with adenine. The two forms are in equilibrium, with nearly all the cytosine existing in the normal form. Only about 1/1,000,000 of the cytosine exists in the tautomeric form at any given moment; a given molecule of cytosine spends about 1/1,000,000 of its time in

the tautomeric form and the rest of its time in the normal form. Suppose, now, that a DNA molecule containing a guanine-cytosine pair happens to replicate at a moment when the cytosine is in its tautomeric form. The cytosine will pair mistakenly with adenine instead of guanine, and at the next replication the adenine will direct the incorporation of a nucleotide containing thymine in place of the cytosine. Point mutations can also be induced chemically, but few of the known chemical mutagens occur naturally in significant concentrations.

  Other known mutagenic agents include ultraviolet light and x-rays. A certain level of x-rays occurs naturally in the form of cosmic rays, and ultraviolet light also occurs naturally at rather low levels. Neither of these naturally occurring sources of radiation is adequate to account for more than a small percentage of naturally occurring mutations.

Several other types of single-gene mutations are known, too. Lewis and John (1963) distinguish the following four types:

1. "Missense" mutations, which result in the substitution of one amino acid for another. This, in turn, results in an altered gene product, with consequences varying from barely discernible in some cases to very drastic (even lethal) in others. Mutations affecting the active sites of enzymes, for example, or altering the three-dimensional structure or net electric charge of the resulting polypeptide, are likely to be of greater consequence than those which substitute similar amino acids for one another in nonessential parts of the molecule.

2. "Samesense" mutations (also called "silent" mutations), in which a codon is replaced by another that codes for the same amino acid (e.g., a change from CUA to CUG, both of which code for lysine).

3. "Nonsense" mutations, in which a codon that codes for an amino acid is replaced by one that codes for chain termination, resulting in the premature cessation of translation prior to completion of a functional gene product.

4. "Frame-shift" mutations by the insertion or deletion of individual base pairs, resulting in a misreading of the genetic code from that point on, and thus generally in a drastically altered and usually nonfunctional gene product.

Single-gene mutations in bacteria and in eucaryotes may confer the ability to use new substrates as nutritional sources or to detoxify harmful substances, thus altering the range of ecological conditions in which the mutants may thrive. The thousands of mutations known in *Drosophila* affect the wings, eyes, body color, bristles, nutrition, temperature tolerance, sex ratio, behavior, and many other characters. In humans, mutations are known that cause various metabolic diseases, or that affect hair distribution, eye and skin pigments, blood groups, abnormalities of finger formation, and so on. Estimates of the number of possible base pairs (thus the number of possible mutable sites) per haploid genome range from about 30 million (3×10^7) to 3 trillion (3×10^{12}) for eucaryotes, with lower estimates for bacteria and still lower ones for viruses. If a typical gene (cistron) is assumed to have about 1000 base pairs, then eucaryotes may have no more than between 30,000 and 3 billion cistrons per haploid genome, and most will have considerably less than this because only a small fraction of the genome is ever transcribed and translated into functional gene products.

The rate at which mutations occur was first measured by H. J. Muller in 1928. The spontaneous mutation rate in *Drosophila* for all sex-linked lethals combined was about one per thousand. By exposing fruit flies to a sublethal dose of x-rays, Muller was able to raise the mutation rate about a hundredfold, to about one in ten (10%). Various spontaneous mutation rates are listed in Box 7–C.

Mutation rates are more easily measured in bacteria. One reason is that bacteria can be cultured in large numbers: a single Petri dish may contain several million of them. The use of chemical agents to detect mutant bacteria is explained in Box 7–D. The medium in the Petri dishes may contain a selective agent, such as an antibiotic that permits the growth of resistant mutations only. Alternatively, the medium may lack some normally essential nutrient, in which case all the colonies die except for those containing a mutation that permits the bacteria to synthesize the nutrient for themselves.

One might well question whether such mutations as these are indeed spontaneous or are instead environmentally induced. By Geoffroy's principle of direct environmental induction (Chapter 3), one might claim that the presence, say, of streptomycin would induce the formation of a streptomycin-resistant strain. Such mutations would be called **postadaptive:** they occur only after exposure to the environment to which they are adaptive. By an ingenious experiment (Box 7–E), Lederberg and Lederberg (1952) were able to show that such mutations are actually **preadaptive:** they occur spontaneously before exposure to the environment, and they are adaptive only fortuitously. The environment, in other words, does not induce their formation; it only selects the preadaptive mutations after they have already occurred.

The rate of point mutations at various positions

BOX 7–C *Spontaneous Mutation Rates in Various Organisms*

Organism	Trait	Mutation rate per gamete (or cell)
Escherichia coli (a bacterium)	Streptomycin resistance	4×10^{-10}
	Phage T1 resistance	3×10^{-8}
Salmonella typhimurium (a bacterium)	Threonine resistance	4.1×10^{-6}
	Tryptophan independence	5×10^{-8}
Diplococcus pneumoniae (a bacterium)	Penicillin resistance	1×10^{-7}
Neurospora crassa (bread mold)	Adenine independence	3×10^{-8}
	Inositol independence	5×10^{-8}
Drosophila melanogaster (fruit fly)	Yellow body	1.2×10^{-4}
	Brown eyes	3×10^{-5}
	Ebony body	2×10^{-5}
	Eyeless	6×10^{-5}
Zea mays (corn)	Shrunken seed	1.2×10^{-6}
	Colorless aleurone	2.3×10^{-6}
	Sugary endosperm	2.4×10^{-6}
Mus musculus (house mouse)	Brown	8.5×10^{-6}
	Pink eye	8.5×10^{-6}
	Piebald	1.7×10^{-5}
Homo sapiens (man)	Epiloia	6×10^{-6}
	Retinoblastoma	$1.2-2.3 \times 10^{-5}$
	Aniridia	5×10^{-6}
	Huntington's Chorea	5×10^{-6}

After M.W. Strickberger, 1976. *Genetics*, 2nd ed. New York: Macmillan.

BOX 7-D *The Use of Chemical Agents to Detect Mutant Bacteria*

The detection of new mutations in bacteria can be simplied greatly by the use of such strong selective agents as antibiotic drugs. Bacteria are first grown, often in a liquid medium, under conditions that permit their numbers to be quantitated. A known volume of medium, containing a large but determin-able number of bacteria, is then spread out thin on Petri dishes in such a way that each colony will be formed by an individual bacterium. The medium in the Petri dishes contains a selective agent, such as an antibiotic, that permits the growth of mutant bacteria only.

The great advantage of this method is that millions of bacteria can be tested within a short period of time, and a rate of mutation can be accurately determined.

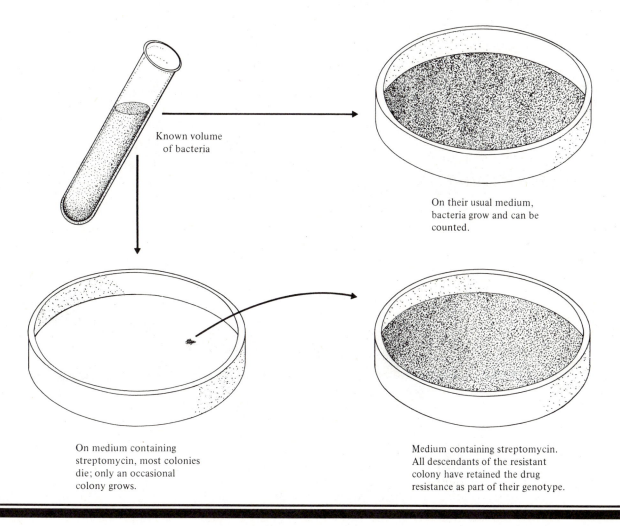

Known volume of bacteria

On their usual medium, bacteria grow and can be counted.

On medium containing streptomycin, most colonies die; only an occasional colony grows.

Medium containing streptomycin. All descendants of the resistant colony have retained the drug resistance as part of their genotype.

BOX 7–E *Lederberg's Replica-Plating Experiment*

This experiment demonstrated the preadaptive nature of mutations. The actual experiment contained many more colonies than are shown here.

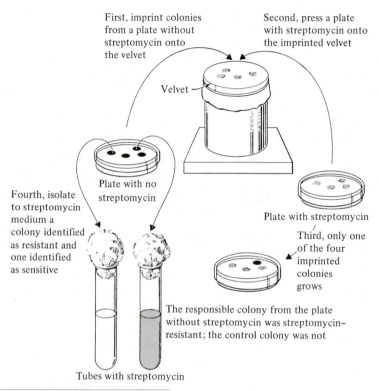

First, imprint colonies from a plate without streptomycin onto the velvet

Second, press a plate with streptomycin onto the imprinted velvet

Velvet

Plate with no streptomycin

Fourth, isolate to streptomycin medium a colony identified as resistant and one identified as sensitive

Plate with streptomycin

Third, only one of the four imprinted colonies grows

The responsible colony from the plate without streptomycin was streptomycin-resistant; the control colony was not

Tubes with streptomycin

From Sager and Ryan, *Cell Heredity*, 1961. New York: John Wiley and Sons.

along a bacterial chromosome can be accurately measured. In one such study, Benzer (1957) discovered that the mutation rate is not at all constant along the length of the chromosome, but that some particularly mutable sites (''hot spots'') are interspersed with sites of much lower mutability (Fig. 7.3).

Chromosomal Aberrations

Many mutational changes take the form of chromosomal aberrations, which may be defined as heritable changes within the structure of single chromosomes. The major types of aberration may be distinguished as shown in Fig. 7.4.

Perhaps the most common of the chromosomal aberrations are those in which a segment of the chromosome turns end-to-end. These changes are known as **inversions**; a change from ABCDEFG to ABFEDCG would be an inversion. Inversions that include the centromere are called pericentric inversions; inversions that do not include the centromere are called paracentric. If an individual is heterozygous for an inversion, its chromosomes will form loops of characteristic form, allowing scientists to detect the inversion and to determine its extent. Para-

Fig. 7.3 The frequency of point mutations at various points along a bacterical chromosome. Each square represents one mutation observed at the indicated point. Note that the frequency of these mutations varies greatly along the length of the chromosome. Reprinted with permission from Jeffrey J. W. Baker and Garland E. Allen, *The Study of Biology,* 3e, © 1977. Reading, Massachusetts: Addison-Wesley (Fig. 14.8), after Benzer.

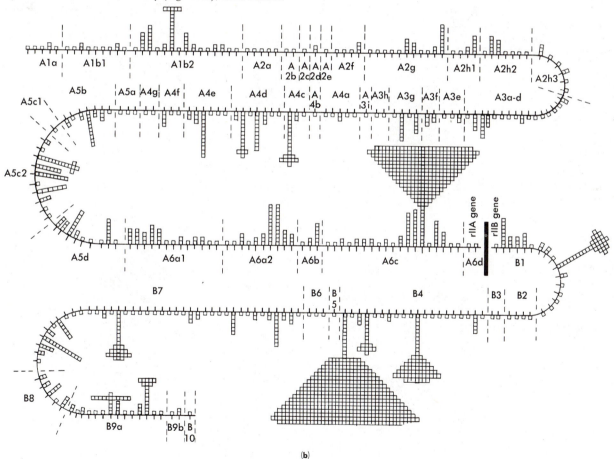

(b)

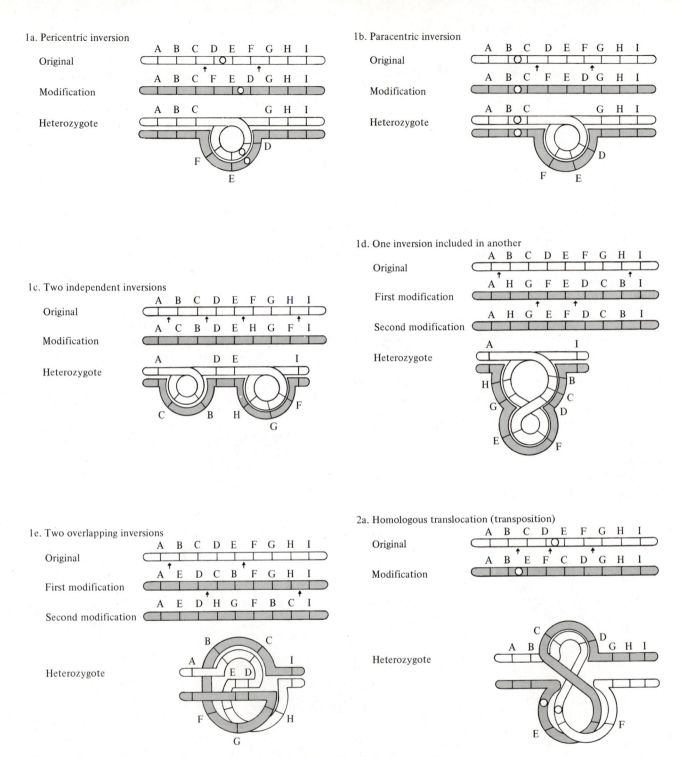

Fig. 7.4 Types of chromosomal rearrangements and the appearance of chromosomes heterozygous for these various rearrangements. Points of chromosome breakage are shown by arrows.

Fig. 7.4 (*continued*)

2b. Heterologous translocation

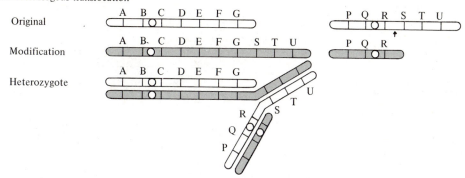

Original

Modification

Heterozygote

2c. Reciprocal translocations

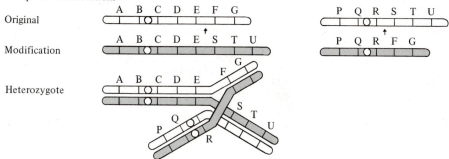

Original

Modification

Heterozygote

2d. Successive reciprocal translocations

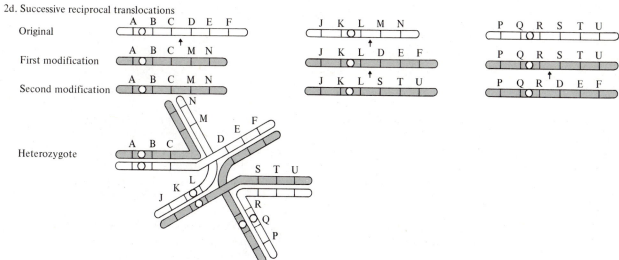

Original

First modification

Second modification

Heterozygote

(*continued*)

3a. Intercalary duplication

Original

Modification

Heterozygote

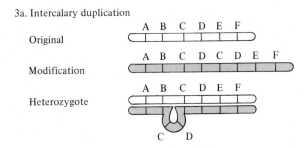

3b. Terminal duplication

Original

Modification

Heterozygote

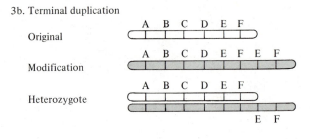

4a. Intercalary deletion

Original

Modification

Heterozygote

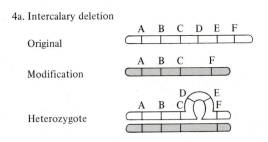

4b. Terminal deletion

Original

Modification

Heterozygote

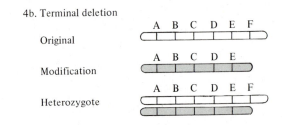

centric inversions have long been known to inhibit crossing over of genes in heterozygotes (Fig. 7.5). These inversions thus preserve in populations the particular sequence of genes that they happen to contain.

Inversions are often studied by evolutionists because they give clues to the differentiation of populations and species. When two inversions partially overlap one another, the stepwise nature of the process can easily be reconstructed. Between a sequence like ABCDEFGHIJK

and one differing from it by two overlapping inversions, like ABGFEIHCDJK, the former existence of the sequence ABGFEDCHIJK can be safely assumed, for no other condition can serve as a suitable intermediate between these two. In fruit flies of the genus *Drosophila*, many such chromosome arrangements are known (Figs. 7.6, 7.7, and 7.8), generally differing from one another by particular inversions. The Standard and Santa Cruz chromosome types differ by two partially overlapping inversions, allowing the former existence of the "hypothetical" chromosome arrangement to be more than just hypothetical.

The transfer of a portion of a chromosome to a different location is known as a **translocation. Heterologous translocations** involve transfer to a different chromosome entirely. **Homologous translocations,** also called transpositions, involve transfer to a different position within the same chromosome. Homologous translocations are detectable cytologically by the presence of paired loops.

Heterologous translocations are detectable during meiosis by the fact that two or more pairs or chromosomes will be brought together by the pairing of their sequences. An extreme case of this is seen in *Oenothera biennis* and certain closely related plant species. A succession of heterologous whole-arm translocations in these species has resulted in all 14 chromosomes' becoming connected during meiosis to form a ringlike configuration known as a Renner complex.

The importance of inversions and translocations in

Fig. 7.5 An explanation for reduced reproductive output of organisms heterozygous for paracentric inversions. (a) Two homologous chromosomes. (b) Pairing at meiosis and crossing-over between two nonsister chromatids. (c) Separation of the chromosomes during early anaphase of the first meiotic division. (d) The resulting chromosomes as meiosis nears completion. The chromosomal segment without a centromere may fail to move toward the poles during meiosis and is usually lost; the chromosome segment with two centromeres usually breaks and, after meiosis II, yields gametes with substantial deletions. Reprinted with permission from Donald I. Patt and Gail R. Patt, *An Introduction to Modern Genetics,* © 1975. Reading, Massachusetts: Addison-Wesley (Fig. 9.19).

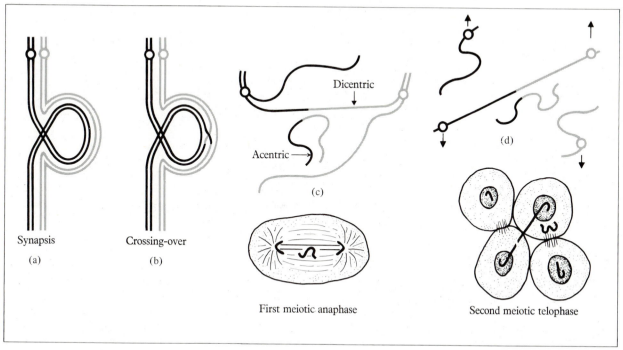

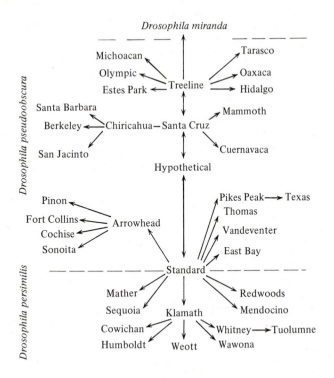

Fig. 7.6 Some of the many chromosomal variations found in three species of *Drosophila*. Chromosome types connected by arrows differ by only a single inversion, translocation, or similar sequence difference. Reprinted with permission from Speiss, *Genes in Populations,* © 1977. New York: John Wiley and Sons, Inc., after Anderson, *et al., Evolution,* 29:26, 1975 and Dobzhansky, *et al., J. Heredity,* 66:204–208, 1975.

Fig. 7.7 Appearance of a portion of the Standard chromosome arrangement of *Drosophila,* as seen in a chromosome taken from the larval salivary gland, where the banding pattern can most easily be seen. The portion marked by the two black lines can become inverted to form the arrangement known as Arrowhead. Adapted with permission from Dobzhansky and Sturtevant, *Genetics,* vol. 23, 1938.

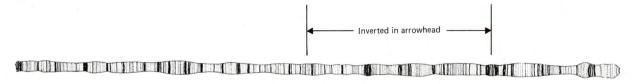

producing genetic variations is partly due to the **position effect:** a given gene may have a somewhat different phenotypic effect depending on what other genes are located next to it (Lewis, 1950). Another important role of chromosomal rearrangement is the formation or breaking up of allelic complexes in which several functionally related alleles may be controlled together as a group. In-

versions and translocations are also important because they rearrange the linkage groups in which genes assort.

Duplications consist of the repetition of a chromosome segment. If an original chromosome sequence is represented as ABCDEFG, then a duplication might be represented as ABCDCDEFG. This would be called an intercalary duplication; a duplication at the end of a

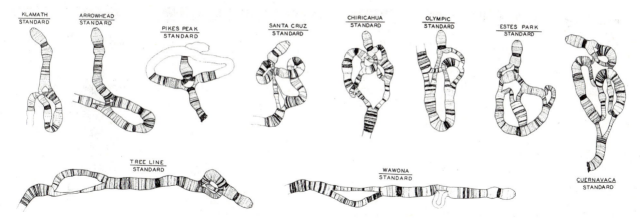

Fig. 7.8 Some chromosome configurations (karyotypes) in the third chromosome of *Drosophila pseudoobscura,* each in heterozygous combination with the karyotypes known as Standard. Adapted with permission from Dobzhansky and Sturtevant, *Genetics,* vol. 23, 1938.

chromosome, such as ABCDEFGFG, would be called a terminal duplication. An organism heterozygous for a chromosomal duplication will have one chromosome of normal length and another chromosome that differs by being longer. In an intercalary duplication, the added portion of the chromosome will tend to form a buckle; in a terminal duplication, the added portion will protrude beyond the end of the normal chromosome. An example of a duplication is the much-studied ''bar eyes'' abnormality in *Drosophila.*

An important advantage of duplications is that they may act as a hedge against future mutations. If a certain gene product is absolutely essential to an organism's existence, a mutational change in the corresponding gene would be likely to prove fatal. But if several copies of the same gene existed at different places within the genome, one copy could undergo mutation to produce a new and perhaps a better product or a product serving an entirely different function, and another copy could continue to serve the original function. Several genes in bacteria exist in multiple copies, showing that duplication has occurred in the past. But investigation of eucaryotic genomes has shown that they contain certain sequences that are repeated thousands of times. These oft-repeated sequences are widely believed to play an important role in control-

ling when and under what circumstances genes will express themselves phenotypically.

Deletions consist of the loss of a part of a chromosome. An intercalary deletion might be represented as a change from ABCDEFG to ABCDG. A terminal deletion might be represented as CDEFG. A chromosome containing a deletion is shorter than normal. An organism heterozygous for a chromosomal deletion will show a buckle in the normal chromosome if the deletion is intercalary, and an extension of the normal chromosome beyond the deleted one if the deletion is terminal. Deletions are likely to be harmful in proportion to the size of the portion deleted—the larger the deletion, the more harmful it is likely to be. But a deletion may not be so harmful if one or more copies of the deleted sequence remain intact in some part of the genome. Single-copy sequences lack this protection, however, and deletions within these sequences are more often very harmful.

D. CHANGES IN CHROMOSOME NUMBER

Changes in the genetic makeup of populations may also occur by changes in the number and arrangement of chromosomes (Box 7–F). Changes in which chromosomes

---

BOX 7–F   *Changes in Chromosome Number*

A. Changes by the addition or loss of single chromosomes (aneuploidy)
   1. Loss of a diploid pair (nullisomy)
   2. Loss of one member of a diploid pair (monosomy)
   3. Addition of an extra chromosome to a diploid pair (trisomy)
   4. Addition of two extra chromosomes to a diploid pair (tetrasomy)
      Etc.
B. Changes by the addition or loss of complete chromosome sets or genomes (euploidy)

1. Loss of an entire haploid set (monoploidy or haploidy, N)
2. Presence of two haploid sets (diploidy or $2N$, the normal condition in most animal and plant species)
3. Presence of more than two sets of chromosomes (polyploidy):

   $3N$—Triploidy
   $4N$—Tetraploidy
   $5N$—Pentaploidy   (See also Chapter 15)
   $6N$—Hexaploidy
   $7N$—Heptaploidy
   $8N$—Octaploidy
   Etc.

---

are lost or added one at a time result in **aneuploidy**. Changes may also occur by the loss or addition of entire genomes, resulting in the halving, doubling, tripling, or quadrupling of the entire chromosome number. Chromosome numbers consisting of complete haploid sets are called **euploid**. If three or more haploid sets are present, the euploid condition is also called polyploid (Chapter 15).

Euploid or aneuploid increases, like duplications, may be viewed as hedges against future mutations. All these changes provide multiple copies of the same genetic material, permitting mutations to occur more readily without harm, and also permitting one gene to be modified in two or more different ways at the same time. There are also disadvantages, especially in the case of aneuploidy, in that previously well-tuned developmental pathways are likely to be disrupted, as in Down's syndrome in humans.

Aneuploidy

Aneuploidy results from a change in chromosome number of less than an entire genome. Many cases of aneuploidy result from the nondisjunction of chromosomes during meiosis.

In a normally diploid species, the number of chromosomes in each pair is two. The loss of a single chromosome results in a reduction from two to one, which is called **monosomy** (*mono-* + chromo*some*). The loss of both chromosomes in a pair results in **nullisomy** (from *nullo,* none). The addition of a chromosome to the original diploid pair results in **trisomy** (Fig. 7.9); the addition of two chromosomes to a diploid pair results in **tetrasomy.**

One of the most familiar examples of aneuploidy is trisomy of chromosome 21 in humans, resulting in Down's syndrome (mongolism). Several other abnormalities of medical interest include monosomy X, or Turner's syndrome (XO), Kleinfelter's syndrome (XXY), and XYY. Turner's syndrome produces an underdeveloped sterile female of short stature; Kleinfelter's syndrome produces a sexually underdeveloped male of tall, thin stature and usually some degree of feminized breast development. Both Turner's and Kleinfelter's syndrome produce some degree of mental impairment or retardation. Such medically important abnormalities as these are so maladapted that they would be of little or no evolutionary consequence under natural conditions, especially since many of them result in partial or total sterility. Aneuploidy is thus uncommon in nature, since it frequently results in a marked reduction in fitness. In species that are already polyploid, however, aneuploidy may be of less drastic consequence, and aneuploids are thus more often viable. Presumably a polyploid species can better tolerate a change in one of its genomes, as long as the others remain intact.

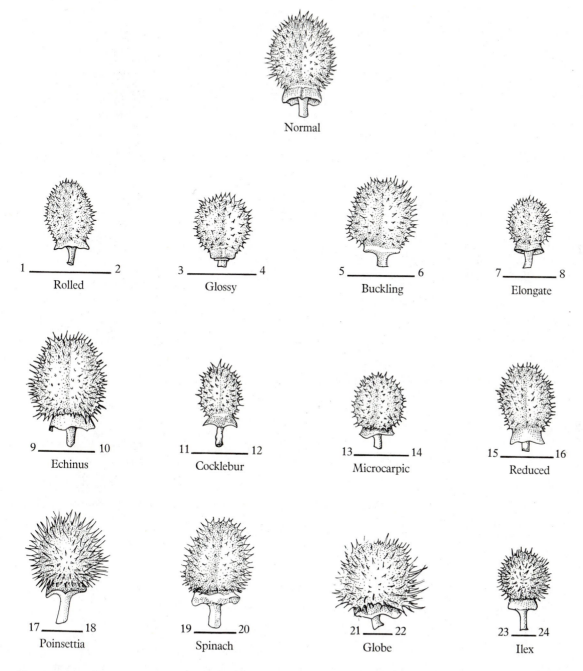

Fig. 7.9 The effects of trisomy for various chromosomes on the seed capsules of the Jimson weed (*Datura*). Each of the mutants shown is trisomic for one of the 12 pairs of chromosomes in this species. Reprinted with permission from Donald I. Patt and Gail R. Patt, *An Introduction to Modern Genetics*, © 1975. Reading, Massachusetts: Addison-Wesley (Fig. 9.23).

Even in diploid species, aneuploidy may occasionally result in changes far less drastic. These may persist in natural populations and may thus serve as the basis for later evolutionary changes. In fruit flies, monosomy 4, resulting from a loss of one of the tiny fourth chromosomes, produces an altered but quite viable phenotype. In humans, the XYY chromosomal type produces a phenotypically normal male, generally of taller than average stature. A number of behavioral tendencies have been noted in XYY males and attributed to the extra Y chromosome; these include a higher level of aggressive and often antisocial behavior. The proportion of XYY males is somewhat higher in prisons than it is in the general population.

The ability of evolutionary changes to take place through aneuploidy is shown by the many cases in which related species differ by relatively small changes in chromosome number. Many small chromosomes in one species may correspond to fewer large chromosomes in the other. According to one possible explanation, the small chromosomes have fused; according to an alternative explanation, the large chromosomes have become fragmented. The number of chromosome *arms,* sometimes known as the *nombre fondamentale* or **NF,** may remain constant within entire families or orders, even while the number of individual chromosomes varies (Fig. 7.10). In cases of polyploidy, both the chromosome number and the NF would increase together.

Euploidy

Unlike aneuploidy, which is uncommon in natural populations, the various types of euploidy, especially polyploidy, are rather common in many groups of organisms. Haploidy, of course, is the normal condition in many species of lower eucaryotes and in all procaryotes. The occurrence of viable haploidy in higher organisms is rare, and it is usually associated with a reversion from sexual to asexual reproduction. Among insects, most males of the order Hymenoptera (bees, wasps, and ants) are haploid, whereas the females (including the sterile workers) are diploid.

Polyploidy is the more common type of euploidy. Polyploidy is treated in Chapter 15, following the consideration of speciation and hybridism, two phenomena frequently associated with polyploidy.

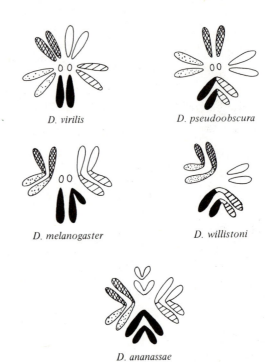

D. virilis   *D. pseudoobscura*

D. melanogaster   *D. willistoni*

D. ananassae

Fig. 7.10 Changes in the chromosome morphology among some species of *Drosophila.* Note that the chromosome number varies, but that the nombre fondamentale (NF) seldom changes as a result. Homologous chromosome arms are distinguished by the shading. The ancestral condition for the genus is believed to be similar to *D. virilis;* other conditions can be derived from this one through chromosome fusions. In *D. ananassae,* the X chromosome (black) differs from that of *D. melanogaster* by a pericentric inversion that has changed the position of the centromere from the end to the middle of the chromosome.

E. GENETIC RECOMBINATION AND BREEDING SYSTEMS

Recombination

Genetic recombination represents the proximate source of by far the greatest number of variations, even though mutations are the ultimate source of genetic variability. No two individuals are exactly alike (except perhaps some identical twins), and this uniqueness is usually the result of recombination rather than mutation. By way of analogy, think of the rich variety of literary works produced by the combination and recombination of the words in our language. We honor Shakespeare and Hemingway and their kind for producing these works by recombining words, but how little we honor people for coining new words, which is the process roughly equivalent to mutation. Because genes interact with one another, and because many characters are controlled by several genes, genetic recombination may result in totally novel phenotypes, as well as in new combinations of previously experienced character states. Look at the people around you and at their many distinguishing traits. Most of these traits have arisen from genetic recombination and from developmental (including environmental) controls, without benefit of new mutations. The recombination of genetic material is thus an important aspect of the evolutionary process. The basic rules governing genetic recombination were covered in Chapter 2. These rules, however, can account for only so much variability without benefit of new mutations.

The rate at which recombination shuffles and redistributes the many genes in the gene pool can itself be influenced by many genetic or environmental factors (Box 7–G). The rate of recombination is always a compromise among many conflicting pressures, but populations adapted to very stable environments are in general selected for lower overall rates of recombination, whereas sexual populations adapted to less predictable environments are generally selected for higher average rates of recombination.

Varieties of Sexual Reproduction

All of us are familiar with the usual type of sexual reproduction among mammals and many other organisms. The principles of genetics covered in Chapter 2 pertain largely to these diploid organisms. Animals in which reproductive cells (gametes) are produced by separate male and female individuals are called **gonochoristic**; the same condition among plants is called **dioecious**. The opposite of gonochorism is **hermaphroditism**, a condition in which organs producing male gametes and organs producing female gametes are housed together in the same individual. (Individual plants bearing both male and female flowers are called monoecious.) In most hermaphroditic species, fertilization of another individual, or cross-fertilization, is the

general rule, and self-fertilization is clearly the exception. The probable reason for this is that the full advantages of sexual reproduction, which include a greatly increased rate of genetic recombination, can be realized in such species only when they cross-fertilize. In some animal species, such as the slipper-limpets (*Crepidula*), cross-fertilization is ensured by the development of male characteristics first (protandry), and by individuals gradually changing into females as they age. The reverse situation, protogyny, is known in parrot-fish (Scaridae). An analogous mechanism operates in those plants whose male and female flower parts mature at different times. Other mechanisms that promote cross-fertilization in plants include heterostyly (Fig. 8.4), and also the various adaptations of insect-pollinated plants which insure that pollen from an incoming insect is deposited before any new pollen is picked up (Fig. 10.19).

Recombination in procaryotes usually does not involve the union of gametes and the formation of diploid cells. In the bacterium *Escherichia coli,* for example, a ''male'' or ''active'' cell injects part of its chromosome into a ''passive'' cell, with crossing-over bringing about the actual recombination of chromosomal material. Many other procaryotes appear to undergo recombination in much the same way.

A peculiar type of reproduction, haplodiploidy, occurs in the insect order Hymenoptera (bees, wasps, and ants). The diploid individuals are all genetically female, though many of them may function only as sterile workers. The males are haploid, produced parthenogenetically from unfertilized eggs. The sex ratio is thus greatly variable and subject to social and ecological control. One consequence of haplodiploidy is that a female shares no more than half of her genotype with her mate, and this puts a limit on the effects of close inbreeding (which might otherwise be a problem in such social species). Other consequences of haplodiploidy are explored in Chapter 24.

Asexual Reproduction and Parthenogenesis

Not all reproduction is sexual. In many organisms, including most procaryotes, the more common form of reproduction is a simple form of asexual reproduction known as binary fission—the simple splitting of the cell in half. The ''budding off'' of new individuals occurs among yeasts and in such lower animals as *Hydra* (phy-

BOX 7–G *Factors That Promote or Inhibit Recombination in Plants*

Several of the factors listed here apply to animals, too.

Regulatory factor	Condition that promotes free recombination	Condition that inhibits free recombination	Condition that leads to a complete absence of recombination
Length of generation	Short	Long	No sexual generations
Chromosome number	High	Low	No meiosis in female line
Crossing over	High chiasma frequency, randomized chiasmata, no structural hybridity	Low chiasma frequency, localized chiasmata, structural hybridity	No meiosis in female line
Breeding system	Entirely or predominantly cross-fertilizing	Partially to predominantly self-fertilizing	No crossing
Pollination system	Widely dispersed pollen, promiscuous pollinators	Pollinating insects with narrow foraging ranges, flower-constant insects	No crossing
Dispersal potential	Diaspores dispersed far from parental plant	Diaspores dispersed close to parental plant	No crossing
Population size	Large continuous populations, margins of smaller populations	Small colonial populations	Asexual clones
Isolating mechanisms	Related species, sections or genera incompletely isolated	Related species strongly isolated	No hybridization
Resultant type of recombination system:	Open	Restricted	Closed

(After V. Grant, 1958, Cold Spring Harbor Symposium Quant. Biol., vol. 23.)

lum Cnidaria or Coelenterata). In certain plants, vegetative propagation may give rise to new individuals by a similar process. In all these cases, an individual's genes are derived from a single source only, with no opportunity for recombination. Evolutionary novelty in these cases consists entirely of mutations.

Some animals and many plants have life cycles in which sexual and asexual reproduction occur alternately

(alternation of phases or generations). This type of life cycle retains both the advantages of sex, namely recombination, and those of asexual reproduction, which include a greater ability to proliferate very rapidly and take advantage of temporary situations.

The asexual production of offspring from unfertilized eggs is called parthenogenesis. Some species undergo parthenogenesis only under ecological conditions that call for rapid proliferation, reproducing sexually at other times. Aphids, for example, produce several asexual generations parthenogenetically in rapid succession each summer when conditions are favorable. But in the fall, the overwintering generation (which faces a more unpredictable future through the winter and into the coming spring) is produced sexually. Certain freshwater crustaceans (*Daphnia* and its relatives) have similar cycles of alternating sexual and parthenogenetic generations. Bdelloid rotifers are unusual: though derived from sexually reproducing ancestors, they are parthenogenetic all of the time, and they consist therefore of only females.

Is Sex Necessary?

The widespread occurrence of sexual reproduction seems to indicate a definite selective advantage; yet the occurrence of asexual reproduction seems to indicate that the advantages of sex are less than universal. Sex is, above all, a costly process in terms of the energy expended in the synthesis of sex organs, in the finding of a mate, in the courtship and mating process, and frequently also in the providing of nourishment and other care for the young. The advantages of sex are shown by the fact that so many species continue to bear this cost; yet the precise selective basis of sex remains a mystery. Several books (e.g., Williams, 1966; Ghiselin, 1974; Maynard-Smith, 1978) have been written on this subject, but many disagreements remain over such issues as group selection (Chapter 10). The selective advantages of sexual reproduction are certainly indirect ones.

One important advantage of sexual reproduction is genetic recombination, with its constant "stirring" of the gene pool, resulting in a greater diversity among the offspring of any individual. In a sexually reproducing species inhabiting a variety of situations, some few individuals are likely to possess the favorable combinations of genes in even the most marginal of those situations. In other words, sex is very efficient at ensuring genetic variability. Asexual reproduction, on the other hand, allows for more rapid production of offspring, and the advantages of this proliferation may in certain situations outweigh the advantages of recombination in each generation. With certain exceptions, asexual reproduction is more common among procaryotes and sexual reproduction among eucaryotes, but the class Rotifera and the insect order Homoptera are major exceptions to this rule.

CHAPTER SUMMARY

Variation is essential to evolution; without it, natural selection would be impotent, and no evolutionary change could possibly take place.

In producing a phenotype, genes must interact with one another and with the environment. Genes are subject to natural selection only through their phenotypic effects. The existence of a single gene with many phenotypic effects is called pleiotropism; the interaction of several genes to produce together a phenotype that none of them could produce alone is called polygeny. An important type of polygeny is the masking of one gene's phenotypic effects by another, a phenomenon known as epistasis.

De Vries called all sudden, heritable changes "mutations," but most biologists would now restrict this definition. The most restrictive definition uses the term mutation only for the changes within single genes, known as single-gene mutations. More drastic are those chromosomal changes that alter the sequence of many genes at once: inversion, in which a chromosome segment is rotated end-to-end; translocation, or the movement of a chromosome segment to a new location; duplication, or the repetition of the same chromosome segment more than once; and deletion, or the loss of a chromosome segment. These changes are especially significant because of the influence that genes may have on other genes nearby—a phenomenon known as the position effect. Genetic recombination will rearrange the genetic material much more rapidly but only within certain limits; the chromosomal aberrations exceed these limits.

Changes in chromosome number may be classed either as aneuploidy, the loss or addition of individual chromosomes, or as euploidy, the loss or addition of entire genomes.

PROBLEMS

1. Why should the rate of mutation vary from point to point along a chromosome? Why, at a given point, might mutations in opposite directions not always occur at the same rates?

2. Assume, as T. H. Morgan initially did, that white eyes were inherited in fruit flies as a simple autosomal recessive. Assume that you have crossed a white-eyed male with

a red-eyed female, as shown in Item 2.7. Where would you first encounter surprising results? Why? Why might you be driven to the conclusion that male flies are heterozygous in the wild?

3. Starting with a standard sequence, ABCDEFGHIJK, explain why two partially overlapping inversions will always

allow the reconstruction of the intermediate step. (Can we always tell which of the two inversions occurred first?)

4. Unequal crossing over is the exchange of material between homologous chromosomes that break at not quite the same points. Show how this phenomenon could lead to both chromosome duplication and deletion.

FOR FURTHER KNOWLEDGE ▐

ITEM 7.1   *Epistasis in the Coat Color of Mice*

In this example, gene B determines the coat color as either "agouti" (dominant) or black (recessive). Gene A may mask this color entirely: mice of genotype AA or Aa are normally colored, but aa mice are albino (white; no pigment produced) regardless of the condition of gene B. The presumed mechanism for this type of epistatic interaction is as follows:

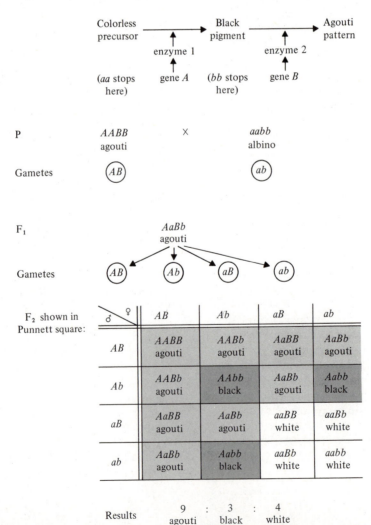

$\male$ \ $\female$	AB	Ab	aB	ab
AB	AABB agouti	AABb agouti	AaBB agouti	AaBb agouti
Ab	AABb agouti	AAbb black	AaBb agouti	Aabb black
aB	AaBB agouti	AaBb agouti	aaBB white	aaBb white
ab	AaBb agouti	Aabb black	aaBb white	aabb white

Results   9 : 3 : 4
          agouti   black   white

ITEM 7.2 *Epistasis in the Color of Poultry Feathers*

In this example, the dominant allele *C* produces a color (*cc* is white). A dominant allele of a different gene, *I*, produces a color inhibitor. The situation may be represented as follows:

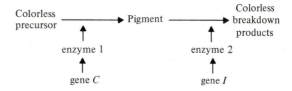

Some breeds of poultry are white because they lack *C*; other breeds are white because they are homozygous for the color inhibitor, *I*.

	P	*iicc* white			*IICC* white

Gametes   (*ic*)   (*IC*)

F₁   *IiCc*
white

Gametes   (*IC*)   (*Ic*)   (*iC*)   (*ic*)

F₂ shown in Punnett square:

♂ \ ♀	*IC*	*Ic*	*iC*	*ic*
IC	*IICC* white	*IICc* white	*IiCc* white	*IiCc* white
Ic	*IICc* white	*IIcc* white	*IiCc* white	*Iicc* white
iC	*IiCC* white	*IiCc* white	*iiCC* colored	*iiCc* colored
ic	*IiCc* white	*Iicc* white	*iiCc* colored	*iicc* white

Results: 13 :   3
             white   colored

8

Genes in Populations

The origin of genetic variations was treated in the previous chapter. But how will these variations act and interact in populations? This chapter examines the simplest model, that of a genetic equilibrium, as well as certain of the simpler types of departure from equilibrium conditions. Extensive discussion of natural selection, the most important type of deviation from this equilibrium, is deferred to a later chapter.

A. THE HARDY-WEINBERG EQUILIBRIUM

Early in the twentieth century, the followers of Mendelian genetics were eager to apply their theories to the human species. Among their discoveries was a trait known as **brachydactyly** (Fig. 8.1), in which the finger bones (*phalanges*) are abnormally short. (The rest of the hand is of normal proportions, and the phalanges are all present, even if shorter than usual.) This trait, inherited as a dominant in simple Mendelian fashion, was hailed as proof that Mendelian laws could be applied to humans as well as to animals and plants.

An objection to this interpretation was soon raised. If the Mendelian interpretation was correct, then a cross between a homozygous dominant (brachydactylous) and a homozygous recessive (normal) should yield three brachydactylous individuals to every normal one in the F_2 generation. Why, then, shouldn't three-fourths of all people be brachydactylous? Since brachydactyly is very rare in human populations, this objection was viewed as a direct challenge to the tenets of Mendelian genetics as applied to humans.

The objection above was answered in the year 1908 by a British mathematician, G. E. Hardy, and a German physician, Wilhelm Weinberg, both working independently. Each discovered the principle that now bears their two names: that under certain specified conditions, genotypic frequencies tend to remain constant (Fig. 8.2). (Spiess, 1977, gives some earlier references in which a genetic equilibrium was predicted for certain restricted cases.) The **frequency** of a genotype refers to its relative abundance in the population, from a minimum of zero to a maximum of 1.00, or 100 percent of the population. Hardy and Weinberg predicted that common traits would remain common, rare traits rare. This has nothing to do with dominance; the terms "dominant" and "re-

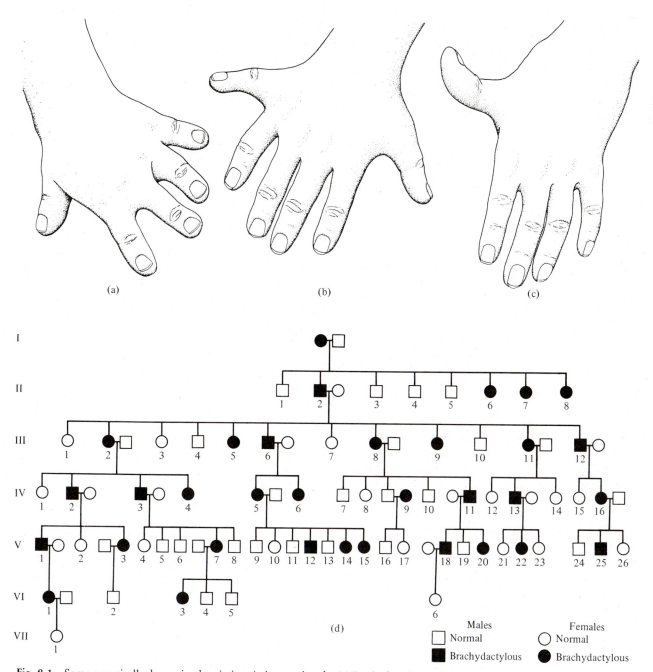

Fig. 8.1 Some genetically determined variations in human hands. (a) Brachydactyly. (b) Poly-
dactyly. (c) Hitchhiker's thumb. (d) A family pedigree showing the incidence of brachydactyly.

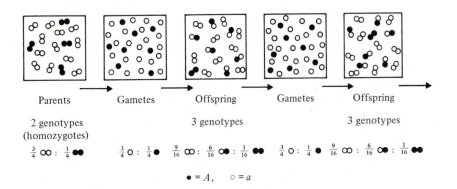

Parents   Gametes   Offspring   Gametes   Offspring

2 genotypes   3 genotypes   3 genotypes
(homozygotes)

$\frac{3}{4}$ ∞ : $\frac{1}{4}$ ●● $\frac{3}{4}$ O : $\frac{1}{4}$ ● $\frac{9}{16}$ ∞ : $\frac{6}{16}$ O● : $\frac{1}{16}$ ●● $\frac{3}{4}$ O : $\frac{1}{4}$ ● $\frac{9}{16}$ ∞ : $\frac{6}{16}$ O● : $\frac{1}{16}$ ●●

● = A,   o = a

Fig. 8.2 The Hardy-Weinberg equilibrium. In this diagram, the open and filled circles represent two alleles with frequencies of 25 % and 75 %. The actual population would have to be much larger than the one shown here. (After Spiess, 1977.)

cessive'' (Chapter 2) refer to the phenotypic appearance of heterozygotes, not to the abundance of any trait in the population.

The **Hardy-Weinberg principle** (also called a ''law'') is here explained in modern notation. Suppose that a new population is founded by homozygous AA and aa individuals. The frequency of AA individuals we shall call p, and the frequency of aa individuals q, where $p + q = 1$, meaning that together they constitute 100 percent of the population. Since the dominants carry only A alleles, and the recessives carry only a alleles, p and q also represent the gene frequencies of these two alleles. Now the AA individuals produce gametes of type A, and the aa individuals produce gametes of type a. These gametes will furthermore be in the same proportion as their parents, p:q. Let us assume that they combine at random, that no gametes are preferentially lost, that no new gametes are introduced, and that all offspring survive and produce new gametes to an equal extent. What proportion of the next generation will be of genotype AA? A male gamete chosen at random has a probability p of being A (and a probability q of being a), and a female gamete chosen at random has the same probability.

The combination of gametes under the assumption of random mating may be shown in a Punnett square. Here the frequencies (p or q) of each gamete are specified, and these are carried into the body of the table by a process of multiplication.

	p A	q a
p A	p^2 AA	pq Aa
q a	pq Aa	q^2 aa

Since the probability of two independent events is equal to the product of the individual probabilities, the outcome AA should occur with probability p times p, or p^2. Similarly, the outcome aa should occur with probability q times q, or q^2. The outcome Aa can occur in either of two ways: male gamete A with female gamete a (probability p times q, or pq), and male gamete a with female gamete A (probability q times p, which is also pq). Thus the total probability of Aa is given by $2pq$. This reasoning can also be shown in a Punnett square or in a geometrical representation. In either case, the new generation consists of three genotypes with frequencies

$$p^2\ AA + 2pq\ Aa + q^2\ aa = 1. \qquad (8.1)$$

(The reader who doubts that these frequencies add up to 1 should write down $p + q = 1$ and square both sides.)

A geometrical representation is also possible. Here each side of the square represents one unit, and it is divided into segments representing p and q, or the proportions of A and a gametes (Fig. 8.3). A randomly chosen point along either side represents a random sampling of gametes. The area of each rectangle represents the frequency of the resulting progeny, meaning the proportion of the next generation having the genotype specified in each cell. The total area represented is 1×1, or 1 square unit, representing 1.00 or 100 percent of the population.

The frequencies of the three genotypes in the next succeeding generation may now be calculated as shown in Box 8–A; these are the exact same frequencies as in Equation 8.1. In other words, an equilibrium has been established that remains constant from one generation to the next. This equilibrium, known as the **Hardy-Weinberg equilibrium,** is described by the equation above. The

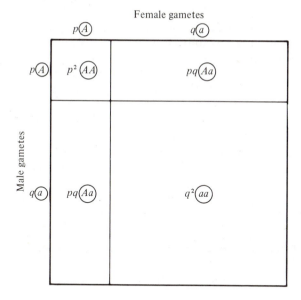

Fig. 8.3 Geometric representation of the Hardy-Weinberg equilibrium.

Hardy-Weinberg principle simply states the conditions under which this equilibrium holds:

> In a large, random-mating population of diploids, in the absence of unbalanced mutation or of unbalanced migration or of selection in any form, the genotypic proportions tend to remain constant.

Note that the Hardy-Weinberg principle specifies that genotypic proportions (therefore also phenotypic and gametic proportions) tend to remain constant; the remainder of the principle merely specifies the conditions under which this is true. We shall soon examine each of these assumptions, but let us first note that natural populations often come close to the predictions of Equation 8.1, even when they fall far short of satisfying the stated conditions. The Hardy-Weinberg principle, in other words, is robust (minimally affected) with respect to many of the departures from equilibrium conditions that commonly occur in natural populations. This fact often makes such departures as selection difficult to substantiate from a simple calculation of gene frequencies. Inbreeding, on the other hand, may cause substantial de-

---

BOX 8–A *The Hardy-Weinberg Equilibrium*

Assume a population with the genotypic proportions

$$p^2\ AA + 2pq\ Aa + q^2\ aa = 1.$$

Dominant homozygotes of genotype AA represent p^2 of this population and contribute p^2 of the gametes, all of them of type A. Similarly, recessive homozygotes of genotype aa represent q^2 of the population and contribute q^2 of the gametes, all of type a. The Aa heterozygotes contribute $2pq$ of the gametes, with two types of gametes present in equal proportions: half of them (pq, which is half of $2pq$) will be A, and the other half (also pq) will be a. All this may be represented as follows:

Parents: $p^2\ AA$ + $2pq\ Aa$ + $q^2\ aa$

Gametes: $p^2\ \textcircled{A}$ $pq\ \textcircled{A}$ $pq\ \textcircled{a}$ $q^2\ \textcircled{a}$

$p\ \textcircled{A}$ $q\ \textcircled{a}$

Combining all terms referring to gametes bearing the dominant

allele, we find the total frequency of A to be

$$p^2 + pq$$
$$= p(p + q)$$
$$= p(1) \quad \text{because } p + q = 1$$
$$= p.$$

Similarly, the total frequency of a alleles is given by

$$pq + q^2$$
$$= q(p + q)$$
$$= q(1) \quad \text{because } p + q = 1$$
$$= q.$$

The frequencies of the gametes are the same as those shown in Fig. 8.3. Either of these may be used to arrive at the genotypic frequencies of the next generation, given by

$$p^2\ AA + 2pq\ Aa + q^2\ aa = 1.$$

These frequencies are the same as those originally specified. They have remained unchanged from one generation to the next, and they will continue to remain unchanged so long as the assumptions made continue to hold.

partures from equilibrium conditions, and it is thus easier to detect.

The astute reader will notice that we have assumed *gametes* to combine at random, although the stated conditions refer only to random mating among organisms. To demonstrate that these will give similar results, we consider the more detailed series of calculations shown in Box 8–B.

Note once more that

$$p^2 AA + 2pq Aa + q^2 aa$$

represents an equilibrium that remains unchanged once it is established.

In our original discussion above, the equilibrium of Equation 8.1 was established initially from a population represented by

$$p AA + q aa = 1$$

with no heterozygotes. Now let us generalize the discussion to include populations in which heterozygotes are present initially. If D stands for homozygous dominants, H for heterozygotes, and R for homozygous recessives, we may let

$$D AA + H Aa + R aa = 1$$

BOX 8–B   *The Hardy-Weinberg Equilibrium*

The following formulation of the Hardy-Weinberg equilibrium gives the same results as in Box 8–A, but they are expressed in terms of the random mating of parents and without the assumption that this necessarily equates to the random union of gametes.

Assume a population with the genotypic proportions

$$p^2 AA + 2pq Aa + q^2 aa = 1.$$

If mating occurs at random, each mating type will have a predictable frequency of occurrence, as well as a predictable number of offspring of each genotype, as shown in the table below.

Summing up the frequencies in the last three columns and factoring out $(p^2 + 2pq + q^2) = 1$ in each case, we arrive at the total figures shown at the bottom of each column. Since these are the same frequencies as were initially present, the genotypic proportions remain unchanged from one generation to the next.

Type of mating	Frequency of matings of this type	Frequency of resultant progeny		
		AA	Aa	aa
$AA \times AA$	$p^2 \times p^2 = p^4$	p^4		
$AA \times Aa$ $Aa \times AA$	$2 \times p^2 \times 2pq = 4p^3q$	$2p^3q$	$2p^3q$	
$AA \times aa$ $aa \times AA$	$2 \times p^2 \times q^2 = 2p^2q^2$		$2p^2q^2$	
$Aa \times Aa$	$2pq \times 2pq = 4p^2q^2$	p^2q^2	$2p^2q^2$	p^2q^2
$Aa \times aa$ $aa \times Aa$	$2 \times 2pq \times q^2 = 4pq^3$		$2pq^3$	$2pq^3$
$aa \times aa$	$q^2 \times q^2 = q^4$			q^4
	Total frequencies:	p^2	$2pq$	q^2

represent any arbitrary set of initial frequencies in which $D + H + R = 1$. In calculating p, the frequency of allele A, we note that all the alleles carried by AA individuals, plus half of the alleles carried by the heterozygotes, are A. Thus p is given by $D + \frac{1}{2}H$, and, by similar reasoning, q, the frequency of a, is $\frac{1}{2}H + R$. If we make these substitutions, Box 8–C shows how an equilibrium is established in a single generation.

The Hardy-Weinberg principle can readily be extended to cover multiple alleles (Box 8–D), haploid organisms, and sex-linked genes (for example, see Li, 1955). It can also be extended to two independent (unlinked) genes or two linked genes under an already estab-

lished equilibrium that will maintain itself. But linkage will slow down the establishment of an equilibrium if one does not yet exist. The eventual outcome is the same, but the closer the linkage, the more slowly the equilibrium is approached asymptotically.

B. DEPARTURE FROM EQUILIBRIUM CONDITIONS

The Hardy-Weinberg equilibrium is typical of many mathematical models: in order to be useful, they run the risk of oversimplification. At their simplest, they require

---

BOX 8–C *Establishment of a Hardy-Weinberg Equilibrium*

Assume an initial population with arbitrary genotypic frequencies.

$$D\ AA + H\ Aa + R\ aa = 1,$$

where $D + H + R$ must total 1. If mating occurs at random,

the frequency of each mating type and the frequency of the resultant progeny are as shown in the table below.

Summing up the frequencies in the last three columns and factoring, we may express the column totals in terms of $(D + \frac{1}{2}H)$ and $(\frac{1}{2}H + R)$, which equal p and q, respectively. Thus a Hardy-Weinberg equilibrium is established in a single generation of random mating, with arbitrary initial frequencies.

Type of mating	Frequency of matings of this type	Frequency of resultant progeny		
		AA	Aa	aa
$AA \times AA$	$D \times D = D^2$	D^2		
$AA \times Aa$ $Aa \times AA$	$2 \times D \times H = 2DH$	DH	DH	
$AA \times aa$ $aa \times AA$	$2 \times D \times R = 2DR$		$2DR$	
$Aa \times Aa$	$H \times H = H^2$	$\frac{1}{4}H^2$	$\frac{1}{2}H^2$	$\frac{1}{4}H^2$
$Aa \times aa$ $aa \times Aa$	$2 \times H \times R = 2HR$		HR	HR
$aa \times aa$	$R \times R = R^2$			R^2
	Total frequencies:	$(D + \frac{1}{2}H)^2$ $= p^2$	$2(D + \frac{1}{2}H)(\frac{1}{2}H + R)$ $= 2pq$	$(\frac{1}{2}H + R)^2$ $= q^2$

BOX 8–D *Genetic Equilibrium with Multiple Alleles*

Assume a system of three alleles, a_1, a_2, and a_3, with frequencies given respectively by p, q, and r, where $p + q + r = 1$. Assume also genotypic proportions given by

$$(p\,a_1 + q\,a_2 + r\,a_3)^2 = 1$$

or

$$p^2\,a_1a_1 + q^2\,a_2a_2 + r^2\,a_3a_3 + 2pq\,a_1a_2 + 2pr\,a_1a_3 + 2qr\,a_2a_3 = 1.$$

Now the gametes produced by each of the six parental types are given in tabular form (at right).

Summing up each column and factoring out $(p + q + r) = 1$ in each case, we arrive at the frequencies of each type of gamete.

Random union of the gametes may be represented algebraically as

$$(p\,a_1 + q\,a_2 + r\,a_3)^2 = 1$$

$$p^2\,a_1a_1 + q^2\,a_2a_2 + r^2\,a_3a_3 + 2pq\,a_1a_2 + 2pr\,a_1a_3 + 2qr\,a_2a_3 = 1.$$

It may also be represented geometrically by the model of the divided square, each of whose sides are divided into segments representing p, q, and r (on facing page).

Parental type and its frequency	Gametes produced		
	a_1	a_2	a_3
$p^2\,a_1a_1$	p^2		
$q^2\,a_2a_2$		q^2	
$r^2\,a_3a_3$			r^2
$2pq\,a_1a_2$	pq	pq	
$2pr\,a_1a_3$	pr		pr
$2qr\,a_2a_3$		qr	qr
Totals:	p	q	r

Here again, each side of the square measures 1 unit, so the table area is $1 \times 1 = 1$, or 100 percent of the population. The area of each rectangle represents the proportion of the next generation having the genotype specified in each case.

Note that the resultant frequencies are the same as those specified initially. Thus the population is in equilibrium, and the genotypic frequencies remain unchanged from one generation to the next.

The calculations above can be readily extended to cover four or more alleles.

---

so many simplifying assumptions that they may in fact be quite unrealistic or unnatural. As the assumptions are relaxed and made more realistic, the conditions covered become less and less unnatural, but the complexity of the model increases greatly.

The Hardy-Weinberg principle states that genotypic frequencies remain the same, provided that certain conditions and simplifying assumptions are met. Some of these assumptions, like the total absence of selection, are quite unrealistic in natural settings. In succeeding sections of this chapter, we shall examine the conditions of the Hardy-Weinberg model more closely and discuss what happens when they are not met. Departures from Hardy-Weinberg equilibrium conditions are discussed in the following order: nonrandom mating, unbalanced

mutation, migration, and genetic drift. A detailed consideration of selection is delayed to Chapter 11.

In this chapter, we assume that reproduction occurs among a random sample of the whole population, and that gamete production also occurs randomly. Relaxation of either of these assumptions leads to a consideration of selection, which is treated at length in Chapter 11. For the moment, we may define selection as a consistent bias in reproduction or a consistent difference in the contributions of the several genotypes to the next generation.

Nonrandom Mating

The stated conditions of the Hardy-Weinberg equilibrium specify a large, random-mating population. Random mating, or **panmixia**, assumes that each individual

BOX 8–D *(continued)*

Female gametes

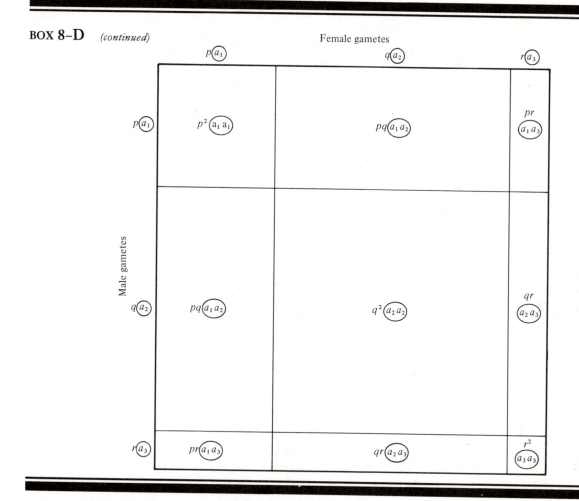

Male gametes

mates without preference: that any individual has an equal probability of being chosen as a mate. One general type of departure permits either sex to show a consistent phenotypic preference in the opposite sex: "gentlemen prefer blondes" would be an example of this (but only if all male phenotypes showed the same degree of preference). In all such cases of *consistent* preference, we are dealing with *sexual selection,* a special type of selection (Chapter 11) in which the opposite sex is the selective agent.

Another type of departure from panmixia involves a mating preference that is dependent on individual phenotype. This is known as **assortative mating,** and it can be of two types: **positive assortment,** in which individuals show preference for their own phenotype (as would be true if blonds preferred blonds), or **negative assortment,** in which individuals tend to avoid their own phenotype. When single genes or multiple independent genes are involved, the effect of assortative mating is largely seen as a change in the equilibrium frequency of heterozygotes. This frequency is increased in negative assortment and decreased in positive assortment, but an equilibrium is maintained nevertheless.

When linked genes are involved, the principal effect of negative assortment is to hasten the attainment of genetic equilibrium. Conversely, positive assortment has the effect of retarding the attainment of genetic equilibrium, especially when linked genes are involved. In such cases, the degree of linkage is also a factor: the closer the linkage, the greater the effect. Changes in gene frequencies under assortative mating are always in the direction of a genetic equilibrium, and once this equilibrium is reached, it tends to be maintained. For an algebraic treatment of assortative mating, the texts by Crow and Kimura (1970) and Spiess (1977) are recommended.

In plants, negative assortment may take the special form of **heterostyly** (Fig. 8.4), in which flowers with high anthers (and short styles) mate preferentially with flowers having long styles (and low anthers). Complete negative assortment may extend to two or more genetically controlled types, acting in extreme cases as self-sterility alleles. Such alleles act in a manner analogous to the determinaton of gender (sexual phenotype) in organisms with separate sexes. Both may be treated as special types of absolute negative assortment: the long-styled form mates only with the short-styled form (and vice versa), just as males mate only with females (and vice versa).

Inbreeding is here understood as a departure from panmixia in which close relatives have an increased probability of mating with one another and a decreased probability of mating with unrelated individuals. As so defined, inbreeding is fairly common in natural populations, and its effects are similar to those of positive assortment. Positive assortment, however, increases the probability of matings between phenotypically similar individuals even if they are unrelated, whereas inbreeding increases the probability of matings between related individuals even if they are phenotypically dissimilar. Inbreeding is especially common in species where individuals tend to remain in the same local area all of their lives, choosing a mate from within this local area. The degree of inbreeding may be measured by the "fixation coefficient" F, developed by Sewall Wright. This coefficient measures the probability that two gametes coming together to form a zygote carry identical alleles. A high value of F thus reflects a high degree of inbreeding. The algebraic treatment of inbreeding under various conditions is covered by Crow and Kimura (1970), Spiess (1977), and Li (1955).

If a population is divided into genetically separate subpopulations that mate randomly within themselves, then each can establish a Hardy-Weinberg equilibrium at a somewhat different gene frequency. If the gene frequencies from all these subpopulations were pooled and treated as if they represented a single population, then an excess of homozygotes would be observed, along with a lower-than-expected number of heterozygotes. This is sometimes called Wahlund's principle, and certain population geneticists believe that it represents a rather common situation, often mistaken for the simpler forms of inbreeding among relatives. The subdivisions of a population need not be completely isolated in order to show the effects of Wahlund's principle, as long as gene exchange with other subpopulations occurs at a reduced rate.

Mutational Equilibrium

We have thus far ignored the effects of unbalanced mutation rates. In introducing mutations, let us consider first the result of mutation in one direction only, say, $A \rightarrow a$. Suppose that in each generation a small proportion, u, of the dominant alleles undergo mutation in this direction. The frequency of a in the next generation would be equal to q, plus a new term, $u\,p$, due to the

Fig. 8.4 Heterostyly in primrose flowers (*Primula officinalis*). (a) The "pin" phenotype, with high stigma and low-placed anthers. (b) The "thrum" phenotype, with low stigma and high-placed anthers. Cross-fertilization (assortative mating) between these two types is effected in two ways: certain insects will enter only the top of the flower (and transfer pollen from thrum to pin), and other insects will penetrate deeper (and transfer pollen from pin to thrum).

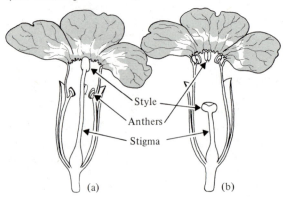

Style

Anthers

Stigma

(a)   (b)

genes just mutated. This term, $u\,p$, may also be called Δq ("delta q") since it represents the change in q. Since both u and p are considered positive, q will continue to increase until it reaches its maximum value of 1, at which point $p = 0$ and the change in q ceases. The entire population would now be homozygous for the recessive allele, a.

Let us now consider the more realistic case of mutation in both directions. Suppose again that a small proportion, u, of the dominant alleles undergo mutation in the direction $A \rightarrow a$, while a small proportion, v, of the recessive alleles undergo mutation in the direction $a \rightarrow A$ in each generation. The proportion, q, of the recessive allele after one generation will be

$$q_1 = q_0 + u\,p - v\,q, \tag{8.2}$$

where q_0 is the initial frequency, $u\,p$ represents an increase due to mutations in the direction $A \rightarrow a$, and $v\,q$ represents a decrease due to mutations in the direction $a \rightarrow A$. The change in q is given by

$$\begin{aligned}
\Delta q &= q_1 - q_0 \\
&= u\,p - v\,q \\
&= u(1 - q) - v\,q \\
&= u - (u + v)q \tag{8.3}
\end{aligned}$$

If an equilibrium is possible, Δq would have to equal zero, or

$$\Delta q = 0 = u - (u + v)q$$

or

$$(u + v)q = u$$

$$q = \frac{u}{u + v} \tag{8.4}$$

Similarly, the equilibrium value of p would occur at

$$p = \frac{v}{u + v} \tag{8.5}$$

In general, both mutation rates u and v will be of similar magnitude, so equilibrium values of p and q will occur when neither A nor a is very rare. In the special case where $u = v$, both p and q have equilibrium values of $\frac{1}{2}$.

As a specific example, consider the effects of mutation alone on the gene causing white eye color in fruit flies (*Drosophila*). Ignoring all but the two most common alleles, we may represent this situation as

$$W \quad \underset{v = 4.2 \times 10^{-5}}{\overset{u = 1.3 \times 10^{-4}}{\rightleftharpoons}} \quad w.$$
$$\text{wild type} \qquad\qquad\qquad \text{white eyes}$$

Ignoring selection and other nonmutational forces, an equilibrium would be established at

$$p = \frac{v}{u + v} = \frac{0.000042}{0.000172} = .244,$$

$$q = \frac{u}{u + v} = \frac{0.000130}{0.000172} = .756.$$

The allele for white eyes would in this case represent about three-fourths of the gene pool, and the "wild type" allele about one-fourth. Since W is far more abundant than w in natural populations, we may safely conclude that some strong nonmutational force is also acting in this instance, presumably some form of natural selection against the recessive allele w.

If we call the equilibrium values $\hat{p}$ and $\hat{q}$ ("p-hat" and "q-hat"), we can rephrase Equation 8.3 as

$$\Delta q = u - (u + v)q$$

$$\frac{\Delta q}{u + v} = \left(\frac{u}{u + v}\right) - \left(\frac{u + v}{u + v}\right)q$$

$$\frac{\Delta q}{u + v} = \hat{q} - q$$

$$\Delta q = (\hat{q} - q)(u + v). \tag{8.6}$$

Thus the change in q is proportional to the combined mutation rate, $u + v$, and also to the departure of q from the equilibrium value $\hat{q}$. The sign of Δq (positive or negative) is determined by the sign of $\hat{q} - q$: when q is below $\hat{q}$, Δq is positive and q increases; when q is above $\hat{q}$, Δq is negative and q decreases. Thus the direction of change is always toward the equilibrium. If a population is ever displaced from equilibrium, the equilibrium will always tend to be restored, and we may therefore refer to it as a *stable equilibrium*. Since this particular equilibrium involves mutation rates only, it is sometimes referred to as a *mutational equilibrium*.

Migration

If a population receives a certain proportion, m, of its genes from a different population every generation, the proportion of the recessive allele, a, after one generation is given by

$$q_1 = q_0(1 - m) + q_\beta m \qquad (8.7)$$

where q_β represents the gene frequency of a among immigrants (assumed here to be constant). The change in q is given by

$$\begin{aligned}
\Delta q &= q_1 - q_0 \\
&= q_0(1 - m) + q_\beta m - q_0 \\
&= -q_0 m + q_\beta m \\
&= m(q_\beta - q_0) \qquad (8.8)
\end{aligned}$$

Note that the change in q is proportional not only to the migration rate but also to the disparity between the gene frequencies in the two populations. The sign of Δq —i.e., the direction of change—is determined by the sign of the factor $q_\beta - q_0$. An equilibrium can occur only when $\Delta q = 0$, which in Equation 8.8 is possible only when $m = 0$ (no migration) or $q_\beta = q_0$ (both populations with identical gene frequencies).

Another type of equilibrium is of course possible if both mutation and migration are operating together. In this case, and drawing on Equations 8.3 and 8.8, we can assume that the total change in q is equal to the change produced by mutation plus the change produced by migration, or

$$\begin{aligned}
\Delta q_{total} &= \Delta q_{mutation} + \Delta q_{migration} \\
&= u - (u + v)q + m(q_\beta - q) \qquad (8.9)
\end{aligned}$$

Setting this $\Delta q = 0$ for equilibrium conditions, we obtain

$$\begin{aligned}
\Delta q_{total} = 0 &= u - (u + v)q + m(q_\beta - q) \\
0 &= u - uq - vq + mq_\beta - mq \\
0 &= u + mq_\beta - (u + v + m)q \\
(u + v + m)q &= u + mq_\beta \\
q = \hat{q} &= \frac{u + mq_\beta}{u + v + m} \qquad (8.10)
\end{aligned}$$

In the discussion above, we have treated only **immigration. Emigration** of a random sample of our population would have no effect under the conditions here postulated, although in real situations it would presumably

have an effect on q_β and on population size as well, the latter contributing to genetic drift (see below). Cases where q_β is subject to change or where more than one group of immigrants is involved are too complex to be discussed here.

Emigration of a nonrandom sample of our population, that is, the differential emigration of the several genotypes, may be treated as a special instance of natural selection. Individuals that emigrate are thereby treated as indistinguishable from individuals that die, though of course in real situations they move to other populations and often contribute to changes in the q_β values.

C. GENETIC DRIFT

The conditions stated for the Hardy-Weinberg equilibrium specify a "large" population. How large? Large enough so that samples (especially breeding samples) drawn from the entire population are representative. Large enough, that is, so that "sampling error" can be overlooked. The mathematical assumption of a zero sampling error in fact requires that the population be infinite! Since no real population is infinite, the assumption can be treated at best as approximate: a small or negligible sampling error. In large populations (say, over 1000 individuals) sampling errors are indeed negligible in most cases, though not in the case of rare genes (see below).

Sampling error is a statistical concept. Suppose I had before me a large bag containing a random mixture of 1000 red balls and 1000 white balls. If I drew balls from the bag indiscriminately (with respect to color) until I had a random sample of 200, or 10 percent of the total, I would expect most often to have drawn a sample of 100 red balls and 100 white balls. I would also expect (a bit less often) samples with 99 red and 101 white, or 99 white and 101 red, and so forth. But I would not expect to have drawn a sample consisting of 200 red balls and no white ones! (If I did draw such a sample, I would probably conclude that I had made an incorrect assumption: either I was not sampling at random, or the bag did not *really* contain equal numbers of red and white balls.) Such an outcome does have a very small but nonzero probability, on the order of 1 in every 10^{30}. Now if I used a bag containing 10 red balls and 10 white ones, and again drew a 10 percent sample, or two balls at a time, I would expect one red ball and one white ball as the most usual outcome, but I could also expect two reds and no whites with a more reasonable probability (9/38, slightly less than ¼) or two whites and no reds with the same probability, 9/38. The difference between these two cases is one of sampling error, definable in rigorous terms as an *error variance* and in more intuitive terms as the results of "chance."

In population genetics, the effects of sampling error may safely be neglected in large populations, but in small or moderate populations its effects are very real. The concept of **genetic drift** was developed by Sewall Wright to treat the effects of sampling error in small and moderate populations. Genetic drift may be defined as *deviations from the Hardy-Weinberg equilibrium due to "chance" (sampling error) in small to moderate populations.*

Genetic drift is of course most noticeable in the smallest of populations, where each individual death abruptly changes the genotypic frequencies (Fig. 8.5). Suppose a particular genotype was present in only two individuals among a population of twenty, that is, in 10% of the population. The death of either of these two individuals would abruptly change the minority genotypic frequency from 10.00% to about 5.26%, whereas the death of any other individual would change this same

frequency from 10.00% to 10.53% with equal abruptness. The most drastic consequences of genetic drift include the total loss of an allele or its total fixation by the loss of other alleles. These changes would normally be irrevocable, unless recurrent mutation and/or gene flow from other populations replaced the allele that had been lost. With rare alleles, even large populations can be affected by genetic drift: the example above of two individuals making up 10% of a population of 20 applies with very little alteration to the case of two individuals making up 0.1% of a population of 2000, or the case of two individuals carrying a rare genotype and making up 0.001% of a population of 200,000. In all these cases, the fortuitous death of either individual would cut the gene frequency approximately in half, and the fortuitous death of both individuals would eliminate the genotype completely.

When the concept of genetic drift was first proposed, some geneticists leaped at the possibility that it might explain the haphazard and supposedly nonadaptive variations often observed among small island populations—each isolated population was assumed to have reached its own assortment of gene frequencies independently and by "random drift." On the other hand, Sewall Wright emphasized that "accidents of sampling . . . may be responsible for nonadaptive differentiation of small island populations but are more likely to lead to

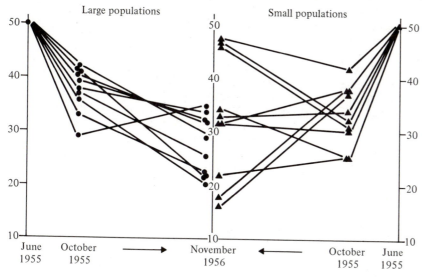

Large populations   Small populations

June 1955   October 1955 → November 1956 ← October 1955   June 1955

Fig. 8.5 Genetic drift in experimental populations of *Drosophila pseudoobscura*. Twenty populations were started with 50% frequencies of the Pike's Peak (PP) chromosome configuration. The left diagram, with time flowing left to right, shows that natural selection has reduced the frequency of PP chromosomes to an average of less than 30% but with a slight variation due to genetic drift. The right diagram, with time flowing right to left, shows the same effect of selection but a much greater and more erratic variation in ten small populations due to genetic drift. From Dobzhansky and Pavlovsky, *Evolution*, 11:315, 1957.

ultimate extinction . . . than to evolutionary advance.'' The debate that ensued over the importance (or even the existence) of genetic drift still persists today under a different guise, the neutralist-selectionist debate (Chapter 12).

Mathematical treatment of genetic drift may be found in the books by Crow and Kimura (1970) and Spiess (1977), and in the many papers of Sewall Wright.

The preceding discussion of genetic drift focuses on a stable population of small but constant size. Another type of genetic drift, already referred to, affects rare genotypes in populations of any size, provided that the number of individuals possessing this genotype is suitably small. Other special types of genetic drift have received special names. For example, the term **founder principle** was proposed by Mayr to cover a type of genetic drift that occurs when a small group of individuals become the founders of a new population. The new population often has genotypic frequencies differing (not in any consistent direction) from the parent population, simply because the founders may have been an unrepresentative sample of the larger parent population. The Dunkers of Pennsylvania, a small religious sect, illustrate this principle. Founded some 100 years ago by a group of 20 to 30 individuals who came from Germany seeking religious freedom, they maintained their genetic isolation by the avoidance of breeding with surrounding populations. Several genetic traits are rather more frequent among Dunkers than they are among other populations living in Pennsylvania, or among the West German populations descended from the same source from which the Dunkers originally came; these traits include blood types A and M. Other traits are much less frequent among Dunkers than among other comparable populations; among these latter traits is the inherited ability to hyperextend the interphalangeal joint in the middle of the thumb, a condition known as ''hitchhiker's thumb,'' and also blood group B, which has all but disappeared from the population. Some of these traits may reflect the genetic composition of the original founder population, and others represent genetic drift that has taken place since the group's origin. A rare trait that just happened to be present in one of these original founders would have an expected frequency of about 3%–5%, whereas any trait that happened not to occur among the few

dozen original founders or in some subsequent generation would have a frequency of zero.

Another special type of genetic drift is called the **bottleneck effect** by Mayr. This operates when a large population becomes temporarily small as the result of its decimation by a severe storm, an epidemic, or other population crisis. When the population regains its former size, its genotypic frequencies may have changed because of genetic drift when the population was temporarily small. The temporarily small population of survivors may well have been unrepresentative of the larger original population. Here, as in other cases, natural selection and genetic drift work side by side. As an illustrative example, assume the existence of several large populations, each with a 10% frequency of a certain genotype. All these populations pass through a ''bottleneck'' (they are decimated by a population crash) from which most of them recover (hardly a certainty in real situations). Now if all these populations have reduced the frequency of the genotype under study to an average value of 7%, this can be attributed to natural selection operating during the population crash. However, if some populations have a 6% or 5% frequency and others have an 8% or 9% frequency, then these unsystematic deviations from the mean value of 7% would be attributed to genetic drift (compare Fig. 8.5).

CHAPTER SUMMARY

In a large, random-mating population of diploids, in the absence of unbalanced mutation, unbalanced migration, or any form of selection, the proportions of the several genotypes tend to remain the same. This phenomenon is known as the Hardy-Weinberg principle, and the equilibrium that it predicts is called a Hardy-Weinberg equilibrium. In the simplest of cases, an equilibrium is rapidly established in a single generation, with genotypes in the proportion

$$p^2 \, AA + 2pq \, Aa + q^2 \, aa = 1,$$

where p represents the frequency of A and q represents the frequency of a. For this equilibrium to be reached, all genotypes must contribute to the next generation in proportion to their numbers; any deviation from this condition would represent a form of selection.

Other factors that may cause departure from the Hardy-Weinberg equilibrium include nonrandom mating, mutation, migration, and genetic drift. Assortative mating, or mating according to phenotype, may influence the proportion of heterozygotes in the equilibrium population, as well as the rate at which such an equilibrium is attained from nonequilibrium conditions. Mutation in one direction causes one allele to increase at the expense of another until the former triumphs and the latter disappears. Mutation in both directions results in a mutational equilibrium, a stable equilibrium determined by the mutation rates only. Gene flow by migration tends to shift the equilibrium toward the gene frequencies of the population from which the newly introduced genes are derived.

In small populations, sampling error ("chance") can cause unsystematic deviations from the Hardy-Weinberg equilibrium. This phenomenon is known as genetic drift. Sampling errors of this kind often lead to the loss or elimination of certain alleles and the fixation of others, thus reducing population variability. In addition to the continual action of genetic drift in small populations, there are also special kinds of genetic drift applicable to temporarily small populations (the "bottleneck effect") or to the founders of new populations (the "founder principle"). Genetic drift may also operate in large populations if a genotype is sufficiently rare that only a small number of individuals possess it.

PROBLEMS

Reminders: 1. Convert percentages into decimal fractions before multiplying them.

2. Observe zeros and decimal points. Remember that 0.3 times itself is 0.09 and that 0.09 times itself is 0.0081.

1. A population is founded by 40% AA and 60% aa individuals. Substitute these values for the symbols p and q, and verify the calculations of Boxes 8–A and 8–B.

2. If a population of fruit flies has a gene frequency of 20% for the recessive allele causing vestigial wings, what proportion of the population shows this trait phenotypically? What proportion of the population is heterozygous?

3. Assume that brown eyes are dominant to blue eyes (an oversimplification). On a certain island, about 9% of the population is blue-eyed. What proportion of the population is heterozygous?

4. The human blood types M and N are determined by codominant genes, so that the MN phenotype is distinguishable from both M (genotype MM) and N (genotype NN). A certain population has the following percentages of these blood groups: M, 15%; N, 24%; MN, 61%. Is this population in equilibrium?

5. The following represent initial population values for D, H, and R, in that order. For each set of values, determine the equilibrium values of p and q and the genotypic frequencies of AA, Aa, and aa at equilibrium.

a) (0.00, 1.00, 0.00)
b) (1/3, 1/3, 1/3)
c) (0.70, 0.30, 0.00)
d) (0.00, 0.70, 0.30)
e) (0.40, 0.30, 0.30)
f) (0.10, 0.30, 0.60)
g) (0.10, 0.70, 0.20)
h) (0.20, 0.20, 0.60)

6. The 25 founders of an isolated island population included one albino (aa). Assuming the remaining 24 founders were homozygous AA, what would be the frequency of the heterozygotes after three generations of random mating? If the population had meanwhile increased to 625 individuals, approximately how many would you expect to be homozygous for albinism? How much would the accidental death of one albino in such a population alter the gene frequencies?

7. In *Drosophila*, vestigial wings are recessive to normal ones. A large island population of *Drosophila* receives 5% of its gene pool in each generation from the adjoining mainland. Normal-winged individuals initially make up 84% of the island population and 75% of the mainland population. What proportion of the island population will have normal wings after three generations? What proportion will be heterozygous?

9

The Ecology of Populations

A. ECOLOGY AND THE NICHE CONCEPT

The word "ecology," first used by Ernst Haeckel (Chapter 5), has now come to have two separate meanings. To many citizens, "ecology" has become a rallying cry for their concern about the environment; ecology in this sense is a social or political movement, some aspects of which are discussed in Chapter 28. **Ecology** as a science is simply the study of the environment and the interactions of various species with their environment, including interactions with each other. It is this meaning of ecology that is emphasized in the present chapter.

One very important concept of ecology is the **niche.** A niche may be defined in one sense as everything that a species needs and does, the sum total of its interactions with its physical environment and with other species. A niche in this sense is the way of life of a species. Some people have tended to confuse the niche with the habitat. The habitat of a species is simply the place where it lives. A frequently used metaphor is that a species's habitat is its address, and its niche is its occupation.

Another definition of niche is given by the so-called hypervolume model. The extremes of temperature, humidity, salinity, and other environmental variables that a species can successfully withstand are measured. Any two such variables will define a space, or set of values for the environments, which that species can tolerate. Three such variables define a volume, and more than three variables define a multidimensional hypervolume. This hypervolume is then defined as the "fundamental niche"; it is the set of all values for the various environmental parameters within which a species is capable of living. A somewhat smaller hypervolume, called the "realized niche," consists of all values at which the species is actually found living in nature.

These meanings of "niche" assume the occupation of the niche by a species. If interpreted literally, the various definitions seem to rule out the possibility of a "vacant" or "unoccupied" niche, since a nonexistent species can have no temperature tolerance, no role in the ecosystem, and no interaction with other species. A "vacant niche" may, however, be defined as an opportunity for a species to occupy a role in the ecosystem. Species may come to occupy such vacant niches by modifying their ecological requirements (e.g., by subdividing or extending an existing niche), or by immigrating. When a species becomes extinct, the niche it occupied may persist in the sense that another species may later be able to invade that niche and successfully occupy it. There are many known instances in which introduced species have been very successful: the American house sparrow, introduced from England, is but one example, and Charles Darwin cited many more in his *Origin of Species*. The success of such an invasion bears testimony to the availability of a vacant niche before the invading species actually occupied it.

B. POPULATION GROWTH AND ITS CONTROL

New populations often arise from few individuals. In the extreme cases, new populations may arise from single individuals that reproduce asexually, from individual pregnant females, from individual hermaphrodites, or from individual pairs in species with separate sexes.

Exponential Growth

The simplest case of population growth is represented by the binary fission of bacteria in an unlimited environment. One bacterium divides to produce 2, then 4, then 8, 16, 32, 64, 128, and so forth. After t generations, the number of bacteria would be represented by 2^t. This assumes discrete, nonoverlapping generations; no account is taken of deaths.

In a case somewhat more applicable to higher organisms, let us assume that in some population of size N, a certain fraction, b, of births take place, and similarly a certain fraction, d, of the population die. We may represent this as

$$\text{births} = bN \quad \text{and} \quad \text{deaths} = dN.$$

If we ignore other causes of change in population size, then the change in N, written as ΔN ("delta N"), may be represented as

$$\Delta N = \text{births} - \text{deaths} = bN - dN$$
$$= (b - d)N. \tag{9.1}$$

If we use the symbol r to represent the quantity $(b - d)$, then the equation simplifies still further, to

$$\Delta N = rN. \tag{9.2}$$

In this equation, the population size, N, is always positive. The rate of change, ΔN, is therefore positive (and the population is increasing) whenever r is positive, which occurs whenever b exceeds d, or when births exceed deaths. On the other hand, ΔN is negative (and the population is decreasing) whenever deaths exceed births, for then d would exceed b, and r would be negative. The quantity r is called either the Malthusian parameter or the "intrinsic rate of natural increase."

Equation 9.2 assumes that the population size changes once each generation at rather discrete intervals. For a population changing continuously, we would use somewhat different notation and write

$$\frac{dN}{dt} = rN. \tag{9.3}$$

Both Equations 9.2 and 9.3 say the same thing, that the rate of change of N is equal to rN. Equation 9.3 can easily be written in the form of a differential equation (see Item 9.1 on page 170), and solved with the aid of very simple calculus. The solution of this equation is usually written as

$$N = N_0 \, e^{rt}. \tag{9.4}$$

This equation is known as the exponential growth equation.

A graph of N as a function of t shows that it increases without limit. This is clearly an unrealistic situation for any natural population, but it does approximate very well the early phases in the growth of real populations.

Logistic Growth

If we graphed $\log_e N$, instead of N, as a function of time, we would expect the equation for exponential growth to graph as a linear function of time, because

$$\log_e N = \log_e N_0 + rt.$$

But most real populations do not keep on growing indefinitely. If we were to graph $\log_e N$ as a function of time

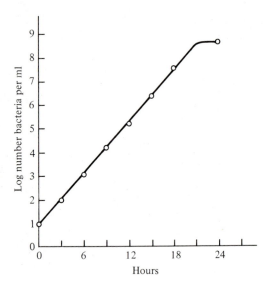

Fig. 9.1 Growth of a bacterial colony under limited resources. Reprinted with permission from Simpson, Roe, and Lewontin, *Quantitative Zoology,* 2e, © 1960. New York: Harcourt, Brace, Jovanovich, Inc. Copyright by Simpson, Roe, and Lewontin. Adapted from Atwood, Schneider, and Ryan.

for the growth of a real population, we would get a curve similar to that shown in Fig. 9.1. Here we see that the population grows until it reaches a definite limit, after which its growth ceases. This limit is known as the carrying capacity of the environment, and it is usually symbolized by K.

The value of K depends on many factors. In the case of bacteria growing in a Petri dish, space is an obvious limitation, and so is the limit imposed by the availability of nutrients. Another serious limitation lies in the production of toxic waste products by the bacteria them-

selves. In many animal species, aggression increases under crowded conditions, and this, too, results in a "negative feedback" that places a limit on population size in any finite area.

If we incorporate into Equation 9.3 a term for the effects of negative feedback when the population approaches its carrying capacity, we obtain

$$\frac{dN}{dt} = rN\left(\frac{K - N}{K}\right). \qquad (9.5)$$

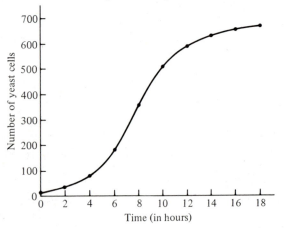

Fig. 9.2 The "logistic curve" of population growth illustrated by the growth of a laboratory colony of yeast cells. (After data from Raymond Pearl.)

This is the usual form of the so-called **logistic growth** equation, discovered independently by P. N. Verhulst and Raymond Pearl. According to Equation 9.5, a growing population should experience the following changes:

a. If N is well below the carrying capacity, then the term $(K - N)/K$ will be nearly equal to 1.0, and dN/dt will be just slightly less than rN. The population will thus be growing *almost* exponentially (cf. Equation 9.3).

b. As N approaches K, the fraction $(K - N)/K$ diminishes to zero. Population growth then slows down, for dN/dt likewise diminishes. At $N = K$, dN/dt will be exactly zero and no change should occur.

c. If N ever exceeded K, the fraction $(K - N)/K$ would be negative, and N would decrease until it again reached K.

A graph of this equation is shown in Fig. 9.2.

Density-Dependent and Density-Independent Controls

The size of a population is ultimately limited by the mortality of its members and by their capacity to reproduce. Any control is called **density-independent** if its effects do not vary with population size or density. The effects of weather are often density-independent. Selection by density-independent factors is sometimes called "hard selection." Lethal genes with 100% penetrance would be an example, since their effects do not vary according to the density, size, or composition of the population.

Density-dependent controls are those that vary with the population size—or more properly, with its density. Many animals lower their reproductive rate, b, in response to crowding. A variety of density-dependent factors also lead to an increase in the death rate, d. Death from the failure to obtain some limited resource is density-dependent: if only a certain amount of food is available, death by starvation will occur more frequently at higher population densities. The same is true of vitamins or mineral nutrients if their supply is limited, and it is also true of living space, foraging areas, nest sites, and the like. Aggression and the deaths resulting from aggression are also density-dependent, for fighting is more frequent under crowded conditions. In general, mortality resulting from competition with other individuals in the species is density-dependent. Selection by density-dependent controls is sometimes called "soft selection," since its intensity will vary according to the size, density, or composition of the population.

Mortality resulting from interaction with other species is also likely to be density-dependent. Predation and parasitism affect dense populations to a greater degree than sparse ones. Disease, too, is often density-dependent. In particular, epidemics are more frequent and more devastating at higher population densities.

A distinction may also be made between **directed mortality** and **random mortality**. Random mortality falls with equal probability on all individuals and does not contribute to evolution. Directed, or nonrandom, mortality is a form of natural selection (Chapter 11).

C. *r*-SELECTION AND *K*-SELECTION

Some populations, especially in stable, dependable environments, succeed by living near the carrying capacity of those environments. The population size is thus limited by the value of K, and selection tends to operate in such a way as to increase K or to favor those individuals most tolerant of population densities near the carrying capacity of the environment. This type of selection is called **K-selection**.

Other populations succeed through their ability to expand rapidly. This is especially important in ephemeral environments, such as the burned-out remnants of forest fires, or the debris left on a tropical island following a hurricane. Populations with rapid growth rates (high r) can exploit these opportunities as soon as they arise and before competing species are able to do so. Success in these species thus depends on a high value of r, and selection in these species tends to operate in such a way as to increase r. This type of selection is called **r-selection**.

Many species characteristics differ according to the type of selection that predominates (Box 9–A). K-selected species tend to achieve increased life spans and stable age distributions (constant proportions of individuals of each age), with parental care often evolving to ensure the survival of offspring. On the other hand, r-selected species tend to have short generation times and leave many offspring, often dispersing them promiscuously. Individuals tend not to survive beyond the time of

BOX 9–A   *Differences between r-Selection and K-Selection*

Correlate	*r*-selection	*K*-selection
Climate	Variable and/or unpredictable: uncertain	Fairly constant and/or predictable: more certain
Mortality	Often catastrophic, nondirected, density-independent	More directed, density-dependent
Survivorship curve	Often type III	Usually types I and II
Population size	Variable in time, nonequilibrium; usually well below carrying capacity of environment; unsaturated communities or portions thereof; ecological vacuums; recolonization each year	Fairly constant in time, equilibrium; at or near carrying capacity of the environment; saturated communities; no recolonization necessary
Intraspecific and interspecific competition	Variable, often lax	Usually strong
Relative abundance	Often does not fit MacArthur's broken-stick model (C. E. King, 1964)	Frequently fits the MacArthur broken-stick model (C. E. King, 1964)
Attributes favored by selection	1. Rapid development 2. High maximum *r* 3. Early reproduction 4. Small body size 5. Semelparity: reproduction only once	1. Slower development, greater competitive ability 2. Lower resource thresholds 3. Delayed reproduction 4. Larger body size 5. Iteroparity: repeated reproductions
Length of life	Short, usually less than 1 year	Longer, usually more than 1 year
Emphasis in energy utilization	Productivity	Efficiency
Colonizing ability	Large	Small
Social behavior	Weak, mostly schools, herds, aggregations	Frequently well developed

reproduction, when they would compete with their own offspring. The *r*-selected species include the many pathogens that cause epidemic outbreaks when they expand rapidly. Such species must necessarily have means of dispersal from one ephemeral environment to another. Other differences between *K*-selected and *r*-selected species are shown in Box 9–A.

D. DEMOGRAPHY

The study of birth, death, reproduction, and their effects on populations constitutes the science of demography. Neither reproduction nor survivorship (the avoidance of death) occurs uniformly—they both depend on age. Letting l_x represent the percentage of individuals that survive from age 0 to age *x,* we may graph a "survivorship curve'' (Fig. 9.3). By plotting $\log_e l$ as a function of age, we can distinguish three types of survivorship curves (Fig. 9.3): Type I, characteristic of many K-selected species, results from deaths occurring mostly at advanced ages. Type II survivorship results from a uniform level of mortality occurring at each age; the probability of surviving from one moment to the next is independent of the age. Type III survivorship is characteristic of all r-selected

species. It is characterized by high mortality at very young ages.

Reproduction is also highly age-dependent. The average number of offspring produced by each individual of age x, or the average number of daughters born to each female of age x, is usually designated m_x. A graph of m_x as a function of age would in most species show a maximum at an age shortly after the attainment of adult-

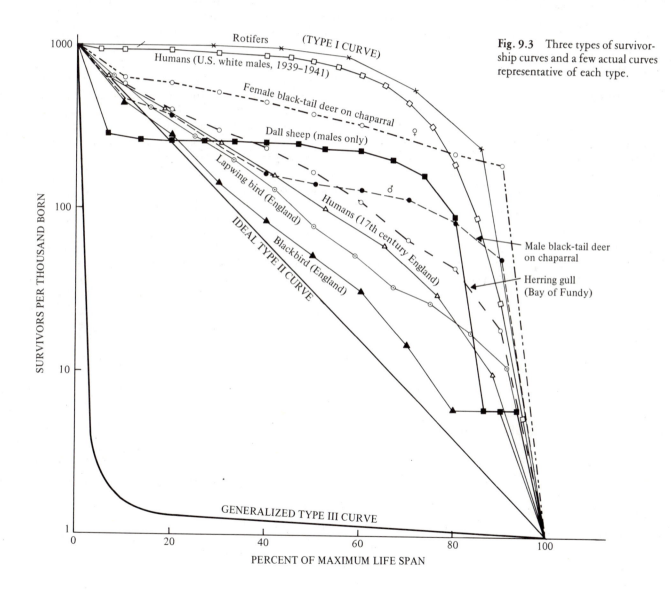

Fig. 9.3 Three types of survivorship curves and a few actual curves representative of each type.

hood. A related concept is the **reproductive value** of an individual or of a class of individuals of the same age. This quantity, v_x, measures the average number of offspring produced by each individual for the remainder of its lifetime, beginning at age x. The essential difference between v_x and m_x is that m_x measures only the number of offspring born to individuals while they are at age x, whereas v_x sums up the number of offspring that individuals of age x will bear over the rest of their lives.

If we multiply l_x by m_x, we may obtain the average number of offspring born to individuals of age x. Summing these values over all ages, we obtain

$$R_0 = \Sigma l_x m_x. \tag{9.6}$$

The quantity R_0, defined by this equation, is sometimes called the **net reproductive rate** of the population. If $R_0 > 1$, the population is growing; if $R_0 < 1$, the population is declining; if $R_0 = 1$, the population size is stable. If the population consists of a continuum of ages, rather than a series of discrete age intervals, R_0 is defined by

$$R_0 = \int_0^\infty l_x m_x \, dx. \tag{9.7}$$

Equations 9.6 and 9.7 are really alternative forms of the same definition.

The net reproductive rate, R_0, is related to the Malthusian parameter, r, and the generation time, T, by the following equation:

$$R_0 = e^{rT}. \tag{9.8}$$

This equation can also be solved for T, giving

$$T = \frac{\log_e R_0}{r}. \tag{9.9}$$

In a species with overlapping generations and reproduction not confined to any single age, the generation time may be thought of as an abstract quantity defined by Equation 9.9.

E. INTERACTIONS BETWEEN SPECIES

Thus far, we have discussed single populations in isolation. But no real population exists in isolation—all populations interact with other populations in ways that we shall now discuss.

The various types of interaction between species are summarized in Box 9–B. These interactions may be favorable (+), unfavorable (−), or neutral (0) to the species concerned. The exact meaning of these three terms is that a slight increase in the numbers of one species causes either an increase (+), a decrease (−), or no change (0) in the population size of another species in the same community. Some of these interactions are more interesting than the others: neutralism (0, 0) requires the least comment. We shall here emphasize competition, predation, and the various forms of symbiosis.

Competition

When two species interact in such a way that an increase in the population size of either causes a decrease in the population size of the other, they are said to be in **competition.** Usually this means that some important limited resource is used by both populations; this resource may be food, water, mineral nutrients, space, or particular sites, such as nest locations, foraging areas, or places of concealment.

The simplest model for competition was given by Lotka (1925) and Volterra (1926, 1931), and it is thus known as the Lotka-Volterra model. It assumes that the effect of each species on the other takes the form of an inhibition that is strictly proportional to the abundance of the other species. Thus, if we represent the two species by the use of the subscripts 1 and 2, we can simply modify the logistic growth equation (Eq. 9.5) by subtracting a term for the inhibiting effect of the other species:

$$\frac{dN_1}{dt} = r_1 N_1 \left(\frac{K_1 - N_1 - \alpha N_2}{K_1} \right), \tag{9.10}$$

and similarly

$$\frac{dN_2}{dt} = r_2 N_2 \left(\frac{K_2 - N_2 - \beta N_1}{K_2} \right). \tag{9.11}$$

These are the so-called Lotka-Volterra equations of population competition, in which α and β are called competition coefficients, and αN_2 and βN_1 represent the proportional effect that each species has on the population size of the other. Taking one species at a time and setting its rate of growth equal to zero, we find that a nonzero population size is stable (neither increasing nor decreasing) only when the term in parentheses is zero, or when

$$N_1 = K_1 - \alpha N_2 \tag{9.12}$$

BOX 9–B *Types of Interaction between Species*

A minus sign in the following schema indicates that the relationship inhibits population growth, meaning that the species in question attains a smaller population size in the presence of the other species than it does alone. A plus sign means that the relationship encourages population growth, and that the species in question reaches a larger population size in the presence of the other species than it does alone. A zero indicates that the species in question is not affected by the presence or absence of the other species. In terms of the Lotka-Volterra equations (Equations 9.10 and 9.11), a minus sign means the subtraction of a term for the interaction, a plus sign indicates an addition of a term, and a zero indicates that no interaction term is either added or subtracted.

Effect on		Interaction
SPECIES 1	SPECIES 2	
−	−	*Competition.* Each species inhibits the other.
−	0	*Amensalism.* The relationship inhibits one species but has no effect on the other.
0	0	*Neutralism.* The relationship has no net effect on either species.
+	−	*Predation;* also *parasitism.* The interaction benefits the predator (or parasite) at the expense of its prey (or host).
+	0	*Commensalism.* One species is benefited, but the host species is unaffected.
+	+	*Mutualism.* Both species are benefited by the interaction.

in the case of species 1, or

$$N_2 = K_2 - \beta N_1 \qquad (9.13)$$

in the case of species 2.

Figure 9.4 shows both Equations 9.12 (black) and 9.13 (gray) on the same set of axes. Any point on this graph represents a pair of values (N_1, N_2) for the population sizes of the two species. At any point along the black line, species 1 is in equilibrium and would tend to remain stable if not for the fact that species 2 might change. To the left of this line, species 1 is below its equilibrium value and tends to increase in numbers, and the point representing population sizes thus moves to the right. To the right of this line, species 1 is above its equilibrium value and tends to decrease, causing the point to move to the left. Species 2 is in equilibrium only for points along the gray line; below this line, species 2 increases and the point moves upward; above this line, species 2 decreases and the point moves down. Combining these two effects simultaneously, we find that, to the right of the dotted line, population sizes tend to move toward the black line, and then they shift downward and

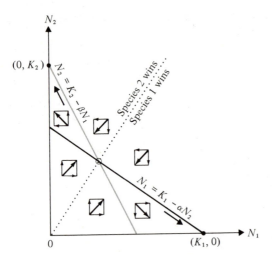

Fig. 9.4 A model of competition between two species, following Lotka and Volterra. Species 1 is in equilibrium at any point along the black line. Species 2 is in equilibrium at any point along the gray line. Both species are at equilibrium where the two lines intersect (open circle), but this equilibrium is unstable. Diagonal arrows are vector sums, which show general direction in which population values will change.

to the right along that line until they reach the point $(K_1, 0)$. At this point, species 1 has reached its carrying capacity, and species 2 has become extinct—it has lost the competition, and species 1 has won. To the left of the dotted line, we find exactly the reverse situation: population sizes tend to move toward the gray line, then shift to the left and upward along that line until they reach the point $(0, K_2)$, in which case species 2 wins and species 1 loses. The only other point where both species are in equilibrium is the point where the two lines cross. But this equilibrium is unstable, because any small, accidental shift of the population values from this point would result in an irrevocable expansion of one species and extinction of the other, depending on whether the shift was initially to the right or to the left of the dotted line. At any other point along the dotted line, the population values will tend to move toward the equilibrium

point, but it is more likely for a small, accidental shift away from this dotted line to occur, in which case one species or the other inevitably wins out.

What we have just discussed, however, is only one of four possible outcomes. All four are shown in Fig. 9.5. In (a) the black line representing species 1 is always above the gray line representing species 2. Population values will tend to move toward the point $(K_1, 0)$, and species 1 always wins. In (b) just the reverse is true: the gray line representing species 2 is always higher, and population values will tend to move toward the point $(0, K_2)$, so species 2 always wins. Case (c) is the one we have already discussed, in which either species may win, depending on their initial numbers. Case (d) is the only one that leads to a stable equilibrium, which is at the point where the two lines cross. Any slight shift from this equilibrium point will result in the restoration of equilibrium.

Controlled experiments on systems where two species are in competition tend to follow the courses predicted by the Lotka-Volterra equations and the accompanying diagrams (Fig. 9.6). For example, two species of flour beetles, *Triboleum castaneum* and *T. confusum,* can be put together in a container with a supply of flour. At high temperature and humidity, *T. castaneum* always wins, but at low temperature and humidity, *T. confusum* is always victorious. Under most temperature-humidity regimes between these extremes, either species may win (Fig. 9.5c), depending on their initial numbers. Long-term coexistence of the two species does not occur in pure flour, but if some environmental heterogeneity is introduced in the form of small glass rods mixed in with the flour, then the two species will coexist indefinitely (Fig. 9.5d).

The Lotka-Volterra model is far from the only possible model of competition. The effects of each species on the other need not be linear. The model also assumes, for simplicity, that the competition coefficients α and β are constants, though in real situations these coefficients may well vary as a function of N_1 and N_2. The model also assumes that each species has an *immediate* effect on the other, though in real situations there would undoubtedly be a time lag for the response of one species to a change in the other.

Amensalism $(-, 0)$ can also be treated under the Lotka-Volterra equations by setting the value of β equal to zero; this situation is shown in Fig. 9.7.

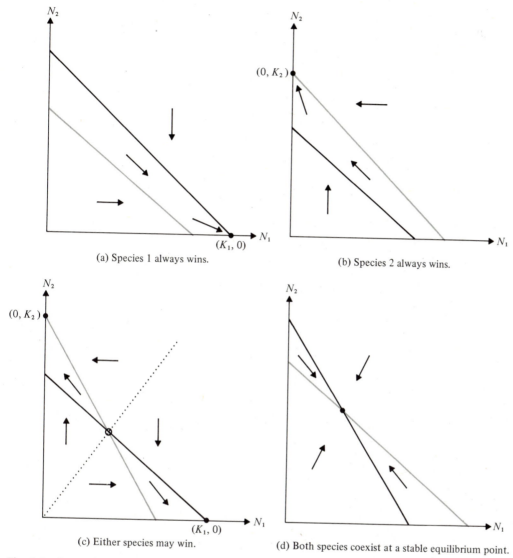

(a) Species 1 always wins.

(b) Species 2 always wins.

(c) Either species may win.

(d) Both species coexist at a stable equilibrium point.

Fig. 9.5 Competition under four sets of conditions. Points of stable equilibrium are indicated by solid circles, unstable equilibria by open circles. See text for further explanation.

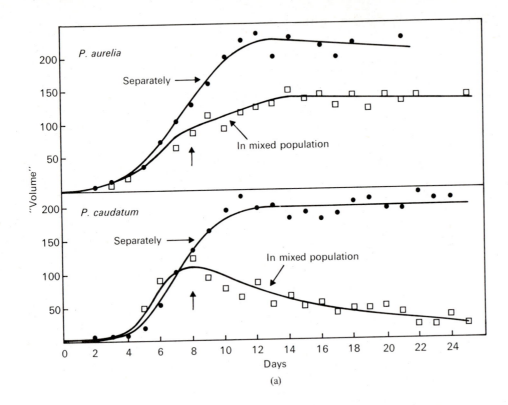

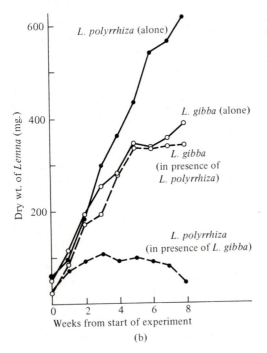

Fig. 9.6 Competition experiments. (a) Competition between two species of *Paramecium* (phylum Ciliata). (b) Competition between two species of the duckweed *Lemna*. (a) is reprinted with permission from G. F. Gause, *The Struggle for Existence,* © 1934. Baltimore: Williams and Wilkins. (b) is reprinted from Harper, 1961, courtesy of the Society for Experimental Biology.

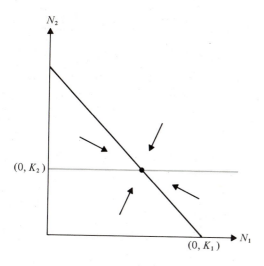

Fig. 9.7 Amensalism, according to a modified version of the model used for competition.

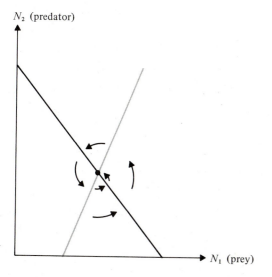

Fig. 9.8 A simple predation model, based on a modified version of the Lotka-Volterra model of competition.

Predation

Predation $(+, -)$ seems a more difficult phenomenon than competition to describe in terms of a mathematical model. By extending the foregoing argument from competition to amensalism and beyond, or by setting β less than zero, one obtains the model of predation shown in Fig. 9.8. This simple model enables us to see that a certain minimum number of prey are needed for the existence of a predator population, and that above this minimum an increase in the prey population would tend to support an increased number of predators. The influence of predators on the prey population is assumed to be the same as in competition. There is a stable equilibrium point in this case, but the population values may tend to spiral or circle around this equilibrium value. Other models of predation are discussed by Wilson and Bossert (1971) and by Pianka (1974); see also Box 9–C.

Rosenzweig and MacArthur (1963) have provided another model of predation. This is shown graphically in Box 9–C. In these diagrams, the prey is assumed to exist at its carrying capacity (K_1) in the absence of predators. A minimum size is also assumed for the prey population below which individuals would be unable to find mates, or below which the effects of genetic drift (Chapter 8) would soon lead to the fixation of too many alleles for the population to remain viable. In this model, there is one value between these limits where the prey population supports as many predators as it can.

The predator population in the Rosenzweig-MacArthur model requires a certain level of prey for its existence. Above this level, the predator population increases rapidly to its carrying capacity, K_2.

Graphs of the Rosenzweig-MacArthur model all contain equilibrium points. But depending on where the two curves intersect, these models predict different outcomes: an overefficient predator (Box 9–C) would kill more prey than could quickly recover, and the population values would tend to spiral outward from the equilibruim point until all the prey were killed, leading to the extinction of both species. Indeed, most attempts to raise predators and their prey under laboratory conditions have met with this fate (compare Fig. 9.9). A more prudent predator would avoid killing too many prey, and the equilibrium would be maintained. In some cases, the population values would spiral inward toward the equilibrium point, producing a damped cycle, or the values

BOX 9–C *Population Modeling of Predator-Prey Interactions*

The simplest model of predator-prey interaction, originally proposed by Lotka and Volterra, assumes that the frequency of predation is strictly proportional to both the frequency of predators and the frequency of prey. Under this model, both predator and prey populations should undergo endless oscillatory cycles that remain exactly the same on each repetition (A):

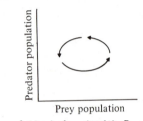

(A)

Rosenzweig and MacArthur (1963), Rosenzweig (1969), May (1972), Emlen (1973), and Roughgarden (1979) discuss other, more sophisticated models. Many of these models assume a prey curve with a "hump"—approximately parabolic in shape. Below this curve, the prey population will grow, and the point representing the population on a graph will move to the right (B). Above the curve, the prey population will diminish, and the point will move to the left. On the curve itself, known as an *isocline,* the prey population is in equilibrium. The left (upward-sloping) portion of the prey curve represents the conditions in which the prey is limited principally by its capacity to reproduce. For a slightly larger prey population, a slightly higher number of predators would be needed to keep the prey population in check (hence the isocline slopes upward). The right (downward-sloping) portion of the prey curve represents conditions in which the prey are limited by mortality. At the lower right end of this curve, the prey reaches its carrying capacity (K) in the absence of predators. An increase in the number of predators diminishes the carrying capacity of the prey population below this value, so the curve has a negative slope.

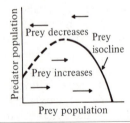

(B)

The diagrams are reprinted with permission from J. Merritt Emlen, *Ecology: An Evolutionary Approach*, © 1973. Reading, Massachusetts: Addison-Wesley (Figs. 11.5 through 11.7).

There is less agreement as to the shape of the predator isocline. At least part of the curve should be upward-sloping (C), because a larger population of prey can in general support a larger number of predators. Under the models we are considering, the exact shape of the predator curve has little influence on the outcome.

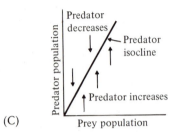

(C)

The outcome of a predator-prey interaction is governed largely by the point where the isoclines meet. If the curves intersect along the right (downward-sloping) limb of the prey isocline, a stable equilibrium is usually reached, as indicated by the dotted arrow in these diagrams (E, F, and G). In certain cases, the population sizes will oscillate around this point as they spiral inward. If, on the other hand, the curves intersect along the left (upward-sloping) limb of the prey isocline, then the equilibrium is unstable (D). The population cycles will "explode" (spiral outward) until one or both species become extinct. Often the prey species becomes extinct first, and the demise of the predator (from starvation) soon follows.

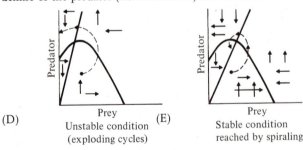

(D) Unstable condition (exploding cycles)

(E) Stable condition reached by spiraling

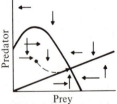

(F) Stable condition reached without spiraling

BOX 9–C *(continued)*

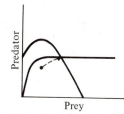

(G)

Predator isocline with satiation limit. Above this limit (horizontal part of predator isocline), the predator population is controlled by factors other than the abundance of prey.

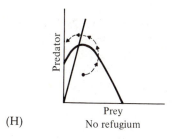

(H)   No refugium

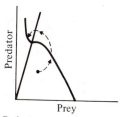

(I)   Refugium available to prey

Laboratory experiments have confirmed that it is difficult to maintain predator and prey populations without both becoming extinct (as in H). But the continuous survival of the prey species might be ensured by a refugium, such as by migration from another locality free of predation (I). If there are enough places for the prey to hide, or if a certain number of new prey are guaranteed each generation as the result of immigration, then a stable equilibrium (or stable cycle, with small oscillations) may result.

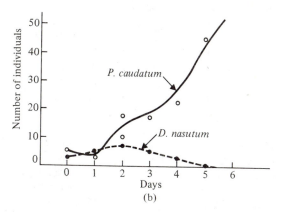

(a)

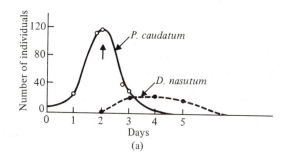

(b)

Fig. 9.9 Laboratory experiments of predation on the protozoan *Didinium nasutum* (phylum Sarcodina), by another protozoan *Paramecium caudatum* (phylum Ciliata). (a) In homogeneous culture. (b) In heterogeneous culture. (c) In homogeneous culture, with one *Paramecium* and one *Didinium* added every three days to simulate "immigration" from another locality. Note that the populations began to cycle in this third case. Reprinted with permission from G. F. Gause, *The Struggle for Existence*, © 1934. Baltimore: Williams and Wilkins.

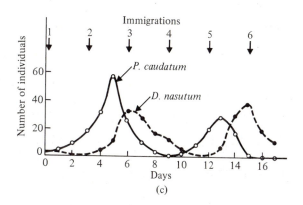

(c)

would circle around the equilibrium point indefinitely, and oscillations around the equilibrium value would persist indefinitely. This explanation has been given in many classic cases of competition, such as that between lynx and varying hare (Fig. 9.10), though population cycles of varying hare have been reported on at least one island devoid of lynx (Keith, 1963). One experiment in which cycles were produced in laboratory populations is shown in Fig. 9.9(c).

The concept of **prudent predation** has recently received much attention. A prudent predator would safeguard the existence of its prey species, while simultaneously maximizing the number of prey (''yield'') that could be eaten without threatening the prey species with extinction. Human populations practicing agriculture or animal husbandry may strive to become prudent predators, but the existence of other prudent predators in nature is still an open question. In particular, a prudent predator species would have to avoid the dangers of overpredation despite the fact that any act of predation works to the selective advantage of the predatory individual. One possible escape from this dilemma is to imagine a nonspecific predator and two or more prey species: when any one prey species falls to low population levels, the predator will naturally tend to consume more of the other, more abundant species, meanwhile giving the threatened prey species a chance to recover. Prudent predators would also be expected to hunt primarily those

prey individuals with lowest reproductive values (v_x), meaning especially the very old and the very young. This type of predation occurs widely, though it probably reflects ease in hunting more than it does conscious predator prudence.

F. SYMBIOSIS

Species interactions in which the two species actually live together are classified as **symbiosis**. The species that live together in this manner are called **symbionts**; in an unequal case of symbiosis, the species that unwittingly provides food and shelter to the other is called the **host**.

Symbiosis must always benefit at least one of the species, for otherwise it would soon dissolve. Symbiosis that is detrimental to the host species is termed **parasitism**. Symbiosis that neither benefits nor harms the host is termed **commensalism**. Symbiosis beneficial to both partners is called **mutualism**, although some authors would restrict this term to the case where the symbiosis is obligatory, meaning that neither species could survive without the other. Authors preferring this more restrictive usage use the term **protocooperation** for those symbioses that benefit both species but are obligatory for neither.

Parasitism

Parasitism is similar to predation in that one species benefits from the association while the other is harmed. Models for predation can also be applied to parasitism. The distinction between the two is rather arbitrary, but it depends on a number of factors: a predator is often larger than its prey, whereas a true parasite is typically many times smaller than its host. A typical predator kills many prey, but a single host individual may harbor many parasites, often thousands. Perhaps a more basic distinction is that predators do not live within or on the bodies of their prey, for death of the prey (or its escape) soon follows the establishment of contact. Parasites, on the other hand, do live within or on the bodies of their hosts. Thus, herbivores, which have been considered plant predators by some ecologists, might better be described as plant parasites.

The distinction between parasites and predators is further strained by the existence of some species of birds and insects that are called **parasitoids**. These are usually within the same general size range as their hosts or prey, though they may sometimes be smaller. The numbers of

Fig. 9.10 Population cycles in the Canada lynx (*Lynx canadensis*, order Carnivora) and varying or snowshore hare (*Lepus americana*, order Lagomorpha), based on the number of pelts received each year by the Hudson Bay Company. (From Solbrig and Solbrig, after MacLulich.)

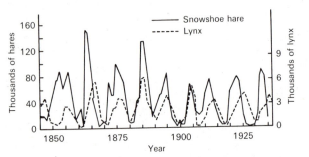

parasitoids relative to their prey depend on the species: cuckoos, which lay their eggs in the nests of their hosts, are in a given nest outnumbered by host individuals. Parasitoid wasps, on the other hand, typically lay many eggs—or one egg that develops into several organisms—within the body of a single victim. Perhaps parasitoids may best be thought of as predators that do not immediately eat their victims but use them for some other purpose, such as taking over their nests or providing food and/or shelter for the predator's offspring.

True parasites, as distinct from predators (including parasitoids), are absolutely dependent on their hosts for food, for shelter, and for the maintenance of a relatively stable environment as long as the host remains alive. The parasite usually depends on the host to provide enough food for both symbionts, and it also frequently depends on the host for water, for protection, and for waste elimination, as well as for the maintenance of relatively even

temperature regimes. The parasite must have its own means to hold on and resist being dislodged.

But the most serious problem by far facing any true parasite (as distinct from a parasitoid or predator) results from the perishability of its host as an individual. All parasites must have a means of dispersal from one host individual to another, and most parasites have thus evolved precautions to ensure against the immediate death of their host, for then the parasite would itself perish. The means evolved by parasites that ensure their dispersal from one host individual to the next are often quite elaborate, and they may involve one or more intermediate hosts for various immature stages of the parasite's life cycle (Figs. 9.11 and 9.12).

Parasites have also evolved numerous means, not all of them clearly understood, by which they avoid causing the immediate death of the host. Parasites have typically evolved in directions that permit them to take what they

Fig. 9.11 Life cycle of the flatworm *Diplostomum baeri* (class Trematoda), an intestinal parasite of ducks. Taken from O. Wilford Olsen, *Animal Parasites*, 1962, Burgess Publishing Company, Minneapolis, Minnesota. Reprinted by permission.

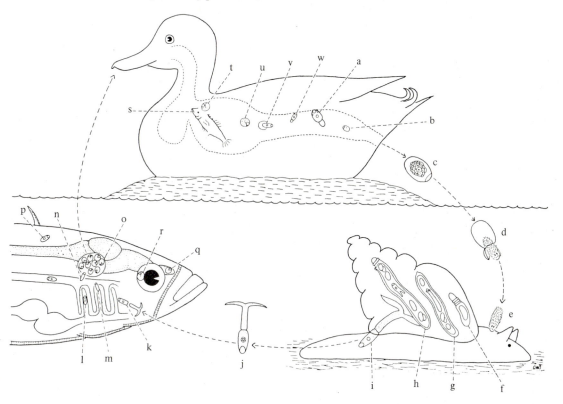

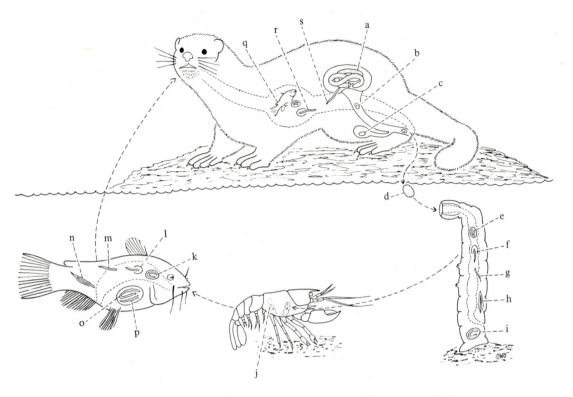

Fig. 9.12 Life cycle of the roundworm *Dicotophyma renale* (class Nematoda), an intestinal parasite of minks and other carnivores, including domestic dogs. Taken from O. Wilford Olsen, *Animal Parasites,* 1962, Burgess Publishing Company, Minneapolis, Minnesota. Reprinted by permission.

need from the host, thereby causing definite harm, but to do so with a minimum of trauma to the host individual. Parasites have thus tended to evolve smaller body sizes, as well as means of entering the host's tissues as painlessly and unobtrusively as possible and causing as little damage as necessary. And in addition to all of these adaptations, parasites tend to be *r*-strategists with short generation times and type III survivorship curves. Like other *r*-strategists, they are adapted for rapid colonization of suitable environments (new host individuals), for rapid expansion, and for rapid dispersal to a new host. Their "hit and run" strategy is yet another way of coping with the problem of the perishability of their hosts.

Commensalism

The ultimate strategy for any parasite is to evolve into a condition that does not harm the host at all, or that returns to the host an equal benefit for whatever harm it inflicts. This condition is called commensalism, and the former parasite is termed a *commensal*. Many if not most commensals are believed to have evolved in this way.

In some cases, it is hard to determine whether a certain species is actually a commensal or not. The bacterium *Escherichia coli,* which inhabits the human intestinal tract, has often been considered a commensal species. It is widespread throughout human populations—all of us have them in our guts. This fact alone has led bacteriologists to argue that it is not a parasite; yet it does occasionally cause disease and even (rarely) death. Recently it was discovered that this bacterium aids in the synthesis of several vitamins that humans might otherwise lack. In these various ways, *Escherichia coli* may confer either benefit or harm on its human host, and it is properly called a commensal only if we are willing to say that the good and the bad balance each other out.

Mutualism

The most highly evolved form of symbiosis is mutualism (Figs. 9.13 and 9.14), although, as noted before, some authors would distinguish between voluntary and obligative interactions and often restrict the term "mutualism" to the latter only.

Mutualism may often evolve from parasitism and commensalism as their ultimate form, in which the former parasite returns to its host a benefit greater than the harm it inflicts. In the long run, this may be the safest strategy of all (even if seldom attained), for the host will then itself tend to evolve characters that make the symbiotic relationship more certain, more steadfast, and more permanent. The famous mutual relationship between the yucca plant and the yucca moth probably evolved in this manner, as did many other cases of mutualism involving insects and plants (e.g., pollination). In this case, the moth lives only on the plant, within which it lays its eggs, and its larvae eat no other species of plant. In return for this use of the plant for the breeding and feeding of its larvae, the moth pollinates the plants, which depend on the moth for this purpose: they can be pollinated in no other way. This obligative form of mutualism presumably evolved first as a case of parasitism by the moth on the plant. The evolution of the moth as a pollinator of the yucca plants was directly beneficial to the moth's own offspring and was therefore selected for among the moths; its long-range benefit to the plant was at first wholly incidental. Another case, studied by Janzen (1966), undoubtedly originated as a case of simple predation on *Acacia* trees by tropical ants. These ants proved quite effective in deterring a wide variety of other herbivores from attacking the acacias, and they actually benefited the acacias more than the equivalent of the harm they had caused by their own feeding. The acacias

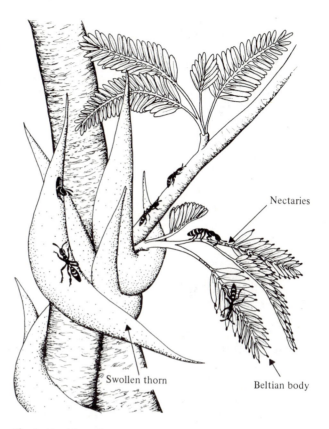

Fig. 9.13 Mutualism between *Acacia cornigera* (order Rosales, family Leguminosae) and ants of the genus *Pseudomyrmex*. The ants live inside specialized "swollen thorns" provided by the acacia, feeding on specially provided nectaries and especially on the protein-rich "Beltian bodies" at the tips of leaflets. In return, the ants will vigorously attack any other herbivore that might feed more destructively.

responded by providing special "swollen thorns" as shelter for the ants, permitting the ants to eat other herbivores under the tree's protection (Fig. 9.13).

There are some cases of mutualism that may not have evolved in this manner. The lichen symbiosis—between an alga and a fungus in each case—is certainly one

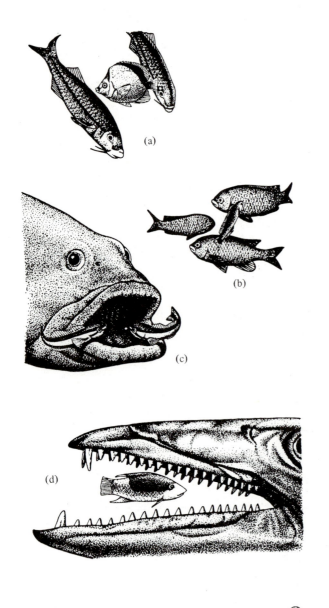

Fig. 9.14   Examples of cleaning symbiosis. (a) A butterfly fish, *Chaetodon nigrirostris,* cleaning two Mexican goatfish, *Pseudopeneus dentatus.* (b) A senorita, *Oxyjulis californica,* cleaning a group of blacksmiths, *Chromis punctipinnis.* (c) Two neon gobies, *Elecatinus oceanops,* cleaning inside the mouth of a grouper, *Epinephelus* sp. (d) A Spanish hogfish, *Bodianus rufus,* cleaning inside the mouth of a barracuda, *Sphyraena barracuda.* (e) The crocodile, *Crocodylus niloticus,* and the Nile plover. (a) through (d) are from *Symbiosis: Organisms Living Together,* by Thomas C. Cheng, © 1970 by the Bobbs-Merrill Co., Inc., Indianapolis, Indiana. (e) is reprinted with permission from John W. Kimball, *Biology,* 4e, © 1978. Reading, Massachusetts: Addison-Wesley (Fig. 43.14).

of the more remarkable examples of mutualism. The relationship is often obligatory and is quite intimate, with the fungal hyphae actually penetrating the individual algal cells. The fungus derives photosynthetically produced nutrition from the alga but does not use so much that it would deprive the alga of its own needed nutrition. The alga, in turn, derives moisture from the fungus, and in this way it is able to subsist in some of the harshest environments on Earth. The first organisms to colonize bare rock surfaces are usually lichens; their use of such environments removes them from competition with other plant species. Lichens have even evolved an armada of chemicals found nowhere else; both alga and fungus must be present together for these chemicals to be produced. Some of them, for example, etch the rock surface to allow the lichen both a holdfast and a source of mineral nutrients. Under favorable conditions, it may be possible to separate the alga from its fungus and culture them separately. In such cases, it is most difficult to persuade the alga and fungus to rejoin and form a lichen once again—unless the environment is made deliberately harsh. Apparently the lichen form of association evolved repeatedly between many different algae and fungi as a response to adverse conditions. The mutual relationship has proceeded so far that fungal spores and algal filaments (trichomes) are often dispersed together as propagules that will establish and perpetuate the symbiosis at a new locality.

G. RESULTS OF INTERACTION

The results of species interaction may be quite diverse. In general, a species benefited by any interaction will evolve so as to intensify the interaction and make it more certain, more permanent, and more obligatory; on the other hand, a species harmed by interaction will evolve so as to avoid the interaction or at least minimize its effects.

Niche Diversification and Specialization

One fundamental principle of ecology has been called the **principle of competitive exclusion,** also known as Gause's principle. This principle states that two species identical in ecological requirements cannot long coexist; in other words, no two species can permanently occupy the very same niche. While Pianka (1974) and others have attacked this "principle" as a mere tautology (true by virtue of its use of terms), it does serve to remind us of several important ecological facts. First, competition between species is likely to be more intense the more similar their ecological requirements (or "niches") are. This may be even more true *within* the species, where individuals having similar ecological requirements suffer the most intense competition. Second, species occupying similar or identical niches will therefore tend to evolve in such a way that their competition is minimized: if neither species succeeds in causing the extinction of the other, they will tend to subdivide the niche, or else one species will succeed in utilizing a different niche altogether. The subdivision of the niche might mean that each species will select a slightly different habitat, a slightly different food, or a slightly different temperature tolerance, or perhaps they will subdivide the niche in some other way. Thus, each species becomes somewhat more of a specialist.

Specialization has its price, however, in the inflexibility of the species in response to adversity or change. A species adapted to a narrow temperature range will suffer far more if the temperature does change than will a more broadly tolerant species. A predator that specializes on a single prey is quite at the mercy of fluctuations in the population size of the prey species. On the other hand, a nonspecific predator may simply switch to a different food source if its usual prey species fails. This has the beneficial effect of allowing the prey species to recover and regain its former numbers, to the ultimate benefit of both predator and prey. Perhaps it is for this reason that most animal predators are generalists to some extent, exploiting a variety of prey species as the availability of prey species permits.

Coevolution

Coevolution is evolution in which one species closely follows, or tracks, the evolution of another. One form of coevolution is that in which a predator tracks its prey's adaptations by evolving means to overcome or circumvent their defensive value. Another form of coevolution is that in which a parasite evolves greater virulence while its host evolves greater resistance. Or a symbiotic relationship may evolve from commensalism to mutualism, and beyond this it may evolve toward an increasing degree of dependence of one species on the other. Once mutualism evolves, both species will tend to evolve adaptations that will benefit the other species and perpetuate if not cement the symbiosis; the swollen thorns and Beltian bodies produced by acacia trees (Fig. 9.13) are examples of this. In such cases, the further evolution of the two species may often proceed together. In particular, speciation of either symbiont may often be accompanied by or followed shortly by speciation of the other.

CHAPTER SUMMARY

Ecology is the study of the interactions between species, and between species and their environment. For a given species, the sum total of its interactions defines its niche.

The simplest model of population growth, that of exponential growth, is unrealistic because every species is limited by the carrying capacity of its environment, symbolized by K. The so-called logistic growth model follows the equation

$$\frac{dN}{dt} = rN\left(\frac{K - N}{K}\right).$$

The controls that operate when the population size, N, approaches the carrying capacity, K, are called density-dependent controls. Populations living in stable, dependable environments often live near the carrying capacity of those environments and are limited by K; they are K-selected. Populations living in ephemeral environments must depend on rapid maturation, rapid and prolific reproduction, and widespread dispersal to continually recolonize patches of suitable environment. These species are limited by their intrinsic rate of natural increase, symbolized by r, and are said to be r-selected.

When two species interact in such a way that each inhibits the other, they are said to be in competition. A frequent outcome of competition is often subdivision of the niche and consequent specialization. The Lotka-Volterra model of competition predicts four possible situations: one in which the first species alone triumphs, one in which the other species triumphs, a third in which either species may triumph, and a fourth in which a stable equilibrium is reached and both species coexist indefinitely. Similar models for the interaction of predators and their prey are not as satisfactory in predicting equilibrium points, overpredation, or the occurrence of population cycles similar to those often observed in nature.

Interaction between two species living together, called symbiosis, is necessarily beneficial to at least one partner. If it is beneficial to both partners, it is called mutualism; if neutral to the host species, it is called commensalism; if harmful to the host species, it is called parasitism. Parasites may coevolve with their hosts, becoming more virulent while the host becomes more resistant. Mutualism, once evolved, tends to become more permanent in those species that become increasingly dependent on each other.

FOR FURTHER KNOWLEDGE

ITEM 9.1 *Derivation of the Formula for Exponential Growth*

To obtain the solution for the equation

$$\frac{dN}{dt} = rN,$$

we multiply both sides by dt to express it in differential form,

$$dN = rN\,dt,$$

then divide both sides by N and integrate:

$$(1/N)\,dN = r\,dt$$
$$\int(1/N)\,dN = \int r\,dt$$
$$\log_e N = rt + \text{constant}.$$

Now, raising e to the power represented by each side, we obtain

$$N = e^{(rt + \text{constant})}$$
$$= e^{rt} \times e^{\text{constant}}.$$

If we represent the quantity e^{constant} by the symbol N_0, we obtain

$$N = N_0 e^{rt},$$

where t = time,
  N = population size (as a function of time), and
  N_0 = initial population size (at time $t = 0$).

10

Adaptation

A. THE NATURE OF ADAPTATION

An **adaptation** may be defined as any characteristic of an organism that makes that organism better suited to its environment. The process of becoming better suited to one's environment may also be termed adaptation. Adaptations may have their basis in structure (**morphological adaptations**), in internal physiological processes (**physiological adaptions**, including biochemical adaptations as an important subset), or in overt behavior. The first two types are considered in this chapter; behavioral adaptations are treated separately in Chapter 24.

Even primitive humans have long realized that organisms possess adaptations by which they are better suited for the exigencies under which they live. Folk mythologies abound with legends attempting to explain why birds have wings or why predators have sharp teeth or beaks, as well as sharp claws or talons. Most of these legends explain such adaptations as the gifts of benevolent deities or supernatural forces, often as a reward for some good deed in the context of a lesson in morals. Some legends even explain the leglessness of snakes, for example, as some sort of divine punishment—again in the context of a moral lesson.

In the eighteenth century and especially in the nineteenth century, interest in adaptation grew almost to an obsession, and numerous adaptations were described in minute detail so that they might more readily be marveled at. The explanation of adaptation became a major problem for nineteenth-century biologists, and Darwin's explanation by means of natural selection (Chapter 11) is testimony to his great genius. No evolutionary theory can be considered successful unless it contains an explanation for adaptation.

The currently accepted explanation of biological adaptations is the Darwinian theory of natural selection. Given an inherent tendency of biological populations to exhibit diversity (Chapters 7 and 12), any characteristic that enables some individuals to leave more offspring than others will become increasingly frequent in succeeding generations. The appearance of such adaptive features by chance, followed by their spread throughout the population by means of natural selection, forms the Darwinian explanation of adaptation, one of the cornerstones of modern evolutionary biology. Unlike some other explanations, this one mentions no foresight, no "purpose," no "goal," nor any inherent tendency toward "progress," all issues to be addressed in Chapter 18. Darwinian evolution proceeds by trial and error, a process of continual groping.

Williams (1966) has argued that the study of adaptations constitutes an important but neglected field of evolutionary studies, for which he proposed the name teleonomy, or the study of function. The rapid proliferation of functional anatomy since Williams wrote bears testimony to the fruitfulness of the functional analysis of adaptations.

B. MORPHOLOGICAL ADAPTATIONS

The most obvious adaptations are those based on structure. They are called morphological adaptations, and their study constitutes the field of functional anatomy or functional morphology, which includes biomechanics.

Many structures are so well suited to the functions and roles they serve that they tempt us to explain them as having been purposefully designed. Even a cursory familiarity with nature teaches us that air-breathing vertebrates have lungs, that animals that paddle through the water often have webbed feet, that anteaters and other animals of similar diet are provided with long, sticky tongues (great for snatching up social insects), and that animals living in cold climates tend to have thick, dense, furry coats. Perhaps more remarkable is the fact that tadpoles acquire lungs during their transformation into air-breathing frogs or salamanders, or that young marsupials are born with precisely those muscles already developed that they need early in life, namely, forelimb muscles with which to drag themselves into the pouch, and forelimb and oral muscles with which to attach themselves to a nipple and begin sucking milk.

In many cases, the structure of organisms reflect good mechanical engineering principles. A columnar support, for example, has maximum strength with a minimum amount of material if it is shaped like a pipe: a hollow cylinder (or better still, a series of such cylinders, arranged pipe-within-pipe). Is it not amazing that the

supporting skeleton of vertebrate limbs has exactly this structure, with the bones shaped like hollow cylinders? Their microstructure follows the same principle, for it is built of a series of cylinders within cylinders, arranged concentrically around a central blood vessel. Nor is this structural adaptation confined to vertebrates, for insects and other arthropods have legs built of hollow cylinders, too. Numerous other examples, similar to this, may be found in the books of D'Arcy Thompson (1942) and Hildebrand (1974), among others.

D'Arcy Thompson and others have repeatedly emphasized the effects of size on many structural adaptations. If an organism increases in size without any change in its shape or structure, it will face a number of problems. The surface areas of any of its absorptive surfaces, the cross-sectional area of its supporting members, or the surface area of its cerebral cortex will increase in proportion to the *square* of linear dimensions; if the animal doubles in length, all areas will increase fourfold. But its volume and therefore its weight will increase eightfold, because volume varies in proportion to the cube of linear dimensions. Thus, if an animal were simply magnified in size, its ability to support its own weight would decrease, its absorptive surfaces (such as those of its intestinal lining) would not be able to supply as much food energy to its enlarged mass, and the ability of its cerebral cortex to control its muscular movements would likewise be impaired. Is it not amazing that animals, though unaware of these structural principles, compensate for them? Animals of increasing size, for example, have limbs that are disproportionately thicker in proportion to the weight they must bear. Their intestines have kept pace with their increasing mass and have become either corrugated or convoluted or both, developing minute absorptive villi on their surfaces. These structural principles place certain limits on organisms that they may not exceed. These principles also determine that certain selective forces must operate when these limits are approached. For instance, given bone as a structural material, there is a physical limit to the size of a terrestrial organism if its supporting limbs are not to be crushed under the weight they must bear—a fact realized by Galileo. Near this limit, we see adaptations for making the most of bone tissue, for example, by disproportionate thickening, or by drawing the bones directly beneath the body and straightening the several joints, as in elephants.

We see other examples of adaptation in the various

modifications of mammalian forefeet in accordance with the purposes they serve. The human hand, used for grasping; the hand of a mole, used for digging; the wing of a bat; the paddle of a porpoise; the foot of a horse—all are different, and each is modified in a manner appropriate for the functions it serves. The horse's foot, for example, has a "spring ligament" that is stretched when it bears the animal's weight and whose elastic rebound helps push the horse off the ground, giving a "bounce" to each step (Fig. 10.1). The mole's hand is broad and flat, and the bony projections to which the various limb muscles attach are lengthened and thickened for added strength. The various modifications of what was once a primitive mammalian forelimb are an example of adaptive radiation, a concept defined in Chapter 18.

Insect mouthparts are likewise adapted to the various functions they serve. The primitive condition probably resembled the chewing mouthparts of the grasshopper (order Orthoptera), which include clearly visible paired mandibles, paired maxillae, and a lower lip (labium). In the true bugs (order Hemiptera), the maxillae and mandibles are modified into a piercing structure (Box 10–A), which is used to penetrate plant tissues and suck up their sugary fluids. Butterflies and moths (order Lepidoptera) have a coiled proboscis, composed of paired maxillae, which enables them to feed on nectar concealed within tubular flowers. Bees (order Hymenoptera) have mouthparts that are adapted to the sipping of liquids. Mosquitoes (order Diptera) have just about the most highly modified mouthparts, in which the upper lip (labrum) and a structure known as the hypopharynx together form a drinking tube, flanked on either side by the piercing stylets, which are borne by the maxillae and mandibles. All these structures are enclosed in a sheath formed by the lower lip, or labium. Like the mammalian forelimbs, these various mouthparts are all built on a common pattern, yet they are variously modified, each in a manner appropriate to its use.

Among plants, those adapted to life in dry habitats (such as deserts) are known as **xerophytes**. Examples include many cacti (family Cactaceae), euphorbs (family Euphorbiaceae), and scattered plants in many other families (Fig. 10.2). Xerophytic plants often have thick, fleshy leaves, very minute, recessed stomata ("breathing pores"), and thick, waxy cuticles, all of which conserve water. Many xerophytes also have thorns or spines, which minimize destruction by desert animals.

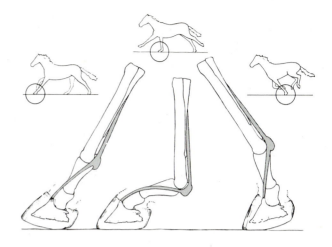

Fig. 10.1   Springing action of the principal suspensory ligament in the foot of the horse. From M. Hildebrand, *Analysis of Vertebrate Structure*, ©1974. New York: John Wiley and Sons, Inc. Reprinted by permission of John Wiley and Sons, Inc.

Fig. 10.2   Three xerophytic plants: *Euphorbia milii, Alluaudia procera,* and *Fourquieria splendens.* They are unrelated but have similar adaptations. Reprinted with permission from L. Cutak and C. Johnston, Missouri Botanical Garden, St. Louis.

BOX 10-A *Insect Mouthparts*

All the mouthparts of insects shown here consist of modifications of the same basic parts.

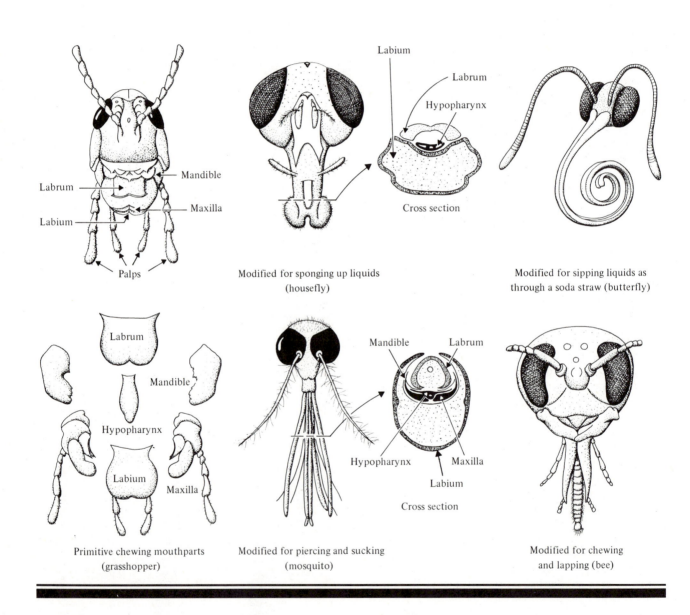

Labrum

Mandible

Labium

Maxilla

Palps

Labium

Labrum

Hypopharynx

Cross section

Modified for sponging up liquids
(housefly)

Modified for sipping liquids as
through a soda straw (butterfly)

Labrum

Mandible

Hypopharynx

Labium

Maxilla

Primitive chewing mouthparts
(grasshopper)

Mandible   Labrum

Hypopharynx   Maxilla

Labium

Cross section

Modified for piercing and sucking
(mosquito)

Modified for chewing
and lapping (bee)

C. PHYSIOLOGICAL (including BIOCHEMICAL) ADAPTATIONS

Adaptations among animals and plants are not confined to matters of structure. Adaptations based on internal physiological (including biochemical) processes are known as **physiological adaptations.** These adaptations, though less visible to the casual observer, may be just as important to the organism's survival. The study of these adaptations constitutes much of physiology and biochemistry but especially the branches of physiology sometimes called comparative physiology and environmental physiology (ecophysiology).

The conservation of water by desert animals is accomplished by a multiplicity of adaptations, many of them physiological. Water loss during breathing is kept minimal by reducing both the breathing rate and the surface area of nasal passages and by exhaling cooler air. The kidneys have the remarkable ability to excrete a very concentrated urine, containing nearly three times the salt and four times the urea of our own urine and correspondingly less water. The feces, too, are much drier, and the loss of moisture through the skin is effected by the reduction of sweat glands. Some of the best-adapted desert rodent species, like the kangaroo rat studied by Schmidt-Nielsen and Schmidt-Nielsen (1951), can also utilize the water produced in the course of their own metabolic processes. Behavioral adaptations, such as spending much of the day in burrows, are also important. Folk (1974) contains further discussion of these adaptations.

The conservation of water by desert plants depends on equally amazing adaptations, some of them mentioned in the previous section of this chapter. Perhaps the most unusual of physiological adaptations is the peculiar "night photosynthesis" of desert cacti. These plants keep their respiratory pores (stomata) closed during the day, minimizing the loss of water by evaporative transpiration. Yet photosynthesis requires carbon dioxide, which cannot be obtained from the air except by opening the stomata. Many cacti have circumvented this difficulty by completing only the light-dependent reactions of photosynthesis (the Hill reactions) during the day, storing an intermediate three-carbon product, and completing the remainder of photosynthesis (including the CO_2-dependent Calvin cycle) at night, when the opening of stomata brings about only a slight water loss.

Plants that are pollinated by insects frequently have adaptations to attract the appropriate pollinating species. Plants pollinated by beetles and other crawling insects often have putrid or fetid odors, similar to the odors of the decaying flesh on which these insects feed. Plants pollinated by nectar-seeking insects, such as bees (order Hymenoptera) or moths (order Lepidoptera), frequently have sweet, perfumy odors, bright colors, and "nectar guides," which are markings near the base of each petal, guiding the insect toward the nectary (Fig. 10.3). Nor are biochemical adaptations for the production of odors exclusive to plants: many animals attract their mates by means of a chemical scent, or pheromone (Chapter 24). Also, both plants and animals may deter predators by

Fig. 10.3 Nectar guides in flowers. Flowers of *Iris* (a) and evening primrose (*Oenothera*) (b) are shown on the left as they appear in normal light and on the right as they appear in ultraviolet light, to which bees and other insects are sensitive.

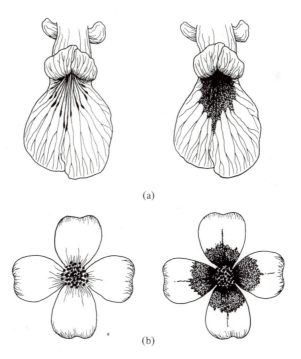

(a)

(b)

secreting distastefully bitter or acrid chemicals. Many of these unpalatable substances, including a large number of the plant products known as alkaloids, are highly specific to the species that secretes them; alkaloids are thus useful in the taxonomy of higher plants.

The mouthparts of mosquitos, described in the previous section, could easily become clogged if the blood of the victim clotted in the mosquito's drinking tube. The mosquito overcomes this difficulty by injecting into its victim's blood an anticoagulant just before it begins to suck the blood. The parasites that cause malaria have adapted their life cycle to this situation: they are carried in the mosquito's saliva and are injected into the human host along with the anticoagulants.

Venomous animals, whether they are snakes or scorpions, jellyfish or zebra-fish, produce highly specific poisons or toxins, many of which have specific action on the nervous systems of their victims. Many of these toxins contain protein-digesting enzymes, which destroy a wide variety of tissues in the bodies of their victims.

Many fungi, lichens, and other simple eucaryotes secrete specific bacteriostatic or bactericidal substances, which act to reduce or prevent bacterial infections. Peni-cillin, manufactured by molds of the genus *Penicillium,* is but one of many such chemicals.

Among the most advanced of biochemical adaptations are the antibodies produced by the immune systems of the higher vertebrates. These protein substances are highly specific to the antigen molecules with which they combine. An antigen, which is generally a foreign protein, will induce a primary reaction the first time it is introduced into the body. This primary response, which is relatively slow, includes the production of an antibody that destroys or sequesters the antigen by combining with it and rendering it harmless. Any subsequent exposure to the same antigen provokes a much more rapid secondary response, in which a much higher level of antibody is achieved.

D. CAMOUFLAGE AND MIMICRY

Camouflage (crypsis)

Protective resemblance, also called either **camou-flage** or **crypsis**, is defined as any resemblance of an organism to some other feature of its environment, by

Fig. 10.4 Cryptic coloration in a ptarmigan, shown here in summer plumage. Courtesy of the American Museum of Natural History.

(a)   (b)

Fig. 10.5 Light-colored "peppered" moths (*Biston betularia*) and melanic (black) individuals of the same species, photographed against a light, moss-covered tree trunk (a) and a dark, soot-covered tree trunk (b). Courtesy of E. B. Ford, *Ecological Genetics*.

means of which resemblance it often escapes detection. Camouflage is widespread throughout the animal kingdom at all levels, and it may involve not only cryptic coloration and patterning but also shape and even behaviorally determined position and orientation. Examples of camouflage are shown in Figs. 10.4 and 10.5.

A special form of camouflage, known as **industrial melanism**, occurs in soot-covered environments near major sources of industrial pollution. Many animals in such environments are dark in color (melanic) and are thus camouflaged against the soot-covered surfaces (Fig. 10.5). The frequency of melanic individuals often varies geographically within species (Fig. 10.6). The adaptive value of industrial melanism is shown by field observations that predators in polluted areas kill mostly nonmelanic individuals, but in nonpolluted areas the same predators kill mostly melanics. Many families of insects and other animals contain species that exhibit this particular form of camouflage.

Batesian Mimicry

Mimicry may be defined as deceptive resemblance to another biological species. There is some overlap between the concepts of mimicry and camouflage in organisms

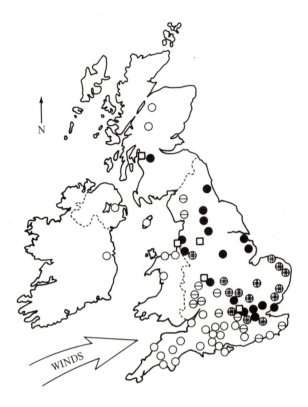

○ 0–30 % melanics
⊖ 30–60 % melanics
⊕ 60–80 % melanics
● 80–100% melanics
□ Major industrial center

Fig. 10.6 Geographic variation in the frequency of dark-colored (melanic) moths of the species *Biston betularia* in the 1950s. Major centers of industrial pollution are shown by squares.

that escape detection by resembling other organisms (instead of inanimate objects).

Mimicry was first studied by the British naturalist Henry Walter Bates (1825–1892), who was also responsible for interesting Alfred Russel Wallace in the study of animal adaptations. From 1848 to 1859, Bates and Wallace conducted a major expedition to the Amazon region of Brazil, collecting thousands of specimens and discover-

Fig. 10.7 Batesian mimicry in the pattern of the African swallowtail butterfly, *Papilio dardanus*. Top row shows nonmimetic female (left) and male (right) *P. dardanus* from Madagascar where no noxious species forms a suitable model. Second to fifth rows show four noxious model species on the left and the corresponding mimetic forms of *P. dardanus* on the right, each occurring in the same geographic area as its model. Reprinted by permission from Ayala, Francisco J. and Valentine, James W., *Evolving*, Menlo Park, California, The Benjamin/Cummings Publishing Company, Inc. 1979, Fig. 4.10.

ing some 8000 species new to science. In the course of his studies, Bates discovered a number of butterflies that deceptively resembled one another. In most cases, one species, the model, was unpalatable, and the other species, the mimic, was not. This type of mimicry is known as Batesian mimicry. Many examples of Batesian mimicry are known, especially among insects; some of these are illustrated in Figs. 10.7 through 10.9.

The adaptive nature of Batesian mimicry has been demonstrated through the experiments of Brower (1958a, 1958b, 1958c, 1960, 1963) and others, using birds as predators. Field observations have shown that birds do avoid both distasteful species and their palatable mimics. In order to test the adaptive nature of mimicry, mealworms (highly palatable larvae of the beetle genus *Tenebrio,* normally colored white) were dipped in green food coloring. Birds soon learned to eat these mealworms despite their unusual color. But when some of the meal-

Monarch butterfly

Viceroy butterfly

Fig. 10.9 Batesian mimicry between the distasteful monarch butterfly (model, above) and the palatable viceroy butterfly (mimic, below).

Fig. 10.8 Batesian mimicry between wasps (models, left column) and harmless flies (mimics, right column) from Borneo. Reprinted with permission from Wickler, *Mimicry in Plants and Animals,* © 1968. New York: McGraw-Hill Book Company.

worms were dipped in quinine (a bitter-tasting alkaloid), the birds soon learned to avoid *all* green mealworms, both distasteful models and their similarly colored but palatable mimics. This was true even if the mimics outnumbered their models by as much as five to one, the actual ratio depending on the distastefulness of the models. If the model was sufficiently unpalatable, the birds would learn on a single encounter to avoid both models and mimics alike. In a natural situation, the mimic would thus escape predation simply by virtue of its adaptive resemblance to the distasteful model.

Warning Coloration

Some animals are characteristically venomous. Many more are distasteful because of the presence of biochemical adaptations that protect them against predation.

Fig. 10.10   Warning coloration in a zebra butterfly.

Predators will learn to recognize and avoid such distasteful or venomous species, but not until several have already been killed. The learning process is made more efficient if the prey species is conspicuously colored. Conspicuous coloration benefits both species: the predators learn with fewer mistakes, and fewer prey are killed during the learning process. Conspicuous coloration of this type is known as **warning coloration** or **aposematic coloration**. Aposematic species are often strikingly colored, in bold, alternating stripes. The colors used are generally those that are visible to the widest variety of predators: yellow and black, white and black, orange and white and black, or yellow and white and black (Figs. 10.10 and 10.11).

Mimics often take advantage of the fact that predators avoid animals with warning coloration. Many cases of mimicry involve palatable animals falsely displaying aposematic colors.

A special type of mimicry may also involve use of warning coloration: the presence of eyespots or conspicuous circular markings that resemble the paired eyes of a predator species, such as an owl. Many moths and other insects normally conceal their hindwings, but upon escape they display on their hindwings a pair of brightly colored eyespots (Fig. 10.12), which startle the predator just long enough for the moth to escape.

Müllerian Mimicry

In the study of mimicry, it is often discovered that two or more species are all unpalatable. None are models—they are all mimics of one another. This type of mimicry, first described by the German zoologist Fritz

Fig. 10.11   Warning coloration in the caterpillar of the monarch butterfly. Reprinted with permission from John W. Kimball, *Biology,* 4e, © 1978. Reading, Massachusetts: Addison–Wesley (Fig. 43.4).

Fig. 10.12   The Brazilian moth *Ophthalmophora*, with eyespots on its hind wings. These eyespots are visible when the moth is flying, but they disappear beneath the cryptically colored forewings when the moth is at rest. Their sudden appearance when the moth senses danger and starts flying away is enough to startle many predators, or frighten them away altogether.

Müller in 1864, is known as Müllerian mimicry (Fig. 10.13). Müllerian mimics are conspicuously colored. Predators soon learn to avoid these colorful mimics, but only after a distasteful experience in which a mimic may perish.

Müllerian mimicry benefits all species involved. The mimics benefit from "spreading around" the risks of attack by a naive predator. For each predator that learns its lesson, a certain number of mimics are killed among all the Müllerian species. But in the absence of mimicry, this same number would be killed in *each* species that had its own individual warning pattern. Not only do the prey species benefit from the mimicry, but the predator also benefits: it has only one pattern of warning coloration to learn, to remember, or to suffer tasting during the learning process.

Perhaps the generally similar color patterns of aposematic species—alternating broad bands of contrasting colors—can be explained as resulting from Müllerian mimicry. The general pattern is easy for a predator to learn and to recognize. Most predators will generalize the pattern to include *all* species with warning coloration, which are therefore Müllerian mimics of one another. Most predators that have been tested will avoid mealworms that have been artificially painted with a warning color pattern.

Self-Mimicry

Some animals have an inconspicuous head and a bold, conspicuous tail that mimics a head in appearance. These animals, which seem to be mimicking themselves, are called **self-mimics.** Self-mimicry is best known among insects and warm-water fishes. Self-mimicking species typically have misleading eyespots on their hind end. Insects exhibiting self-mimicry often have imitation antennae and other structures on their imitation heads (Figs. 10.14 and 10.15). Some self-mimicking species even compound the deception by jumping backwards.

Self-mimicry may be explained adaptively by appealing to the effects of predation. Most predators attack other animals where they are most vulnerable, usually at their heads. A self-mimicking species would redirect the predator's attack to a false head at the much less vital hind end, and would thus be far more likely to survive the attack.

Fig. 10.13 Batesian and Müllerian mimicry in various butterflies, with unpalatable models above the horizontal bars and their palatable mimics below. Resemblance between models and mimics is Batesian mimicry; resemblance among models is Müllerian mimicry. Each butterfly shown belongs to a different species. Reprinted with permission from Wickler, *Mimicry in Plants and Animals,* © 1968. New York: McGraw-Hill Book Company.

Fig. 10.14 Self-mimicry of the butterfly *Thecla togarna,* with head facing to the right. Reprinted with permission from Wickler, *Mimicry in Plants and Animals,* © 1968. New York: McGraw-Hill Book Company.

Fig. 10.15 Self-mimicry in the Thailand lantern-fly, with head concealed toward the right, tucked very low. The false eyes, antennae, and "beak" are all modified parts of the wing tips. Reprinted with permission from Wickler, *Mimicry in Plants and Animals,* © 1968. New York: McGraw-Hill Book Company.

Aggressive Mimicry

Most animals recognize and avoid potential predators. A few predators circumvent this avoidance by mimicking a harmless species, especially one whose presence the prey species normally welcomes. The females of the firefly genus *Photuris,* for example, mimic the flashing patterns by which fireflies of the genus *Photinus* find their mates. Male *Photinus* are attracted to these flashes, which they mistake for the mating signal of their own species, only to find themselves caught in a death trap set

Fig. 10.16 Reproductive mimicry in the freshwater clam *Lampsilis ventricosa.* The body of this decoy "fish" is constructed of the clam's marsupium (brood pouch), and the "fins" are constructed from the clam's mantle flaps. When an unwary real fish comes to investigate, the clam shoots its parasitic larvae toward the unsuspecting host, whose gills will serve as their new home. The decoy "fish" even mimics the swimming motions of a real fish. Photo by John H. Welsh.

by the *Photuris* females. Another case (Eisner *et al,* 1978) concerns ant lions that disguise themselves as aphids by covering themselves with pieces of wax, which they pluck from the aphids that abound in many ant colonies. Since the ants tolerate the aphids, the ant lion can enter the ant nest unnoticed, only to surprise an unsuspecting ant that should happen to come too close to its jaws.

Reproductive Mimicry

In some species, mimicry serves the function of reproduction instead of individual survival. A remarkable case of this occurs in the clam *Lampsilis ventricosa,* whose larvae live as temporary parasites on the gills of fishes. The female *Lampsilis* sports a brood-pouch and mantle flaps that form an imitation fish (Fig. 10.16). When a real fish approaches this decoy, the female sprays the young larvae upwards. Many of these larvae reach the gills of the host fish in this way.

Some insect-pollinated species serve as "copulation dummies" by mimicking the females of the pollinating insect species. A male comes to copulate with the dummy, and in so doing he picks up pollen and pollinates the plant. Several orchids of the genus *Ophrys* mimic female bees, especially those of the genus *Eucera* (Fig. 10.17). The plant is pollinated by the male *Eucera,* which come to copulate with the dummy females. The mimicry is complete to such details as the pattern of minute hairs on the labium of the flower, which closely resemble the distribution of hairs on the back of the female *Eucera.*

E. COADAPTATION

The most remarkable of adaptations are often those by which species are adapted to each other in cases of mutualism (Chapter 9). The reciprocal adaptations of symbiotic species to one another are often called **coadaptations**, and their evolution together or in consort is a most important form of coevolution (Chapter 9). A coadaptation may be defined as an adaptation that benefits its possessor only indirectly, by means of its more direct benefit to another, usually symbiotic, species.

Some fish that are normally predatory will allow certain species of smaller fish to enter their mouths unharmed and clean food fragments from between their teeth. This type of "cleaning symbiosis" (Fig. 9.14)

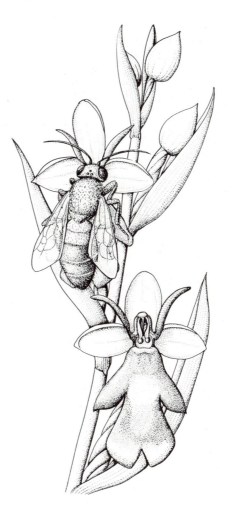

Fig. 10.17 Reproductive mimicry in an orchid, *Ophrys insectifera,* which forms a "copulation dummy" that mimics female bees of the genus *Eucera.* The flowers are pollinated by male *Eucera* bees, which are attracted to the flowers. The bee attempts to copulate with the flower, thus bringing its head into contact with the pollen sacs. The pollen sacs adhere to the bee's head and are transferred to the next flower visited. Reprinted with permission from Donald D. Ritchie and Robert Carola, *Biology,* © 1979. Reading, Massachusetts: Addison-Wesley (Fig. 15.34).

Fig. 10.18 Cleaning symbiosis and aggressive mimicry in fishes. Two cleaning wrasses (*Labroides dimidiatus*) are shown cleaning the gills of a red snapper (*Lutianus sebae*), and an aggressive mimic, the saber-toothed blenny (*Aspidontus taeniatus*) attacks the tail fin of another red snapper (note bites already taken from tail fin). The cleaning wrass (above) and its mimic (below) are also shown separately for closer comparison. Reprinted with permission from Wickler, *Mimicry in Plants and Animals,* © 1968. New York: McGraw-Hill Book Company.

benefits both species: the smaller fish receives a ready supply of food, and the larger fish gets its teeth cleaned. Other forms of cleaning symbiosis include those in which small fish (or small shrimp) clean the gills or the external scales of larger fish. There is also a bird, the Nile plover, that cleans food particles from between the teeth of the Nile crocodile. Some predators take advantage of the cleaning symbiosis: they mimic a cleaning species and are thus permitted to get close to their prey (Fig. 10.18). Some corals will allow certain species of fish to hide among their tentacles. The fish receives protection from predators, since the coral's stinging tentacles are armed with a poison. In exchange for this protection, the corals receive an occasional food fish, which the symbionts catch for them and deposit within their tentacles. Some aggressive fishes also mimic the symbionts and thus pass unmolested amongst the tentacles of the coral, only to take an unexpected bite out of one of the coral's tentacles.

Coadaptations between plants and insects are especially common. The nectaries, swollen thorns, and Beltian bodies provided by certain acacia trees to symbiotic ants (Fig. 9.13) are remarkable examples of coadaptation. The ants, for their part, attack and often destroy any other vegetation (including nonfood species) that encroaches on their shade-intolerant hosts.

Many of the coadaptations of plants and insects have as their effect the more efficient transfer of pollen and thus the growth of new plants. Most insect-pollinated plants secrete nectar that attracts the pollinating insects, and many bees have pollen-carrying baskets on their legs. In flowers of the genus *Salvia,* the feeding activities of bees automatically daub them with pollen (Fig. 10.19), and cross-fertilization is ensured by the maturing of male and female flower parts at different times. In some plants, the nectar and pollen are so arranged in the flower that an entering insect will first brush off any pollen it already has onto the female flower parts (stigmata). As the insect reaches into the nectary, the structure of the flower forces it to pick up new pollen from the male flower parts (anthers). In some flowers, the pollinating insect must enter one way and leave by another. The female parts are located near the "entrance" and the male parts near the "exit," ensuring cross- rather than self-fertilization.

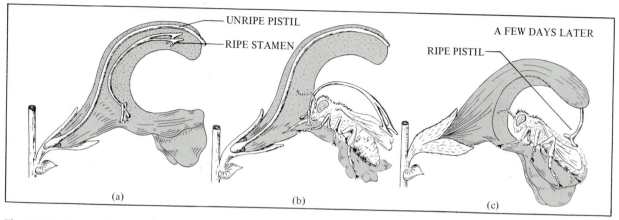

Fig. 10.19 Coevolution of *Salvia* flowers for pollination by bees. Cross-pollination is ensured by (a) ripening of the stamen before the pistil, (b) a lever-action trigger mechanism that dusts pollen on a bee that probes inside the flower for nectar, and (c) later growth of the pistil so that it picks up pollen from bees as they enter the flower for more nectar. Reprinted with permission from John W. Kimball, *Biology*, 4e, ©1978. Reading, Massachusetts: Addison-Wesley (Fig. 20.7).

F. THE EXPLANATION OF ADAPTATION

Among the greatest problems of nineteenth-century biology was the explanation of adaptation. At the beginning of that century, Paley's *Natural Theology* (Chapter 3) offered the omnipotence of a benevolent and omniscient God as an explanation for all adaptation. Indeed, Paley insisted that adaptation was so perfect that it was explainable by no other agency.

Lamarck offered the theory of use and disuse (Chapter 3) as an explanation for adaptations. The strengthening of any organ through use or its atrophy through disuse would, according to Lamarck, be passed on to the next generation. The adaptation that resulted would have been determined by the animal's voluntary use or disuse. Structural adaptations were more easily explained by this mechanism than camouflage or mimicry. Ultimately, Lamarck's mechanism proved faulty, for it depended on the inheritance of acquired characteristics, but a better explanation for adaptations had meanwhile

been offered. This was Darwin's theory of natural selection, which we shall examine carefully in the next chapter.

CHAPTER SUMMARY

No theory of evolution is complete without an explanation of adaptation. An adaptation is any feature that makes an organism better suited to its environment. Locomotor organs, feeding organs, and other structural or morphological adaptations are widespread throughout the living world, as are biochemical adaptations, some of which render their possessors distasteful and so reduce predation.

Camouflage is protective resemblance of an organism to its background, by which the organism escapes detection. Distasteful or harmful species are often warningly colored instead. Mimicry is a situation in which one species gains an advantage by resembling another. In Batesian mimicry, a palatable mimic resembles a distasteful model. In Müllerian mimicry, several distasteful species resemble one another, and all are mimics. In self-mimicry, an animal's hind end resembles its head, causing predators to attack at the less vulnerable end. In aggressive mimicry, the predator is benefited by being allowed to come closer to its prey without causing alarm.

Coadaptations are adaptations in which symbiotic species benefit one another and are reciprocally benefited. Such coadaptations may exist between flowers and insects or between other pairs of symbiotic species.

Historical explanations for adaptation have included Paley's Natural Theology and Lamarck's theory of use and disuse. The modern explanation of adaptation is based on Darwin's theory of natural selection.

11

Natural Selection

A. THE CONCEPT OF NATURAL SELECTION

The theory of natural selection was Charles Darwin's greatest contribution to evolutionary biology. The concept was as simple as it was revolutionary: among numerous individual variations (i.e., genotypes), death would fall on most, and those that were best suited to the conditions of life would survive and reproduce in greatest proportion. If the selected variation had a genetic basis, evolution would result from the inheritance of this variation by successively larger proportions of future generations.

The concept of natural selection was alien to many minds. It was accused of being mindless, Godless, and blindly mechanistic. Above all, it was a statistical or population concept, and population thinking was a very new and even upsetting way of thought to many people. Instead of the reassuring constancy of the Platonic *eidos* (Chapter 3), population thinking insisted that each individual was unique, and that the average was just a statistical abstraction that had no independent existence. Malthus and other British economists had introduced population thinking to Britain, and it is noteworthy that both Darwin and Wallace claimed that Malthus had inspired their independent theories of natural selection, though a few dissenting historians have expressed disbelief in these claims.

Selection, whether natural or artificial, may be defined as *consistent differences in the contribution of various genotypes to the next generation.* Dobzhansky's definition is that "the chance of contributing to the gene pool of the next generation is a statistically predictable function of the genotype." As Darwin and others have emphasized repeatedly, the differential contributions may result from death of certain types of individuals from whatever cause, death by predation, starvation, or disease being in most cases more frequent than death by any literal "struggle." But differential contributions could also result, with equal effectiveness, from differences in reproductive ability, including both fertility (ability to reproduce) and fecundity (number of offspring).

From Malthus, Darwin drew the observation that any species, if left unchecked, would soon increase without limit and outstrip its resources. Since no species actually increases in this manner, Darwin concluded that death and destruction were falling on the population in each generation. In nature, this destruction often fell on the seeds, the eggs, or the newly hatched larvae; in domestication, it usually came in the form of harvest or slaughter for the marketplace. The result, said Darwin, was intraspecific competition, which he called the "struggle for existence," though he cautioned that it was a struggle only in the "large and metaphorical sense." In this competition, those kinds of organisms that had an advantage at any stage in their life cycles would, as a rule, leave more offspring, whereas those suffering any disadvantage would leave fewer offspring or none at all.

Many of Darwin's readers were not sufficiently impressed by the nature of population thinking. Some misunderstood Spencer's phrase "survival of the fittest" to imply that some individuals were guaranteed survival. Others took "survival of the fittest" to be a circular definition of natural selection, since the "fittest" were identified by the mere fact that they had survived. Yet, in the discussion above and in Darwin's own discussions in his first four editions, the term "fittest" never appears.

The potential power of selection is amply demonstrated by the variety of domesticated animals and plants that humans have produced over the past few centuries of methodically practiced artificial selection. Charles Darwin frequently compared natural selection with such artificial selection. Though both operated in the same way, artificial selection could be documented more readily, and the evidence for its existence was uncontestable. Yet, compared with natural selection, artificial selection was always considered a distinctly inferior force: "Natural Selection," declared Darwin (1859: 61), ". . . is as immeasurably superior to man's feeble efforts, as the works of Nature are to those of Art."

The existence of selection in nature has been repeatedly documented by field studies on a variety of vertebrates, insects, and plants. Several of these studies will be cited later in this chapter.

Today Darwin's theory of natural selection has won almost universal acceptance among scientists, its power most convincingly demonstrated by its efficacy as an explanation of adaptations (Chapter 10). Natural selection has helped explain adaptations as diverse as the flower's fragrance, the eagle's eye, or the human hand, not to mention mimicry, camouflage, and bacterial resistance to antibiotics.

B. SELECTION FOR SINGLE-GENE TRAITS

In this section we consider the theory of selection as it applies to a trait controlled by a single gene. The symbols we shall use are defined in Box 11–A. The survival rate (λ) of each genotype is defined as the fraction of that genotype that contributes to the next generation. Individuals that die or fail to reproduce are not counted, and those that reproduce more than once are counted more than once. For simplicity, we usually assume that all selection acts at once, but in reality it acts all the time, and there is never any "population before selection" or any "population after selection."

The relative fitness (W) of each genotype is defined as its survival rate expressed as a fraction of the maximal survival rate. The relative fitness (W) of the optimal genotype is therefore defined as equal to 1.00, and the relative fitnesses of other genotypes are calculated as proportionate fractions or percentages. The selection coefficient (s) of each genotype is defined as 1 minus its relative fitness, i.e., its relative disadvantage as compared with the optimum genotype. Please consult Box 11–A and be certain that these various concepts are clear to you before proceeding.

Selection against Dominants, Favoring Recessives (λ_{aa} Highest)

We will first consider the simplest case, that of complete selection against all individuals with the dominant genotype. The survival rate of genotype aa is here assumed to be the highest, and the survival rate of the other two genotypes is assumed to be zero (see Item 11.1 on page 200). The result is the complete fixation of a and the elimination of AA in a single generation. We may summarize this as follows: *dominant lethal traits are eliminated in a single generation of selection.*

Consider next the case in which there is only partial selection against dominants. The survival rate of genotype aa is still assumed to be the highest, but the survival rate of the other two genotypes is no longer assumed to be zero. For simplicity, we are assuming complete dominance, in which AA and Aa are indistinguishable in both phenotype and in relative fitness. In this instance (Item 11.2), the gene frequency of A is reduced in each generation by a factor equal to W_{AA}. The inevitable result, in

BOX 11–A   *Numerical Example Illustrating the Calculations of Survival Rate, Relative Fitness, and Selection Coefficient for Each of Three Diploid Genotypes. (Fictitious Data.)*

Genotype	AA	Aa	aa
Population before selection	4000	6400	3200
Population after selection	3600	5400	2000
Survival rate, λ	$\lambda_{AA} = \dfrac{3600}{4000}$ $= .900$	$\lambda_{Aa} = \dfrac{5400}{6400}$ $= .844$	$\lambda_{aa} = \dfrac{2000}{3200}$ $= .625$
Relative fitness, W (compared with that of AA)	$W_{AA} = \dfrac{\lambda_{AA}}{\lambda_{max}}$ $= \dfrac{0.900}{0.900} = 1.00$	$W_{Aa} = \dfrac{\lambda_{Aa}}{\lambda_{max}}$ $= \dfrac{0.844}{0.900} = 0.94$	$W_{aa} = \dfrac{\lambda_{aa}}{\lambda_{max}}$ $= \dfrac{0.625}{0.900} = 0.69$
Selection coefficient, $s = 1 - W$	$s_{AA} = 1.00 - 1.00$ $= 0.00$	$s_{Aa} = 1.00 - 0.94$ $= 0.06$	$s_{aa} = 1.00 - 0.69$ $= 0.31$

the absence of such contravening forces as recurrent mutation, is the continual reduction of p and the elimination of A. The case of incomplete dominance would be only slightly more complicated, but it would still have the same general result: *selection against dominants leads inexorably to the elimination of the dominant allele and the complete fixation of the recessive one.*

Selection against Recessives, Favoring Dominants (λ_{aa} Highest)

The simplest case of selection against recessives is represented in Item 11.3. We have here assumed both complete dominance ($\lambda_{AA} = \lambda_{Aa}$) and complete selection ($\lambda_{aa} = 0$). The result is a very small reduction of the value of q in each generation, the rate of reduction slowing

down as q gets smaller and smaller. Thus, even the most vigorous selection against recessives would eliminate them from the population with the utmost slowness. The complete elimination of recessives is made all the more difficult by the fact that selection against them becomes very ineffective when the recessive allele becomes rare.

To illustrate this last point, let us take the equation

$$q_n = \frac{q_{n-1}}{1 + q_{n-1}} = \frac{q_0}{1 + n\,q_0} \qquad (11.1)$$

and solve for n, giving

$$n = \frac{1}{q_n} - \frac{1}{q_0}. \qquad (11.2)$$

Now let us consider the case where we have reduced the gene frequency by one half. In this instance, we may substitute $q_n = \frac{1}{2} q_0$ in Equation 11.2, which then simplifies to

$$n = 1/q_0. \qquad (11.3)$$

In others words, if q had an initial value of $1/100$ (and $q^2 = 1/10,000$, meaning that one individual in every 10,000 was a recessive), then it would take 100 generations of relentless and unmitigated selection against these recessives simply to reduce the gene frequency to $1/200$ or the frequency of recessive individuals to $1/40,000$. The consequences for artificially imposed selection are clear: if one individual in every 20,000 is, say, an albino ($q = 1/141$, approximately), then it would take 141 generations (about four thousand years in human terms) to reduce this frequency to a value of one in every 80,000. And many of the most commonly studied genetic defects in humans (albinism, phenylketonuria, G6PD deficiency, etc.) are much less frequent than this, meaning that selection against any of them would take even longer. The reason for this is that most of the recessive alleles are hidden unnoticed in the bodies of heterozygotes, and only those that occur in homozygotes are subject to selection.

What, now, if selection is less than totally efficient? The computations, shown in Item 11.4, may be summarized as follows: *selection against recessives always proceeds very slowly*, and *any* level of survival among *aa* homozygotes will slow down this selection even further.

What will happen over the great expanse of geologic time? Selection, even inefficient selection, against recessive traits will lead (however slowly) to the near total elimination of these traits from the population, given enough time. "Enough time" is here conceived of as much greater than any human civilization has at its disposal for a program of applied selection, but as possible over the course of geologic time. "Near total elimination" is here specified for a number of reasons. First, selection against recessive traits in an infinite population can reduce but can theoretically never eliminate completely any recessive trait under the conditions here specified. Second, once a recessive trait becomes sufficiently rare, the effects of even the most vigorous selection become less important than the effects of recurrent mutation on the one hand and the effects of genetic drift on the other (see Chapter 8). Rare traits will in the long run be subject to accidental loss through genetic drift (which *can* act on heterozygotes as well as homozygotes); yet

they always will tend to reappear through recurrent mutation. We should thus expect deleterious recessive traits, even lethal ones, to be exceedingly rare in natural populations—perhaps absent—but this absence would be only temporary in evolutionary terms.

Combined Effects of Selection and Mutation

To make this last point somewhat more explicit, let us consider the joint action of mutation and selection against recessives on the frequency, q, of a rare recessive allele. The assumption of a rare allele ensures that the effects of mutation will be significant, and it also permits us to ignore the effects of reverse mutation. The computations, shown in Item 11.5, result in an equilibrium value that will generally be small, because u is usually much smaller than s, typically by several orders of magnitude.

Selection against Homozygotes, Favoring Heterozygotes (λ_{Aa} Highest)

The selective superiority of heterozygotes leads to a stable equilibrium in which both alleles persist in the population indefinitely (Item 11.6). This situation, known as balanced polymorphism, is discussed further in Chapter 12. Examination of natural populations reveals that this particular form of selection is very common; it constitutes the major exception to the rule that natural selection generally tends to reduce population variability in the long run.

Selection against Heterozygotes, Favoring Both Homozygotes

This case, treated in Item 11.7, results in an equilibrium which, however, is unstable. Any small deviation from this equilibrium (as by genetic drift) leads to an increase in one allele at the expense of the other and eventually in the complete fixation of the former and the complete elimination of the latter. Here natural selection tends to magnify the effects of drift: the elimination of either allele and the fixation of the other are clearly the result of selection, yet it was sampling error that made the initial choice that determined which allele was to prevail.

BOX 11–B *Summary of the Outcome of Various Cases of Selection*

	$W_{Aa} < W_{aa}$	$W_{Aa} = W_{aa}$	$W_{Aa} > W_{aa}$
$W_{AA} < W_{Aa}$	Fixation of a	Slow fixation of a	Balanced polymorphism (Items 11.0, 12.1)
$W_{AA} = W_{Aa}$	Rapid fixation of a (Items 11.1, 11.2)	Hardy-Weinberg equilibrium (Chapter 8)	Slow fixation of A (Items 11.3, 11.4)
$W_{AA} > W_{Aa}$	Unstable equilibrium (Item 11.7)	Rapid fixation of A	Fixation of A

The various results given in the preceding sections are all summarized in Box 11–B. Note that most of the forms of selection included in this table lead eventually to genetic uniformity, even if very slowly.

C. SELECTION FOR MULTIGENE TRAITS

For traits determined by the interaction of two or three genes, selection can often be treated as if it applied to each gene separately. For traits whose genetic basis is somewhat different, different calculations along the same general guidelines are carried out. As the number of genes controlling a trait increases, the complexity of the calculations increases very rapidly.

In the treatment of more complex situations the phenotype rather than the genotype is used. This procedure is valid because natural selection acts on phenotypes, and genotypes that do not achieve phenotypic expression cannot be selected.

Nondirectional Selection

Natural selection operating on traits controlled by polygenes can be thought of as resolvable into directional and nondirectional components (Fig. 11.1). **Nondirectional selection** may be defined as selection for polygenic traits that causes no shift in the mean value. Two types of

Fig. 11.1 Different forms of selection for a polygenic trait.

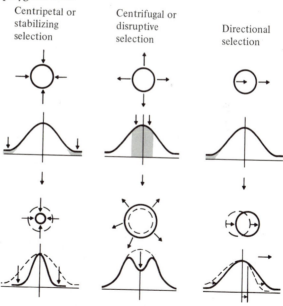

Centripetal or stabilizing selection

Centrifugal or disruptive selection

Directional selection

"Pure" directional selection

Directional selection together with centripetal selection

Directional selection together with centrifugal selection

BOX 11–C *A Case of Compromise Selection in Birds*

In the winter of 1898, after an "uncommonly severe storm of snow, rain, and sleet," 136 starving English sparrows were brought to the laboratory of H. C. Bumpus at Brown University in Providence, Rhode Island. Of these, 72 survived and 64 died. Bumpus had meanwhile taken certain standard measurements on all the birds and published a study showing the differences between the measurements of the survivors and those that died.

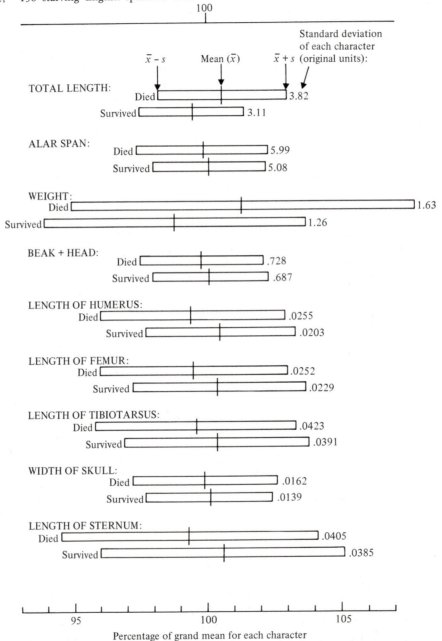

nondirectional selection are possible. The first type, known either as **stabilizing** or **centripetal selection,** is assumed to be more common. In a now classic study by Bumpus, various measurements were taken of sparrows that survived a severe New England winter storm and of sparrows that succumbed to the same storm. In each of the many characters studied, the mean measurements for survivors and nonsurvivors were fairly close (indicating the virtual absence of directional selection), but the variance for nonsurvivors was much greater than that of survivors in every case (Box 11–C). Other studies have also supported the belief that stabilizing selection, especially by the elimination of both extremes, is a very important force in maintaining the adaptiveness of natural populations. Note that this may be true even for characters (such as many of those used by Bumpus) whose relation to selective forces is hardly obvious. It is of historical interest to note that Charles Lyell believed in the ability of stabilizing selection to hold species constant (Chapter 3) long before Charles Darwin proposed natural selection as a mechanism of evolutionary change.

Disruptive selection, also called **centrifugal** or **diversifying selection,** is a much-debated type of nondirectional selection, favoring both extremes of a continuous distribution at the expense of the midrange values. Since it selects most strongly against phenotypic values representing high levels of heterozygosity, disruptive selection may be reducible to many simultaneous cases of selection against heterozygotes (see treatment above). This would lead to the ultimate elimination of one or the other extreme, the initial choice being made by deviations due to sampling error from an unstable equilibrium. In a series of isolated populations, this might well lead to the fixation of low phenotypic values in some and of high phenotypic values in others, but any given population would reach only one of these extremes, never both.

Some early geneticists believed that disruptive selection could lead to the partitioning of a niche, the fractioning of a single population into two separate populations, or perhaps ultimately to sympatric speciation (Chapter 14). Fisher (1930) and Mayr (1963) argued strongly against this. The experimental results of Thoday and Gibson (1962) tend to suggest that disruptive selection might indeed subdivide the population into separate breeding units, given a sufficient level of positive assortment. In plants, Antonovics and Bradshaw have studied similar effects resulting from sharp differences between the selective regimes of closely adjacent populations (Chapter 13). Positive assortment acts in these several cases as an incipient mechanism of reproductive isolation (Chapter 14).

Directional Selection

Directional selection in the case of polygenic traits is defined as selection that causes a shift in the mean value for the population. "Pure" directional selection would cause no change in the variance; directional selection that involves a change in variance may be thought of as having a nondirectional component (either stabilizing or disruptive).

Directional selection in present-day organisms has been demonstrated by Klauber (1956) in rattlesnakes, by Hecht (1952) in the lizard genus *Aristelliger* (Fig. 11.2), and by Cain and Sheppard (1950, 1954) in the snail genus *Cepaea* (Fig. 11.3). It might come as a surprise that directional selection has been conclusively demonstrated in so few cases, and that even in these cases the directional component of selection has turned out to be rather weak. The slightness of directional selection would seem to explain why stabilizing selection is in general easier to substantiate. The importance of directional selection is better demonstrated by such long-range effects as evolutionary trends (Chapter 17) and adaptive radiation (Chapter 19).

The force of directional selection has also been demonstrated in experimental populations of *Drosophila* (Fig. 11.4), corn (Fig. 11.5), and mice (Fig. 11.6). In these experiments, selection for either high or low phenotypic extremes results in an artificial "trend," with resulting mean values far beyond the extremes of the original population. In some of these experiments, the direction of selection was reversed after a certain number of generations, and the ensuing reversal of the trend showed that little if any genetic variability had been lost.

Evolutionary trends in the fossil record are in general many times more gradual than in these artificial selection experiments (Chapter 20). This presumably indicates that directional selection in natural populations proceeds much more slowly (but no less potently) than in artificial experiments; the studies cited above on present-day populations also seem to indicate this.

Wright's model of an adaptive landscape with "peaks" and "valleys" incorporates all the forms of selection listed above: stabilizing selection in the vicinity

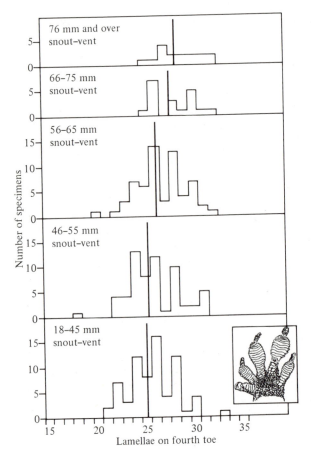

Fig. 11.2 Natural selection in the lizard genus *Aristelliger*. Since lizards continue growing throughout life, their size (as measured by snout-to-vent length) gives an indication of age. The number of lamellae (specialized scales) on the clinging surface of the fourth toe, however, does not change with age. Differences in the scale counts of small (young) and large (old) lizards therefore reflect differential survival. Note that the older lizards have a somewhat higher mean and reduced variance. Adapted with permission from *Evolution*, vol. 6, 1952.

Fig. 11.3 Natural selection in the land snail *Cepaea*. Three genetically determined color patterns are shown. The differences in the frequency of these color variants in different habitats reflects selection by thrushes and other birds that locate the shells visually. In low vegetation (grass fields and hedgerows), the banded light-colored shells are most difficult to see because of shadows cast by vegetation. In dark woodlands, the dark unbanded shells are most difficult to see. Adapted with permission from Cain and Sheppard, *Genetics, 54*, 1954.

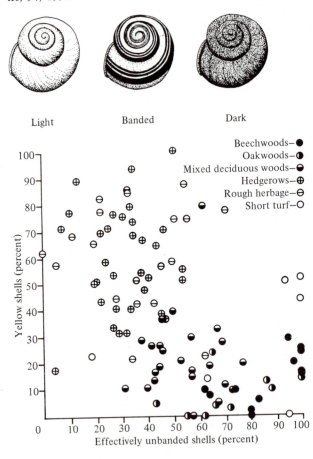

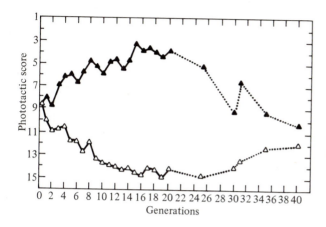

Fig. 11.4 Results of a selection experiment in *Drosophila*. These fruit flies usually fly toward the light (phototaxis), but the behavior is partly under genetic control and can therefore respond to selection. The open triangles represent a line of flies selected for high phototactic score. The closed triangles represent flies selected for low phototactic score. The dotted lines after 20 generations indicate that selection in both lines has been relaxed. Reprinted with permission from Spiess, *Genes in Populations*, © 1977. New York: John Wiley and Sons (after Dobzhansky and Spassky).

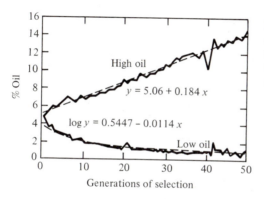

Fig. 11.5 Results of selection experiments in corn (*Zea mays*), selected for high oil content, low oil content, high protein content, or low protein content. Reprinted from *Agronomy Journal*, volume 44, 1952, page 61 by permission of the American Society of Agronomy, Inc., Madison, Wisc.

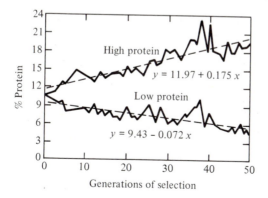

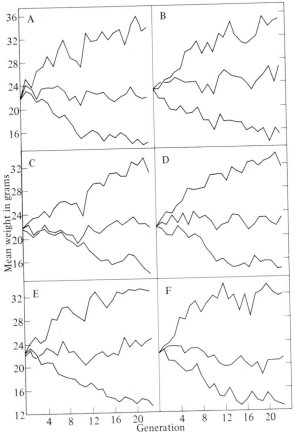

Fig. 11.6 Selection for body weight (measured at age six weeks) in mice. Six separate experiments were carried out. In each experiment, one line was selected for greater weight, a second for lesser weight, and a third was left unselected. Reprinted by permission from Ayala, Francisco J. and Valentine, James W., *Evolving,* Menlo Park, California, The Benjamin/Cummings Company, Inc., 1979, Fig. 3.6.

of the "peaks," disruptive selection along the "valleys" between peaks, and directional selection along the valley's sloping walls.

D. SEXUAL SELECTION

Sexual selection may be defined as selection of one sex in which the opposite sex serves as the selective agent. Particularly noteworthy is the fact that sexual selection operates usually by determining reproductive success, without sacrificing the lives of the losers.

Sexual selection frequently acts on traits that are phenotypically expressed in one sex only. The result is a morphological distinction between the sexes known as **sexual dimorphism** (Fig. 11.7). Some examples of sexual dimorphism are size differences between the sexes, differences in coloration, the presence of the mane in lions, the antlers in deer, the brightly colored plumes of peacocks and lyrebirds, and the hooked jaw in salmon. Sexual dimorphism might also extend to the secretion of chemical scents (including pheromones) by one sex and not the other, or the production of sounds by one sex only. The organs of sound production may also be confined to one sex only, as are the vocal sacs in frogs, the enlarged larynx in howler monkeys, and the comblike stridulating (chirping) organs of grasshoppers and crickets. Of course, the sounds or scents of one sex would have to be recognized by the other sex for sexual selection to occur.

Fig. 11.7 Some examples of sexual dimorphism: fallow deer (*Dama dama*), California quail (*Lophortyx californicus*), and lions (*Panthera leo*).

Sexual selection operates by giving greater reproductive success to those having certain traits. Very frequently, the structure thus selected appears in a rudimentary form in the opposite sex. The rudimentary nipples of the human male's breast may serve as an example: they serve no useful purpose, but the genes that control their development in the young of both sexes are adaptive in females, where they become part of an organ that serves to nourish the young. The female's breast, incidentally, contains a much larger amount of fat tissue than glandular tissue, a phenomenon peculiar to humans. The outward shape of the human female breast has been attributed to various causes, including sexual selection by the tastes of their mates, sexual selection because of their supposed resemblance to the buttocks, selection as fat storage depots for possible use during pregnancy, and selection as a consequence of upright posture in nursing mothers (Cant, 1981).

E. GROUP AND KIN SELECTION

Natural selection may work either within groups (**intragroup** or **individual selection**) or between groups (**intergroup selection**, or simply **group selection**). Darwin himself made mention of this possibility, but the theory that selection takes place between groups as well as within them has been championed principally by Wynne-Edwards (1962). According to his view, social organization, age distributions, and other population traits may be subject to group selection, in which groups having an adaptive social structure persist and flourish, but those with inadaptive social structures perish. Indeed, claims Wynne-Edwards, they can be explained *only* by group selection, for selection between individuals can never explain such social characteristics as territories, sex ratios, or age distributions, which often sacrifice individual reproductive success.

Among biologists, Wynne-Edwards has attracted far more criticism than support. Not since Ernst Haeckel's proposal that "ontogeny recapitulates phylogeny" has a biological idea provoked so much new thought and research in an effort to prove it wrong. The opponents of group selection have quite successfully shown how selection within groups can account for nearly all group characteristics. Sex ratios, for example, can be explained quite well as the outcome of individual selection between the sexes. In any biparentally reproducing species, the next generation's genes will come half from the males and half from the females of the present generation. If one sex is more plentiful than the other, individuals of the minority sex usually have a selective advantage, for each will leave, on the average, a proportionately greater share to the next generation. (For example, if the population has 800 males but only 600 females, an average male's contribution to the gene pool is half of 1/800, or 1/1600, but an average female's contribution is half of 1/600, or 1/1200, which is proportionately larger.) In a nonequilibrium population containing (for example) more males, a given individual will leave more *grandchildren* by giving birth to more females than males. If a sex ratio among offspring could be inherited, it would be subject to selection, and there is good evidence that sex ratios can indeed be selected for in *Drosophila*. The selective advantage to the minority sex should result in its increase to the point were the sexes are produced in equal numbers. Thus, the sex ratios in such a species tends to stabilize at an equal number of males and females.

The deterioration of phenotype with age (**senescence**) and ultimately death itself can also be explained without recourse to group selection. One explanation, at least in nonsocial species, is that an individual will leave more grandchildren if it dies after reproducing, rather than remain alive to compete ecologically with its own offspring. (In social species, it might also, of course, contribute some advantage to the reproductive success of its offspring sufficient to offset such competition.) A second factor (Williams, 1966, 1975) is that genes producing favorable effects early in life will be favored by selection even if they produce unfavorable effects later, after reproduction has already occurred. A third factor (Gould, 1977) is that genes with harmful effects can be selected to delay the expression of these effects until long past reproductive age. In wild populations, few individuals ever reach old age, and a gene whose effects are thus postponed to an age of senility may be considered lost, just as if it had been eliminated entirely. This is especially important if the gene is pleiotropic and has some beneficial effect at another stage of life, for postponement of the harmful effects would still allow the gene and its beneficial effects to persist.

Changes in the timing of gene activity can also account for such demographic features as survivorship curves, age-specific mortality rates, and reproductive values (Chapter 9). Selection within the population can certainly control the timing of expression of the many genes that control these population characteristics. Similarly, the various spacing mechanisms that limit population density, including territoriality (Chapter 24), can also be accounted for by individual selection. Lack (1954) has shown that individual fitness in birds decreases to either side of the optimal population density. In particular, when the population density is too high, mortality is also high, and individual fitness is therefore lowered. An individual finding itself surrounded by too many neighbors would increase its fitness more by emigrating than by reproducing under crowded conditions. This is most obvious in territorial species: individuals not having a territory or having too small a territory face a lowered fitness if they reproduce, and they would do better to migrate to a region of lower population density in search of a better territory.

The most persistent claim on the part of group selection is that **altruism**, or self-sacrifice for the benefit of others (or of the group as a whole), could never be accounted for by any other means, because the altruistic individual sacrifices its own fitness. A way out of this difficulty has been shown by Hamilton (1972), who distinguishes between exclusive and inclusive fitness: my **exclusive fitness** is determined only by the number of my offspring, but my **inclusive fitness** is the sum total of reproductive success in all individuals sharing my genotype. Altruism among relatives is thus easily explained, for an individual might leave more grandchildren if it sacrificed its own life to save the lives of its own children, who bear its genotype. The same is true among relatives other than offspring: each of your brothers and sisters shares, on the average, about half of your genotype, and you could therefore increase the inclusive fitness of your genotype by sacrificing your own life to save the lives of more than two siblings. With first cousins, the degree of relationship is one-eighth instead of a half, and the British evolutionist J. B. S. Haldane once remarked, "I would lay down my life for two brothers or eight cousins." **Kin selection** may be defined as selection that sacrifices the individual but favors its genotype by conferring advantage on its near relatives. Remember, too, that among sedentary species inbreeding ratios may be very high, and all individuals in the population may be related to one another.

What, then, of altruism among unrelated individuals? The incidence of such altruism in natural populations is hard to assess, nor has it ever been shown, for that matter, that there is any such thing as a "gene for altruism." In social species, even altruism among nonrelatives can be explained on the basis that the favor may someday be returned (**reciprocal altruism**, Chapter 24). If I risk my life to save your family, you may feel obligated to risk your life someday to save mine. Even if I die saving your family, you may thereafter feel obligated to care for my orphaned children, thus increasing their fitness. Even less extreme sacrifices can be explained in this way. If an individual is known to be kind and generous and is valued as a friend by his or her peers, that individual is more likely to be helped and cared for in time of ill health or other adversity. It is not necessary, by the way, that the performer of a good deed expect any form of repayment for the favor, either personally or on behalf of his or her offspring; all that is needed is that children and other relatives eventually benefit, whether this fact was realized at the time of sacrifice or not.

Charles Darwin realized the importance of kin selection. He used kin selection among social insects as an explanation for the instincts of neuter ant workers and as an argument against Lamarckism. More recently, Trivers and Hare (1976) have used kin selection to explain the prevalence of sociality in the insect order Hymenoptera as a consequence of their peculiar mating system. Female hymenopterans are diploid, whereas males are haploid, a system known as either haplodiploidy or arrhenotoky. In this strange system, a female worker shares an average of half her genes with her mother or her daughters, 3/4 of her genes with her sisters, and only 1/4 of her genes with her haploid brothers. A female worker would therefore increase her own inclusive fitness more by contributing to the raising of her sisters than by raising her own off-spring! Selection has therefore favored the evolution of worker females, themselves sterile, who care for the safety and survival of their sisters, who are all offspring of the same queen. Little wonder that social behavior of this sort has evolved nearly a dozen times independently in the Hymenoptera but only once in all the other insects. The one group of truly social insects outside the Hymenoptera are the termites (order Isoptera). Here both sexes are diploid, and workers may be either male or female, instead of exclusively female as they are in the Hymenoptera. Even the ratio of sizes between the sexes of social insects can be explained by the theory of kin selection: in most of the social Hymenoptera, the workers are related three times as much to their sisters as to their brothers (3/4 as compared to 1/4), and their ratio of energy invested in feeding them is also in this proportion, so males weigh only about one-third as much as females on the average. But in termites or in those species of ants that use the workers of other species to care for the young (the so-called slave-making ants), no such inequality exists, and the males and females are of approximately equal size.

CHAPTER SUMMARY

Natural selection is the differential capacity of various genotypes for contributing to the gene pool of the next generation.

Selection for a single-gene trait operates rapidly if selection is against dominants, but more slowly if it operates against recessives. Except in the case of heterozygote superiority, natural selection tends to reduce the variability of natural populations in the long run, leading in extreme cases to genetic uniformity. Balanced polymorphism is a special type of equilibrium that arises from the selective superiority of heterozygotes, a situation that ensures the persistence of both alleles, even if one is lethal.

In the case of selection for traits controlled by polygenes, stabilizing selection tends to favor midrange values at the expense of both extremes; disruptive selection, which favors the extremes, is much less common. Selection for polygenic traits may also have a directional component, but the stabilizing component is usually also present and is typically stronger.

Sexual selection operates because one sex may operate as the selective agent in determining the reproductive success of the other sex. Sexual selection usually leads to sexual dimorphism, in which the two sexes are distinguishable by morphological differences. Selection between groups, often called group selection, may favor groups with favorable social structures over those with less well-adapted social structures. Altruistic behavior has been attributed to group selection, but more evolutionists would now explain it as resulting from kin selection, in which a self-sacrificing individual favors its own genotype by contributing to the fitness of its near relatives.

FOR FURTHER KNOWLEDGE

ITEM 11.1 Complete Selection against Dominants (λ_{aa} Highest, $\lambda_{AA} = \lambda_{Aa} = 0$)

Genotype	AA	Aa	aa	Frequency of a
Frequency before selection	p^2	$2pq$	q^2	$q_0 = q$
Survival rate	$\lambda_{AA} = 0$	$\lambda_{Aa} = 0$	$\lambda_{aa} \neq 0$	
Relative fitness	$W_{AA} = 0$	$W_{Aa} = 0$	$W_{aa} = \dfrac{\lambda_{aa}}{\lambda_{aa}} = 1$	
Selection coefficient	$s_{AA} = 1$	$s_{Aa} = 1$	$s_{aa} = 0$	
Frequency after selection	0	0	1	1

ITEM 11.2 Partial Selection against Dominants (λ_{aa} Highest, $\lambda_{AA} = \lambda_{Aa}$)

Genotype	AA	Aa	aa	Frequency of a
Frequency before selection	p^2	$2pq$	q^2	$q_0 = q$
Relative fitness	$1 - s$	$1 - s$	1	
Frequency after selection	$(1 - s)p^2$	$(1 - s)2pq$	q^2	$q_1 =$

$$q_1 = \frac{q}{(1 - s)(1 - q^2) + q^2} = \frac{q}{1 - s + sq^2}$$

The new frequency of a will always be greater than the previous frequency because the denominator, $1 - s + sq^2$, is always less than 1 as long as q remains below 1. This continues until A disappears and a is fixed ($q = 1$).

ITEM 11.3   *Complete Selection against Recessives ($\lambda_{aa} = 0$)*

Genotype	AA	Aa	aa	Frequency of a
Frequency before selection	p^2	$2pq$	q^2	$q_0 = q$
Relative fitness	1	1	0	
Frequency after selection	p^2	$2pq$	0	$q_1 = pq/(1 - q^2)$ $= q/(1 + q)$

The relation between each value of q and the next is recursive:

$$q_1 = \frac{q_0}{1 + q_0}$$

$$q_2 = \frac{q_1}{1 + q_1} = \frac{q_0}{1 + 2q_0}$$

$$q_3 = \frac{q_2}{1 + q_2} = \frac{q_0}{1 + 3q_0}$$

$$\dots\dots\dots\dots\dots\dots\dots\dots$$

$$q_n = \frac{q_{n-1}}{1 + q_{n-1}} = \frac{q_0}{1 + nq_0}$$

ITEM 11.4   *Partial Selection against Recessives (λ_{aa} Lowest)*

Genotype	AA	Aa	aa	Frequency of a
Frequency before selection	p^2	$2pq$	q^2	$q_0 = q$
Relative fitness	1	1	$1 - s$	
Frequency after selection	p^2	$2pq$	$q^2(1 - s)$	(see below)

The frequency of a after selection will be

$$q_1 = \frac{pq + q^2(1 - s)}{p^2 + 2pq + q^2(1 - s)}$$

$$= \frac{(1 - q)q + q^2 - sq^2}{(1 - q^2) + q^2 - sq^2}$$

$$= \frac{q - sq^2}{1 - sq^2}$$

$$= q\left(\frac{1 - sq}{1 - sq^2}\right).$$

In those cases where q is already very small (near zero), the quantity sq^2 is negligible, and the denominator is sufficiently close to 1 that we may write

$$q_1 \approx q(1 - sq).$$

We may also calculate the change in q from one generation to the next as

$$\Delta q = q_1 - q_0$$

$$= q\left(\frac{1 - sq}{1 - sq^2}\right) - q$$

$$= \frac{q(1 - sq) - q(1 - sq^2)}{1 - sq^2}$$

$$= \frac{q - sq^2 - q + sq^3}{1 - sq^2}$$

$$= \frac{-sq^2(1 - q)}{1 - sq^2}$$

or

$$\Delta q \approx - sq^2(1 - q).$$

ITEM 11.5 *Combined Effects of Selection and Mutation*

This case is similar to that of Item 11.4, with the added effect of mutation. The value of q is assumed to be very small, so terms containing q^2 can be ignored and mutation can be assumed to take place in one direction only $(A \rightarrow a)$.

If we assume the change in q to be additive,

$$\Delta q_{total} = \Delta q_{selection} + \Delta q_{mutation},$$

then, combining results obtained in Item 11.4 with a term for mutation,

$$\Delta q_{total} \approx -sq^2(1-q) + u(1-q).$$

This quantity can equal zero (i.e., no change in q) only when

$$0 = (1-q)(-sq^2 + u),$$

which occurs only when $q = 1$ (a trivial solution) or when

$$u = sq^2,$$
$$q^2 = u/s,$$
$$q = \sqrt{u/s}.$$

This is the equilibrium value of q, often called $\hat{q}$, under the conditions stated above, and it is equal to the square root of the ratio u/s. An equivalent statement is that the ratio of u to s must equal the frequency of recessive homozygotes, q^2.

ITEM 11.6 *Heterosis, or Selection Favoring the Heterozygotes over Both Homozygotes*

Genotype	AA	Aa	aa	Frequency of a
Frequency before selection	p^2	$2pq$	q^2	$q_0 = q$
Relative fitness	$W_{AA} = 1 - s$	1	$W_{aa} = 1 - t$	
Frequency after selection	$p^2(1-s) = p^2 - p^2 s$	$2pq$	$q^2(1-t) = q^2 - q^2 t$	(see below)

The frequency of a after one generation of selection can be calculated as

$$q_1 = \frac{pq + q^2(1-t)}{p^2 - p^2 s + 2pq + q^2 - q^2 t}$$

$$= \frac{(1-q)q + q^2(1-t)}{(p^2 + 2pq + q^2) - p^2 s - q^2 t}$$

$$= \frac{q - q^2 + q^2 - q^2 t}{1 - p^2 s - q^2 t}$$

$$= \frac{q(1-q)t}{1 - p^2 s - q^2 t}.$$

The change in q can now be calculated as

$$\Delta q = q_1 - q_0 = \frac{q(1-q)t}{1 - p^2 s - q^2 t} - q$$

$$= \frac{q - q^2 t - q(1 - p^2 s - q^2 t)}{1 - p^2 s - q^2 t}$$

$$= \frac{q - q^2 t - q + qp^2 s + q^3 t}{1 - p^2 s - q^2 t}$$

$$= \frac{p^2 qs - (q^2 - q^3)t}{1 - p^2 s - q^2 t}$$

$$= \frac{p^2 qs - q^2(1-q)t}{1 - p^2 s - q^2 t}$$

$$= \frac{p^2 qs - pq^2 t}{1 - p^2 s - q^2 t}$$

$$= \frac{pq(ps - qt)}{1 - p^2 s - q^2 t},$$

which is zero for $p = 0$ or $q = 0$ or for $ps = qt$.

ITEM 11.6   *(continued)*

Solving the last equation for p and for q, we obtain

$$ps = (1 - p)t \qquad (1 - q)s = qt$$
$$ps = t - tp \qquad s - qs = qt$$
$$p(s + t) = t \qquad s = qs + qt$$

$$\hat{p} = \frac{t}{s + t} \qquad \hat{q} = \frac{s}{s + t},$$

which are the equilibrium values.

Also, for $0 < q < 1$, the denominator of Δq is always positive, and the sign of the numerator depends on the term

$ps - qt$, which equals

$$ps - qt = (1 - q)s - qt = s - qs - qt = s - q(s + t)$$

$$= \frac{s}{(s + t)}(s + t) - q(s + t) = (\hat{q} - q)(s + t).$$

Thus Δq is always positive (and q increases) when q falls below its equilibrium value, and Δq is negative (and q decreases) when q is above its equilibrium value. The equilibrium is therefore stable, because deviations from equilibrium values tend to be restored to equilibrium (except for the extreme values $q = 0$ and $q = 1$, which are also equilibria).

ITEM 11.7   *Selection against Heterozygotes, Favoring Both Homozygotes*

Genotype	AA	Aa	aa	Frequency of a
Frequency before selection	p^2	$2pq$	q^2	$q_0 = q$
Relative fitness	1	$1 - s$	1	
Selection coefficient	0	$s_{Aa} = s$	0	
Frequency after selection	p^2	$2pq(1 - s)$	q^2	$q_1 =$

$$\frac{q^2 + pq(1 - s)}{1 - 2pq(s)}$$

Calculating the change per generation, we obtain

$$\Delta q = q_1 - q_0$$

$$= \frac{q^2 + pq - pqs}{1 - 2pqs} - q$$

$$= \frac{q^2 + pq - pqs - q(1 - 2pqs)}{1 - 2\,pqs}$$

$$= \frac{q^2 + q(1 - q) - (1 - q)qs - q + 2(1 - q)q^2s}{1 - 2\,pqs}$$

$$= \frac{q^2 + q - q^2 - qs + q^2s - q + 2q^2s - 2q^3s}{1 - 2\,pqs}$$

$$= \frac{-qs + 3q^2s - 2q^3s}{1 - 2\,pqs}$$

$$= \frac{-qs(2q^2 - 3q + 1)}{1 - 2\,pqs}$$

$$= \frac{-qs(2q - 1)(q - 1)}{1 - 2\,pqs}.$$

This quantity can equal 0 (no change in q) only when the numerator is zero, which occurs only at the values

$$q = 0, \qquad q = \tfrac{1}{2}, \qquad q = 1.$$

Except for the extreme values of zero and one, an equilibrium can only exist at a single intermediate value, $q = \tfrac{1}{2}$. But the equilibrium at this point is unstable, because the fraction

$$\frac{-qs(2q - 1)(q - 1)}{1 - 2pqs}$$

has a denominator that is always positive as long as $p + q = 1$. The numerator is equivalent to

$$sq(1 - q)(2q - 1),$$

in which every permissible value of q between 0 and 1 makes the first three terms positive, and the last term is positive for $q > \frac{1}{2}$ and negative for $q < \frac{1}{2}$. We can interpret this to mean that for any nonequilibrium value of q, the value of q increases when it is already high ($q > \frac{1}{2}$) and decreases when it is already low ($q < \frac{1}{2}$). This makes the equilibrium unstable, for if q accidentally changed from 0.5000 to 0.5001 (say, by genetic drift), it would continue to increase until $q = 1$ and A disappeared, or if it accidentally changed to 0.4999, it would continue to decrease until $q = 0$ and a disappeared.

In the more general case, where the relative fitness of AA and aa is not necessarily the same, the same general rule holds, though the calculations are lengthier. A single unstable equilibrium point exists at some intermediate point (not always $q = \frac{1}{2}$), and any change from this equilibrium will soon result in the elimination of one allele and the fixation of the other.

12

The Persistence of Variability Within Populations

Chapter Outline

A. THE IMPORTANCE OF VARIABILITY

Early in the twentieth century, many Mendelian geneticists asserted that natural populations should all be uniform and consist of a single ubiquitous "wild type"; yet now we know that most natural populations are highly variable. The Russian naturalist-turned-geneticist Chetverikov was among the first to realize the importance of genetic variability in natural populations. His studies were very influential on Fisher (1930) and on another Russian-born naturalist-turned-geneticist, Theodosius Dobzhansky. Through studies on natural populations of *Drosophila,* Dobzhansky and his students brought to light an immense degree of genetic variability existing in natural populations at all levels. The "mutations" that some scientists took to be mere laboratory curiosities turned out to be naturally present in nearly all populations. Hardly a mutation was described in the *Drosophila* laboratories whose existence in natural populations was not soon confirmed, usually in the heterozygous condition. Even at the level of chromosomal translocations and inversions, a great deal of natural variability was found to exist in all wild populations.

The existence of such natural variability raises an important problem, because the simpler forms of both natural selection and genetic drift would tend always to reduce genetic variability and in the long run to eliminate it. In Chapter 7, we saw that variation originates as the result of mutations, recombinations, chromosomal changes, and changes in gene expression. In the present chapter we shall examine the problem of how this variability manages to persist in the face of such forces as natural selection and genetic drift, forces that have often been seen as tending to produce changes in the direction of greater population uniformity.

There probably never was any such thing as a genetically uniform natural population of any significant size; at any rate, nobody has ever found one. The fact that natural populations are genetically highly variable has been firmly and repeatedly established in many species. But why isn't the population uniformly homozygous for the best-adapted or most "fit" genotype? The simplest (but not necessarily adequate) explanations would rely on the concept of group selection (Chapter 11) to show that variable populations have a greater evolutionary potential than more uniform ones, or that variability is the key to adaptability. Fisher (1930), for example, called this the "**fundamental theorem of natural selection**: . . . the rate of increase in fitness within any population is proportional to the variance in fitness at that time." In other words, evolutionary change can occur more rapidly in variable than in uniform populations. Experimental demonstration of this phenomenon was provided by Ayala (1965, 1968), who showed that experimental *Drosophila* populations derived from a mixture of natural populations adapted about twice as fast to experimental temperature regimes as did similar populations derived from single sources.

Population variability may also be viewed as a hedge against future contingencies. If the environment were to change, a variable population would be far more likely to harbor one or more genotypes better suited to the new environment. In fact, group selection might even eliminate uniform populations under such conditions, since they would be unable to adapt to new conditions unless the "right" mutation just happened to occur at the "right" time. By a corollary of Fisher's fundamental theorem, a variable population has an increased chance of continuing to evolve under changing environmental conditions.

The astute reader will notice that the group-selectionist explanations above do not really show *how* variability manages to persist, but only that such persistent variability would be advantageous. This has the appearance of a teleological explanation, an explanation like "corn grows in order that people may eat." What is needed here is an explanation in terms of an efficient cause, like "corn grows because the farmer has planted its seed in good, rich soil." The search for an efficient cause of persistent population variability based on individual selection is one of the major problems to which the rest of this chapter is addressed.

B. POLYMORPHISM FOR VISIBLE TRAITS

Polymorphism may be defined as *the existence in a natural population of two or more alleles in frequencies too large to be explained by recurrent mutation alone.* The final qualifying phrase in this definition is inserted specifically to exclude alleles with no permanent or continuous existence in the population, such as recur in each generation as the result of new mutations. The level of polymorphism in a population is simply the fraction or percentage of loci that are polymorphic.

Transient Polymorphism

Two kinds of polymorphism, transient and balanced, are generally recognized. **Transient polymorphism** exists as a temporary condition only, while one allele is in the process of replacing another. Transient polymorphism is often presumed to be both preceded and followed by conditions of genetic uniformity at that locus within the population, though this fact has seldom been adequately documented. Difficulties for the interpretation of transient polymorphism, including Haldane's dilemma, are discussed in a later section of this chapter.

Balanced Polymorphism

Balanced polymorphism may be defined as persistent polymorphism, perpetuated by selection favoring heterozygotes and by other mechanisms that we will soon discuss. The discovery of balanced polymorphism, together with its explanation in terms of heterozygote superiority, was one of the major advances of evolutionary biology in the present century. The geographic variation of many types of balanced polymorphism provided one of the most convincing arguments for the genetic basis of geographic variation (Chapter 13), thus resolving one of the major differences between the early twentieth-century geneticists and naturalists (Chapter 6). The study of balanced polymorphism was thus instrumental in the advent of the modern evolutionary synthesis.

One of the classic cases of balanced polymorphism was studied by Dobzhansky and his students among the *Drosophila* populations of the southwestern United States. These populations contain a variety of chromosomal aberrations (inversions, translocations, etc.), and nearly every population studied was polymorphic. Another classic case of polymorphism was the industrial melanism among British peppered moths studied by Ford and by Kettlewell (Chapter 10). This case involves both geographical variation and variation through time.

The dark, melanic moths, first noticed in the 1890s, became so ubiquitous throughout Britain as a result of industrial pollution that the original peppered form could hardly be found. Now, however, efforts to control pollution have allowed the peppered form to make a comeback. Transient and balanced polymorphism may in such cases occur together, a situation that may indeed be fairly common. In such cases, the balanced polymorphism may slow down the replacement of one allele by another.

Sickle-Cell Anemia

The most well-known case of polymorphism is that of sickle-cell anemia, an uncommon genetic disorder that reduces the oxygen-carrying capacity of the blood and impairs the blood supply to various organs. Secondary effects resulting from the oxygen deficiency include impaired mental and physical development, towering of the skull, kidney failure, brain damage, heart failure, and certain death (Box 12–A). The disorder, which is inherited as a Mendelian recessive, is more frequent among American blacks than among American whites. Studies on African populations showed a surprisingly high frequency for the responsible allele, with maximum values as high as 23%. This result was immediately turned into an argument against natural selection: here was a harmful trait (and there is none so harmful as a lethal) that was remarkably frequent, even though natural selection would have been expected to rid the population of such harmful traits long ago. The paradox remained until maps were published (Fig. 12.1) showing the incidence of the disease or the responsible allele throughout Africa. It was partly on the basis of such maps that Allison (1955, 1961) first showed the connection between sickle-cell anemia and malaria, for both diseases tend to be rare in the same places, and the areas of greatest incidence also frequently coincide.

A simple blood test can distinguish heterozygotes from normal homozygotes. A few drops of blood are placed on a microscope slide in a specially made depression from which one can exclude oxygen by sealing the slide with a cover glass. The red blood cells of a normal person will remain round, but half of those from heterozygotes and all the red blood cells from sickle-cell homozygotes will assume a characteristic "sickle" shape, which gives the disease its name. Allison identified the various genotypes by this procedure and tested their resistance to malaria. When *Plasmodium falciparum* (Fig. 12.2), the parasite that causes the most serious form of malaria, was injected into volunteer patients, the heterozygotes contracted a far milder form of the disease and were much more likely to recover. One reason is that the parasite causes the red blood cells of these heterozygotes to become sticky and to adhere to the walls of very small blood vessels, where they assume their sickled shape. The body's defensive cells then kill the sickled cells, as well as many of the parasites along with them. Also, the mosquitos that normally carry the malarial parasite seem to bite homozygous normal individuals in preference to heterozygotes.

The type of selection exhibited here is selection favoring the heterozygotes (see Item 12.1 on page 222 and Item 11.6 at the end of Chapter 11). In this particular example, the heterozygotes are resistant to malaria and enjoy the highest survival rate. The dominant homozygotes are susceptible to malaria and thus have a lower survival rate; the decrease in their relative fitness depends largely on the (geographically variable) incidence of *Anopheles*-borne malaria. The recessives invariably die from sickle-cell anemia, and so they have a survival rate that is virtually zero. (Current research in the United States is rapidly improving this situation, which affects persons of African descent almost exclusively.) In cases like this, where the heterozygotes are maximally fit, both alleles will be maintained indefinitely in the population, even if one of them is lethal. The equilibrium reached is a stable one (any deviation from it tends to be corrected in the next generation), and the equilibrium frequencies are determined by the values of the selection coefficients against the two homozygous classes.

All of the known symptoms of sickle-cell anemia can be traced to an abnormal type of hemoglobin molecule, known as hemoglobin S in contrast to the normal hemoglobin A. Each hemoglobin molecule is made of four long-chain proteins plus an iron-containing "heme" group. Hemoglobin S differs from hemoglobin A in just a single amino acid residue in position 6 of the β protein chain. Knowledge of the genetic code shows that this amino acid difference can be accounted for by a change in only a single base pair in the nucleic acid sequence.

BOX 12–A   *Some Consequences of Sickle-Cell Disease*

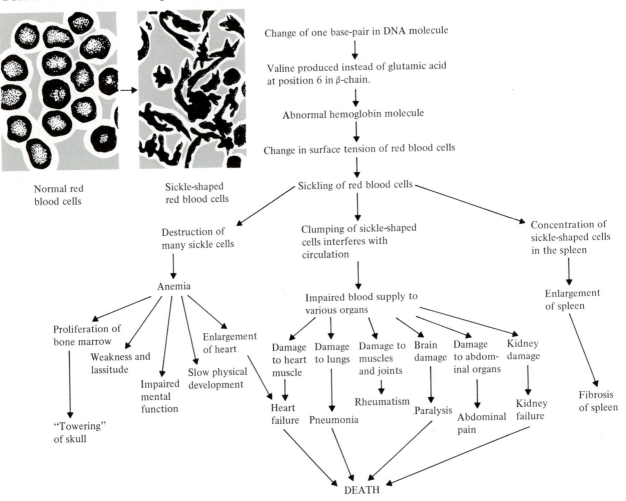

Normal red
blood cells

Sickle-shaped
red blood cells

Change of one base-pair in DNA molecule

Valine produced instead of glutamic acid
at position 6 in β-chain.

Abnormal hemoglobin molecule

Change in surface tension of red blood cells

Sickling of red blood cells

Destruction of
many sickle cells

Clumping of sickle-shaped
cells interferes with
circulation

Concentration of
sickle-shaped cells
in the spleen

Anemia

Impaired blood supply to
various organs

Enlargement
of spleen

Proliferation of
bone marrow

Weakness and
lassitude

Impaired
mental
function

Slow physical
development

Enlargement
of heart

Damage
to heart
muscle

Damage
to lungs

Damage to
muscles
and joints

Brain
damage

Damage
to abdom-
inal organs

Kidney
damage

Fibrosis
of spleen

"Towering"
of skull

Heart
failure

Pneumonia

Rheumatism

Paralysis

Abdominal
pain

Kidney
failure

DEATH

(a)

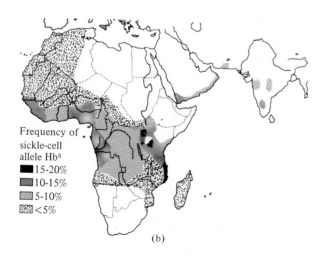

Frequency of
sickle-cell
allele Hb^s

■ 15-20%
▨ 10-15%
▧ 5-10%
▨ <5%

(b)

Fig. 12.1 Maps showing (a) the occurrence of falciparum malaria and (b) the frequency of the Hb^s allele, which causes sickle-cell anemia, both in the Old World. Part (b) is reprinted by permission from Ayala, Francisco J., and Kiger, John A. Jr., *Modern Genetics,* Menlo Park, California: The Benjamin/Cummings Publishing Co, Inc., 1980, Fig. 20.6, p. 683.

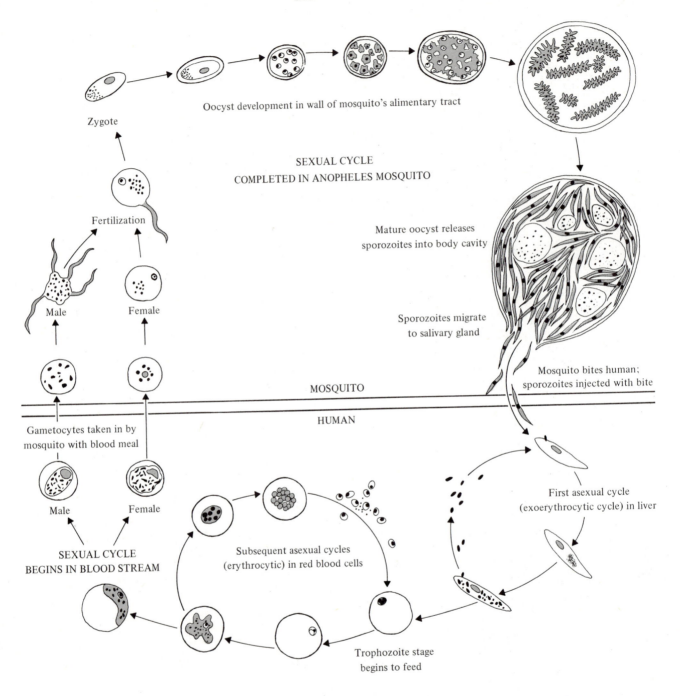

Fig. 12.2 The life cycle of the malarial parasite, *Plasmodium falciparum*.

C. ELECTROPHORETIC POLYMORPHISM

In the last two decades, the development of new bio-chemical techniques has permitted the rapid analysis of long-chain macromolecules, especially proteins, revealing differences in their structure within biological populations. Most important for the study of evolution is a procedure in which long-chain proteins are first extracted, then either broken up by enzymes into shorter fragments or analyzed whole. The analysis itself is carried out by **electrophoresis,** a technique that sorts out the proteins or protein fragments according to their speed of migration through an electrical field (Box 12–B). The im-

---

BOX 12–B *Electrophoresis Technique*

Starch-gel and acrylamide-gel electrophoresis are powerful techniques for surveying the genetic variation of natural populations. Tissue samples are first homogenized and purified (e.g., by centrifugation). Liquid, protein-rich samples may be further purified, and the proteins may be either broken into smaller fragments or analyzed whole. Samples to be analyzed are placed in slots, one for the material derived from each individual. The gel is then exposed to an electric field (a), and the enzymes or other protein molecules migrate through the gel at a rate determined by their molecular weight and electrical charge. The key to the technique is that different molecular varieties of the same enzyme often migrate at different rates. The electric field is then taken away (b), and the gel is exposed to chemicals that will change color and reveal the positions to which the enzyme samples have migrated. Many individual genotypes, one for each sample slot, can be determined in this way. Homozygous individuals will produce enzymes that migrate to a single location. Heterozygous individuals will produce two types of enzyme molecules, and in most cases this will show up as two distinct bands (c). If the enzymes form dimers, however, three bands will often show: one for dimers made of the first type of enzyme molecule, one for dimers made of the other type; and a third band, halfway between the first two, for dimers made of one molecule of each type (d).

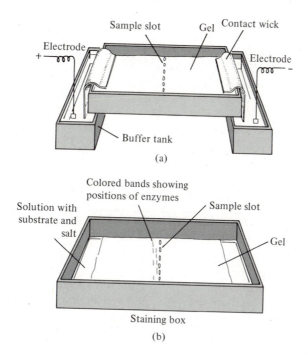

(a)

(b)

(*continued*)

BOX 12-B   *(continued)*

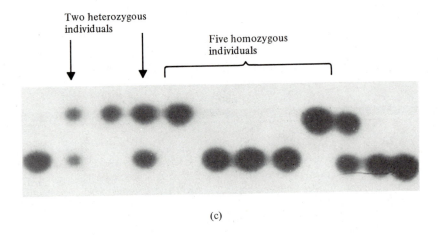

Two heterozygous individuals

Five homozygous individuals

(c)

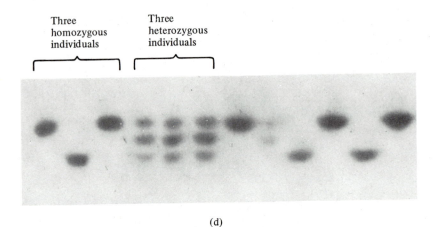

Three homozygous individuals

Three heterozygous individuals

(d)

Illustrations are reprinted with permission from Ayala, Francisco J., and Kiger, John A. Jr., *Modern Genetics*. Menlo Park, California: The Benjamin/Cummings Co., Inc., 1980, Figs. 18.7, 18.8, 18.9, pp. 614–615.

portance of this technique lies in the fact that it permits rapid analysis of slight variations in the protein structure of many individuals at a time and thus enables the rapid determination of frequencies for each electrophoretic variant ("electromorph") within the population. Electrophoretic studies have been carried out on an ever-increasing list of animal, plant, and microbial species.

Studies on the genus *Drosophila* alone already number more than a hundred. Most of the proteins analyzed by this technique are enzymes, but human hemoglobins and certain other nonenzymatic proteins have also been analyzed.

Not every genetic change will result in a protein variation detectable by electrophoresis. First, the redundancy of the genetic code ensures that certain mutational changes will result only in the substitution of one equivalent codon for another, such as CGU for CGC or CGA for CGG, all of which code for arginine. Second, mutations that do result in amino acid substitution may still go undetected unless an alteration in electrical charge also results. Sixteen of the 20 amino acids commonly found in proteins are electrically neutral; the substitution of these amino acids for one another would have little if any effect on the electrophoretic mobility of a protein. Only two amino acids (lysine and arginine) are basic (positively charged at neutral pH), and two others (aspartic acid and glutamic acid) are acidic (negatively charged at neutral pH). Only those changes that resulted in the substitution of an amino acid *carrying a different charge from the original one* would be detected by electrophoresis. Lewontin (1974) has estimated that 32% or less of the DNA base substitutions that result in the substitution of one amino acid for another would be electrophoretically detectable; he also presents arguments for the belief that these detectable variations should form a nearly random sample of all single-gene mutations. The analysis of pedigrees can detect only those genes whose products differ conspicuously from one another, but Lewontin emphasizes that electrophoresis can detect variation at a much finer level and can also identify those genetic loci whose protein products do not vary electrophoretically.

The ability to identify both variant and invariant loci allows estimations to be made of their relative frequencies. Studies on humans, rodents, and the phylogenetically relict horseshoe crab *Limulus* (Chapter 20) reveal that 20–30% of the loci studied show polymorphisms within populations, though several of these studies are based on relatively few populations (5 or less). The somewhat more extensive studies of the genus *Drosophila* have revealed anywhere from 25% to 86% polymorphism, depending on the particular species studied, and to a lesser extent on the populations and loci sampled. *Drosophila willistoni* appears to be the most highly polymorphic (80–86%). At a randomly selected locus, the proportion of individuals that are heterozygous varies from near 5% in certain mice to over 18% in *Drosophila willistoni*. Estimates of heterozygosity, however, are subject to statistical uncertainties (standard errors) of ±1 to 5%, meaning that differences of this magnitude would not be considered statistically significant.

D. THE NEUTRALIST-SELECTIONIST CONTROVERSY

The discovery of substantial levels of variability in natural populations has brought on a debate between opposing schools over the interpretation of this variability. We will now examine this debate.

The "classical" theory of population genetics, as exemplified by the writings of H. J. Muller (and with philosophical roots going back to Darwin or de Candolle), envisioned a situation in which well-evolved populations were expected to be genetically uniform for the "wild type." Every genetic variation was assumed to have some effect on fitness, however small, and natural selection was seen almost exclusively as a "purifying" force that would rid the population of the less fit variants and preserve only the most fit. Mutations would continue to arise, but most of them would be harmful, and the eternal vigilance of natural selection would be necessary to keep them in check and prevent the deterioration of the population from its "load of mutations." (Such was the origin of the term "genetic load," as well as of Muller's sound of alarm at the reduction of selective forces in human populations through medical practice.)

The discovery of widespread polymorphism in natural populations, and especially (since 1966) of fine-level electrophoretic variation, has made the "classical" theory untenable in its original form. Lewontin and Hubby (1966, also Hubby and Lewontin, 1966) were able, using electrophoresis, to detect and document high levels of both heterozygosity and polymorphism in natural populations of *Drosophila pseudoobscura*, and other workers have made similar findings on other species (Box 12–C). According to Lewontin and Hubby's calculations, a *minimum* of 30 percent of all loci were polymorphic, and each individual was heterozygous for an average of somewhat more than 12 percent of all its loci. Assuming (as Lewontin and Hubby themselves first did) that this unexpectedly high level of heterozygosity was maintained by balanced polymorphism, the genetic load necessary to maintain this polymorphism would at first seem intolerably high. But if the genes in question are assumed to manifest their phenotypic effects principally or exclusively in the form of polygenic traits (controlled by the additive contributions of many genes at a time), then the calculated values of genetic load due to polymorphism are reduced considerably.

BOX 12–C   *Levels of Genic Variation in Selected Natural Populations*

Group	Number of species	Mean number of loci per species	Proportion of loci	
			POLYMORPHIC PER POPULATION (P) MEAN (S.D.)	HETEROZYGOUS PER INDIVIDUAL (H) MEAN (S.D.)
Drosophila	28	24	0.529 (.030)	0.150 (.010)
Other insects	4	18	0.531	0.151
Haplodiploid wasps	6	15	0.243 (0.039)	0.062 (.007)
Marine invertebrates	9	26	0.587 (0.084)	0.147 (.019)
Snails				
Land	5	18	0.437	0.150
Marine	5	17	0.175	0.083
Total invertebrates	57	Mean = 21.8	Mean = 0.469	Mean = 0.135
Fish	14	21	0.306 (.047)	0.078 (.012)
Amphibians	11	22	0.336 (.034)	0.082 (.008)
Reptiles	9	21	0.231 (.032)	0.047 (.008)
Birds	4	19	0.145	0.042
Rodents	26	26	0.202 (.015)	0.043 (.005)
Large mammals	4	40	0.233	0.037
Total vertebrates	68	Mean = 24.1	Mean = 0.247	Mean = 0.061
Annual plants	33	11	0.446 (.296)	0.142 (.107)
Biennial plants	11	18	0.220 (.250)	0.079 (.081)
Herbaceous perennials	14	12	0.250 (.312)	0.116 (.118)
Woody trees and shrubs	7	12	0.652 (.314)	0.359 (.102)
Total plants	65	Mean = 12.6	Mean = 0.391	Mean = 0.149

From J. L. Hamrick, 1969. Genetic variation and longevity. In O. T. Solbrig *et al.* (eds.), *Topics in Plant Population Biology*. New York: Columbia University Press. Reprinted by permission of Columbia University Press (for North America) and of Macmillan, London and Basingstoke (elsewhere).

Since 1966, the further development and use of electrophoresis and other biochemical techniques has resulted in the detection of large amounts of genetic variation (in nucleic acids) and epigenetic variation (in proteins) *for which no phenotypic consequences have as yet been demonstrated.* This must mean one of two things: either the phenotypic consequences do not exist, or they do exist but await discovery. Proponents of both viewpoints can be found, and a lively debate has arisen between their camps. Each group has rejected a different assumption of the "classical" school in their attempt to explain the available data (Box 12–D).

One school, led by Crow, Kimura, Ohta, and Lewontin, is variously called the "neutralist," "panneutralist," or "neoclassical" school; excellent summaries of their position can be found in Kimura and Ohta (1971, 1974), and especially Lewontin (1974). This school has retained the explicit assumption that selection is usually "purifying" and also the implicit emphasis on "fitness" as an average or overall measure. What they have rejected is the assumption that variation at all levels has phenotypic consequences (and therefore effects on fitness, however small). They have instead postulated that most electrophoretic variants are selectively neutral and of no phenotypic consequence. In this they have allied themselves with the earlier proponents of genetic drift as a determining factor in evolution. Genetic drift among selectively neutral mutations is held responsible for genetic variation at the protein level both within and between populations, and natural selection is relegated to the role of "purifying" the population by ridding it of harmful mutations (the vast majority), and only with utmost rarity to the role of substituting a new and superior allele for a former one. "If we take it as given that balancing selection is rare and that natural selection is nearly always directional and 'purifying,'" writes Lewontin (1974:197–198),

"how can we explain the observed polymorphism for electrophoretic variants at so many loci? We can do so by claiming that the variation is only apparent and not real . . . [and] in most cases not detectable by the organism [and therefore of no phenotypic consequence]."

BOX 12–D *Different Interpretations of Polymorphism in Natural Populations*

"Neutralist" or *"neoclassical"* school (Crow, Kimura, Lewontin)	*"Classical"* school (Early geneticists; Fisher, Muller)	*"Balanced"* school (Dobzhansky, Wallace)
Neutralist position: Most electrophoretic variants viewed as selectively neutral (with no phenotypic consequences); drift of neutral characters very important in determining frequencies.	*Selectionist position:* Most electrophoretic variants assumed to have phenotypic consequences (including effects on fitness), even if these effects are small; selection viewed as more important than drift in determining gene frequencies.	
Belief in "purifying selection," in which natural selection leads to greater uniformity of populations.		*Belief in "balanced selection,"* in which natural selection leads to balanced polymorphism.
(Variation of phenotypic effects admitted but never emphasized; overall *mean fitness* emphasized instead.)	Phenotypic effects of particular genotypes viewed as invariant; "wild type" usually best adapted.	Phenotypic effects on fitness vary with time, place, and genetic background; "wild type" only an arbitrary designation.
Variation viewed as generally neutral	*Variation viewed as generally harmful*	*Variation viewed as often beneficial*

. . . The variations detected by electrophoresis . . . may be completely indifferent to the action of natural selection. . . . From the standpoint of natural selection they are *neutral mutations.* . . . The so-called neutral mutation theory is, in reality, the classical Darwin-Muller hypothesis about population structure and evolution, brought up-to-date. It asserts that when natural selection occurs it is almost always purifying, but that there is a class of subliminal mutations which are irrelevant to adaptation and natural selection. This latter class . . . is what is observed . . . when the tools of electrophoresis and immunology are applied to individual and species differences.''

Opposed to this school is the ''balance'' or ''selectionist'' school, led by Dobzhansky, Wallace, G. B. Johnson, and others. Waving the flag of opposition to the Platonic *eidos,* these geneticists have rejected the concept of the ''wild type,'' and with it the assumption that natural selection is a ''purifying'' force that leads to genetic uniformity. The selectionists have retained the assumption that most genetic variants have phenotypic consequences (however small), and that selection is a much more important force than drift in determining gene frequencies. Instead of emphasizing the overall or ''average'' fitness of a given genotype, the selectionists emphasize the *variation* in fitness with time, place, and genetic background. Above all, the selectionists assume that if genetic variation is so common, it *must* be favored by selection, and the factors that reduce variation (including both genetic drift and several forms of selection) are opposed by other factors that tend to maintain variability. Balanced polymorphism is obviously one such mechanism. We will examine several other mechanisms in the next section.

Several researchers have supported the selectionist hypothesis by showing that enzyme polymorphisms vary ecologically and geographically, thus suggesting a selective basis even in the absence of a determinable phenotype. (The parameters of fitness often vary geographically, and genes may well affect fitness without any overt phenotypic effects.) For example, Somero and Soulé (1974) found that freshwater fish species in temperate regions have higher levels of polymorphism than in the tropics. Vigue and Johnson (1973) found that the alcohol dehydrogenase (ADH) enzyme in *Drosophila melanogaster* contains but a single electromorph in Florida, with

the level of heterozygosity increasing steadily northward to a maximum in New York State. Merritt (1972) has shown that different electromorphs of the lactate dehydrogenase (LDH) enzyme predominate at different water temperatures in freshwater fish. Not only do polymorphisms vary with certain environmental conditions, but they are sometimes so sensitive to these conditions that Nevo *et al* (1978, 1980) have suggested their possible use as indicators of pollution!

G. B. Johnson (1976 and earlier work cited there) has repeatedly shown in both *Colias* butterflies and *Drosophila* that high levels of heterozygosity for the enzyme α-glycerophosphate dehydrogenase (αGPD) correspond to environments with more variable temperature regimes, whereas lower levels of heterozygosity correspond to more uniform temperature regimes. Johnson has also shown that in more than a dozen *Drosophila* species, levels of polymorphism are consistently higher for enzymes whose substrates, obtained from the external environment, are likely to vary more, but lower for enzymes acting on substances obtained from within the cell. As further evidence of adaptive polymorphism, Johnson notes that regulatory, rate-limiting enzymes are, in general, more highly polymorphic and thus more highly buffered against chemical or temperature variation.

We mentioned earlier in this chapter that a number of base changes in the genetic code (the so-called silent mutations) would have no effect whatever on the initial gene product (epigenotype) and therefore no possibility of any effect on the ultimate phenotype. It is also possible that a certain minority of amino acid substitutions in the noncritical regions of enzymes might leave these enzymes minimally affected, though it is doubtful that they would be completely unaffected. Amino acid substitutions affecting the three-dimensional configuration of the enzyme or altering its ''active site'' will of course have profound effects. Nonenzymatic proteins (like hemoglobin) may prove to be less sensitive to amino acid substitutions. Their direct synthesis from nucleic acids raises another large possibility for ''silent'' or at least ''quiet'' mutations to exist.

Beyond the level of immediate gene products, the possibility has been debated that enzymes with detectable alterations in their own sequences might nevertheless produce indistinguishable phenotypes. It is certainly possible when single characters are considered: any of several distinct red pigments might produce the same shade of red, and if this were the only character under scrutiny by natural selection, any change resulting in the substitution of one red pigment for another would in theory be selectively neutral. On the other hand, the pleiotropic effects of these red pigments (or of the enzymes that produced them) might well have either positive or negative selective value by affecting some other part of the phenotype. Remember that the phenotypic characters most often studied are the size, shape, and color of externally visible parts, and that resistance to disease or to environmental extremes often plays a greater selective role.

The debate between neutralists and selectionists continues. Important early papers and criticisms include those of Lewontin and Hubby (1966), Hubby and Lewontin (1966), King and Jukes (1969), Dobzhansky (1970), and Crow and Kimura (1970). More recent summaries are contained in Lewontin (1974), Johnson (1976), and Wallace (1975).

E. HOW DOES VARIABILITY PERSIST?

We began this chapter with a group-selectionist view of population variability. If a population is to survive, it must possess not only the variations that are adaptive to present conditions but also those variations that are more suitable to possible future conditions. In the long run, genetically variable populations will survive at the expense of genetically uniform ones, especially in times of environmental stress or environmental change. It may therefore benefit the population as a whole to produce a small proportion of unfit individuals in each generation, if these individuals would have a better chance of survival under unusual circumstances, because in a time of crisis it may well be these currently unfit individuals who pull the population through. But there is a serious difficulty to this type of explanation: regardless of any potential future advantage, no disadvantageous trait *could* be maintained in a population for any length of time without certain mechanisms for the storage and protection of genetic variability. We will now examine such mechanisms.

Internal Mechanisms That Preserve Variability

Natural selection was traditionally viewed as a force that would channel populations into a greater degree of uniformity. Of the several mechanisms that protect genetic variation against this type of selection, we shall first consider the internal mechanisms. These are the mechanisms that result from the way the genetic material is transmitted chromosomally, the way that genes are expressed, and the way that genes tend to modify the expression of other genes.

Developmental control of phenotype (phenotypic suppression). Since natural selection can act only on those genes that are expressed phenotypically, any mechanism that prevents phenotypic expression thereby conceals the unfavored genes from the ravages of selection. Remember: natural selection can act only on phenotypes. Phenotypic suppression can work at several levels, as follows:

RECESSIVENESS. The simplest form of phenotypic suppression is recessiveness, in which the gene cannot produce a phenotype unless it is homozygous. The recessive allele in a heterozygote is thus not expressed phenotypically, and it is not exposed to natural selection.

EPISTASIS. Many genes control, modify, or even suppress totally the phenotypic effects of other genes. In these cases, the gene whose expression is modified (the hypostatic gene) is either not exposed to natural selection or is exposed in an altered (usually weakened) form.

INCOMPLETE PENETRANCE AND LIMITED EXPRESSIVITY. Certain genes are expressed—or expressed fully—only a certain percentage of the time (incomplete penetrance), or when not fully expressed, they may be expressed only partially and to varying degrees (limited expressivity). In either or both of these cases, the complete expression of the gene may depend on other genes (as in epistasis) or on certain environmental conditions. For example, the genes controlling susceptibility or resistance to certain diseases are expressed only when those diseases occur. In these cases, again, the gene in question is exposed to natural selection only when expressed; at other times it is either not exposed at all or is exposed only partially.

Cytogenetic mechanisms. The cytogenetic mechanisms discussed here distort the effects of natural selection, thus favoring a different set of genotypic proportions than would be the case if simple selection were acting alone. In the special case where selection would tend to eliminate an unfavored allele from the population, these cytogenetic mechanisms would tend to thwart or at least retard this elimination.

LINKAGE DISEQUILIBRIUM. If two genes are linked, it will take longer for their alleles to assort independently in the population and be distributed according to the Hardy-Weinberg equilibrium. The closer the linkage, the stronger this effect. Moreover, if epistasis is also operating, selection will proceed still more slowly because the optimal genetic combinations do not occur as readily; this is especially true with linked polygenes.

UNEQUAL SEGREGATION ("MEIOTIC DRIVE"). The strange and uncommon phenomenon of unequal segregation is represented by the T-locus in house mice. The recessive t-alleles are lethal when homozygous, but this selection against them is compensated by the production in heterozygous individuals of 90–99% t-bearing gametes and only 10% or less T-bearing ones, instead of the expected 50/50 ratio. This may be treated as a special type of gametic selection against T.

Selective Superiority of Heterozygotes (Heterosis). If heterozygotes have a selective advantage, however slight, over both homozygous conditions, then both alleles will persist in the population indefinitely, even if one is lethal when homozygous. This phenomenon seems quite widespread and might be nearly universal. Several mechanisms exist:

BALANCED DISADVANTAGES. In the situation of balanced disadvantages, each homozygous condition lowers fitness, as in sickle-cell anemia.

HETEROLOGOUS ADVANTAGE. Certain enzymes function only when two or more nearly identical subunits come together to form the functional enzyme, which is thus a dimer or polymer. In some cases, dimers formed of two nonidentical subunits (heterologous dimers) may have more enzymatic activity than dimers composed of identical subunits (Johnson, 1976). Such heterodimer or heteropolymer enzymes are produced by heterozygous genotypes, thus conferring on such genotypes a heterolo-

gous advantage, i.e., an advantage resulting from internal variation.

ALLOSTERIC CONTROL OF ENZYMES. In order for an enzyme to function properly, part of the enzyme (the "active site") must be able to undergo a particular change in shape while the rest of the enzyme molecule keeps its shape to hold the active site in place. This type of enzyme control, called allosteric control, is possible only over a restricted range of temperature, pH, and other conditions. At low temperatures, the entire enzyme molecule is rigid and cannot change shape, but at high temperatures the entire molecule moves too freely and uncontrollably. Similar arguments apply to pH and other conditions.

Different electrophoretic variants of the same enzyme often differ in the range of temperatures and other conditions over which allosteric control is possible. A heterozygous individual, possessing two different gene products (enzyme variants), may therefore be capable of enzyme activity over a wider range of temperatures and other conditions. This gives to heterozygotes a selective advantage because they effectively have two enzymes (e.g., for higher and lower temperature ranges) instead of only one, and they are thus more effectively buffered against variations in temperature, substrate concentration, etc.

Ecological Mechanisms That Preserve Variability

In addition to the mechanisms above, which operate internally, a larger and more diverse number of mechanisms operate at the population level to protect the genetic variability of populations. In the absence of these mechanisms, the population might become genetically more uniform, to its own peril if selective values changed.

Inefficiency of selection. Natural selection is hardly ever an all-or-none process, eliminating the unfit without exception and preserving all other individuals. It is far more common for the less optimal genotypes to survive in *some* proportion, though necessarily in smaller proportion than do the more optimal genotypes. At the other extreme, it is important to realize that even the most optimal genotype realizes some level of mortality, morbidity, and reduction in fecundity: no genotype can guarantee survival; it can only make survival more probable. "The race is not always to the swift, nor the battle to the strong . . . , but time and chance happeneth to them all" (Eccles. 9:11).

Opposing selection pressures (compromise selection). Often a particular phenotype will be a compromise between opposing forces of selection, each acting in the opposite direction. The size of the human head, for example, is controlled by a compromise involving the selective advantages of increased intelligence (which has acted throughout primate history to produce bigger brains and bigger heads), and the selective disadvantage, once a certain size is reached, of not being able to fit through the birth canal (Fig. 12.3). (Thus large fetuses

and their mothers often died in childbirth before the practice of cesarean section became widespread.) Actually, many other selective forces are at work (i.e., the compromise is between many forces and not just two); one of these has already given to the human female a pelvis differing in shape from that of the male, and another has resulted in prolonged infancy, i.e., earlier birth (at a smaller head size). In the past, rather less attention has been paid to this fairly obvious mechanism than it deserves; it is one of the most important of the ecological mechanisms for the preservation of genetic variability. Compromise selection will in general favor an optimum phenotype not at either extreme of the range of variation. In a continuously variable character controlled by polygenes, this would favor any of several equally optimal genotypes (*AAbbcc, AaBbcc, AabbCc, aaBBcc, aaBbCc, aabbCC*, for example), many of them highly heterozygous.

Changes in selection pressure with time (temporal mosaicism). In any case where different selection pressures operate at different times, the results are much the same as in compromise selection with the opposing selection forces acting at the same time.

ONTOGENETIC CHANGES. The selective forces acting on the young or larval stages often differ from those acting on adults.

SEASONAL CHANGES (Fig. 12.4). Different selective forces may act in different seasons. The partial loss of one allele in winter may compensate for the partial loss of another allele in summer.

TEMPORARY, SHORT-TERM CHANGES. Changes in weather may likewise bring about fluctuations in selective forces, with different alleles favored at different times. Fluctuations in population size may have a similar effect, especially if selection is density-dependent.

TIMES OF CRISIS. An important type of change in selective forces may result from the combined effects of seasonal and temporary changes (e.g., severe droughts, severe winters, hurricanes), or from epidemics (not always seasonal), volcanic eruptions, forest fires, or earthquakes.

LONG-TERM CHANGES. Long-term changes include climatic trends, the disruption of environments by human colonization and industry, and other changes whose causes may be obscure.

Fig. 12.3 A case of compromise selection. The head of this infant rhesus monkey was too large to fit through its mother's pelvic girdle. Both mother and infant died in childbirth. Reprinted with permission from Schultz, in S. L. Washburn, *Classification and Human Evolution*. Chicago: Aldine Publishing Co., 1963.

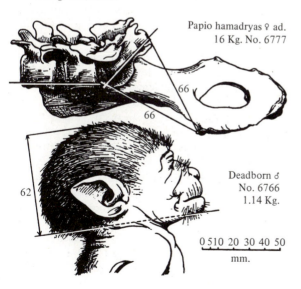

Papio hamadryas ♀ ad.
16 Kg. No. 6777

66

66

Deadborn ♂
No. 6766
1.14 Kg.

62

0 5 10 20 30 40 50
mm.

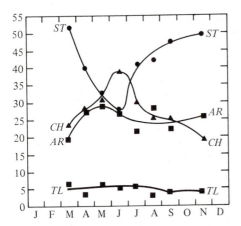

Fig. 12.4 Seasonal variation in the frequency of four chromosome arrangements in a population of *Drosophila pseudoobscura* at San Jacinto, California.

Mosaicism of the environment in space (the "Ludwig effect"). In different habitats, different selective forces will act. A species acquires a selective advantage when it diversifies to fill many subniches (Ludwig's theorem), since it is unlikely to be simultaneously exterminated in all of them. This also makes the population as a whole more resistant to such crises as floods, droughts, violent storms, and epidemics, which may be local in their occurrence or which may affect different environmental niches unequally.

Inverse assortative mating (heterogamy, including heterostyly in plants). If dominants (*AA* or *Aa*) mate preferentially with recessives (*aa*) and vice versa, then genotype *AA* will be selectively eliminated (because it can arise only from crosses between two individuals of dominant phenotype). In the extreme case of complete negative assortment, genotype *AA* will be eliminated in a single generation. All matings thereafter will be *Aa* × *aa* (or *aa* × *Aa*), tending to produce equal numbers of *Aa* and *aa* offspring, and perpetuating both alleles in the population. The case of *Primula* (Fig. 8.4) is a good example of this.

Inherent advantages of rareness. At least three separate mechanisms are known that tend to favor rare genes just because of their rarity, preventing or at least retarding their complete elimination from the population.

MUTATION PRESSURE. If mutations occur at random, the more abundant allele will be affected more frequently and the rare allele only rarely. If a gene pool contains 90% *A* and 10% *a*, then approximately 90% of all mutations at this locus will be from *A* to *a* (or to a novel allele), and only 10% from *a* to *A* or a new allele. (This assumes equal mutation rates in both directions, a condition realized in most cases only approximately. The principle explained here requires only that the disparity between mutation rates, seldom differing from one another by a factor of more than ten, be substantially less than the disparity between gene frequencies.) The long-range effect of this phenomenon is to increase the number of alleles and to increase the frequency of any alleles that happen to be rare in the population.

FREQUENCY-DEPENDENT SELECTION. Thus far we have assumed (for the sake of simplicity) that the selective values of genotypes are independent of genotypic frequencies, but this is not always so. Frequency-dependent selection often favors rare genes simply because of their rareness. The inherent advantages of rareness may be due to a number of causes. Many genes, for instance, produce characters to which parasites and predators may be sensitive. Most predators will evolve search images and behavior patterns that seek out the most frequent prey phenotype. (For example, if two prey colors are equally visible, the predator would do better to seek out the more *frequent* color; if different prey odors are produced, the predator would do better to seek out the more frequent scent.) Parasites and pathogens, likewise, will evolve and adapt to the most frequent or most prevalent phenotype, thus sparing the rare type simply because it is rare. Selection by antibodies, for example, favors pathogens that "adapt" to the most frequently encountered antibody types; an epidemic will thus favor host individuals having rare antibody types to which the pathogen has not become adapted.

Sexual selection (Chapter 11) may also be frequency-dependent. Ehrman and Petit (1968) have shown that *Drosophila* females prefer males of rare genotypes and mate with them preferentially. Since most females mate several times, this "rare male effect" greatly reduces the chances of a female inadvertently mating with the same male twice. It also confers on the rare male an advantage in mating that is a direct consequence of his rarity.

DENSITY-DEPENDENT SELECTION. Density-dependent selection, like frequency-dependent selection, also tends to favor rare genotypes. If a species produces, say, a wet-soil and a dry-soil genotype, density-dependent selection will operate more severely against whichever type is more crowded, thus favoring the less crowded type.

F. GENETIC LOAD

Genetic variability, emphasized in the last section, is purchased at a price; that price is called the **genetic load**. More precisely, the genetic load is defined as

> the relative difference between the actual mean fitness of a population and the fitness that would result if the optimal genotype were ubiquitous.

Genetic load is thus the cost of evolutionary potential. Since the concept was introduced by H. J. Muller in 1950, it has been the subject of intensive study, especially by J. F. Crow and Bruce Wallace.

The calculation of a genetic load is shown in Item 12.2.

Three types of genetic load are here distinguished; this arrangement has been modified after one proposed by J. F. Crow.

Input load. This is the "cost" (reduction in fitness) due to the continual introduction into the population of new (and potentially harmful) genotypes. This cost is tolerated, within limits, for the sake of the usually less frequent beneficial genotypes that are also introduced.

MUTATIONAL LOAD. New genotypes arising through mutation are usually harmful, and tend to reduce the fitness of the population.

IMMIGRATION LOAD. New genotypes introduced from other populations by immigration (gene flow) also tend to reduce fitness.

Balanced load. This is the "cost" (or reduction in fitness) of the various mechanisms that store and protect genetic variation, especially the cost of balanced polymorphism. It is the "cost" of maintaining diversity.

HOMOZYGOUS-DISADVANTAGE LOAD ("SEGREGATIONAL LOAD"). This is the "cost" of producing in each generation homozygous types of reduced fitness (e.g., having sickle-cell anemia) for the sake of heterozygote superiority (in this case, because of their resistance to malaria).

INCOMPATABILITY LOAD. In mammals, antigenic incompatability of fetal and maternal genotypes will reduce fitness. The "cost" of diversity is seen here in the case of Rh-antigens. (Human Rh-positive babies born to Rh-negative mothers tend to die unless they receive special medical attention. If the population were genotypically uniform for either condition, this situation would never occur.)

ENVIRONMENTAL HETEROGENEITY LOAD. This is the "cost" of occupying several different niches simultaneously, or the cost of environmental mosaicism. The reduction in fitness results from gene flow, which introduces genotypes from one environment into another, where they may be disadvantageous.

Substitutional load. This is the "cost" of transient polymorphism, or the cost of replacing one allele with another. In order to completely substitute a new allele for an old one, the number of deaths required is about 5 to 15 times the size of the population at any given time, and the process takes about 300 generations per substitution. This has also been called "the cost of evolution."

British evolutionist J. B. S. Haldane, taking note of the high cost of replacing one allele by another, has questioned how evolution can take place at all without destroying the population altogether. Much has since been written about "Haldane's dilemma," as it has come to be called. Alice Brues (1964), for example, has suggested that the "cost" of evolving be compared with the cost of *not* evolving (keeping the less favorable allele in the population): the advantage is clearly to the evolving rather than the stagnant population. A somewhat different explanation of Haldane's dilemma has also been suggested by Van Valen (1963).

CHAPTER SUMMARY

Compared with the Hardy-Weinberg equilibrium discussed in Chapter 8, natural selection and genetic drift may act to reduce genetic variability and may even result in the complete elimination of one or more alleles from a population. In the face of such potent forces tending to reduce population variability, it is important that this variability be maintained and protected; otherwise the population is in danger of extinction if conditions were to change. Intrinsic mechanisms such as recessiveness and epistasis may conceal certain alleles by repressing their phenotypic effects, thus shielding them from loss through selection. Balanced polymorphism, resulting from the superiority of heterozygotes, represents another important mechanism for the maintenance of genetic variability. One well-known example of this is sickle-cell anemia, a disease perpetuated in many African populations by selection against normal homozygotes. The homozygotes are more susceptible to malaria, whereas the heterozygotes are more malaria-resistant. In malaria-infested regions, this selection has perpetuated sickle-cell anemia in many human populations. Other important mechanisms for the preservation of variability include spatial and temporal variations in selection, density- and frequency-dependent selection, compromise selection, and inverse assortative mating.

The protection and maintenance of genetic variability comes at a cost known as the genetic load. Genetic load is defined as a percentage reduction in overall population fitness, compared with the situation in which the entire population is homozygous for the most favored allele. The importance of the genetic load has been much debated, as has also the relative importance of random versus deterministic factors in evolution.

The technique of electrophoresis detects small differences in the electrical charge on proteins. Using this technique, evolutionary geneticists have discovered large amounts of genetic variation with no obvious phenotypic consequences. Some geneticists believe that much of this variation is selectively neutral, representing the effects of "silent mutations" and genetic drift. Others doubt whether any such variation can be wholly neutral, and they have instead advanced many arguments to show a selective basis for electrophoretic variation and for balanced polymorphism.

FOR FURTHER KNOWLEDGE

ITEM 12.1 *Selection Favoring Heterozygotes: The Case of Sickle-Cell Anemia*

The two forms of hemoglobin known as hemoglobin A and hemoglobin S are perpetuated in malaria-infested regions of the world by a selection which favors the AS heterozygotes. The SS condition (sickle-cell anemia) has always been fatal ($W_{SS} = 0$). The actual mean fitness of AA homozygotes, W_{AA}, varies geographically with the incidence of malaria. The value chosen here is a realistic one.

Genotype	*AA*	*AS*	*SS*	Frequency of *S*
Frequency before selection	p^2	$2pq$	q^2	q
Relative fitness	$W_{AA} = 0.85$	$W_{AS} = 1.00$	$W_{SS} = 0.00$	
Frequency after selection	$0.85p^2$	$2pq$	0	$\dfrac{pq}{0.85p^2 + 2pq}$

ITEM 12.1   *(continued)*

$$\Delta q = \frac{pq}{(.85p + 2q)p} - q$$

$$= \frac{q}{0.85p + 2q} - \frac{(0.85p + 2q)q}{0.85p + 2q}$$

$$= \frac{q - 0.85q(1 - q) - 2q^2}{0.85(1 - q) + 2q}$$

$$= \frac{q - 0.85q + 0.85q^2 - 2q^2}{0.85 - 0.85q + 2q}$$

$$= \frac{0.15q - 1.15q^2}{0.85 + 1.15q}$$

The denominator of this fraction is always positive, and the numerator is zero (no change in q) only when $q = 0$ or $q = 0.15/1.15 = 0.13$. For values of q above 0.13, the numerator is negative (meaning that q decreases); for values of q below 0.13 the numerator is positive and q increases. There is thus a stable equilibrium point at $q = 0.13$; the value of q always changes in the direction of this equilibrium, decreasing when q exceeds 0.13 and increasing when q falls below this value.

ITEM 12.2   *Calculation of the Genetic Load in the Case of Sickle-Cell Anemia*

The data for the calculation of the genetic load in this case are taken from Item 12.1.

Genotype	AA	AS	SS
Frequency before selection	p^2	$2pq$	q^2
Relative fitness	$W_{AA} = 0.85$	$W_{AS} = 1.00$	$W_{SS} = 0.00$

The actual mean fitness of this population is calculated as

$$\overline{W} = (0.85)p^2 + (1.00)2pq + 0$$
$$= 0.85p^2 + 2pq.$$

The genetic load is defined as

$$\frac{W_{max} - \overline{W}}{W_{max}} = \frac{W_{AS} - \overline{W}}{W_{AS}}$$

$$= \frac{1 - 0.85p^2 - 2pq}{1}.$$

For the population at equilibrium under these conditions, we have $q = 0.13$ (see Item 12.1), and $p = 0.87$. Substituting these values in the formula above, we have

$$1 - 0.85(0.87)^2 - 2(0.87)(0.13) = 0.13$$

In other words, compared with a theoretical (and genetically impossible) population composed of 100 percent heterozygotes, a population at equilibrium under these conditions suffers 13 percent mortality each generation, of which $p^2(1 - W_{AA}) = 0.017$, or 1.7 percent, is attributable to selection (by malaria) against AA homozygotes and $q^2(1 - W_{SS}) = 0.114$, or 11.4 percent, is attributable to selection by sickle-cell anemia against SS homozygotes. This is the price of genetic diversity at this locus.

13

Geographic Variation

A. THE NATURE AND IMPORTANCE OF GEOGRAPHIC VARIATION

The phenomenon of geographical variation exists in most species—fish or fowl, vertebrate or invertebrate, animal or plant. Nearly all species that have been adequately investigated show geographic variation, whether they live on continents or islands, in streams or oceans. There are some exceptions: species with very restricted ranges, confined to a single island, lake, or forest, for example, often show no such variation. Species with wide-ranging, migratory populations may often show very little geographic variation, suggesting a constant mixing of their gene pools.

Nearly any character may show geographic variation at any scale or level, either within a species or between related species. Variations in size and in external color markings are most conspicuous to human observers, but variations in other features have been found equally prevalent whenever these have been adequately studied. Body proportions, food preferences, habitat and other ecological preferences, seasons of mating and of flowering, temperature tolerance and similar physiological tolerances, behavioral traits—all these may vary geographically. Nor is geographic variation restricted to visible phenotypic traits, for gene frequencies and protein polymorphisms commonly vary geographically, and the same is often true of chromosomal polymorphisms as well.

Earlier in this century, the study of geographic variation provided an important basis for the reconciliation of the issues that had divided the naturalists and geneticists of the previous era (Chapter 6). The discovery that geographic variation had a genetic basis and the recognition of its important role in speciation were two of the major advances that made the study of geographic variation one of the cornerstones of the modern synthetic theory of evolution. We now realize that geographic variation is an essential prelude to species formation in most if not all cases. Species that do not vary geographically rarely (some would say never) give rise to new species, for the usual modes of speciation require geographical variation as the first step. The role of geographic variation in species formation is discussed in the next chapter.

Some of the early twentieth-century geneticists challenged the importance of geographic variation on the grounds that heritable variation had to be discrete, and that geographic variation, being usually continuous, was nonheritable and of no lasting consequence. The strongly heritable component of continuous geographic variation was first established through transplant studies. In a now classic study by Clausen, Keck, and Hiesey (1940, 1948), plants of the species *Potentilla glandulosa* from three localities in California were each grown at the other two localities (Fig. 6.6). As might be expected, the plants from Stanford and from Mather thrived best in their own localities. But plants taken from Timberline thrived best at Mather. The natural absence of these Timberline plants at Mather results from their lack of success in competition with other plants. At Timberline, however, the harsh environmental conditions exclude all competitors, and the Timberline plants are the only ones capable of surviving under these conditions. Evolutionary success at Stanford and Mather depends on success in competition with other plants of the same species, but at Timberline, evolutionary success depends more on the ability to eke out an existence in this marginal habitat.

More direct evidence for the genetic basis of geographic variation was provided by Dobzhansky and his students, who identified several inversions and other chromosomal variations in the genus *Drosophila*, which they were then able to study in the field. Most natural populations were found to be highly polymorphic for these variations, but even more striking was the fact that the polymorphism itself varied geographically (Fig. 7.1). Here was geographic variation of a trait with an obvious genetic basis: a gradual variation in the frequency of discrete chromosomal rearrangements. Another classic example of geographic variation in a trait with a known genetic basis was provided by Kettlewell's studies of industrial melanism in the peppered moth, *Biston betularia* (Chapter 10).

Still further evidence for the genetic basis of geographic variation has recently been provided through a growing number of studies using the electrophoresis technique. For example, Johnson (summarized in Johnson, 1976) has repeatedly shown that electrophoretic polymorphism for the enzyme α-glycerophosphate dehydrogenase varies geographically in *Colias* butterflies. Populations inhabiting the thermally more variable montane habitats have a much greater degree of electrophoretic polymorphism than those inhabiting either al-pine or lowland habitats, where variation in temperature is much less. Also, where a population straddles the boundary between alpine and montane habitats, the degree of polymorphism varies dramatically, even within the limits of a single population.

B. GENETIC VARIATION BETWEEN POPULATIONS

The modern study of geographic variations begins with the relatively simple case of variation among a series of adjacent local populations. Each local population, also called a **deme**, consists of a group of organisms belonging to the same species and inhabiting the same locality. The limits of a deme are in most cases poorly defined, but it is generally understood that each deme is small enough to permit relatively free interbreeding (if not total panmixia) among its members. In general, species of more sedentary individuals have demes encompassing a smaller area than species composed of more wide-ranging individuals. Williams (1975) defines *vagility* as the mean distance, in a straight line, between the place where an individual comes into being by fertilization and the place where it gives rise to its offspring. The higher the vagility, the greater the area covered by the deme.

The number of individuals in a deme depends not only on the area covered but also on the population density. Most demes have between 100 and 1,000,000 sexually reproducing individuals. Smaller demes cannot long be maintained without undergoing genetic drift and consequent loss of genetic potential, leading in many cases to local extinction.

The differentiation of demes according to geography depends on a number of factors, which fall into two broad categories. First, a greater difference among the selective regimes of different localities generally favors an increase in the extent of geographic variation among populations. Second, anything that restricts gene flow among populations favors an increase in geographic variation, whereas an increase in gene flow diminishes the chances of geographic differentiation. The relative importance of selection versus restricted gene flow is a crucial issue on which opinion is divided, with neutralists (Chapter 12) often favoring restricted gene flow and selectionists continuing to emphasize the importance of natural selection.

The amount of gene flow among populations is one important factor in the origin of geographic variation. Using some of the equations developed in Chapter 8, we can examine the simplest case in which selection against a recessive allele is balanced by introduction of that allele through gene flow (see Item 13.1 on page 237). The difference in gene frequency that can be maintained between two demes under these conditions is proportional to the quantity $q^2 - q^3$, which can never exceed $4/27$, and to the ratio, s/m, between the selection and migration coefficients. If m exceeds the quantity $4s/27$, then the effects of selection will be overcome by the effects of gene flow, and the two populations will be unable to maintain different gene frequencies.

The effects of migration are even greater if we consider gene flow in both directions, or among more than two populations. Lewontin (1974) calculates that, for gene flow in both directions between two populations, the maximum gene frequency difference that can be maintained between them is

$$d = 2 \sqrt{\frac{\bar{p}(1 - \bar{p})}{(1 + 4Nm)}},$$

where $\bar{p}$ is the average gene frequency, m is the average rate of migration per generation, and N is the size of each population. Lewontin points out that the product, Nm, represents the absolute number of individuals migrating. For $Nm = 10$ migrant individuals exchanged per generation (regardless of population size), the maximum value of d (at $\bar{p} = 0.5$) is 0.15. For $Nm = 100$ migrants exchanged per generation, the maximum value of d is only 0.05. Thus, by itself, even a modest amount of migration would bring about a great reduction in the differences between populations. Without some restriction on gene flow, gene frequency differences between populations are difficult to maintain.

The reduction of gene flow among populations would help explain why mountainous regions and islands in archipelagoes are far more productive of geographic variation and ultimately of new species than are vast expanses of grassland, and why lakes and streams contain more geographic variation than ocean waters. Wide-ranging species of broad ecological tolerance will generally show less geographic variation than species in which the ecological tolerance of their populations is more restricted. Species whose distribution is spotty are likely to show much more geographic variation than species whose distribution is continuous. Over the same area, species composed of sedentary organisms with small home ranges and little opportunity for dispersal will show much more geographic variation than species having efficient dispersal mechanisms or larger individual home ranges. Increased isolation of populations from one another will allow accumulated differences between these populations to persist more easily, whether these differences originated by selection or by genetic drift. The extent to which genetic drift contributes to geographic variation was once controversial, but most evolutionists would now acknowledge that it does play a minor role in causing gene frequencies among isolated or semi-isolated populations to diverge. (Section D of the present chapter treats this subject further.)

The isolation of populations greatly increases the occurrence and also the intensity of geographic variation. But is this restriction of gene flow entirely necessary? In the 1940s and 1950s it seemed that a certain impediment to free migration and gene flow was needed to prevent a single, continuous gene pool from developing, even in the face of environmental differences. Computer models (Endler, 1977) have demonstrated the theoretical ability of geographic variation to exist without any restriction of gene flow, provided the selective differences between successive environments are sufficiently great to overcome the effects of gene flow. Endler also lists many studies of natural populations in which geographic variation is maintained, despite extensive gene flow, by selective differences among geographically separated populations.

In one such study, Antonovics and Bradshaw (1970) and Antonovics (1971) compared populations of grasses growing on abandoned mine tailings with populations growing on more natural soil only one or two meters away. Here the intense differences in selection are sufficient to maintain genetically distinct populations, despite the close proximity. These differences in heavy metal tolerance and other characters of obvious selective importance are truly genetic, not mere phenotypic responses. This fact was demonstrated by growing the progeny from seed under experimentally controlled conditions and showing that the differences were maintained. Sharp boundaries of this sort between different selective regimes are known as **ecotones.** Ecotones may produce so-called step clines, which are abrupt changes in phenotypic values or genotypic frequencies. But Endler (1977) and others have shown that step clines may also arise without ecotones under certain conditions.

Selective differences between adjacent populations need not be so abrupt as in the case of ecotones. Even a less drastic ecological variation within the range of a species will generally result in different selective optima and different selection intensities from place to place. These conditions strongly favor the origin of geographic variation. Species whose ranges are ecologically uniform seldom vary geographically. Indeed, the finding of geographic variation is usually taken as strong evidence for the existence of ecological differences among the various populations. More uniform environments are generally characterized by a lesser amount of geographic variation among their inhabitants, and more variegated environments by a greater amount. In the United States, for example, many more mammalian species reach the limits of their range in either Utah or southern California than in Illinois or Kansas, and many more species have within the former two states populations that are placed in different subspecies. The apparent reason for this is that Utah and southern California are much more varied ecologically than either Kansas or Illinois.

C. PATTERNS OF GEOGRAPHIC VARIATION

A thorough description of a geographically variable species might begin with many maps of the species range, each depicting the variation in a different gene frequency or phenotypic character. Most descriptions fall far short of this, and in their attempt to summarize complex variation in simple terms they tend to follow either of two models, depending on whether the variation is described as continuous or discrete.

Discrete Variation and Polytypic Species

Botanists and zoologists of past centuries often began their careers by describing the fauna and/or flora of their own districts. Without exception, these local naturalists came to notice that the animals and plants of their local district fell into discrete types, or *species.* The Platonic theory of forms explained these discrete species by saying that each had a separate and distinct *eidos* (Chapter 3). But as local naturalists ventured to districts beyond their own, problems soon arose. Animals or plants in a neighboring district differed, but only slightly. But animals and plants of distant countries sometimes differed by so much that the differences simply could not be ignored, and each was then described as a separate species.

A serious problem arose when two supposedly distinct species were found to be connected by a series of intermediate populations. Were they *really* two distinct species? A solution proposed with increasing frequency since the early nineteenth century was the description of **subspecies.** A subspecies may be defined as a spatially or temporally circumscribed subdivision of a species characterized by reduced gene flow with other subspecies. Mayr has defined a subspecies as "a group of populations within a species that differ taxonomically from the rest of the species." Mayr's definition more closely reflects current taxonomic practice, but the first definition given above has some theoretical merit: it emphasizes that subspecies are subdivisions of a species, and that there is reduced gene flow between them. Phenotypic characters that vary geographically are certainly used as evidence for reduced gene flow, but they are not in my view the essential feature of subspecies.

Species that are subdivided into subspecies are called **polytypic species.** Polytypic species are simply species with a sufficient degree of geographic variation—and most species are geographically variable to at least some extent. It is a historical fact that since the 1930s many formerly described species have been redesignated as subspecies within a larger, polytypic species.

Figures 13.1 through 13.4 illustrate some of the features of subspecies maps. Species isolated on a series of islands, lakes, or mountain peaks tend to form many subspecies, each with a rather confined range. Sedentary species also form numerous subspecies; this appears to be a large part of the explanation of the numerous subspecies in the genus *Thomomys,* or pocket gophers, in which each individual normally spends its entire life within a radius of less than 100 meters. Broad-ranging continental species, on the other hand, may have a few subspecies,

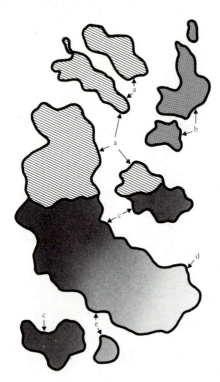

Fig. 13.1 The distribution of a hypothetical species, showing some features of distribution maps. Supspecies **a** contains several disjunct populations, some of which are in conjunction with subspecies **c**. Subspecies **b** is disjunct and differentiated from the rest of the species, and it contains several disjunct but not differentiated populations. Subspecies **c** is in conjunction with **a** but in gradation with subspecies **d**, because a cline goes from **c** to **d**. Subspecies **c** also has disjunct populations, as does the center of the **c**–**d** cline, which is designated as subspecies **e**. From John A. Endler, *Geographic Variation, Speciation, and Clines* (copyright © 1977 by Princeton University Press): Fig. 1–1 p. 5. Reprinted by permission of Princeton University Press.

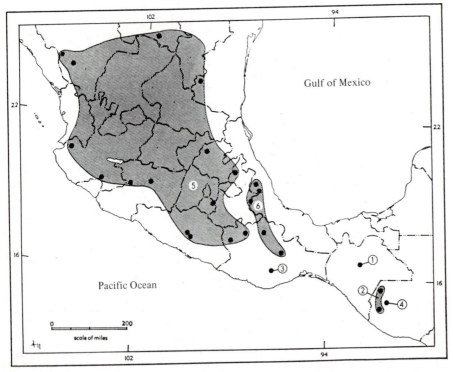

Fig. 13.2 *Sorex saussurei*, a polytypic species inhabiting the mountainous regions of Mexico and Guatemala. Each of the six subspecies is disjunct, being separated from the other high-altitude subspecies by intervening regions of lower altitude. From Hall and Kelson, *The Mammals of North America*. New York: Ronald Press, 1959, p. 48, map 26. Reprinted by permission of John Wiley and Sons, Inc.

1. *S. s. cristobalensis*
2. *S. s. godmani*
3. *S. s. oaxacae*
4. *S. s. salvini*
5. *S. s. saussurei*
6. *S. s. veraecrucis*

228

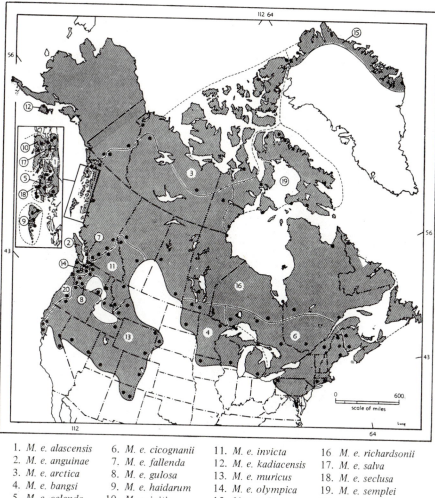

1. *M. e. alascensis*	6. *M. e. cicognanii*	11. *M. e. invicta*	16. *M. e. richardsonii*
2. *M. e. anguinae*	7. *M. e. fallenda*	12. *M. e. kadiacensis*	17. *M. e. salva*
3. *M. e. arctica*	8. *M. e. gulosa*	13. *M. e. muricus*	18. *M. e. seclusa*
4. *M. e. bangsi*	9. *M. e. haidarum*	14. *M. e. olympica*	19. *M. e. semplei*
5. *M. e. celenda*	10. *M. e. initis*	15. *M. e. polaris*	20. *M. e. streatori*

Fig. 13.3 The ermine, *Mustela erminea*, a polytypic species with a broad continental distribution across North America (and Eurasia). Several of the North American subspecies have a broad distribution, but note that along the western periphery of the range there are many subspecies confined to individual islands or to the Coast Range province. From Hall and Kelson, *The Mammals of North America*. New York: Ronald Press, 1959, p. 905, map 463. Reprinted by permission of John Wiley and Sons, Inc.

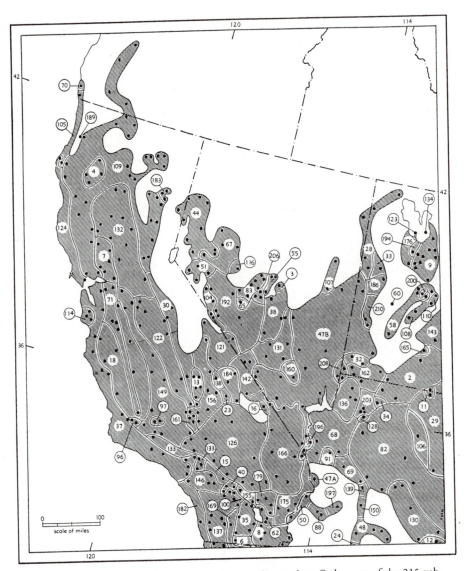

Fig. 13.4 *Thomomys umbrinus,* the southern pocket gopher. Only some of the 215 subspecies listed by Hall and Kelson are shown here. The species ranges over the southwestern United States and adjoining parts of Mexico. The sedentary habits of pocket gophers have probably contributed to this great amount of subspecific differentiation. Adapted from Hall and Kelson, *The Mammals of North America.* New York: Ronald Press, p. 414, map 262. Reprinted by permission of John Wiley and Sons, Inc.

each covering a wide area, although the periphery of the species range may also be inhabited by subspecies with more restricted ranges. For example, the long-tailed weasel, *Mustela frenata,* has five subspecies covering the eastern United States east of the Mississippi River, but in Oregon alone it has six subspecies, and in California it has nine. A related species, the ermine (*Mustela erminea*) has six subspecies along the coast of southernmost Alaska and the nearby Queen Charlotte Islands of Canada, and another seven subspecies within a 100-kilometer (about 60-mile) radius of Seattle, Washington; yet one subspecies (*M. erminea richardsonii*) ranges across mainland Canada from British Columbia to the Atlantic.

Parasites represent a somewhat special case, because they may form "host races," which are more or less equivalent to subspecies, except that they are each characterized by host distribution rather than by geographic distribution in the strict sense. Host races may be considered a peculiar form of subspecies whose boundaries are defined by spatial distribution within their hosts, rather than spatial distribution on a mappable portion of the Earth's surface.

Students should be warned that there *is* a considerable amount of variation even within a subspecies, as everyone who has studied them well knows. In broad-ranging continental species, the boundaries between mapped subspecies are often rather arbitrary in location, since there is really no sharp boundary but only a slow intergradation. The impression one might get from a map, of sharply distinct but internally uniform subspecies, is wholly mistaken, and at best it would fit only a few insular subspecies of restricted geographic range.

Continuous Variation

In view of what has just been said, it became apparent to many naturalists and population geneticists that the geographical variation of a given character in a continental species could better be described as a gradient or continuum between extremes, rather than a discrete number of subdivisions. In the 1930s, this viewpoint came to a head, and in 1939 Julian Huxley coined the term **cline** to refer to such character gradients.

A cline, defined simply, as a character gradient, exists whenever there is continuous geographic variation (Figs. 13.5 through 13.7), with extreme values typically

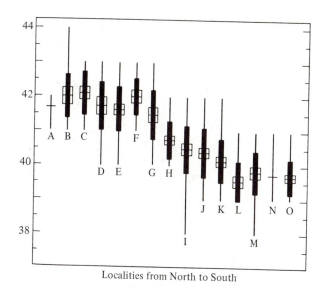

Localities from North to South

Fig. 13.5 Clinal variation in the anchovy, *Anchoviella mitchilli* (class Osteichthyes). The localities are arranged from North to South. The vertical scale gives the number of vertebrae. In each sample, the vertical scale gives the total observed range, the black rectangle gives an interval of one standard deviation to either side of the mean, and the hollow rectangle represents twice the standard error to either side of the mean; the mean is shown as a horizontal crossbar. Reprinted with permission from Mayr, *Principles of Systematic Zoology.* New York: McGraw-Hill, 1969.

coinciding with the extremes of a species range. The recognition of clinal variation has greatly reduced the importance of subspecies descriptions. However, one should keep in mind that a given population belongs simultaneously to as many character gradients (clines) as it has characters.

Clines may be uniformly smooth, or they may have "steps"—regions of more sudden or even abrupt change. A uniform cline cannot be divided except arbitrarily. But a cline with "steps" can be divided at places where the change is most abrupt and the character gradient therefore steepest. This solution would be simplest in

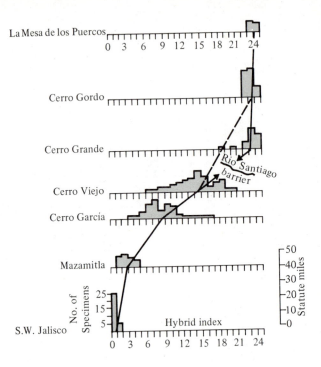

Fig. 13.6 Clinal variation in populations of red-eyed towhees from Mexico. A hybrid index of 0 indicates "pure" *Pipilo ocai* characteristics; a hybrid index of 24 indicates "pure" *P. erythrophthalmus*. Solid lines connect means of adjacent populations. The dashed line indicates the cline that would be expected in the absence of a geographic barrier formed by the Rio Santiago. Reprinted with permission from Simpson, G. G., A. Roe, and R. C. Lewontin, *Quantitative Zoology*, 2e. New York: Harcourt, Brace, Jovanovich, 1960. Modified from a figure by C. G. Sibley. Present version copyright by G. G. Simpson, A. Roe, and R. C. Lewontin.

Fig. 13.7 Clinal variation in the height of the plant *Achillea lanulosa* along a transect from West to East across East Central California. Adapted with permission from Clausen, J., David D. Keck, and William M. Hiesey, *Experimental Studies on the Nature of Species. III. Environmental Responses of Climatic Races of Achillea*. Carnegie Institute of Washington, publication 581, 1948.

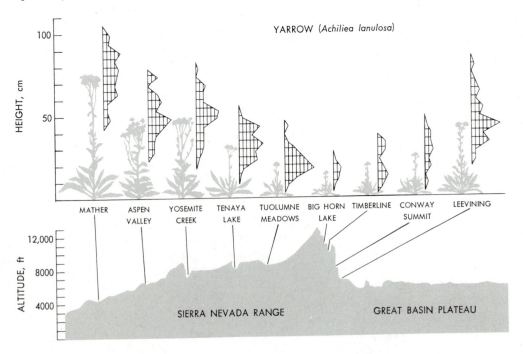

species having a long, narrow range crossing several climatic zones and varying only in this one direction. But in practice it is often hard to divide a species into subspecies using "step" clines, because steps that are steep in one place may become much more gradual elsewhere. Different characters, furthermore, do not always vary together (concordantly). They may have steps in different places, or they may vary according to totally different patterns, forming what are called discordant clines (Dillon, 1962). One promising new approach to the study of clines uses the frequencies of electromorphic alleles to compare geographically separated populations and to estimate rates of gene flow among them (see, for example, Rogers, 1972, Nei, 1972, Ayala et al., 1974, and White, 1978).

D. THE ADAPTIVENESS OF GEOGRAPHIC VARIATION

The causes of geographic variation have been widely debated for many years. Three general explanations have been offered: nonheritable phenotypic variation, adaptive genetic variation, and nonadaptive "random genetic drift." The early twentieth-century geneticists often favored the first of these explanations, for they thought that heritable variation had to be discrete and discontinuous. Geographic variation, since it was continuous, was thought to be phenotypic only and of no evolutionary consequence. Soon the geneticists themselves succeeded in showing how discrete genes could give rise to continuous variation, but it remained for the naturalists to demonstrate clearly that the overwhelming majority of geographic variation was both continuous and heritable.

A more serious challenge was raised against the adaptiveness of geographic variation by certain supporters of the concept of genetic drift, mostly during the 1950s. According to these scientists, certain instances of geographic variation, especially among isolated island populations, were the result of "random drift" and were not necessarily adaptive. The relative importance of genetic drift versus natural selection became the focus of a debate, which continues in somewhat different form today (see Chapter 12, Section F). It is now generally conceded that both natural selection and genetic drift are operating to some extent in all populations, and that neither alone can account for 100 percent of the varia-

tion. Sewall Wright, author of the concept of genetic drift (sometimes called the "Sewall Wright effect"), declared that "accidents of sampling . . . may be responsible for nonadaptive differentiation of small island populations but are more likely to lead to ultimate extinction . . . than to evolutionary advance."

At least four independent lines of evidence are now used to support the hypothesis of adaptive genetic variation as the principal cause of geographic variation. The hypothesis of adaptive genetic variation explains the following evidence better than the hypothesis of genetic drift among selectively neutral characters:

Ecotypic variation. Many botanists, working in such places as California, have documented a number of plant species that vary according to altitude, humidity, and other ecological factors. Clausen and his co-workers have analyzed such variation in *Achillea, Potentilla,* and other plant genera. Through transplant studies, they were able to demonstrate that this type of variation was truly heritable, not merely a phenotypic response to a variable environment. The variation in *Potentilla* is discussed earlier in this chapter.

Substrate variation. Many animals protectively resemble their backgrounds, against which they appear camouflaged. In these species, any abrupt change in the color of the substrate should be accompanied by a fairly abrupt change in the color of the inhabitants. This is exactly what happens in a number of cases, some of them quite striking. In some parts of Arizona, white sands, derived from light-colored rocks (gypsum, quartzite), are occasionally interrupted by virtually black substrates derived from basaltic lava flows. The contrast between light and dark sands is both striking and sudden, yet the climate and other ecological factors are otherwise nearly identical. Several species of desert rodents have developed dark-colored substrate races on the dark sands, whereas the races inhabiting the light-colored sands are themselves very light in color. There is also some evidence that the animals are aware of this difference, since they show a definite preference for backgrounds against which they are camouflaged. The adaptive significance of substrate-dependent variation is clear: individuals who protectively resemble their backgrounds have a better chance of escaping detection by predators. The pattern of color variation, dark where the sand is black and light where the sand is white, shows clearly that the variation is influenced by substrate color and not by any of several other possible variables.

Industrial melanism (Chapter 10) provides another case of substrate-dependent variation, in which many moth species exhibit a much higher frequency of dark individuals when living in soot-covered woods near major centers of industrial pollution.

Variation in physiological traits. Physiological traits, though not obvious to the casual observer, nevertheless have a great effect on survival and reproductive success. Temperature and humidity tolerances, rates of development, and times of flowering may all vary geographically. Rates of water and food intake also vary geographically, even when individuals drawn from different populations are measured under controlled laboratory conditions; the same is true of rates of activity and water loss. In frogs, for example, the following physiological features were found to vary clinally along a north-to-south gradient across eastern North America: rate of water loss, resistance to desiccation, metabolic rates under controlled laboratory temperatures, developmental rate of embryos at controlled temperatures, optimal developmental temperature for embryos, and the age at which the growth of embryos begins to slow down (Moore, 1944, 1950). These findings are unaffected by subsequent studies showing that the frogs placed by Moore in the species *Rana pipiens* belong in reality to two closely similar species (sibling species) with closely adjacent (parapatric) ranges.

Ecogeographic rules. Important evidence for the adaptiveness of geographic variation arises from cases where clines run parallel to major climatic gradients. The same type of cline, responding to the same climatic variable, may exist in a large number of species and be recognizable in the form of an **ecogeographic rule**. These rules pertain primarily to intraspecific variation, particularly to clines (character gradients within a species), although they are sometimes applied across species boundaries as well. The best-known ecogeographic rules are those of Bergmann, Allen, and Gloger.

BERGMANN'S RULE. Bergmann's rule relates body size to average environmental temperature: within a geographically variable species of warm-blooded animals, the populations inhabiting the colder parts of the range will tend to have a larger average body size, and the populations inhabiting the warmer parts of the range will tend to have a smaller average body size (Fig. 13.8). The reason is clear: heat loss is related to surface area, which varies in proportion to the square of linear dimensions.

Fig. 13.8 Bergmann's rule in the geographic variation of house sparrows (*Passer domesticus*) in North America. On this map, dark areas indicate large-sized birds, and light shading indicates small birds, based on a composite of 16 skeletal characters. Reproduced from *Ever Since Darwin, Reflections in Natural History,* by Stephen Jay Gould, by permission of the publisher, W. W. Norton & Company, Inc. Copyright © 1977 by Stephen Jay Gould. Copyright © 1973, 1974, 1975, 1976, 1977 by the American Museum of Natural History.

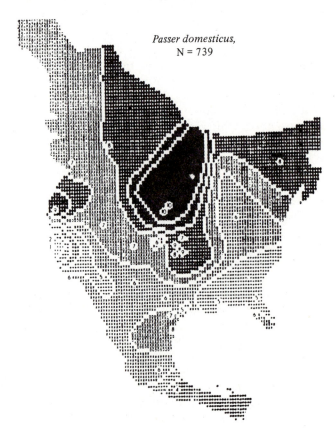

Passer domesticus,
N = 739

The biomass cooled by this heat loss increases as the cube of linear dimensions. The surface-to-volume ratio thus diminishes with increasing size, and the efficiency of heat retention increases. This, in turn, places a selective premium on larger size in colder populations.

ALLEN'S RULE.   Allen's rule deals with protruding body parts, such as legs, tails, snouts, and ears, again in relation to temperature. The rule states that within a geographically variable species of warm-blooded animals, the populations inhabiting the colder parts of the range will have shorter and stubbier protruding parts, and populations in the warmer parts of the range have, on the average, longer and more slender body parts (Fig. 13.9). Again, heat loss is the controlling variable.

GLOGER'S RULE.   A somewhat more complex rule than the previous two, Gloger's rule relates body color to both humidity and temperature. Within a geographically variable species of warm-blooded animals, the populations inhabiting the warmer, arid parts of the range tend to have greater average levels of red and yellow pigments, the populations inhabiting the warmer and more humid parts of the range tend to have high levels of melanin pigment and thus to be much darker, and the populations inhabiting the colder and humid parts of the range tend to be white or largely devoid of pigment (Fig. 13.10).

OTHER ECOGEOGRAPHIC RULES.   Although Bergmann's, Allen's, and Gloger's rules are perhaps better known, they are certainly not the only ecogeographic rules. Bernhard Rensch, who has made a study of these rules, found several others. In birds, for instance, the number of eggs laid together in a nest, the so-called clutch size, tends to increase with increasing latitude in many species, reaching its maximum in the colder parts of the range and its minimum in the warmer parts. This same rule is followed also by a number of cold-blooded animals, and even among mammals litter size varies according to the same pattern. A partial explanation for this is that density-independent mortality (in cold places) favors *r*-selection, whereas density-dependent mortality, which predominates in the tropics, favors *K*-selection.

There are also several additional lines of evidence that corroborate the hypothesis of adaptive geographic variation, but that are equally consistent with the alternative hypothesis of genetic drift among selectively neutral characters:

Fig. 13.9   Allen's rule, illustrated by four closely related species of the genus *Lepus*. Note that the ears, tail, and legs are shortest in the Arctic hare, *L. arcticus* (a), which inhabits the coldest regions, intermediate in the snowshoe hare, *L. americanus* (b), which inhabits boreal forests, and longest in the black-tailed jackrabbit, *L. californicus* (c), and the antelope jackrabbit, *L. alleni* (d), which inhabit the warmest regions in northwestern Mexico and the western United States. Reprinted by permission from Luria, S. E., S. J. Gould, and S. Singer, *A View of Life*. Menlo Park, California: The Benjamin/Cummings Publishing Co., Inc., 1981, Fig. 26.11, p. 616.

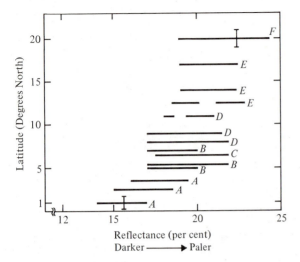

Fig. 13.10 Gloger's rule, illustrated by variation in the coat color of mainland populations of *Tupaia glis* in Southeast Asia. The darkest populations (in the south) live in regions of highest annual rainfall, and the paler population (in the north) live in regions of somewhat less rainfall. Reprinted with permission from Buettner-Janusch, *Origins of Man*. New York: John Wiley and Sons, 1966.

somal rearrangements in *Drosophila,* which they were then able to study in the field. Most natural populations were found to be highly polymorphic for these variations, but even more striking was the fact that the polymorphism itself varied geographically (Fig. 7.1). Among the human genetic polymorphisms that vary geographically (including racially) we may list the blood groups (ABO, MN, Rh, etc.), sickle-cell anemia (Chapter 12), thalassemia, lactose intolerance, and the ability to taste phenylthiocarbamide (Chapter 27). In fact, genetically controlled traits almost always vary geographically if they vary at all. It was this type of evidence that proved most decisive in convincing the geneticists of an earlier generation of the importance of geographic variation to evolution. In certain cases (including sickle-cell anemia, thalassemia, and lactose intolerance in humans; industrial melanism in moths), the selective basis for the geographic variation is known. In many other cases, such as the human ABO blood groups, weak selective differences are known among the several genotypes, but no causal link with the pattern of geographic variation has as yet been established. Genetic drift and selective neutrality (or near neutrality) have also been invoked as possible explanations for much of the geographic variation of genetic traits.

Microgeographic variation. Even within a subspecies, there is a good deal of geographic variation, sometimes known as microgeographic variation. The reason is that every local population is to some extent unique, differing at least slightly from every other local population. This type of variation dispels the illusion of discrete, homogeneous subspecies, each internally constant despite variations in the local environment. It corroborates the suggestion of adaptive variation, for small differences in microclimate should be expected to result in small differences between closely adjacent populations, as we in fact observe. The same effects could also be explained as resulting from genetic drift among neutral characters.

Variation in gene frequencies and chromosomal rearrangements. Theodosius Dobzhansky and his students identified several inversions and other chromo-

Variation in protein polymorphisms. More evidence for the adaptive nature of geographic variation comes from recent studies of protein polymorphisms using the electrophoresis technique. Geographic variation in these proteins, which are direct manifestations of the epigenotype, confirms that the underlying genes also vary in frequency geographically. An increasing number of proteins are being analyzed each year in an increasing number of species. The cumulative evidence for the geographic variation of these polymorphisms is now substantial (Chapter 12); yet the selective significance of this type of variation is still debated. The final resolution of this debate may well depend on the study of such geographically varying polymorphisms.

Various theories concerning the nature of geographic variation and its origins have in past centuries modified our thinking on the nature of species. In our own century, the modern synthetic theory of evolution first crystallized around the study of geographic variation and the realization that geographic variation played an essential role in the origin of nearly all species. This, the formation of new species, is the subject of the next chapter.

CHAPTER SUMMARY

Most species of animals and plants vary geographically. Our understanding of geographic variation has progressed hand in hand with our understanding of the nature of species. It was the study of geographic variation that led to our modern understanding of speciation around 1940, and on this basis the modern synthetic theory of evolution was founded.

Geographic variation within a species may often be described by subdividing the species into discrete units called subspecies; any species so subdivided is called a polytypic species. Another way of describing geographic variation is to consider separately the continuous variation in each character; each character gradient is then known as a cline.

Environmental differences within a species range accompany most cases of geographic variation. These environmental differences result in differing selective regimes throughout the species range, and these differing selective regimes are in turn responsible for the observed geographic variation. Greater environmental variation is generally reflected in a greater amount of geographic variation, whereas more uniform environments are generally inhabited by geographically less variable species. A reduction in gene flow between populations, and their consequent genetic isolation, greatly increases both the occurrence and the extent of geographic variation. Recent evidence suggests, however, that geographic variation can also occur in the absence of such impediments to gene flow.

The adaptiveness of geographic variation is shown by the fact that such variation characterizes even the microgeographic level and by the fact that characters with known selective value and/or known genetic basis vary geographically, too. Among such characters are resemblances of animals to their substrates, physiological rates and other physiological characteristics, gene frequencies, chromosomal polymorphisms, and protein polymorphisms revealed by electrophoresis. The major climatic gradients of temperature and humidity have given rise to certain ecogeographic rules that characterize many independently evolved clines. Bergmann's rule states that a geographically variable species will reach larger average body sizes in colder parts of its range and smaller average body sizes in warmer regions. Allen's rule states that in a geographically variable species the protruding parts will be shorter on the average in cold parts of the range and longer on the average in warmer parts. Gloger's rule predicts darkest average coloration in those parts of a species range that are both warm and moist, the color grading off to paler red and yellow hues as the humidity decreases, or to lighter pigments and eventually white as temperature decreases climatically within the species range.

FOR FURTHER KNOWLEDGE

ITEM 13.1 *The Joint Effects of Gene Flow and Natural Selection*

If both natural selection (Box 11–E) and gene flow (Chapter 8, Eq. 8) are operating together, we may add their effects as if they were worked independently:

total change in gene frequency	=	change due to gene flow	+	change due to selection

or,

$$\Delta q = -m(q - q_\beta) + -sq^2(1 - q)$$

At equilibrium, Δq would be zero (no change in q), so that

$$0 = -m(q - q_\beta) - sq^2(1 - q)$$

$$m(q - q_\beta) = -sq^2(1 - q)$$

$$\frac{m}{s} = \frac{-q^2(1 - q)}{q - q_\beta} = \frac{q^2(q - 1)}{q - q_\beta} = \frac{q^3 - q^2}{q - q_\beta}$$

$$\boxed{\frac{m}{s} = \frac{q^2 - q^3}{q_\beta - q}.}$$

Solving for the difference, $q_\beta - q$, we obtain

$$(q_\beta - q)\frac{m}{s} = q^2 - q^3$$

$$\boxed{(q_\beta - q) = \frac{s}{m}(q^2 - q^3).}$$

The quantity $q_\beta - q$ represents the difference between the gene frequencies of two populations. The largest such difference that can be maintained at equilibrium is limited by the quantity $q^2 - q^3$, which reaches its maximum value of $4/27$ at $q = 2/3$, and by the ratio s/m. This last ratio is large only when selection is intense (large s) or when migration is restricted (small m) or both.

14

The Origin of New Species

Chapter Outline

A. SPECIES DEFINITIONS AND THEIR IMPORTANCE

In order to understand the origin of something, we must first have a clear understanding of what it is whose origin we wish to know. In the case of species, the problem of definition takes on added importance because of the many famous naturalists who failed to solve the problem of species origins through their own misconceptions of the nature of species.

The reproduction of animals and plants, each after its own kind, is mentioned in the Bible (Genesis 1:11–12, 1:24–25) and may be taken as evidence for the belief in species in ancient times. From Aristotle to the modern period, species have always been accorded special importance. Linnaeus emphasized both the reality and the fixity of species. It is no surprise that Darwin used the term "origin of *species*" in the title of his famous book, for it was precisely this possibility that had been denied by supporters of the species concept, especially Linnaeus.

The species concept defined by Aristotle and especially by Linnaeus was a **typological species concept.** Each species was real and immutable because of its *eidos.* The *eidos,* or "type," was considered more perfect and more real than any of the individual members of the species. Variation from the "type" was considered abnormal, if not immoral. Constancy and uniformity were considered most important to every species.

Species descriptions under the typological species concept were supposedly descriptions of the *eidos,* or "type." Often they were simply descriptions of the specimen chosen to represent the type—usually on the grounds that it was "most perfect." Thus the typological species concept was, through species descriptions, also a **morphological species concept.** *Individual specimens were considered to be members of the species if they agreed morphologically with the "type."*

The **biological species definition** replaces the Platonic *eidos* with a view of the species as a group of related populations. Instead of being defined on the basis of their individual morphology, all species are defined on the basis of a single criterion: populations belonging to the same species interbreed with one another whenever they can; populations belonging to different species cannot. The most frequently quoted wording of the biological species definition is that of Ernst Mayr (1942):

''Species are groups of actually or potentially interbreeding natural populations reproductively isolated from other such groups.''

The usual test of distinct species in nature is their ability to coexist over the same area without interbreeding. Populations or species that together inhabit the same area are called **sympatric**; populations or species that inhabit nonoverlapping areas are called **allopatric**. Sympatric populations clearly belong to separate and distinct species if they do not interbreed, but allopatric populations cannot be tested by this criterion.

Those who neglect the population thinking that underlies the biological species definition are certain to misinterpret the definition itself. For example, the definition says nothing of the ability of captive individuals to interbreed—only the interbreeding of natural populations is dealt with. An example may clarify matters. Eastern North America is inhabited by two closely related species of ducks, the mallard, *Anas platyrhynchos,* and the pintail, *Anas acuta* (Fig. 14.1). Captive mallards will mate with captive pintails, and yet the two natural populations coexist sympatrically over much of eastern North America without interbreeding. Given a choice, a bird of either species will mate preferentially with a member of its own species and avoid those of the other species. The behavioral rituals of the two species are different, and this normally precludes their mating. Under natural conditions, a bird of either species is likely to have many potential partners of its own species. But if it does not, and it mates instead with the other species, the hybrids produced have a mixed-up ritual that uses disconnected elements of both mallard and pintail mating rituals. The hybrids are for this reason shunned as mates by mallards and pintails alike, and even by each other. So despite the fact that captive mallards and pintails can be made to mate, the reproductive isolation between them is sufficient to prevent the populations from interbreeding in nature.

The difference between the morphological (typological) species definition and the biological species definition is accentuated in the case of sibling species. **Sibling species** are defined as reproductively isolated species whose morphological differences are cryptic (not immediately evident, though always discernible when sought). Despite their morphological similarity, they are true species because of their reproductive isolation. The limitations of a morphological species definition are shown by such cases as these.

Sibling species are usually discovered when a study is made of some nonmorphological feature, such as behavior or mating calls. A study of crickets, for example, revealed that two distinct types of mating call were given by members of what previously had been thought to be a single species. In this case, as in many others, the discovery of this one difference led to the discovery of other differences (Box 14–A), including minute morphological distinctions that had escaped notice before.

Sibling species show that the amount of morphological differentiation between related species can in some cases be very slight. On the other hand, extreme cases of sexual dimorphism (Chapter 11) illustrate that a great deal of morphological variation can be contained within what is unquestionably a single species. Certain further problems in the characterization of species differences are treated later in this chapter.

Fig. 14.1 Mallard (above) and pintail (below) ducks, two reproductively isolated species of ducks that coexist sympatrically over much of North America. Both species show sexual dimorphism. The cryptically colored females are similar morphologically; males can distinguish among females largely on the basis of courtship behavior. From *A Field Guide to Birds* by Roger Tory Peterson; copyright © 1980 by Roger Tory Peterson; reproduced by permission of Houghton Mifflin Company.

BOX 14–A *Characters Distinguishing Three Sibling Species of Crickets (Class Insecta, Order Orthoptera)*

Morphological differences of the sound file and characteristics of the song in the *Nemobius fasciatus* group of crickets

Measurement	*N. allardi*	*N. fasciatus*	*N. tinnulus*
Average number of teeth in file (right wing)	192 (165–220)	118 (101–126)	214 (196–218)
Average file length (mm)	1.438 (1.32–1.50)	0.992 (0.81–1.12)	1.600 (1.5–1.74)
Duration of pulse (sec)	0.002	0.006–0.010	0.02
Number of teeth struck per pulse	162 (84%)	56 (47.5%)	126 (58%)
Frequency (cy/sec)	7500	7740	6300
Number of pulses or chirps per sec	14–20 pulses per sec, lasting 8 sec	4–12 pulses per chirp, 1.4–5.0 chirps per sec	5–10 single-pulse chirps per sec
Nature of song	Series of separate and distinct pulses	Series of discrete chirps or trains of pulses	High-pitched bell-like note

After Pierce, from E. Mayr, *Animal Species and Evolution*. Cambridge, Mass.: Harvard University Press, 1963. Reprinted by permission.

B. REPRODUCTIVE ISOLATING MECHANISMS

The existence of separate species depends on the existence of **reproductive isolation** between them. Reproductive isolation, at least in the form of hybrid sterility, has been known since antiquity. Yet it was the only important evolutionary force to which Darwin was totally oblivious, presumably because of his strong nominalist position on the matter of species (Chapter 5). Reproductive isolation may be defined as the existence of intrinsic barriers to the interbreeding of natural populations. Each of these intrinsic barriers is called a **reproductive isolating mechanism.** The importance of these mechanisms has been repeatedly emphasized by Mayr, who defines them as "biological properties of individuals which prevent the interbreeding of naturally sympatric populations."

According to the time at which they act, reproductive isolating mechanisms may be classified as either **premating** or **postmating,** a distinction that I believe has greater meaning than the more customary distinction of prezygotic from postzygotic mechanisms. Box 14–B lists these various isolating mechanisms.

BOX 14–B *Reproductive Isolating Mechanisms. Modified after Mayr and other sources*

Premating isolating mechanisms These mechanisms act prior to mating, preventing both the interbreeding of populations and the waste of their gametes in any futile attempts at interbreeding.

1. *Ecological isolation.* This is reproductive isolation in which potential mates never meet, either because they occupy different habitats within the same geographical area (habitat isolation), or because they are reproductively active at different times of the day (secular or temporal isolation) or at different times of the year (seasonal isolation). The importance of these types of reproductive isolation lies in their great efficiency: little if any energy is wasted in unfruitful mating attempts. Ecological isolation is important in both plants and animals. Many species of plants, for example, open their flowers only at certain times of day or flower only in a particularly short season. Birds may develop breeding plumage only during their brief mating season, and the gonads of very many animals are functional only seasonally. In tropical rain forests, many animal species are restricted to a particular stratum in the forest canopy, and they are thus isolated from their near relatives occupying a different stratum.

2. *Behavioral isolation,* also called ethological isolation. This is reproductive isolation in which potential mates do encounter each other but never attempt to mate. The mallard and pintail ducks, discussed earlier in this chapter, exemplify this type of isolation, which occurs in animals only. Secular isolation (above), which some might consider a form of behavioral isolation, is here treated as a form of ecological isolation.

3. *Mechanical isolation.* This is isolation in which copulation or pollen transfer is attempted but does not succeed because of mechanical difficulty. Mechanical isolation among animals often takes the form of a ''lock-and-key'' arrangement of male and female genitalia, which prevents their fitting together properly except with the matching genitalia of the same species. It is an important isolating mechanism among animals with rigid genital armatures, principally insects and other arthropods. Numerous species-level distinctions among insects are based for this reason on the structure of the male genitalia. Mechanical isolation among plants may occur if pollen transfer, whether by insects or by other means, is prevented between species whose stamens and styles differ in height or are inaccessible to pollinators for similar mechanical reasons.

Postmating isolating mechanisms. These mechanisms act subsequent to mating, preventing the interbreeding of populations by removing the hybrids or potential hybrids from the gene pool, a process that wastes many gametes.

4. *Gametic mortality.* This destroys the potential hybrid by destroying the gametes before fertilization can be effected. In plants, it may act through the failure of the pollen to form a pollen tube. In sea urchins, the eggs of different species have receptors that are sensitive only to proteins carried by the sperm of the same species. The death of sperm carrying the lethal gene T in mice is a frequently cited case of gametic mortality, though it has not been proved to act as an isolating mechanism.

5. *Zygotic mortality.* Here fertilization takes place, and a zygote is produced, but the zygote dies, presumably from an inability of its chromosomes to pair during the first mitosis. This is a frequent isolating mechanism between those animals or plant species that are distinguished by their chromosome numbers.

6. *Embryonic or larval mortality.* A zygote is produced, but the hybrid dies in a larval or embryonic stage, presumably because of the developmental anomalies caused by disharmonious genotypes and unequal rates of development.

7. *Hybrid inviability.* The F_1 hybrid is formed, but it fails to reach reproductive age. This case goes one step beyond the previous case because the hybird, during its brief life, competes with both parent species for ecological resources.

8. *Hybrid sterility.* The F_1 hybrid is viable but sterile. The mule, a sterile hybrid between horse (*Equus caballus*) and donkey (*Equus asinus*), is a familiar example of hybrid sterility.

9. *F_2 breakdown.* The F_1 hybrid is both viable and fertile, but the F_2 generation is incapable of proper cell division and dies in an early stage of development.

Premating isolating mechanisms act prior to mating, preventing both the interbreeding of populations and the waste of their gametes in any futile attempts at interbreeding. Postmating isolating mechanisms act subsequent to mating, preventing the interbreeding of populations by removing the hybrids or potential hybrids from the gene pool. Unfortunately, the late action of these mechanisms is accompanied by the waste of many gametes. Natural selection therefore tends to favor organisms that exhibit premating isolation (and do not waste their gametes in unproductive unions) over those that have only a postzygotic isolation, for the latter will waste many gametes (and often much energy as well) that might have been spent in productive unions with compatible mates. Natural selection also favors the earlier action of premating mechanisms and the reinforcement of existing isolating mechanisms by the evolution of additional ones. No isolating mechanism is foolproof, and the existence of one or more "backup" mechanisms ensures further against gene flow between separate species.

One common feature of reproductive isolating mechanisms is that they display **character displacement** (Brown and Wilson 1956), which may be defined as the enhancement of differences between two species in their region of sympatric overlap. Such character displacement may often have an ecological basis in the principle of competitive exclusion. In Fig. 14.2, compare the values given for the bill length of each species in those localities where it occurs by itself. The frequency distributions of these values are broadly overlapping, meaning that the character in question cannot be used by the organisms as a reproductive isolating mechanism or by scientists as a means of distinguishing the two species. But in the area of sympatric overlap, where both species occur together, there is a character displacement such that the two species are readily distinguishable from each other on the basis of a single trait. Character displacement may therefore lead to niche subdivision in the region of sympatric overlap. Another adaptive advantage of character displacement is that individuals of either species can more readily distinguish potential mates, with less waste of energy and with less chance of error. In the region where each species occurs by itself, there is no need for character displacement, and the character reaches at equilibrium a mean value determined by other selective forces.

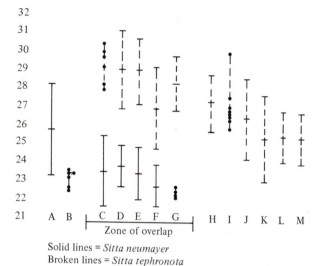

Solid lines = *Sitta neumayer*
Broken lines = *Sitta tephronota*

Fig. 14.2 Character displacement in the bill length of two partially sympatric species of nuthatches, *Sitta neumayer* and *S. tephronota* (class Aves), along a series of localities from west (Greece, A) to east (China, M). The differences between the two species are greatest in the region of sympatric overlap (in Iran). From Vaurie, *Amer. Mus. Novitates*, 1472 (1950).

C. PHYLETIC VERSUS TRUE SPECIATION

Speciation—the origin of new species—has been the subject of much controversy and a certain amount of misunderstanding as well. Several distinct phenomena are involved. The fusion of two previously existing species through a breakdown in their reproductive isolation leads to the establishment of a new species, distinct from either parent species. This rare occurrence will be discussed in Chapter 15.

Much of the confusion centers on the anagenetic process (Chapter 17) often called "phyletic speciation," in which one species gradually changes and becomes sufficiently different to be recognized as distinct from its ancestors. This process, more accurately called **phyletic transformation**, comprises several types of change through time: changes in gene frequencies, changes in the frequencies of chromosomal inversions and other rearrangements, changes in behavior, and most conspicuously, a variety of changes in morphology. No subdivision of species occurs; no reproductive isolating mechanisms evolve; all that occurs is the gradual transformation of one species into another. Most of Darwin's discussions on the ''origin of species'' focus on this process rather than true speciation. Darwin had sought to explain the ''origin of species,'' but he succeeded in explaining only the process of successional change (anagenesis, Chapter 17). By confusing phyletic transformation with true speciation, Darwin ultimately failed to solve the problem he had posed for himself.

True speciation, on the other hand, involves something more: the *splitting* of species and the origin of two or more species where only one had previously existed. No great degree of morphological change need occur, as the case of sibling species shows. But in the light of the biological species concept, true speciation requires another essential change: subdivision of the gene pool, or the origin of two or more separate gene pools where only a single gene pool had previously existed. True speciation may thus be defined as the splitting or multiplication of species by subdivision of the gene pool, including the requisite acquisition of reproductive isolation.

D. SUDDEN SPECIATION (BY SALTATION)

Several possible mechanisms (Boxes 14–C and 14–D) have been proposed to account for true speciation. True speciation could occur either suddenly, through innovating individuals, or gradually, through innovations that spread through populations.

Early in the twentieth century, theories of sudden (if not instantaneous) speciation were in vogue. The ''bridgeless gap'' that was seen to separate species could hardly be crossed, so it seemed, by any gradual process. The predominant frame of mind was to look for sudden, discrete changes, such as those postulated by De Vries. In his mutation theory, doubtlessly influenced by the unusual manner of chromosomal variation in *Oenothera* (Chapter 7), De Vries stated that new species could arise

---

BOX 14–C *Possible Modes of Speciation. Modified after Mayr (1963) and Endler (1977).*

PHYLETIC SUCCESSION: The gradual replacement of one species by another, without branching.

FUSION OF SPECIES by the complete breakdown of reproductive isolation.

TRUE SPECIATION: The splitting of species, with concomitant evolution of one or more reproductive isolating mechanisms.

   Sudden speciation, through individuals (speciation by saltation).

      Genetically, through ''macromutation'' (''macrogenesis'').

      Chromosomally, through hybridization and/or polyploidy.

   Gradual speciation, through populations (see also Fig. 14.3).

      Sympatric speciation, in which new species arise within the same population.

      Nonsympatric speciation, in which new species arise within different populations.

        Allopatric or geographical speciation, in which new species arise through spatially separated (isolated, allopatric) populations.

        Parapatric speciation, in which new species arise through populations that maintain genetic contact wherever they are contiguous (parapatric).

        Alloparapatric speciation, in which new species arise through populations that are at first allopatric but later become parapatric before a completely effective reproductive isolating mechanism has evolved.

BOX 14–D *An Alternative Classification of Modes of Speciation (after White, 1978a).*

1. Strict allopatry without a narrow population bottleneck
2. Strict allopatry with a narrow population bottleneck ("founder principle")
3. Extinction of intermediate populations in a chain of races
4. Clinal speciation
5. Area-effect speciation (primarily genic)
6. Stasipatric speciation (primarily chromosomal)
7. Sympatric speciation

Models 1–3 of the scheme above would be considered forms of allopatric (or geographic) speciation by Mayr (1963) or Endler (1977). Models 4 and 5 clearly fall into the category that Endler called "parapatric," or that Mayr called "semi-geographic." According to White, model 6 also falls into this category, though it clearly differs from any of the forms of speciation that Endler or Mayr recognize. In analogy with Endler's term "allo-parapatric" for populations that undergo one form of change, White's stasipatric speciation model would have to be called "symparapatric" to fit into Endler's scheme, for White envisions that the new population would be characterized by a chromosomal rearrangement "originating somewhere within the area occupied by the ancestral species," i.e., sympatrically, but would then lead to a parapatric distribution pattern.

suddenly through a single large mutation, or "macromutation." The term "saltation," from a Latin word meaning "jump," is often applied to such theories of sudden speciation; the word "macrogenesis" is similarly used.

The sudden creation of a new species by a single individual is fraught with problems. With whom would this individual mate? If the new individual truly belongs to a new species, any back-crossing to its parent species is by definition precluded. Second, most species are estimated to differ from one another by thousands of genetic differences. A "macromutation" of sufficient magnitude to create a species distinction would have to involve a large amount of change, and large, sudden mutational changes are nearly always very harmful. Finally, to maintain a harmonious genotype despite sudden change, a very large number of genes would have to mutate simultaneously in a well-planned manner—a wholesale "mass mutation." Richard Goldschmidt, one of the last proponents of sudden speciation, saw a "hopeful monster" arising by a process of mass mutation and producing a new species all at once. Goldschmidt's many critics, including Mayr, have emphasized that the probability of a harmonious genotype resulting from a mass mutation of this sort is vanishingly small, and that a hopeless monster, rather than a hopeful one, would inevitably result.

Yet some evolutionists now searching for alternative mechanisms of sudden speciation have begun to view some of Goldschmidt's ideas more sympathetically.

Of the possible mechanisms for sudden speciation, chromosomal changes have always seemed more plausible than macromutation. Speciation by polyploidy has been documented in several plant species, as has species fusion (or partial fusion) by introgressive hybridization. Both of these mechanisms are treated in Chapter 15. Stasipatric speciation, treated in the next section, is sometimes considered a form of sudden speciation.

E. MODES OF GRADUAL SPECIATION

It has now been definitely established that nearly all speciation is gradual. The only mechanism of sudden speciation that has received widespread contemporary support is polyploidy (Chapter 15). (White's stasipatric model is here considered "gradual.")

Various possible modes of gradual speciation are listed in Boxes 14–C and 14–D and diagrammed in Fig. 14.3. The major distinction is between **sympatric speciation,** in which the resultant daughter species originated in the same population, and all the other possibilities,

THE INTERRELATIONSHIPS AMONG
PATTERNS OF GEOGRAPHIC VARIATION AND MODES OF SPECIATION

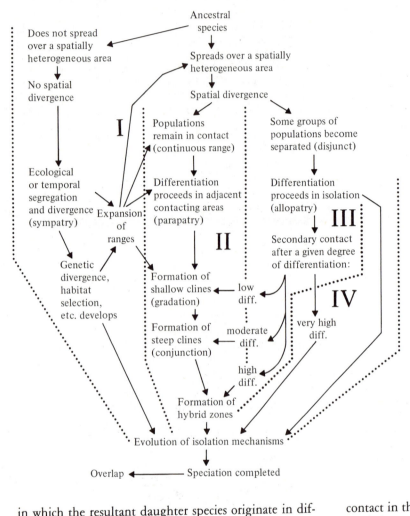

Fig. 14.3 Comparison of four modes of speciation: I, sympatric speciation; II, Parapatric ("semigeographical") speciation; III, alloparapatric speciation; IV, allopatric (geographical) speciation. From John A. Endler, *Geographic Variation, Speciation, and Clines* (copyright © 1977 by Princeton University Press): Fig. 1–2, p. 13. Reprinted by permission of Princeton University Press.

in which the resultant daughter species originate in different populations. These different populations may inhabit discontiguous ranges truly isolated from one another (allopatric populations), or they may inhabit adjacent or contiguous ranges in genetic contact with one another (**parapatric** populations). Speciation through extrinsically separated, allopatric populations is called either **geographic speciation** or **allopatric speciation,** and it is considered to be the major if not the exclusive means of speciation for the vast majority of species. Speciation through parapatric populations, without loss of genetic

contact in the formative stages, is called either **parapatric** or **semigeographic speciation.** These two modes of speciation are furthermore the ends of a continuum, for in some cases speciation may undergo its early stages allopatrically and its later stages parapatrically; this intermediate type of speciation has been termed alloparapatric speciation by Endler (1977). One additional form of speciation, the stasipatric speciation model of White (1968, 1978a) and his co-workers (e.g., Key, 1968), does not fit conveniently into the scheme above and will thus be considered separately.

Gradual, Sympatric Speciation

Once it became apparent that sudden speciation was unlikely to work, attention was turned to models of gradual, sympatric speciation. Darwin had himself been a gradualist, and some of the early population geneticists were searching for a simple solution consistent with Darwin's ideas. Gradual speciation was more consistent with Darwinism, and speciation from a single population seemed at first a simpler model than allopatric speciation.

Disruptive selection (Fig. 11.1) seemed to be a likely mechanism to account for the splitting of a species and the subdivision of its gene pool. Fisher (1930) suggested that such a model of speciation by disruptive selection would not work, because any genetic interchange between the extremes of a frequency distribution would restore the middle values, even in the face of strong disruptive selection against them. In other words, any interbreeding whatever between the extremes of a distribution (the incipient species) would tend to make them more similar and more connected by intermediate values. Only if the two populations were reproductively isolated and did not interbreed, that is, only if two separate species were present initially, could two separate species result.

Despite these theoretical objections, the theory of sympatric speciation through disruptive selection continues to find occasional supporters. Contemporary support for the possibility of sympatric speciation is summarized especially by White (1978a), who recognizes three lines of supportive evidence: ecological and host "races"; speciation in small, isolated habitats; and experiments simulating the speciation process.

Many species have two or more genetically distinct strains living in different habitats or on different hosts within the same broad geographical range. These "ecological races" (or "host races") interbreed whenever they come in contact, but their habitat or host preferences usually ensure that such contacts remain infrequent. The ability of these ecological and host races to evolve into separate species has been disputed. Reproductive isolation, the usual criterion of separate species, has been demonstrated in very few instances. In one of these instances, Bush (1975 and works cited there) has studied the formation of host races in flies of the genus *Rhagoletis*. These flies have narrow host preferences; Bush describes them as plant parasites. Some of the host races of *Rhagoletis* have formed on introduced fruit plants (e.g., apple) within historic times. Some have achieved at least a partial degree of reproductive isolation, as shown by the reduced viability of hybrid eggs. Many other claims of sympatric speciation have failed to demonstrate either (1) that presently sympatric host races arose sympatrically, or (2) that reproductive isolation has been achieved. Bush (1975) believes that sympatric speciation "appears to be limited to...parasites and parasitoids."

Small, isolated islands and similarly isolated freshwater lakes often support faunas in which a few families or genera are unusually well represented. Rapa Island in the Pacific supports over 40 endemic species of the weevil genus *Microcryptorhynchus*. Lake Baikal in the Soviet Union supports many endemic species, including nearly 240 endemic species of amphipod crustaceans (Kozhov, 1963). In Africa, Lake Victoria, Lake Tanganyika, and Lake Malawi each supports a large fish fauna. Over 200 species of fishes of the family Cichlidae live in Lake Malawi, 126 in Lake Tanganyika, and 170 in Lake Victoria, though very few of these species (4 in Lake Malawi, 6 in Lake Victoria, and none in Lake Tanganyika) occur elsewhere (Fryer and Isles, 1972). Supporters of sympatric speciation claim that these facts can be explained only by sympatric speciation, since the localities are so small and isolated as to preclude either allopatric speciation or the geographic variation that allopatric speciation would require. But champions of allopatric speciation (e.g., Mayr, 1963) insist that such localities are very old and ecologically varied, thus affording ample opportunity for allopatric speciation, and that sympatric speciation still lacks a plausible mechanism.

In a series of experiments often cited in support of sympatric speciation, Thoday and Gibson (1962, 1970) and Thoday (1972) achieved partial success in maintaining population cages of fruit flies at different temperatures, though still connected in such a way that interbreeding was not precluded. A good degree of morphological distinctness was produced in these experiments, though subsequent workers (e.g., Robertson, 1970) have been unable to duplicate the results. What Thoday's experiments have shown, furthermore, is a model more for geographic variation and subspecies formation than for sympatric speciation, because the crucial final stage, reproductive isolation, was never achieved.

The Geographical Theory of Speciation

The theory that species can multiply only through allopatric populations is known as the **geographical theory of speciation,** or sometimes as the allopatric or geographical speciation model. According to this theory, the subdivision of the gene pool takes place only through spatial (especially geographical) isolation. Geographical speciation, in other words, is by overwhelming odds the principal mode of species multiplication.

Recall, now, that reproductive isolation requires intrinsic barriers to the interbreeding of natural populations, and that these intrinsic barriers, called reproductive isolating mechanisms, allow two or more natural populations to live sympatrically without interbreeding. One of the cornerstones of the geographical theory of speciation is the belief that *reproductive isolation generally arises during geographic isolation,* i.e., that intrinsic barriers evolve only during the existence of extrinsic ones.

Geographic isolation is the separation of populations by extrinsic barriers to dispersal. It is generally a fortuitous consequence of inhabiting a range that is or becomes discontinuous. An extrinsic barrier may consist of an obvious geographic feature, like a mountain, or it may simply be an ecologically unsuitable habitat. Of course, a barrier that is formidable to one species may be easily penetrated by another. The dispersal abilities of each species determine what constitutes a barrier. Not all extrinsic barriers to the dispersal of populations are geographical in the strict sense. In the case of parasites, isolation of ''host races'' from one another can be equally effective and serves as an extrinsic barrier in much the same way as does geographic separation.

The geographical theory of speciation invokes several means by which one species can give rise to two. According to the most frequently cited model, the species must first be geographically variable. Second, it must become subdivided into units that are *geographically isolated* from each other. Third, the geographically isolated units must later become *reproductively isolated* from each other. Finally, the two resultant species may now come into contact with each other, and their failure to interbreed in nature serves as proof of their distinctness as species. *Sympatric overlap without interbreeding* is thus the test for the completion of this process (Fig. 14.4). White (1978a) refers to this as the ''dumbell diagram''

model, in which geographic variation arises before geographic isolation. On islands and other discontiguous ranges it is more common for geographic isolation to come first, and geographic variation to evolve subsequently as the result of differing selective regimes; White (1978a) calls this the ''founder effect'' model. Yet a third model of geographical speciation is explained by Mayr (1963) as ''extinction of the intermediate links in a chain of populations of which the terminal ones had already acquired reproductive isolation.''

Adaptive radiation and other long-range consequences of allopatric speciation will be discussed in later chapters.

Parapatric and Alloparapatric Speciation

The possibility of speciation without total loss of genetic contact until the final stage has been warmly debated for nearly 40 years. The founders of the modern synthetic theory of evolution were also supporters of the theory of geographical speciation, though both Julian Huxley and Ernst Mayr admitted the possibility of ''semigeographic'' or parapatric speciation of populations separated by distance but never completely isolated from one another. In order to maintain their differences and to evolve reproductive isolating mechanisms, such populations would have to experience selection regimes that differed even more than in the case of geographic speciation. Endler (1977) had developed a model to predict the conditions favorable to parapatric, as opposed to allopatric, speciation.

The annual plant genus *Clarkia* contains several well-studied instances of parapatric speciation (Lewis and Roberts, 1956; Lewis, 1966, 1972). *Clarkia lingulata,* with tongue-shaped petals and a chromosome number of $2N = 18$, appears to have formed parapatrically on the periphery of the range of *Clarkia biloba,* a more wide-ranging species with heart-shaped petals and $2N = 16$. A chromosomal sterility barrier separates these two species. Two chromosomal translocations account for most of the sterility barrier, for the aneuploid increase in chromosome number, and for the change in petal shape. There are also ecological differences between the two species.

Endler (1977) has proposed the term ''alloparapatric'' for another possible mode of speciation, in which the early stages take place in allopatric populations, but

Stage 1. A uniform species with a large range

Followed by:
Process 1. Differentiation into subspecies

Resulting in:
Stage 2. A geographically variable species with a more or less continuous array of similar subspecies

Followed by:
Process 2. a) Isolating action of geographic barriers between some of the populations;
also b) development of isolating mechanisms in the isolated and differentiating subspecies

Resulting in:
Stage 3. A geographically variable species with many subspecies completely isolated, particularly near the borders of the range, and some of them morphologically as different as good species

Followed by:
Process 3. Expansion of range of such isolated populations into the territory of the representative forms

Resulting in either
Stage 4. Noncrossing, that is, new species with restricted range
or

Stage 5. Interbreeding, that is, the establishment of a hybrid zone (zone of secondary intergradation)

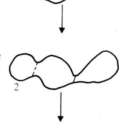
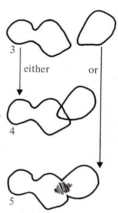

either or

(a)

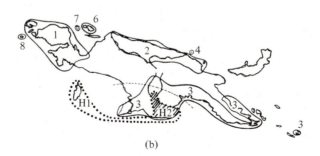

(b)

Fig. 14.4 The geographic model of speciation. (a) Mayr's original figure and explanation. (b) A case of geographic speciation (stage 5) in New Guinea kingfishers, *Tanysiptera* (area of sympatric overlap shaded). Reprinted with permission from Ernst Mayr, *Systematics and the Origin of Species.* New York: Columbia University Press, 1942. Copyright by Ernst Mayr.

the later stages take place in parapatric ones. Although this mode of speciation has long been recognized by Mayr and others as theoretically possible, in practice it is hard to distinguish cases of alloparapatric speciation from truly allopatric speciation on the one hand, and from parapatric speciation on the other.

Stasipatric Speciation

White (1968, 1978a, 1978b) and co-workers (e.g., Key, 1968) have advanced what they call a "stasipatric" model of speciation. According to their model, a new species arises first by means of a chromosomal rearrangement within the geographic range of a parent species (i.e., sympatrically). The chromosomal rearrangement must have reduced fitness when heterozygous, so that selection against heterozygotes serves from the start as an incipient means of reproductive isolation. The population carrying the new chromosomal arrangement spreads from its origin within the range of the parent species and establishes a parapatric distribution. In some localities, the new arrangement will be favored, and the ancestral arrangement is favored in others. Competition between the two incipient species ensures their nonoverlapping (parapatric) distribution, and selection against heterozygotes results in perfection of further reproductive isolating mechanisms between them.

The morabine grasshoppers of the Australian genus *Vandiemenella* (Fig. 14.5) were the original basis for the stasipatric model, and they remain one of the most intensively studied cases. According to White (1978a), there are seven species in this group, four of them with chromosomal races. Most of the species and all of the chromosomal races have parapatric distributions, with occasional narrow zones of hybridization about 200–300 meters wide. All twelve races and species are chromosomally distinguishable, and White and his co-workers believe that it was originally these chromosomal changes that brought about the present taxonomic differences. Nevertheless, the precise geographic origin of these chromosomally defined taxa is uncertain: Key (1968) postulates a parapatric origin of most chromosomal variations, whereas White (1978a and elsewhere) believes that most of these variations arose sympatrically within the range of the original species, *V. viatica*.

The importance of chromosomal rearrangements in speciation has been noted by many previous workers who

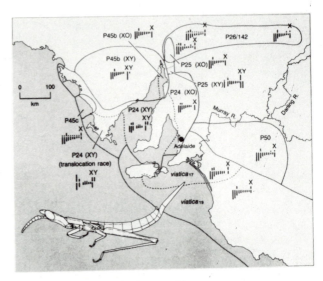

Fig. 14.5 Speciation in the morabine grasshoppers (genus *Vandiemenella*) of Southern Australia, according to White (1978). Seven species ranges are shown by solid boundaries; chromosomal races within these species are shown by dotted boundaries within these ranges. This is one of the examples used by White and his co-workers to support their model of stasipatric speciation. Reprinted with permission from *Modes of Speciation* by Michael J. D. White. W. H. Freeman and Company. Copyright © 1978.

do not necessarily accept the stasipatric model. Indeed, a large majority of new species seem to differ chromosomally from their parent species in some major respect, though a number of *Drosophila* species are chromosomally indistinguishable from one another. The possibility that chromosomal heterozygosity is at a selective disadvantage is also supported by many workers, though it is far from universal—many chromosomal arrangements (e.g., in *Drosophila*) show heterozygote superiority.

The most controversial aspect of the stasipatric model is that the new chromosomal arrangement is supposed to characterize a new population from the start, yet be disadvantageous in the heterozygous condition. However, most new chromosomal arrangements arise first in single individuals and are initially heterozygous.

The objections to most models of sudden speciation can also be invoked against the initial stage of stasipatric speciation, which is supposed to occur suddenly. If the new chromosomal arrangement arose first in a single individual, it would be able to mate only with its parent species and would thus produce heterozygotes only. If the heterozygotes were favored, their numbers might increase to the point where matings between them were sufficiently common for several homozygotes to be produced. But the model postulates just the reverse situation, in which heterozygotes are selected against! The new arrangement, however superior it may be in homozygous condition, seems foredoomed if it arises in a single individual and is disadvantageous when heterozygous, unless the founder individual is somehow capable of either self-fertilization or asexual reproduction. We might therefore expect more stasipatric speciation in organisms capable of uniparental reproduction, as in *Clarkia*. The case of *Clarkia lingulata* has in fact been considered an instance of stasipatric speciation by some workers.

Now, instead of assuming that the new chromosomal arrangement arises first in a single individual, what if we assume that it occurs simultaneously in many individuals at once? White himself (1978a) argues strongly against this possibility, insisting that each chromosomal rearrangement must be an event so rare as to be unique. A single-gene mutation involves a change at a single location, but any chromosomal change (except a terminal deletion) takes place at a minimum of two locations —frequently more. Given the very low probability of a chromosome break at any one particular location, White calculates that the probability of an independent recurrence of any two-break chromosomal change should be at most 1 in every 12.5×10^9 for the polytene chromosomes of *Drosophila*. The independent recurrence of such a chromosomal change in more than two individuals is vanishingly small.

F. INCOMPLETE SPECIATION AND THE "SPECIES PROBLEM"

The geographical theory of speciation has been quite successful in explaining the origin of most species. Few biologists would now quarrel with the origin of species through geographic speciation, yet many would take ex-

ception to the universal applicability of this model to the practical matter of classifying particular species. For one thing, the geographical theory of speciation pertains primarily to sexually reproducing species. In the case of species that always reproduce asexually, many biologists have even questioned whether the concept of "species" has any true meaning. This difficulty may be more apparent than real, however, for many asexual species do reproduce sexually (or parasexually) at least on occasion. Bacteria, for example, have a number of parasexual processes: conjugation (including the exchange of plasmids), transduction, and transformation.

Various difficulties are encountered in applying the biological species concept to particular species. These difficulties include asexual species, poorly known species (incomplete data), cases of incomplete speciation, and species extended through space and time (allopatric and diachronic species). Critics of the biological species concept (e.g., Ehrlich, 1961, Sokal and Crovello, 1970) have long used these difficulties to argue against the applicability, even against the reality, of the biological species concept. Many of their arguments are similar to those used in the eighteenth century by Buffon and Adanson against the then typological species concept of Linnaeus (Chapter 3).

Some of the knottiest problems surrounding the biological species concept concern those cases in which speciation is incomplete. These cases are as salutary to the evolutionist as they are annoying to the taxonomist, for their existence is quite predictable from geographical speciation theory. Recall that during geographical isolation, a reproductive isolating mechanism *may* evolve. If such an isolating mechanism has had sufficient time to evolve before sympatry is reestablished, the effectiveness of this isolating mechanism precludes interbreeding and necessitates treating the resultant populations as separate species. But it should also be apparent from this model that an isolating mechanism might in some cases remain imperfectly evolved, such that a certain (low) level of interbreeding is still possible. All cases of geographical speciation should pass through such a stage; it is only a matter of chance that sympatry should in some cases be reestablished at precisely this stage. The possibility of incomplete speciation is predictable under the geographic theory of speciation, and supporters of the theory properly take delight in such occurrences. But these cases of

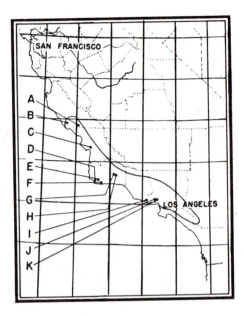

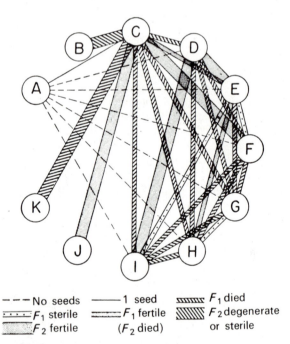

Fig. 14.6 Fertility relationships among *Clarkia deflexa* populations in California. Reprinted with permission from H. Lewis, *Evolution,* 7:10, 1953.

incomplete speciation are annoying to taxonomists because the taxonomic status of imperfectly isolated groups of populations is unclear: should they be placed in two species or in one? Practical solutions to this problem may be hard to find. A consensus may be difficult to achieve in particular cases, especially where data are also lacking. A theoretical answer may be found, however, in the very instability of this situation. The partial reproductive isolation that characterizes incomplete speciation is usually temporary only: *either* the reproductive isolation is strong enough that the species differences will persist, *or* these differences will continue to diminish through interbreeding until there exists only a single species again. Then we may say, in retrospect only, that there had existed two separate species in the first case but only one species in the second. In higher plants, but far less often in animals, a third possibility may exist in the formation of a stable hybrid zone (Chapter 15).

Another problem concerns ring-shaped clines with circular overlap. Suppose, in a series of populations, that each could interbreed with the adjacent population, but that distant populations were reproductively isolated from one another. The grass frogs (*Rana pipiens*) of eastern North America were originally described by Moore (1946) as illustrating such a case, though his results have since been disputed by others (e.g., Littlejohn and Oldham, 1968). According to Moore, attempted crosses between adjacent populations in his north-to-south transect (Georgia × North Carolina, North Carolina × Virginia, Virginia × Maryland, Maryland × Pennsylvania, Pennsylvania × New York, New York × Vermont, Vermont × Quebec) always produced viable hybrids, but crosses between distant populations (New York × Georgia, Virginia × Quebec) did not, with the hybrids usually dying as early embryos. The Georgia and Quebec populations are reproductively isolated, although they are still connected to one another, according to Moore, by a long series of intermediates. A similar situation in the plant genus *Clarkia* is shown in Fig. 14.6, and another, in *Drosophila,* is described by Thompson and Woodruff (1980).

A few cases exist where such a cline is bent into the form of a circle, often with the ends overlapping sympatrically without interbreeding. Such cases of circular overlap present a problem, not so much to the evolutionist to explain their cause, but rather to the taxonomist, who must classify species. In the region of sympatric overlap, two distinct species appear to exist, but around the circle itself these two apparent species are connected by a series of intermediates, with no good place to draw a taxonomic boundary. These are usually classified as a single species, but if the circle were to become broken at any point by the extinction of intermediate populations, two species would be formed. Again, all sorts of intermediate conditions are possible, for the two ends of such a circle may meet each other at any stage of the evolution of reproductive isolating mechanisms.

The case of the great tit, *Parus major* (Fig. 14.7) may be illustrative of these several points. Of the three recognized subspecies, two confirmed cases of intergradation are known; one is in Iran, and the other is in South China. At first, the meeting of the two remaining subspecies in Manchuria was reported as a case of sympatric overlap, making this a circular cline with reproductively isolated ends. A later study of this same case re-

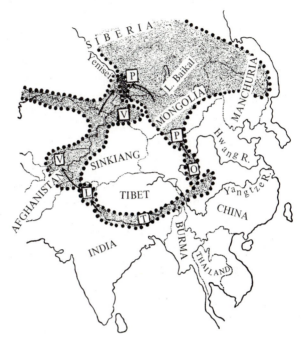

Fig. 14.8 Circular overlap in a warbler, *Phylloscopus trochiloides* (class Aves). The zone of sympatric overlap occurs to the east of the Yenisei River in Siberia. Reprinted with permission from Ernst Mayr, *Systematics and the Origin of Species*. New York: Columbia University Press, 1942. Copyright by Ernst Mayr.

Fig. 14.7 Circular overlap (?) in the great tit, *Parus major*. The three subspecies designated as A, B, and C form hybrid zones (D) in Iran and in South China. Populations in the Amur region (E) of easternmost Asia were first described as sympatrically coexisting, but a certain amount of hybridization has since been reported to occur there. Reprinted by permission from E. Mayr, *Animal Species and Evolution*. Cambridge, Mass.: Harvard University Press, 1963.

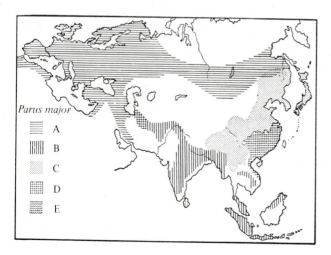

ports the Manchurian populations to hybridize, just as in the other two zones of intergradation. Perhaps it is just possible that both studies were correct, and that the populations had been sympatric in Manchuria but imperfectly isolated reproductively. In the intervening years, the reproductive isolation might have broken down, and a zone of secondary intergradation might have been established. Another, generally similar case is shown in Fig. 14.8.

G. ALLOPATRIC AND DIACHRONIC SPECIES

Two of the more formidable difficulties in application of the biological species concept are those of allopatric and diachronic species. The first of these difficulties concerns two or more geographically separate, or allopatric, sets of populations. Are they all one species, or are several species present? In the absence of sympatric overlap, it is indeed hard to tell. The populations never come into contact, so their ability to interbreed in nature is never tested. It is desirable in such cases to give the populations a name that is noncommittal as to species status. For this reason, the allopatric groups of interbreeding populations are called **semispecies** (Fig. 14.9), and the larger group to which all these semispecies belong is called a **superspecies**. Again, the theory of geographical speciation would predict the existence of precisely such cases as these, annoying as they are to the taxonomist, who must decide whether in a particular case to classify semispecies either as species or as subspecies.

Fig. 14.9 Geographic distribution of the semispecies of the *Drosophila paulistorum* species group. Dobzhansky and Spassky have described this as a cluster of species *in statu nascendi* (in the process of being born).

▲ Centro-American
▼ Amazonian
■ Orinocan
● Andean-Brazilian
△ Transitional
◆ Interior

The case of the diachronic species, that is, the species evolving through time, is equally vexing to the taxonomist. Even in the simplest case, that of phyletic transformation without true speciation, there is no way to draw species boundaries across an evolving continuum except arbitrarily. In practice, these boundaries are usually drawn where the fossil record is relatively sparse and the gap between known forms is therefore large. But if we had a complete record of all individuals that had lived in a series of populations that underwent phyletic transformation, there would be no way of drawing a horizontal boundary without separating some parents on one side from their children on the other.

When true speciation and phyletic transformation take place together, the problem is even more complex. A species, *A*, splits into two separate species, *B* and *C*. Now, if *B* and *C* are reproductively isolated, they are separate from each other. But is either one of them, or both, separate at the species level from *A*? As a matter of practical convenience—and no more—the following rule of thumb is generally followed: if *B* and *C* are about equal in numbers, then *A* is considered distinct from either and is assumed to have gone out of existence when the speciation occurred. If *B* is decidedly larger than *C*, however, it is often considered to represent the same species as *A*. Hennig (see Chapter 22) believes this practice to be inconsistent. He insists that *any* species should be considered to have ceased its existence when it undergoes speciation, and that both the resultant species are always new. He goes still further in denying phyletic transformation except through true speciation: no matter how much a species may change morphologically, it should still be considered the same species until it becomes subdivided, at which point its existence ceases.

Eldredge and Gould deny the possibility of phyletic transformation in a somewhat different way. The fossil record, they insist, never contains a record of gradual transformation from one species to another. Old species are always replaced by new ones, often their immediate descendents. These new species develop on the periphery of the species range, where they are less likely to fossilize, and they later become successful enough to invade the main portion of the species range. This controversial model is further treated in Chapter 21.

CHAPTER SUMMARY

Species are groups of interbreeding natural populations that are reproductively isolated from other species. Interbreeding takes place freely within species but not between them. The intrinsic mechanisms that prevent interbreeding between species are known as reproductive isolating mechanisms. Those mechanisms that act before mating (premating isolating mechanisms) prevent the waste of gametes; mechanisms that act after mating (postmating isolating mechanisms) prevent the interbreeding of species but do not prevent the waste of gametes. Selection tends to favor multiple isolating mechanisms and to favor mechanisms that act earlier over those that act later.

True speciation is the multiplication or splitting of species, rather than the mere succession of one species after another. Theories to account for the sudden origin of new species by saltation were popular in the early twentieth century, but it now seems that speciation cannot occur suddenly except by polyploidy. There is wide disagreement over the possibility of gradual, sympatric speciation, but even those who claim that this type of speciation is possible admit that it occurs only infrequently. The vast majority of new species originate in fully isolated, allopatric populations, according to the geographical theory of speciation. According to this model, intrinsic reproductive isolating mechanisms generally evolve during geographic isolation, when the populations are separated by some extrinsic barrier. Other models for speciation include the evolution of reproductive isolation in populations that are never completely separated (parapatric speciation), and the evolution of reproductive isolation in populations that are initially separate but later become contiguous (alloparapatric speciation). Even if these other models account for a few new species, it is now generally acknowledged that the vast majority of new species arise through the geographical speciation of isolated, allopatric populations.

Incomplete speciation results from the coming together of populations before completely effective reproductive isolation has evolved between them. Although this is in accordance with the expectations of evolutionary theory, it causes problems for the taxonomist seeking to classify species. Taxonomic difficulties at the species level may also arise from lack of data, from asexuality, from gradual transitions within a phyletic series, from circular overlap, or from allopatry. In the case of allopatric populations, it is often not possible to determine whether or not they belong to the same species, and the term "semispecies" has been applied to such allopatric populations.

15

Hybridism and Polyploidy

Chapter Outline

A. THE MEANINGS OF "HYBRIDIZATION"

In the past, much confusion has arisen in the discussion of hybridism, simply because such varied meanings have been given to the term by various authors. The broadest and most general definition of hybridization is the crossing of genotypically distinct individuals or populations. It is in this sense that Mendel's *Experiments in Plant Hybridization* refers to individuals differing in only a single gene, calling their offspring "hybrid" and the cross a "monohybrid cross." Charles Darwin's discussion of hybridization was also thoroughly confused by the lack of any distinction between individual hybrids and hybrids between populations or species.

Modern biologists usually prefer a stricter definition of **hybridization**: the "crossing between individuals belonging to separate populations which have different adaptive norms" (Stebbins, 1950). The strictest definition of all would restrict the term to the case of interspecific hybrids (hybrids between species).

B. HYBRIDISM AS A SOURCE OF NEW VARIABILITY

In animals as well as plants, many novel genetic combinations may arise first as the result of interpopulational hybrids within the same species. Hybridization in which genes from one population or species are introduced and incorporated into another is known as **introgressive hybridization**. In plants, the introgression of genes from other species by interspecific hybridization is a significant evolutionary force, and it appears to be one of the principal ways in which plant populations acquire their variability. With introgressive hybridization, the possible sources for new variations are expanded beyond the limits of even a single species to include an entire hybrid complex or "syngameon" (Grant, 1971).

Many hybrid plants are hardier than their nonhybrid relatives, and they grow much more vigorously and in more marginal situations. Entire hybrid populations are sometimes more vigorous than their nonhybrid counterparts. Their store of genetic variability is certainly far greater, an advantage that often enables them to colonize new and varied habitats more readily than nonhybrids. Perhaps this is why hybrids occur so often in disturbed habitats.

A case in point is provided by the irises of the Mississippi Delta of the United States (Anderson, 1949; Stebbins, 1950; Grant, 1971). *Iris fulva* and *Iris giganticaerulea* are isolated from one another by partial hybrid sterility and by preference for different habitats: *I. fulva* prefers clay soils along the partially shaded banks of rivers and drainage ditches, but *I. giganticaerulea* (formerly considered a variety of *I. hexagona*) prefers mucky tidal marsh soils in open sunlight. Where the habitat is undisturbed, the two species occur sympatrically over much of the Gulf Coast region. But human activity (clearing forests, draining swamps, etc.) has created several mixed or intermediate habitats in which hybrid populations thrive. Careful study has revealed that these populations contain some F_1 hybrids between the two species, plus many intermediates formed by back-crossing with *I. giganticaerulea* (but apparently not with *I. fulva*). Natural selection appears to favor hybrid plants with a preponderance of *I. giganticaerulea* characteristics. The circumstances of hybridization plus selection has resulted in the introgression of *I. fulva* genes into the hybrid populations and perhaps into *I. giganticaerulea* itself. The hybrid populations grade into the latter species, but not into *I. fulva*. The frequency of hybridization between these two species varies geographically as well as ecologically.

C. HYBRIDIZATION BETWEEN SPECIES

Interspecific hybrids, meaning hybrids between species, occur rather often in plants but are comparatively rare in animals. Some zoologists have altogether doubted the existence of such hybrids in entire large groups of animals, but careful inquiry seems to confirm their occasional presence in nearly every group, though nowhere to the extent of their occurrence in plants. The relative infrequency of interspecific hybrids among animals (as compared to plants) may reflect the greater reliance of animals upon premating isolating mechanisms, especially ethological isolation (see Box 14–B). Another possible explanation is that sex is determined chromosomally among most animal species, and a disruption of the delicate balance between the sex chromosomes and autosomes (such as might occur during hybridization) will very likely lead to sterility—the mule is a classic example

of this. The somewhat more frequent occurrence of hybrids among fishes and amphibians than among reptiles, birds, mammals, or insects has been attributed to the fact that fertilization among fishes and amphibians is usually external, while it is generally internal among the other groups listed. This, in turn, may reflect the fact that internally fertilizing species generally have more elaborate mating rituals than externally fertilizing species have, which relates back to the possibility of ethological isolation.

Among animals, many more interspecific hybrids are known to occur in captivity than in nature. No known animal species is of certain hybrid origin, but there are several known cases in which reproductive isolating mechanisms have broken down and resulted in the production, especially the local production, of **hybrid swarms**. A hybrid swarm may be defined as a population or series of populations consisting of nothing but hybrids. The two species of sparrows shown in Fig. 15.1 coexist sympatrically, and without interbreeding, over a broad area. Partial breakdown of reproductive isolation has resulted in the hybridization of the two species in Italy and on the island of Crete. A complete breakdown of reproductive isolation in parts of Tunisia has resulted in the production of a hybrid swarm. This again illustrates that the degee of reproductive isolation between two species, as measured by the frequency of hybridization between them, may vary geographically.

In plants, hybridism is a somewhat more regular occurrence (Fig. 15.2); even intergeneric hybrids are known to occur. Not only are plant hybrids more common, but quite a few plant species are either known or strongly suspected to have originated by hybridization. If the two parent populations are adapted to somewhat different habitats, then hybridization will be more frequent whenever "hybridization of the habitat" has occurred. This is another reason why hybridism is more frequent in disturbed habitats, as where forests have been cleared for human settlement (see the *Iris* example above).

Hybridization between species often occurs as the result of incomplete speciation (Chapter 14). When a reproductive isolating mechanism has not quite been perfected, hybridization between the species will occur, and the frequency of such hybridization may gradually increase. But only a certain level of hybridization is consistent with the maintenance of a partial reproductive isolating mechanism. Once this level of hybridization is

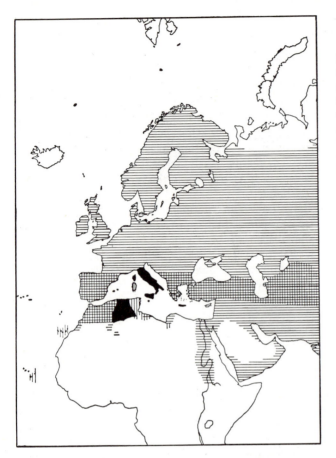

Fig. 15.1 Hybridization through local breakdown in reproductive isolation between two species of birds, the house sparrow (*Passer domesticus,* horizontal shading) and the willow sparrow (*P. hispaniolensis,* vertical shading). Note that the two species are sympatric without interbreeding over an extensive region (cross-hatched), but that hybridization has occurred in the places indicated by dark shading. Reprinted by permission of E. Mayr, *Animal Species and Evolution,* 1963. Cambridge, Mass.: Harvard University Press.

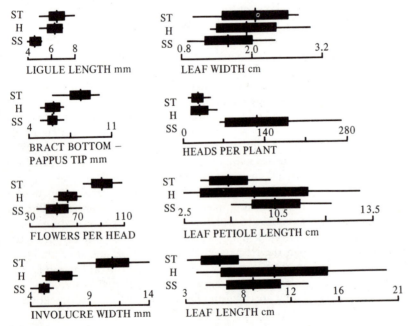

Fig. 15.2 Variation of quantitative characters in the plants *Senecio tomentosum* (ST), *Senecio smalli* (SS), and their hybrids (H). From G. C. Chapman and S. B. Jones, *Brittonia,* vol. 23, 1971. Reprinted with permission of the authors and The New York Botanical Garden.

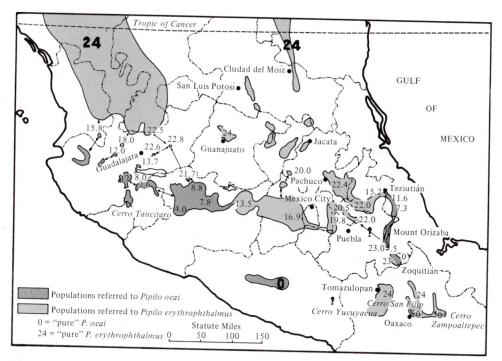

Fig. 15.3 Hybridization between two Mexican birds, the red-eyed towhee (*Pipilo erythrophthalmus*) and a related species, *Pipilo ocai*. The numbers indicate a hybrid "score" based on many characters and arranged so that "pure" *P. ocai* populations have a score of 0 and "pure" *P. erythrophthalmus* have a score of 24. Mean scores for several populations are shown. This case of hybridization is believed to have arisen as the result of agricultural disturbances of the birds' normal habitat during the last few centuries. (See also Fig. 13.6.) Reprinted by permission of E. Mayr, *Animal Species and Evolution*, 1963. Cambridge, Mass.: Harvard University Press.

exceeded, the reproductive isolating mechanism may disappear entirely, and a hybrid swarm will result. In other cases, the level of hybridization may remain low, and a continuum of hybrid populations may connect the non-hybrid extremes (Fig. 15.3).

Hybridization between species may occasionally result in the production of a new species, a process known as **hybrid speciation** (Grant, 1971). When it does occur, hybrid speciation usually takes place very rapidly. In such cases, the new reproductive isolating mechanisms are usually the direct consequence of chromosomal rearrangements, and their origin therefore accompanies the founding of the new hybrid populations. Hybrid speciation often involves polyploidy too, but Grant (1971) has insisted that such is not always the case. The genomes of hybrid populations are highly heterozygous because they combine large genomic fractions derived from different source populations. Under free sexual recombination, they would rapidly dissociate and become lost, but Grant (1971) lists various mechanisms by which such hybrid genomes may be stabilized. Of these mechanisms, asexual reproduction (apomixis, including vegetative propagation) avoids the dissociation of favorable but highly heterozygous gene complexes that sexual reproduction would normally bring about. Two further but unusual mechanisms (permanently odd polyploidy and permanent translocation heterozygosity) allow sexual reproduction to take place without free recombination—the adult genome is divided into a few complementary portions that easily reassemble during zygote formation. The somewhat uncommon process of recombinational speciation (Stebbins, 1950; Grant, 1971) involves the origin of a hybrid genome that is homozygous for some chromosomes derived from each parental species. The chromosomes derived from each parental species may carry sterility factors isolating the hybrids from the other parental species, while the largely homozygous genome can undergo free recombination without risk of permanent dissociation. The mechanism that Grant (1971) designates as "hybrid speciation with external barriers" involves two interfertile parental species that produce a hybrid species which is ecologically isolated from both.

Through the several mechanisms explained by Grant (1971), new species may arise through hybridization, occasionally supplanting one or both parental species entirely through subsequent competition. Repeated hybrid speciation gives rise to a pattern of **reticulate evolution,** in which phylogenetic diagrams look more like nets than like trees. Several botanists have identified such patterns as characteristic of the evolution of many angiosperm families.

Many plant hybrids achieve the best of both parental genomes by becoming polyploid. The occurrence of polyploidy following hybridization is discussed next.

D. POLYPLOIDY AND ITS OCCURRENCE

Polyploidy, defined as an increase in chromosome number by the addition of entire genomes, is a relatively frequent occurrence in most families of higher plants. Many of our domesticated grains are polyploid, including wheat (tetraploid), oats (hexaploid), and rye (also hexaploid). Polyploidy among animals is far less common than among plants and is far less likely to produce a viable genotype. Grant (1971) has surveyed the frequency of occurrence of polyploidy among eucaryotes and reports that it is rare among animals, fungi, and most groups of gymnosperms (conifers, cycads, and ginkgos), though it occurs regularly among angiosperms,

pteridophytes (ferns and other vascular plants lacking true seeds), and one group of gymnosperms (*Gnetum, Ephedra,* and *Welwitschia*). The comparative rarity of polyploidy among animals has received various explanations that are not mutually exclusive but may reinforce one another. One explanation is that sex is determined chromosomally among most animal species and that changes in chromosome number are more likely to disrupt sex determination and thus result in sterility. In those rare cases where polyploidy has been recorded in *Drosophila,* sterility has always resulted. (Aneuploidy involving the sex chromosomes also leads to sterility with much greater certainty than does aneuploidy involving the other chromosomes, in humans as well as in *Drosophila.*)

After haploidy or N (from *haplos,* single) and diploidy or 2N (from *diplos,* double), the levels of polyploidy up to eight are designated by Greek prefixes: triploid (3N), tetraploid (4N), pentaploid (5N), hexaploid (6N), heptaploid (7N), octaploid (8N). Beyond octaploidy, the levels of polyploidy are usually designated by Arabic numerals, e.g., 12-ploid, 16-ploid, or 32-ploid.

The two major types of polyploidy may be distinguished as **autopolyploidy,** or reduplication of a single genome, and **allopolyploidy,** the multiple combination of several genomes. If the haploid genome of one species is represented by A, then AA would be the normal diploid condition, and AAA, AAAA, AAAAA, and so forth would represent autopolyploids. If the genome of another species is represented as B, then AABB, AAAABB, and AAABBB, would represent three types of allopolyploidy. Designations of autopolyploidy or allopolyploidy may be combined with numerical designations: AAA would be called autotriploid, BBBB is autotetraploid, and AABBBB, AAABBB, and AAAABB would all be considered allohexaploid.

Even levels of polyploidy, such as tetraploidy, hexaploidy, and octaploidy, are far more frequent than odd levels such as triploidy and pentaploidy. Presumably this is because of the difficulty that unpaired genomes experience during prophase I of meiosis. Experimentally produced triploids and pentaploids are inviable or sterile in many species. In those species (e.g., the *Rosa canina* group) where triploids and pentaploids occur naturally and are viable, they usually avoid the problems of meiosis by reproducing asexually.

E. AUTOPOLYPLOIDY

Autopolyploidy is polyploidy in which a single genome occurs repeatedly. Autopolyploidy arises through duplication of the chromosome complement, often by the disruption of spindle fibers during mitosis. Somatic polyploidy, or polyploidy of somatic cells, results in polyploid tissue within an otherwise diploid individual. Your liver cells, for instance, are tetraploid.

Autopolyploids occur naturally in a number of plant genera, including *Sedum, Galax,* and others. Several cases previously reported as instances of autopolyploidy have since been determined to have resulted from allopolyploidy between two closely related genomes. Autopolyploids have been produced artificially in many species of domestic plants, usually as autotetraploids. The morphological effects of this and other types of polyploidy are well known to botanists: the primary effect is an overall increase in cell size. Other consequences of polyploidy commonly include an increased thickness of leaves, increased retention of water, slower growth, delayed flowering, and flowering over a longer season.

Autopolyploidy often occurs in a species already polyploid, usually an allopolyploid. The resulting polyploid may be designated an autoallopolyploid.

F. ALLOPOLYPLOIDY

Allopolyploidy is polyploidy in which two or more distinct genomes occur together. Allopolyploidy usually arises in hybrid material through the duplication of one or more of the genomic constituents.

In the first prophase of meiosis, hybrids frequently encounter a difficulty in chromosome pairing. An easy solution to this difficulty is to become polyploid, for in this case each set of chromosomes will have at least one identical set of chromosomes with which to pair. Polyploidy arising following hybridization has been called **amphiploidy,** and many instances of this are known among higher plants. Nearly half of all angiosperm species are thought to be polyploid; Grant (1971) reports that "the vast majority" of well-analyzed polyploid plant species are amphiploids. Grant (1971) calls amphiploidy "a type of permanent hybridity" because it allows the genomic contributions of both parent species to be permanently perpetuated in the hybrid state. None of

the advantages that the hybrid enjoys from combining both parental genomes need be lost during sexual reproduction, since the amphiploids possess multiple copies of each parental genome, usually either in the form AABB (allotetraploid) or AAAABB (or AABBBB, or AABBCC, all allohexaploid).

Many natural and artificial plant hybrids are amphiploids. Most commercial wheat is hexaploid, while durum and certain other varieties are tetraploid. Other polyploid crops include oats, rye, alfalfa, sugar cane, cotton, tobacco, potato, banana, coffee, sweet potato, apple, pear, and strawberry. Human preference for enlarged ("plump") plant parts with larger cells and a higher water content ("juicy") has augmented natural selection favoring the more drought-resistant polyploids. Selection for polyploid characteristics is thought to have played a major role in the origin of many important cultivated plants, including cereals.

Artificial hybrids between cabbages and radishes have been produced as allotetraploids under the name "raphanobrassica." These plants unfortunately have the inedible leaves of the radish and the useless roots of the cabbage, instead of having both edible vegetables combined into a single plant.

Both polyploidy and aneuploidy occur in the jimson weed (*Datura*). Specimens of varying genetic constitution are shown in Fig. 15.4.

Fig. 15.4 Polyploidy and other chromosomal rearrangements in the Jimson weed, *Datura*. From *Principles of Genetics* by Sinnott, Dunn, and Dobzhansky. Copyright © 1958, McGraw-Hill Book Company. Used with the permission of McGraw-Hill Book Company.

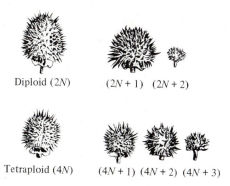

Diploid (2*N*)   (2*N* + 1)   (2*N* + 2)

Tetraploid (4*N*)   (4*N* + 1)   (4*N* + 2)   (4*N* + 3)

G. SPECIATION BY POLYPLOIDY

Speciation by polyploidy occurs at times among angiosperms and in certain other plant groups. Among animals and fungi, it is exceedingly rare and occurs only in some unusual circumstances, such as the parthenogenetic hybrid species of *Cnemidophorus* lizards studied by Fritts (1969) or Lowe et al. (1970a, 1970b).

Polyploidy is the one sudden mechanism of speciation that modern evolutionists generally accept. Crosses between the original parent species and a derivative polyploid often produce sterile hybrids, owing to the difficulties of meiosis with unbalanced chromosome sets. In particular, the most frequent case would involve a diploid parent species and a tetraploid daughter species; hybrids between these would initially be triploid and therefore usually sterile. The immediate difficulties of back-crossing to either parent species means that *most polyploid species are reproductively isolated from their diploid parents from the time of their origin.*

Speciation by autopolyploidy is possible by this mechanism, and several genera are known in which species with different multiples of the same basic chromosome number are thought to have arisen by autopolyploidy. Far more important, however, is a form of speciation that involves both hybridization and allopolyploidy. Two reproductively isolated species must first hybridize; the hybrids, if viable, may be sterile, or capable of asexual reproduction only. (If the two parent species have incompatible sets of chromosomes, then the hybrid may be incapable of meiosis.) But if the hybrid becomes polyploid, it acquires duplicate chromosome sets capable of meiotic pairing. If the resultant allopolyploid cannot readily hybridize with either parent, then it immediately forms a new species. Many plant species are believed to have arisen in this manner (Figs. 15.5 and 15.6).

Fig. 15.5 Family tree of the plant genus *Clarkia*, showing the origin of several of the species as interspecific hybrids. From Lewis and Lewis, *The Genus Clarkia* (U. Calif. Publ. Botany, vol. 20, no. 4), 1955. Reprinted by permission of the University of California Press.

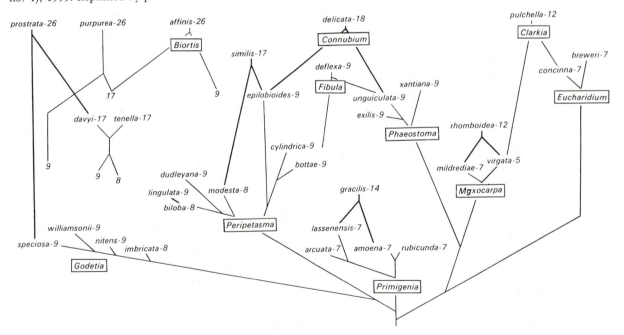

Fig. 15.6 This hemp nettle, *Galeopsis tetrahit,* was hypothesized to be a natural hybrid between the two related species *G. pubescens* and *G. speciosa.* The hypothesis received strong support when these two species were crossed, for the hybrids matched *G. tetrahit* in its visible features as well as in chromosome morphology. Reprinted by permission from Donald D. Ritchie and Robert Carola, *Biology,* © 1979. Reading, Massachusetts: Addison-Wesley (Fig. 25.16).

CHAPTER SUMMARY

Hybridization is the crossing of genetically dissimilar individuals, especially if derived from genetically distinct populations. Interpopulational hybridization occurs in both animals and plants, but interspecific hybridization is far more frequent among plants than among animals.

Polyploidy is an increase in chromosome number by the addition of entire genomes; it occurs more often in the higher plants. Polyploids that consist of reduplications of a single genome are called autopolyploids; polyploids in which two or more distinct genomes are combined are called allopolyploids. Allopolyploids, especially allotetraploids and allohexaploids, are more common than other forms of polyploidy. Allopolyploidy often follows hybridization between populations; such allopolyploidy is known as amphiploidy. All types of polyploidy are rare among animals.

Speciation may occur by polyploidy if the polyploids are from the time of their origin incapable of breeding successfully with either of their parent species. Hybridization followed by allopolyploidy is the most frequent mechanism of such speciation, and many plant species are thought to have originated in this way.

16

The Geologic Record

INTRODUCTION TO UNIT III

The past several chapters have dealt with microevolutionary mechanisms operating at the level of individual populations. The remaining chapters will examine the long-range and large-scale phenomena studied by comparative morphologists and paleontologists. This is evolution above the species level, also called **transspecific evolution** or simply **macroevolution**.

The examples used in Unit III are drawn from studies on particular groups of organisms. These groups of organisms are the subject of Unit IV, but the reader who may wish to have a brief overview of them is invited to turn to the classification at the end of the book, both now and repeatedly throughout the reading of Units III and IV.

A. THE VASTNESS OF GEOLOGIC TIME

When we think of time, we normally think of the hours and minutes shown on our watches, or the days and weeks on our calendars. We also have a concept of historical time and are thus aware that certain events took place in 1066 or 1492. But the Earth's time is too long to be measured in years or even centuries—it is measured in many millions of years. A million years is equal to ten thousand centuries, or about 14,000 lifetimes, but it is a very short time to a geologist. The oldest known rocks are about 3.8 *billion* years old, which equals 3,800 million years, or 38,000,000 centuries, or about 53 million human lifetimes. And the Earth is considerably older than even this; current estimates place the Earth's age at somewhere near 4.7 billion years! (Note to British Commonwealth readers: the American billion, used in this book, is 1,000,000,000, or 10^9.)

Such immense units of time are not readily grasped. Piaget has shown that childern develop their concept of time gradually. Young children will confuse age with size, or time travelled with distance travelled. Appreciation of clock-time develops much sooner than appreciation of calendar-time. Appreciation of historic time develops still later, which is why many children find dates like 1066 or 1492 meaningless. It is therefore no wonder that many adults have difficulty in conceiving geologic time.

An analogy, inspired by Rettie (1950), may develop a proper feeling of perspective. Let us compress the Earth's history into the scope of a normal calendar year of 365 days. To do this, imagine a picture of our planet taken once each year, and these pictures run as frames in a motion-picture projector at the rate of 144 frames per second, six times the usual speed. Each second that our imaginary movie is shown, 144 years of the Earth's history flashes by. To show the entire 4.7 billion year history, we must keep the projector running continuously, 24 hours a day, from midnight on New Year's Eve until the next New Year's Eve. Throughout January, February, March, and much of April, the history of the Earth is unrecorded, and our film is blank. On April 12, for a few seconds only, one corner of the screen is visible, and shows us a picture of the oldest known rocks. On April 20, life first appears, but again only for a few seconds and only in a small corner of the screen. May, June, July, and August go by, and the film is nearly as blank as before, with only fleeting glimpses here and there of a bleak and desolate scenery, still containing no multicellular organisms. The picture improves during September and October, but only after November 16, representing the beginning of the Cambrian period some 600 million years ago, are we watching anything even approaching a continuously visible motion picture. The Devonian period, or "Age of Fishes," begins on December 1. The "Age of Reptiles," or Mesozoic era, begins on the morning of December 14—for 12½ days the dinosaurs and other large reptiles dominate the scene. The extinction of the dinosaurs ushers in the Age of Mammals on December 26 at about 11 P.M. The 3.75 million year old genus *Homo,* to which we belong, first appears at 4:45 on New Year's Eve, and the species *Homo sapiens* at about 23 minutes to midnight. The entire history of human civilization, since the start of agriculture, is shown on the screen in less than a minute.

B. HOW THE GEOLOGICAL TIME SCALE IS ESTABLISHED

We have so far assumed that certain past events took place at known dates. But where are these events recorded, and how do we know their dates?

Relative Dating Methods

Sedimentary rocks are rocks that form from accumulated material under the conditions of temperature and pressure found near the Earth's surface or at shallow depths. These rocks usually occur in layers, and in a given region these layers can be arranged in a local time sequence according to the "law of superposition" (Chapter 3): the older rocks (those deposited first) are on the bottom, and the younger rocks (deposited later) are on top.

The local time sequences established at different localities can be compared lithostratigraphically by following a distinctive layer over the countryside between them. Such comparison is not always possible, however, and so comparisons between widely separated localities are usually made biostratigraphically on the basis of fossil assemblages. **Correlation by fossils** is possible because rocks of similar age have many of the same fossils embedded within them. Correlation among overlapping local sequences allows a **relative time scale** to be established—one in which dates are "earlier" or "later," though not generally known in years. Long before absolute ages (in years) were determinable, geologists had established a sequence of names for **geological time periods** (Box 16–A). The sequence is read from bottom to top (Cambrian, Ordovician, etc.), because this is the sequence in which the rocks lie. The numbers to the right of the vertical line indicate how many millions of years ago each time period was. The geological time periods are defined with reference to "type sections," which are formations in which rocks of that period are well represented. A designation such as "Permian" is an implicit reference to certain rocks exposed in the Russian province of Perm, with which the rocks so designated are believed to be contemporaneous. Similarly, rocks believed to be contemporaneous with the White Cliffs of Dover are called "Cretaceous," a name derived from the Latin word for chalk. The Mississippian and Pennsylvanian periods are named after rocks exposed in the states of Mississippi and Pennsylvania, but outside North America the Mississippian is usually called Lower Carboniferous and the Pennsylvanian is called Upper Carboniferous.

The geological periods are grouped into **eras: Paleozoic,** meaning "first life"; **Mesozoic,** meaning "middle life"; and **Cenozoic,** meaning "recent life." These three eras are sometimes grouped together as the **Phanerozoic**

BOX 16–A   *The Geological Time Scale*

Eras		Periods	Epochs	Recorded life
CENOZOIC (65 MILLION YEARS DURATION)	2	Quaternary	Recent Pleistocene	Human civilizations Ice age glaciations
	65	Tertiary	Pliocene Miocene Oligocene Eocene Paleocene	Age of Mammals
MESOZOIC (145 MILLION YEARS DURATION)	130	Cretaceous		More dinosaurs and other large reptiles; first angiosperm floras
	160	Jurassic		Dinosaurs and other large reptiles
	210	Triassic		Abundant reptiles, including therapsids and thecodonts
	235	Permian		Therapsids dominant
	255	Pennsylvanian (Upper Carboniferous)		Early reptiles, including pelycosaurs
	275	Mississippian (Lower Carboniferous)		Amphibians abundant
	315	Devonian		Diversity of fishes (especially placoderms) and land plants
	350	Silurian		Oldest land plants
PALEOZOIC (390 MILLION YEARS DURATION)	440	Ordovician		Oldest vertebrate remains; most modern invertebrate phyla represented
	600	Cambrian		Trilobites dominant; very few modern invertebrates
PRECAMBRIAN				Soft-bodied forms only

Eon ("evident life"), contrasted with the **Cryptozoic Eon** ("hidden life"), which is also called Precambrian. Geologic periods may be divided into *epochs;* only the Tertiary and Quaternary periods will be so divided in this book. Any time period may be subdivided into early, middle, and late portions; the corresponding rock units are called lower, middle, and upper, respectively.

Absolute Dating Methods

We may learn through relative dating that the Permian followed the Pennsylvanian and preceded the Triassic. But how long did the Permian last, and when did it begin, and was it longer or shorter than the Cretaceous? Questions of this sort can be answered only by **absolute dating** methods that give us numerical dates in years.

Several techniques are available for the calculation of absolute ages in years. Many of these methods are of limited applicability. The book by Zeuner (1970), though now out of date, is still a valuable reference on these various methods.

The most important method of absolute dating is based on the rate of radioactive disintegration. **Radiometric dating** is based on the fact that each radioactive isotope decays with a characteristic **half-life.** The decay is exponential because the atoms are statistically independent of one another, and the average number decaying during any given time interval is proportional to the number of atoms present. After one half-life, half of the original material remains; after two half-lives, one-quarter remains; after three half-lives, one-eighth, then one-sixteenth, and so forth. The decay proceeds according to the equation

$$\frac{X}{X_0} = \left(\frac{1}{2}\right)^{T/T_{1/2}},$$

where X_0 is the initial quantity of radioactive material, X is the present quantity, T the elapsed time, and $T_{1/2}$ the half-life. This equation can be solved for T, the elapsed time, if all other quantities are known. The half-life, $T_{1/2}$, is a known constant for each isotope, and X can be measured directly. A frequent difficulty lies in the determination of the initial quantity, X_0.

Selected radioactive isotopes of long half-life are listed in Box 16–B. Some of these isotopes are unsuitable for general use because they are rare or because they occur only as trace amounts—too small to be measured accurately. The decay of uranium or thorium into lead can be used only with certain assumptions as to the quantities of lead derived from each of several independent decay processes, as well as the quantity of non-radiogenic lead that was present from the start.

For absolute ages within the last million years, a radiometric clock based on the rate of decay of carbon-14 is used. The method can be used on any organic material

---

BOX 16–B *Selected Radioactive Isotopes of Geochronological Interest*

Parent element, isotope		Daughter element, isotope		Disintegration mode	Half-life (in billions of years)
uranium,	^{238}U	lead,	^{206}Pb	$8\alpha + 6\beta$	4.51
uranium,	^{235}U	lead,	^{207}Pb	$7\alpha + 4\beta$	0.71
thorium,	^{232}Th	lead,	^{208}Pb	$6\alpha + 4\beta$	13.9
rubidium,	^{87}Rb	strontium,	^{87}Sr	β	47.0
potassium,	^{40}K	argon,	^{40}Ar	electron capture	12.4
carbon,	^{14}C	nitrogen,	^{14}N	β	5730 years

containing carbon. Assuming that the ratio of atmospheric ^{12}C to ^{14}C has remained the same over the last million years, a ratio of ^{12}C to ^{14}C is calculated for the sample to be dated. Any excess over the present atmospheric ratio of $^{12}C/^{14}C$ is then attributed to the radioactive decay of ^{14}C into ^{12}C. Archaeological and other materials less than a million years old are often dated by this method. For older materials, however, the level of ^{14}C is too small to be measured accurately, and the assumption of atmospheric constancy is less secure.

Perhaps the best radiometric clock for much of geological time is that based on the decay of ^{40}K into ^{40}Ar. Potassium-argon (K/Ar) dates may easily be determined because potassium is a common constituent of many rocks and because argon is not usually trapped inside rocks except as a product of radioactive decay. Argon is a gas. Any rock that has been heated during its formation readily loses any argon that it may contain, thus "setting the clock." Any argon accumulating thereafter may be assumed to result from the radioactive decay of potassium-40, assuming that no argon escapes. Thus, any potassium-containing rock that is heated during its formation can be dated by this method.

Rocks that were melted during their formation are called **igneous rocks.** (A few other rock types associated with volcanoes are also considered igneous.) Igneous rocks are usually formed under conditions that destroy any fossils that may be present. But many igneous rocks contain potassium or other elements that can be dated radiometrically. When such igneous rocks occur within a layered sedimentary sequence, or cutting across such a sequence, the relative ages of the rock units can be established by geological methods. In practice, most dating is now done this way: a series of relative dates is established from sedimentary sequences, and absolute dates are given to this sequence from sporadically associated igneous rocks that are dated radiometrically.

C. FOSSILS AND FOSSILIZATION

A **fossil** is defined as any remains or other direct evidence of past life. Fossils may contain some original organic material, or else none at all. They may accurately reflect the shape of the object fossilized, or they may be flattened to varying degrees. They may be fossils of the organism itself, or of part of an organism, or of the work of an organism. The latter are called **trace fossils;** they may include tracks, trails, burrows, and fossil dung. The various types of fossils are listed in Item 16.1 on page 276.

D. SURVEY OF GEOLOGICAL PERIODS

In the following brief survey of geological time, each period is described in terms of its characteristic fauna and flora. Some of these organisms are familiar to the average reader; others may not be. All are described in Unit IV (especially Chapter 26). Readers unfamiliar with animal and plant diversity may either defer this section until they have read Chapter 26, or they may make repeated reference to the classification that appears at the end of the book.

Precambrian or Cryptozoic Eon

Most Precambrian sedimentary rocks have been converted by high temperatures and pressures into **metamorphic rocks,** which only rarely contain fossils. Precambrian fossils do occur, but not in sufficient numbers to establish a time scale by traditional means. The only reliable ages for Precambrian rocks are radiometric ages.

The Precambrian represents about six-sevenths of the earth's geological history. Various terms have been suggested for subdivisions of Precambrian time, though none has achieved widespread acceptance. The oldest Precambrian rocks, containing no remains of life, are sometimes called Archean or Azoic, in contrast to the later Precambrian rocks, which are called Proterozoic. Some geologists have suggested that the name "Precambrian" be abandoned, and that the time before the base of the Cambrian be referred to as the Cryptozoic Eon, meaning literally "hidden life." The time since the base of the Cambrian, that is, the last one-seventh of the Earth's history, would then be called the Phanerozoic, which means "evident life."

The Onverwacht Shale of South Africa, one of the Earth's oldest rock formations, has been dated radiometrically at 3.6 billion years old using K/Ar. The oldest fossils, contained in the Fig Tree formation immediately overlying the Onverwacht, can be dated at 3.5 billion years. Most of the remainder of Precambrian time may be called the ''Age of Blue-Green Algae'' (Schopf, 1974), after the dominant forms of life of that era. Important fossil-bearing formations of the upper Precambrian include the Gunflint Iron Formation of the Lake Superior region, dated at 1.9 billion years, and the Bitter Springs Formation of Australia, dated at 1.0 billion years.

Paleozoic Era

The base of the Cambrian marks a great biological revolution or breakthrough, for it is at this time that innumerable forms of animal life first acquired hard skeletons capable of being readily fossilized. One current theory explains this acquisition as a protection against predators. The same theory explains the sudden expansion of species numbers in Cambrian times as the consequence of the richer faunal diversity that predation makes possible. Most major animal phyla—except the vertebrates and their relatives—are already represented in Cambrian rocks. The most advanced of invertebrate groups were already present: the highly characteristic trilobites belonged to the phylum Arthropoda. Other forms of Cambrian life (Fig. 16.1) included sponges, brachio-

pods, snails, primitive echinoderms, and occasional worms, all further described in Chapter 26.

By Ordovician times, all the principal groups not already present in the Cambrian had made their entrance. These included the corals, bryozoans, graptolites, and vertebrates. In addition, some groups poorly represented in Cambrian deposits now came to the fore; these included the various echinoderms and the cephalopod molluscs. In fact, all the phyla of the animal kingdom were already present in Ordovician times, with the possible exception of a few phyla of ''worms''—soft-bodied animals that have left no usable fossil record at all. The trilobites reached their peak in Ordovician times, and began a slow decline that lasted through the rest of the Paleozoic.

The Silurian (Fig. 16.2) was in some respects a transitional period. At the beginning of the period, vascular land plants were unknown, and vertebrates were represented by some fragmentary remains only. By the end of the Silurian, vascular plants had begun to invade the land, and several groups of vertebrates were represented in the seas. Except for the insects, most major invertebrate classes that have any usable fossil record were

Fig. 16.1 Life of the Cambrian. Prominently featured are the tubular, spongelike *Archaeocyathus* and a variety of primitive arthropods, including trilobites. Courtesy of the Field Museum of Natural History, Chicago.

Fig. 16.2 Life of the Silurian. Prominently featured are several types of coral (center and left foreground), cystoid echinoderms (the flowerlike stalks on either side), brachiopods (right foreground), trilobites, and several nautiloid cephalopods (one to the right of center, another atop the leftmost coral). Courtesy of the Field Museum of Natural History, Chicago.

Fig. 16.3 A Devonian landscape, featuring some of the earliest land plants. Landscape courtesy of the Brooklyn Botanic Gardens.

already present. The corals were at their peak, and the invertebrates were still the dominant forms of animal life.

By Devonian times (Fig. 16.3), some dramatic changes had taken place. Vascular plants were spreading over the land surfaces of the Earth, while fishes proliferated in the seas. The so-called Age of Fishes witnessed the diversification of these fishes into a number of orders and classes, both jawed and jawless. Among the invertebrates, the brachiopods were at their peak of diversity.

Many important groups declined at the end of the Devonian or in early Mississippian times; these included the corals, brachiopods, jawless fish, and the armored fish called placoderms. These declines were either accompanied or soon followed by the development of some new or previously insignificant groups, including the ammonoids and the sharks. Among echinoderms, both the crinoids and blastoids reached their peak in the Mississippian. On land, the Mississippian witnessed the first few amphibians and the expansion of land floras; both ferns and lycopods grew to large size.

Pennsylvanian faunas and floras were generally similar to Mississippian ones, but with a great increase in diversity, due largely to the expansion of gastropod, bivalve, and ammonoid molluscs, and bony as well as sharklike fishes. Seed ferns, lycopods, and the earliest conifers were among the plants of the period, with the fernlike forms occurring in dense, swampy forests (Fig. 16.4). Compression fossils and carbonized remains of these plants form the extensive coal deposits of many countries.

Beginning first in these coal swamps, the land animals began to proliferate. The first insects and myriapods date from this period, showing that the arthropods had successfully invaded the land. The amphibians diversified considerably, and the first reptiles appeared; their expansion may have been directly effected by the expansion of land floras, or they may have resulted secondarily from the appearance of the first insects.

The Permian period was marked by a decline in many of the older groups, and the ascendance of others that were later to rise to dominance in the Mesozoic. Sev-

Fig. 16.4 A Pennsylvanian coal-swamp, featuring many seed-ferns, lycopods, and some primitive insects. Courtesy of the Field Museum of Natural History, Chicago.

eral invertebrates became extinct, including the last of the trilobites; several groups of fishes declined, too. Insects, reptiles, and ammonoids expanded in diversity, and several new orders appeared, including the first mammal-like reptiles. Ferns and conifers expanded, largely at the expense of the more primitive vascular plants; cycads made their first appearance. In the Southern Hemisphere, a period of extensive glaciation occurred, and this was followed everywhere by a flora in which the seed-fern *Glossopteris* predominated. The simultaneity of glaciation, followed in each case by the

Glossopteris flora, on several disconnected land masses (Africa, India, Australia, Antarctica, South America) may be cited as evidence that these continents were once combined into a single land mass called Gondwanaland.

Mesozoic Era

The Mesozoic was truly the Age of Reptiles, on land and in the seas. Many of the dominant reptile orders had already appeared before the Mesozoic began, but now they proliferated, and new orders rose to join them. In the seas, the ammonites were most characteristic of the invertebrate life of the era.

The Triassic period saw the rise of the great reptilian subclass Archosauria, beginning with the order Thecodontia, which was exclusively Triassic. Turtles, lizards, crocodiles, and dinosaurs all made their first appearance in the Triassic, as did such marine reptiles as the ichthyosaurs and plesiosaurs. Among the Permian groups that persisted into the Triassic are the therapsids, or mammal-like reptiles. The mammals themselves made their first appearance toward the end of Triassic times. In the seas, brachiopods declined further as the bivalve molluscs rose to replace them in importance. Both gastropod and cephalopod molluscs underwent considerable turnover, with many older orders and families being replaced by more modern ones. Sharks were at their low point, as the bony fishes successfully invaded marine waters in large numbers.

The Jurassic period (Fig. 16.5) saw the rise of the great dinosaurs, including all of the major dinosaur suborders except the Ceratopsia. The thecodonts themselves became extinct, but their Jurassic descendants included, in addition to dinosaurs, the first flying reptiles and also the first birds. The mammal-like reptiles of earlier periods became largely extinct and were replaced by their mammalian descendants. In the seas, the bivalves, cephalopods, and marine reptiles continued to prosper, while modern bony fishes, including the first teleosts, increased further. Many ancient types of plants became extinct in the Jurassic, and the cycads and ginkgos were at their peak.

The Cretaceous period (Fig. 16.6) was marked by many changes. In the seas, the ammonoids and the various marine reptiles all rose to their maxima and then declined to extinction. Bivalves and teleost fishes increased in diversity, as did planktonic Foraminifera.

Fig. 16.5 Life of the Jurassic, featuring the dinosaurs *Stegosaurus* in the foreground, *Allosaurus* in the middle distance, and *Brontosaurus* half-submerged in the background. Courtesy of the American Museum of Natural History.

Fig. 16.6 Life of the Cretaceous. Dinosaurs featured, left to right, include the crested duckbills *Corythosaurus* (in the water) and *Parasaurolophus* (feeding in the background), the armored *Ankylosaurus* (in the foreground), the ostrichlike *Ornithomimus* (in the background), and the large duckbill *Anatosaurus* (= *Trachodon*) (far right). From a mural by Charles R. Knight. Courtesy of the Field Museum of Natural History, Chicago.

On land, the dinosaur suborders Sauropoda and Stegosauria became extinct, while the Ceratopsia or horned dinosuars rose to prominence. Marsupials and placental mammals both made their first appearance. The flowering plants, or angiosperms, arose and diversified during Cretaceous times; by the end of the period all floras were dominated by angiosperms. With the extinction of the dinosaurs, marine reptiles, and ammonoids, the Cretaceous period was brought to a close.

Cenozoic Era

During the Cenozoic era, the bivalves and teleost fishes, and later also the marine mammals, expanded into the niches previously occupied by the ammonoids and marine reptiles, possibly contributing to the extinction of these two groups. On land, however, the void created by the extinction of the dinosaurs and flying reptiles was much greater, and so the greatest changes during the Cenozoic were those brought about by the evolution of mammals and birds to fill this void. Plant life was already quite modern; the major change during the Cenozoic was the great expansion of grasslands during the Miocene.

The Cenozoic era is divided into seven epochs: Paleocene, Eocene, Oligocene, Miocene, Pliocene, Pleistocene, and Holocene (or Recent). The Tertiary period includes the first five of these epochs; the Quaternary period includes the last two.

The Paleocene epoch marks the first great expansion of the placental mammals, largely at the expense of the marsupials. Many orders of placentals made their first

(a)

(b)

Fig. 16.7 Two early Tertiary scenes. (a) The primitive ungulates *Coryphodon* (left, also background) and *Barylambda* (foreground) roam a Paleocene landscape. (b) An Eocene landscape, featuring the large, grotesque *Uintatherium* (left and center) and a contemporaneous early horse, probably *Epihippus* (right). Courtesy of the Denver Museum of Natural History.

appearance, including the rodents, carnivores, and edentates. The primates, insectivores, and condylarths, already present by the end of the Cretaceous, diversified greatly with the origin of many new families. Many archaic mammals, belonging to completely extinct orders, also originated during the Paleocene (Fig. 16.7a).

The Eocene epoch (Fig. 16.7b) marks the greatest diversity of the archaic mammals, as well as the origin of nearly all the modern orders not already present. Modern types of hoofed mammals originated in the Eocene, including the first horses, but they were all rather small and were dominated by the much larger archaic mammals.

The Oligocene epoch saw the extinction of many archaic forms, and the origin of the last few major orders and suborders, all of them of African ancestry and pos-sibly present though undiscovered in earlier times. Many modern families appeared, and much of the fauna took on a modern cast. Old World monkeys and apes appeared during the Oligocene for the first time.

The Miocene epoch witnessed a great expansion of grasslands, though grasses themselves had already been present since the Cretaceous. Monkeys and modern rodents expanded and proliferated, as did the horses and the ruminant artiodactyls. Both horses and artiodactyls became adapted to running over the new grasslands and eating the grittier and more fibrous grasses themselves.

The Pliocene epoch saw the further expansion of grasslands and of modern artiodactyls and rodents. Except for the persistence of certain families and orders that became extinct in the Pleistocene, Pliocene faunas are almost entirely modern.

(a)

(b)

Fig. 16.8 Two Pleistocene scenes. (a) Ground sloths and glyptodonts on the pampas of Argentina, with *Toxodon* (left) and *Macrauchenia* (right) in the background. (b) Sabertooth and California condor at the Rancho La Brea tar pits in Los Angeles, California, with imperial mammoths in the background and vultures overhead. Courtesy of the American Museum of Natural History.

The end of the Tertiary was marked by a drop in sea level and a corresponding rise or emergence of both the Bering and Panama land bridges. Faunal interchange between North America and Asia was now possible on a greater scale, and, perhaps more important, South America became connected to North America and ceased to be an island continent. As a result of the competition of invading forms, many groups rapidly became extinct, but others persisted and expanded into new areas.

The Pleistocene epoch was marked by a succession of glacial advances and retreats in the Northern Hemisphere, the so-called Ice Age. Many types of mammals grew to large size, though most of these later became extinct (Fig. 16.8). Most significant to ourselves was the rapid evolution of humans during the Pleistocene, as well as the advent of modern one-toed horses.

Much of human evolution took place in the Pleistocene epoch. During four successive cold episodes within the Pleistocene, polar ice caps and Alpine glaciers expanded until massive continental ice sheets were formed. In central Europe, these four episodes of glaciation are called Gunz, Mindel, Riss, and Würm, in that order. Pleistocene faunas that precede the first, or Gunz, glaciation are referred to a preglacial portion of the Pleistocene called the Villafranchian.

The Recent epoch hardly differs from the Pleistocene, except for the effects of human civilization. The retreat of the Wisconsin glaciers some 15,000 years ago is usually taken to mark the end of the Pleistocene and the onset of the Recent epoch. Many geologists, however, believe we are now living in an interglacial age.

CHAPTER SUMMARY

The Earth is about 4.7 billion years old, but the well-documented portion of the fossil record covers less than one-seventh of this time, or about 600 million years. Geologic time may be measured in relative units, such as periods and epochs, established through the study of fossil sequences in sedimentary rocks. Absolute dating, in years, is also possible; most absolute dates are determined from the known rate of decay of certain radioactive substances.

A fossil is any evidence of an ancient organism or its activities. Fossils may include actual organic remains, casts and molds, tracks and trails, minerals that have replaced organic structures, and other records of past organisms or of organic activity.

Once life originated, about 3.5 billion years ago, the remainder of the Precambrian was dominated by the blue-green algae and bacteria. Multicellular animals with readily fossilized hard parts first diversified in early Cambrian times. The Paleozoic faunas were dominated first by invertebrates, then by fishes and amphibians. Vertebrates and land plants first became numerous during the Silurian. Insects and land vertebrates (amphibians) first appeared during the Carboniferous. Reptiles first became abundant during Permian times, and during the Mesozoic they dominated the land faunas, while the ammonoids dominated the seas. The Tertiary period was marked by the rise to prominence of mammals, birds, teleost fishes, and the modern families of invertebrates.

FOR FURTHER KNOWLEDGE

ITEM 16.1 *Types of Fossils, Classified by Their Methods of Preservation*

I. FOSSILS OF ORGANISMS OR THEIR PARTS

  A. Fossils consisting largely of the original organic material

    1. *Unaltered remains* In colder climates, carcasses can remain frozen and essentially intact for many thousands of years. Even cellular details are preserved, including the banding patterns of individual muscle fibers. Frozen mammoths are the most spectacular examples of this type of fossilization; they range in age up to about 40,000 years.

    2. *Compressions* These are the flattened remains of animals or plants, often dehydrated, but otherwise largely intact. Leaves are commonly preserved in this way. Cellular details, especially surface details, are often preserved; the individual stomata (breathing-

ITEM 16.1 *(continued)*

pores) on the undersides of leaves are often distinctly visible, and the waxy cuticle of the leaf can often be recovered intact.

B. Fossils in which the original material has been partly or wholly replaced (*replacement fossils*)

1. Fossils in which the original material is gradually replaced, so that cellular details are preserved

a. *Permineralization*   This is the gradual filling of interstitial spaces by minerals provided by ground water. The organic material remains, but minerals are added. Bones are often preserved in this way. The name *petrifaction* is given to those cases where silica (SiO_2) or calcite ($CaCO_3$) is the mineral added.

b. *Impregnation or embedding*   This is like permineralization, except that the impregnating material surrounds the fossil completely as well as filling in any interstitial spaces.

2. Fossils in which the original material is destroyed

a. *Carbonization or distillation*   This is preservation in which volatile components are lost, leaving only a tarlike or carbonized skeleton that is usually compressed as well. Insect fossils preserved in amber are carbonized.

b. *Mineralization*   This is the replacement of the original material by minerals usually provided by ground water. No organic material remains, and the finer details of structure are not preserved.

C. Casts and molds

1. *Natural molds* are "negative" fossils in that surface relief is represented in reverse—depressions are represented by projections and projections as depressions. Molds may preserve the three-dimensional shape of the original or they may be flattened. An *impression* is a special type of mold—it is the negative imprint of a compression.

2. *Natural casts* are "positive" images that usually result from the filling of natural molds. A *core* or *steinkern*, also called an *endocast*, is an internal cast resulting from the filling in of a naturally hollow structure such as a shell.

II. TRACE FOSSILS

A. Fossils of organic material derived from biological activities

1. *Amber*   Fossil resin.

2. *Coprolites*   Fossil dung.

B. Inorganic material indicating the "work of an organism"

1. *Tracks and trails*   Worm trails, dinosaur footprints, and other traces of locomotor activity.

2. *Burrows and tubes*   These may be preserved if filled with sediment different from the surrounding material.

3. *Castings*   Sand or other sediment, often with some organic matter added, which has passed through the body of a detritus feeder and been discarded.

17

Evolutionary Lineages and Trends

A. LINEAGES AND RELATED CONCEPTS

Lineages

Central to much of macroevolutionary theory is the study of sequences of species, which we call lineages. A **lineage** is defined as a succession of species arranged in a continuous ancestor-to-descendant sequence. Strictly speaking, a lineage extends from the very origin of life to either a species living today or one which became extinct at some time in the past. What we usually call a lineage is in fact only part of a lineage. Thus, the sequence of species from the ancestral horse *Hyracotherium angustidens* to the modern horse *Equus caballus* represents a lineage. The concept of a lineage can be found in many nineteenth-century writings; the definition above is due largely to Simpson (1953, 1961, 1967) and to Minkoff (1974).

Cladogenesis and Anagenesis

Evolution, as depicted in family trees, includes two very distinct processes: the branching of lineages and the evolution of lineages between branching points. The shortcomings of many early evolutionary theories can be traced to the failure to make this distinction. Although implicit in the writings of a few earlier workers, the distinction was never emphasized until Rensch (1947) coined the terms *cladogenesis* ("Kladogenesis" in the original) and *anagenesis*.* **Cladogenesis** (from the Greek words *klados,* a branch, plus *genesis,* development or becoming) is defined as the branching or splitting of lineages. **Anagenesis** (from the Greek words *ana-,* upwards, plus *genesis*) is defined as the evolution of lineages between branching points (Fig. 17.1). It is also sometimes called "phyletic evolution."

Charles Darwin was one of many evolutionists who failed to see the distinction between anagenesis and cladogenesis: he believed that his theory of evolution accounted for the origin of species, while in reality he had provided a direct explanation only of anagenesis. The currently received theory of anagenesis is essentially

---

*The term *anagenesis* was used previously in a different sense by Cope, who contrasted "upward" or "progressive" evolution (*anagenesis*) with "regressive" or "degenerative" evolution (*catagenesis*). In modern usage, anagenesis refers to evolution between branching points regardless of whether it is considered progressive or retrogressive.

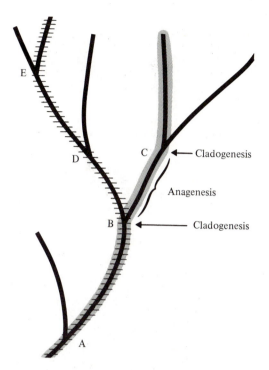

Fig. 17.1 A hypothetical family tree, illustrating both cladogenesis (at points A, B, C, etc.) and anagenesis (between these points). The hatched and shaded portions of the diagram represent two lineages, splitting from each other at point B by cladogenesis and diverging thereafter by anagenesis.

Darwin's, but the currently received theory of cladogenesis or speciation (Unit II, especially Chapter 14) did not become generally accepted until the 1940s (Chapter 6). The relation of cladogenesis to anagenesis—whether they are interrelated or independent—stands today as one of the major unresolved issues of macroevolution.

The term **phylogenesis** may be used to refer to the evolutionary processes of anagenesis and cladogenesis together. The study of phylogenesis is called **phylogenetics**, and the phylogenetics of a particular group of organisms is called its **phylogeny**. This and the next few chapters deal specifically with phylogenetics.

Evolutionary Trends

The continued change of a character within an evolving lineage is known as an evolutionary **trend**. The number of trends in any lineage is therefore the same as the number of characters that are evolving. Different characters often exhibit different trends within the same lineage, varying in either the direction or rapidity of change, and the same character may or may not exhibit the same trend in different lineages. It is a common mistake to assume that an evolutionary trend represents "progress," for a trend may equally well be retrogressive, as when an organ degenerates and is lost.

Of course, no lineage leaves behind a perfect fossil record, and every observation of a trend is to some extent an inference based on the filling in of gaps. The reality of these gaps, and the smoothness of evolutionary trends, are currently the subject of much debate (Chapter 21).

Most evolutionists believe that the direction and intensity of evolutionary trends can be explained as the adaptive results of natural selection, as we shall see shortly.

In a lineage that exhibits a trend, the original or ancestral condition is called **primitive** (or plesiomorphous), and the resultant or descendent condition is called either **derived** or **advanced** (or apomorphous). The use of these terms does not imply any distinction between progressive and retrogressive trends, nor does it imply a causal explanation for trends. See Hennig (1966), Hull (1970), Mayr (1969), Minkoff (1974), and Cracraft and Eldredge (1979) for discussion on the meaning of these terms and their importance in cladistic analysis (Chapter 22).

A condition is called **generalized** (or unspecialized) if a variety of alternative conditions have been or could be derived from it. A **specialized** condition is one from which either no alternative conditions, or only a few (and these more specialized), can be derived. The limbs of the first land vertebrates were generalized; from them have evolved the jumping legs of frogs, the digging legs of turtles and moles, the wings of bats and birds, the flippers of whales, the running legs of horses and deer, and the hands of humans, each specialized in a different direction. Specialized conditions often arise as irreversible adaptations to particular environmental needs. The emphasis here is on the word "irreversible," for no change is considered a specialization unless it is supposed that some capacity for further change or for reversion has been lost. It is, of course, difficult to prove claims about future adaptive potentials, and so these terms are most meaningful when used in retrospect.

Because different characters evolve in different directions and at different rates (the principle of mosaic evolution, Chapter 18), each organism is a conglomeration of primitive and advanced, and of generalized and specialized characteristics. Each of the four terms applies to individual characteristics, not to whole organisms. The platypus (*Ornithorhynchus*) is considered a primitive mammal because it lays eggs and regulates its body temperature rather poorly. It also exhibits many derived ("advanced") features that are also specialized: its duck-like bill and webbed feet are among these. Yet it is incorrect (though common) to contrast the terms "primitive" and "specialized" as if they were opposites. The terms generalized and specialized refer to the number and variety of directions in which future modification is possible; primitive and advanced refer to the extent of previous modification along one of these directions. The vestigial and functionless limbs occasionally found in some snakes are primitive in that they represent an ancestral condition, but they are also specialized, since further modification is restricted to the loss of these limbs and can then proceed no further.

B. EXAMPLES OF EVOLUTIONARY TRENDS

Examples of evolutionary trends are not hard to find, though they may be less gradual than is commonly assumed (Eldredge and Gould, 1972, 1974). Trends can often be inferred from the study of living species (Chapter 21), but documentation for these trends exists only in those cases where a good fossil record is available. The following examples are therefore drawn from among organisms with (relatively) well-preserved fossil records.

Our first example concerns the larger Foraminifera (Protozoa, Sarcodina) of the family Fusulinidae. These one-celled organisms have a spirally coiled shell or test made of calcium carbonate. In *Millerella,* one of the earliest genera, the axis of coiling is short, and the shell as a whole is disc-shaped (Fig. 17.2). As we ascend the Pennsylvanian series, the axis of coiling lengthens relative

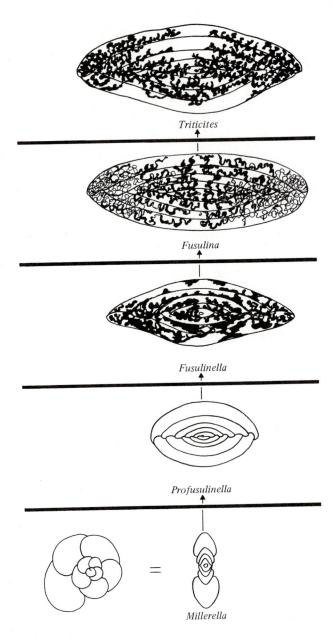

Triticites

Fusulina

Fusulinella

Profusulinella

=

Millerella

Fig. 17.2 Evolutionary trends in fusulinid Foraminifera (phylum Sarcodina). Reprinted with permission from Moore, Lalicker, and Fischer, *Invertebrate Fossils,* © 1952. New York: McGraw-Hill Book Company.

to the diameter of the coil, and the shell becomes progressively football-shaped (in *Fusulinella*) and even cigar-shaped (in *Fusulina*). In the latest members of this subfamily (*Tricitites,* etc.) the trend to axial elongation reverses somewhat. Other trends are exhibited by this group of Foraminifera: the septa, for example, which divide the shell into chambers, are planar in *Millerella* and become progressively more and more fluted as we follow the lineage through time. In *Fusulinella,* the septa are planar throughout most of the shell, but are fluted near the axis of coiling, especially at the poles. In *Fusulina,* the septa are fluted throughout.

Our second example concerns the cephalopod molluscs of the order Nautiloidea. Here, a series of trends affect the shape of the shell, which may be uncoiled (in the earliest genera), slightly coiled, or fully coiled (Fig. 17.3). The coils may be so loose that they do not touch one another, or they may touch and even overlap, partially or wholly enclosing the earlier coils. In some genera, the degree of coiling changes with age, so that the early turns may be more tightly coiled than the later ones, or the degree of coiling may change several times during life to produce varied shapes.

The extinct graptolites (subphylum or phylum Hemichordata, class Graptolithina) exhibit a number of interesting trends, shown in Fig. 17.4. The earliest graptolites had colonies that were branched, the number of branches (or stipes) often exceeding 40. The number of branches was reduced to 16, then 8, 4, 2, and 1. Meanwhile, other trends can be traced in a number of graptolite lineages, some of them independently in the four-branched and two-branched forms. The earliest of both two-branched and four-branched forms had these branches hanging down in a *pendent* position. Successively later forms raised these branches upwards to varying degrees, with both two-branched and four-branched forms ultimately (and independently) reaching the fully upright (scandent) position. Among the fully scandent forms, one or more rows (branches) tend to be lost, the two-branched or biserial forms giving rise to uniserial forms.

The primates (Chapter 27) exhibit a number of evolutionary trends of which the progressive increase in brain size is perhaps the most familiar. This same trend occurs independently in many primate lineages, including the one leading to our species. But if a series of living

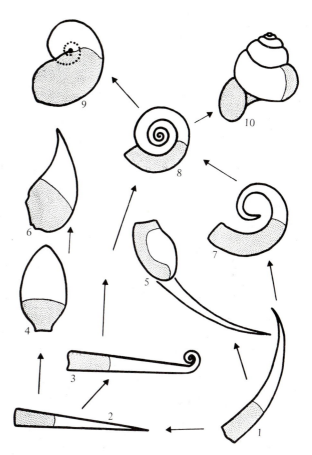

Fig. 17.3 Evolutionary trends in nautiloids (phylum Mollusca, class Cephalopoda). Reprinted with permission from Moore, Lalicker, and Fischer, *Invertebrate Fossils,* © 1952. New York: McGraw-Hill Book Company.

primates is arranged in order of increasing brain size they would not constitute a lineage, because none is ancestral to any of the others; the increase in brain size seen among these contemporary forms is therefore not a trend (Minkoff, 1974).

The family Elephantidae, to which our living elephants belong, shows in its well-documented fossil record a number of interesting trends; these have been

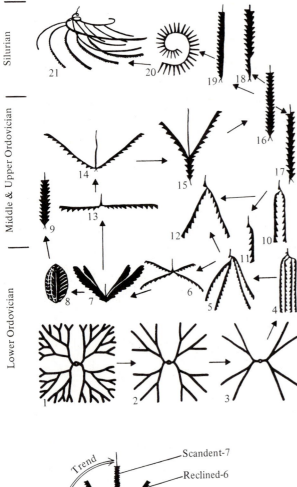

Silurian

Middle & Upper Ordovician

Lower Ordovician

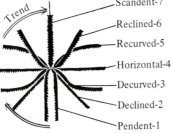

Trend

Scandent-7

Reclined-6

Recurved-5

Horizontal-4

Decurved-3

Declined-2

Pendent-1

Fig. 17.4 Evolutionary trends among graptolites (phylum Hemichordata). Reprinted with permission from Moore, Lalicker, and Fischer, *Invertebrate Fossils,* © 1952. New York: McGraw-Hill Book Company.

studied by Osborn and others, and most recently by Maglio (1973). The lower jaw underwent progressive shortening, the reversal of an earlier trend toward lengthening of both upper and lower jaws in mid-Tertiary Proboscidea. The number of molar teeth in use at any one time decreased, but the number of transverse ridges on each increased. The sequence of molar eruption became progressively slower, so that among the living genera there is always a pair of teeth undergoing eruption, thus exposing new and unworn enamel ridges, at all times during life. The enamel ridges themselves became progressively thinner, and the rear of the skull was reoriented so as to give more favorable leverage to the muscles used in grinding the food. A number of these trends are shown in Fig. 17.5.

C. EVOLUTIONARY TRENDS IN HORSES

Evolutionary trends within the horse family (Equidae) have long been the subject of evolutionary studies (Fig. 17.6). Several theories concerning the nature or course of all evolution have been suggested or tested by the history of horses. It is for this reason that we shall examine them so closely; in later chapters we will also refer back to this extended example in order to test various hypotheses about evolution. The following account is based on the research of scores of workers, as summarized by Stirton (1940) and Simpson (1951). Quinn (1955) dissents from the majority viewpoint, but Forsten (1975) and other recent workers have reaffirmed that viewpoint.

The earliest horses belong to the genus *Hyracotherium,* also commonly but incorrectly known as "Eohippus." A small animal, about the size of a fox terrier, *Hyracotherium* was a browser of leaves. It had 44 teeth, as in primitive placental mammals. The molar teeth had a simple, four-cusped pattern, and the individual cusps were distinct from each other and low-crowned. On each of its front feet, *Hyracotherium* had four toes, and on each of its hind feet it had five, though only three of these were functional; the first and last were too small to reach the ground. Small hoofs were present on each of the toes, but the weight of the body rested largely on soft, cushioned pads, similar to those of dogs or cats.

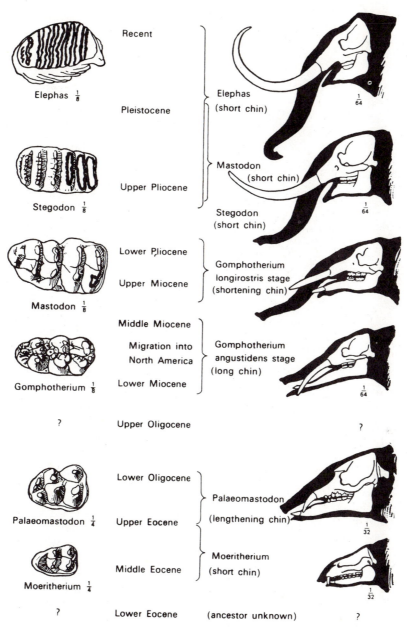

Fig. 17.5 Evolutionary trends among elephants (order Proboscidea). Note that the lower jaw became first longer and then much shorter, that the tusks grew longer rather consistently, that the molar cusps increased in number and gave way to a series of transverse ridges, and that these ridges finally became more numerous and pressed closer together. The impression of a single lineage is an oversimplification; there were actually several lineages in which these changes took place. Adapted with permission from W. B. Scott, *A History of Land Mammals in the Western Hemisphere*, 1913. New York: Macmillan.

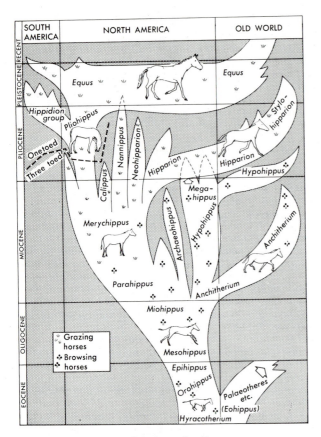

Fig. 17.6 Phylogeny of the horse family (Equidae). From *Horses: The Story of the Horse Family in the Modern World and Through Sixty Million Years of History* by George Gaylord Simpson. Copyright © 1951, 1979 by Oxford University Press, Inc. Reprinted by permission.

Evolutionary trends among horses were not always universal throughout the family. Still, a number of important trends can be identified, including a general increase (but occasionally a decrease) in size, a progressive loss of toes, lengthening of the toes that were retained, lengthening of the limbs generally, enlargement of the brain (especially the cerebral hemispheres), an increase in the height and complexity of the molar teeth, and an enlargement of the last two and eventually of the last three premolars until they came to resemble the molars.

From the 1870s well into the present century, studies of horse evolution tended to emphasize, if not overemphasize, the origins of the genus *Equus*. The lineage that led to modern horses was singled out, and other lineages were given less attention (Box 17–A). Defenders of the theory of evolution emphasized evolutionary trends within this lineage: the loss of toes, the increase in size, the increase in molar complexity and crown height, and the molarization of the premolars.

A closer look at horse evolution reveals that the lineage leading to *Equus* was but one of many, nor was it even the most direct (Box 17–B). Simpson's interpretation of horse evolution as involving a series of adaptive radiations and several adaptive shifts is probably much closer to the truth.

There were indeed evolutionary trends within the Equidae, but none has the steadily progressive character, nor the uniformity and universality, that have been attributed to evolutionary trends in the past. Increases in size, for example, occurred frequently within the horse family, but in at least three cases there was a reduction in size, and among Eocene horses there was hardly any change at all. The Miocene genus *Merychippus* had at least three direct descendants that grew larger, but also two others that became smaller.

A loss of toes indeed took place, but not gradually, nor in all lineages. The reduction from five toes (the primitive mammalian number) to four had already taken place in the forefeet of the earliest horses. In the hind feet, there was indeed a gradual diminution of the first and last toes throughout the Eocene, but the four toes on the front feet persisted without any trend at all toward their reduction. The transition to the fully three-toed condition in the earliest Oligocene horses was quite abrupt.* Despite this, the three-toed condition was maintained with very little change throughout much of the remaining history of horses. Only in one of the several lines descended from *Merychippus* was there a shift to the one-toed condition.

Other evolutionary trends in horses show similar characteristics. The lengthening of the limbs, and the

---

*On a geological time scale, that is. In the ensuing discussion, it should be borne in mind that changes described as "sudden" or "rapid" may have taken a few million years.

BOX 17–A *Evolutionary Trends in Horses*

A.   Evolution of horses' skulls

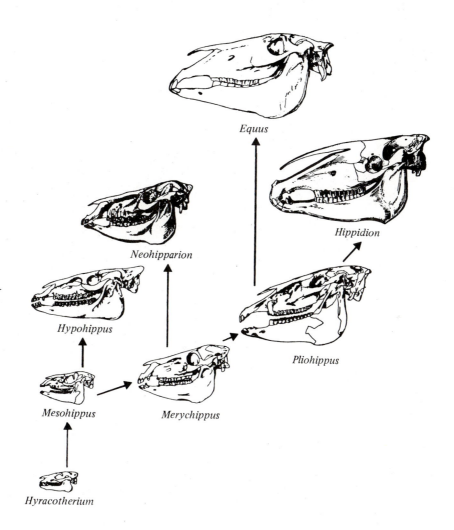

Equus

Neohipparion

Hippidion

Hypohippus

Pliohippus

Mesohippus   Merychippus

Hyracotherium

BOX **17–A**   *(continued)*

B.   Evolution of horses' feet

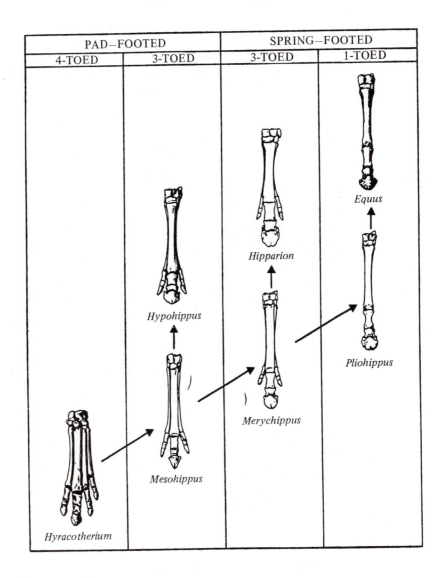

BOX 17–A *(continued)*

C. Evolution of horses' teeth

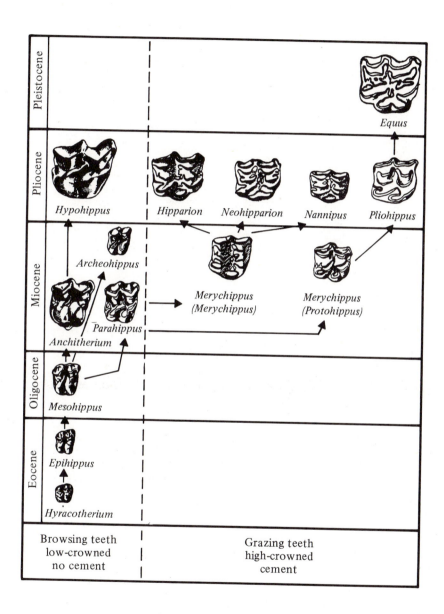

(continued)

BOX 17–A   *(continued)*

D. Evolution of horses' brains

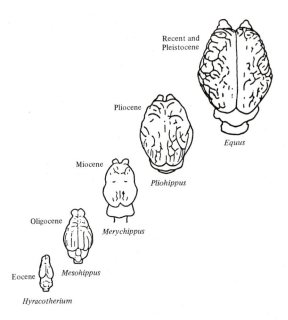

From *Horses: The Story of the Horse Family in the Modern World and through Sixty Million Years of History,* by George Gaylord Simpson. Copyright © 1951, 1979 by Oxford University Press, Inc. Reprinted by permission.

attainment of modern limb proportions, was essentially complete by the Oligocene, though this trend was reversed in the genus *Hippidion* and its descendants. The evolution of spring ligaments took place rapidly during the Miocene, giving to the feet a bounciness similar to that of a pogo stick. The molarization of the last two premolars took place gradually throughout the Eocene, but the premolar immediately in front of these was molarized quite suddenly, and in the earliest Oligocene horses, the process of molarization was essentially complete. Some trends were only temporary: the molars curved inward in many species of *Merychippus* and certain of their descendants, but among modern horses the molars are straight once more.

One of the most significant changes in horse evolution took place in the Miocene lineage that includes *Parahippus* and its descendant *Merychippus*. It is in this lineage, and this alone, that horses first adapted to a diet of grasses, taking advantage of the great expansion of grasslands that apparently occurred at this time. The teeth were greatly modified as an adaptation to this new diet; they became high-crowned (hypsodont), with the enamel folded into a nearly vertical orientation. Yet, the increase in hypsodonty was not at all uniform through time, nor was it the same in different lineages. The increase was most rapid beginning with *Merychippus* (see Box 19–B), and was in fact more rapid in the *Hipparion* lineage than in the lineage leading to modern horses.

BOX 17–B *Three Interpretations of Horse Evolution*

1. Evolution in a straight line (''orthogenesis'').

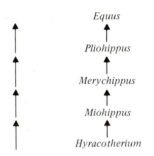

2. A straight ''main line,'' with side branches.

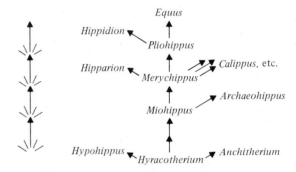

3. A series of adaptive shifts, with side lines often outlasting their ''main line'' cousins.

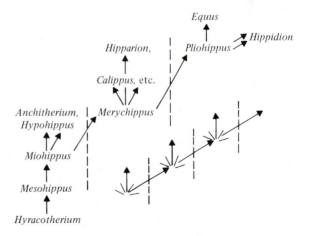

D. PARALLEL EVOLUTION

Some evolving lineages exhibit the same trend as other lineages, a phenomenon known as **parallelism.** The similarity of the end results is not crucial here, as in the case of convergence (Chapter 19); it is rather the similarity of sequence or process that defines parallelism. Of course, many cases of parallelism also result in convergence, but it would be a mistake to equate the two, or to treat either one as a special case of the other (both of which have been suggested a number of times).

Parallel trends, like parallel roads, may have different starting points, and they may or may not continue for the same time or distance (or at the same rate). Thus, there is no assurance of reaching the same destination or results through parallelism. Convergence, on the other hand, always leads to the same destination or to similar results, even if usually by different paths. Convergent descendants usually have distantly related or unrelated ancestors, but parallel trends usually develop in closely related groups, although it is not part of the definition that they must necessarily do so. Simpson and others have remarked that parallel trends demonstrate the relatedness of the ancestral forms, because only if their hereditary endowments were similar to begin with could the same sequence of mutational changes have resulted.

As an example of parallelism, consider the evolution of two closely related, extinct families (Notohippidae and Proterotheriidae) of South American mammals belonging to the order Litopterna. These plant-eating ungulates were superficially similar to horses, and were adapted, like horses, to running rapidly over hard ground for great distances. In the course of their evolution, these animals paralleled the horses in a number of evolutionary trends (Fig. 17.7): they increased in size, their limbs lengthened (with the distal segments lengthening the most, as in horses), their brains grew in size (though far less than in horses), the grinding surfaces of their teeth became more complex, and their toes became reduced in number. One family, the Notohippidae, reduced the number of toes to three, but stopped there. Their achievement of hypsodont molars, with cementum filling in the spaces between the molar cusps, was not only independent of the similar development among horses, but was achieved earlier: Oligocene notohippids in fact were more hypsodont than the most advanced of Miocene horses. The

family Proterotheriidae never evolved the hypsodont molars that permitted a shift from browsing to grazing, but in the reduction of toes from five to three to one, the proterotheres went further than any of the horses, losing all traces of the second and fourth toes, including the metatarsals still retained by horses as "splint bones." When North and South America were joined across

Fig. 17.7 Parallel trends in the horse family, Equidae (order Perissodactyla), and in extinct horselike mammals of the family Toatheriidae (order Litopterna). Adapted with permission from W. B. Scott, *A History of Land Mammals in the Western Hemisphere,* 1913. New York: Macmillan.

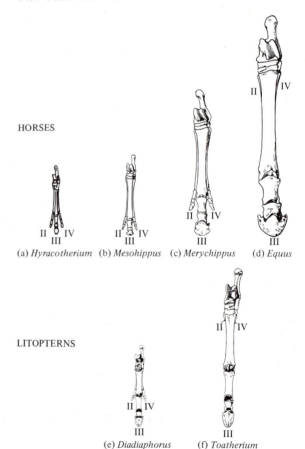

HORSES

(a) *Hyracotherium* (b) *Mesohippus* (c) *Merychippus* (d) *Equus*

LITOPTERNS

(e) *Diadiaphorus* (f) *Toatherium*

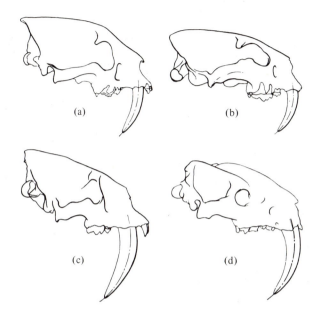

(a)

(b)

(c)

(d)

Fig. 17.8 Parallelism between sabertooth cats (order Carnivora, family Felidae) and South American sabertooth marsupials (order Marsupicarnivora, family Borhyaenidae). (a) The early Oligocene sabertooth cat *Hoplophoneus*. (b) The Pliocene sabertooth cat *Machairodus*. (c) The Pleistocene sabertooth cat *Smilodon*. (d) The sabertooth marsupial *Thylacosmilus*. From Simpson, American Museum Novitates, no. 1130, 1941.

Panama at the close of the Tertiary, both these families succumbed to intense competition from more efficient herbivores (including deer, camels, and others, not just horses), and eventually they became extinct.

As another example, consider the Borhyaenidae, an extinct family of South American marsupials. The canine teeth in these marsupials underwent an enlargement similar to that found among sabertooth cats (order Carnivora, family Felidae). Here, the enlargement of the tooth was carried further among the borhyaenids than among the sabertooth cats, and the most advanced genus (*Thylacosmilus*, Fig. 17.8) had a descending flange on its lower jaw to an extent not equalled among the cats. The purpose of this flange was presumably protection of the enlarged tooth, including protection of its tip when the animal rested its chin on the ground.

Other examples of parallelism between South American mammals and mammals on other continents include the peccaries of the family Tayassuidae (Fig. 17.9), which paralleled the Old World pigs of the family Suidae, and the ceboid monkeys (Chapter 27), which paralleled the cercopithecoid monkeys of Asia and Africa.

Like South America, Australia had been an island continent throughout the Tertiary, and it remains so today. Its mammalian fauna is unique, and shows the results (but, owing to a poor fossil record, very little of the process itself) of much parallelism with the placental mammals of other continents. Molelike, catlike, doglike, and squirrel-like forms have all evolved, some of them hardly distinguishable from their placental counterparts except in those features diagnostic of marsupials (see Fig. 18.9).

As the above examples illustrate, parallelism and convergence often occur together, but not always. Lineages exhibiting parallelism are usually related, but need not always be so. A case of parallelism in brain evolution among several related lineages of carnivores is shown in Fig. 17.10.

The existence of parallel trends has been used as an argument for the adaptive nature of trends, a point that we shall examine shortly. Parallel trends have also been used to argue for the upward or progressive nature of trends. Those who view trends as demonstrating advancement or "progress" often use parallelism to bolster their case.

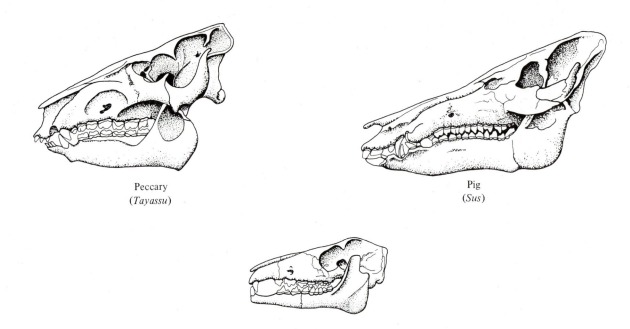

Peccary
(*Tayassu*)

Pig
(*Sus*)

Primitive artiodactyl
(*Stibarus*)

Fig. 17.9 Parallelism in the evolution of Old World pigs (order Artiodactyla, family Suidae) and New World peccaries (order Artiodactyla, family Tayassuidae).

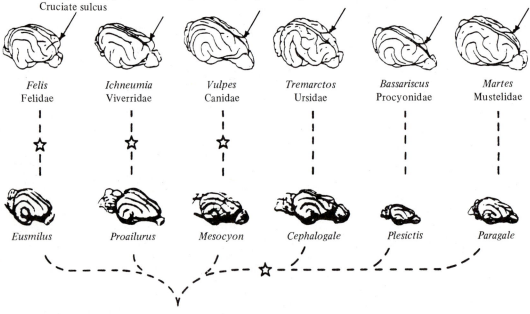

Cruciate sulcus

Felis
Felidae

Ichneumia
Viverridae

Vulpes
Canidae

Tremarctos
Ursidae

Bassariscus
Procyonidae

Martes
Mustelidae

Eusmilus

Proailurus

Mesocyon

Cephalogale

Plesictis

Paragale

Fig. 17.10 Parallelism in brain evolution among carnivores (order Carnivora). The six modern genera in the top row all have a cruciate sulcus (shown by arrow) within the frontal lobe. The Eocene or Oligocene ancestors of each (bottom row) show that the cruciate sulcus evolved independently at least four times, as indicated by stars. Reprinted with permission from Philip D. Gingerich, *Patterns of Evolution in the Mammalian Fossil Record*, 1977.

E. PROGRESSIVE AND REGRESSIVE TRENDS

Progressive Trends

The existence of evolutionary trends has suggested to many persons that evolution is on the whole a directional process, and that later organisms are "improvements" upon earlier ones. The concept of progress had ancient roots, but came to fruition with the philosophical movement known as the Enlightenment and with the French Revolution. Belief in progress inspired many early evolutionary theories and some modern ones, too. Supporters of scientific theories of "progress" have included biologists like Georges Buffon, Charles Bonnett, and Lamarck (Chapter 3), and geological and paleontological writers like Cuvier, Louis Agassiz, and Robert Chambers (Chapter 4). Philosophers supporting theories of "progress" have included Kant, Schopenhauer, Hegel, Marx, Engels, Spencer, and Teilhard de Chardin. Many modern evolutionists, on the other hand, see little if any meaning in the claim that evolution is a form of progress.

Distinctions between progressive and regressive evolution were first emphasized by the American paleontologist E. D. Cope (1886) and by the American botanist Bessey (1915). Most evolutionists today would agree that different trends may sometimes proceed in opposite directions, but few would see any purpose in calling some of these "progressive" and others "regressive." We can only justify calling a certain change "progress" if we can explain why its opposite would be regressive. Adaptation to the changing conditions of life may be termed an improvement in efficiency or in fitness, and evolution is in this limited sense progressive. But the adaptations may only be temporary, and as conditions change, adaptation must either keep pace or fall behind. A long-term trend does allow us to distinguish between a further continuation ("progress") and a reversal. In this limited sense, the continued increase in size among horses was progressive, and the occasional decreases in size were retrogressive.

Regressive Trends

If the term "progress" applies to any change whatever, then the claim that evolutionary change represents progress is a tautology that only tells us that evolutionary change represents change. If, on the other hand, only certain changes are considered progress, then the claim that all evolutionary change represents progress may be taken as a falsifiable hypothesis denying the possibility of regressive or degenerative change. Various theories of "progress" in evolution may thus be tested by a search for degenerative or regressive change.

Regressive evolution may be observed whenever a trend reverses its course, or whenever a organ degenerates and is lost. The degeneration of unused organs is quite frequent among parasites or cave dwellers. Vestigial organs in humans are shown in Fig. 17.11.

One might imagine that a useless structure, such as an eye in a cave-dwelling species, would be selectively neutral, and would thus evolve at random, or persist unchanged at the same level. In fact, such unused organs tend as a rule to degenerate and become lost. At least

Fig. 17.11 Vestigial organs in humans.

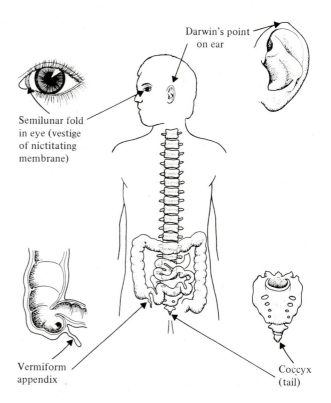

three types of explanation have been proposed to account for such cases: the several "nutritional" theories, the "clockwork" theory, and the Lamarckian theory of use and disuse.

Lamarck's theory of disuse, which Darwin accepted as a possible evolutionary mechanism (along with natural selection), went quite far toward explaining the reduction of unused organs to vestiges. According to this theory, any unused organ tends to atrophy, and its muscles in particular grow weaker. If this diminution is passed on to the offspring, then succeeding generations will be born with the organ somewhat reduced, and over a large number of such generations, the organ would be reduced to a mere vestige or disappear entirely. Darwin repeatedly fell back on this theory as an explanation for the loss of organs through regressive or degenerative evolution. In particular, he noted that muscles not exercised by domestic animals, such as those of the wings in domestic fowl, or those of the ears in certain breeds of dogs (which thus have drooping ears), tend to weaken greatly and also to diminish.

The crucial deficiency of this theory lies in its mechanism. The atrophy of unused organs is *not* passed on to the offspring, and thus has no effect on future generations. The reduced wings of domestic fowl could certainly be explained by one of the "nutritional" theories discussed next.

Several related "nutritional" or "economic" explanations for regressive evolution argue that an unused organ is somehow *not* selectively neutral, but is slightly disadvantageous because it diverts nutrition, biochemical energy, or some other limited resource from other, more useful functions. Darwin recognized this in the following words: "natural selection is continually trying to economise in every part of the organisation. If under changed conditions of life a structure before useful becomes less useful, any diminution, however slight, in its development, will be seized on by natural selection, for it will profit the individual not to have its nutriment wasted in building up an useless structure" (Darwin, 1859, pp. 147–148). Natural selection would thus tend to eliminate all organs for which there was not some positive adaptive value; hence the disappearance of eyes in cave animals, and of nearly all sense organs in highly evolved parasites.

The "clockwork" theory is a statistical explanation of regressive evolution. A complex and highly evolved organ, such as the vertebrate eye, exists as a well-tuned mechanism only because selection has brought about the most favorable combination of responsible genes. In the absence of selection, random changes would tend to persist. A random change in a well-tuned clockwork, or a finely tuned racing car, or any other complex structure, would only by rare chance be an improvement; most often it would be a change for the worse. So it is with evolution: complex structures, such as vertebrate eyes, would deteriorate greatly if they could undergo random changes in the absence of selection. Thermodynamically, an unused organ would tend to return to the condition of maximum entropy by random rearrangement of its component parts, or of the responsible genes.

F. THE ADAPTIVENESS OF TRENDS

Modern evolutionists acknowledge that one of the foremost characteristics of evolutionary trends is that they are nearly always adaptive. No theory of evolutionary trends would be satisfactory if it did not somewhere include an explanation of the adaptive nature of trends.

Evidence for the adaptive nature of trends comes from many sources: the adaptations of living organisms, functional analysis of adaptations in both living and extinct species, the persistence and independent occurrence of trends, mosaic evolution (to be defined shortly), and variations in evolutionary rates.

The strongest evidence for adaptive trends is of course the finding of many adaptations among the living organisms resulting from these trends. These various adaptations are often remarkable in their efficiency of design. They include anatomical structure, camouflage and mimicry, and the various coadaptations that exist between mutually dependent species. The adaptations of living organisms are treated more extensively in Chapter 10. Some of these same adaptations are also discernible in extinct organisms, lending support to the theory that adaptation is not merely a latter-day phenomenon.

Function, in the case of fossils, cannot be directly observed, but in many cases it may be obvious to us from comparison with living forms. Through the building and testing of physical models, or through the analysis of mathematical models, functional analysis may reveal how well suited a particular structure was to a specified task. If a structure seems optimally suited to a particular task, it is often assumed to have actually performed the task, especially if the structure varies in a particularly adaptive way across life stages (from young to old), adult size ranges, or taxa. In other instances, functional analysis may reveal the extent to which structures might have served several functions at once, or the extent to which they were preadapted to functions not yet actualized.

The mere persistence of many trends for many millions of years, with or without branching, is itself evidence that the trends are at least not inadaptive. If they were inadaptive, the trends would surely have led to rapid extinction. Persistence alone is therefore evidence of a certain minimum level of adaptiveness.

Parallelism (discussed earlier) and convergence (Chapter 22) are also used as arguments for adaptation under the same logic, but more persuasively: if the same condition evolved more than once, it is far less likely to have done so by chance. Therefore, its persistence was favored, which means that it must have had positive selective value; in other words, it was adaptive.

Variations in Evolutionary Rates

Some of the strongest evidence for the adaptive nature of trends comes from studies of variation in evolutionary rates. As physical and biotic conditions change, the direction and intensity of selection changes with them. The rate of evolution responds to these changes by slowing down or accelerating, possibly by coming to a halt, and even occasionally by reversing itself. Rates of anagenesis in long-continued trends do indeed vary with time, slowing down when a certain plateau is neared, accelerating with each ''breakthrough'' or improvement, occasionally coming to a complete halt, and even reversing themselves in a few cases. Horse evolution shows several instances of this: brain development was evolving at its maximal rate during the Eocene, but has since slowed down. Meanwhile, the evolution of hypsodonty has continued to accelerate, especially since the Miocene. The reduction in toes occurred in two or more brief shifts rather than in any continuous fashion. Other organisms

show trends with variable rates, too: coiling in nautiloids, septal fluting in Foraminifera, reduction in the number of dermal skull bones from fish to mammals. All of these trends exhibit phases of more rapid acceleration alternating with a generally lower if barely perceptible rate in between. The general rule is clear: evolutionary trends do not usually continue at a constant rate for more than a few million years. The observed changes in the rates of evolutionary trends are best explained by the hypothesis of adaptive control.

Mosaic Evolution

An important corollary to the variation of evolutionary rates with time is the principle that these rates may vary independently of one another. Different characters may evolve at different rates, at different times, and in different patterns. Thus, at any one time, any species may include a mosaic of both highly evolved and little evolved characters, characters that are just beginning to evolve rapidly, characters that have just finished evolving rapidly, characters that have not changed at all, and so forth. In a group of organisms, advancement in one character may or may not be accompanied by advancement in another. The mosaicism of characters among the green algae is shown in Box 17–C.

The Eocene browsing horses (*Hyracotherium, Orohippus,* and *Epihippus*) illustrate the principle of mosaic evolution. Following this lineage through time, we find an increase in the size of the brain, the progressive strengthening of the middle (third) toe, and the progressive molarization of the last two premolar teeth in each jaw. The molar teeth show the development of a small cusp (the mesostyle) on the outside margin, and the gradual connection of all cusps on the outside margin of the tooth into a W-shaped ridge called the ectoloph. But while these features show conspicuous, ongoing change, other features changed very little: the four toes on the front feet persisted during the Eocene without any reduction (though a reduction to three toes soon followed), while overall body size remained the same or decreased slightly. The gap (diastema) between the first two premolars varied without showing any consistent trend. Clearly, not all characters were evolving together at the same rate or at the same time.

BOX 17–C *Mosaic Evolution among Green Algae (Chlorophyceae)**

Order	Flagella	Sexual reproduction†	Vegetative structure	Other notable achievements
UNICELLULAR VOLVOCALES	persistent	isogamous	unicellular	—
COLONIAL VOLVOCALES	persistent	isogamous to oogamous	colonial	—
CHLOROCOCCALES	may be temporary only	isogamous to anisogamous	colonial	—
TETRASPORALES	temporary only	isogamous to anisogamous	simple colonial	—
ZYGNEMATALES	amoeboid instead	conjugation in some	unicellular to filamentous	chloroplasts may have unusual shapes
CLADOPHORALES	four	isogamous to anisogamous	multinucleate; some diploid	—
SCHIZOGONIALES (= PRASIOLALES)	none	oogamous	filamentous or sheet-like	often marine
DASYCLADALES	two	usually isogamous	multinucleate; often diploid	lime-secreting, often marine
SIPHONALES	two to four	isogamous to oogamous	multinucleate; often diploid	novel pigments
SIPHONOCLADALES	four	usually isogamous; may be oogamous	multinucleate	"segregated" cell division, often marine
OEDOGONIALES	many, forming crown-ring	anisogamous to oogamous	cylindrical cells, forming filaments	—
ULOTRICHALES	two to four	isogamous to oogamous	branching thallus	some are partly erect (most advanced morphologically)
ULVALES	two to four	usually isogamous; may be anisogamous	both haploid and diploid stages sheetlike or cylindrical, often two-layered.	sometimes marine

*Data are from various sources. The conditions indicated are common, but not necessarily universal, in each group.
†See Fig. 26.9 for an explanation of the terms used in this column.

The principle of mosaicism predicts that transitions between major taxonomic groups should not take place all at once, nor even in all characters at the same rate. There should be many character shifts, some taking place sooner than others. *Archaeopteryx* (Box 17–D) was reptilian in most respects, but Owen classified it as a bird because it had feathers—a fairly advanced condition. Rather than being half bird and half reptilian in each of its characters, *Archaeopteryx* was instead birdlike in a few and persistently reptilian in most others. Here, as in many other cases, a single character, feathers, can be singled out as the **key character** whose origin is taken to mark the transition from one major taxon to another. A key character is one that evolves early in the history of a group, and influences the evolution of a number of other characters. In the ideal case, the key character unlocks new opportunities previously closed to the organisms, enabling them to invade new continents or new adaptive zones or niches. In the origin of rodents, the key character was the chisel-like incisor tooth that grew continuously and remained sharp throughout life. In the case of artiodactyls, the key character was the possession of an astragalus bone with two pulleylike facets, enabling them to run faster and with less fatigue. In the origin of the Hominidae, the attainment of upright posture was a key character.

Causal Explanation of Trends

Five major causal explanations—and various versions or combinations of them—have been advanced over the years to account for evolutionary trends. Three of these, orthogenesis, finalism, and Geoffroyism, are now usually rejected because they fail to account for the adaptiveness of trends. Orthogenesis and finalism will be discussed in Chapter 18.

The possibility of direct environmental induction was inferred by the French naturalist Isadore Geoffroy Saint-Hilaire (Chapter 3) from his studies of abnormal embryos. Geoffroy thought that environmental changes could directly result in organic changes that were not necessarily adaptive. Embryos reared at high temperatures, for example, or with certain chemicals, develop characteristic abnormalities. Geoffroy thought that these changes, if administered repeatedly from one generation to the next, might become permanent, and the abnormal become normal. This theory, Geoffroyism, requires the inheritance of acquired characteristics, which the overwhelming majority of modern biologists reject as a significant evolutionary mechanism. Belief in direct environmental induction has often been associated mistakenly with Lamarckism, but Lamarck himself rejected this theory, and it is therefore incorrect to refer to it as Lamarckian.

Of all suggested mechanisms for the control of trends, two have emphasized the essential role of adaptation. These two are Lamarckian use and disuse (Chapter 3), and natural selection (Chapter 5).

Lamarck is often said to have believed in volition as a causative factor in evolution, with use and disuse as its method of action. Lamarck believed that animals would always act to fulfill their own "needs" (*besoins*), including the need for food or water. In fulfilling these needs, an animal would repeatedly use certain organs, at the expense of others, thereby strengthening those used while allowing the unused organs to atrophy. If characters so acquired were inherited, the augmented or diminished organs would then be passed on to the offspring, only to be augmented or diminished further in the next generation. Thus, through use and disuse, organisms would in a sense control their own evolution. The kernel of truth in this argument is that voluntary behavior often does influence the direction of evolution (Chapters 18, 24). But considerable evidence exists against the inheritance of acquired characteristics, and thus the theory loses its mechanism. Mayr has moreover questioned whether Lamarck truly believed in volition as an evolutionary force to the extent that has usually been assumed. One major point to Lamarck's credit is that his theory did predict the adaptive nature of evolutionary change.

The control of adaptive trends by natural selection is the only theory that explains all the facts that we have outlined: the generally adaptive nature of trends, the variations in evolutionary rates, parallelism, convergence, mosaic evolution, and so forth. Both the rate and the direction of change are controlled by natural selection. To be sure, certain minimal rates of mutation and recombination are needed, too. But these factors are seldom limiting or rate-determining, as is shown by artificial selection experiments (Chapter 11) in which changes can be made to occur at rates far greater than those of any well-documented evolutionary trend.

BOX 17-D   Archaeopteryx, *The First Bird*

(a) **As preserved.** The teeth, claws, and long tail were persistently reptilian, but the feathers were already birdlike, showing that *Archaeopteryx* was a warm-blooded flier.

(b) **As compared to a modern pigeon.** Comparable regions of the skeleton are shaded black. Notice the long tail, unmodified hand, weak sternum (breastbone), small brain case, weak pelvis, and teeth. These features had evolved very little in *Archaeopteryx* (over its reptilian forebears), but the feathers had evolved rapidly.

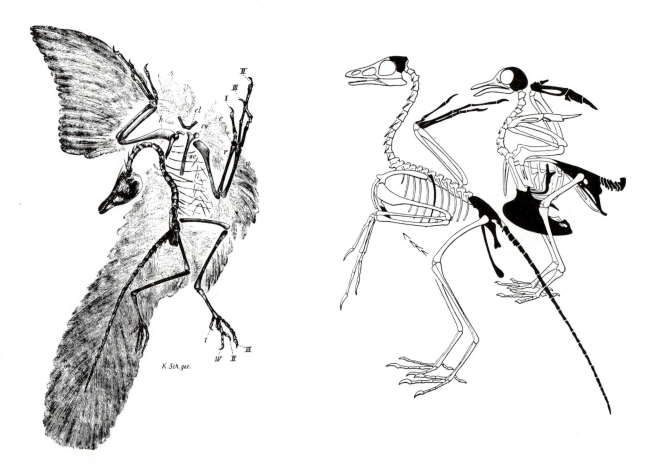

Part (b) is reprinted with permission from Colbert, *Evolution of the Vertebrates,* 1980. New York: John Wiley and Sons.

G. ARE THERE ANY INADAPTIVE TRENDS?

We have reviewed evidence of the theory that most evolutionary trends are adaptive. But is this necessarily true of *all* trends? Or are any trends demonstrably inadaptive? The inadaptiveness of a continued trend is not at issue: all evolutionists would acknowledge a long-continued trend as an adaptive one. Instead, two separate issues confront us: the possibility of inadaptiveness in the early stages of a trend, and the possibility of inadaptiveness in the later stages.

The Early Stages of Trends

The adaptiveness of the early stages of trends was questioned in 1872 by St. George Mivart, one of Darwin's critics. Mivart argued that the evolutionary changes postulated by Darwin would not necessarily be adaptive in their early stages, especially in the case of organs that needed to achieve a certain level of perfection in order to be functional at all. Mivart cited the eye and the wings of birds as two cases in point: how could an eye originate by gradual steps, if any half-formed eye would be incapable of seeing? How could a bird's wing originate, if a half-formed wing would be of no aid in flying?

Darwin's reply to this type of criticism was to an extent given in advance, since he anticipated that the issue would be raised. The evolution of such structures, said Darwin, would indeed be adaptive in all stages. The eye, for instance, could have originated as a simple photoreceptor, not capable of forming an image. A lens could later have been added as a simple improvement, increasing the sensitivity of the eye in dim light. The improvement of the lens' focusing ability enabled both greater sensitivity to the detection of movements, and a greater ability to discriminate between different types of images. The eye thus evolved in small steps, several of which have been preserved to us, at least approximately, among living animals (Box 18–A). The argument outlined over 100 years ago by Darwin is still followed today in its essentials; the evidence for this viewpoint has been reviewed by Simpson (1967) and especially Salvini-Plawen and Mayr (1977).

Other cases seem to bear the same treatment. The bird's wing, for example, is now commonly explained as an adaptation that originally retarded the descent of a reptile that jumped from one branch to another. Any increase in the surface area and maneuverability of the forelimb would have aided in gliding, even long before sustained flight was possible. The ability to flap the wings forcefully aided in reaching just a little further than was previously possible. With favorable atmospheric conditions, true flight became possible, and subsequent improvements were under the influence of natural selection favoring better and more reliable flying ability. Note that in this explanation, preadaptation and functional shifts play a prominent role: the first adaptation was climbing behavior (arboreal locomotion), then behavior that included jumping from limb to limb. This behavior preadapted the forelimbs to serve as organs that could be used to increase air resistance and retard descent, resulting in their selection for a winglike form. The attainment of winglike form in a maneuverable forelimb preadapted the wing for active, flapping flight. Subsequent evolution strengthened the wing and increased its efficiency as a flying structure.

Preadaptation and functional shift are characteristic of the early stages of many other trends as well. To cite just one more example, Simpson (1949) showed that the early evolution of the strange horns of the extinct titanotheres (Fig. 17.12) was probably the result of their behavior of butting with the head. Behavior of this sort, common to most ungulates, places a selective premium on the evolution of any thickenings or protuberances that would render the head more effective as a butting organ. The selective advantage of ever more prominent protuberances would, through natural selection, lead to the evolution of various types of hornlike structures in various early ungulate groups. Even in advanced ungulates (see Fig. 18.3), the diversity of horn shapes betrays the fact that they are multiple solutions to a common problem, and that almost any type of horn was selectively advantageous and so was preserved in the course of evolution.

The Later Stages of Trends

The grotesque or extreme conditions that occur in the late stages of evolutionary trends have been cited in support of the claim that adaptive trends, once under way, could continue beyond their adaptive optima and ultimately lead to extinction. This explanation depends

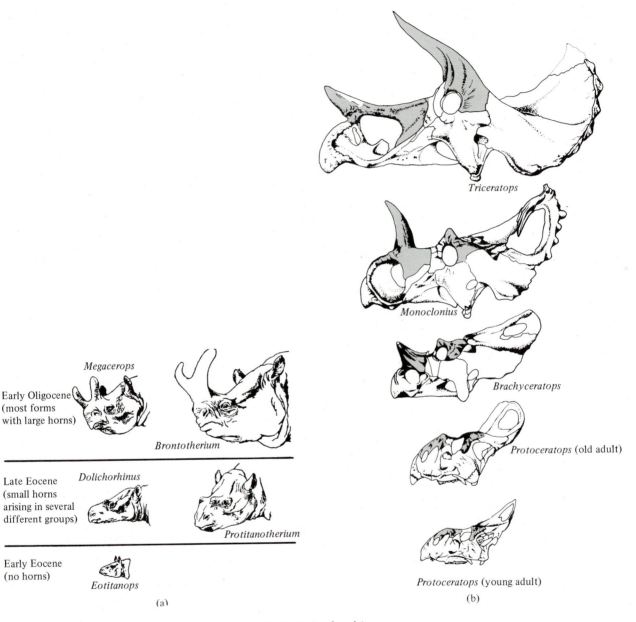

Early Oligocene
(most forms
with large horns)

Megacerops

Brontotherium

Late Eocene
(small horns
arising in several
different groups)

Dolichorhinus

Protitanotherium

Early Eocene
(no horns)

Eotitanops

(a)

Triceratops

Monoclonius

Brachyceratops

Protoceratops (old adult)

Protoceratops (young adult)

(b)

Fig. 17.12 (a) Evolution of horns among titanotheres (order Perissodactyla). The trend for the enlargement of the horns was begun with the earliest protuberances. (b) The parallel evolution of horns in ceratopsian dinosaurs (order Ornithischia, suborder Ceratopsia). Part (a) is reprinted with permission from G. G. Simpson, *The Meaning of Evolution,* © 1967. New Haven, Conn.: Yale University Press.

upon the existence of some sort of momentum in evolutionary trends, a phenomenon for which there is no known mechanism. Belief in evolutionary momentum has been a frequent characteristic of orthogenetic and finalistic theories of evolution (Chapter 18), but the majority of practicing evolutionists now reject such theories.

Three classic cases in particular have been cited by paleontologists as possibly indicating that adaptive trends may continue beyond the point where they cease to be adaptive. These three are the Irish elk, the sabertooth cats, and the oysters of the genus *Gryphaea*. All three cases involve trends that were ended by the extinction of highly evolved, specialized forms.

The so-called Irish elk, *Megaloceras* (Fig. 17.13) has been the subject of thorough reviews by Gould (1973, 1974). This magnificent beast had enormous antlers that were shed each winter and regrown with each successive spring, presumably at considerable expense nutritionally

Fig. 17.13   The Irish elk, *Megaloceras*.

and biosynthetically. The weight of the antlers must have been considerable, and the ability of the neck to support the head in its movements has been questioned from time to time. The claim has often been voiced that the antlers of the Irish elk grew so big and cumbersome that the poor creature could no longer support their weight. The extinction of the genus has often been attributed to this supposedly inadaptive terminal stage in the evolution of antlers.

According to Gould (1973, 1974) and to other evolutionists who have studied the Irish elk, its antlers were adaptive to the very end, and were in no way responsible for this great deer's extinction. Allometry of the antlers as a function of body size within the deer family shows that the Irish elk had antlers exactly as large as a deer its overall size might be expected to have (Fig. 17.14). The antlers were also palmate, meaning that the individual tines were connected at their base by a flat surface. Any male deer with palmated antlers displays his dominance status by showing his antlers off to maximum advantage, turning the maximum surface area toward its rival. According to Gould, this ability reached its height in the Irish elk, which had only to look at its rival, without need of lowering its head in the manner of many other deer.

The sabertooth cats (also called sabertooth ''tigers'') of the Pleistocene were the end members of a lineage that had its origins in the Eocene. In this lineage, the enlarged canines for which the sabertooth cats are so famous developed early, but then remained rather constant in proportion to the rest of the skull as overall size continued to increase (see Fig. 17.8). The sabertooths have been depicted as incapable of closing their jaws because of their overgrown teeth, but this view is fallacious. The teeth were quite functional at all stages, and were undoubtedly formidable as weapons. Perhaps most revealing in this case is that a parallel trend for the evolution of enlarged, ''sabertooth'' canines existed in a South American marsupial, *Thylacosmilus* (family Borhyaenidae, see Fig. 17.8). The parallel development of this structure in two distantly related orders of mammals on different continents strongly supports the belief that these structures were highly adaptive.

The oysters of the genus *Gryphaea* differed from the more typical genus *Ostrea* principally in the greater degree of coiling. The coiling apparently increased

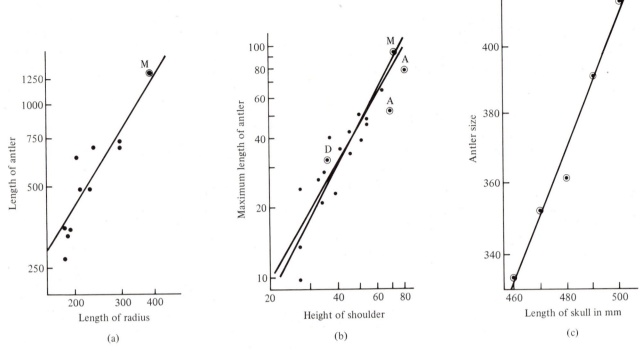

Fig. 17.14   Allometry of antler size versus other size dimensions within the deer family (Cervidae). Comparison with (a) Pleistocene cervids and (b) recent cervids shows that the Irish elk (M) had antlers whose size could have been predicted from its body size. The same adaptive relationship also holds within the genus *Megaloceras* (c). Reprinted with permission from Gould, *Evolution*, vol. 28, 1974.

throughout the evolution of the genus, until the latest forms coiled so far that they supposedly caused their own death by self-immurement. Simpson (1967), however, has pointed out that this was true of the most aged and postreproductive individuals only, the younger individuals being considerably less coiled. There is also indirect evidence to suggest that the extinction of *Gryphaea* was at least partially the result of predation by bottom-feeding crabs, a cause unrelated to the degree of coiling. It has furthermore been suggested that the genus *Gryphaea* is unnatural, that the specimens assigned to this genus are in fact the most highly coiled individuals in several related species, and that the "extinction" of *Gryphaea* is thus a taxonomic artifact.

CHAPTER SUMMARY

A lineage is a succession of species arranged in a continuous ancestral-descendent sequence. The branching of lineages is called cladogenesis. Development within a lineage between branching points is called anagenesis. Within any evolving lineage, the continued change in any character is called a trend. The early stages of a trend are called primitive, the later ones advanced. A generalized condition may be used as the starting point of diverse trends in several directions. A specialized condition occurs at or near the end of a trend, and involves the loss or restriction of evolutionary potential.

Evolutionary trends may be found in most groups with adequate fossil records. Careful studies of horses, a rather well-documented family, show that the direction and speed of these trends are subject to much variation. Several trends are restricted to one lineage among many within the family.

Parallel evolution occurs whenever two or more lineages exhibit the same trend. This always means the same sequence of changes, usually but not always with the same end results.

Evolution is always a change, but not every change represents ''progress.'' No trend can be considered ''progressive'' unless the opposite trend would be considered regressive. Lamarck's theory of disuse was intended to account for such regressive changes as the loss of unused organs, but the theory depends upon the inheritance of acquired characteristics. Two other theories are more in accord with modern evolutionary thought; one is basically a statistical argument, that random changes in a well-tuned mechanism lead generally to a reduced efficiency or diminution of function. The remaining family of theories is that useless organs are somewhat harmful, rather than selectively neutral, because they deprive other more useful structures of energy and nutrition.

The adaptiveness of trends is demonstrated by the adaptations of living organisms, by the persistence of many trends over long periods of time, by parallelism and convergence, by adaptive variations in evolutionary rates, and by mosaic evolution. Mosaic evolution is the evolution of different characters at different rates, each according to its own pattern. At any given time, any species contains a mosaic of advanced traits, primitive traits, slowly evolving traits, and rapidly evolving traits in all stages of evolution.

Several possible explanations of evolutionary trends have been put forth. Of these, orthogenesis, orientation toward finalistic goals, and direct environmental induction are now usually rejected because they fail to account for the adaptive nature of trends. Of the two explanations that emphasize adaptation, Lamarckism is rejected by most modern evolutionists because of its reliance on the inheritance of acquired characteristics. Natural selection, the accepted explanation, accounts for all the observed characteristics of trends, including their adaptiveness and their generally opportunistic nature.

Evolutionary trends are adaptive in both their early and their later stages. The early stages of many trends involve preadaptation of structures to their eventual function, even while fulfilling some prior function. The supposedly inadaptive trends can all be explained adaptively.

18

Directional Versus Opportunistic Evolution

Chapter Outline

Does evolution have a destination or a goal? Is it constrained to proceed along predetermined paths, or in predetermined directions? Or is evolution a groping process, following whatever paths it can, in as many simultaneous directions as it can? These are the basic questions we shall examine in this chapter.

A. DIRECTIONALISM

Several theories of evolution have attributed to the evolutionary process some sort of overall direction, with or without a specific goal. We may call these theories directionalist, while those that presuppose a goal can be called finalist.

Orthogenesis

A few evolutionists have discussed under the name of ''orthogenesis'' a number of interrelated evolutionary concepts. Succinctly, we can enumerate three: evolution in a straight line, evolutionary momentum, and the cause of evolutionary trends.

Simplest of these concepts is **unilineal evolution,** or evolution in a simple, unbranching progression. It is not surprising that evolution *sometimes* proceeds in straight, unbranching lineages (Fig. 18.1), or that evolutionary trends *sometimes* persist for long intervals of time. Close examination of the fossil record reveals that some lineages are orthogenetic in this sense, while others are not. The horses (see Box 17–A) and graptolites (see Fig. 17.4) are two groups containing branched sequences documented by fossils.

A second orthogenetic concept is the idea that evolutionary trends have some kind of momentum that causes them to continue unless stopped by a contravening force. There is no evidence to support this hypothesis; all evidence in fact points to the opposite: evolutionary trends, left to themselves, will come to a stop unless continuing selection in the same direction keeps them going.

Many orthogeneticists have claimed that orthogenesis is also the cause of evolutionary trends. Some are even guilty of the ''naming fallacy'': ''orthogenesis'' is *defined* as the cause of evolutionary trends, and trends are explained by appealing to orthogenesis. Thus the statement, ''orthogenesis is the cause of trends'' simply means that the cause of trends is the cause of trends.

Defining orthogenesis in this way defines nothing—it merely gives a name to "the cause of trends." Supporters of orthogenesis have claimed that the concept somehow sheds light on evolution in general, but most modern evolutionists reject this claim.

Teleology

One of the most often recurring fallacies in evolutionary biology is the fallacy of teleology, or the assumption that natural events can be explained as purposefully determined by future goals.

Much of human behavior can be explained as purposeful. We all justify many of our actions in terms of some desired outcome: "I will study hard *in order to* get a good grade on the next test," or "I am buying this record *so that* I might enjoy listening to it." Explanations such as these, which explain events in terms of a purpose or goal, are called **teleological explanations.** Teleological explanations tend to explain present events in terms of future ones: the purchase of the record is explained in terms of future enjoyment in its use. Yet the explanation remains unchanged even if the record should break on the way home.

Physiological activities are often so beneficial that we are tempted to invoke teleological explanations for them: the rat shivered "in order to" raise its body temperature; the plant deposited starch "so that" it could have a reserve supply of energy to draw upon. Vitalists since Aristotle have used the seemingly goal-directed activities of living organisms to argue for the existence of vital forces. Most modern biologists would explain these apparently teleological processes by invoking evolution through natural selection: plants that stored starch built up a reserve supply upon which they could draw, and thus had a competitive advantage over plants that did not store starch. Over long periods of time, this resulted in the persistence of starch-forming plants and the elimination of many plants that lacked this ability.

In general, we may translate "A occurs *so that* B will occur" to mean that among past occurrences of A, those followed by B increased the chances of A's recurrence. Some have claimed that this is a "new," mechanistic form of teleology, in contrast to the vitalistic and often anthropomorphic teleology of Aristotle.

In all forms of teleology, but more obviously in the traditional form, events are explained with reference to

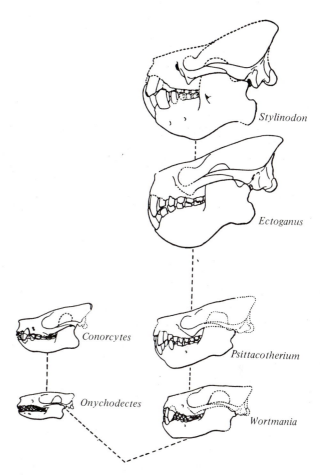

Fig. 18.1 Straight-line evolution in a group of extinct mammals (order Taeniodonta). One branching occurred very early in the history of this order, but no subsequent branching is recorded, despite considerable morphological change in one of the resulting lineages, leading to the genus *Stylinodon.* From Bryan Patterson, "Rates of Evolution in Taeniodonts," in *Genetics, Paleontology, and Evolution,* edited by Glen Jepsen, Ernst Mayr, and George Gaylord Simpson (copyright 1949, © 1977 by Princeton University Press), Fig. 2, p. 246. Reprinted by permission of Princeton University Press.

their consequences. A process may be described as teleological if it is *goal-oriented,* that is, if the goal (*telos*) influences the process itself.

Finalism

It is difficult for us to forget living descendants when we think of their fossil ancestors. Many trends seem to reach their limit in present-day organisms, suggesting to some people that the result might have been predetermined or preordained. This is **finalism,** the theory that evolution is directed by its goals.

Finalistic theories of evolution may be distinguished as either organic or superorganic. **Organic finalism** involves the voluntary striving of the organisms themselves after some goal; **superorganic finalism** is the belief that evolution is directed toward the goal by a supernatural or divine force. Very few evolutionists currently support either form of finalism. Several theories that have taken their inspiration from Lamarck have been finalistic, but Lamarck himself was not a finalist.

Organic finalism is a belief in goal-seeking evolution directed by the evolving organisms themselves, as when crossopterygian fishes are depicted as "striving" to evolve into amphibians capable of walking on land. As a possible explanation of trends, organic finalism presupposes that organisms possess cognitive powers that have never been demonstrated, such as the ability to determine their own distant future needs and to direct their evolution accordingly. Some vitalists attribute exactly these abilities to their "vital force"; one of them even defined the vital force as a force that causes evolutionary trends to follow an undeviating path from one species to the next. Yet, nobody has explained (for instance) how a crossopterygian fish, even one with human intelligence, might have known that there were land environments to colonize, or that its descendants would ever be capable of colonizing them.

Superorganic finalism is a belief in goal-directed evolution under supernatural or divine influence. Among such theories, none has attracted more attention in recent years than that of Pierre Teilhard de Chardin. A devout priest belonging to the Jesuit order, Teilhard tried very hard to reconcile his strong belief in evolution with his strongly held religious beliefs. His best known work is *The Phenomenon of Man,* published in 1955, shortly after his death. The strongly metaphorical style of Teilhard's books stands in marked contrast to the lucidity of his earlier geological writings. In his books, Teilhard often coined new words without defining them. He seldom presented reasoned arguments against any conflicting views. Indeed, because of the vagueness of Teilhard's writing, it is often unclear exactly what those conflicting views would be. Perhaps this is why G. G. Simpson called Teilhard a "religious mystic."

Teilhard saw the human species as the outcome, if not the original goal, of a long evolutionary progression that began with the origin of life from lifeless matter. Evolution, according to Teilhard, is not the blind, groping process that some would see it to be; rather, it is a progression to higher degrees of perfection. Evolutionary trends have a definite direction, and this is to the improvement of every life form on Earth. Life, said Teilhard, creates a shell or envelope, the Biosphere, around the lifeless globe, which is the Lithosphere. Human thought creates around this yet another layer, or Noösphere. Human evolution is as different from organic evolution as life was from the nonliving world in which it arose.

The finalist conception of goal-directed or teleological evolution uses the existence of evolutionary trends as one of its major pillars of support. A retrospective glance at any such trend shows clearly how each successive improvement led one step closer to the final outcome, as if the result had been preordained. Predestination is exactly the thesis that many superorganic finalists seek to prove, and *Homo sapiens* is the goal toward which most finalists see evolution as striving. These ideas, and others, are contained, for example, in the writings of Du Noüy (1940) or those of Teilhard de Chardin.

Of course, evolutionary trends terminating in living species can always be seen as leading toward those species. Perhaps it is relevant that long-continued trends seldom progress toward their apparent "goals" in a uniform direction or at a uniform rate—there are many detours, pauses, and even reversals. The frequency with which trends may arrest themselves or reverse their direction would alone seem to discredit finalism. Also, since evolution is a branching process, it is equally valid to point to pigs, worms, or mushrooms as the "goals" of evolution as it is to single out humans—they have certainly resulted from trends that have continued as long as any that gave rise to our own species. Also, if grasses (or orchids) are the culmination of plant evolution, or honeybees the pinnacle of arthropod evolution, then how does one explain the evolution of the various other plants or other arthropods? Or if humans were evolved as evolution's goal, then how does one explain malarial parasites, polio viruses, or tuberculosis bacteria, to name just a few of our natural enemies?

The fossil record contains considerable evidence against both orthogenesis and finalism. The theory of opportunism, explained next, is far more consistent with the evidence at hand.

B. OPPORTUNISM

George Gaylord Simpson has used the term "**opportunism**" to cover several related phenomena in evolution. As Simpson has stated (1967), opportunism means (1) "that what can happen usually does happen," (2) that "changes occur as they may and not as would be hypothetically best," and (3) that "the course of evolution follows opportunity rather than plan," proceeding simultaneously in as many directions as it can.

The above senses of "opportunism" are related to one another, and all represent a rejection of finalism. Evolution, in other words, is determined by opportunities of the moment, rather than by conscious or unconscious precognition of future or ideal possibilities. Opportunism in the first sense means that any existing possibility will sooner or later be seized, and any opportunity exploited. We see this phenomenon in the invasion of new continents or islands by species that soon exploit whatever opportunity they find. Darwin cited many instances of this in his *Origin of Species*.

Opportunity, however, is limited. Only certain species are given certain opportunities. Their morphology is never completely plastic; they can only evolve new structures out of old ones. Even then, the old structure must be suitable to its new function, and moreover it must either be capable of relinquishing its former function with no harmful effects, or else be capable of subserving both old and new functions simultaneously. The course of evolution, again, is determined only one step at a time. Every step must be adaptive in its own right; there is no deferral of adaptation to some future time. Here is where the line is drawn between finalistic and opportu-

nistic theories: finalism would predict evolution according to plan, with occasional deferral of adaptation in order that a future, more perfectly adapted state might later evolve; opportunism denies this possibility, and insists that each new stage must be adaptive.

We shall therefore examine the evidence that will enable us to decide between opportunistic and finalistic theories of evolution.

Organic Diversity and Adaptive Radiation

One of the simplest, yet most striking of evolutionary phenomena is the increase in organic diversity through branching evolution or cladogenesis. The fossil record clearly shows that, with only occasional or localized exceptions, life as a whole has always tended to increase in both numbers and in diversity. The same tendency also applies to most larger taxonomic groups (phyla, classes), but with occasional exceptions (lycopods, cycads, pteridosperms, trilobites, cephalopods, and crinoids among them).

The branching process of evolution is difficult to explain from either an orthogenetic or finalistic viewpoint. We can, however, explain it opportunistically: diversity within and between populations can expose a species to varying opportunities and different selection pressures. Competition (Chapter 9) may force species to exploit these new opportunities, thus increasing the number of available niches by subdividing existing ones. This process, if continued, will produce branching evolution, and ultimately an increase in the diversity of life, though not necessarily in each and every taxon.

When an evolutionary "breakthrough" opens up a variety of opportunities, rather than only one, diversification often results. The diverse adaptations among the descendants of a single form is called **adaptive radiation**. Examples of this phenomenon include the diversification of "Darwin's finches" on the Galapagos Islands (Fig. 5.3), of Australian marsupials (see Fig. 18.9), of Hawaiian birds (Fig. 18.2) and Hawaiian *Drosophila*, of antelopes in Zaire (Fig. 18.3), or of lemurlike primates on the island of Madagascar (Chapter 27). The diversification of Devonian fishes, Cretaceous angiosperms, or Tertiary bivalves may also be cited as examples. The placental mammals are yet another example: not long after

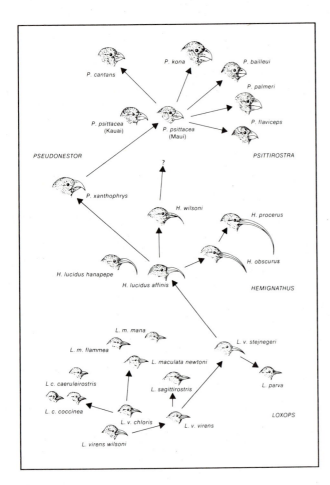

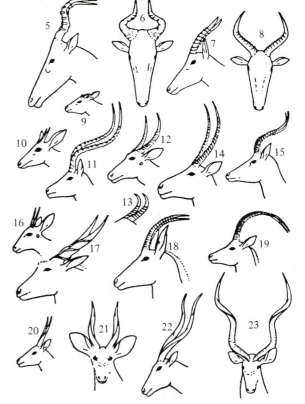

Fig. 18.2   Adaptive radiation among Hawaiian honeycreepers (family Drepanididae). The different types of bills are each adapted to a different diet. Reprinted by permission from Ayala, Francisco J. and Valentine, James W, *Evolving: The Theory and Processes of Organic Evolution*, Menlo Park, California, The Benjamin/Cummings Co., Inc., 1979, Fig. 8.3, p. 263.

Fig. 18.3   Adaptive radiation among the horns of antelopes from Zaire. Despite the variety of shapes represented, there is no evidence that any of these horns are any less effective than the rest. Reprinted by permission from Schouteden, *De Zoogdieren van Belgisch-Congo en van Ruanda-Urundi,* 1947, p. 166. Copyright Koninklijk Museum Voor Midden-Afrika, Tervuren, Belgium.

the evolution of the placenta (Chapter 26), these mammals radiated out into various modes of life, giving rise to more than two dozen orders within a relatively short span of time.

Essential to any adaptive radiation is the existence of a large number of opportunities (vacant niches) within easy reach, and subject to less competition than the former niches. Opportunities for a flying arthropod, though theoretically available, were not within easy reach until insects first evolved wings. A number of new opportunities became at once available to them, and their adaptive radiation soon followed. Their relatively great freedom from competition was insured by their rapid diversification, and by the fact that they had been first to evolve wings in an otherwise wingless world. To use another example, opportunities for a hummingbird-finch, a woodpecker-finch, or a flycatcher-finch do not normally exist in the presence of true hummingbirds, woodpeckers, or flycatchers. These opportunities did, however, exist on the Galapagos islands, where they awaited the first land bird capable of exploiting them. The arrival of the first finches on these islands made these opportunities readily available, and soon the finches underwent an adaptive radiation in which they seized opportunities normally unavailable to finches living elsewhere.

Opportunism involves a strong element of chance. Among all land birds, the finches may *not* have been the group maximally suited for colonizing of the Galapagos, either in their adaptability to the islands themselves or in their possibility of transoceanic dispersal. Of course, they were sufficiently adaptable, and sufficiently capable of transport, but this does not mean that other species might not have been equally or even more suitable in both these respects. The colonization of the Galapagos by finches, instead of some other family, was thus to some extent a matter of chance. Likewise, among cynodont reptiles, several families might have been capable of evolving into mammals, and yet only one is believed to have done so. The choice of this family from among several possibilities may have been a matter of selective advantage, but chance may also have been partially responsible. Again, among all known forms of life, amino acids are almost exclusively of the L- ("left-handed") variety. Chemical evolution (Chapter 25) of-

fers an explanation for this kind of uniformity, but not for any preference of L- over D-amino acids. It may only have been a matter of chance that some early and successful form of life had one kind of amino acid and not the other.

In all of the above cases involving chance, the subsequent outcome was strongly influenced by the "**first arriver**" principle, or "**king of the mountain**" principle, as I prefer to call it. This principle merely states that the first species to exploit a new opportunity, such as a new environment or niche, gains thereby a selective advantage against all future competitors simply because it got there first and established a foothold. The successful establishment of winged insects so completely preoccupied all available niches that the evolution of some second group of winged arthropods was thereby precluded, for the newcomers would have had to compete against the already established insects. On the Galapagos islands, the successful establishment of finches precluded the establishment of a second group of land birds, just because the finches had arrived first.

Multiple Solutions

Many evolving species have independently encountered similar limitations. These "problems" may have many solutions, which may all be equally suitable, but more often the various "solutions" are unequally successful in different situations. One might suppose that, if evolution were controlled by a benevolent Deity, the optimal solution to such problems would always be preferred. Alternatively, if evolution were controlled by an experimenting Deity (the "divine Tinker" hypothesis), one might expect *every* possible solution to be tried. Neither of these predictions holds; evolutionary "problems" are met by differing "solutions," but not all possibilities are tried everywhere. This is what one would expect on the basis of opportunism, for each group encountering the same problem independently would evolve the "solution" most readily available *to them*, without regard to the choices made by other groups encountering the same problem. Evolution, in other words, takes place along the path of least resistance.

Whenever a new structure evolves, it evolves out of whatever previously existing structures happen to be

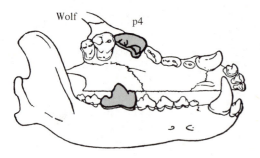

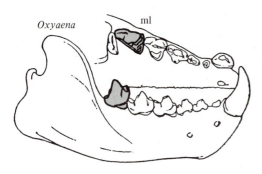

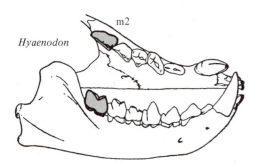

Fig. 18.4 Multiple solutions among the meat-shearing (carnassial) teeth of various carnivorous mammals. All three developed these shearing teeth, but a different pair of teeth was used in each case. Reprinted with permission from G. G. Simpson, *The Meaning of Evolution,* © 1967. New Haven, Conn.: Yale University Press.

available. Thus the bird's wing evolved from the modification of a forelimb, as did also the bat's wing and the wing of the extinct flying reptiles known as pterosaurs. But insect wings evolved from a different source: extensions of the dorsal "roof" of the body's exoskeleton became modified into wings, while the walking legs were never modified for this purpose.

Multiple solutions to the same problem can be seen in the evolution of respiratory adaptations among land animals. Land vertebrates rely upon lungs, insects rely upon a "tracheal system" of branching air-filled tubes, and spiders and land snails each rely upon a different form of modified gill structure. Swimming tails have evolved independently many times, e.g., among fishes, whales, and a group of extinct marine reptiles known as ichthyosaurs. It happens, however, that both fishes and ichthyosaurs evolved vertical tails that undulate from side to side, while whales, like lobsters, evolved a horizontal form of tail that beats up and down. The various locomotor adaptations among the Protista (Chapter 26) are yet another example of multiple solutions. Two further examples of multiple solutions are shown in Figs. 18.4 and 18.5.

External parasites of land vertebrates are better adapted if they are very flat, probably because they are more resistant to being dislodged. Mites, ticks, and two different orders of lice have convergently evolved bodies that are flattened dorsoventrally (top to bottom), but fleas have achieved comparable results through the evolution of a body shape that is flattened from side to side.

One of the most striking cases of multiple solutions concerns the evolution of photoreceptor organs (eyes), a subject last reviewed by Salvini-Plawen and Mayr (1977). Throughout the animal kingdom, a great diversity of photoreceptive organs is found, some unicellular and others multicellular. Some can merely detect overall differences in light intensity, while others can determine the direction of light sources, and still others can form complex images. Image-forming eyes may use various physical principles. Four such principles are known: a convex lens, a concave reflector, a pinhole, and a series of individual photoreceptors (Box 18–A). The principle of a convex lens is used in the eyes of both vertebrates and cephalopods, though in one case the retina is everse while in the other it is inverse. The pinhole principle is used in the eyes of starfish and some cephalopods. The com-

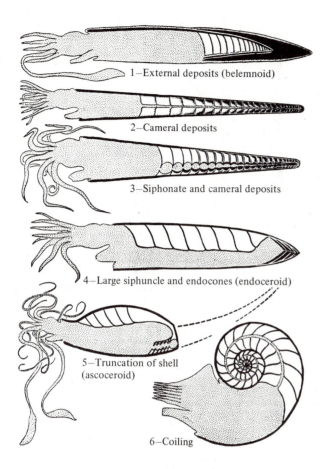

1—External deposits (belemnoid)

2—Cameral deposits

3—Siphonate and cameral deposits

4—Large siphuncle and endocones (endoceroid)

5—Truncation of shell (ascoceroid)

6—Coiling

Fig. 18.5 Multiple solutions among cephalopod molluscs to the problem of maintaining a constant and stable orientation in the water. Light, gas-filled chambers (white) and heavy shell material (black) were used in various ways, but all achieved the same general result. Reprinted with permission from Moore, Lalicker, and Fischer, *Invertebrate Fossils,* © 1952. New York: McGraw-Hill Book Company.

pound eye principle is used by insects and other arthropods: each photoreceptor is sensitive to varying intensities of light, and the image is formed by the pattern of these differences, just as the image on a television screen is formed by a pattern of individual dots. No known organism has yet exploited the possibility of using a concave reflector to build an image-forming eye. This is an instance where evolution is limited by possibilities of the moment: solutions that may theoretically exist are often out of reach.

Photoreceptors may arise from the body surface (epidermis) or from deeper-lying structures; those arising from the epidermis may have the cell membrane unfolded (unplicated), but more often these membranes have their surface areas greatly increased either by numerous folds and microvilli (the rhabdomere type) or by cilia (the ciliated type). The light-sensitive parts may be closer to the surface than are the cells that they stimulate, so that the stimulus travels away from the surface (everse or direct organization). Alternatively, the light-sensitive parts may lie deeper than the cells that they stimulate, with the stimulus traveling back toward the surface (inverse organization). The otherwise similar eyes of mammals and squids differ in that mammalian eyes are inverse while squid eyes are direct. By classifying eyes according to the above criteria, Salvini-Plawen and Mayr (1977) conclude that eyes have evolved independently at least 40 times, or perhaps more than 60 times. In a lineage in which photoreception is favored by selection, a photoreceptor may be built out of whatever materials are at hand, according to any of several physical principles and using any of several alternative designs. The multiple

BOX 18–A   *Multiple Solutions among Animal Photoreceptors*

The following simple types of photoreceptors are: (a) light-sensitive pigment spot (stigma) in a protist (*Pouchetia*); (b) light-sensitive cells in the skin of an earthworm; (c) three grades of complexity in the photoreceptors of starfish; (d) photoreceptor of the flatworm *Dugesia;* (e) three types of eyes in gastropods; (f) highly differentiated eye of a clam. Also shown are several types of advanced, image-forming eyes: (g) the eye of a cuttlefish (class Cephalopoda), with lens and cup-shaped retina; (h) the human eye, also with lens and cup-shaped retina, but with a different arrangement of cells in the retina; (i) the pinhole eye of the chambered nautilus (class Cephalopoda); (j) the compound eye of an insect with multiple tubes.

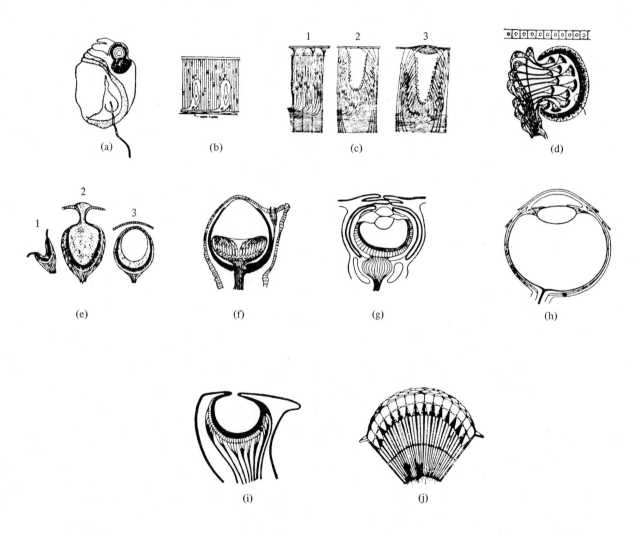

From Simpson, 1967. *The Meaning of Evolution.* New Haven: Yale University Press.

solutions to the problems of photoreception show that these materials, designs, and physical principles have been different in different lineages, and moreover have varied independently of one another, so that eyes that resemble one another in several respects (as between mammals and squids) may nevertheless differ from one another in detail.

Fig. 18.6 An example of iterative evolution among trilobites of the genus *Olenus*. Four separate lineages underwent similar changes in their pygidia (terminal portions), in each case from a similar ancestry, indicated by the broken arrows, and in each case terminated by unfavorable local conditions (heavy solid lines). Reprinted with permission from Simpson, *Major Features of Evolution,* © 1953. New York: Columbia University Press.

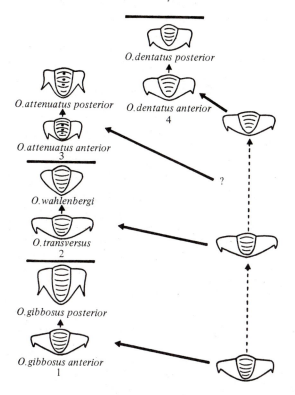

O.dentatus posterior

O.attenuatus posterior   *O.dentatus anterior*
                                         4

O.attenuatus anterior
            3

                                    ?

O.wahlenbergi

O.transversus
        2

O.gibbosus posterior

O.gibbosus anterior
          1

Iterative Evolution

The exploitation of a new opportunity by one successful group may preclude the later exploitation of this same opportunity by another group. But what if the first group is no longer successful? The opportunity now becomes available to any other group capable of seizing it, and the newcomers would in this case have just as much freedom from competition as did the unsuccessful pioneers of the earlier group. This situation, if repeated, leads to the phenomenon of **iterative evolution** (Fig. 18.6), or repeated evolution of the same or similar adaptations independently.

Examples of iterative evolution include the repeated evolution among Foraminifera of "large," bottom-dwelling genera from "small," pelagic ancestors, or the repeated invasion of Eurasia by various horse genera of North American ancestry (see Fig. 17.6).

Opportunism and Convergence

Strong evidence for opportunistically adaptive evolution comes from the phenomenon of **convergence,** which we may define as the *independent evolution of similar adaptations in unrelated organisms.* The independent evolution of the fishlike streamlined body shape in sharks, ichthyosaurs, and cetaceans (whales, porpoises, and dolphins) is a frequently cited example of convergence (Fig. 18.7).

It is a mistake to consider convergence as a special case of parallelism, or *vice versa.* Parallelism produces similar trends; convergence produces similar results. If the starting points are dissimilar, the results of parallelism are equally dissimilar, while the results of convergence are more closely similar. Also, parallelism typically occurs between groups of organisms somewhat related to begin with, while convergence seldom occurs between close relatives.

Convergent resemblance, indicative of similar adaptation, is also called resemblance by **analogy.** Another kind of resemblance, called homology (Chapter 22), is indicative of common ancestry, and may or may not be present at the same time. The winglike shape of insect wings, bird wings, and bat wings reflect similar adaptation for flying, and is thus a resemblance by analogy. The bird's wing and the bat's wing share both kinds of resemblance: their winglike shape, an analogous resemblance, results from convergence, while their common possession of a humerus, radius, and pectoral muscu-

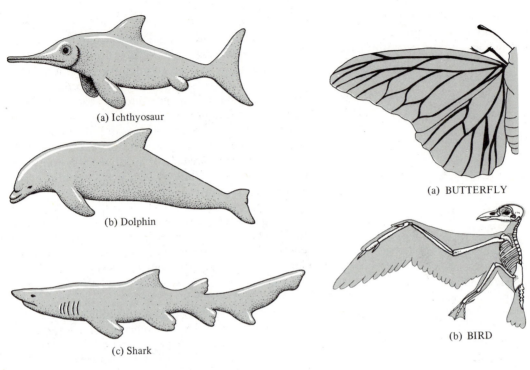

(a) Ichthyosaur

(b) Dolphin

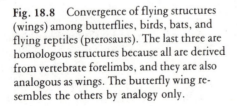

(c) Shark

Fig. 18.7   A striking example of convergence: (a) a marine reptile (ichthyosaur) and (b) a marine mammal (dolphin), both of which evolved a streamlined body, bilobed tail, dorsal fins, and pectoral flippers functionally similar to those of (c) sharks and other fishes.

(a) BUTTERFLY

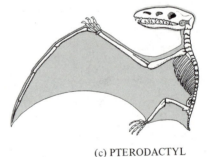

(b) BIRD

(c) PTERODACTYL

(d) BAT

Fig. 18.8   Convergence of flying structures (wings) among butterflies, birds, bats, and flying reptiles (pterosaurs). The last three are homologous structures because all are derived from vertebrate forelimbs, and they are also analogous as wings. The butterfly wing resembles the others by analogy only.

lature is homologous resemblance and results from their remote common ancestry (Fig. 18.8). Neither type of resemblance is perfect: the bird has lost certain bones that the bat still possesses, presumably because of their differing evolutionary histories. ·

Convergence is never perfect either—this fact alone provides strong evidence for the opportunistic basis of convergence. The marsupials of Australia and the placentals of other continents (Fig. 18.9) exhibit many cases of convergence, but in no case have the marsupials lost their characteristic pouch or other distinctive features. Whales and sharks have convergently evolved similar body shapes, but with a vertical tail in one case and a horizontal tail in the other. Neither has lost the characteristics that reflect their independent ancestries; the whale, for instance, has not re-evolved gills, but has retained the lungs of its terrestrial ancestors. Multiple solutions can also be seen in cases of convergence: the wings of bats, birds, and insects (and airplanes, too) have similar shape because of aerodynamic considerations. Yet they are constructed very differently: the flight surface of a bird's wing is made of feathers, while the other flying animals use membranes (and the airplane uses a rigid wing of metal and other materials). Even the membrane wings are not all the same. The membranes of insect wings are made largely of chitin and reinforced by hollow "veins," and the membrane wings of bats are made largely of skin and reinforced by bones. The extinct flying reptiles (pterosaurs) had a membrane wing made of skin stretched over bone, as in bats, but the membrane was supported almost entirely by the enlarged fourth finger, instead of by the second through fifth fingers as in bats.

Functional Shifts
and Preadaptation

Retrospectively, we can point to certain useful structures that were already present before they acquired their present functions. Crossopterygian fishes, for example, had at least three traits that served their amphibian descendants very well: lungs, internal nostrils, and lobe-shaped fleshy fins (Fig. 18.10). Characteristics that are suitable for future use are called preadaptive characteristics, and the possession of such characteristics is called **preadaptation.** (A slightly different concept, that of preadaptive mutations, was covered in Chapter 7.)

Finalists have pointed to preadaptations as unexplainable by natural selection. How, they ask, can a

character evolve before it is of any benefit to its possessor, if not by some finalistic or goal-directed process? If lungs are of no use to fish, then their presence in the ancestors of the land vertebrates indicates some planning with foresight. The fallacy in this argument is the assumption that a structure that has not yet assumed its present function is of no use at all. All preadaptive structures appear on close examination to be quite useful to their possessors, often serving quite a different function.

Lungs are by no means useless to freshwater fishes. Modern lungfishes live in environments subject to seasonal desiccation, and there is evidence that many Devonian fishes lived in similar environments. During the dry season, many streams in such environments contract into a series of unconnected ponds of varying size; these ponds soon become stagnant, and their oxygen supply quickly diminishes. A fish would survive these seasonal conditions better if it developed the habit of rising to the surface and gulping air. It could also burrow into the mud and aestivate, thereby diminishing greatly its need for oxygen—this might allow the fish to last just a little bit longer until the rains finally returned. Of course, the rains might not return in time, and the fish might become fossilized, encased in its muddy tomb. The number of fossil fish preserved in their aestivation burrows bears testimony to the dryness of their climate.

In many early fishes, the floor of the pharynx (the chamber just behind the mouth) possessed a rich blood supply; several living bony fishes have blood-rich patches or depressions there. The gulping of air or the "biting off" of air bubbles would enable the fish to aerate the blood through the vascularized floor of the pharynx and by other means. (Functionally similar adaptations occur in the family Anabatidae, whose members occasionally venture forth from the water.) Fishes with a larger or more highly vascularized pharyngeal floor would have a selective advantage, and by natural selection the pharyngeal floor would grow out and proliferate until it became a lung. As evidence for this scenario, we may point to the presence of lungs not only in lungfishes (Dipnoi) but also in *Polypterus, Erythrinus, Umbrina,* and *Gymnarchus,* and even a Devonian freshwater fish (*Bothriolepis,* a placoderm, Fig. 18.11, though at least one paleontologist has questioned the usual interpretation of these structures as lungs). We may also point out the absence of lungs in the sharklike fishes (Chondrichthyes), the one major group that has been (with occasional excep-

PLACENTALS

Wolf
(*Canis*)

Ocelot
(*Felis*)

Ground hog
(*Marmota*)

Flying squirrel
(*Glaucomys*)

Anteater
(*Myrmecophaga*)

Mole
(*Talpa*)

Mouse
(*Mus*)

MARSUPIALS

Tasmanian "wolf"
(*Thylacinus*)

Native "cat"
(*Dasyurus*)

Flying phalanger
(*Petaurus*)

Wombat
(*Phascolomys*)

Marsupial "anteater"
(*Myrmecobius*)

Marsupial "mole"
(*Notoryctes*)

Marsupial "mouse"
(*Dasycerus*)

Fig. 18.9 Examples of convergence between the marsupials of Australia (right) and the placental mammals of other continents (left). The marsupials all diverged from a common ancestor, and the diversity among them results from adaptive radiation. Reprinted with permission from Jeffry J. W. Baker and Garland E. Allen, *Study of Biology*, 4e, © 1982. Reading, Massachusetts: Addison-Wesley (Fig. 24.7).

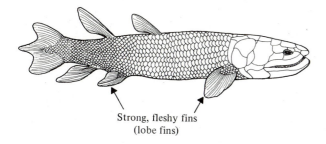

Strong, fleshy fins
(lobe fins)

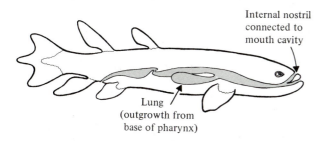

Internal nostril
connected to
mouth cavity

Lung
(outgrowth from
base of pharynx)

Fig. 18.10 Preadaptations to the land in the crossopterygian fish *Eusthenopteron*.

tions only) characteristically marine throughout its history.

Gulping for air, however, would dry out the mouth. The existence of a passageway leading from the external nostrils to the roof of the mouth would be a selective advantage, for it would allow more exchange of air with less loss of humidity. Nostrils would also allow the fish to obtain air with only the tip of its snout exposed, as opposed to its whole head in the case of a fish that gulped air with its mouth. Nostrils would also be advantageous in the phylogenetically older function of smelling, for the nose could receive samples of odor-filled water more rapidly and more continuously if it had an exit as well as an entrance. These several selective advantages led to the evolution of internal nostrils opening into the roof of the mouth from above.

A small pond in the dry season may diminish into a puddle, and eventually dry up entirely. Larger ponds will undergo the same changes, only more slowly. A fish, stranded by the drying up of its pond, would tend to wriggle, and thereby propel itself a short distance up or

down the dried-up stream bed. The most vigorous of these fishes would occasionally wriggle their way into the next pond, which might perchance be somewhat larger and therefore still watery. This behavior placed a selective premium on the possession of strong, fleshy fins that could aid in propelling the fish along the stream bed, and in the amphibian descendents of these fishes, the fleshy fins eventually became transformed into walking legs. It is somewhat ironic that these organs of loco-

Fig. 18.11 Presence of lunglike organs in the Devonian placoderm *Bothriolepis*.

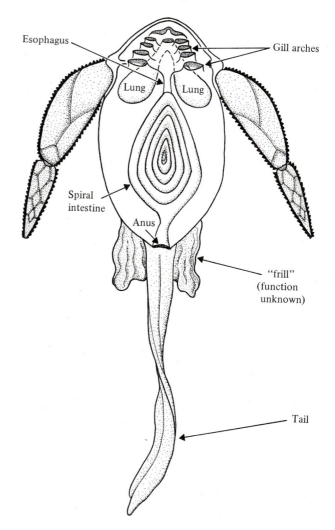

Esophagus

Gill arches

Lung

Lung

Spiral
intestine

Anus

"frill"
(function
unknown)

Tail

motion on land seem to have evolved first in fishes, and that their original function enabled the fish to find their way back into the water!

Preadaptation often involves functional shifts. Structures evolved in connection with one function may happen to be usable in connection with an entirely different function. Lungs, for example, were in most fishes no longer needed in connection with their original function, and they presumably would have been reduced to mere vestiges had they not served a totally unrelated purpose quite by accident. Any air-filled structure increases buoyancy in a swimming animal, and by controlling the amount of air in the lung, the fish could control the depth at which it swam. The most successful group of fishes converted the lungs, by a series of small steps, into a swim bladder capable of controlling buoyancy.

Among protists of the group known as Foraminifera, early members had thin shells of chitin surrounded by a gelatinous mucus. Selection by predators favored those individuals that were able to agglutinate sand grains and other foreign particles into this gelatinous mucus. The sand grains could be cemented into a still firmer coating by the addition of a cementing substance such as calcium carbonate. (A related group evolved silica shells by a similar process.) The calcium carbonate, initially serving as a cement, proved able to serve as the entire shell, and the presence of sand grains became superfluous. The foraminiferans that agglutinated sand grains into their mucous coats were preadapted to evolving shells made of calcium carbonate, and the adaptive shift was made possible in part by a change in the role of calcium carbonate from that of a cement to that of a major shell component.

Other cases of functional shifts included the conversion of the reptilian jaw suspension into a pair of mammalian ear bones, or the conversion of the sticky ''ovigerous frena'' of barnacles, from structures keeping the eggs from washing away, into gills used for breathing (Chapter 5). In all these cases, it is important that there is never a functionless stage intervening when function changes: the transition always includes at least one stage where the old function and the new are both served by the same structure.

CHAPTER SUMMARY

Evolution may be viewed either as having a direction (directionalism) or as branching out in all possible directions at once (opportunism).

Evolution in a straight line is sometimes called ''orthogenesis.'' This is a property of some lineages, but not of all. By no means is ''orthogenesis'' a cause of trends.

Some theorists have seen evolution as goal-seeking or purposive (finalistic or teleological); many among these have singled out humans as the culmination or goal of evolution. Yet evolution is a branching rather than a unilineal process overall, and it makes little sense to single out just one branch and ignore all others. There is no evidence that trends are guided by their outcomes, for their outcomes are numerous, and are knowable only in retrospect.

Evolution is basically opportunistic, meaning that things happen as they may, that available opportunities are eventually seized, and that evolution follows opportunity rather than plan. When opportunities are many and varied, branching evolution results in adaptive radiation, in which each descendant is differently adapted. Evidence for opportunism, and against finalism, comes from many sources. Multiple solutions to the same problem are often varied, but do not always exhaust all possibilities, because each group follows the solution most readily available to it. Repeated evolution of similar forms from the same ancestral lineage may result in iterative evolution. Common solutions to the same problem in unrelated groups may result in convergence, in which unrelated species resemble one another. This type of resemblance is called analogy, and reflects similar adaptations only, while homology is a more basic resemblance, reflecting common ancestry.

A structure that serves one function may simultaneously be preadapted to serve another function as well. A functional shift may take place, but only if the structure can abandon its former function without harm, or if both functions can be subserved simultaneously. Most cases of functional change have a transitional stage in which both old and new functions are subserved simultaneously.

19

Rates of Evolution

Evolution is a process of change. Rates of evolution measure how fast these changes take place. Like all other rates, evolutionary rates are defined as so much change per unit time, that is, the amount of change divided by some measure of time.

A. WHY MEASURE EVOLUTIONARY RATES?

We measure evolutionary rates for several reasons. In science, quantification is usually the first step in any abstract or theoretical treatment. Lord Kelvin, the British physicist, once said that if you can't measure something, then your knowledge is of a very meager and inferior sort.

But rates of evolution are themselves of further interest. Why is evolution fast or slow, and what controls its rate? Is the rate of evolution constant or changeable? In response to what forces might it change? Are its changes predictable in any manner? Do different groups of organisms evolve at the same rate? Have insects evolved faster than mammals, or mammals faster than bacteria? Finally, evolutionary rates may allow us to choose among conflicting explanations of evolutionary trends, or of evolution in general, since different theories often predict differences in evolutionary rates.

B. TYPES OF EVOLUTIONARY RATES AND DIFFICULTIES IN MEASURING THEM

Various problems are encountered in the measuring of evolutionary rates. Since different questions require different answers, the proper choice among the major kinds of evolutionary rates (Box 19–A) is important. Most essential is the distinction between rates of anagenesis and rates of cladogenesis. Anagenetic rates measure change within a single evolving lineage; cladogenetic rates measure the branching or extinction of lineages or the origin and cessation of taxa. Another important distinction may be made between morphological rates, measured in terms of individual characters, and taxonomic rates, measured in terms of genera or families.

Taxonomic rates are quite sensitive to the opinion of the taxonomist who determined the number of families, genera, or species, a phenomenon known as **taxonomic**

BOX 19–A   *Types of Evolutionary Rates*

I. ANAGENETIC RATES (within a single lineage)

A. **Morphological rates of anagenesis**

1. Unit character rates (measurable in darwins). These can be further subdivided by the scale against which they are measured, as follows:

a. Absolute temporal scale (in years, millions of years, etc.)

b. Correlative geological scales

i. By periods, epochs, etc.

ii. By thickness of strata (stratigraphic scale)

iii. By cyclical changes (varves, cyclothems, etc.)

c. Correlative biological scales

i. By generations

ii. By relative changes in other organs (allometric rates)

B. **Taxonomic rates of anagenesis**

1. Rate of true phyletic replacement in a single lineage, at a given taxonomic level (genus, species, etc.).

C. **Genetic rates of anagenesis**

1. Rates of replacement of individual genes, etc.

D. **Ecological rates of anagenesis**

1. Rate of change in population size.

2. Rate of change in population density.

3. Rate of change in diversity of habitats or localities occupied.

II. CLADOGENETIC RATES

A. **Pure cladogenetic rates**

1. Rate of true speciation (splitting of lineages).

a. Along a single lineage

b. Averaged among several lineages

2. Rate of true extinction of lineages. This and the following three types of rates refer to all-or-none phenomena that must be averaged amongst several lineages.

3. Rate of persistence of lineages (total diversity minus extinctions).

4. Rate of change in diversity (speciations minus extinctions).

5. Rate of turnover (speciations plus extinctions).

B. **Mixed taxonomic rates**

1. Rate of first appearances (''taxonomic origins''); equals sum of phyletic replacement rate (I.B1, anagenetic) plus 2 × rate of true speciation (II.A1, cladogenetic); see Box 19–E.

2. Rate of last appearances (''taxonomic extinctions''); equals (approximately) sum of true extinction rate (II.A1, cladogenetic), and rate of phyletic replacement (I.B1, anagenetic); see Box 19–E.

3. Rate of recurrence or continuity; equals total diversity minus taxonomic extinctions.

4. Rate of change in diversity; equals first appearances minus last appearances.

5. Rate of taxonomic turnover; equals II.B2, or may be expressed as a percentage of preexisting diversity.

Classification modified after Simpson, 1953, and further expanded.

subjectivity. Simpson (1953), for example, calculated that the lineage from *Hyracotherium* to *Equus* included eight genera in about 65 million years, or roughly one genus every 8 million years. These results are based in part on a classification that recognizes eight genera in this particular lineage. A more finely split classification might recognize more than eight genera, while a more ''lumped'' classification would recognize seven or fewer, with consequent changes in the calculated rates. Variation from this source may tell us more about the different kinds of taxonomists than about different kinds of horses. This is especially true in comparing, say, insects with mammals or flowers. Suppose it were determined that in a certain lineage of beetles a new genus is produced, on the average, every four million years. Does this mean that beetles evolved twice as fast as horses, or rather that beetle taxonomy is twice as split? Obviously, it may mean either of these, or some combination of them.

Other taxonomic artifacts are possible, too. Suppose we are studying the rate of origin of new genera within a given taxon. We may find that fossils from certain geologic periods are more richly preserved than at other periods, or have been more adequately studied, or have been studied by a taxonomic splitter; the apparently rapid origin of many genera within a short time may reflect any or all of these artifacts. Numerous Australian marsupials are known, for example, from the Pleistocene, and excellent monographs have been written about them. Very few Tertiary marsupials are known, however, and only one genus (*Wynyardia*) has received much attention. The apparent increase in marsupial genera for the Pleistocene may thus reflect the lack of preservation and discovery of numerous earlier faunas, plus a healthy degree of splitting by the classifiers of Pleistocene forms.

The usual definition of a rate is an amount of change divided by some measure of elapsed time. The scale against which evolution is measured thus appears in the denominator of a fraction. Scales of measurement may be in absolute time units such as years, or they may appear as a correlative scale, that is, a scale correlated with time. Correlative geological scales may be established in terms of periods or epochs, in terms of stratigraphic thickness (measuring change per meter of sedimentary rock), or in terms of recurring cyclical units such as varves or cyclothems. Correlative biological scales may be in terms of generations or in terms of other characters. For example, if body size is known to have increased, the amount of change in some other character may be measured against the scale of increasing size, giving what is called an allometric rate of increase.

Inaccuracy in the determination of absolute geologic time can affect the denominator of any rate-determining fraction, too. This difficulty is potentially great in poorly dated material, but where a fossil sequence is sufficiently well known to establish a lineage it is usually less serious. The increasing number of dates firmly established by radiometric means has also diminished this problem considerably.

C. ANAGENETIC RATES

Morphological Rates of Anagenesis

The simplest and most commonly calculated rates are the **morphological rates of anagenesis**. Some character, like the length or volume of a particular structure, is measured in both ancestor and descendant, and the rate of its evolution is expressed, say, in centimeters per unit time. In the last million years of human evolution, for example, brain size has increased from a cranial capacity of approximately 900 cm^3 to about 1400 cm^3. Hence a rate of 500 cm^3 per million years may be calculated for this character. This may also be expressed as 0.0125 cm^3 per generation, taking 25 years as the average length of a human generation (though in the past it was undoubtedly lower).

Rates of evolution may also be expressed in relative or percentage terms. The above rate, for example, represents a change of 500/900 or about 55.6% per million years. The British evolutionist J. B. S. Haldane has proposed that a relative change by a factor of e (the base of natural logarithms) per million years be known as a **darwin**, and that the rate in darwins would then be given by

$$\frac{\log_e (\text{final value}) - \log_e (\text{initial value})}{T_2 - T_1}$$

Using this formula, we learn that the human brain size has evolved at the rate of about 0.442 darwins, or 442 millidarwins. More rapid rates of change may be expressed in kilodarwins (10^3 darwins) or megadarwins (10^6 darwins); slower rates of change may be expressed in millidarwins (10^{-3} darwins) or microdarwins (10^{-6} darwins).

Some additional rates of anagenesis are calculated in Box 19–B. The example shows clearly that the rate of anagenesis may change markedly within a lineage and that two lineages diverging from a common source may have very different anagenetic rates. This is a strong argument for the adaptive control of evolutionary trends (Chapter 17).

We have thus far been discussing single character rates, which measure change in only one character at a time. It is also possible to measure many characters at

BOX 19-B *Rates of Anagenesis in Horses (Order Perissodactyla, Family Equidae)*

The character measured here is an index of hypsodonty. Specifically, it is

$$\frac{\text{Height of paracone}}{\text{Length of ectoloph}} \times 100 \text{ on the third upper molar } (M^3)$$

This is a dimensionless fraction, expressed as a percent. Notice the great variation in rate, both between lineages, and within the same lineage at different times.

Hyracotherium to *Mesohippus* (L.Eoc.–M.Olig.)	0.56 per m.y.
Mesohippus to *Hypohippus* (M.Olig.–U.Mioc.	0.89 per m.y.
Mesohippus to *Merychippus* (M.Olig.–U.Mioc.)	5.58 per m.y.
Merychippus to *Neohipparion* (U.Mioc.–M.Plioc.)	11.65 per m.y.

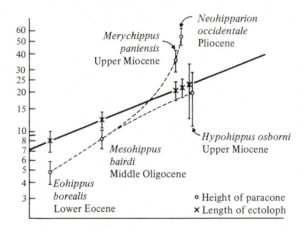

Height of unworn paracone and length of ectoloph of M^5 in five genera of Equidae. Vertical lines are standard ranges, statistical estimates of variation in a population of 1000 individuals.

After Simpson, 1953. *Major Features of Evolution.* New York: Columbia University Press.

once, and calculate a composite index based on the entire character complex. An example of such a composite rate of anagenesis is shown in Box 19–C, which deals with the order Dipnoi, or lungfishes. Here we see that the lungfishes, an ancient group, evolved fairly rapidly at first, reaching a nearly modern condition rather early in their history. Since then, however, their rate of evolution has slowed considerably. There are other ancient groups that display the same pattern of moderate to rapid evolution at first, followed by a much longer period of very slow evolution. Horseshoe crabs (Xiphosura) and inarticulate brachiopods, for example, both fit this pattern.

Biochemical Rates of Anagenesis

Modern biochemistry has given us at least three other ways of quantifying the differences among organisms. Large, complex proteins may be analyzed and their exact sequence of amino acids determined. The number of amino acid differences, or the percentage of such differences, may then be used as a measure of evolutionary change. Such calculations have been made on a number of large protein molecules, including cytochrome *c*, serum albumins, and hemoglobin and myoglobin proteins. A second method is used to compare immunoglobins, which are also proteins. A distantly related animal is immunized: a rabbit, for example, may be immunized against human serum. The level of reaction of this rabbit's serum against human serum is designated as 100%, and is used as a standard. Reaction of this rabbit's serum against the sera of various apes, monkeys, and other primates can now be determined, and expressed as a percentage of the reaction to human serum. In this way, the **immunological distance** of the other primates from humans can be calculated. Notice that the technique requires the production of immune serum by an animal very distantly related to the species being compared.

While the above techniques compare proteins, a third technique directly measures the nucleic acid differences upon which these protein differences are based.

BOX 19–C *Rates of Anagenesis in Lungfishes, Order Dipnoi*

The rate of "modernization" in lungfishes is measured here by a 26-character complex. Each character was scored from 0, the most primitive condition, up to 2, 3, 4, 5, 6, or 7 for the most "modern" condition, and the values for all 26 characters were added together. The earliest lungfishes have a "modernization index" of zero, or 0% modern; the living lungfish have a "modernization index" of 100, or 100% modern. Notice that this morphological rate of anagenesis has slowed to near zero in the last 150 million years.

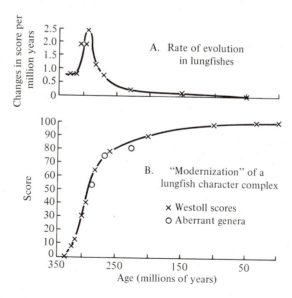

From Simpson, 1953. *Major Features of Evolution*. New York: Columbia University Press; data after Westoll.

Nucleic acid (usually DNA) is obtained from several sources and purified, then it is heated so that double-stranded molecules are separated into single strands. A heated mixture of DNA from two sources is allowed to cool slowly, and the rate at which it renatures into the double-stranded form is measured. The percentage of homologous sequence can then be determined, since DNA strands with identical sequences will reunite more rapidly, and fragments with differing sequences more slowly, or not at all, depending upon the fraction of sequence that is common to the two sources. See Wilson, Carson, and White (1977) for further details on the biochemical techniques used to measure evolutionary divergence.

All of the above techniques can be used to compare species differences. If one could compare an ancestor and its descendant by any of these techniques, one would only have to divide by the elapsed time between them in order to arrive at an evolutionary rate. Unfortunately, these techniques can be used only on living species, and thus can never be strictly anagenetic.

If the time of divergence between two surviving lineages is known from the fossil record, then we can measure the degree of divergence between the protein sequences of two modern species, and calculate a minimal rate of anagenetic divergence between them averaged over geologic time (Fig. 19.1). Using this technique, one study showed that divergence between humans and apes took place faster than divergence among frogs. But such rates are only minimum estimates (or underestimates) for several reasons: a single amino acid replacement (e.g., proline to alanine) usually cannot be distinguished from a double or multiple change (such as proline to valine to alanine); the same amino acid may have evolved twice convergently, in which case no change would be apparent; or a later change may reverse an earlier one, and again no change would be apparent. These difficulties would be greater among unrelated organisms than among close relatives. Biochemical measurements of the divergence between apes and humans are discussed in Chapter 27.

Fig. 19.1 A, Evolutionary tree of mammals, based on myoglobin sequences. B, Evolutionary tree of mammals based on α-globin sequences. Reprinted with permission from M. Goodman, *et al.*, *Journal of Molecular Evolution*, 3, 1974.

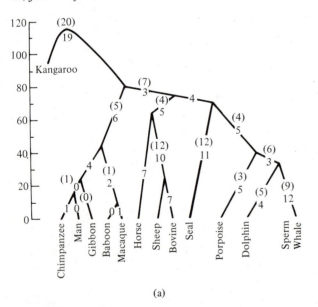

(a)

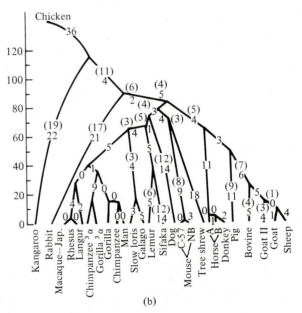

(b)

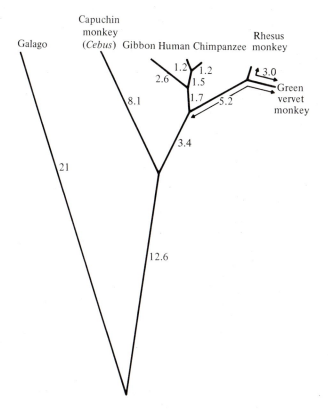

Fig. 19.2 Similarity in the DNA of a number of species of primates based on DNA hybridization techniques. The numbers on the branches estimate the percentage of nucleotide-pair substitutions that have occurred during evolution. The pattern of similarity suggests a family tree.

On the other hand, the phylogeny may not be known, and the biochemical differences may be used to reconstruct it (Figs. 19.2 and 19.3). If the rate of divergence is assumed to be relatively uniform over time, then by comparing a number of such different pairs, a relative time-sequence of divergences (hence, a phylogeny) can be reconstructed, subject always to the challenge that the rate of divergence may not really have been uniform. A large number of evolutionary biochemists are willing to accept the assumption of generally uniform rates; others are not. In the descent of humans from common ancestry

with pongids, different proteins (globulins, albuminoids) have evolved at decidedly unequal rates. Goodman *et al.* (1974) report marked variations in the rate of evolution among beta-globins. Avise *et al.* (1980) studied 16 proteins among New World wood warblers (Parulidae), and found the rate of biochemical evolution to be much lower than in their own similar study of the rodent subfamily Cricetinae (Fig. 19.4). Several evolutionists have argued that biochemical traits should be equally subject to convergence or mosaic evolution as are morphological traits. Other evolutionists claim that these phenomena are rare and their effects minimal. The major issues are reviewed by Fitch (1976), Dayhoff (1972), Goodman (1976), and Maeda and Fitch (1981).

Taxonomic Rates of Anagenesis

Taxonomic rates of anagenesis measure changes in taxonomic units such as families, genera, or species. In theory at least, these units are based on the totality of all characteristics (Chapter 22), and yet it is not necessary to investigate all these characteristics firsthand: once a reliable classification exists, the evolutionist can use it to determine evolutionary rates. Moreover, all taxonomic rates expressed, say, in genera per million years can be compared with one another. Yet it cannot be over-emphasized that all taxonomic rates are subject to the problems of taxonomic subjectivity mentioned earlier.

Box 19–D shows examples of taxonomic rates of anagenesis in three taxa, two of which are closely related.

Rapid and Slow Anagenesis

Some of the best evidence for the adaptive control of anagenesis comes from the now well-established observation that different groups of organisms may evolve at different rates. Nowhere is this more evident than in the extreme cases of very rapid or very slow anagenesis. These cases bear close examination because of what they might tell us about the causes of anagenesis, or of evolution in general.

Rapid anagenesis, also called **tachytely,** may occur following the invasion of a new ecological niche or the origin of a new adaptive type, often representing a new major taxon. Tachytely of this sort is often inferred from the abruptness of adaptive shifts, for the origin of a new major taxon often occurs so rapidly as not to be preserved in the fossil record. Paleocene bats, for example, had

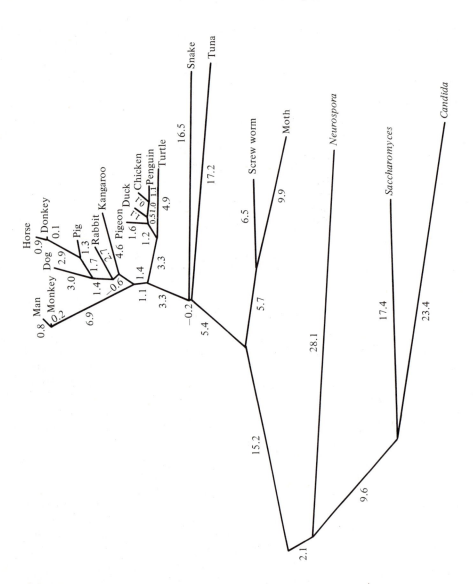

Fig. 19.3 Phylogeny inferred from pattern of differences in cytochrome *c* protein sequences. After data from Fitch and Margoliash.

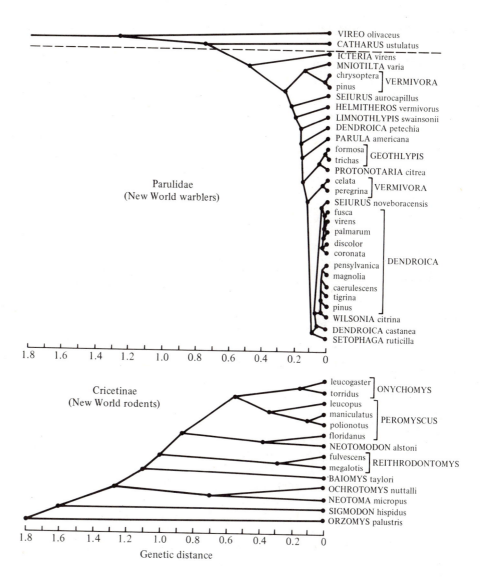

Fig. 19.4 Comparison of evolutionary rates in warblers and cricetine rodents. The warblers have undergone more speciation (cladogenesis) with less genetic change, whereas the cricetine rodents have experienced more anagenesis with less speciation. Reprinted with permission from J. C. Avise, *et al., Journal of Heredity,* 71:303–310, 1980.

BOX 19–D   *Comparison of Three Taxonomic Rates of Anagenesis Based on Genera*

	No. of genera	Million years	Genera per m.y.	M. yrs. per genus
EQUIDAE (*HYRACOTHERIUM* TO *EQUUS*)	8	65	0.13	8
CHALICOTHERIIDAE	5	40	0.13	8
TRIASSIC AND EARLIER AMMONITES.	8	160	0.05	20

From Simpson, 1953. *Major Features of Evolution.* New York: Columbia University Press.

fully formed wings, though none of the possible candidates for their ancestry had even the slightest hint of limb modification in this direction. Tachytely may also occur following the extinction of a competitive group. For example, the true carnivores (cats, dogs, etc.) evolved more rapidly following the extinction of such competitors as amphicyonids, mesonychids, oxyaenids, and especially hyaenodontids.

Slow anagenesis, or **bradytely,** is marked by persistence with little or no change for long periods of time. Such lineages often terminate in **living fossils,** modern species that closely resemble their ancestors of remote time periods (Fig. 19.5). The opossum has changed very little in 70 million years, and the horseshoe crab (*Limulus*) has changed very little for over 500 million years. *Lingula,* an inarticulate brachiopod, belongs to the same genus as its Cambrian ancestors of some 600 million years ago.

Rates of anagenesis are limited in part by the rapidity of environmental change, and in part by the rate of mutation, which limits the rate of response to this environmental change. In cases of bradytely, it has been suggested that lack of sufficient mutations has limited the rate of evolution. But Selander *et al.* (1970) have shown in *Limulus* a sufficient store of genetic variation to effect a severalfold higher rate of anagenesis. The occurrence of new variation is seldom the rate-limiting process. Compromise between conflicting selection pressures may more often limit the anagenetic rate.

As an example of one way in which the rate of anagenesis may be limited, consider the trend toward enlarged brain size (hence also enlarged head size) among the higher primates and particularly in our own ancestry. This rapid enlargement of the brain was presumably adaptive; the greater intelligence of bigger-brained animals was presumably important as a selecting force. Now consider the mother monkey shown in Fig. 12.3 with her stillborn infant; both died because the infant's head was too large to fit through the birth canal. This selective force operates against larger-headed infants, imposing some limit on the maximum size of the head or the brain, and placing a new selective premium on babies born in earlier stages or on mothers with larger birth canals. The actual rate of brain enlargement through evolution may thus be limited more by this type of counter-selection than by the rate of selection for bigger brain size.

What would happen if there were no counter-selection? The unopposed selection would surely cause a rate of anagenesis far more rapid than normal, a rate limited only by the speed at which new favorable genotypes arise, hence ultimately by the rate of mutation. In selection

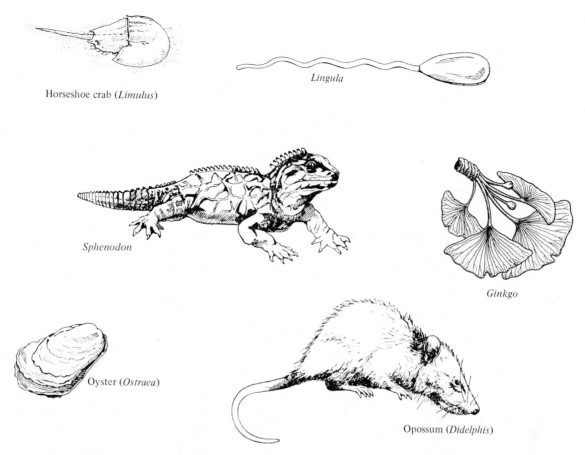

Horseshoe crab (*Limulus*)

Lingula

Sphenodon

Ginkgo

Oyster (*Ostraea*)

Opossum (*Didelphis*)

Fig. 19.5 "Living fossils": organisms that have changed very little for many millions of years.

experiments, such as those shown in Figs. 11.4 through 11.6, this is exactly what occurs. It is not known how frequent this type of tachytely may be. Simpson has said that phyletic evolution (anagenesis) proceeds in "spurts and pauses." The spurts could possibly be caused by unopposed selection, while the much longer pauses and gradual changes may be limited in their rate of change by compromise selection (Chapter 12). I have given the name of spasmodic evolution to this phenomenon.

The "punctuationalist" theory that most anagenetic change occurs during speciation is discussed later in this chapter. Under this theory, the rate of anagenesis would be determined by the rate of speciation, a cladogenetic rate. Other aspects of punctuationalism are examined in Chapter 21.

D. CLADOGENETIC RATES

Types of Cladogenetic Rates

Rates of cladogenesis depend not upon changes within single lineages, but upon processes that increase or decrease the number of lineages. Since the splitting of lineages requires speciation, cladogenetic rates are necessarily taxonomic.

Two types of cladogenetic rates may be distinguished. Rates that involve taxa only may be termed **mixed** or **taxonomic rates**; these rates may be determined from a list of taxa giving the time range of each, without knowing the exact phylogeny. **Pure cladogenetic rates,** or **lineage rates,** are based on true lineages rather than taxa, and can be determined only from a knowledge of the phylogeny. The true rate of speciation, for example, is a pure cladogenetic rate, and measures the frequency of branching lineages. The corresponding mixed rate is a rate of first appearances, measuring the frequency with which new taxa appear in the fossil record. This is a ''mixed'' rate because taxa may originate by either of two processes: phyletic replacement of earlier taxa without branching (an anagenetic process), or the splitting of a lineage into two (a cladogenetic process). Box 19–E illustrates the differences between these types of cladogenetic rates.

Rapid and Slow Cladogenesis

Terms for rapid and slow cladogenesis are not in common usage, but may formed from the Greek root *schizo-,* meaning a splitting. The rapid splitting of lineages would be called **tachyschizia,** moderate rates of splitting would be called **horoschizia,** and long persistence without splitting would be called **bradyschizia.**

While thorough studies of the factors responsible for tachyschizia and bradyschizia remain to be done, a few generalizations can be made. From studies on the speciation of present-day organisms, one could infer that tachyschizia, or rapid cladogenesis, would be more likely in mountainous regions or in island regions, where the opportunities for geographic isolation are greater. Tachyschizia occurs early in the history of many groups, but it may also occur at other times and in some cases occurs repeatedly. Episodes of tachyschizia often follow the invasion of a new continent, the exploitation of a new habitat or other ecological opportunity, or the extinction of an important group of competitors. Examples of tachyschizia include the Darwin's finches (Geospizinae) of the Galapagos Islands, the fruit flies (*Drosophila*) of the Hawaiian Islands, or the rapid expansion of mammals following the extinction of Mesozoic reptiles.

The conditions associated with bradyschizia are much more varied. Bradyschizia may begin at almost any time in the history of a group, it may be associated with either rapid or slow rates of anagenesis, and it may occur in both short-lived and long-lived groups. The phylum Onychophora are a case in which bradyschizia and bradytely both occurred throughout a very long history, reaching back some 550 million years. The phylum, which shows many transitional features between the Annelida and the Arthropoda, has remained small throughout its history and has changed very little. The reptilian order Rhynchocephalia present a different case, for their early history includes a period of diversification and at least two major adaptive radiations. But the early members of this order underwent much extinction, and the lone survivor, *Sphenodon,* stands at the end of a lineage that was both bradytelic and bradyschizic.

Are Cladogenesis and Anagenesis Related?

Determining evolutionary rates and their causes represents a major unsolved problem in macroevolutionary biology. In particular, the relationship between cladogenesis and anagenesis needs to be investigated more thoroughly. Orthogeneticists have always taught that anagenesis usually occurred in unbranching lineages. Eldredge and Gould, at the other extreme, have argued that anagenesis hardly ever occurs without speciation; their ''punctuationalist'' theory is examined in Chapter 21.

The existence of sibling species (Chapter 14) is itself evidence that cladogenesis (speciation) can take place more rapidly than anagenesis (morphological change). In some cases, tachyschizia (rapid and repeated speciation) can take place with very little anagenesis, as in the case of Hawaiian *Drosophila*. The occurrence of anagenesis without cladogenesis has been documented in one of two lineages of taeniodonts studied by Patterson (1949; see Fig. 18.1), and in the early Tertiary primates and condylarths studied by Gingerich (1976, 1977). Such isolated, peculiar species as the rodentlike primate *Daubentonia* or the strange plant *Welwitschia* may also mark the existence of lineages that have undergone much anagenesis but little or no cladogenesis. Independence of anagenesis

BOX 19–E *An Example Illustrating the Calculation of Several Types of Evolutionary Rates.*

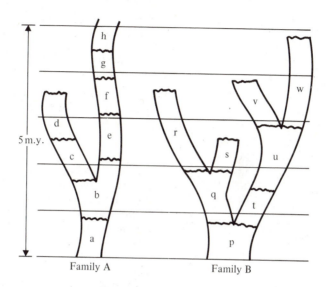

Family A   Family B

	Family A (8 species)		Family B (8 species)	
EVOLUTIONARY RATES (IN SPECIES PER MILLION YEARS):	SPECIES COUNTED	RATE	SPECIES COUNTED	RATE
I.B1. RATE OF TRUE PHYLETIC REPLACEMENT (EXCLUDING BRANCHING):	origin of b,d,e,f,g,h	6/5 = 1.2	origin of u	1/5 = .2
II.A1. RATE OF TRUE SPECIATION:	b only	1/5 = .2	p,q,u	3/5 = .6
II.A2. RATE OF TRUE EXTINCTION:	d only	1/5 = .2	r,s,v,w	4/5 = .8
II.B1. RATE OF FIRST APPEARANCES (TAXONOMIC ORIGINS):	a–h	8/5 = 1.6	p–w	8/5 = 1.6
II.B2. RATE OF LAST APPEARANCES (TAXONOMIC EXTINCTIONS):	a–g	7/5 = 1.4	p–w	8/5 = 1.6

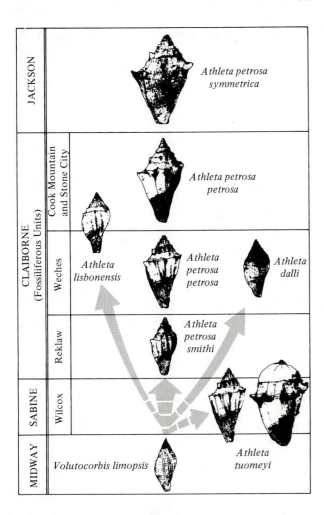

Fig. 19.6 Independence of cladogenesis and anagenesis among Eocene marine gastropods. Cladogenesis among the forms shown here was confined to the brief upper Midway—lower Sabine interval, whereas anagenesis occurred unequally among the resultant lineages: in the lineage of *A. dalii* there was little morphological change, whereas the lineage of *A. petrosa* continued to change throughout Claiborne and into Jackson time. Reprinted with permission from W. L. Fisher, P. U. Rodda, and J. W. Dietrich, *Evolution of Athleta petrosa stock (Eocene, Gastropoda) of Texas.* Courtesy of the Bureau of Economic Geology, University of Texas at Austin.

and cladogenesis is also shown in gastropods of the genus *Athleta* (Fig. 19.6) and in certain fishes (Fig. 19.7).

The broader question is whether anagenesis and cladogenesis have any necessary relationship, or whether the two are independent phenomena with unrelated causes. This is hardly a new question. New approaches to answering it have already begun to appear.

Stochastic Modeling of Cladogenesis

One of the newer approaches to the study of cladogenesis is the comparison of actual phylogenies with those generated at random by a computer. In each inter-

Fig. 19.7 A comparison of anagenetic and mixed cladogenetic rates in coelacanth and dipnoan fishes (after Schaeffer). The greatest morphological change took place during the Devonian in both groups, coinciding with a maximum in the rate of origin in dipnoan (lungfish) genera, but the maximum origination rate for coelacanth genera occurred much later, in the Triassic.

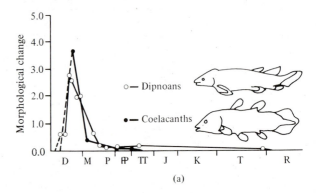

(a)

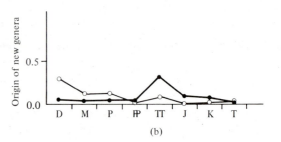

(b)

val of time, the species in the model are subjected to a certain probability of becoming extinct, or of splitting into two species, or of persisting without splitting. If it persists without splitting, it may or may not undergo enough anagenetic change to warrant placing it in a separate species. If it undergoes speciation (splitting), the resultant species may or may not become the founders of a new major taxon. With the aid of a computer, one can specify these probabilities and then use "pseudorandom" numbers to select which species will survive, split, persist without splitting, etc. A model of this sort is called **stochastic,** because the overall outcome is determined by numerous small, random events of specified probability: extinction of a major branch of the "family tree," for instance, can occur only as the result of the extinction of all of its species.

If the causes of rapid and slow cladogenesis are unique to each case, and are randomly distributed across geologic time and across the various taxa, then we should expect such stochastic models to mimic well the appearance of actual phylogenies. On the other hand, if certain phyla or classes, or certain geologic periods, had characteristically higher or lower rates of speciation, extinction, and so forth, then we should expect to find systematic deviations between real patterns of phylogeny and the patterns generated for us by stochastic models. Studies by Raup *et al.* (1973) and Gould *et al.* (1977) show that stochastic models can mimic actual phylogenies quite well once the ecological "world" has been "filled" with species. The great diversification of life in the lower Cambrian, on the other hand, shows that the rate of speciation in nearly all major groups was characteristically higher then and has subsequently diminished. Further studies may reveal whether there are systematic differences in speciation rates among major taxa.

The control of cladogenetic rates is poorly understood. The fact that cladogenesis can be mimicked to a great extent by stochastic models suggests that many of the factors that influence cladogenetic rates are distributed randomly across major taxonomic boundaries and across geologic time periods.

CHAPTER SUMMARY

The quantification of evolutionary rates may aid our understanding of evolutionary processes. Rates may be measured against absolute time, or against some other variable correlated with time. Rates involving taxonomic units such as genera are sensitive to the subjective assessment of these units by taxonomists, particularly with regard to splitting and lumping.

Rates of anagenesis measure change within a single lineage, ignoring any branches. These rates may be morphological, biochemical, or taxonomic. Morphological changes may be based on single characters or on character complexes. These rates often vary between lineages or within the same lineage at different times. These rates may be expressed in darwins, one darwin being an increase by a factor of *e* per million years.

Biochemical rates of change are subject to many of the same evolutionary phenomena as influence morphological rates. In particular, the rates are subject to variation and similarity may in some cases reflect convergence. Taxonomic rates of anagenesis are most useful in comparing groups of unrelated organisms.

Rates of cladogenesis are of two types. Taxonomic or mixed rates include anagenetic as well as cladogenetic components and may be determined from a simple list of taxa and their time ranges. Pure cladogenetic rates are based on lineages and can be determined only from a knowledge of the phylogeny.

Rapid anagenesis, or tachytely, may occur following the invasion of a new ecologic niche or the seizing of some other new opportunity. Slow anagenesis, or bradytely, results in "living fossils." Constancy of environment, hence also of selection, is more likely responsible for bradytely than inadequacy of mutations. Compromise selection may control anagenetic rates between these limits. The control of cladogenesis is poorly understood. The causes of rapid cladogenesis may vary from group to group, and can thus be mimicked by stochastic models to a great extent.

20

Extinction and Its Causes

A. PHYLETIC VERSUS TRUE EXTINCTION

True extinction may be defined as the termination of a lineage without issue. It is the dying out of a form of life, the permanent elimination of all their genes from the face of the Earth. Little else in common experience goes out of existence in this way; societies, governments, political parties, corporate bodies, musical groups, and fraternal organizations all go out of existence by disbanding and by becoming assimilated into other groups of the same kind. Family groups or royal dynasties may undergo true extinction if their living members all die without issue, a process that does take place bit by bit, just as in the extinction of a lineage.

Another kind of extinction is properly called **pseudoextinction,** though it has more often been called phyletic extinction or phyletic transformation. The genus *Hyracotherium* (''Eohippus'') contains no living members and is therefore ''extinct'' in one important sense. It is extinct simply because it has become transformed into something else, namely the living genus *Equus* (and also several truly extinct genera). Some of the genes of *Hyracotherium* are still with us; they are now among the genes of *Equus*. We might therefore say that since *Hyracotherium* has living descendants, it has not become truly extinct—it has simply become transformed into something else, and survives under a different name. In like fashion the ancient Persian Empire never really became extinct; it survives today as the modern nation of Iran.

Taxonomic groups such as families and genera may undergo either type of extinction or both. The ammonoids and the dinosaurs have undergone true extinction—they have left no living descendants, unless, as Bakker (1971) suggests, the living birds are derived from dinosaurs. The therapsid reptiles, on the other hand, underwent phyletic transformation only, for they have given rise to the mammals. The pteridosperms or seed ferns have likewise undergone phyletic transformation into the diverse angiosperms that are alive today. In the case of certain taxa, it is not yet clear which type of extinction has taken place: the Trilobita are all extinct, but it is not certain whether any or all living arthropods are descended from them. The Placodermi are likewise extinct, but the group may include the ancestors of both the bony fishes and the sharks. In either case, we may refer to taxonomic extinction, or the termination of the geological (stratigraphic) range of a taxon. A taxon is extinct in this sense simply if it contains no living members.

In seeking the causes for extinction, our primary concern will be with the true extinction of lineages. Pseudoextinction is simply a form of anagenesis, and has other causes that have been treated in previous chapters.

B. PROGRAMMED EXTINCTION AND RACIAL LONGEVITY

When a lineage becomes extinct, especially after a long and seemingly successful career on this Earth, it may be tempting to speculate that its end has been fated or predetermined. This belief in **programmed extinction** is part of a larger view that evolution as a whole is internally programmed and relatively unaffected by the vicissitudes of environmental change.

Supporters of programmed extinction divide the evolution of major taxa into three phases: an initial phase of rapid diversification and adaptive radiation, an intermediate phase of much anagenesis but little or no cladogenesis, and then finally a phase of "racial old age," during which the lineages become extinct, one by one, until none is left. This pattern may apply to the ammonoids, which inspired the theory, though their maximum diversity came late in their history. But periods of rapid cladogenesis and maximum diversity may occur early, late, or in the middle of a group's history. They may occur either once or repeatedly, and in the case of bradyschizic lineages (Chapter 19) they may not occur at all. Rapid cladogenesis and maximum diversity occurred early in the history of trilobites and amphibians, but late in the history of rodents, artiodactyls, and teleost fishes, and somewhere in the middle of the history of brachiopods. The sharklike fishes (Chondrichthyes) and mammal-like reptiles (Synapsida) each enjoyed two such periods. Sharklike fishes dwindled during the Triassic and appeared to be on the verge of extinction, but this "racial old age" was followed by a "rejuvenation" in the Jurassic and Cretaceous, and the diversity of sharks has continued strong to the present day.

Champions of programmed extinction believe that lineages, like individuals, are born with certain maximum life expectancies. But no mechanism to explain racial life expectancies has ever been described, except for Hyatt's original "old age theory" (described by Gould, 1977b), which could not be reconciled with twentieth-century theories of embryology and genetics. For want of a suitable mechanism, theories of programmed extinction have few if any modern supporters.

C. RANDOM EXTINCTION

When a form of life becomes completely extinct, we marvel at the occurrence and wonder what its causes were. But is this reaction justified? How likely is it, or how improbable, that a certain number of deaths should result in the extinction of a species? If extinction is simply the sum of so many deaths, then we should be able to calculate the probability of species extinction from the demography of its constituent populations. We should also be able to calculate the probability of extinction of a major taxon from the speciation and extinction probabilities of its constituent species. If this probability is moderate to large, then perhaps we need not marvel at extinction when it takes place.

Through the use of computer models (such as the one described in Chapter 19), evolutionists can now examine the probability of extinction for large groups as well as small ones, distinguishing which events in the fossil record are so unlikely as to require a special explanation. Raup (1978b), for example, compared such a model to several actual extinction curves (Fig. 20.1). Raup also questions whether the extinction of certain early classes of the phylum Echinodermata requires a special explanation—it may simply be a consequence of the high average rate of extinction for the group as a whole.

The assumption of a constant rate of taxonomic extinctions (which is not really the same as a constant probability of extinction for lineages) is supported by Van Valen (1973). Van Valen, following MacArthur, has shown that the rate of extinction within major taxonomic groups remains remarkably constant over long spans of time despite other changes. This is even true regardless of changes in the size of the taxonomic group. Van Valen's

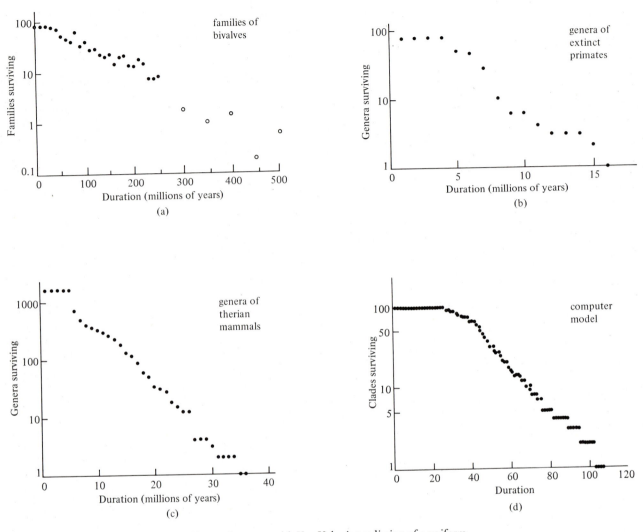

Fig. 20.1 Comparison of some actual longevity curves with Van Valen's prediction of a uniform rate of extinction. Perfect uniformity in the rate of extinction would result in a straight-line plot of descending slope, and there is a high level of agreement with this prediction. Reprinted with permission from David Raup, *Paleobiology,* vol. 1, 1976.

critics have pointed out certain exceptions to this rule, especially in those cases where taxa within newly diversifying groups are subject to a higher rate of extinction than their later relatives. Raup (1978a) found that average extinction rates among Phanerozoic invertebrates have remained rather constant across geologic time, except for a notable increase in extinction during the Permian–Triassic transition. Raup also calculated a value of 11.1 million years for the average species duration among fossil invertebrates.

D. PREVALENCE AND PATTERN OF EXTINCTION

Study of the fossil record reveals few patterns that characteristically are associated with extinction. Major groups of herbivorous vertebrates are a bit more susceptible to extinction than major groups of carnivorous vertebrates, and large organisms are more susceptible than small ones (Minkoff, 1979), but there are many exceptions. Extinction of a major group is often accompanied or followed by the proliferation and adaptive radiation of another group occupying the same broad ecological zone, but not always. The extinction of Mesozoic reptiles was soon followed by a great proliferation of placental mammals. The extinction of several major cephalopod groups at about the same time was followed or accompanied by the proliferation of the bivalves, and an earlier expansion of bivalves followed close on the heels of the extinction of many brachiopods. The extinction of the Multituberculata may have been related to the rise of the adaptively similar Rodentia. McKenna (1960) found that the frequency of rodents was inversely related to the frequency of multituberculates among the lower Eocene local faunas that he sampled in northwest Colorado. This strengthens the hypothesis that the two groups were in competition, and that the success of rodents led to the demise of multituberculates.

Van Valen (1973) has documented a constancy through geologic time of the rate of extinction in a number of major taxa. Van Valen refers to this as "MacArthur's law," and explains it on the basis of the "Red Queen's hypothesis": every new adaptation that increases the fitness of its possessor by a certain amount also decreases the fitness of all other ecologically related species by a small fraction of that amount. The continued, small decreases in fitness will ultimately cause the extinction of any form that fails to undergo enough change to compensate for this, or, as the Red Queen said, "you have to keep running pretty fast, just in order to stay in the same place."

Stanley (1979) believes that the average rate of extinction is usually equal to the average rate of speciation in each major taxon, and that both these rates vary together from one major taxon to the next. Stanley also believes that these rates should be higher for morphologically complex animals (such as vertebrates and cephalopods). Morphologically complex animals, argues Stanley, have more complex behavior patterns and thus more opportunity to build up reproductive isolation by differences in mating behavior. The higher rate of speciation then brings about increasing competition and thus more extinction. Stanley has amassed many graphs summarizing the evidence to support these several hypotheses among animals. Similar patterns may also apply to plants, though the argument about behavioral complexity would not apply in the plant kingdom.

E. CAUSES OF EXTINCTION

A stubbornly persistent question in evolutionary biology is that of extinction. Do extinctions have intrinsic causes, dependent only upon the characteristics of the species at the time, or is an environmental (extrinsic) causation always a necessary factor? Gould (1977a) considers this issue to be one of the "eternal metaphors" of paleontology, unresolved since the issue was first raised in the late eighteenth century. On the side of intrinsic causes, we may list the supporters of the various theories of programmed extinction, as well as those who would claim that statistically determined forces (similar to those assumed in various computer models) are responsible for all extinction irrespective of external environmental influences. On the side of extrinsic or environmental causation we have most modern ecologists and population biologists, as well as a large number of respected evolutionists both past and present: Darwin, Haldane, Simpson, and many others. Certainly an extrinsic cause can usually be found (or imagined) for any particular case of extinction. The real question, then, is whether there are any *general* causes applicable to all situations.

Adaptive failure, meaning usually an insufficiency of adaptation, can certainly lead to extinction. But what are the underlying causes of such insufficiency? The possibility of maladaptation resulting from inadaptive trends was considered in Chapter 17. A more plausible hypothesis is that the species simply fails to adapt rapidly enough to changing conditions. In most cases in which the cause of extinction has actively been sought, an ecological change in the conditions of life can usually be found. Climatic changes and mountain-building episodes (orogenies) may sometimes be responsible, but even these usually act through the changes that they effect in the biotic environment. Of the many species that became extinct during the Pleistocene, few seem to have perished from the cold, while many suffered greatly from the competition by invading species that crossed land bridges between the continents. In Chapter 9 we saw that competition may frequently lead to the extinction of one species and the establishment of another. Extinction is of course one possible outcome of such a competition.

As conditions change, organisms must either change with them, migrate in search of more favorable conditions, or become extinct. If adaptation is the major cause of trends, then adaptive failure is the major cause of extinction. Yet adaptive failure is a relative term: few organisms are maladapted in the sense of having zero fitness. All that is really necessary is that some other species have a competitive advantage.

The "Red Queen's hypothesis" figures here again: any adaptive advance by one species decreases the fitness of all other species slightly. Thus the mere failure to continually improve ultimately dooms a species to extinction, especially since environments are bound to change as well. This hypothesis is testable, for it predicts that extinction rates will be higher when other taxa are undergoing rapid evolutionary advances. Unfortunately, no adequate study has yet been done to test this prediction.

In most cases of extinction, an environmental change, and a concomitant failure to adapt, can usually be found. This observation suggests that each case of extinction has its own unique causes, which differ from group to group. Insufficient adaptation can often be shown, but the reasons for the insufficiency are diverse.

In rare cases, even very well adapted species can become extinct: a parasite confined to a single host may be very well adapted, but if the host becomes extinct the parasite will, too. The tambalacoque tree (*Calvaria*) on the island of Mauritius almost met with a similar fate, for its seeds were disseminated principally by the dodo, and it was thus seldom propagated since the dodo became extinct.

Another possibility that has been invoked is that extinction is the result of overspecialization. This may indeed be the reason why larger animals, which tend to be more specialized, are more often affected by extinction than are small ones, or why herbivores often exhibit higher rates of extinction than closely related carnivorous groups (Minkoff, 1979). It is hard to define "overspecialization" precisely, but one criterion is morphological peculiarity. Lipps (1970) reports that planktonic species that survived the late Cretaceous extinctions tended to be the morphologically simpler and ecologically more widespread ("generalized") forms. But morphological oddity is not necessarily a good indicator of the life expectancy of a group. Flessa, Powers, and Cisne (1975) report that the morphologically more specialized arthropods—those whose several appendages were very different from one another—have persisted just as long as those whose appendages were more uniform ("generalized"). Simpson (1953) has emphasized that many strange and bizarre creatures survive today (Fig. 20.2), despite the extinction of their more "normal" relatives. Simpson also points out that morphologically more generalized dinosaurs and ammonites persisted side by side with their more bizarre ("specialized") descendants, and succumbed to extinction at about the same time (Fig. 20.3).

If we think of specialization not as a structural change, but as a narrowing of ecological tolerance, it becomes evident that overspecialization, an excessive narrowing of ecological tolerance, can lead to the extinction of a species if the environment changes. In the case of a parasite, or an insect confined to a single type of host plant, the narrowness of host tolerance can jeopardize the continued existence of a species in case its host becomes extinct. Some environmental changes occur slowly enough for the species to evolve new and more suitable genotypes. But other changes occur too suddenly to permit this—the species must either have the resistant genotypes already present in its population or else fall victim to extinction. Populations living at the limits of a species'

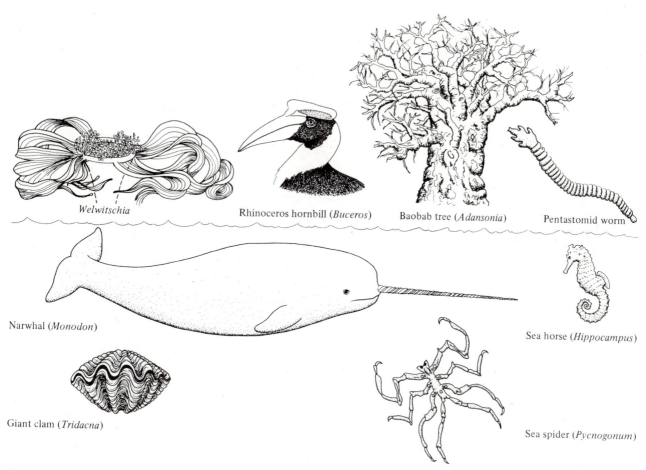

Welwitschia

Rhinoceros hornbill (*Buceros*)

Baobab tree (*Adansonia*)

Pentastomid worm

Narwhal (*Monodon*)

Sea horse (*Hippocampus*)

Giant clam (*Tridacna*)

Sea spider (*Pycnogonum*)

Fig. 20.2 Some bizarre forms still living today.

ecological tolerance are especially prone to this form of extinction. Suppose, for example, that a population was limited by the availability of drinking water during late summer dryness. Its failure to extend its range into still drier country stands as evidence of its lack of sufficiently drought-resistant genotypes. The genetic resources of this species would be inadequate to adapt to increasing dryness, and if the environment changed in this direction, the species would become extinct. In this case, the dying species would leave no indication of any attempt to adapt, nor would the fossilized corpses leave any indication, even under the best of conditions, that the reasons for mortality had changed (for they hadn't!). Ranges of ecological tolerance or host preference are seldom adequately recorded in the fossil record, and so the "reasons" for the extinction of this species or that may never be known. Simpson has stressed that neither environmental change nor the failure of a species to respond adaptively is an adequate explanation of extinction —both factors are always necessary.

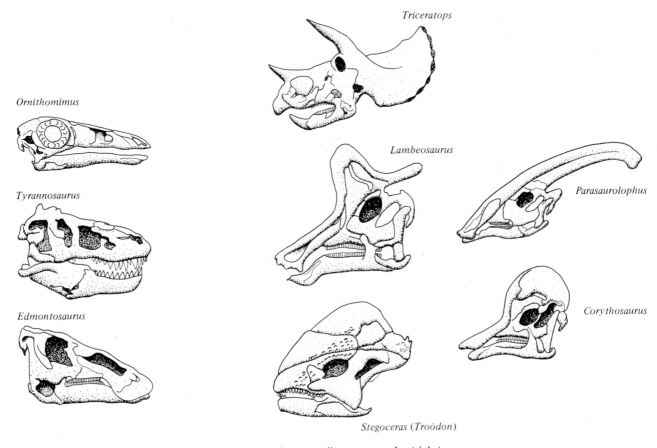

Fig. 20.3   A comparison of structurally simple (left) and structurally more complex (right) Cretaceous dinosaurs, all of which became extinct at or near the close of Cretaceous time.

F. SPECIES THAT HAVE AVOIDED EXTINCTION

Each species now living stands at the end of a lineage that stretches back in an unbroken sequence to the very origin of life. Each of these lineages has avoided extinction for the entire duration (thus far) of life on Earth. An equal or greater number of lineages has become extinct prior to reaching the present time. But how have the surviving lineages managed to avoid extinction so long? They have simply had the capacity (and the good fortune) to adapt sufficiently well to those situations that have allowed them continued success.

If there is any way to avoid extinction, surely the "living fossils" described in Chapter 19 have already found it. These are species that have persisted for very long periods of time, without any great morphological change and without extinction. They live in diverse environments, and except for their great longevity they have very little in common. One fact common to all living fossils is that they are adapted to environments that are very stable. These environments are not subject to seasonal decimation, nor is it likely that any future events would so disrupt the environment as to cause extinction. The food upon which they depend may be of several kinds, but is never confined to only one or a few food species. On a global scale, "living fossils" may be quite rare, yet their populations are often locally dense. They have avoided many of the specializations that have placed other species at high risk of extinction. Beyond this, the key to their amazing longevity is elusive.

G. GREAT EXTINCTIONS

Most puzzling in the history of life are those cases in which extinction came suddenly, without apparent cause, or in which many unrelated animals perished within a short period of time. The extinction of most major reptile groups at the end of the Mesozoic was a case in point, all the more so because many other animal groups, both marine and terrestrial, died out together with them. Yet the extinctions that occurred among many unrelated vertebrate and invertebrate groups at the end of Permian times were perhaps more devastating of world faunas, and the extinctions that occurred during the Pleistocene were no less dramatic.

The causal explanations that have been offered for these widespread changes fall into four groups: climatic, tectonic, biological, and extraterrestrial.

Postulated climatic changes include the warming or cooling of the entire Earth, increases in the extent of daily or yearly temperature fluctuations, and changes in the circulation of warm and cold ocean currents. Some theories invoke tectonic or astronomical factors to explain these climatic changes. Changes in daily or annual temperature fluctuations would have a greater effect on land than in the open sea, because the ocean acts as a heat reservoir. Yet the major Permian and Cretaceous extinctions affected sea life at least as much as life on land.

Changes in the Earth's average temperature are among the few known causes that could affect both marine and terrestrial habitats on a world-wide basis. A few dramatic changes in the Earth's climate are documented in the geologic record: an Ice Age during the Pleistocene, and another at the base of the Permian. But these ice ages do not coincide with the major extinctions at the close of Permian and Cretaceous times. Since the Cretaceous, we have evidence of the Earth's ocean temperatures preserved to us in the ratio of oxygen isotopes in the shells of certain Foraminifera. This evidence suggests a gradual cooling of the Earth throughout the Tertiary, and four episodes of dramatic cooling during the Pleistocene, corresponding to the four known glacial periods. No large or abrupt change is indicated at the end of Cretaceous time.

The evidence of Pleistocene extinction shows that widespread and well-documented climatic change was indeed accompanied by major extinctions, but the extinctions affected terrestrial organisms primarily, and especially the larger and more conspicuous among them. Detailed stratigraphic analysis shows that the extinctions of the Pleistocene generally did not coincide with the extremes of temperature or of glaciation, nor were they confined to the colder regions of high latitude. Many Pleistocene extinctions can be attributed to the exchange of North American and South American faunas across the newly emergent Panama land bridge, or between North American and Eurasian faunas across the Bering land bridge between Alaska and Siberia. Most of the extinctions resulting from these invasions took place as a result of increased competition brought about by the introduction of new faunal elements. The largest number of Pleistocene extinctions, in fact, took place coincident in time and place with invasions by a single species, *Homo sapiens* (Martin and Wright, 1967). Cave paintings depict the hunting of mammoth, reindeer, and other species. Arrowheads have been found associated with mammoth and bison bones. In one case, bison were stampeded into a dead-end canyon and slaughtered en masse. The extinction of the large, flightless elephant-bird in Madagascar, and of certain large marsupials in Australia, coincided approximately with the arrival of humans. In the Western Hemisphere, horses became extinct at the close of the Pleistocene, and the most likely explanation is that they were hunted to extinction by humans. Possible climatic or floral changes as major causal factors in this extinction seem ruled out by the fact that horses became extinct in various climates throughout the Hemisphere and by the fact that horses reintroduced by Spanish conquistadors have flourished quite successfully.

Within historic time, we can document the human slaughter of the now-extinct moa of New Zealand, the dodo, and the passenger pigeon. The threatened extinction of the bison, gorilla, orang-utan, Indian lion, and American alligator are well known. Many other species

—typically those of large body size—are threatened by human encroachment into their habitats. We hope that there is still time to save those species that are threatened with extinction today, although for a few it may already be too late.

Possible tectonic explanations for major extinctions include known episodes of mountain building (orogeny), as well as various crustal movements associated with plate tectonics. Newell (1952) has studied the possibility that extinctions resulted from major mountain-building events. Stanley (1979) has suggested that the more varied, mountainous topography would increase opportunities for speciation (see Chapter 14), and that the increased competition would lead to more extinctions. Some geological evidence is indeed suggestive: the Laramide orogeny, which was the major mountain-building episode in the Rocky Mountain region of North America, occurred at the end of the Cretaceous, coincident in time with many extinctions. Yet no such orogeny accompanied the extinctions at the end of Permian time, and the major Appalachian orogenies, earlier in the Paleozoic, were not accompanied by any dramatic extinctions. Most of all, these major mountain-building episodes were never world-wide, but were localized to a single continent, or half a continent, only.

According to the theory of plate tectonics (Chapter 23), the Earth's continents are slowly moving. This theory has suggested other possible explanations for major extinctions; for instance, the currently accepted theory describes the Permian as the time when all the world's continents were coalesced into a single continent called Pangaea. The crustal movements that brought about the ''break-up of Pangaea'' have continued to the present day, and the continents have been drifting apart ever since. Now, most of the organisms that became extinct at the end of the Permian were bottom-dwelling inhabitants of shallow inland seas (epeiric seas). T. J. M. Schopf (1974) argues that the coalescence of all the world's continents into a single supercontinent would have meant great reduction in the extent of these epeiric seas, and it was this reduction that he believes responsible for the ''great dying'' of later Permian times.

Gartner and McGuirk (1979) use plate tectonics to explain late Cretaceous extinctions. They believe that the Arctic Ocean became separated from the rest of the

world's oceans, with consequent changes in ocean currents and wind patterns. When, as the result of continental drift, the continuity among the world's oceans was reestablished, the sudden redistribution of ocean water temperatures caused drastic changes in ocean currents, prevailing winds, and climate, with disastrous consequences for dinosaurs, marine reptiles, ammonites, and other groups.

Extinctions may also have been caused by biological changes in the fauna and flora. Possibly these changes would greatly magnify the effect of some physical change in the global ecosystem. The changes in fauna and flora would not necessarily occur at the same time in both terrestrial and marine environments.

The extinction of the dinosaurs and other reptiles at the end of the Cretaceous has been the object of intensive study. The general climatic cooling that continued throughout Tertiary times and eventuated in the Pleistocene Ice Age had just begun. A major orogeny was taking place in western North America. Perhaps more significantly, the world floras were changing. The rise of the angiosperms had already taken place in earlier Cretaceous times, and these angiosperm floras were replacing the older floras of fernlike plants and gymnosperms (Chapter 26). The insects, which are now highly dependent upon angiosperms, must have undergone a great but undocumented adaptive radiation at about this time. Mammals were a small and insignificant part of the fauna, but they became increasingly important as the dinosaurs and other reptiles became extinct. The rise of mammals or of insects may explain the extinction of the dinosaurs on land, but not the simultaneous extinction of the marine reptiles and ammonoids in the seas. The major changes that can be documented in the geologic record were faunal and floral. The mammals may have been the first vertebrates to take advantage of the insects and other terrestrial invertebrates as a major food source. In the seas, the major changes were the great expansion of the smaller Foraminifera, the partial replacement of the ammonoids and marine reptiles by the expansion of the bivalves, and the evolution of many epifaunal predators among the Crustacea.

The extinction of the major reptile groups occurred piecemeal, one family or genus at a time, and not as suddenly as is sometimes imagined. Axelrod and Bailey

(1968) have documented the frequently misunderstood case of late Cretaceous dinosaurs, showing that the various species died out one at a time, over many millions of years. The pattern of piecemeal extinction argues strongly against any sudden, catastrophic cause.

The extraterrestrial causes that have been suggested as explanations for major extinctions include changes in the Earth's orbit (perhaps inducing climatic changes), explosion of a nearby supernova (Whitten *et al.*, 1976), increases in the cosmic radiation falling upon Earth, or collision of the Earth with a large (about 10 km diameter) asteroid (Alvarez *et al.*, 1980). Most of these explanations involve sudden, catastrophic events.

Increases in cosmic radiation have been suggested as possible causes of greatly increased mutation and extinction rates. No such increases are documented in the geologic record, however. Cosmic radiation would also become somewhat attenuated upon passage through ocean water, and should therefore have affected terrestrial and near-shore marine organisms more than those in deep marine waters. Yet, the victims of Cretaceous extinction included the ammonites and the bottom-feeding placodonts. Also, the bottom-living trilobites and many groups of sessile, bottom-living brachiopods became extinct during the extinctions that ended the Paleozoic Era.

Newest among the astronomical explanations is the suggestion by Alvarez *et al.* (1980) that the Earth collided with an asteroid of about 10 km diameter. (Earlier workers had suggested collisions with comets or meteors, objects that are far too small to be of any significance.) According to Alvarez's hypothesis, a 10-km asteroid would throw up sufficient dust to obscure significant amounts of sunlight for many years, thus affecting many photosynthetic plants on both land and sea. Changes in these plants would have repercussions throughout the Earth's food chains, and could easily have led to the extinction of many important animal groups. Support for the asteroid hypothesis has been found in the sediments that span the Cretaceous–Tertiary boundary, which show a brief but dramatic increase in the level of iridium (Ir). Iridium is very rare on Earth, but is much more abundant in asteroid material. But extraterrestrial material should also be enriched in certain other rare elements, and the enrichment of these elements has not been reported thus far.

CHAPTER SUMMARY

True extinction is the termination of a lineage without issue. Phyletic replacement of one species by another results only in pseudoextinction. Termination in the stratigraphic range of a taxon by either means may be called taxonomic extinction.

Various theories have explained extinction as resulting from "racial old age" or other intrinsic causes. But limited cladogenesis and extreme morphological specialization do not always herald extinction. A growing number of evolutionists have recently turned instead to the statistical prediction of extinction from random models. The general aim of these models is to predict the patterns of extinction for major taxa from the random extinction of their constituent species.

Extinction results frequently from competition with other species. Overspecialization, including an excessive narrowing of ecological tolerance, may likewise figure prominently in extinction. "Living fossils," which have resisted extinction for many millions of years, are characterized by locally abundant populations and adaptation to very stable environments.

The widespread extinctions of Permian and Cretaceous times are difficult to explain, especially since they affected unrelated species on both land and sea. Climatic changes, tectonic changes (such as mountain-building episodes), astronomical events, and faunal and floral changes have all been suggested as possible causes. Changes in faunas and floras can explain the extinction of such land reptiles as dinosaurs, but the extinction of many marine animals during the same time period cannot be explained as easily. Another explanation offered in recent years for the Permian extinctions is that they were related to the reduction in the extent of shallow inland seas resulting from the presence of the supercontinent known as Pangaea. A recently proposed explanation for late Cretaceous extinctions involves collision of the Earth with an asteroid. During the Pleistocene, most extinctions of land animals can be explained as resulting from the competition that followed the establishment of contact between previously isolated faunas. The largest and most conspicuous animals were driven to extinction by humans, a process that has continued into historic times.

21

Patterns of Regularity in Macroevolution

A. REGULARITIES OF NATURE: ARE THEY "LAWS"?

Progress in many sciences (those described in Chapter 1 as *nomothetic*) often involves the discovery of laws that are regularities of nature, permitting no exceptions. One such "law" might predict that hydrogen and oxygen will combine under certain specified conditions to make water. There may be circumstances under which this reaction would not occur, but such "exceptions" are treated by suitably restricting the specified conditions under which the law holds.

The realization that evolution is a science has led many scientists to search for regularities of this type, which are quickly dubbed "laws." But evolution is one of the historical sciences, which are largely *ideographic* (Chapter 1) because they deal with particular events or particular individual phenomena. Sometimes, these individual phenomena show recurring patterns. To what extent may these patterns of nature be called "laws"? The Grand Canyon, the genus *Hyracotherium*, the presidency of Andrew Jackson, and the paintings of Michelangelo were all unique events in the history of the world. Regular patterns may describe how the Grand Canyon resembles certain other canyons, or how *Hyracotherium* resembles certain other mammalian herbivores of similar size, but these patterns cannot explain the unique aspects of particular canyons or particular mammalian genera.

Historical regularities, unlike the laws of physics, must admit to a certain percentage of exceptions, and it is largely for this reason that many scholars prefer not to call such regularities "laws." The regularities ("laws") of evolution are more like those of history than like the laws of physics. They are generalizations governing very complex and incompletely understood phenomena, often demanding the use of qualifying expressions like "tend to" or "all other things being equal." Many of these generalizations are probabilistic—they cannot be used to predict the outcome in individual instances. The finding that rapid cladogenesis often follows rapid anagenesis has about the same validity as the historical generalization that economic depressions often follow wars.

Generalizations concerning the mechanisms of evolution include Malthus's "principle of population," Mendel's laws of heredity, and a number of Darwinian laws including the principle of natural selection. Such generalizations have been treated in Unit I of this book.

The generalizations of macroevolution, treated in this chapter, are of three kinds, as follows:

1. Dollo's law of irreversibility, a general law applicable to *all* historical or ideographic sciences.
2. Generalizations concerning the overall patterns of phylogeny.
3. Empirical generalizations based on common evolutionary trends.

Dollo's law is singled out for special attention because it is applicable to *all* the historical or idiographic sciences, while the others are specific to evolutionary biology only.

B. DOLLO'S LAW OF IRREVERSIBILITY

Of all the laws of evolutionary biology, none has such profound importance as the law of irreversibility associated with the name of Louis Dollo (1857–1931), and often known as Dollo's law. Other evolutionists, to be sure, were aware of this principle before Dollo, including

Charles Darwin (1859:315). In brief, Dollo's law states that evolution, like historical processes generally, never returns to a previous condition. *History,* in short, *never repeats itself exactly.* Unfortunately, Dollo himself created a certain amount of confusion by using the term, ''Loi d'Irreversibilité,'' in at least three distinct senses (Box 21–A).

The Presumption of Methodology

The strictest meaning of irreversibility is that *entire species can never revert exactly to previous conditions,* just as in history entire civilizations can never exactly revert to a previous state of affairs. This presumption is a necessary part of the methodology of paleontology and other historical sciences. A paleontologist who finds a well-preserved fossil having all the characteristics of the genus *Brontosaurus* makes use of this presumption in assigning the fossil to *Brontosaurus,* instead of proclaiming that another genus has independently evolved all the same characteristics. An archaeologist uses the same presumption in assigning a newly found ruin to the Myce-

---

BOX 21–A *Three Meanings of Irreversibility (Dollo's Law)*

Nature of statement	Statement	Basis for statement
A necessary methodological presumption	1. ''An entire organism (or an entire civilization) can never return exactly to any previous, ancestral condition.''	Evolution is a complex historical process, subject to statistical phenomena when many independent events are involved. (True)
An empirical hypothesis (testable, refutable, falsifiable)	2. ''A structure (i.e., a part of an organism) can never return exactly to any previous ancestral condition.''	
	3. ''Evolutionary trends (cf. historical trends) tend to remain constant in direction and rate unless acted upon by some external force'' (i.e., evolution has a momentum).	Evolution is a preprogrammed process; its direction is predetermined. (False)

naean civilization because it has all the characteristics associated with that civilization. In these cases, the law of irreversibility assures the paleontologist that no other genus has all the characteristics of *Brontosaurus,* just as it assures the archaeologist that no other civilization ever had all the characteristics of the Mycenaean. The Mycenaean civilization, like *Brontosaurus,* was historically unique, and can never again be recreated in exactly its prior form. If the conditions suitable to the evolution of the genus *Brontosaurus* were again to recur, the genus *Brontosaurus* would never re-evolve, because its ancestors are no longer present. And if a brontosaurlike creature *did* evolve under these conditions, it would perforce evolve from different ancestors, and would bear some telltale trace of its different ancestry.

We may therefore restate the law of irreversibility in this sense by saying that in all historical processes, independently produced phenomena are always distinguishable. Convergence, in other words, is never so perfect as to be undetectable. Dollo himself pointed to the failure of whales to re-evolve gills as an instance of this type of irreversibility. Irreversibility of this kind is so basic to the historical sciences that there could indeed be no historical inquiry without it!

The Empirical Generalization

The most frequently used meaning of irreversibility is the statement that complex phenotypic traits, or complex characteristics of individual historical phenomena, can never exactly recur. This is a testable, falsifiable hypothesis (unlike the methodological presumption, which cannot be falsified). Irreversibility in this sense tells us that Mycenaean pottery could never independently be produced outside Mycenae, or that the teeth of *Hyracotherium* could never evolve independently in another genus. These are falsifiable statements: the finding of a tooth indistinguishable from *Hyracotherium* in the body of another beast, or the finding of Mycenaean pottery in a Mayan ruin, would falsify these statements. Convergence, even of component structures, is still never perfect: the identical mammalian molar tooth is never found in the jaws of nonidentical species.

The fact that no two individual genotypes are alike (excluding monozygous or "identical" twins) is an im-

portant special case of this law. The reason why this is so provides insight into the basis for irreversibility in both this and the previous sense. Each individual has many thousands of individual genes, if not millions. The probability of two individuals independently having identical genotypes is vanishingly small—it could theoretically occur, but with such low probability that it might just as well be treated as an impossibility. It is exactly this same reasoning that allows us to state that a monkey sitting at the keyboard of a typewriter could never produce Shakespeare's *Hamlet* simply by depressing the keys at random. Again, there is a vanishingly small probability, but we can safely ignore it.

Now what if we asked that a monkey striking the typewriter keys at random produce only a single line from *Hamlet,* rather than the entire play? The probability is still rather small, but we can no longer ignore it altogether. Suppose we required only that the monkey produce a single word from *Hamlet,* such as the word "to"? The probability would now become so high, that given enough time at the typewriter, the appearance of such a word is a virtual certainty.

Returning to a discussion of genotypes, we can see that a complex structure, such as a molar tooth, would have a sufficiently complex genetic basis that it would never independently recur in two separate species. Yet, if we asked the same question of a less complex but still highly polygenic trait, say, a head length of 15 centimeters, we would be forced to admit that the same head length *could* occur independently in unrelated animals, though it might not have the identical genetic basis in each. If we chose a trait controlled by a single gene, such as white versus red eye color in *Drosophila,* the case is even simpler: mutational changes from one allele to the other can and do occur, and furthermore they occur reversibly! Thus, a trait with a very simple genetic basis *could* reverse itself in the course of evolution, but the reversibility of a more complex character becomes successively less probable with increasing complexity. In fact, the probability of many genes reverting together varies exponentially with the number of genes, so that any increase in the number of genes greatly reduces the probability. Evolutionary processes are, in other words, capable of reversing themselves only if simple characters (with simple genetic bases) are concerned. As the complexity of a character increases, the probability of its reversion to a former exact state becomes vanishingly small. Irreversibility may thus be seen as a consequence of the statistical nature of evolution. Like the second law of thermodynamics, it is true only because of the large number of individual events summarized in macroscopic processes.

Nopsca and Abel, among others, have suggested that Dollo's law be restricted to a special case only: that an organ, once lost, could never be regained. Dollo, of course, included this in his conception of irreversibility, but he also included much more. In fact, some of the examples used by Nopsca to argue against a broader meaning of irreversibility are the very examples that Dollo had used to argue *for* irreversibility. Nopsca, for example, used the whale as an argument against irreversibility, noting that its ancestors were terrestrial mammals that were ultimately descended from aquatic fishes. But Dollo himself used this example to illustrate his principle that despite such superficial reversals, the convergence was never complete: the whale had not re-evolved gills.

The Inherent Causation

In addition to the two previous senses, which modern evolutionists accept, there is a third meaning that Louis Dollo gave to the concept of irreversibility, a meaning that most modern evolutionists would now reject. A friend of mine claims that this meaning of irreversibility was represented by a handwritten sign affixed to a malfunctioning vending machine: "you can't get your money back by pouring hot coffee into this machine." Why not? Because the coffee machine is a programmed mechanism, designed to act only in one direction, and incapable of working in the other direction. The third meaning of irreversibility states that evolution is just such a programmed mechanism, capable of operating in one direction only.

Our previous discussion of traits with simple genetic bases will serve to demonstrate that this form of irreversibility is a false conception. Evolutionary trends are not programmed to operate in one direction only, for when conditions reverse themselves, the trends may themselves reverse, and characters with relatively simple genetic bases may indeed revert to an antecedent condition.

C. PATTERNS OF PHYLOGENY

Various evolutionists have occasionally noted regularities in the overall patterns of phylogeny. Some of Bessey's (1915) dicta (see Box 21–E) fall into this category, such as the rule that evolution can be both progressive and retrogressive, proceeding both "up" and "down." The principle of mosaic evolution is another example, as are such classic dicta as "evolution is a bush, not a ladder," or "Natura non facit saltus" (Chapter 3), though the latter has recently been questioned (see below). Other principles of this type, including orthogenesis (Chapter 18), programmed extinction (Chapter 20), or belief in the momentum of evolutionary trends (Chapter 17), have been rejected by most modern evolutionists.

Cope's "Law of the Unspecialized"

The American paleontologist Edward Drinker Cope (1840–1897) noted that mammals had evolved not from advanced, highly evolved reptiles, but from reptiles of a rather primitive sort. Reptiles, likewise, evolved from a small group of early amphibians, not from the more successful amphibians that dominated the Pennsylvanian coal swamps. So, too, the amphibians evolved from an early and only moderately successful group of Devonian fishes, not from the more advanced and more successful fish groups of later geologic periods. Among plants, monocots are now believed to have evolved from rather primitive dicots, and dicots from seed ferns rather than from the more specialized conifers or gnetophytes (Chapter 26).

Cope originally proposed his "law of the unspecialized" to apply to cases such as these: the origin of new major taxa from the less evolved members of ancestral taxa. Other possible meanings of the "survival of the unspecialized" soon suggested themselves (Box 21–B), and as a result quite a number of distinct meanings have been given to this concept.

One important meaning is that organisms with a number of generalized, multipurpose structures tend to have higher chances of survival than do organisms with highly specialized, single-purpose structures. "Overspecialized" organisms tend to have low ecological tolerance

BOX 21–B *Cope's "Law of the Unspecialized"*

To the phrase "survival of the unspecialized" many possible meanings can be (and indeed have been) given. Discussions on the validity of this "law" are plagued by confusion over the meaning of the law. Unfortunately, neither "survival" nor "unspecialized" has an unambiguous meaning in these discussions. I propose that the various meanings be distinguished as shown in the table.

Meanings of "survival":

A. Taxonomic continuance to the present (taxonomic survival).
B. Survival with change; ancestry to another living taxon.
C. Survival with or without change; possession of living descendants (phyletic survival, the *union* of A and B).

Meanings of "unspecialized":

	EVIDENCE:
Past status	
1. Early origin	1. Early appearance in fossil record.
Present status	
2. Broad ecological tolerance; broad adaptations.	2. Ability to withstand environmental fluctuations; variety of habitats occupied, etc.
3. Broad structural adaptation; possession of multipurpose structures	3. Functional analysis to reveal multiple functions
Potential future status	
4. Broad taxonomic potential	4. Taxonomic variety of descendants
5. Broad adaptive or structural potential	5. Adaptive variety of descendants; variety of derived homologues

By combining the above possibilities (or others), we may list a large number of possible meanings for the "survival of the unspecialized," as follows:

A1. Long-lived taxa are those that originate early.
A2. Long-lived taxa are those with broad ecological tolerance.
A3. Long-lived taxa are those with broad structural adaptations.
A4. Long-lived taxa are those whose descendants are taxonomically varied.
A5. Long-lived taxa are those whose descendants are adapted to a variety of environmental situations.

B1. New taxa originate from ancestral taxa that originate early.

B2. New taxa originate from ancestral taxa with broad ecological tolerance.
B3. New taxa originate from ancestral taxa with broad structural adaptations.
B4. New taxa originate from ancestral taxa whose descendants are taxonomically varied.
B5. New taxa originate from ancestral taxa whose descendants have become adapted to a variety of environmental situations.

C1. Living descendants are more often found among taxa of early origin.
C2. Living descendants are more often found among taxa of broad ecological tolerance.

BOX 21-B *(continued)*

C3. Living descendants are more often found among taxa with broad structural adaptations.

C4. Living descendants are more often found among taxa whose descendants are taxonomically varied.

C5. Living descendants are more often found among taxa whose descendants are varied in their ecological adaptations.

Of the above meanings (which by no means exhaust all the possibilities), B1 is probably closest to Cope's original meaning, though B2 through B5 were also suggested by many of Cope's own writings. Many of these statements (including A2, B1, B2, B3, B4, B5, C3, C4, and C5) are probably true, though B4 and B5 are at least in part tautological. Some statements (A1, A4, A5, C1) are perhaps of questionable validity, the characteristics of descendent taxa being usually independent of the time of origin or the longevity of their ancestral taxa.

---

and tend to succumb more easily to ecological stress or ecological change. On the other hand, Alfred S. Romer has cautioned that "no organism can make a living simply by being an unspecialized ancestor." True to Romer's statement, many ancestors are found to have at least a few specialized features. Specialization may decrease a species's chances of survival in the long run, but a certain level of specialization may well be necessary to survival.

Gradualism versus Punctuated Equilibria

New interest in general patterns of phylogeny has recently been sparked by the theory of **punctuated equilibria,** enunciated by Eldredge (1971), Eldredge and Gould (1972, 1974), Gould and Eldredge (1977), and Stanley (1979). According to this theory, speciation takes place allopatrically (Chapter 14) in isolated, peripheral populations, which are far less likely to be preserved in the fossil record. A new species enters the fossil record not when it originates as an isolated peripheral population, but when it successfully replaces the prior species in the major portion of its range (Fig. 21.1). The fossil record will then show a series of species which suddenly replace one another, rather than a gradual transition from each species to the next.

The more traditional viewpoint has been designated as the theory of **gradualism.** Darwin's *Origin of Species* predicted a smooth transition from one species to the next and rejected the abrupt replacement of one species by another as an artifact resulting from "The Imperfection of the Geologic Record." Modern theories inspired by Darwin have emphasized evolutionary trends as re-

sulting from the continued, gradual replacement of each species by the next. According to the gradualist school of thought, *most morphological change takes place during anagenesis.* In fact, an ongoing lineage is viewed as continually changing and divided into a series of arbitrary species for convenience only. Such species are sometimes called either paleospecies or chronospecies.

The theory of punctuated equilibria teaches that species do not succeed one another by such a process of gradual phyletic replacement. Very little morphological change takes place during the life of a species; *most morphological change takes place during speciation (cladogenesis).* On the basis of this last assumption, Stanley (1979) has predicted that rates of morphological change should be proportional to rates of speciation. Furthermore, since behavior is the basis for so much reproductive isolation among animal species, Stanley predicts that the behaviorally more complex groups should exhibit higher rates of speciation and thus higher rates of morphological change as well. Also, since the rate of speciation is faster for behaviorally more complex groups ("higher" animals), the average duration of species or genera should be much shorter than in the case of simpler and behaviorally less complex organisms.

The patterns of phylogeny predicted by gradualists and by punctuationalists are contrasted in Fig. 21.2. Gingerich (1976, 1977) and Gingerich and Schoeninger (1977) have studied several relatively well-documented species transitions among early Tertiary mammals (Figs. 21.3 and 21.4) and conclude that they support the gradualist model. Eldredge and Gould's (1977) defense of the punctuational model contains an attempt to fit Gingerich's data to a punctuational model.

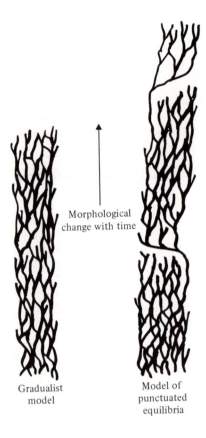

Morphological change with time

Gradualist model

Model of punctuated equilibria

Fig. 21.1 Two models of an evolving lineage. In the gradualist model, the species evolving through time is an anastomosing network of populations that split and fuse. Any boundary (such as a species boundary) drawn across this evolving continuum is wholly arbitrary. In the punctuational model, species transitions are marked by the abrupt replacement of most central populations by the descendants of a rapidly evolving, peripherally isolated population. Boundaries between successional species are therefore interpreted as non-arbitrary, and abrupt transitions between such species are expected.

Fig. 21.2 Patterns of phylogeny assumed by (a) an extreme punctuationalist who interprets all morphological change as speciational, (b) a more moderate punctuationalist who admits some anagenetic change as well, (c) a moderate gradualist who recognizes most morphological change as anagenetic, and (d) an extreme gradualist who recognizes all morphologic change as anagenetic. Adapted with permission from Steven Stanley, *Macroevolution: Pattern and Process*, copyright © 1979. San Francisco: W. H. Freeman.

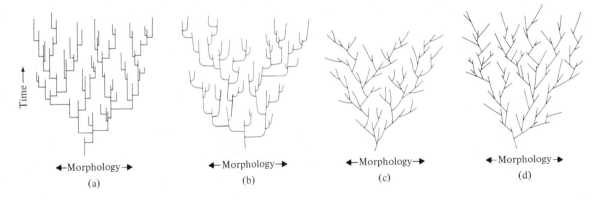

Time →

←Morphology→
(a)

←Morphology→
(b)

←Morphology→
(c)

←Morphology→
(d)

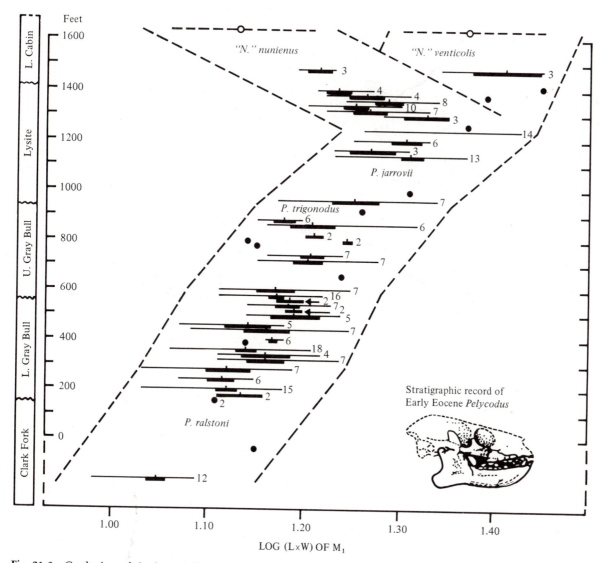

Fig. 21.3 Gradual trends in the early Tertiary primate *Pelycodus*. Reprinted with permission from P. D. Gingerich and M. Schoeninger, The fossil record and primate phylogeny. *J. Human Evol.* 6: 483–505, 1977.

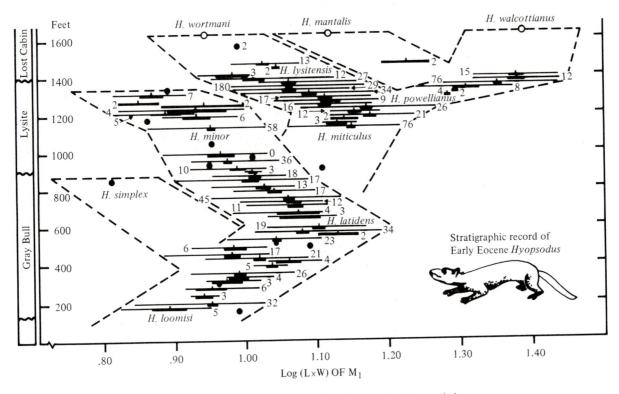

Fig. 21.4 Gradual trends in the early Tertiary mammal *Hyopsodus*. Reprinted with permission from P. D. Gingerich, 1977, Patterns of evolution in mammalian fossil record. In A. Hallam, ed. *Patterns of Evolution as Illustrated by the Fossil Record.* Amsterdam: Elsevier.

The major assumptions of the gradualist and punctuationalist models are contrasted in Box 21–C. Some of these assumptions cannot be tested without a nearly continuous fossil record of species durations or of species transitions. One pair of contrasting assumptions that can be tested concerns the relationship between anagenesis and cladogenesis: gradualists believe that these two processes are independent, while punctuationalists insist that anagenesis does not occur without cladogenesis. Examination of many groups shows that anagenesis and cladogenesis commonly occur together, but not always. Cladogenesis with little or no concomitant anagenesis has occurred in most cases of sibling species (Chapter 14), and among the Hawaiian *Drosophila*. Much anagenesis with very little cladogenesis has occurred among whales, pterosaurs, and taeniodonts, and perhaps among the earliest bats (but not the later ones, which diversified

greatly). Cases of no cladogenesis whatsoever, in the face of anagenetic change, would be exceedingly difficult to document to the satisfaction of most paleontologists. A few species transitions among Tertiary mammals may have occurred without cladogenesis; the best-documented cases include some of those recorded by Gingerich, as well as the transition between the horse genera *Parahippus* and *Merychippus*. It will be an important task for the next generation of paleontologists to document in detail as many species transitions as they can, as a test of the gradualist and punctuational models.

The controversy between gradualists and punctuationalists has done much to spark a renewed interest in evolutionary biology, and particularly in evolutionary theory. A number of invertebrate paleontologists have been won over to the punctuationalist viewpoint. But many workers on Recent organisms have been more reluctant to give up their belief in Darwinian gradualism. Stebbins and Ayala (1981) argue that both viewpoints are consistent with modern microevolutionary theory, and that the resolution of the controversy must therefore come from macroevolutionary rather than microevolutionary studies. Macroevolution, they argue, is therefore an independent field of study.

D. COMMON EVOLUTIONARY TRENDS

These several rules are empirical generalizations concerning events that occur more frequently than might at first be assumed. As generalizations, they hold ''more often than not,'' or ''all other things being equal'' (*ceteris*

paribus), rather than invariably. Misunderstandings of this simple fact have led to undue emphasis on ''exceptions'' to such rules, which ought not to be thought very surprising if the true character of the generalizations is kept in mind.

Williston's Rule

Many animal and plant structures contain parts that are serially repeated: the several petals of a flower, or the several legs of a many-legged animal are familiar examples. Other examples include the serial repetition of vertebrae, of gill slits, of body segments in annelids and many other animals, or of chromosomes. The American paleontologist Samuel Wendell Williston (1852–1918) called much attention to the fact that such repeated structures often decreased in number during the course

---

BOX 21-C *Contrasting Viewpoints of Gradualist and Punctuational Models*

Gradualist model	Punctuational model
Most morphological change occurs during anagenesis.	Most morphological change occurs during cladogenesis.
Morphological change is gradual and occurs continuously.	Morphological change is rapid but occurs only rarely.
Paleospecies are arbitrary subdivisions of a continuous anagenetic sequence.	Paleospecies are real, discrete, and nonarbitrary.
Transitions between paleospecies are gradual and anagenetic.	Transitions between paleospecies are abrupt and cladogenetic.
New species originate gradually. The seemingly abrupt origin of any paleospecies is an artifact resulting from the imperfections of the fossil record.	New species originate abruptly. The abrupt origin of a paleospecies occurs by speciation (cladogenesis), plus sudden invasion of its ancestral range and replacement of its immediate ancestor.
The duration of a paleospecies is marked by gradual but continual change.	The duration of a paleospecies is marked by evolutionary stasis with no morphological change.
Gaps between paleospecies represent the imperfections of the geological record; a complete record would show continuous transitions from one paleospecies to the next.	Gaps between paleospecies are real and represent abrupt replacement of one paleospecies by its immediate descendant.
Phyletic replacement does not involve the splitting of any species.	Phyletic replacement involves the splitting off of a new species, which then replaces its ancestor abruptly.
Anagenesis and cladogenesis are independent processes; either can occur without the other.	Anagenesis never occurs in the absence of cladogenesis; usually they occur together.

of evolution, while the structures that persisted tended to undergo regional specialization. This generalization has been called ''Williston's law,'' though others before Williston seem to have been aware of it, including even Charles Darwin. This rule may be stated as follows:

> In the evolution of organisms with serially repeated parts, there is a frequent tendency for the parts to diminish in number, while those remaining tend to undergo regional specialization and differentiation from one another.

Williston's rule does not deny the possibility of repeated structures ever becoming more numerous, as has indeed happened in the vertebrae of snakes and in some other cases, too. The rule merely states that trends more often run in the direction of a decrease in number and regional specialization. The trilobite and crayfish (Box 21–D) show an instance of Williston's rule: trilobite appendages (legs) were numerous, and each was a simple all-purpose organ (a ''biramous appendage'') resembling all the others. Crayfish appendages are far less numerous, and they are also regionally specialized for use as mouthparts, claws, walking legs, water-propelling ''swimmerettes,'' and so forth.

The selective advantages of regional specialization can be used to explain Williston's rule, but to this we must also add the disadvantages of possessing many repeated parts when a few will do. By their mere position, some repeated parts may be used more often for one purpose, others for another purpose. In ancestral trilobites, for instance, the anterior appendages were used more often in directing food toward the mouth, while the posterior ones were used more often in fanning the water, even before any structural difference between them became apparent. The division of labor created selection pressures favoring anterior appendages more efficient in food handling, while also favoring posterior appendages more efficient in fanning the water. In this manner, the division of labor led ultimately to regional specialization of the repeated appendages. Meanwhile, some appendages were hardly ever used, and these tended to disappear for the reasons given in Chapter 17. The numerous similar appendages of the trilobite thus evolved into the fewer, more specialized appendages of the crayfish. In other independent but related lineages, the same uniform series of biramous appendages evolved into the differentiated appendages of insects, of spiders and scorpions, and of several other groups of arthropods. Most insects, for instance, have one pair of maxillae, one pair of mandibles, and three pairs of walking legs (themselves frequently differentiated), derived ultimately from trilobite appendages. Between the mouthparts and walking legs, and also posterior to the walking legs, a number of appendages have been lost, and are represented by legless segments.

Cope's Rule of Size Increase

It is frequently the case that descendents are much larger than their ancestors; it is less often the case that descendants are smaller than their ancestors. This generalization has been called ''Cope's law,'' after the American paleontologist Edward Drinker Cope (1840–1897), though it was known to other evolutionists before Cope. It is perhaps better considered a rule than a law, for evolutionary trends toward a decrease in size occur with low frequency in nearly all taxonomic groups and are more frequent in some. There is a further confusion in that the ''law of the unspecialized,'' treated earlier in this chapter, has also at times been called ''Cope's law.''

Early attempts to explain Cope's rule usually emphasized the selective advantages of large over small body size. Heat loss and various other metabolic factors are frequently more favorable at larger body sizes. The areas of internal surfaces, including brain surfaces, absorptive intestinal surfaces, and respiratory surfaces

---

BOX 21–D *Williston's Rule*

The variety of arthropod appendages (including mouthparts, etc.) are all derived from the simple biramous appendages possessed by trilobites. Shown here are the appendages of an insect, a crustacean, and a trilobite, as illustrating Williston's rule. Note that the ancestral condition (as represented by the trilobite) has numerous serially repeated parts of multiple function, and that the less numerous appendages of specialized arthropods (as represented here by the crayfish and the grasshopper) are more differentiated from one another, each serving a somewhat distinct function in relation to the rest.

BOX **21-D** *(continued)*

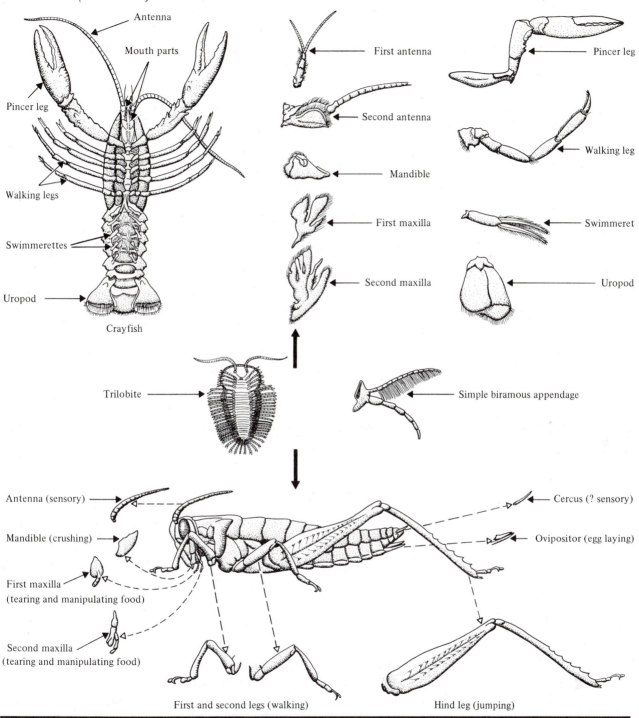

Antenna

Mouth parts

Pincer leg

Walking legs

Swimmerettes

Uropod

Crayfish

First antenna

Second antenna

Mandible

First maxilla

Second maxilla

Pincer leg

Walking leg

Swimmeret

Uropod

Trilobite

Simple biramous appendage

Antenna (sensory)

Mandible (crushing)

First maxilla
(tearing and manipulating food)

Second maxilla
(tearing and manipulating food)

Cercus (? sensory)

Ovipositor (egg laying)

First and second legs (walking)

Hind leg (jumping)

(gills or lungs) must increase at a rate in keeping with the animal's volume or mass. These surfaces are often coiled, folded, or subdivided as a result. In larger animals, weight-bearing structures (legs) must increase in proportion to body mass, according to the principle of scaling (Chapter 10). Plants, too, run into problems of support (more need for woody tissues) and of fluid transport (more need for vascular transport) as size increases. Larger organisms thus tend, for all these reasons, to be more specialized than their smaller relatives.

Stanley (1970) has offered a different explanation for Cope's rule, pointing out that smaller animals tend for the above reasons to be less specialized, and thus more suitable as ancestors for their more specialized descendants. According to Stanley, Cope's rule results from the greater probability of small animals giving rise to new taxa, and the greater probability of large animals becoming extinct through overspecialization. If there is an optimally efficient body size for a given taxonomic group, the origin of the group tends to occur at a body size well below this optimum value, and the more frequent evolutionary trends are therefore toward increasing body size.

Other Common Trends

In addition to Williston's rule and Cope's rule of size increase, other regularities have from time to time been observed in the study of life's evolutionary history. The American botanist Charles Bessey (1845–1915) published in 1915 a number of ''dicta,'' some of which included precisely such regularities among higher plants: that numerous petals generally evolve into fewer petals, that superior ovaries generally evolve into inferior ones, and so forth (Box 21–E). We may now add several more to this list, for example the generalization that chromosome numbers tend to increase among higher plants by doubling or quadrupling through polyploidy, while they tend also to decrease more gradually through the loss of individual chromosomes one at a time.

Among animals, we may note that parasites tend frequently to undergo regressive evolution and to lose many structures, especially those of the sense organs and nervous system. Animals that move actively in a given direction tend to undergo cephalization, which is the concentration of the sense organs and brain at the anterior end. The most frequently used sense organs may even tend to move furthest forward in many but not all cases.

BOX 21–E *Bessey's Dicta (1915)*

The evolutionary classification of the higher plants proposed by the American botanist Charles Bessey in 1915 forms the basis for all later classifications, including those of Takhtajan, Cronquist, and Stebbins (Chapter 26). In explaining the evolutionary basis of his classification, Bessey put forth 28 ''dicta.'' Some of these (nos. 1–5, 7) are general statements about evolution; the remainder are empirical generalizations concerning the evolution of plants, especially angiosperms. In some cases these generalizations serve as statements regarding the primitiveness of certain character states (Chapter 22). Of the more general statements, no. 3 may be recognized as the principle of mosaic evolution (Chapter 17), no. 2 as Williston's rule, and no. 5 as a cautious version of Dollo's law. The listing that follows is taken directly from Bessey (1915).

A. General Dicta

1. Evolution is not always upward, but often it involves degradation and degeneration.

2. In general, homogeneous structures (with many and similar parts) are lower, and heterogeneous structures (with fewer and dissimilar parts) are higher.

3. Evolution does not necessarily involve all organs of the plant equally in any particular period, and one organ may be advancing while another is retrograding.

4. Upward development is sometimes through an increase in complexity, and sometimes by a simplification of an organ or a set of organs.

5. Evolution has generally been consistent, and when a particular progression or retrogression has set in, it is persisted in to the end of the phylum.

6. In any phylum the holophytic (chlorophyll-green) plants precede the colorless (hysterophytic) plants, and the latter are derived from the former.

7. Plant relationships are *up and down* the genetic lines, and these must constitute the framework of phylogenetic taxonomy.

BOX 21–E (continued)

B. Dicta Having Special Reference to the General Structure of the Flowering Plants

8. The stem structure with collateral vascular bundles arranged in a cylinder is more primitive than that with scattered bundles, and the latter are to be regarded as derived from the former.

9. Woody stems (as of trees) are more primitive than herbaceous stems, and herbs are held to have been derived from trees.

10. The simple, unbranched stem is an earlier type, from which branching stems have been derived.

11. Historically the arrangement of leaves in pairs on the stem is held to have preceded the spiral arrangement in which the leaves are solitary at the nodes.

12. Historically simple leaves preceded branched ("compound") leaves.

13. Historically leaves were first persistent ("evergreen") and later deciduous.

14. The reticulated venation of leaves is the normal structure, and the parallel venation of some leaves is a special modification derived from it.

C. Dicta Having Reference to the Flowers of Flowering Plants

15. The polymerous flower structure precedes, and the oligomerous structure follows from it, and this is accompanied by a progressive sterilization of sporophylls.

Quoted matter courtesy of the Missouri Botanical Garden, St. Louis.

16. Petaly is the normal perianth structure, and apetaly is the result of perianth reduction (aphanisis).

17. The apochlamydeous perianth is earlier and the gamochlamydeous perianth is derived from it by a symphysis of the members of perianth whorls.

18. Actinomorphy is an earlier structure than zygomorphy, and the latter results from a change from a similar to a dissimilar growth of the members of the perianth whorls.

19. Hypogyny is the more primitive structure, and from it epigyny was derived later.

20. Apocarpy is the primitive structure, and from it syncarpy was derived later.

21. Polycarpy is the earlier condition, and oligocarpy was derived from it later.

22. The endospermous seed is primitive and lower, while the seed without endosperm is derived and higher.

23. Consequently, the seed with a small embryo (in endosperm) is more primitive than the seed with a large embryo (in scanty or no endosperm).

24. In earlier (primitive) flowers there are many stamens (polystemonous) while in later flowers there are fewer stamens (oligostemonous).

25. The stamens of primitive flowers are separate (apostemonous), while those of derived flowers are often united (synstemonous).

26. The condition of powdery pollen is more primitive than that with coherent or massed pollen.

27. Flowers with both stamens and carpels (monoclinous) precede those in which these occur on separate flowers (diclinous).

28. In diclinous plants the monoecious condition is the earlier, and the dioecious later.

CHAPTER SUMMARY

The generalities of nature that are encountered in the historical sciences are more like the rules of history than like the laws of physics. In particular, they reflect numerous uncontrolled, interacting conditions, and are thus true in a majority of cases only, "other things being equal."

Dollo's law of irreversibility has a special status, for it is the only evolutionary law that is applicable to all fields of historical investigation. Dollo's law reflects the statistical nature of historical processes. Complex structures or complex phenomena in general are historically unique, and cannot recur exactly. The probability of reversion in a complex trait is nonzero, but so vanishingly small that it can be safely ignored. Another interpretation, that evolution is a process inherently programmed to work only in one direction, is now rejected by most evolutionists.

Regularities that deal with the general nature of phylogeny include the "survival of the unspecialized," and several of Bessey's dicta, such as the statement that evolution can be either progressive or regressive. The past decade has seen a new controversy developing between the advocates of Darwinian gradualism and the advocates of "punctuated equilibria." The gradualist model assumes that most morphological change is anagenetic, that paleospecies are arbitrary subdivisions of a lineage, and that anagenesis and cladogenesis are independent. The punctuational model assumes that most morphological change is cladogenetic, that paleospecies succeed one another through cladogenesis plus geographic replace-ment, and that anagenesis does not occur without clado-genesis.

Several of the regularities that have been called evolutionary "laws" are generalizations based upon common trends. Among these are Cope's rule of size in-crease, which states that size tends to increase more often than not in the course of evolution. Williston's rule is also of this kind; it states that in the evolution of serially repeated parts, the number of these parts tends to dimin-ish more often than to increase, and that the remaining parts tend to become increasingly specialized and increas-ingly differentiated from one another. Other common trends include the loss of sense organs in parasites, or the increase in the chromosome numbers of higher plants by polyploidy.

22

Morphology, Phylogeny, and Classification

A. MORPHOLOGY AND EVOLUTION

Morphology is the study of form; one important aim of morphology is to explain the relations between form and function.

An influential pre-evolutionary school of morphology was the so-called **idealistic morphology** that was part of the German tradition of Naturphilosophie (Chapter 3). In their search for a common structural plan (**Bauplan**) that would unite related species, the idealistic morphologists made great advances in the study of form. They found many instances where diverse structures could be interpreted as constructed on a common structural plan, which they interpreted sometimes as a Platonic archetype and sometimes as a divine Plan, or a theme upon which God played many variations. From this school arose the important finding that two distinct types of similarity existed in nature: similarity resulting from the sharing of a common Bauplan, and similarity resulting from other causes, which usually meant adaptive similarity. Sir Richard Owen designated the former by the term homology (defined below in its modern meaning), and the latter by the term analogy (Chapter 19). Classification under the influence of idealistic morphology united those species that shared a common Bauplan, as shown by their possession of homologous structures. The discovery of homologies at all levels was therefore an aim of taxonomy as well as of morphology.

Evolutionary morphology continued many of the same practices as idealistic morphology, but changed their theoretical basis of justification. Homologies were still sought after; indeed, they were sought after with renewed vigor. But homology was now explained by appealing to common ancestry rather than to a common Bauplan.

In its modern, evolutionary meaning, **homology** may be defined as *similarity resulting from common ancestry*. The human arm, the whale's flipper, the horse's foreleg, the bat's wing, the bird's wing, and the pectoral fins of certain fishes are all homologous—their possession of many similar bones, muscles, nerves, and blood vessels, in similar positions and in similar relation to one another, indicates their common ancestry (Fig. 22.1). The wings of bats and birds are also analogous as flying struc-

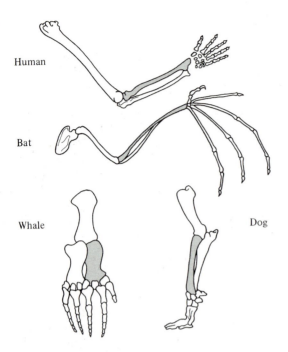

Fig. 22.1 The forelimbs of vertebrates are examples of homologous structures. The radius, for example, has been shaded on all these forelimbs; other nearby bones are homologous among the forelimbs, too.

tures (Chapter 18), a convergent resemblance that they share also with insect wings (see Fig. 18.8), reflecting the similar adaptations of all three.

Homologies are detected by various means: study of internal structure, study of finer details of structure at increasingly microscopic levels, study of embryological development, or study of intermediate forms whenever these are available. These various studies provide *evidence* for the determination of homologies, but the *definition* of homology is on the basis of a common ancestry that is not observed directly. A somewhat broader definition would be that homology is similarity or resemblance that reflects derivation from any common source. It is in this broader sense that we can say that insect mouthparts and insect legs are serially homologous (they are both derived from the same type of primitive appendages) or

that the ovaries and testes of vertebrates are sexually homologous (they are both derived from sexually indifferent gonads).

B. ONTOGENY AND PHYLOGENY

The sequence of an individual's development from ovum to embryo to adult is called its **ontogeny**; the sequence of ancestors in its evolutionary lineage is called its **phylogeny**. A similarity between these two sequences has often been noticed. The fascinating history of this entire subject has been extensively reviewed in an excellent book by Gould (1977), and much of what follows is based on his account.

Even before the theory of evolution became accepted, a similarity was often noticed between the sequence of human ontogeny and the static series or "natural order" specified by the *scala naturae* (Chapter 3). Aristotle and other ancient writers saw this as an indication that man, the microcosm, obeyed the same laws as Nature, the macrocosm. The scientific concept of a parallel between the natural order of the *scala naturae* and the order of development in ontogeny reached its first fruition within the largely German school of Naturphilosophie (Chapter 3).

Karl Ernst von Baer (1792–1876), a disciple of Cuvier, was opposed to any direct parallelism of this sort. An opponent of Naturphilosophie, and later also of Darwinism, von Baer denied any close parallelism between ontogeny and phylogeny, proposing instead four principles of development that are still held to be true today:

1. During embryological development, the general characteristics of any group of animals develop first.
2. From these more general characters the less general and finally the special characters develop. These first two principles indicate that during the embryonic development of any given species, the characters of its phylum become evident first, then those of its class, its order, its family, its genus, and finally its species, in proper hierarchical or taxonomic order.
3. The developmental stages of any given species do not pass through the corresponding developmental

stages of other species, but depart more and more from them. Thus, if the ontogenies of two species were followed side by side, the differences between them emerge gradually and progressively, becoming more and more pronounced as development proceeds.

4. A "higher" species in an embryonic stage of development does not resemble the adult forms of any "lower" species, but only the embryos of that species at a comparable stage of development (Fig. 22.2).

These principles predict that the embryos of two species should resemble one another more than the adults, or, in evolutionary terms, that the embryonic characters should evolve more conservatively than those of the adult. The use of embryonic characters in forming "natural" classifications had been known since at least 1682, when John Ray introduced the modern distinction between monocotyledonous and dicotyledonous plants. Charles Darwin gave his strong support to von Baer's views, and Robert Chambers and Herbert Spencer were among von Baer's other early supporters. Gould (1977) even believes that the first two of von Baer's principles were the original source of Herbert Spencer's idea that all change proceeds from the more general to the more specialized, or, in Spencer's terms, from homogeneity to heterogeneity.

Despite Darwin's agreement with von Baer's perceptions, the more forceful expression of a relation between ontogeny and phylogeny within an evolutionary context was to be Ernst Haeckel's (Chapter 5). Unfortunately, Haeckel rejected von Baer's more correct theory and instead restored the earlier belief in a direct parallelism. An inveterate coiner of words (including *ontogeny*, *phylogeny*, and *ecology*), Haeckel gave the name **recapitulation** to this parallelism, which he also explained in what he called the "**biogenetic law**":

> Ontogeny recapitulates phylogeny. The history of the individual is a brief recapitulation of the history of its race. Ontogeny is the direct consequence of phylogeny, and phylogeny is the mechanical cause of ontogeny.

No modern biologist supports Haeckel's theory in quite this extreme form. In particular, no modern biologist would now support the claim that phylogeny is the mechanical cause of ontogeny.

But Haeckel was less interested in discovering the mechanical cause of ontogeny than he was in reconstructing phylogenies. Despite the "Biogenetic law," Haeckel recognized that there were certain exceptions, where a literal interpretation of an ontogeny as an account of phylogeny would lead the observer astray: larval adaptations to larval conditions (which he called **caenogenesis**), and unequal rates of development (which he called **heterochrony**). The mammalian placenta, for example, is caenogenetic, as are the embryonic membranes of other vertebrates, or the larval adaptations of caterpillars. As for heterochrony, or the acceleration and retardation of development of one characteristic relative to another, Alpheus Hyatt and Edward D. Cope suggested that development was in general accelerated, that this acceleration was responsible for the recapitulation of phylogeny, and that the unequal acceleration of different characters resulted in heterochrony of one character relative to another.

According to Gould (1977), it was the coming of Mendelian genetics and experimental embryology that robbed Haeckel's recapitulation theory of any claim to truth by depriving it of a mechanical basis. Also, retarded development and accelerated development could be explained with equal facility, and there was therefore no reason for giving any primacy to either one. The embryologists therefore led a march away from Haeckel's theory, meanwhile coining dozens of names for various possible relationships between ontogeny and phylogeny. Gould has suggested that these terms can be reduced to a bare minimum as described in Box 22–A. The *results* of such evolution include recapitulation, in which the early stages of descendants resemble the adult stages of their ancestors, and the opposite phenomenon of **paedomorphosis**, in which adult descendants resemble the youthful stages of their ancestors. The *causative processes* are **acceleration**, the speeding up of development, and **retardation**, the slowing down of development in the course of evolution. But the causative processes and their results do not always correspond. Paedomorphosis, for example, can result either from accelerated sexual development (**progenesis**) or from retarded somatic development (**neoteny**). Recapitulation can also result either from accelerated somatic development or from retarded sexual development (delayed maturity).

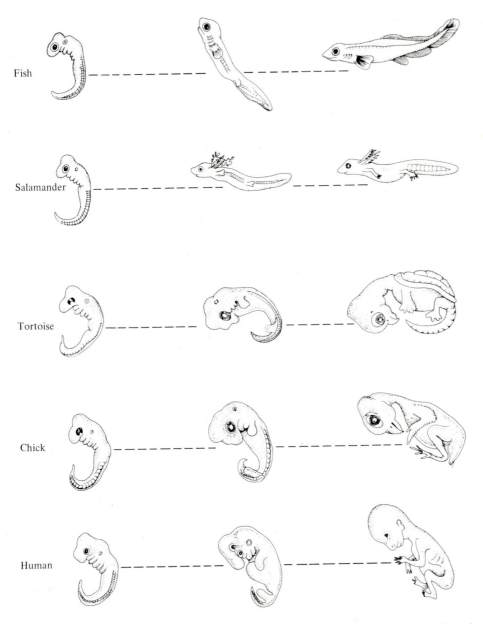

Fish

Salamander

Tortoise

Chick

Human

Fig. 22.2 Comparison of embryological development in selected vertebrates. Note that the early stages of all these species possess gill slits and a tail. Reprinted with permission from Aaron O. Wasserman, *Biology*, 2e, © 1975. Reading, Massachusetts: Addison-Wesley (Fig. 16.17).

BOX 22–A *The Evolution of Ontogeny: Processes and Their Results, Based on the Terminology of Gould (1977)*

There are two major *processes* in the evolution of ontogeny:

a. The origin of novel characteristics, which may take place at any time during the life cycle as the result of adaptation through variation and natural selection. The origin of purely larval or embryonic adaptations to purely larval or embryonic conditions is called **caenogenesis**, and forms one special instance.

b. **Heterochrony**, or alteration in the rate of development of any organ. The two types of heterochrony are: (1) **Acceleration**, or speeding up of development; and (2) **Retardation**, or slowing down of development.

The *results* of the above processes should carefully be distinguished from the processes themselves. For example, by the speeding up or slowing down of sexual development and of

"somatic" (all other) development independently, we may distinguish four separate possible results:

Causative change	Morphological result
Acceleration of somatic development	Recapitulation (by acceleration)
Acceleration of sexual development	Paedomorphosis by truncation (**progenesis**)
Retardation of somatic development	Paedomorphosis by retardation (**neoteny**)
Retardation of sexual development	Recapitulation by prolongation (hypermorphosis)

Organisms are subject to natural selection at all stages of their life cycle. The embryonic or larval adaptations may evolve by processes that are identical to the evolution of adaptation in adult forms. Since the action of genes during development is often cumulative, genes that act earlier generally produce more drastic changes, including changes in the action of many later-acting genes. Since most of these changes are harmful, it follows that changes in earlier stages of development should in general be selected against more strongly than changes in later ones. It is for this reason that the evolution of characters expressed early in development has been more conservative than the evolution of characters whose expression appears later in development. Conservation of early-acting genes in the course of evolution thus explains von Baer's principles as the consequence of successive gene actions.

But the timing of development is also subject to variation and hence to natural selection. Characters that benefit the organism as soon as they appear in ontogeny are thus selected to appear earlier and earlier, resulting in the acceleration of a character and a possibly recapitulatory sequence. On the other hand, characters that are in-

adaptive cause less reduction in fitness if they appear later, and thus natural selection will cause such characters to be increasingly delayed (retarded) in their development. The resultant appearance of numerous maladaptations in aged and postreproductive individuals leads to their senility and eventually their death. Aging may thus be the consequence of the delayed action of harmful genes or the accumulated effects of genes that have beneficial effects early in life in return for harmful effects later, after reproduction has already taken place.

C. MODERN THEORIES OF CLASSIFICATION

In recent years, the philosophical basis of classification has been thrown into question. Evolutionary taxonomists have been called upon to justify each of their beliefs, and taxonomists in general have been asked to justify many of their practices, some of which had previously been justified only by custom or intuition. The attacks have been launched from two opposing camps that have been called phenetic and cladistic, two terms that are defined below. The debate has been carried out in several books, including those by Hennig (1966), Sokal and Sneath (1963), Sneath and Sokal (1973), Mayr (1969), Jardine and Sibson (1971), and Crowson (1970). There have also been numerous journal articles on the subject, especially in the pages of *Systematic Zoology*. Particularly good reviews of the major issues may be found in Mayr (1965) and in Hull (1970, 1979).

Debates concerning the theoretical basis of classification are often couched in a strange language in which the names of certain taxonomic concepts are given specially restricted meanings. Some of these terms and their definitions are as follows:

DEFINITIONS

Taxon: A group of species or populations at any hierarchical level. Examples include Mammalia (a class), *Drosophila* (a genus), Saxifragales (an order), or *Homo sapiens* (a species).

Category: A rank or hierarchical level to which taxa may be assigned. Examples include the class, the order, the family, the genus, and the species.

Hierarchy: An arrangement of categories in their proper sequence: Kingdom, phylum (or division), class, order, family, genus, species.

Classification: An orderly arrangement of organisms into taxa, including the assignment of each taxon to a category. Also, the process of arriving at such a classification.

Taxonomy: The study of classification, including its theoretical basis.

Systematics: The study both of classification and of the evolutionary processes that give rise to new taxa or produce divergence between existing taxa.

Phenetics: The study of phenotypic resemblances and differences among organisms.

Phenetic classification: A classification based entirely on phenetics.

Phenetic (or numerical) taxonomy: A taxonomic theory that insists that all classifications should be strictly phenetic.

OTU (Operational Taxonomic Unit): A taxon or unit at the finest level recognized in a particular classification. The units whose grouping into successively larger taxa is desired, and whose classification is therefore under discussion.

Character: Any variable characteristic of the organisms in question. Examples include color, number of toes, presence or absence of horns, length of tarsus.

Character state: A particular condition or value of a character, such as red, three toes, horns absent, or 7 mm.

Primitive (or plesiomorphous): The ancestral character state for a given taxon and character; the condition of that character in the ancestor of the entire taxon.

Derived (or apomorphous): Any character state not ancestral within a given taxon.

Monophyletic taxon: A taxon that includes the common ancestor of all species included in the taxon. A taxon with a single evolutionary origin (Fig. 22.3).

Polyphyletic taxon: A taxon that does not include the common ancestor of all species included in the

Fig. 22.3 Diagrams illustrating the meaning of the terms holophyly, paraphyly, and polyphyly, for comparison with the definitions presented in the text. Ashlock (1968) and most other workers (but not Hennig) would use the term monophyly to cover both holophyly and paraphyly.

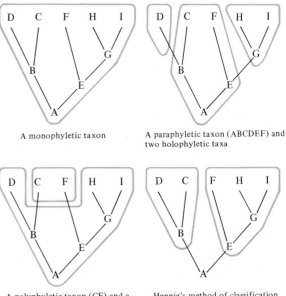

A monophyletic taxon

A paraphyletic taxon (ABCDEF) and two holophyletic taxa

A polyphyletic taxon (CF) and a paraphyletic taxon

Hennig's method of classification into two holophyletic taxa, with the ancestral species (A) included in neither group

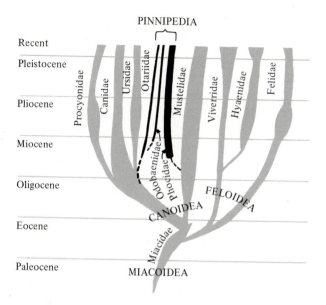

Fig. 22.4 A possibly polyphyletic taxon, the Pinnipedia (seals, sea lions, and walruses) is recognized as a suborder of the order Carnivora by most mammalian taxonomists. The true ancestry of these aquatic mammals within the order Carnivora is uncertain, but if the phylogeny shown here turns out to be correct, phylogenetic taxonomists would be forced to abandon the taxon Pinnipedia as unnatural, but phenetic taxonomists might continue to recognize it.

taxon. A taxon with multiple or heterogeneous origins (Figs. 22.3, 22.4).

Holophyletic taxon: A monophyletic taxon that includes the common ancestor and all of its descendants without exception (Fig. 22.3). *Important:* Hennig and certain other authors have suggested that the term ''monophyletic'' be restricted to this case only.

Paraphyletic taxon: A monophyletic taxon that is not holophyletic. An ancestral species and some but not all of its descendants (Fig. 22.3).

Cladistics: The study of the branching sequences of phylogenies, including the reconstruction of these phylogenies on the basis of the comparative distribution of character states among the descendants.

Cladistic classification: A classification based entirely on cladistics, ignoring the degree of morphological change (anagenesis) occurring between branching points.

Cladistic taxonomy: A taxonomic theory that insists upon cladistic classifications.

Dendrogram: Any treelike diagram.

Phenogram: A dendrogram depicting phenetic similarities only.

Cladogram: A dendrogram depicting cladistic relationships only (Fig. 22.5).

Please read the definitions above and make sure that you understand them before proceeding; also review

Fig. 22.5 A portion of a phylogeny and its classification according to Hennig's cladistic system. Reprinted with permission from Malcolm C. McKenna *in* Luckett and Szalay, eds. *Phylogeny of the Primates*, 1975. New York: Plenum Publishing.

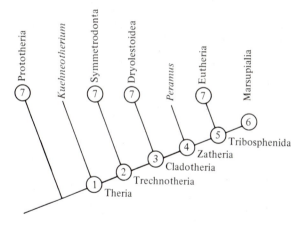

Subclass Theria
  Superlegion Kuehneotheria
  Superlegion Trechnotheria
    Legion Symmetrodonta
    Legion Cladotheria
      Sublegion Dryolestoidea
      Sublegion Zatheria
        Infraclass Peramura
        Infraclass Tribosphenida
          Supercohort Marsupialia
          Supercohort Eutheria

from Chapter 17 the meaning of anagenesis and cladogenesis, and from Chapter 18 the meanings of analogy and convergence.

Some of the divergent viewpoints of the three current schools of taxonomy are summarized in Box 22–B.

Phenetic Taxonomy

Phenetic classification, though simpler in theory, has presented the greater philosophical challenge, and represents a more distinct break with evolutionary classification as it has been practiced since Darwin's time. Despairing of attempts to reconstruct phylogenies in the absence of a complete fossil record, or to base classifications on any but perfectly and completely known phylogenies´ (which, of course, do not exist), the pheneticists have sought to ignore phylogeny entirely, and to base classifications upon resemblance alone. Furthermore, since all modern taxonomists of whatever school acknowledge that classifications based on single characters are undesirable, the pheneticists have taken inspiration from the late eighteenth-century French nominalists such as Georges Buffon (1707–1788, in his early writings) and Michel Adanson (1727–1806). The pheneticists seek to base their classifications on numerous characters haphazardly chosen without regard to their preconceived importance, a program that Adanson had long advocated, and which Buffon advocated in his early writings. This program was difficult to carry out in the eighteenth century, but the advent of modern, high-speed computers has made the comparison of numerous characters at a time a workable possibility.

Phenetic taxonomy is also called "numerical taxonomy." Two of its strongest advocates, Robert Sokal and Peter H. A. Sneath, have written two books and many articles advocating their methods and urging them upon taxonomists of all groups of organisms.

It is the expressed aim of the pheneticists to remove all matters of subjective judgment from taxonomy, and to base classifications on purely objective and quantifiable principles. Yet, phenetic taxonomists use many methods, and the resulting classifications are not always the same. The choice among methods is unfortunately subjective at the present state of the art, and more methodological research is needed.

A common fault of phenetic classifications is that convergence is not distinguished from resemblance inherited through common ancestry. Polyphyletic taxa, de-

fined earlier in this section and illustrated in Figs. 22.3 and 22.4, may therefore occur in phenetic classifications. Another problem of phenetic classifications is that larval and adult characteristics may produce widely differing results. When immature and adult stages are included in the analysis separately, the youthful stages may appear closer, for they resemble each other more than do the adult forms, according to von Baer's principles (see the previous section of this chapter, and also Fig. 22.2).

Phylogenetic Taxonomy

The two remaining schools of thought in taxonomy share several viewpoints in common, especially with regard to the importance of phylogeny in classification. We may therefore designate these two schools together by the term **phylogenetic taxonomy**, using that term rather broadly. In fact, each of the two schools of phylogenetic taxonomy claims that it alone has proper claim to that title, and each claims Charles Darwin as an advocate of their position (on which point see Ghiselin and Jaffe, 1973).

Phylogenetic taxonomists of both schools agree that phylogenies can be reconstructed in many cases, that phylogenies should be used in classification wherever possible, that taxonomic statements are scientific hypotheses subject to falsification by later findings, that convergent characters have no place in taxonomy, that polyphyletic taxa are inadmissible, and that characters should be weighted phylogenetically. This last issue is one of the most explosive among those that divide the phenetic and phylogenetic taxonomists.

On the issue of whether characters should be "weighted" in favor of those characters least likely to be convergent, phenetic taxonomists have generally taken the position that all characters are of equal value, and that all should therefore be weighted equally. The **equal weighting of characters**, they argue, is necessary if subjectivity in taxonomy is to be avoided. (Buffon raised a similar argument against Linnaeus' sexual system of plant classification about 200 years ago.) The phylogenetic taxonomists have usually replied that some characters evolve more conservatively, and therefore carry a higher phylogenetic information content, while certain other characters are more liable to convergence and other misleading phenomena, and some are so easily changeable as to be taxonomically worthless. The **phylogenetic weighting of characters**, on which these phylogenetic taxono-

BOX 22–B *Comparison of the Viewpoints of Three Schools of Taxonomy*

Phenetic taxonomy	Phylogenetic taxonomy	
	ECLECTIC (EVOLUTIONARY) TAXONOMY	CLADISTIC TAXONOMY
Classification should reflect resemblance only.	Classification should reflect both phylogeny and degree of morphological difference.	Classification should reflect phylogeny only.
All characters should be weighted equally.	Characters should be weighted unequally according to their phylogenetic content.	
Convergent resemblance should enter into classification with the same status as any other resemblance.	Convergent resemblance should be clearly distinguished from resemblance by homology and should be excluded from classifications.	
All categories are of arbitrary rank.	The species category is of nonarbitrary rank, and deserves a special place in taxonomy; other categories are of arbitrary rank. The special status of species is important to both these schools.	
(This issue is irrelevant because phylogeny has no business in classification.)	If a common ancestor gives rise over a short period of time to several taxa, it is still permissible to rank these taxa in different categories if they are of unequal size, or if one was more successful than the others.	If a common ancestor gives rise over a short period of time to several taxa, then these taxa must all be of equal rank regardless of their size or later success.
(This issue is irrelevant in theory; practice largely reaches results similar to evolutionary taxonomists.)	An unbranching lineage can be arbitrarily divided into several species that differ from one another about as much as do contemporaneous, sympatric species.	An unbranching lineage is all one species, however much morphological change occurs.
Age of taxa is irrelevant because classification should be based on resemblance only, not phylogenetic history.	Taxa originating in a certain time period need not be at same rank. Of two taxa originating together, one may have produced only a single species, while the other differentiated into an entire phylum or class.	All taxa originating in a single geological period (e.g., Permian) must be placed at same rank; if an order originated in the Permian, then all taxa originating in the Permian are orders.
Taxa may be either monophyletic or polyphyletic because phylogeny is irrelevant to classification.	Taxa should not be polyphyletic but may be paraphyletic; both types of monophyletic taxa are permissible.	Taxa should be holophyletic only.
Taxonomic statements are measurements, subject only to refinement.	Taxonomic statements are hypotheses, subject to falsification by additional evidence.	

mists insist, would give added importance to characters of high phylogenetic significance, and correspondingly low importance in classification to characters that are phylogenetically insignificant, inconsistent, or misleading. Fine, say the pheneticists, but we are not interested in phylogenetic reconstructions, only in assessing similarity among organisms, which makes phylogenetic weighting irrelevant.

Cladistic Taxonomy

Of the two phylogenetic schools of taxonomy, the more extreme school is properly known as **cladism. Cladistic taxonomy** was promulgated by the German entomologist Willi Hennig (1913–1976), who used the term "phylogenetische Systematik," or phylogenetic systematics. Cladism has since won the support of many entomologists and paleontologists, as well as that of scattered workers in other fields. The basic principles of cladism

are that primitive character states can usually (or always) be identified, that characters can be used to reconstruct phylogenies, and that phylogenies so reconstructed can serve as the direct basis for classification without any other input. Any treelike diagram depicting this type of phylogeny is called a **cladogram;** cladistic taxonomists often draw the lines of their cladograms to have the same slope (in either direction), indicating that they carry no information about time duration, degree of morphological divergence, or the like.

In Hennig's original system of cladistic taxonomy, each lineage that undergoes phyletic evolution (anagenesis) without branching is considered a single species. Upon branching or cladogenesis, the species will give rise to two monophyletic taxa. A cladistic classification will therefore divide each taxon into a single species, which is left unclassified at lower ranks, and two coordinate "sister group" taxa at the next lower categorical level (Fig.

BOX 22-C *The 22 Possible Phylogenetic Trees Involving Three Species*

To a strict cladist, only the diagrams in the top row would be acceptable as cladograms. If only A, B, and C (but no intermediates) were known, a cladist would interpret all the family trees in each vertical row to look like the cladogram at the top of that row. The same cladogram can therefore result from several topologically distinct phylogenies.

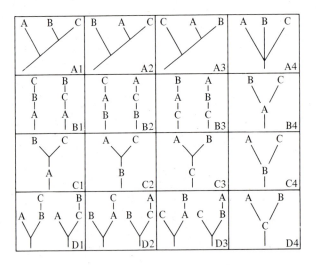

After Platnick, 1977. Courtesy of Society of Systematic Zoology.

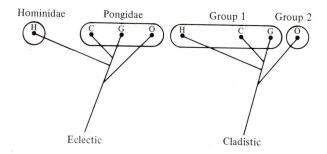

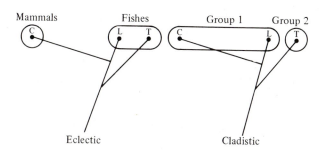

Fig. 22.6   Two examples of the differences between eclectic and cladistic practices in classification, with time on the vertical axis and morphological divergence shown horizontally. Cladists consider only the order of branching points, but eclectic (evolutionary) taxonomists also take into account the amount of morphological divergence. Upper diagram: the cladist would classify humans (H), gorillas (G), and chimpanzees (C) in one taxon and orang-utans (O) in another, instead of the more traditional families Pondigae (apes) and Hominidae (humans). Lower diagram: the cladist would classify cows (C) and lungfish (L) in one taxon and trout (T) in another. Reprinted by permission from Luria, S. E., Gould, S. J., and Singer, S., *A View of Life,* Menlo Park, California: The Benjamin/Cummings Publishing Company, Inc., 1981, p. 681, Fig. 28.12.

22.5). A strictly cladistic classification according to Hennig's principles could just as easily be written as a phylogeny, since there is an exact one-to-one correspondence between phylogeny and classification under Hennig's system if all portions of the phylogeny are known, or if both known and unknown taxa are classified. A given phylogeny has only one possible representation as a strictly cladistic classification under Hennig's system. But if portions of a phylogeny are missing, and only the known portions classified, then a given classification may correspond to more than a single phylogeny (Box 22–C).

Eclectic (Evolutionary) Taxonomy

Eclectic or evolutionary taxonomy has the support of well-established taxonomists in nearly all fields. In fact, prior to 1960, the other two schools of taxonomy were hardly known at all, except to a few pioneering specialists.

Though evolutionary taxonomists agree with cladists on a number of important issues, the two differ on several points, illustrated in Fig. 22.6. Evolutionary taxonomists have always insisted that classifications must be a compromise between phenetics and cladistics, and between theory and convenience. Classification, say the

evolutionary taxonomists, must be consistent with phylogeny, but it may also reflect morphological divergence, evolutionary success, or the size of taxa. If a common ancestor gives rise to several taxa over a short period of time, these taxa may be ranked in different categories if they are of unequal size, or if one was more successful than the others. The birds are legitimate as a class of equal rank as the reptiles, even though birds evolved from one group of reptiles only (Box 22–D). The fact that a given classification may reflect more than a single phylogeny is illustrated in Box 22–E.

D. PHYLOGENETIC ANALYSIS: THE ART OF RECONSTRUCTING PHYLOGENIES

Two of the three contemporary schools of taxonomy believe that classification should reflect phylogeny; thus classifications can only be as good as the phylogenies upon which they are based. The dissension of the phenetic school is based principally upon their belief that phylogeny can never be known with certainty, and that it is therefore folly to base any classifications upon it. Phylogenetic analysis is thus of crucial importance to two of the

BOX 22-D *The Phylogeny of Reptiles and Their Descendants*

In (a), the phylogeny is used to illustrate a traditional evolutionary classification. The class Reptilia is divided into six subclasses (indicated by abbreviations), while the birds and mammals each represent a distinct class. In (b), the same phylogeny is used to illustrate a strictly cladistic classification, in which the traditional Reptilia is broken up into six taxa of equal rank. Under this scheme, the mammals would become a small part of the Class Synapsida, and the birds would become part of the Class Archosauria.

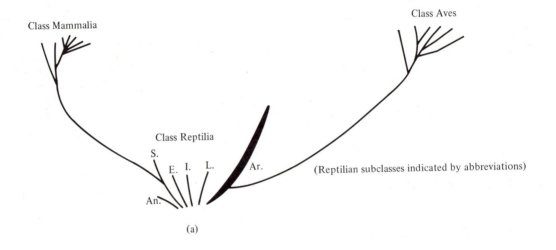

(a)

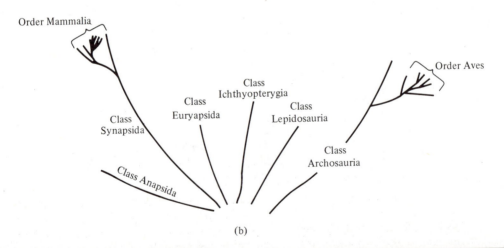

(b)

BOX 22–E *Possible Representations of the Taxonomic Position of the Fossil Genus* Oreopithecus

Some Possible Opinions as to the Phylogeny and Distinctiveness of *Oreopithecus*, and Some Classifications Consistent with those Opinions

Opinions as to phylogeny of *Oreopithecus*	Opinions as to distinctiveness of *Oreopithecus*	
	a. LESSER	b. GREATER
I. In or near hominine ancestry.	A. Hominoidea 　Hominidae 　　Homininae 　　　*Oreopithecus*	B. Hominoidea 　Hominidae 　　Oreopithecinae 　　　*Oreopithecus*
II. Divergent from early hominids after separation from pongids.	B. Hominoidea 　Hominidae 　　Oreopithecinae 　　　*Oreopithecus*	C. Hominoidea 　Oreopithecidae 　　*Oreopithecus*
III. Divergent from early pongids after separation from hominids.	D. Hominoidea 　Pongidae 　　Oreopithecinae 　　　*Oreopithecus*	C. Hominoidea 　Oreopithecidae 　　*Oreopithecus*
IV. Divergent from common stem of pongids and hominids.	C. Hominoidea 　Oreopithecidae 　　*Oreopithecus*	E. Oreopithecoidea 　Oreopithecidae 　　*Oreopithecus*

From Simpson, *in* Washburn, ed., 1963. *Classification and Human Evolution*. Chicago: Aldine Publishing Company.

---

major schools of contemporary taxonomy, while the third school holds the impossibility of phylogenetic analysis as one of its fundamental beliefs. To an evolutionist, phylogenetic analysis may be considered as an end in itself, independent of any use in classification.

Two of the most persistent problems in phylogenetic analysis are the detection of convergence and the determination of which of several character states is most primitive.

The Role of Paleontology

Paleontology plays an essential role in phylogenetic analysis by the discovery of fossil forms. These may simply add more species to be classified, thus adding to the complexity of phylogenetic analysis. But when many fossil species become known in any group of organisms, at least some of them usually begin to give clues as to the ancestry of one or more previously recognized groups. Of course, a complete fossil series would in fact constitute a phylogeny without much need for analysis, but few if any fossil series are in fact so complete as this.

Those who do not themselves practice paleontology are often surprised to discover that fossil species connecting previously known taxa are not perfectly intermediate in all their characters. The principle of mosaic evolution (Chapter 17) explains why a transitional form between two groups may have some of its characters still in a primitive and others in an advanced state. In the case of actual common ancestors, there may frequently be a mixture of derivative character states resembling one of their descendant groups, other derivative character states resembling other descendants, and still other characters that are more primitive than in any descendant group, and therefore different from them all. Most surprising are those cases where a common ancestor seems to possess specializations, and when these are discovered it is often tempting to suggest that their possessor was not in fact the ancestor, but represented some sterile side branch. Several paleontologists, notably A. S. Romer, have cautioned against making this assumption too rashly: no animal, said Romer, could make a living by simply being an unspecialized ancestor.

The Role of Comparative Morphology

Though paleontology may provide the essential time element, the further interpretation of fossils is usually based on a study of their morphology. Indeed, in the absence of fossils, phylogenetic analysis must perforce be based on a comparative study of living species.

One of the most important tasks of the comparative morphologist is the careful examination of structures that are similar in appearance, in an attempt to discover whether they are in fact homologous or merely analogous. If the same character state is independently (convergently) derived in several groups, it is then considered of low utility in classification. If, on the other hand, a complex internal pattern of homology is discovered, then the common possession of this derived character state may be used to define a monophyletic group. The implicit assumption here is that a complex structural change is not likely to have taken place more than once independently in exactly the same way.

These same methods may be applied to the great problem of the detection of convergence. In the absence of direct fossil evidence, convergence is usually detected through the study of comparative morphology, including comparative embryology. Homologous character states, derived in common, may be expected to be very similar in their internal structure, their relation to nearby structures, and their embryonic development. Convergent structures, on the other hand, should possess none of these detailed similarities, or should possess them only by chance, or as the result of the convergence itself.

On Distinguishing the Primitive from the Advanced

Often in the study of contemporaneous organisms there are two or more character states displayed, of which either may be considered the more primitive. Deprived of a time dimension, are there any rules that may be followed in reconstructing the phylogeny as a sequence of events, or in establishing "time's arrow"? To this, the most important problem in phylogenetic reconstruction, we cannot provide infallible answers, for morphology is a naturalistic science and Nature does not always test our hypotheses in the way that we might wish. Yet, there are several guidelines that may aid us in establishing which of several characters is more primitive, and which is derivative:

1. If a sufficient fossil record is known to establish the relative antiquity of character states, the older state may be assumed to be more primitive. Of course, one's confidence in this criterion depends upon one's confidence in the completeness of the fossil record.

2. If one character state represents a greater homogeneity (e.g., many similar parts) and another a greater heterogeneity (e.g., differences among the parts), the more homogeneous state may be assumed to be primitive and the more heterogeneous state advanced (Williston's rule; compare also von Baer's principles earlier in this chapter). The known exceptions to this principle can usually be explained by simplification through the loss of parts.

3. If one character state appears early in ontogeny, only to become transformed during embryonic development into a different state, then the earlier-appearing state is usually the more primitive and the later-appearing state the more derivative. This is Haeckel's principle of recapitulation, to which there certainly are exceptions, but on the whole the principle works for the majority of characters examined in this manner.

4. If two or more states of the same character are found within a given taxon, and one of these states has widespread distribution in coordinate related taxa, that widely distributed character state is likely to be more primitive. Thus, among primates or rodents, the character state that is widely distributed among other mammals may be assumed to be more primitive, and within any particular angiosperm order, the most primitive character is that which achieves the widest distribution among

other angiosperms. There are in each class a very small number of orders (the most primitive ones) where this principle would not hold, but these orders can usually be identified by other criteria, and they would in any event be the same orders regardless of which character was being discussed.

5. If a given taxon contains species possessing two or more character states, a_1, a_2, etc., and if one of these occurs together with a character state known to be primitive in a different character (b_1), then that state is also primitive. Thus, having determined among angiosperms that a certain type of flower is primitive, one can also conclude

that the most primitive type of stem anatomy or leaf anatomy is the type that occurs in those groups having the most primitive flowers. (Mosaic evolution, however, creates exceptions to this rule.)

6. As an extension of this last principle, if a set of uniquely derived character states forms a nested series (Wilson, 1965), then the innermost character state of the nest is the most highly derived, and the successively more inclusive character states (working out from the center of the nest) are increasingly primitive (Fig. 22.7).

7. Hennig recognizes an additional principle, that if the location of the ancestral home of a taxon is known,

Fig. 22.7 Wilson's consistency test, based on unique, unreversed character states. One may establish many hypotheses declaring certain character states to be unique and unreversed, therefore defining holophyletic taxa. But the taxa defined in this way must form a nested series for the hypotheses to be mutually consistent. In this example, the seven taxa of marsupials show three character states forming a nested series. From this, one can infer that each of these three character states is unique and unreversed (i.e., each evolved only once and was not subsequently lost). One can also infer the phylogeny shown here, in which each of these character states defines a holophyletic group, and in which Australian occurrences evolved first, conjoined toes second, and reduction in the number of incisor teeth last.

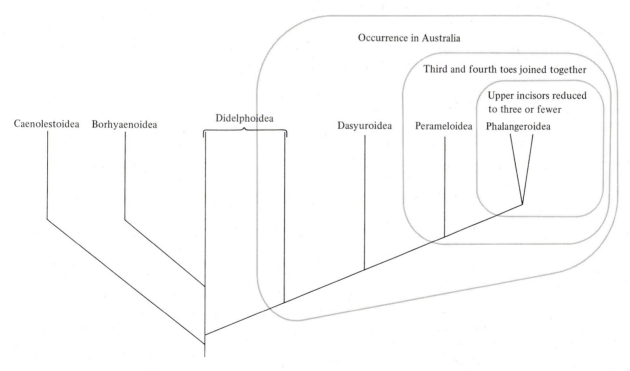

then those character states that are found among species still inhabiting the ancestral homeland are more primitive. I would consider this to be a particular case of principle 5 (in which character *b* is the geographic distribution). There are some striking exceptions to this principle in ancient groups, and the principle is particularly inapplicable to taxa that have long ago abandoned their ancestral homeland, or whose ancestral homeland can no longer be ascertained with any certainty.

All of the foregoing principles admit exceptions; they hold only *ceteris paribus*—all other things being equal. Thus, it becomes important to confirm the primitiveness of a certain character state using as many independent criteria as possible.

In particular groups of organisms, hypotheses asserting one or another character state to be primitive are of long standing. Bessey's dicta (Box 21–E) are perhaps the best known example of such hypotheses.

The Role of Molecular Biology

To those who seek independent confirmation of phylogenies determined by any of the foregoing methods, molecular biology has provided some new methods of its own. All of these methods involve comparisons between complex macromolecules, and in each case the data gathered are in the form of distances or similarity coefficients between species. The great advantage of these methods lies in the fact that each determination summarizes hundreds of individual comparisons at once, producing a figure that represents the degree of similarity between the macromolecules compared.

There are three such methods currently in use:

1. Immunological comparisons (comparative serology). In this method, which compares proteins, a distantly related species is used to produce antibodies against one of the species being compared. A chicken, for example, might be immunized against a human protein in an attempt to compare various primate species. The reaction of the chicken antiserum against human protein is quantified and set at 100%; the reactions of chicken antiserum against proteins of other primate species is then expressed as percentages of this value. For example, the reaction to chimpanzee protein may be 99% as strong as the reaction to human protein, while the reaction to protein from a squirrel monkey might be only 85% as

strong. From this comparison, one could conclude that the protein tested was far more similar in structure in humans and chimpanzees than it was between humans and squirrel monkeys. In one of the earliest applications of this technique, Moody *et al.* (1944) immunized a chicken against rabbit serum (a mixture of blood proteins), and compared the reactions of the chick antiserum against a variety of mammalian sera. Their findings showed that rabbits were no more related to rodents than they were to many other mammals, and that they were closest to sheep and other members of the order Artiodactyla (among the species tested). Additional methods of immunological comparison have been developed in recent years, including immunodiffusion, immunoelectrophoresis, and others, in addition to the more traditional "precipitin" method described above.

2. Sequence analysis of proteins. This method again compares proteins, usually nonenzyme proteins such as hemoglobin, myoglobin, or cytochrome *c*. Conceptually, the method is very simple: the amino acid sequence of a widespread protein is determined for a variety of species, and the sequences thus obtained are compared. Closely related species will in most cases differ from one another by fewer amino acid residues than will distantly related species.

3. Nucleic acid hybridization. Nucleic acid (generally DNA) is obtained from various species, and the double-stranded molecules are in each case separated into single strands by heating them. A mixture of single-stranded DNA from two species is then allowed to cool slowly, and from the speed of formation of double-stranded molecules one can calculate the percentage of base pairs that are the same in the two species.

In all the above methods, we should expect phylogenetically related species to have closely similar macromolecular structures, which in fact has been observed in a number of cases. There are, however, some known exceptions: higher primate globulins and higher primate albumins, for example, have evolved according to two different patterns.

Macromolecular comparisons between species are useful in assessing the relative biochemical similarities among them. These should be expected to fall into patterns established on the basis of phylogenetic analysis: phylogenetically close species should also be biochemically close. This provides an independent check on the results of phylogenetic analysis.

If the fossil record of a group permits their phylogeny to be known with a high degree of certainty, then an independent check on the methods of macromolecular systematics can be obtained. Molecular-based phylogenies that have thus far been published (e.g., Fig. 19.3) show a high degree of correspondence with phylogenetically based classifications. But the early claims of some biochemical taxonomists, that their macromolecules evolve at constant rates, seem to be overstated (Chapter 19). The neutralist–selectionist controversy (Chapter 12) is being argued again at the molecular level.

CHAPTER SUMMARY

Morphology, the study of form, formerly explained homology as resulting from a common structural plan. Homology is now defined as similarity resulting from common ancestry.

Different species often resemble one another more as embryos than they do as adults because the embryonic characters are more conservative in evolution. Von Baer explained that, as development proceeds, the characters appear in proper taxonomic sequence, and the embryos of higher forms resemble the corresponding embryonic stages of lower forms rather than their adult stages. Haeckel, on the other hand, asserted a direct parallelism: embryologic development, or ontogeny, was a brief recapitulation of phylogeny. Later scientists have explained this recapitulation, and various other evolutionary alterations of embryonic development, as resulting from

heterochrony, the speeding up or slowing down of one character's development relative to another's.

Of the three contemporary schools of taxonomy, the phenetic school stands farthest apart in insisting that phylogeny should be ignored and classification should be based on resemblance only. The other two schools are both phylogenetic—they seek to have classification reflect phylogeny. Of these, the cladistic school believes that only the branching sequences of phylogenies should be used in classification, but not the degree of morphological divergence that follows branching, nor the subsequent evolutionary success of a group. Cladistic taxonomists insist that classifications and phylogenies should be topologically identical. Evolutionary or eclectic taxonomists believe that classification should reflect all known information about taxa, including their time of divergence from one another, their rate of morphological change, and their subsequent evolutionary success. Classification should thus reflect morphological distinctness as well as phylogenetic origin.

In the analysis of phylogenies, two recurrent problems are the detecting of convergence and the determination of which character state is most primitive. Comparative morphology, including comparative embryology, provides much of the evidence upon which such determinations are based; to this must be added important evidence gained from the knowledge of the geological and geographical history of the group in question. Once the primitive state of a particular character is determined, the primitive states of other characters can often be deduced more easily. An independent check on phylogenetic analysis can be provided via molecular biology, since closely related species should be expected to have a greater degree of similarity in the molecular structure of their nucleic acids and proteins.

23

Biogeography

A. PATTERNS OF GEOGRAPHICAL DISTRIBUTION

Biogeography is the study of the geographic distribution of species. The traditional emphasis of biogeography was on the description of animal and plant distributions and on the summarizing of these descriptions in terms of biomes (or life zones) and zoogeographic or phytogeographic regions. Traditional explanations of these distribution patterns have tended to emphasize historical events of migration and extinction. In recent decades, increasing emphasis has been placed on elucidating and quantifying the causal factors responsible for both present and past distributions.

Biomes

Geographers and ecologists since Alexander Humboldt have used vegetation types to characterize the life zones of different climatic regions. The major ecological zones, or **biomes** (Fig. 23.1) are each characterized by a set of climatic conditions, principally temperature, rainfall (or snowfall), and the extent to which these vary seasonally. These conditions in large measure determine the type of vegetation that dominates the landscape and thus the type of habitat provided for animals and other organisms.

Northernmost of the world's major biomes is the **tundra,** a region of low temperature and rainfall characterized by the absence of trees and the presence of permanently frozen soil (permafrost) not far beneath the surface. In latitudes just south of the tundra lies the **northern conifer forest** biome, or **taiga,** dominated by cone-bearing evergreen trees such as spruce and fir. Here, the winters are milder and rainier. **Temperate deciduous forests** have milder temperatures in both winter and summer; they are dominated by broadleaf trees that shed their leaves in winter, including maples, oaks, hickories, beeches, ashes, elms, and many others. **Temperate grasslands,** also called **prairies** or **steppes,** have slightly less rainfall than temperate deciduous forests, with a similar range of temperatures. These grasslands have few trees (usually confined to the banks of streams) and are dominated instead by grassy (herbaceous) plants.

In low temperate latitudes on the west coasts of continents, a **broad sclerophyll biome** is characterized by mild, wet winters, warm, dry summers, and short to

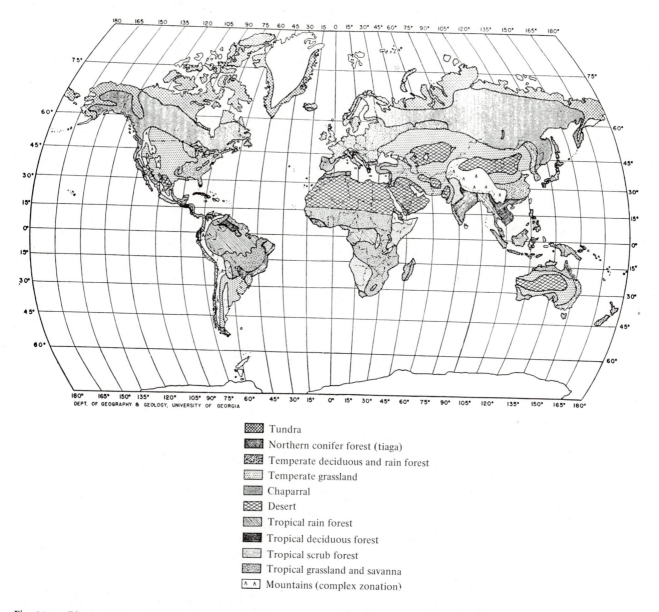

Tundra
Northern conifer forest (tiaga)
Temperate deciduous and rain forest
Temperate grassland
Chaparral
Desert
Tropical rain forest
Tropical deciduous forest
Tropical scrub forest
Tropical grassland and savanna
Mountains (complex zonation)

Fig. 23.1 Biomes of the world. From Eugene P. Odum, *Fundamentals of Ecology*, 2e, copyright © 1953 and 1959, W. B. Saunders Company. Reprinted with permission of Holt, Rinehart and Winston.

moderate-sized, drought-resistant evergreen shrubs. This biome occurs around the Mediterranean Sea, in Chile, in southwest Africa, in western Australia, and in southern California, where it is called ''chaparral.'' At slightly lower latitudes, characteristically 30° north or south of the Equator, regions of very low and intermittent rainfall may support a **desert biome** characterized by desert shrubs (sage, creosote) and cactuslike succulents, usually with bare ground between them.

Tropical biomes are distinguished from one another largely by the amount of rainfall. The **tropical rain forests** of equatorial regions have the most rainfall of all and the most luxuriant vegetation, with the largest array of species, especially of broadleaf evergreens. Trees in these rain forests are often tall, and their upper, leafy stories typically form a continuous or nearly continuous canopy that excludes much of the sunlight from reaching ground level. The **tropical deciduous forests** of Asia and other continents have much more seasonal variation in rainfall;

bamboos and other deciduous species commonly occur here. In regions of somewhat more moderate (but seasonally distributed) rainfall, a **tropical thorn forest** or **scrub forest** biome is characterized by thorny and often twisted, low- to moderate-sized trees. The greatest seasonal variation among tropical biomes is seen in **tropical grassland** or **savannah** biomes, where dry and rainy seasons differ markedly. These grasslands have few widely spaced trees but many grasses, and support mixed faunas containing many herbivore species.

From one continent to another, the biomes of the world are largely discontiguous (except in high northern latitudes). Areas of the same biome on different continents are inhabited by ecologically equivalent but often unrelated species. For example, the cactuslike succulents occurring in desert biomes may be true cacti, euphorbs, or members of another family, but they are always similar in appearance and in their ecological niches. The same is true of the flightless birds that inhabit certain grass-

Fig. 23.2 Parallelism of latitudinal and altitudinal life-zones. Reprinted with permission from Charles K. Levy, *Elements of Biology,* © 1978. Reading, Massachusetts: Addison-Wesley (Fig. 22.7).

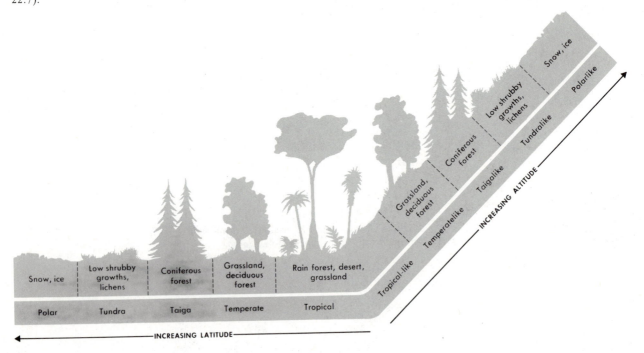

lands: they may be true ostriches in Africa, rheas in South America, or emus in Australia, a fact that much impressed Charles Darwin on his visits to these places. Convergence between ecologically equivalent species, or between entire faunas and floras on widely separated continents, has been the subject of repeated investigations.

The mountains of the world have a complex pattern of zonation, with altitudinal life zones or biomes roughly analogous to the latitudinal life zones or biomes of non-mountainous regions (Fig. 23.2).

Zoogeographic and Phytogeographic Regions

In contrast to the life zones, which are regions of similar climate on separate continents, it is often useful to characterize the fauna or flora of certain broad geographical regions such as continents. Charles Darwin was deeply impressed by the resemblance of South American species to one another despite differences in climate within the continent. In fact, this was probably a major determinant of his belief in evolution, while the observation of strikingly different faunas inhabiting the same life zones of different continents caused him to reject the environmental determinism that had given rise to the only other evolutionary theories with which he was familiar at the time. Equally impressed with these same facts was Alfred R. Wallace, the codiscoverer of natural selection, who may also be considered the father of systematic zoogeography. With his contemporaries, principally P. N. Sclater, Wallace devised a series of **zoogeographic regions** and provinces based largely on the distribution of mammals and birds. The zoogeographic regions recognized today (Fig. 23.3) are based on their system.

The northern temperate biota of the New World constitute the **Nearctic Region,** including Canada, the United States, and most of Mexico. The northern tem-

Fig. 23.3 Zoogeographic regions of the world, based largely on the distribution of mammals and land birds. Reprinted with permission from Charles K. Levy, *Elements of Biology,* © 1978. Reading, Massachusetts: Addison-Wesley (Fig. 22.11).

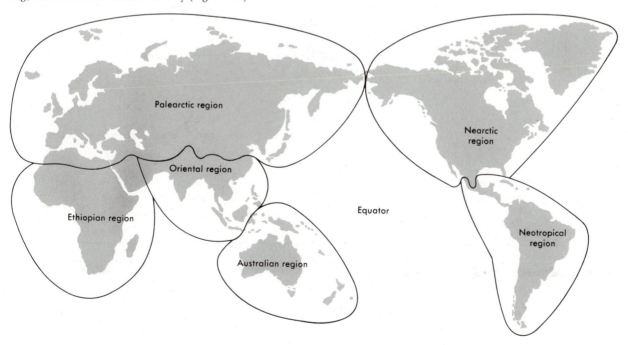

BOX 23-A *Some Taxa Endemic (or nearly so) to Certain Zoogeographic Regions*

Nearctic Region
bowfins (Amiidae)
moon-eyes (Hiodontidae)
ameiurid catfishes (Ameiuridae)
turkeys (Meleagrididae)
moles (Talpidae, also in Palaearctic region)
pronghorns (Antilocapridae)
mountain beaver (Aplodontidae)
pocket mice (Heteromyidae)
pocket gophers (Geomyidae)
Solenodontidae

Neotropical Region
gymnotid eels (Gymnotidae)
xantusiid lizards (Xantusiidae)
rheas (Rheiformes)
tinamous (Tinamiformes)
new world vultures
screamers
cracids
hoatzins
trumpeters
sun bitterns
oilbirds
motmots
toucans (Toucanidae)
Furnarioidea
opossum rats (Caenolestidae)
platyrrhine monkeys (Platyrrhina)
sloths, anteaters, etc. (Edentata; one
  species reaches Texas)
peccaries (Tayassuidae)

caviomorph rodents (Caviomorpha; one
  extends into North America)
noctilionid bats (Noctilionidae)
leaf-nosed bats (Phyllostomatidae)
vampire bats (Desmodontidae)
disk-winged bats
furipterid bats (Furipteridae)

Palaearctic Region
pandas (*Ailuropoda*)
moles (Talpidae, also Nearctic)
unique genera of deer, etc.

Ethiopian Region
bichir fish (*Polypterus*)
chameleons (Chameleonidae)
ostriches (Struthionidae)
secretary bird (Secretariidae)
turacos (Turacidae)
golden moles (Chrysochloridae)
elephant-shrews (Macroscelididae)
galagos (Galaginae)
lemurs (Lemuridae, Indrisidae)
aye-aye (Daubentoniidae)
aardvarks (Tubulidentata)
hyraxes (Hyracoidea)
hippopotamus (Hippopotamidae)
giraffes (Giraffidae)
tenrecs (Tenrecidae)
unique genera of antelopes, monkeys,
  apes, etc.

unique families of rodents (Pedetidae,
  Anomaluridae,
  Barythergidae, Ctenodactylidae)
Welwitschia (see Fig. 20.2)

Oriental Region
uropeltid snakes (Uropeltidae)
xenopeltid snakes (Xenopeltidae)
sea snakes (Hydrophidae)
flying lemurs (Dermoptera)
tree shrews (Tupaiidae)
tarsiers (Tarsiidae)
unique genera of monkeys
Ginkgophyta

Australian Region
ceratodontid lungfish (Ceratodontidae)
cassowaries
emus
megapodes
owlet-frogmouths
lyre-birds
honey eaters
birds of paradise
bowerbirds
egg-laying mammals (Monotremata)
dasyures (Dasyuridae)
kangaroos (Macropodidae)
phalangers, koalas (Phalangeridae)
marsupial ''mole'' (Notoryctidae)
bandicoots (Peramelidae)
wombats (Phascolomidae)

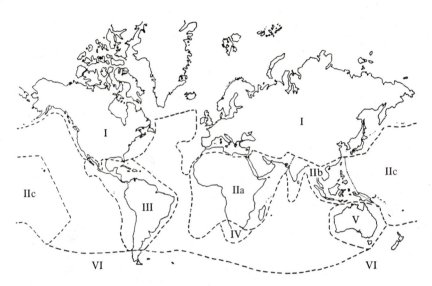

Fig. 23.4 Good's phytogeographic regions of the world, based largely on the distribution of flowering plants. Floristic "kingdoms" recognized by Good are I, Boreal; II, Paleotropical (with a, African; b, Indo-Malaysian; c, Polynesian); III, Neotropical; IV, South African; V, Australian; VI, Antarctic. From Eugene P. Odum, *Fundamentals of Ecology*, 2e, copyright © 1953 and 1959, W. B. Saunders. Reprinted by permission of Holt, Rinehart and Winston.

perate biota of the Old World constitute the **Palaearctic region,** including Europe, extreme north Africa, and Asia north of the Himalayas. These two regions are very similar in many respects, and are often grouped together under the term **Holarctic.**

The tropical and south temperate biota of the New World, from Cape Horn to central Mexico, constitute the **Neotropical region.** The tropical and south temperate biotas of the Old World are divided among three zoogeographical regions: the **Ethiopian region,** including Africa south of the Sahara plus the Arabian peninsula (which was once part of Africa), the **Oriental region,** including the Indian subcontinent and Southeast Asia, and the **Australian region,** including Australia, New Zealand, and the Pacific islands.

Each major zoogeographic region can be characterized by the presence or absence of certain major families or orders (Box 23–A). Most characteristic are those taxa that are confined, or **endemic,** to a particular region.

The zoogeographic regions just outlined are based on the distribution of mammals and birds, but have proven useful in describing the distribution of other land animals such as reptiles and even freshwater fish. Botanists have long preferred a somewhat different set of **phytogeographic** regions for plants, shown in Fig. 23.4. The

distribution of insects accords much better with the phytogeographic than with the zoogeographic regions, apparently reflecting a general dependence of insects upon plants ecologically.

B. THE EVOLUTION OF GEOGRAPHICAL DISTRIBUTIONS

Dispersal and Colonization

The migration of species from one region to another, or from a continent to an island, depends upon the **dispersal** ability of the species. The smallest unit by which a species can establish a new population is known as a **propagule;** some species are capable of dispersing their propagules far and wide, while others are more narrowly confined. A propagule may consist of a single seed or fruit, a single hermaphrodite, a pregnant female, or a pair capable of mating. The organisms dispersed as propagules are not necessarily adults; in many species the eggs or young larval stages are far more easily transported or dispersed.

Dispersal of terrestrial species is most difficult across broad stretches of salt water. Species that fly even short distances can be carried by strong winds once they are airborne and carried over much greater distances in unusual cases. Species that live in or near trees can sometimes be carried out to sea on floating islands of vegetation that occasionally break off the banks of large rivers. The significance of these ''rafts'' as a dispersal mechanism has been debated for over a century. Most of them certainly sink, but a number have been spotted far out at sea, complete with a varied fauna and flora; Darwin observed one such ''floating island'' while aboard the *Beagle*. The seeds of many plants can be dispersed in the digestive tracts of birds, or clinging to the feet of birds. Some groups of animals disperse very poorly across salt water: large mammals, especially ungulates, are too heavy to survive on floating islands, and amphibians in all stages of their life cycles are soon killed by contact with sea water.

The occasional dispersal of organisms across formidable barriers has been called **sweepstakes dispersal** by Simpson. Sweepstakes dispersal characteristically is a chance phenomenon: arboreal or flying species are far more likely to be transported in this way, while ungulates and amphibians are never transported in this manner. In Simpson's analogy, many species hold ''tickets'' in a sweepstakes drawing for which each ticket carries a low probability of dispersal. Some species hold more tickets than others, while some hold none at all. Those species holding more tickets have a greater chance of ''winning,'' but occasionally an unlikely species (holding fewer tickets) will cross before a more likely one. It is characteristic of sweepstakes dispersal that only a small percentage of the fauna is transported, that the colonizers arrive at varying times, and that the transported species represent a biased sample of the fauna from which they are derived, with a higher proportion of small and/or arboreal species and few if any ungulates.

Continents may also be connected by **land bridges,** such as the isthmus of Panama that currently connects North America and South America. It is characteristic of land bridges that they transport whole faunas or major faunal samples, that nearly the entire fauna crosses soon after the land bridge is made available to them, that dispersal occurs in both directions, and that the sample of species dispersed is a representative rather than a biased sample. In Pleistocene times, a land bridge connected Asia to Alaska across the Bering Strait. Camels and horses were among the species that crossed this bridge from North America to Asia; mammoths were among the species that migrated from Asia to North America.

Successful colonization depends not only upon dispersal, but also on hardiness and ecological flexibility, both of which will determine success or failure in establishing a colony. Species with a broad range of ecological tolerance are more likely to establish new colonies than are species with more narrowly defined ranges of tolerance.

Plate Tectonics

In the last two decades, it has become increasingly evident that the slow movement of continental land masses relative to one another has influenced the ease or difficulty of intercontinental dispersal. The majority of today's geologists support the theory of **plate tectonics,** which teaches that the Earth's crust is fragmented into less than a dozen large plates (Fig. 23.5), moving slowly across the outer layer of the Earth's core. Evidence for the relative motion of the plates comes from the apparent spreading of the sea floor: the mid-ocean ridges contain very young rock, and the age of the rocks increases steadily as one progresses further from such ridges. Other evidence of continental movement is paleomagnetic: many rocks contain a remnant magnetism that can serve as a paleomagnetic compass, pointing to the former location of the North Pole. Paleomagnetic compasses from different continents do not always point to the same location, however, and the data are best explained by assuming that the continents have drifted relative to one another. The remarkable correspondence between the Atlantic coastlines of Africa and South America, as if they were parts of a giant jigsaw puzzle, provides further evidence, and the correspondence is even better if you match the continental shelves instead of the present coastlines.

The theory of **continental drift,** as it was originally called, was first proposed by the German geologist Alfred Wegener in 1910. Central to Wegener's theory was the belief that Africa, India, Australia, South America, and Antarctica had once been united into a single land mass known as **Gondwanaland,** while the continents of North America and Eurasia (excuding India) were similarly united into a land mass called **Laurasia.** The geophysicists, knowing no mechanism to account for such continental drifting, attacked the theory during its early years,

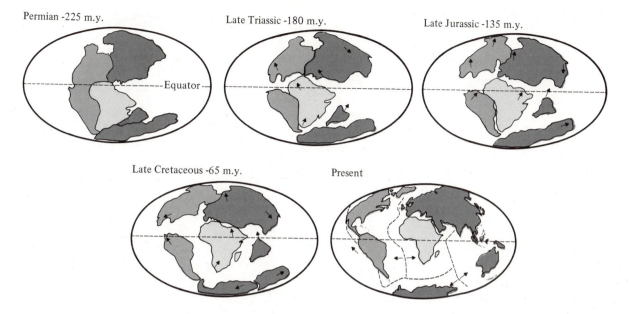

Fig. 23.5   Relative movement of the Earth's continents through time. Adapted with permission from Aaron O. Wasserman, *Biology,* 2e, © 1975. Reading, Massachusetts: Addison-Wesley (Fig. 17.7).

though a few paleontologists and one South African geophysicist did support it.

A symposium on the subject was held in 1928, in which geophysicists were united in their condemnation of Wegener's theory, all the more because Wegener had claimed that continental drift had persisted into the Tertiary and even the Pleistocene. Only with the discovery of paleomagnetic compasses, especially in the 1950s, and the discovery of the spreading of the sea floor (Fig. 23.6), also in the 1950s, did the geophysicists alter their position and come to support the theory (though not Wegener's original time frame). Convection movements within the Earth's mantle are now believed responsible for the relative movements of continental plates.

The fossil evidence supporting the theory of plate tectonics is rather varied. Faunal evidence, especially from the Paleozoic Era, suggests that the several southern continents may once have been united or interconnected. The peculiar fossil reptile, *Mesosaurus,* is found in freshwater deposits only in Africa and South America. The mammal-like reptiles of the Permian in Africa are nearly identical to other mammal-like reptiles recovered from South America and even from Antarctica. A distinctive flora, containing the fossil seed fern *Glossopteris,* is present in Paleozoic deposits of several southern continents, including South America, Africa, India, Australia, and Antarctica. These are only some of the kinds of fossil evidence that support the theory of plate tectonics; more such evidence is being found each year, as the phylogenetic history of one group of organisms after another is reconciled to the theory.

According to a widely accepted version of plate tectonic theory, the Paleozoic Era was marked by the movement of continent-sized plates toward one another, until they were all united to form a supercontinent known as Pangaea. At or near the close of Permian time, Pangaea broke in half, forming a northern supercontinent (Laurasia) and a southern one (Gondwanaland). Laurasia later broke apart to form the modern continents of North America and Eurasia (excluding the Indian subcontinent); Gondwanaland broke up into the modern continents of South America, Africa (including Arabia), Australia, Antarctica, and India. India drifted northward

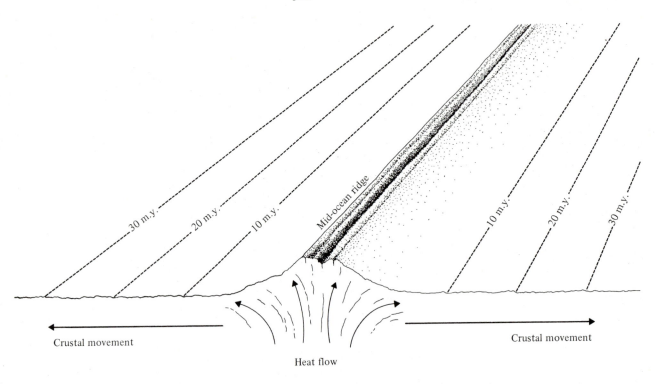

Crustal movement

Heat flow

Crustal movement

Fig. 23.6 Evidence for plate tectonics in the spreading of the ocean floor. The youngest rocks on the ocean floor are found along the mid-ocean ridge, and successively older rocks are found at greater distances from this ridge.

until it collided with Eurasia, forming the Himalayan mountain range where the two met. In fact, most of the world's major mountain ranges are believed to have formed in similar fashion, under the pressure of two continental plates that had come together. Each major mountain range represents a seam joining two plates that have been pressed together. The boundaries of ancient continental plates are thus marked by modern mountain ranges, and the mountain-building episodes (orogenies) should coincide in age with the further movement of these plates toward one another following their collision. In this one coherent theory are explained many varied and seemingly unrelated observations concerning animal and plant distributions, ocean floor spreading, mountain buildings, and many other geological and paleontological phenomena.

Cenozoic Faunas and Continents

There is now sufficient evidence from the distribution of Cenozoic mammalian faunas to suggest a somewhat different relationship among the world's continents than we see at present. Here is a case where our knowledge of past faunas preceded our knowledge of the actual geology. In fact, reconstructions of Cenozoic continental geography have always relied heavily on the evidence provided by the distribution of fossil mammals.

The northern continents of North America and Eurasia have been much more thoroughly explored by paleontologists than the southern continents. The Eocene faunas of Europe and North America have been found strikingly similar, with hardly a family known from either continent that is not also present in the other. In fact, many families (and even some genera) known at first from only one of these continents were subsequently discovered in the other. The faunas are so similar as to argue strongly for a land connection between these two continents, even though the exact location and extent of this connection have never been firmly established. Throughout the Tertiary, many new mammalian families arose, and several of these (generally the more successful ones) rapidly spread across both continents. The horses, for example, evolved principally in North America, but they invaded Eurasia on at least three different occasions (Chapter 17). Many other families, however, failed to cross between North America and Europe, and gradually the faunas of these two land areas diverged.

The land faunas of Asia are less perfectly known than those of either Europe or North America, but great progress has been made in recent years as the result of the Polish-American Museum Expeditions. The degree of faunal similarity between Asia and other northern continents was initially less than that between Europe and North America, with fewer genera common to both Asia and either Europe or North America. The uniqueness of Asian faunas increased from Cretaceous to Paleocene to Oligocene times.

During the Pleistocene, much of the Earth's water became frozen into glaciers, and the sea level correspondingly fell. The Bering and Panama land bridges became exposed, allowing a greater degree of faunal interchange between North America and adjacent continents. To the north, the exposure of additional land allowed such Old World species as mammoths to migrate to North America, while an even greater number, including horses and camels, crossed in the opposite direction from North America to Asia.

The southern continents have shown a much greater degree of faunal isolation throughout Tertiary time. Australia, for example, has apparently always been an island continent. Its unique mammalian fauna contains an endemic order (Monotremata, the egg-laying mammals), as well as a wider variety of marsupials (pouched mammals) than are present on any other continent. Except for bats, which can fly or be blown across many ocean barriers, the native placentals of Australia include only a number of rats and a single carnivore, the dingo, which was probably introduced by the early human inhabitants. The Pleistocene faunas of Australia were equally unique, with

marsupials predominating, just as they now do. The few Tertiary faunas known to us also indicate a preponderance of marsupials, and there is no evidence of Cenozoic faunal interchange with any other continent.

South America was likewise an island continent during most of the Tertiary. Six orders of mammals (Edentata, Notoungulata, Litopterna, Astrapotheria, Pyrotheria, Xenungulata) were confined, or nearly confined, to South America during Tertiary times, and two more orders (Marsupialia and Condylarthra) were represented in South America by endemic families in the earliest Tertiary faunas. Two further orders (Primates and Rodentia) invaded South America during the Tertiary, probably by sweepstakes dispersal, and developed into endemic suborders in both cases. Among all mammalian families known during the Tertiary in South America, only one (the marsupial family Didelphidae) was also known with any certainty on any other continent prior to the Pleistocene. Perhaps equally impressive was the total absence of entire orders, including the very successful Carnivora, Insectivora, Artiodactyla, and Perissodactyla, from South American faunas. This faunal evidence alone strongly suggests that South America was an island during Tertiary times, a hypothesis further corroborated by studies of Central American geology, including the finding of many marine fossils. The close of the Tertiary brought with it the exposure of the land bridge across Panama, and a great mixing of North American and South American faunas ensued, with migrations in both directions. Horses, mastodonts, carnivores (jaguars, etc.), rabbits, and artiodactyls (llamas, alpaca, etc.; also deer) were among the southward migrants, while the northward migrants included armadillos, porcupines, ground sloths, and perhaps the familiar opossum. Competition resulting from this faunal mixing brought about the extinction of many families and even entire orders. Much more of the South American fauna succumbed to extinction than did the North American fauna.

African faunas prior to the upper Eocene are known from fragmentary remains only. In the upper Eocene and lower Oligocene, we encounter the first large faunal sampling, in the Fayûm deposits of Egypt. The rich faunal assemblage preserved in these deposits suggests that Africa, too, had been faunally isolated from the rest of the world for one or more epochs, for several endemic families and orders are present. Four entire orders

(Sirenia, Hyracoidea, Proboscidea, Embrithopoda) were at the time endemic to Africa, and a fifth (Cetacea) was nearly so. Surely all had evolved in Africa during some long period of isolation from the rest of the world. Several other orders of mammals (Insectivora, Primates, Rodentia, Artiodactyla among them) were represented by endemic families or suborders. Many entire orders (notably Perissodactyla and Carnivora) and suborders (the more typical rodents, advanced artiodactyls, and prosimian primates) were absent, despite their faunal diversity elsewhere in the world. The Fayûm faunas suggest that Africa had developed as an island continent during early Tertiary times. The few faunal fragments of earlier epochs in Africa offer no evidence against this hypothesis.

Miocene and later faunas of Africa continue to show heavy representation of the groups present in the Fayûm, but with increasing admixture of Palaearctic faunal elements, notably the advanced Artiodactyla, beginning in the Pliocene. Certain groups of African ancestry, notably the Proboscidea (elephants and their relatives) and the catarrhine primates, meanwhile spread to Eurasia, and in a few cases to other continents as well. The African fauna of today shows a mixture of truly native elements (monkeys, apes, elephants, hyraxes, elephant-shrews, hippopotami, etc.) and various descendants of groups that invaded from the north during the second half of the Cenozoic (including zebras, lions, leopards, giraffes, and a host of antelopes).

Relict and Disjunct Distributions

In the study of the geographic distribution of living taxa, we may encounter taxa with a **disjunct distribution** composed of two or more widely separated occurrences. The mammalian family Tapiridae, for example, occurs today in Southeast Asia (Sumatra, Malaya) and in South and Central America. This particular disjunct distribution is also a **relict distribution**—the remains of a once more extensive continuous distribution, which in this case included both North America and Europe. A similar example is provided by the giant salamanders (family Cryptobranchidae), now living in eastern Asia and eastern North America, but known as fossils also from Europe and Nebraska. The lizard family Iguanidae is common throughout the Western Hemisphere and also occurs in Madagascar and a few Pacific islands; this is

surely a relict distribution, even though former occurrence on Eastern Hemisphere continents has never been conclusively proven.

A relict distribution need not be disjunct. The once geographically extensive reptilian order Rhynchocephalia occurs today only in New Zealand. The most primitive suborder of rodents occurs today only in mountainous regions of western North America. Relicts often occur on islands and similar havens, but not always. They may occur on the periphery of a formerly more extensive range, as in the camel family (formerly North American, but now confined to high altitudes in South America and arid portions of Southern Asia and Northern Africa). In some cases (the camels or rhynchocephalians, for example), the group may be extinct in its ancestral homeland, but this is not always true. Elephants (Elephantidae) and apes (Pongidae), both formerly more widespread across Eurasia, are today confined to relict distributions that include their ancestral homeland of Africa as well as portions of tropical Asia.

In certain cases, an apparently disjunct distribution may, upon taxonomic reexamination, prove instead to be an indication that the geographically separated groups are in fact only distantly related. Porcupines, for example, occur in both Africa and South America (and have extended their ranges northward into Europe and North America, respectively), but examination of their anatomy has convinced Woods (1972) and many other mammalogists that the African and South American forms are unrelated. Similar remarks can be made for the Old World and New World monkeys, and for a group of supposedly related insectivores that includes the tenrecs of Madagascar, the genus *Potamogale* in the rivers of West Africa, and the two peculiar genera *Nesophontes* and *Solenodon* in the Caribbean. In this last case, as in many others, there has been much disagreement as to whether the disjunct populations are truly related or not.

Of Cradles, Museums, and Vicariance

In attempting to reconstruct past phylogenetic or faunal history, it often is necessary to ask where a particular taxon originated. Each taxon is, of course, a unique situation. The major question here is whether or not certain generalizations can be made: whether certain places are more likely to serve as places of origin, or as havens for geographic relicts, and whether from a distribution alone we can ever determine the direction in which migration took place.

Some taxa with well-known fossil records have an identifiable place of origin; others do not. It is the fossil record, for instance, that identifies Africa as the ancestral home of the elephants and their relatives. Widespread, cosmopolitan taxa, on the other hand, offer few clues to their places of origin: the common bats (Vespertilionidae), fruit flies (Drosophilidae), or colon bacteria (*Escherichia*) could well have originated almost anywhere (but see one qualification of this below), and the rats and mice (Muridae) might have originated in nearly any portion of the Old World. The oldest fossil occurrence may indicate a place of origin, as, for example, Africa in the case of elephants and their relatives. Among larger taxa, however, the earliest occurrence may simply be an accident of discovery: the oldest known primate comes from Montana, and the oldest known bat from Wyoming, but these are not necessarily the places where the bats or primates originated.

In certain relatively young taxa, the area of greatest diversity is also the area of origin, while the geographically peripheral forms are also the most divergent anatomically. In older taxa, a greater diversity may evolve on the periphery of the former range. In taxa with relict distributions, the area of origin is usually hard to reconstruct, as it need not bear any particular relationship to the present distribution. An endemic taxon may have originated in the place where it now occurs (probably true of many but not necessarily all Australian marsupials), or it may be a geographic relict that originated elsewhere (as in the case of the reptilian order Rhynchocephalia, formerly more widespread but now confined to New Zealand).

Faunal diversity tends to be greater on larger land masses (e.g., continents) than on smaller ones (e.g., islands), and it also tends to be greater in the tropics than in colder regions. Migration more often occurs from continent to island, or from large area to small, than in the opposite direction. It is partly for this reason that we may reject the Hawaiian Islands as the ancestral homeland of the Drosophilidae (fruit flies), though their greatest present diversity is there. Most taxa, especially of land vertebrates, are poorly represented in Hawaii, as on most islands.

The botanist G. Ledyard Stebbins (1974) has raised the question as to whether the tropics, especially the Old World tropics, are a "cradle" of origin for many taxa, or merely a "museum" in which the most primitive representatives of each taxon are preserved as relicts. Charles Darwin (1859) and William Diller Matthew (1915) believed strongly in the origin of major taxa in temperate regions, and their subsequent migration toward the tropics. Darlington (1957), on the other hand, favors the Old World Tropics as the richest source area for new taxa, with subsequent migration outward toward either pole. The question may never be resolved, but a majority of zoologists seem to favor Darlington's position. Valentine (1973), meanwhile, has shown that the rich faunal diversity of tropical as compared to temperate faunas has persisted at least since the Paleozoic.

Croizat (1958) and his followers have rejected the practice of assuming that taxa occurring on southern continents (like Africa or South America) must have originated elsewhere. According to their theory of "vicariance," present geographical distributions result largely from the differentiation of taxa *in situ*, with migrations playing a minor role only. As a corollary of this theory, families resembling one another, but inhabiting different continents, are assumed to be unrelated, and their resemblances analogous only. Debates concerning the vicariance theory have therefore tended to center about the taxonomy of particular families and other taxa having this type of geographic distribution. One example of such taxa are the Old World and New World porcupines (Hystricidae and Erethizontidae), which were formerly judged to be closely related. But Woods (1972) has reexamined the anatomy of these rodents, and has found many deep-seated differences. Presumably this shows that the ancestry of these groups was independent, and, in particular, that neither was descended from the other. There is thus no need to ask whether porcupines originated in the Old World or the New, for they originated independently in both places.

But the theory of vicariance also provides a methodological approach to zoogeography that challenges the traditional reliance on dispersal as a means of explaining distributions. The first step in this approach is cladistic analysis (Chapter 22), and the identification of monophyletic groups. Next, for each monophyletic group that occurs on two different land masses, a "tract" is drawn connecting those land masses. Some tracts, reflecting relict distributions, may be unique to a particular groups—tapirs, for example, are among the few taxa common to tropical South America and tropical Southeast Asia. But certain land masses, like North and South America, are common to many monophyletic taxa; these land masses will be connected by numerous tracts, which are said to constitute a "generalized tract." The theory of vicariance postulates that the generalized tracts determined by this procedure will identify land masses that were formerly connected to one another. The theory of plate tectonics provides an important confirmation of the vicariance theory, in that the land masses identified by the theory of plate tectonics as formerly connected are the same as those now connected by generalized tracts. See Nelson and Rosen (1981) for an excellent symposium on vicariance theory.

Island Biogeography

The study of islands has always fascinated evolutionists and zoogeographers alike. Islands have often played a critical role in the development of both evolutionary and ecological theories. Darwin's belief in evolution has often been attributed to the observations he made on the Galapagos Islands, while it is equally if not more clear that Alfred Russell Wallace was profoundly influenced by his studies among the islands of Southeast Asia (especially those which now constitute Indonesia) and by his earlier trip to Brazil with Henry Walter Bates. Darwin was most impressed with the observation that island faunas tend to resemble those of the nearest continent, indicating that the continent had served as a source from which the island fauna was derived. More recently, islands have provided ecologists with "natural experiments" of various kinds. For example, the "ecological release" of a prey species—its ecological reaction to the absence of predators—can be tested simply by finding it on an island where its predators do not occur. The concepts of r- and K-selection (Chapter 11) were first developed by population biologists studying the distribution of r-selected versus K-selected species on islands.

Islands like Great Britain or New Guinea would be connected to a continental mainland if the sea level were dropped less than 100 meters, as it was during the Pleistocene or Ice Age. These islands are called **continental islands,** and their faunas and floras tend to be closely similar to those of the adjacent mainland, but perhaps a bit less varied. Endemic species, found nowhere else, tend to be very few.

In contrast to these are the truly **oceanic islands,** which have never been connected to any mainland. Geologically, these islands are beyond the limits of any continental shelf. In the past, some overenthusiastic "bridge builders" advocated the former existence of land bridges, which subsequently sank beneath the sea, to account for the biota of these islands. There is however no mechanism to account for the sinking of a land bridge of the required dimensions, not to mention its utter disappearance by vanishing into the deep ocean without a trace. The small and unrepresentative faunas and floras of such islands are furthermore inconsistent with such a hypothesis, and are instead much more consistent with dispersal of species to these islands by the repeated action of rare events, i.e., by "sweepstakes" dispersal. The frequency with which various species reach oceanic islands, and the greatest distance from any mainland that each attains, are both in direct proportion to the ability of propagules to be transported great distances across salt water by occasional means. Species that cannot tolerate transoceanic dispersal (notably frogs and ungulates) do not occur on oceanic islands unless they have been introduced by human activities. The faunas of oceanic islands are characteristically impoverished in inverse proportion to the dispersal abilities of each taxon.

Reptiles, freshwater fish, mammals other than ungulates, and nonoceanic birds may occur on oceanic islands, but they will represent only a small and unrepresentative sample of those occurring on continents at the same latitude. Those families and genera that do occur will frequently diverge from their relatives elsewhere. Adaptive radiation is common among those groups that do reach oceanic islands, particularly when the islands occur in archipelagos, or when they are sufficiently large to permit ecological diversity—the finches of the Galapagos, the *Drosophila* species of Hawaii, or the lemurs of Madagascar may serve as examples. In the absence of competitors or of major carnivores, many species (or entire families) may survive on islands long after their mainland relatives suffer extinction. For these several reasons, the proportion of endemic taxa on oceanic islands, especially on archipelagos or large islands, is characteristically high.

A new era in the study of island biogeography was initiated by MacArthur and Wilson (1963, 1967), who have approached the diversity of island biotas as an exercise in mathematical model-building. The springboard of their theory is the so-called species-area curve, which can be plotted as a straight line on fully logarithmic coordinates (Fig. 23.7). Within a given taxonomic group, the number of species (S) on an island is (approximately) related to the size of the island by the equation

$$S = CA^z$$

where A represents the area of the island and C and z are constants. The value of C depends on the units chosen, on the particular taxonomic group, and on the islands chosen for study. The value of z, on the other hand, is generally in the range 0.20 to 0.35, regardless of the islands or the taxonomic group chosen.

The number of species present on any island at equilibrium depends rather obviously on the rate at which new species arrive on the island (or arise by speci-

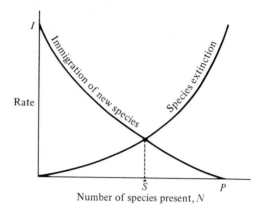

Fig. 23.8 Immigration and extinction curves. A stable equilibrium is achieved where these two curves intersect, for at that point the rate of immigration and the rate of extinction just balance each other. From Robert H. MacArthur and Edward O. Wilson, *The Theory of Island Biogeography* (copyright © 1967 by Princeton University Press), Fig. 7, p. 21. Reprinted by permission of Princeton University Press.

Fig. 23.7 The species-area curve for the herpetological fauna (amphibians and reptiles) of the West Indies. Curves of this type usually follow equations of the form $S = CA^z$. From Robert H. MacArthur and Edward O. Wilson, *The Theory of Island Biogeography* (copyright © 1967 by Princeton University Press), Fig. 2, p. 8. Reprinted by permission of Princeton University Press.

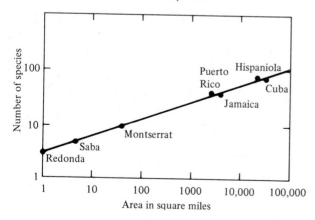

ation), and on the rate at which species become extinct on that island (Fig. 23.8). The rate of arrival depends on the dispersal abilities of the species in question, and on the distance of the island from the mainland (Fig. 23.9). The rate of extinction depends on many factors, but the ecological diversity within the island, which in turn depends on the island's total area, is undoubtedly one of them (Fig. 23.10). For islands of a given area, those nearest the mainland should have a higher rate of new arrivals and should thus support a larger species diversity (Fig. 23.11). At the same time, islands of differing size but equal distance from the mainland should have approximately equal rates of arrival, but the rate of extinction should be higher on the smaller and less diverse islands.

If a number of islands are arranged so that they can be used as "stepping stones," the probability of colonization is thereby increased. MacArthur and Wilson (1967) have shown that, of two possible source areas ("mainlands") competing for the colonization of an

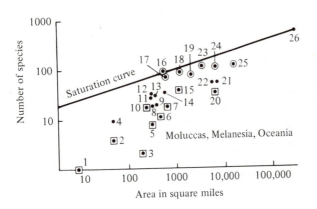

Fig. 23.9 Species-area relationship for the bird fauna of the Indo-Pacific. The largest island (26) is New Guinea. Islands within 500 miles of New Guinea (indicated by circles) fall along a line of the form $S = CA^z$ or just below it. Islands 500–2000 miles from New Guinea (indicated by unenclosed symbols) and those more than 2000 miles from New Guinea (enclosed in squares) fall below this theoretical line in approximate relation to their distance from New Guinea, the presumed source area. A possible interpretation is that islands falling along the theoretical line are at or near equilibrium, whereas remote islands, which fall well below the line, are below equilibrium capacity because of an insufficient supply of immigrants. Reprinted with permission from Robert H. MacArthur and Edward O. Wilson, *Evolution,* vol. 17, 1963.

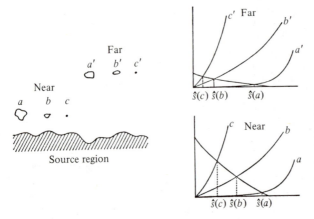

Fig. 23.10 The effect of area on near and far islands of different sizes. Larger islands tend to have more species, but this effect is greater for far islands than for near ones. From Robert H. MacArthur and Edward O. Wilson, *The Theory of Island Biogeography,* copyright © 1967 by Princeton University Press, Fig. 12, p. 28. Reprinted by permission of Princeton University Press.

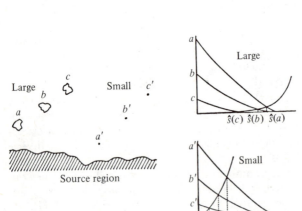

Fig. 23.11 The effect of distance on large and small islands of varying distance. Farther islands tend to have smaller faunas. This "distance effect" is greater among smaller islands, provided there is no "stepping-stone" migration from island to island. From Robert H. MacArthur and Edward O. Wilson, *The Theory of Island Biogeography,* copyright © 1967 by Princeton University Press, Fig. 13, p. 29. Reprinted by permission of Princeton University Press.

island, the source area that has a "fringing archipelago" of possible stepping stones will generally have an increased ability to colonize the island. Furthermore, this "fringing archipelago" effect also explains the predominant direction of species colonization *between* the two source areas, using the islands as stepping stones. The presence of a single island "stepping stone" increases the probability of colonization in both directions, but does not favor either one over the other. But if one source area acquires a "fringing archipelago" of several islands, it also becomes more efficient as a source of colonizing species, and the direction of migration becomes largely a flow from this source area to the other. Of course, different species will differ in their ability to colonize islands.

One technique in the study of island biogeography, favored by Wilson and Simberloff, consists of the complete denudation of a small island, followed by repeated faunal sampling of that island at selected intervals following denudation. The rate at which islands of varying characteristics (area, distance from source area, etc.) are repopulated, and the fate of the early colonizers, can then be studied experimentally. These experiments are in part inspired by the earlier study (Dammerman, 1948) of the repopulation of Krakatau island (now part of Indonesia) following a volcanic eruption that completely denuded the island in 1883. The colonizing species in some cases persist, but in other cases they may become extinct on the island as they are replaced by other, more successful species. Wilson has in fact shown that some species may succeed on a global scale, despite competetive inferiority, simply by being rapid colonizers. These species depend for survival on their ability to be the first colonizers of islands denuded by natural catastrophes (hurricanes, etc.) or of totally new islands. Here they expand rapidly, with a high natural rate of increase (r), and hopefully survive long enough to send propagules to colonize another "empty" island, before competition from later arrivals causes their local extinction. There is yet another aspect to this "taxon cycle": the rapid colonizers may speciate, giving rise to new species that inhabit rain forests and other interior habitats. The rain forest species are generally poor colonizers, and are thus subject to total extinction should the island ever become denuded. The rapid colonizers, meanwhile, survive on the other islands they have colonized.

CHAPTER SUMMARY

The world's environments may be classified into ecological zones or biomes, each characterized by a set of climatic conditions such as seasonal patterns of temperature and rainfall. The major biomes are the tundra, taiga (northern coniferous forest), temperate deciduous forest, temperate grassland (prairie, steppe), chaparral, desert, tropical rain forest, tropical deciduous forest, tropical thorn forest, and tropical grassland (savannah). Areas of the same biome on different continents are in most cases isolated from one another.

The faunas of the world can also be grouped into zoogeographic regions based on the continental distributions of major groups of animals; analogous phytogeographic regions exist for plants. The major zoogeographic regions are the Neotropical and Nearctic in the New World, and the Australian, Ethiopian, Oriental, and Palaearctic regions of the Old World. These regions correspond roughly to the major continents, and are connected to one another either by broad corridors or by land bridges.

The geographic dispersal of land animals across "land bridges" usually occurs as soon as the opportunity arises, and a large and representative sample of the entire fauna is thereby transported. Dispersal by sporadic, occasional means ("sweepstakes dispersal") transports small and unbalanced faunal samples only. Reconstructions of past faunas show that South America was isolated from other continents during much of the Tertiary, that Australia's isolation as an island is of long standing, and that Africa may also have been isolated in early Tertiary times. The movement of continental masses relative to one another, brought about by plate tectonics, has separated North America from Eurasia (formerly they were united) and has broken up the former supercontinent of Gondwanaland into South America, Africa, India, Australia, and Antarctica. The survival of species in inaccessible places may result in a relict distribution, but it is not always easy to distinguish between such relict distributions and distributions that have always been restricted. Places of geographic origin are often difficult to ascertain in the absence of fossils.

Islands have always assumed a special importance to biogeographers. Compared to continental environments of equal size, islands can boast a much higher proportion of endemic species and genera, found nowhere else on Earth. The species-area curve, $S = CA^z$, indicates that larger islands generally have more diverse faunas. The number of species on an island reflects a balance between immigration, speciation, and extinction. Immigration to an island is generally reduced with increasing distance of that island from the mainland, unless "stepping stone" islands or a fringing archipelago are present in between. With increasing ecological diversity on an island (approximately proportional to the island's area), speciation generally increases, and extinction generally decreases. Some r-selected species may succeed largely by their ability to colonize new islands ahead of competitively superior species.

24

Behavior and Sociobiology

A.   FORMS OF BEHAVIOR

Behavior may be defined as the reaction of a living organism to stimuli. The evolution of behavioral characters figured prominently in the writings of Lamarck, Charles Darwin, and many subsequent evolutionists. Stanley (1979) and Wilson (1975) are but two of the many modern evolutionists whose theories have emphasized the importance of behavior.

Different kinds of behavior vary in their degree of complexity, and in the extent to which they are based on prior experience. The behavior forms described below as taxes and tropisms are simple in that they are not further divisible into sequential steps and can thus be described as a single action. Complex behavior, on the other hand, is generally described as a series of sequential steps that may form an entire pattern. Patterned or complex behavior has the capability of being either species-specific or situation-specific or both. It may be capable of modification to suit the situation at hand, and may also enable members of a particular species, sex, caste, age class, or status to be recognized and to recognize each other.

Another important distinction is between behavior that is learned and behavior that does not require learning. Behavior that does not require learning is called **innate** behavior. Innate behavior need not be present from birth, but is "inborn" in the sense that no prior learning experience is required.

Learned behavior generally requires a trial-and-error period for its correct performance, it improves with practice, and its exact pattern of performance may vary with the individual or from one occasion to the next. Innate behavior requires no trial-and-error period, it improves only minimally with practice, and it varies very little with the individual or from one occasion to the next. Innate behavior may also be thought of as genetically determined or genetically programmed behavior, modified very little (if at all) by experience. Learned behavior is environmentally determined behavior, including behavior learned by the imitation of conspecific individuals in one's environment. In theory, a purely learned behavior pattern would have a zero genetic component, but no such behavior has yet been discovered. Even maze-running, that most classical of all learned behavior patterns, has a genetic component that can be improved by selective breeding. The limits upon learning are also genetically determined: the types of stimuli responded to, the

BOX 24-A *Plant Tropisms*

Phototropism: Oriented growth either toward light (positive phototropism) or away from light (negative phototropism).

Geotropism: Oriented growth either toward the Earth's center of gravity (positive geotropism) or away from it (negative geotropism).

Chemotropism: Oriented growth either toward higher concentration of a chemical substance (positive chemotropism) or toward a lower concentration (negative chemotropism).

Two kinds of chemotropism are distinguished by their own names: hydrotropism, an orientation to water, and halotropism, an orientation to salt.

Anemotropism: Oriented growth either toward the source of prevailing winds (positive anemotropism) or away from that source (negative anemotropism).

Thigmotropism: Oriented growth toward a surface in response to touch (positive thigmotropism); negative thigmotropism is uncommon.

Magnetotropism: Oriented growth parallel to magnetic lines of force.

---

variety of possible responses, the rapidity at which learning takes place, and the complexity of a pattern that can be learned—all these factors are to a great extent under genetic control.

Simple Behavior Patterns

Tropisms are the directionally oriented growth movements of plants, sessile animals, or fungi. Tropisms may be classified by the nature of the orienting stimulus (Box 24–A); thus, orientation in response to light would be called phototropism, and orientation in response to the Earth's gravity would be called geotropism. Tropisms may also be classified as either positive or negative, according to whether the behavior is oriented toward or away from the stimulus.

The adaptive importance of tropisms is quite evident. Plants exhibiting either positive phototropism or negative geotropism (the two reinforcing each other in most natural situations) acquire an advantage in being closer to the unobstructed light that affords maximal opportunity for photosynthesis. Leaves and stems are often positively phototropic and negatively geotropic for this reason. Roots, on the other hand, are often positively geotropic and positively hydrotropic (water-seeking). In addition, positive chemotropism brings the roots closer to the source of a beneficial chemical, while negative chemotropism minimizes contact with a harmful one. Many tropisms are adaptively programmed to avoid ex-

cesses: plants that normally show positive chemotropism or phototropism will instead exhibit a negative tropism when the light is too intense or the chemical too highly concentrated, stimuli that could lead to tissue damage. Any inadaptive tropisms (for instance a positive chemotropism to a harmful chemical) would tend to be eliminated quite rapidly by natural selection.

The oriented locomotor patterns of animals are called **taxes** (singular, *taxis*). Movement toward a stimulus is a positive taxis, and movement away from a stimulus is a negative taxis. Common taxes are listed in Box 24–B.

The adaptive significance of animal taxes is often easy to assess. Most animals are attracted to dissolved nutrients in low concentrations (positive chemotaxis), but repelled by higher and potentially toxic concentrations of nearly any substance (negative chemotaxis). Marine organisms are within certain limits positively halotaxic (salt-seeking). Freshwater fish exhibit both negative halotaxis (within limits) and positive rheotaxis (swimming upstream). Many flying insects are positively phototaxic, especially when threatened: they escape by flying toward the sun. Negative phototaxis is quite commonly exhibited by a variety of animals, including most burrowers, cave animals, crawling insects, and desert animals, nearly all parasites, and virtually any animal if the light stimulus is too intense. A wide assortment of male animals find their mates by positive chemotaxis to a chemical substance (called a *pheromone*) secreted by the females. Male silkworm moths find their mates by a combination of positive chemotaxis to such a pheromone, and positive anemotaxis (flying upwind) toward the source of the wind-dispersed scent.

Complex Behavior: Instincts

Complex behavior that does not require learning is known as **instinct**. Instinct is thus complex innate behavior.

The usual test of an instinct is that it be performed experimentally by an individual reared in isolation from other members of its species or from any other potential learning source. Bird songs, for example, are usually instinctive, and are displayed by individuals reared in isolation from other birds or from any auditory stimuli other than their own. This is all the more remarkable because many of these songs are both sex- and species-specific; that is, an isolation-reared bird displays not just any song, but the song characteristic of its species and its sex.

Some songs do also have learned components: a deafened bird often sings imperfectly, or not at all, for want of the opportunity to hear itself sing, or to correct its early mistakes.

The study of instinctive behavior under natural conditions usually requires hours of patient waiting, and in some cases careful analysis of sound recordings or motion pictures. Sometimes, however, the behavior has as its outcome a visible product, such as a nest, which may be studied directly. Burrows, nests, and other types of homes may be very elaborate, and in some cases (termites, certain birds) may aid in species identification. Among the most striking of these products are the webs and traps constructed by spiders for the capture of their insect prey. The webs are usually very distinctive, permitting experts to identify the correct species in many cases merely by studying its web.

A remarkable instinct in honeybees is the method by which they communicate to one another information about a source of food. In this highly social species, a bee

---

BOX 24–B *Animal Taxes*

	Positive	Negative
PHOTOTAXIS	Oriented movement toward light	Oriented movement away from light
GEOTAXIS	Oriented movement toward the Earth's center of gravity (down)	Oriented movement away from the Earth's center of gravity (up)
CHEMOTAXIS	Oriented movement toward higher concentration of a chemical substance	Oriented movement toward lower concentration of a chemical substance
HYDROTAXIS	Oriented movement toward water	Oriented movement away from water
HALOTAXIS	Oriented movement toward salt, toward greater salinity	Oriented movement away from salt, toward lesser salinity
ANEMOTAXIS	Oriented movement toward source of wind (upwind)	Oriented movement away from source of wind (downwind)
RHEOTAXIS	Oriented movement toward source of running water (upstream)	Oriented movement away from source of running water (downstream)
MAGNETOTAXIS	Oriented movements that bring about alignment with magnetic lines of force.	

returning from a food-gathering flight is surrounded by other bees, to which it feeds small samples of the nectar or other food it has gathered. The returning bee also communicates to its hive-mates information on the distance, direction, and "quality" (or intensity) of the food source.

Many instincts are remarkably invariant: the pitch and duration of each note, the shape of each web or nest, or the amplitude and duration of each head-flick are all remarkably constant from individual to individual or from occasion to occasion. Behavior that is this constant is called **stereotyped** behavior. Instinctive behavior patterns are very often stereotyped.

The stereotypy of instinctive behavior, as well as the perfection with which it is performed the first time, even in isolation-reared individuals, makes it ideal for species-recognition behavior, courtship, and mating. The price would be high if these were learned, or performed imperfectly by untrained individuals: mating opportunities would be lost, and reproductive isolation between species would break down. The instinctive nature of this behavior insures an unvarying perfection that permits correct species recognition and mate recognition with very little chance of error. It is largely for this reason that courtship and mating behavior are so often instinctive, even among mammals, where learned behavior is at its highest. Only among humans and apes does mating behavior have an appreciable learned component. In animals with short life spans and little opportunity to learn anything, the unvarying perfection of instinctive behavior assumes a selective importance, and any behavior that is necessary to their survival is usually instinctive.

Complex Behavior:
Learned Behavior

Learned behavior may be defined as behavior that is altered to fit the environmental situation and that improves with practice. Learned behavior patterns are much less rigid and more flexible than instincts; they may be modified to suit the situation or the experience of the individual. This is possible only if the individual has some sort of memory, which we may define as the capacity to modify behavior on the basis of past experience. Compared with instincts, which are generally complex, learned behavior patterns present a far wider spectrum:

they may be as simple as many taxes, or they may be far more complex than any instinct. Most psychology textbooks treat learned behavior in much more detail than is appropriate here.

The advantages of learned behavior are many; foremost among them is the ability to modify the behavior to adapt more fully to the immediate situation. But learning comes at a price: there must always be a time in the life of every individual when the proper behavior has not yet been learned. Naive individuals are thus more vulnerable if a learned behavior pattern is necessary to their survival. Learning is therefore accompanied in many species by parental care, and the survival of naive individuals may thus be insured by their parents' behavior, not just their own.

There is another price paid: learning experiences are never quite the same, and learned behavior patterns therefore differ from one individual to the next. This is one reason why handwriting patterns are so distinctive in humans. In most cases, this price is unimportant, but in the case of mating behavior or species-recognition behavior generally, the price may be high enough to tip the balance in favor of instinct as the more advantageous type of behavior.

Learning has been demonstrated in animals as simple as flatworms, which can be trained to recognize a flash of light given before a mild electric shock. (The trained worms will assume a characteristic "defensive" posture whenever the light is turned on.) Nearly all animals capable of responding to stimuli can learn to avoid noxious stimuli. Among vertebrates, learning becomes increasingly important, especially among mammals. The greatest complexity of learned behavior is developed among the primates.

If several individual acts have already been learned, it may sooner or later happen that these acts are spontaneously performed in a particular sequence. If these acts produce a desirable result, the entire sequence is reinforced. With repeated occurrence of the same sequence, a new, more complex behavior pattern may be learned. The role of behavioral variations in this process is partially comparable to the role of mutations in the evolution of genetically determined traits. Such a variation may affect one step only in an already established sequence, or it may bring about a new succession of acts that had not previously occurred together.

Learning to distinguish between a desirable and an undesirable behavior pattern may require a certain intelligence, and it also brings about an increase in intelligence in the course of evolution. For example, if the sequence $ABCD$ leads to a desirable result, but the sequence $ABFD$ to an undesirable one, a selective premium is thereby placed on animals that can learn to distinguish C from F in the proper context with the fewest number of errors in the learning process. This may be a test of sequence as well, for example if $ABCD$ has a desirable outcome, while $ACBD$ does not. There may also be an environmental context if this behavior produces its result only under certain environmental conditions; a selective premium is thereby placed on those animals with sufficient intelligence or sensory abilities to distinguish proper circumstances from improper ones.

Another type of learned behavior, sometimes called "insight," differs in that it is perfected all of a sudden rather than step by step. A chimpanzee unable to reach a banana suspended from the ceiling may turn his attention to playing with some boxes. After some time, the chimpanzee may suddenly realize the solution to the previously unsolved problem: he stacks one box upon the other and is thus able to reach the banana. This type of learning is also easily imitated: another chimpanzee who views the solution is then able to repeat it at will. The components of the final solution are often learned through exploratory behavior such as play. Play is especially important among the higher primates. Play also serves other functions, e.g., it assures others that no harm or danger is present.

B. THE ROLE OF BEHAVIOR IN EVOLUTION

The evolution of behavior has always figured prominently in discussions about evolution, from Darwin and his predecessors to many present-day discussions. There are at least three important ways in which behavior can influence evolution: behavior itself is an important form of adaptation, behavior can function as a reproductive isolating mechanism, and behavior can alter the functional use of an organ and thus redirect its evolution through an adaptive shift.

The functional relevance of behavior is often very obvious. It is therefore easy to slip into anthropomorphic description, ascribing human qualities to nonhumans, or teleological description, explaining activities by the results achieved or sought after. Animals can be provoked into attack, or into retreat—they then behave *as if* they were angry or fearful. Animals deprived of water will behave *as if* they experienced thirst—they search for drinking water, and will go to some difficulty if necessary in order to obtain some. Birds build nests in which they later lay their eggs—they behave *as if* they were consciously providing for the care of their future offspring. We cannot know what these animals are thinking, except through their overt behavior; even then, we must be careful to say that they behave *as if* a certain emotion or purpose was guiding their actions, which is quite different from saying that they do in fact follow such emotions or purposes. The functional relevance of behavior follows from the fact that so much behavior can be described in these terms.

It is but a short step from functional relevance to actual adaptive values. Seldom, however, have adaptive values actually been calculated for animals both with and without certain behavior patterns under natural or near-natural conditions. Yet the widespread existence of behavior that seems purposeful argues strongly for such behavior to have evolved adaptively through natural selection. In fact, some of the most elaborate adaptations in the animal world are behavioral, as Darwin well recognized. The "dances" of honeybees, for example, help the colony exploit sources of nectar more efficiently by communicating to other colony members information about the location and content of the nectar source.

In animals, one of the most prevalent forms of reproductive isolation is behavioral or ethological isolation (Chapter 14). Members of either sex recognize the proper behavioral repertoire of the opposite sex, where "proper" behavior really means behavior characteristic of that particular species. Mating calls, such as the chirping of crickets or the singing of birds, almost always differ between closely similar sympatric species (Fig. 24.1). The result is that potential mates belonging to the same species will recognize each other's behavior and mate, while members of a different species will be recognized as different by their behavior, and interbreeding will thus be avoided.

Mating behavior often takes the form of a complex ritual. One sex performs step 1, then the other responds

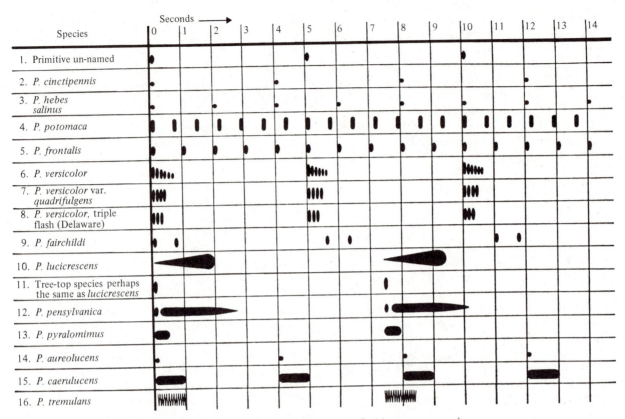

Fig. 24.1 Flashing patterns of fireflies (*Photuris*). Species differences in flashing patterns are instinctive, and they are used as reproductive isolating mechanisms. Reprinted with permission from E. Mayr, *Animal Species and Evolution*, 1963. Cambridge, Massachusetts: Harvard University Press.

with step 2, then step 3 and so on; in a sequence of a few steps or a dozen, one sex performs all the odd-numbered steps, while the other performs all the even-numbered ones (Fig. 24.2). If either partner makes a mistake, the ritual is broken off and must usually resume again from step 1. After several repeated failures, the courtship may be broken off permanently; if this happens between members of two distinct but related species, it serves as a premating reproductive isolating mechanism (Chapter 14). The complexity of courtship and mating behavior thus helps insure correct species identification. It is much like spies exchanging secret code words.

Another way that behavior may influence evolution is by changing the way in which an organ is used. The way that an organ is used determines to a great extent the selective forces acting upon it. Changes in function, and the consequent changes in selection pressures, often lead to the restructuring of anatomical form. Most organs have at least one function, often several, and these functions may change with the passage of time, some being added while others disappear. Functionless organs tend to degenerate into vestiges, and eventually to disappear (Chapter 17).

Behavioral changes are often instrumental in the early stages of functional shifts. Recall, for instance, the explanation given in Chapter 18 for the evolution of lungs. A vascularized area on the floor of the pharynx had the potential to promote oxygen exchange, but was

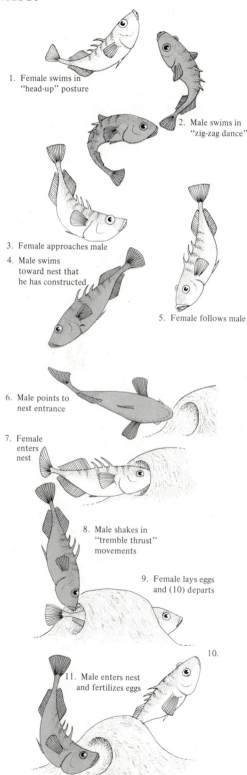

1. Female swims in "head-up" posture

2. Male swims in "zig-zag dance"

3. Female approaches male

4. Male swims toward nest that he has constructed

5. Female follows male

6. Male points to nest entrance

7. Female enters nest

8. Male shakes in "tremble thrust" movements

9. Female lays eggs and (10) departs

10.

11. Male enters nest and fertilizes eggs

Fig. 24.2   A courtship sequence between stickleback fishes. Reprinted by permission from Donald D. Ritchie and Robert Carola, *Biology,* © 1979. Reading, Massachusetts: Addison-Wesley (Fig. 28.19).

not selected for this function until a behavioral change occurred. The habit of rising to the surface was learned as a response to the low oxygen tension in stagnant water; this behavior was immediately functional because it promoted oxygen exchange across the gills. The behavior also altered selection pressures on the vascularized pharyngeal floor, favoring the downward expansion of this area until eventually lungs were formed. The habit of rising to the surface, which determined the selective forces, evolved long before any structurally definable lung was present. This habit was thus essential in utilizing the structural preadaptation of a well-vascularized pharyngeal floor. Behavior, in this case as in others, has a potentially great effect on evolution through its ability to modify selective forces or to create new ones. During the early establishment of new functional complexes, and during change in the function of an organ, these behaviorally induced evolutionary changes are especially important. Gans (1979) suggests that behavior may also function at other times to test the limits of phenotypic flexibility or plasticity, thereby creating the selective regimes that extend these limits.

C. SOCIOBIOLOGY

Most animals occur in social groups, rather than as solitary individuals. The study of these groups and of the behavior of animals within them is a field known as **sociobiology.**

Social behavior may be defined as behavior whose importance lies in the interaction among members of the same species. Courtship and mating behavior are included in this definition, as is all aggressive or submissive behavior between members of the same species. Also included are the many behavior patterns that determine or regulate group size, group membership, group defense, and group harmony.

Social Groups

The social groups formed by animals vary greatly in size, in cohesiveness, in permanence, and in the degree of cooperation between group members. A pack of wolves may be a small group, but its members hunt together, share food, and cooperate in other important ways, and the group may remain together for a long period of time. A herd of antelopes may be extremely large, but cooperation among group members is rather limited. If one antelope detects a predator, they may all flee together, but at other times they may feed independently side by side and interact very little. Some flocks of birds split apart soon after they are formed, and have only an ephemeral existence, with individuals coming and going much as they please. Penguins, on the other hand, form very large, more stable flocks. Many bird species flock together only in certain seasons, during which they may mate, and in some cases migrate.

The size of social groups may vary greatly, even among related species. Often, the size of social groups is related to ecology: animals feeding on very large prey, compared to their own size, tend to hunt in large groups, and herbivores foraging on evenly distributed food sources (like grass) tend also to form large herds. Animals feeding on food that has a patchy distribution will tend to do so alone if the patches are small enough to be monopolized by one individual, but when the patches are very large the animals will tend to form larger groups that often cooperate in the search for these patches. Nocturnal feeding and reliance upon smell often favor solitary or small-group behavior; feeding by day and reliance upon vision often favor larger social groupings. Defense by camouflage or concealment, or by being strongly distasteful or venomous, often works best in solitary species; animals that live in large social groups tend more often to defend themselves by scattering and fleeing, or by standing ground and fighting back.

There are several ways in which group size is controlled. Density-dependent selection is one obvious mechanism in which the fitness of individuals decreases whenever the population becomes too large. Beyond a certain population size, the cohesion of a social group may also diminish as population increases, thus favoring the fragmentation of large groups into smaller units that then become separate groups.

The largest and most cohesive social groupings are found among the social insects, including termites of the order Isoptera, and bees, wasps, and ants of the order Hymenoptera. It is among the social Hymenoptera that the greatest degree of cooperative behavior is found, for reasons to be explained shortly.

The Evolution of Altruism

One of the central issues in modern population biology and sociobiology is the question of the evolution of altruism. **Altruism** may be defined as behavior that decreases individual fitness, but that increases the overall fitness of the population. The seeming paradox is that any individual exhibiting altruism would decrease its own fitness, and whatever genetic basis determined this behavior would therefore be selected against. The other side of the coin, less often discussed, is equally intriguing: in a socially cooperative species, any individual who selfishly violated the rules might thereby increase its own fitness, even at the expense of other group members or of the group as a whole. In other words, selfish, antisocial behavior might increase individual fitness and lead by selection to the destruction of social groupings. (An analogy with the criminal or other antisocial elements in human societies is tempting.) The very prevalence of social organization within the animal kingdom shows that somehow altruism is favored by selection and antisocial behavior weeded out. But how is this possible? Three major attempts to explain the evolution of altruism are the theory of group selection, the theory of inclusive fitness with kin selection, and the theory of reciprocal altruism.

The theory of **group selection** was championed especially by Wynne-Edwards (1962), who postulated that selection operates at the level of social groups as well as at the individual level. According to this theory, groups in which individuals are altruistic are favored by selection, while groups containing selfish individuals are selected against. Many other population biologists, however, have been loath to accept this kind of group selection if there is any other way to explain altruism as resulting from selection among individuals.

Certain opponents of group selection have proposed as an alternative the theory of **kin selection**. Natural selection, say the kin selectionists, works not on individuals, but on their genotypes. In many cases, these are one and the same, but if the members of a group include a number of close relatives, then an **inclusive fitness** must be calculated for each member. In diploid populations, an individual will share an average of 50% of its genotype with each sibling, 50% with each child, 25% with each niece or nephew, and 12½% with each first cousin. If an altruist risks its own life to save the lives of two children and a niece, it has increased the fitness of the genes it shares with these other individuals. Even if the altruist is killed, it has saved the lives of individuals carrying an estimated total of 125% of its genotype. The concept of inclusive fitness means that the altruist in our example would perpetuate more of its own genotype by the seemingly unselfish act of sacrificing its life to save its kin, than by selfishly saving its own neck. It is in this spirit that J. B. S. Haldane is once said to have exclaimed, ''I would lay down my life for two brothers or eight cousins.'' Many populations have a moderate to high rate of inbreeding, which makes most any other individual a relative. The evolution of altruism is favored in this case, for the beneficiaries of any altruistic act are likely to be related to the altruist. An altruist, in other words, would increase the selective value or fitness of the genotype that it shares with many of its relatives, even at the expense of its own individual fitness.

The concept of kin selection also explains the evolution of social behavior among the Hymenoptera. Most insect species in this order are haplodiploid, meaning that males are haploid and females are diploid. In this strange mating system, a diploid female shares at least 50% of her genotype with each sister (more if her mother and father had been related), and 50% with each daughter, but only 25% with each son. If her mother, the queen, lays mostly fertilized eggs (females), then the female we are examining would increase her inclusive fitness more by helping her mother raise new sisters (who are 50% related) than by raising her own children (who would, on the average, be only 37½% related!). She can accomplish this by developing into a sterile worker. Trivers and Hare (1976) and Wilson (1975) have expanded upon this idea to explain other characteristics of hymenopteran societies.

One final explanation is offered by the theory of reciprocal altruism. Under this theory, I may perform an al-truistic act today, but only under the expectation that I may someday be the beneficiary of an altruistic act performed by another individual. Axelrod and Hamilton (1981) explain the evolution of this type of cooperation by means of a strategy that they call TIT FOR TAT. According to this strategy, I would begin by cooperating with you, even though I could gain a small, temporary advantage by acting selfishly. On subsequent encounters, I would cooperate with you if you had cooperated with me, or try to take advantage of you if you had tried to take advantage of me. In short, I would return your actions TIT FOR TAT. Axelrod and Hamilton discuss this and other strategies, and show that TIT FOR TAT can successfully insinuate itself into a population of completely selfish individuals. It can successfully compete with other strategies. It is also evolutionarily stable, meaning that it cannot be replaced by any other strategy. Once the entire population was using the TIT FOR TAT strategy, all would be cooperating as reciprocal altruists. There is even an interesting exception: if I were very old or sickly, and unlikely to survive long enough to reciprocate to your action, you might achieve greater advantage by acting selfishly *just this once*. It is interesting that we find these exceptions in nature, just as the theory predicts we should. Another interesting phenomenon may be called either behavioral mimicry or hypocrisy, in which I try to act selfishly and take advantage of you, but in such a way that you think I have acted altruistically and are thus likely to reciprocate. This theory, in short, explains both the good and bad aspects of human nature (altruism and hypocrisy) as the outcome of a long-range genetic selfishness.

CHAPTER SUMMARY

Behavior is defined as any reaction of a living organism to a stimulus present in its environment. Behavior is itself an important form of adaptation. Behavior can also influence evolution by serving as a reproductive isolating mechanism, or by altering the function of an organ and thus redirecting its evolution through means of an adaptive shift.

Simple (single-step) behavior patterns include the oriented growth movements of plants (tropisms) and the oriented locomotor movements of animals (taxes). These simple forms of behavior provide organisms with quick, automatic, adaptive responses to environmental conditions.

Complex, multi-step behavior may either be learned or innate. Innate behavior can be displayed by animals reared in isolation. Innate behavior that is complex is known as instinct. Instinctive behavior is often remarkably uniform (stereotyped) and frequently also species-specific. These features enable instincts to be used in mate recognition and courtship, often serving as reproductive isolating mechanisms.

Learned behavior is behavior that improves upon repetition. It is more flexible than instinct and may be molded to fit the environmental situation. Species that depend heavily on learning usually have some degree of parental care to protect naive individuals.

Sociobiology is the study of social organization and social behavior in biological terms. Social groupings may vary in size, in permanence, or in the amount of cooperation practiced between its members.

Altruistic behavior, benefitting the group to the detriment of the individual, has received several evolutionary explanations. One is selection between groups (group selection). Another explanation is kin selection, in which the altruist may benefit related individuals that share a certain proportion of its own genotype. The altruist, in this theory, appears to be doing something for the good of others, but is also meanwhile increasing its own inclusive fitness, or the fitness of its genotype throughout the population. A third explanation for altruism is the theory of reciprocal altruism, in which there is a certain expectation that the altruist will someday reap the benefits of another individual's altruism.

25

The Origin and Early Evolution of Life

INTRODUCTION TO UNIT IV

The next three chapters contain an account of the evolution of the various forms of life from the simplest to the most complex, beginning with life's very origin. The evolutionary history of the major groups will be sketched briefly. The principles of units II and III are thus applied to particular groups of organisms. In many cases, the particular phylogenies were elucidated first, and the general principles abstracted from them. Progress in our understanding of particular phylogenies has gone hand in hand with advances in our understanding of general principles, and each major advance in one endeavor has stimulated important advances in the other.

The simplest organisms of all are the bacteria, which have one of the poorest of fossil records. We must in this case rely upon a reconstructed phylogeny based on a study of living diversity. Just one generation ago, biologists would often do no more than survey the diversity among these lower organisms. Today, our understanding of simple organisms has progressed phenomenally, and provisional phylogenies have been proposed for most major groups. Of course, the most crucial datum of any phylogeny is its starting point, which ultimately means the origin of life. The past generation of theorists and experimenters has built up an impressive array of evidence concerning this most momentous step in the history of our universe. It is with life's origin that the present chapter begins.

A. PAST AND PRESENT THEORIES

The Theory of Spontaneous Generation

The problem of life's origins has always fascinated people. Many ancient legends deal with the origins of particular species: worms, for instance, were thought to have originated from horse's hairs that had fallen into the water. The origin of life from nonlife was in these legends usually accomplished in a single generation: plants sprang forth from the Earth itself, or fishes and worms from the mud.

The theory that life originated from nonliving material is now called **abiogenesis.** An early version of this theory was **spontaneous generation,** or the origin of life without apparent cause and without divine intervention. Belief in spontaneous generation was widespread in ancient Greece: Thales, Anaxagoras, Empedocles, Democritus, and Aristotle all supported the concept. Saint Augustine declared spontaneous generation to be God's will, and thus adapted the theory to a Christian setting. Philosophers such as Descartes and biologists such as Harvey and Leeuwenhoek supported the theory, a support which continued well into the nineteenth century.

The occasional dissenters included Francesco Redi (1626?–?1697), who experimented with maggots, the white, wormlike larvae of flies. The spontaneous generation of maggots from rotting meat was widely believed, but Redi began to doubt this when he saw flies attracted to the rotting meat before any maggots appeared. He tested his hypothesis by sealing meat inside four closed flasks, while leaving four other meat-filled flasks open (Fig. 25.1). Soon the meat in the open vessels was full of maggots, "and flies were seen entering and leaving at will; but in the closed flasks," wrote Redi, "I did not see a worm, though many days had passed. . . . Outside on the paper cover there was now and then a deposit, or a maggot that eagerly sought some crevice by which to enter and obtain nourishment." Redi had disproved the spontaneous generation of maggots, but he went considerably beyond this to suggest "that the Earth, after having brought forth the first plants and animals at the beginning by order of the Supreme and Omnipotent Creator, has never since produced any kinds of plants or animals, either perfect or imperfect; and everything

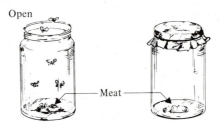

Open

Meat

Fig. 25.1 Redi's experiment. Maggots appeared on the rotting meat in the open jar only, but none appeared in the closed jar. Reprinted with permission from Jeffrey J. W. Baker and Garland E. Allen, *Study of Biology*, 3e, © 1977. Reading, Massachusetts: Addison-Wesley (Fig. 21.1).

which we know in past or present times that she has produced, came solely from the true seeds of the plants and animals themselves, which thus, through means of their own, preserve their species."

The invention of the microscope in the late 1600s revealed a whole new world of remarkably tiny organisms in undreamed variety. Adherents of spontaneous generation easily retreated to the newly discovered world of microbes: although the higher animals did not arise spontaneously, perhaps these much simpler microbes did. The spontaneous generation of microbes was vigorously supported by the Scottish clergyman-naturalist John Needham (1713–1781). He introduced mutton broth into a glass vial, which he sealed tightly with a cork, and heated it (but not sufficiently) on hot coals; it soon teemed with microbial life introduced from the surface of the unsterilized cork. Lazzaro Spallanzani (1729–1799) repeated Needham's experiment with much greater care: he sterilized various animal and vegetable broths by prolonged boiling, and sealed the vials hermetically while they were still hot. Under his microscope he could demonstrate that the broth remained sterile. But if he broke one of the sealed vials open, it immediately teemed with microbial life, proving that conditions for the growth of microbes had been suitable all along. Needham challenged Spallanzani's experiment, claiming that the air in the vial had been "spoiled" and was no longer capable of supporting life. This claim received the support of the chemist Joseph Gay-Lussac (1778–1850), who demonstrated that when liquids were boiled in flasks and the flasks then sealed, the vapor sealed above the liquid contained no detectable oxygen. This was a fatal blow, because fermentation was not yet understood, and oxygen was assumed to be essential for all forms of life. Belief in spontaneous generation continued, especially in France; Buffon, Lamarck, and many other prominent scientists supported the theory.

Louis Pasteur and the Theory of Biogenesis

In 1859, a French chemist named Felix Pouchet wrote a book supporting spontaneous generation. Another French chemist, Louis Pasteur (1822–1895), took up the cause against spontaneous generatioon. A lively debate ensued, most of it conducted before the *Academie des Sciences.* The *Academie* even offered a prize for anyone who could devise an experiment to settle the matter; the prize was awarded in 1862, to Pasteur.

Pasteur had already distinguished himself as a chemist by his discovery of **optical rotation,** a phenomenon exhibited by asymmetrical molecules. He showed that many compounds derived from biological sources were ''optically active,'' meaning that they would rotate a beam of plane-polarized light either right or left. Biological molecules could thus be distinguished from inorganic molecules, a distinction whose origin we shall discuss later in this chapter. To Pasteur, optical rotation was somewhat of a vital principle (see Chapter 1). His belief in a great, unbridgeable gap between the living and the nonliving caused him to reject spontaneous generation: life was for Pasteur (as for most modern scientists) far too complex to have originated in a single step from the nonliving. The ''spark of life'' could not be created where it did not already exist. This is the theory of **biogenesis,** the origin of life from preexisting life only.

Pasteur supported his belief in biogenesis through many experiments designed to refute all claims of spontaneous generation. For example, Leeuwenhoek had claimed that microorganisms in rain barrels had been spontaneously generated from the rainwater. Pasteur collected rain in sterile wine goblets, both near Paris (already a major industrial city) and on Mont Blanc in the Alps. The air near Paris soon contaminated the water with microorganisms, but on Mont Blanc 19 of the 20 flasks remained sterile.

Pasteur then began repeating Spallanzani's experiments in a way that would meet Needham's objections. First he showed that air could be sterilized by heating, without destroying its oxygen or its ability to support life, but abolishing its ability to contaminate sterile broth. He also left sterilized broth connected to the atmosphere through a long tube that was continually heated to sterilize the air. To each experiment Pouchet would object whenever he could, and his objections drove Pasteur to ever more rigorous proofs. In Pasteur's most convincing experiment, some nutrient medium was boiled in a flask whose neck was then drawn out into a long curve, resembling a swan's neck (Fig. 25.2). The flask thus remained open to the atmosphere, but dust and airborne contaminants were trapped in the lower bend. The medium remained sterile—no microorganisms could be seen under the microscope—as long as the flask remained intact. But if Pasteur introduced a contaminant—sterile cotton through which air had been drawn to trap airborne particles—or if he merely cut off the bent neck of the flask, it soon swarmed with microbial life. Pasteur kept some of these flasks for years, and exhibited them before the *Academie des Sciences;* none of them became contaminated. In a much-heralded lecture before the Sorbonne, Pasteur exhibited one of these sterile swan's neck flasks, and declared: ''I have taken my drop of

Fig. 25.2 Pasteur's swan's neck flask. The nutrient medium inside remained sterile because airborne microbes were trapped at the lower bend, B. Reprinted with permission from Jeffrey J. W. Baker and Garland E. Allen, *Study of Biology,* 3e, © 1977. Reading, Massachusetts: Addison-Wesley (Fig. 21.2).

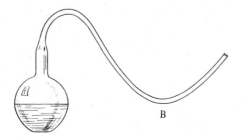

water from the immensity of creation . . . , full of the elements appropriated to the development of inferior beings. And I wait, I watch, I question it, begging it to recommence for me the beautiful spectacle of the first creation. But it is dumb, dumb since these experiments were begun . . . , because I have kept it from the only thing man cannot produce, from the germs which float in the air, from Life, for Life is a germ and a germ is Life. Never will the doctrine of spontaneous generation recover from the mortal blow of this simple experiment.'' Pasteur then concluded, ''there is now no circumstance known in which it can be affirmed that microscopic beings came into the world without germs, without parents similar to themselves.''

The physicist John Tyndall soon improved upon Pasteur's experiment by using an airtight box that allowed fine colloidal particles, usually suspended in the air, to settle out. He fitted the cover of the box with a pair of bent tubes to allow air to enter but dust to be trapped. Once the colloidal particles had settled out, Tyndall added a nutrient medium to some test tubes lodged in the bottom of the box, and sterilized them by boiling. The nutrient medium remained sterile, but if Tyndall put some settled dust from the floor of the box into one of the tubes, then that tube soon teemed with microbial life, while the others remained sterile.

With Pasteur's and Tyndall's experiments, the spontaneous generation of present-day organisms had finally been disproven. Pasteur declared, ''I have been looking for spontaneous generation during twenty years without discovering it. No, I do not judge it impossible.'' But despite Pasteur's caution, the British physicist Lord Kelvin (1824–1907) held that Pasteur had proven that spontaneous generation was impossible at any place and time, under any conditions. Kelvin was not alone in this assumption, which was apparently widespread throughout scientific circles. Ernst Haeckel (1834–1919, Chapter 5) was one of the few who foresaw what Oparin would later champion: that spontaneous generation had occurred in the remote past, under conditions far different from those which prevail in the world today.

What were the scientists who rejected spontaneous generation left to believe? If life had not been spontaneously generated, perhaps it had been eternal.

Wilhelm Preyer, who supported this idea, even went so far as to assert that life had existed on Earth when our planet was still molten. Another possibility was to seek an extraterrestrial origin of life. Under such names as ''Cosmozoa'' and ''Panspermia,'' hypotheses of interplanetary dispersal were seriously entertained as explanations of how life had reached the Earth. Today, such explanations are virtually ruled out by the hazards now known to exist in interplanetary space: extremely low temperatures, lack of any atmosphere, utter dryness, and a very high flux of both cosmic rays and ultraviolet radiation. (Yet, renewed support for this idea has recently come from Francis Crick.) It has also been repeatedly observed that the assumption of an extraterrestrial origin of life failed to solve the problem; it only transferred it to a different planet and an earlier point in time.

If asked about the first origin of life, anywhere in the Universe, the many scientists who rejected abiogenesis were left with only two alternatives: eternal life, or miraculous creation, and even the first alternative was becoming precluded by the growing evidence that the Earth had once been uninhabitable. Here was a dilemma that scientists were often loath to face. Many scientists simply stopped discussing the origin of life, or declared the problem totally insoluble. Darwin himself wrote, ''It is mere rubbish to talk about the origin of life; one might as well talk about the origin of matter.''

Alexander Oparin and the Theory of Primary Abiogenesis

In 1924, a young Russian chemist, Alexander I. Oparin (1897–   ), published a short paper advancing a new theory, which he elaborated on further in his book, *The Origin of Life,* published in 1936 (English translation, 1938). Meanwhile J. B. S. Haldane, a British scientist known largely for his contributions to genetics, independently advanced many of Oparin's views. The theory advanced by Oparin and Haldane has now become accepted by nearly all practicing scientists.

Oparin's theory assumes that life originated on Earth at some time in the remote past, and under conditions no longer present. This is often called the theory of **primary abiogenesis,** and may be summed up in the phrase, *abiogenesis at first, but biogenesis ever since.* The theory of biogenesis had depended not only on Pasteur's experiments, which Oparin fully accepted, but also in part on Kelvin's overly strict uniformitarianism, which is what Oparin really rejected.

Oparin ruled out any form of miraculous creation. He was also aware that conditions in space precluded the interplanetary transmission of life. He concluded that the origin of earthly life had to be sought on Earth. Since it was then believed that the Earth had formed from a molten mass, Oparin concluded that life could not be eternal, but must have originated at a definite time in the Earth's history after sufficient cooling had taken place. Yet Pasteur had shown that spontaneous generation was impossible under present conditions; therefore the conditions prevailing when life originated must have been very different. Free oxygen, in particular, was not present, for this gas is a product of photosynthesizing organisms.

According to Oparin, conditions on the Earth today are no longer suitable to the production of life from non-living matter. The present atmosphere is strongly oxidizing, a condition almost wholly attributable to the actions of green plants. Simple organic compounds, under present conditions, are unstable over long periods because they are readily oxidized to CO_2. Furthermore, any organic compound, if formed under present conditions, would instantly be devoured by some organism, thus precluding a large number of such molecules ever coming together to form even the first rudiments of a living system. It is the *"king of the mountain" principle* (Chapter 17) applied to the simplest forms of life: whatever first became alive used up so many available nutrients that a second origin of life soon became impossible. Darwin had anticipated this when he wrote,

> "It is often said that all the conditions for the first production of a living organism are now present, which could ever have been present. But if (and oh what a big *if*) we could conceive in some warm little pond with all sorts of ammonia and phosphoric salts,—light, heat, electricity etc. present, that a protein compound was chemically formed, ready to undergo still more complex changes, at the present day such matter would be instantly devoured, or absorbed, which would not have been the case before living creatures were formed."

Further details of Oparin's theory are given in the next section.

One possible predecessor to Oparin was Friedrich Engels (1820–1895). An ardent supporter of evolution, Engels believed in the inherent tendency of matter to develop into higher and higher levels of organization, from the physical to the chemical, then to the biological, and finally to the social. This doctrine of historical emergence was called **dialectical materialism.** Oparin (1936) cited Engels' ideas in support of his theory, but there is good evidence that Oparin had developed the theory without the benefit of Engels' inspiration. Oparin's initial paper appeared in 1924; Engels' principal contribution to science, the posthumous *Dialectics of Nature,* was not published in any language until 1925. Graham (1972) notes that Oparin's 1924 paper was reductionist in tone and aimed largely against vitalism, while his longer book (1936), which embraced Engels' dialectics, was more double-edged in its criticism of reductionist as well as vitalist ideas. Curiously, Haldane, like Oparin, was also a Marxist. It was Haldane who collaborated in the first translation of Engels' *Dialectics of Nature* into English! Yet Haldane claimed that he had formulated his ideas on the origin of life before he became a Marxist, and had nearly completed his 1929 book on the subject before he first read Engels' posthumous work.

It is strange that Oparin's book was so little noticed by Western scientists. Few biologists seem to have been familiar with it before 1953, despite its translation to English in 1938. H. J. Muller was one of the few biologists who ever mentioned it in any context. One reason perhaps was that Oparin was known as a chemist, and that chemistry and evolutionary biology were in those decades worlds apart from one another.

The first adequate recognition of Oparin's work outside Russia was accorded in the 1950s by the American astronomer, Harold C. Urey, who had encountered Oparin's book while writing his own book, *The Planets,* and was won over to its thesis. Urey asked one of his graduate students, a biochemist named Stanley L. Miller, to try to replicate the primitive earth conditions that Oparin had postulated, in an attempt to study the consequences of abiotic synthesis in such a world.

Miller sealed in a flask a mixture of hydrogen (H_2), methane (CH_4), ammonia (NH_3), and water (H_2O), in the belief that these four substances had been the principal constituents of the primordial atmosphere. At room

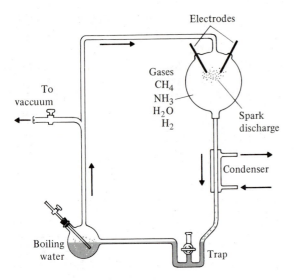

Fig. 25.3   Miller's experimental apparatus (one of several models). Most chemical reactions take place in the spark chamber, and the reaction products are condensed. The boiling flask of water keeps the system circulating in the direction shown by the arrows. Reprinted with permission from Jeffrey J. W. Baker and Garland E. Allen, *Study of Biology*, 3e, © 1977. Reading, Massachusetts: Addison-Wesley (Fig. 21.3).

temperature, the water would settle out as a liquid, but at higher temperatures all four components were gaseous. Miller's apparatus (Fig. 25.3) allowed the water to be evaporated and condensed, simulating rain. Electrical sparks could be created, simulating lightning. Finally, an outlet was provided for samples to be withdrawn for chemical analysis.

Miller circulated the materials in his apparatus for several days. Simple organic compounds began to appear, and they became dissolved in the water to form a dilute solution. After 18 days, the solution became discolored. Upon analysis, it proved to contain quite a variety of organic compounds, including several amino acids and even simple peptide chains. Under conditions as they might have existed on the primitive Earth, Miller had synthesized abiotically a number of important compounds (Fig. 25.4).

Other workers, using slightly different starting materials (different presumed "primitive atmospheres"), performed experiments similar to Miller's (Box 25–A). In some of these experiments, ultraviolet light or some other energy source was used instead of an electrical spark gap. In all cases, a variety of interesting organic compounds were formed, usually including several amino acids. With hydrogen cyanide included among the original constituents, even adenine and other nitrogenous bases were formed. These compounds are important constituents of the nucleic acids DNA and RNA. Thus the major "building blocks" of life have now been produced experimentally under conditions that simulate those of the primitive Earth.

Fig. 25.4   Some of the organic molecules formed in Miller's experiment.

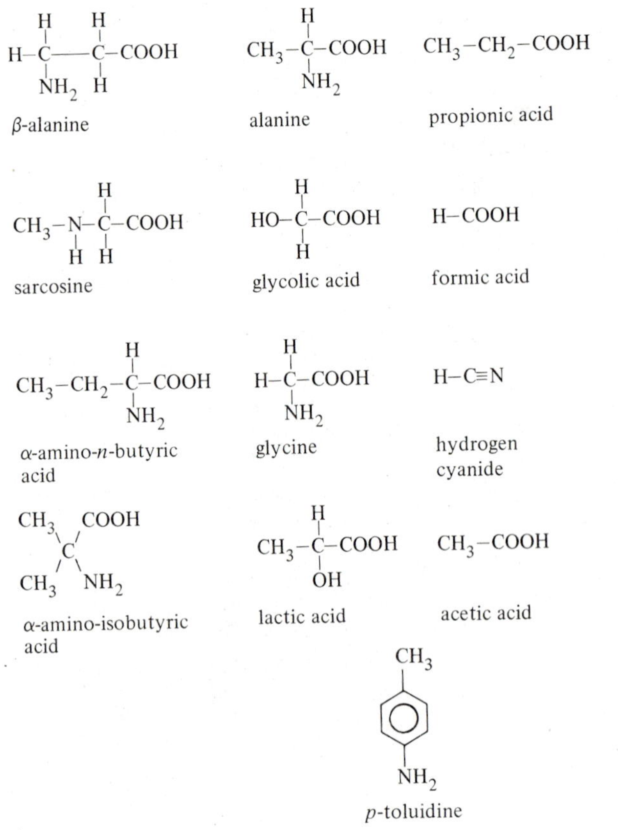

$$\underset{\beta\text{-alanine}}{\overset{\overset{\displaystyle H}{|}\ \ \overset{\displaystyle H}{|}}{H-C\!-\!\!-\!C-COOH}}$$

β-alanine

alanine

propionic acid

sarcosine

glycolic acid

formic acid

α-amino-*n*-butyric acid

glycine

hydrogen cyanide

α-amino-isobutyric acid

lactic acid

acetic acid

p-toluidine

BOX 25-A   *A Summary of Some Primitive-Earth Experiments*

Table 1   **Synthesis of Amino Acids and Related Compounds**

Date	Investigators	Starting conditions	Energy sources	Significant products
1953	Miller	Methane; ammonia; hydrogen; water	Electric discharge	Glycine, alanine; β-alanine, glutamic acid
1955–1956	Abelson	Hydrogen, methane, carbon monoxide or carbon dioxide; ammonia or nitrogen; water; oxygen	Electric discharge	Alanine, glycine, sarcosine, serine, aspartic acid (only when conditions are nonoxidizing)
1957	Walter and Mayer	Methane; ammonia; water; hydrogen sulfide	Helium spark	NH_4SCN (ammonium thiocyanate)
1959	Oró	Formaldehyde; hydrogen cyanide	Room temperatures (spontaneous)	Several common amino acids
1959	Pavlovskaya and Pasynskii	Methane; ammonia; water; carbon monoxide	Electric discharge	Common amino acids (depletion of hydrogen from system enhances amino acid synthesis); aminobutyric acid; lysine
1960	Groth and Von Weysenhoff	Methane; ethane; ammonia; water	Ultraviolet light: 407 nm; 219.6 nm	Glycine; alanine; aminobutyric acid
1959–1965	Becker, et al.; Hanafusa et al.; Ponnamperuma et al.; Steinman, et al.	n-carboxyamino acid anhydrides; aminoacetonitrile sulfate; kaolin; dicyandiamide; HCl and corresponding amino acids	Heat to 100°C	Aspartic acid, threonine, serine, glutamic acid; proline; glycine; alanine; valine; isoleucine; leucine; tyrosine; phenylalanine
1963	Oró	Ethane; methane; ammonia; water.	—	Several common amino acids; proline; valine; leucine; asparagine
1965	Grossenbacher; Knight	Methane; ammonia bubbled over ammonium hydroxide	—	Several common amino acids plus threonine; serine; glutamic acid; isoleucine
1965	Oró	Methane; ammonia; water	1300°K heat	Aspartic acid; threonine; serine; alloisoleucine; isoleucine; phenylalanine; β-alanine; glutamic acid; etc.

(continued)

BOX 25–A   *(continued)*

Table 1   Synthesis of Amino Acids and Related Compounds

Date	Investigators	Starting conditions	Energy sources	Significant products
1957–1966	Schram et al.; Fox et al.; Ponnamperuma	Tripeptides; polyphosphoric acid esters; 18 amino acids	Heat 50–60°C	Polypeptides; proteinoids
1966	Ponnamperuma	Methane; ammonia; water	Electric discharge	Polypeptides
1966	Abelson	Ammonium formate; ammonium hydroxide; sodium cyanide; ferrous sulfate	Ultraviolet light	Hydrogen cyanide intermediate (glycinonitrile glycine)
1966	Ponnamperuma and Flores	Methane; ammonia; and water to hydrogen cyanide, then hydrolysis of HCN	—	Common amino acids, etc.; threonine; valine; phenylalanine
1968	Steinman, Smith, and Silver	NH_4SCN; hydrolyzed $6N$ HCl; N_2 gas phase	Ultraviolet light, 260 nm for 3 hr	Methionine

Table 2   Synthesis of Nucleic Acid Derivatives

Date	Investigators	Starting conditions	Energy sources	Significant products
1960	Oró	Ammonium cyanide	Spontaneous at high concentrations	Adenine
1961	Oró	Aminoimidazole carboximide	Heat (100–140°C)	Guanine; xanthine
1961	Fox and Harada	Malic acid; urea	Heat	Uracil
1963	Ponnamperuma et al.	Methane; ammonia; water; hydrogen	Ultraviolet light	Adenine (enhanced by removal of hydrogen)
1963	Oró	Acrylonitrile; aminopropylmalonamide semialdehydimine (found in comets) in aqueous ammonium chloride solutions	Heat (135°C)	Uracil
1963	Ponnamperuma et al.	Ribose, adenine, and phosphate in dilute aqueous solution	Ultraviolet light	Adenosine

Table 2 (continued)

Date	Investigators	Starting conditions	Energy sources	Significant products
1963	Ponnamperuma and Sagan	Adenine; adenosine; adenosine monophosphate (AMP); adenosine diphosphate (ADP); ethyl phosphate	—	Adenosine; AMP; ADP; ATP
1964	Ponnamperuma and Kirk	Deoxyribose; adenine; HCN	Ultraviolet light	Deoxyadenosine
1964	Miller and Parris	Hydroxyapatite and cyanate salts (reaction takes place on surface of mineral)	—	Pyrophosphate
1965	Ponnamperuma	HCN (dilute)	Ultraviolet light	Adenine; guanine; urea
1968	J. Ferris, et al.	Cyanoacetyline; cyanate; cyanogen	Electric discharge (cyclical temperature changes to freeze water lead to eutectic concentration techniques)	Pyrimidines; cytosine; uracil
1974–	Orgel and others	Bases, ribose, sea salt	Heat cycles	Adenosine, guanosine, inosine, xanthosine

Adapted from Margulis, Lynn, 1981. *Symbiosis in Cell Evolution*. San Francisco: W. H. Freeman and Company.

B. CHEMICAL EVOLUTION

The modern theory of chemical evolution is based on the theory first formulated by Oparin and Haldane, and on experimental evidence provided by Miller, Fox, Ponnamperuma, and others.

Cosmic Evolution

Most astronomers now accept the theory that the Universe originated in a "big bang." All matter was created in this "big bang," including the matter of which stars and planets are composed.

Star formation begins wherever matter happens to be sufficiently dense for the mutual attraction of atoms to result in "gravitational collapse." As the young star increases in density, mounting pressures in its interior cause it to heat up. When a certain critical temperature is reached, thermonuclear reactions begin. Astronomers believe that our Sun was formed in this way.

To account for our Solar System and its planets, most astronomers now support a modified version of the **nebular hypothesis,** originally proposed by Immanuel Kant (1724–1804) and independently by Pierre Simon de Laplace (1749–1827). According to this hypothesis, the Solar System first formed from a swirling nebula whose rotation caused it to flatten into the shape of a disc. The center of this swirling mass became the Sun; the planets and their satellites formed from the more peripheral material. An opposing hypothesis invokes a star passing just close enough to the Sun to pull out some of its mass by gravitation into an elongated extension, and the condensing of this extension into a series of discrete planetary bodies. Many astronomers believe the Earth to have formed from the Solar Nebula as a molten mass that solidified as it cooled. Other astronomers assume a ''cold'' origin of the Earth by the aggregation or accumulation of many smaller solid bodies.

As the Earth formed, astronomers believe that the heavier elements and their compounds were drawn to the center, while the lighter ones floated to the surface. The lightest elements, such as hydrogen (H), oxygen (O), nitrogen (N), and carbon (C), were more abundant in the primordial atmosphere than in the solidifying mass. A crust of medium-weight materials such as silicon (Si), calcium (Ca), and aluminum (Al) solidified at the surface, insulating the core against further heat losses. Today, the Earth derives its interior heat from radioactive decay within its core.

The Primordial Earth
and Its Atmosphere

The atmosphere of the primitive Earth was moderately to strongly reducing, meaning that an abundance of hydrogen guaranteed the availability of electrons. The assumption of a primitively reducing environment is based on the reducing character of other planetary atmospheres, on the overwhelming cosmic abundance of hydrogen, and on the fact that meteorites, lunar samples, and the oldest terrestrial rocks all seem to have been formed under strongly reducing conditions. Hydrogen was so abundant that many other elements were combined with it: oxygen was present largely as water vapor (H_2O), nitrogen as ammonia (NH_3), carbon as methane (CH_4) and other hydrocarbons, and sulfur as hydrogen sulfide (H_2S). No free oxygen (O_2) was present, and the common oxides, such as those of carbon (CO_2) and silicon (SiO_2), may not have been present either.

The Earth's gravitational field was sufficient to hold onto methane, ammonia, and many other gasses, but not hydrogen. The escape of hydrogen from Earth's gravitational field gradually made the atmosphere less strongly reducing. On the Earth's surface, conditions for the origin of life became more favorable. Water vapor condensed from the atmosphere to form liquid pools and eventually oceans. Rain tended to wash substances from the atmosphere into these pools, where they were further concentrated by evaporation and entered into still more

chemical reactions with each other. The oceans accumulated more and more of these organic compounds, and came to resemble what Haldane called a ''hot, dilute soup,'' similar to the aqueous solutions formed by prebiotic chemists such as Miller.

The most abundant constituent (about 79%) of our present atmosphere is nitrogen gas, producd from ammonia according to the reaction

$$2\,NH_3 \quad \rightleftharpoons \quad N_2 \quad + \quad 3\,H_2 \uparrow .$$
$$\text{ammonia} \qquad \text{nitrogen} \qquad \text{hydrogen}$$

The gravitational escape of hydrogen favored the breakdown of ammonia, and also the production of hydrocarbon chains:

$$CH_4 + CH_4 + CH_4 + CH_4 + CH_4 \qquad \rightleftharpoons$$
$$\text{methane}$$

$$CH_3{-}CH_3 + CH_3{-}CH_2{-}CH_3 + \quad 3H_2 \uparrow \rightleftharpoons$$
$$\text{ethane} \qquad \text{propane} \qquad \text{hydrogen}$$
$$\text{(escapes)}$$

$$CH_3{-}CH_2{-}CH_2{-}CH_2{-}CH_3 + H_2 \uparrow .$$
$$\text{pentane}$$

Reaction of methane with ammonia yielded hydrogen cyanide and aminonitrile, precursors of many nitrogen-containing organic compounds:

$$CH_4 + NH_3 \rightleftharpoons HCN + 3\,H_2 \uparrow$$
$$\text{methane} \quad \text{ammonia} \quad \text{hydrogen} \quad \text{hydrogen}$$
$$\text{cyanide}$$

$$\begin{array}{c} \qquad\qquad\qquad H \\ 2\,HCN \rightleftharpoons N{\equiv}C{-}C{=}N{-}H . \end{array}$$
$$\text{hydrogen} \qquad\qquad \text{aminonitrile}$$
$$\text{cyanide}$$

Eventually, the oxides of carbon were produced on the primitive Earth by reactions such as the following:

$$CH_4 + H_2O \rightleftharpoons HCHO + 2\,H_2 \uparrow ,$$
$$\text{methane} \quad \text{water} \qquad \text{formaldehyde} \quad \text{hydrogen}$$

$$HCHO \rightleftharpoons CO + H_2 \uparrow ,$$
$$\text{formaldehyde} \qquad \text{carbon} \qquad \text{hydrogen}$$
$$\text{monoxide}$$

$$CO + H_2O \rightleftharpoons CO_2 + H_2 \uparrow .$$
$$\text{carbon} \qquad \text{water} \qquad \text{carbon} \qquad \text{hydrogen}$$
$$\text{monoxide} \qquad\qquad \text{dioxide}$$

Carbon dioxide may also have been released by volcanic activity, along with hydrogen sulfide and water. Some scientists believe that much of our atmosphere was formed in this way, a process known as "outgassing."

Polymers and the Problem of Concentration

The most important of biomolecules, namely proteins and nucleic acids, are polymers, meaning that they consist of many smaller units linked together in long chains or other shapes. As a rule, polymers tend to be more stable than their subunits (monomers), and this very stability lends permanence to polymer-containing systems. Recall that amino acids, the subunits of proteins, were among the first molecules formed, and that many constituents of nucleic acids had also been formed abiotically.

Polymers are far more likely to form in situations where their subunits exist in greater concentration. But as chemical evolution continued, each compound eventually reached a steady-state chemical concentration, at which point further synthesis was balanced out by spontaneous decomposition. In many if not most cases, this steady-state concentration was low, so that the "hot, dilute soup" never became a "hot, concentrated soup."

This placed a limit on chemical complexity: if compounds of modest complexity could at most be present in a very dilute solution, then they would have little opportunity to combine with one another to form compounds of greater complexity. A mechanism to concentrate monomers, and thus favor polymerization, would have constituted an evolutionary "breakthrough" at this point. Evaporation of water in tidal pools may have increased the concentration of dissolved substances. Synthesizing reactions may have been speeded up (catalyzed) on the surface of asymmetrical apatite crystals $(Ca_3(PO_4)_2)$, which may account for the presence of phosphorus in living systems and for the synthesis of asymmetrical molecules. Fox and his co-workers have championed a theory based on proteinoid "microspheres," and have shown that experimentally produced microspheres have many of the properties expected of prebiotic systems, especially the ability to catalyze self-perpetuating chemical reactions. Several other concentration mechanisms (Box 25–B) have also been suggested.

The most widely favored theory, that of **coacervation,** was championed principally by Oparin. Coacervation is the separation of tiny droplets, bubbles, or bilayers from colloidal mixtures of two or more compounds. Each coacervate droplet has a definite boundary,

---

BOX 25–B *Concentration Mechanisms Suggested by Various Authors*

Coacervates and protobionts	DeJong, Oparin (Bernal 1967), Oparin 1969
Apatite surface (also solves low phosphorus problem)	Bernal 1967 Hutchinson 1965
"Ocean line"	Weyl 1968
Microspheres (made from driving water off by heating organic compounds)	Fox and McCauley 1968
Excitable membrane	Gabel (Ponnamperuma and Gabel 1968)
Drying fringes of the sea, salt pools, environments rich in sulfur and iron	Ehrensvaard 1960
Eutectic concentration mechanisms	Orgel 1970

From Margulis, 1970, p. 82; courtesy of Lynn Margulis, Boston University.

and the concentrations of dissolved substances within the droplet may differ from those prevailing outside in the surrounding medium. Life may first have acquired individuality in these droplets, since each was potentially different from the next. Within some of these coacervate droplets, certain substances may have become more concentrated than in the surrounding environment, and the synthesis of still larger molecules, especially proteins, was thus favored.

The formation of proteins was most important because many of these spontaneously formed polymers were capable of acting as enzymes, organic compounds that speed up the rate of chemical reactions. Reactions favored by enzymes could now proceed at rates often several orders of magnitude higher than before, and the diversity of enzymes (which depends upon their amino acid sequence) now enabled the synthesis of far more complex substances. Of course, many of these spontaneously formed enzyme systems were to no avail, or they made their droplets less stable. These last enzymes were selected against, since they led ultimately to the destruction of their systems. But at least a few enzymes favored those reactions that increased the stability of the systems of which they were part, and these systems were favored by selection. In addition to making their systems more stable, the enzymes might also have favored the ''growth'' of their systems by favoring reactions in which some part of the external environment was incorporated into the system. Some enzymes, moreover, may have catalyzed energy-producing reactions, which could now become coupled to other, energy-requiring reactions (up to a point), thereby enabling a bewildering variety of chemical reactions not previously possible. We are here approaching the complexity of simple organisms, whose very complexity depends largely on enzyme-controlled reactions, and thus on the existence of proteins. The most diverse kind of biological compounds today are proteins, most of them enzymes. There are more different proteins in a living cell than all the different types of nonprotein compounds combined.

The Origin of Asymmetrical Molecules

Any molecule that cannot be made to coincide with its mirror image is **asymmetrical,** or **chiral** (literally, *handed*). In solution, such molecules tend to rotate plane-polarized light in either a clockwise direction (dextral rotation, D) or a counterclockwise one (levorotation, L). Abiotic syntheses of asymmetrical compounds from symmetrical precursors tend to produce equal amounts of dextrorotatory and levorotatory molecules; such a mixture, termed a **racemic** mixture, has no rotatory effect on polarized light and is said to be **optically inactive.**

Louis Pasteur's fame as a chemist predates his interest in the disproof of spontaneous generation. As a chemist, Pasteur had investigated the behavior of a number of asymmetrical, or **optically active** molecules. He found that inorganic syntheses from symmetrical reactants always produced optically inactive racemic mixtures, but that asymmetrical molecules isolated from biological sources were invariably optically active. For example, most naturally occurring amino acids (except glycine, which is symmetrical) are of the levorotatory rather than the dextrorotatory form. The reason for this phenomenon is that biological syntheses take place with the aid of asymmetrical enzymes, and these enzymes favor one optically active form over the other. Pasteur, a vitalist, concluded that this was an important distinction (if not *the* important distinction) between biologically synthesized and abiotic molecules.

The origin of asymmetrical molecules has been the subject of much speculation. Asymmetrical enzymes depend on asymmetrically coiled protein chains, which in turn depend on asymmetrical amino acids. The asymmetry of tRNAs and other nucleic acids also looms large here, but these were synthesized using protein enzymes, so the asymmetry of amino acids and proteins is of prime concern. The fact that all forms of life tend to have the same type of asymmetry suggests that life on Earth originated only once, not repeatedly.

The initial choice of levorotatory over dextrorotatory amino acids may have been a matter of historical accident, but once this choice had been made, subsequent syntheses were generally asymmetrical, owing either to asymmetrical precursors or asymmetrical enzymes or both. If the earliest successful, self-replicating droplets happened to have a preponderance of levorotatory amino acids, merely by chance, then these levorotatory amino acids gained the upper hand simply by being there first. The presence of a successful system with levorotatory amino acids filled the seas with so many levorotatory molecules that the evolution of a form with dextrorotatory amino acids became nearly impossible, an application of the ''king of the mountain'' principle (Chapter 18). The selection that strongly favored conformity to the preponderant direction of molecular asymmetry would have granted further and continued success to whichever type had gained the initial advantage.

The Origin of Membranes and of Individuality

The same breakthrough that overcame the limits on chemical complexity imposed by very low equilibrium concentrations may also, according to Oparin, have accounted for the origin of membranes. These first membranes need not have been as highly structured or as complex as the plasma membranes of existing cells, but they must have been capable of forming a perimeter that distinguished an ''outside'' from an ''inside.'' Such primordial membrane-bounded systems, called ''protobionts'' by Oparin, were probably the first ''individuals'' of the biological world, for each was unique and no two were long identical. Chemical substances within these protobionts could now reach concentrations different from, and in particular higher than, those prevailing outside. A selective process, called ''protoselection'' by Oparin, caused certain fortuitous protobionts to persist longer than others, perhaps even to grow. Those that were internally balanced and capable of incorporating more of their environmental surroundings into themselves, grew larger and may have divided spontaneously in half, perhaps with the aid of wave action or other mechanical forces, thus in effect ''reproducing'' themselves. Nearly all the characteristics of life listed in Chapter 1 were present in elementary form. Metabolism and self-regulation took the form of chemical reactions that perpetuated the system and incorporated more into the system from the environment.

Within the membranes of the early protobionts, chemical concentrations reached the point that permitted the polymerization of amino acids to form proteins, and possibly also of nucleic acid bases to form nucleotides and then nucleic acids. Some of the spontaneously formed proteins were capable of functioning as enzymes. Those that happened, by chance, to increase the stability of their protobionts, or to increase the ability of these protobionts to incorporate and derive energy from energy-rich molecules in the environment, were selected for and

increased so rapidly that they soon became ubiquitous. Most important were those enzymes that not only conferred advantage on the protobiont as a whole, but also directed their own molecular synthesis, a process known as autocatalysis. Autocatalytic enzymes, once formed fortuitously, furthered their own synthesis and were thus self-perpetuating.

Oparin, Fox, and others have experimented with simple model systems thought to possess many of the traits of the original protobionts. Fox has produced in his laboratory many coacervate ''microspheres''; when viewed under the microscope, they can easily be mistaken for living organisms.

The Evolution of Protein Synthesis

All modern proteins are synthesized from nucleic acid templates, though some are subsequently modified into their final form. Nucleic acids in modern systems can replicate or be transcribed only with the aid of enzymes, which are proteins. Could proteins have evolved before nucleic acids, or *vice versa*? (Did the chicken come before the egg?) Research on this point is currently very active: Fox and his co-workers have studied several protein-containing systems produced without the aid of nucleic acids.

It is now widely believed that the earliest proteins probably formed by the random polymerization of abiotically produced amino acids. But modern organisms all synthesize proteins by a far more complex mechanism, typically involving DNA, messenger RNA, transfer RNA, ribosomes, and a variety of enzymes, including those used in the synthesis and replication of the nucleic acids themselves. When did this change come about? Did it come about gradually, or all at once? This is now a fruitful area for research, but the problems are many. Protein synthesis, in even the simplest of organisms, is far more complex, by several orders of magnitude, than any step-by-step chemical synthesis. In particular, cellular protein synthesis depends on the presence of ribosomes, and the structure of even the simplest of ribosomes is as yet incompletely known.

Many workers in this field believe that protein synthesis in the earliest biological systems was entirely protein-dependent and very inefficient. The exact sequence of a particular protein was nowhere specified or was at best specified imperfectly by proportionate mixtures of various enzymes and amino acids. The origin of a "genetic code"—a correspondence between protein and nucleic acid sequences—was certainly a major breakthrough, and one that enabled definite sequences of proteins to be produced reliably again and again. Perhaps the earliest genetic code worked in both directions, synthesizing nucleic acid sequences from proteins and then using the nucleic acid sequences to direct the synthesis of new protein molecules that would duplicate the original sequence. If this were true, perhaps messenger RNA evolved first, and the DNA that now determines its sequence evolved much later.

The modern genetic code, whatever its origin, must have conferred a very high selective advantage upon its first possessor. That first modern system must have outmultiplied any competing systems so that other possible genetic codes had no opportunity to become established. Again, the "king of the mountain" principle applies: the first organic system to evolve a genetic code gained such an advantage that it was able to preclude other organic systems from evolving different but equally plausible codes. This is presumably why *all living organisms have the same genetic code.* Yet the initial correspondence, say between the nucleotide sequence AGA and the amino acid arginine may initially have been a chance occurrence. (It is not at all obvious why a code in which AGA codes for arginine should have an advantage over other possible codes.) Some researchers believe that the original code may have been a "two-letter" code, using only two nucleotides (instead of the current three) to code for each amino acid. A maximum of 16 amino acids could be specified by such a code, and the evolution of a "three-letter" code may have been favored initially by its ability to code for additional amino acids.

At what point in the origin of life would the term "living" first be applicable? If the origin of life had been sudden, as the proponents of spontaneous generation had assumed, then the demarcation of life from nonlife would be easy. But scientists since Oparin have hypothesized instead a gradual origin of life, one in which the boundary between life and nonlife is harder to draw. As we learn more about life's gradual unfolding, we may someday be able to select a "key character" (Chapter 17)—a major breakthrough—as the arbitrary starting point of life. The origin of proteins, of membranes, or of

asymmetrical synthesis are all possible criteria. The perfection of the nucleic acid-based type of protein synthesis is a much more stringent criterion.

C. EXOBIOLOGY: THE SEARCH FOR EXTRATERRESTRIAL LIFE

If life originated on the Earth, might it not also have originated elsewhere? The search for (and, hopefully, the study of) extraterrestrial life is a new field known as **exobiology.**

The origin of life was certainly an event of very low probability. Yet among the millions and billions of stars and their planets, might it not have occurred several times, if not hundreds of times? Suppose that the origin of life had in any given year a probability of one in a billion, or .000000001. Then the probability that life would originate at least once in a period of a billion years is $1 - (.999999999)^{1000000000}$, or about 0.63. Over two billion years, the probability of life's originating at least once rises to 0.86, and over four billion years the probability becomes 0.98, which is nearly certain. And the number of planets with at least this much probability for life may indeed be great.

Our prospects of ever seeing or studying life beyond our own Solar System are nevertheless very poor. To reach Alpha Centauri, the nearest star to our Sun, would take over 8700 years, even at speeds 100 times faster than the maximum attainable by our present rockets. For the foreseeable future, we shall have to be content with the exploration of our own Solar System.

Prospects for Life within Our Solar System

Thus far, the Moon is the only celestial body that has been directly sampled, and it has been shown to contain neither life nor conditions suitable to life. There is virtually no atmosphere on the Moon, nor any liquid water. The smallness of the Moon's mass would explain its lack of an atmosphere, for its gravitational field is not strong enough to retain one. This would also explain the Moon's inability to hold liquid water, for without an atmosphere pressing down upon it, liquid water would soon evaporate and pass off into space. Chemical evolution could hardly be expected on a planetary body retaining neither liquid water nor an atmosphere. This also precludes the origin of life on the planet Mercury, and on all but one of the satellites in our solar system. None of the asteroids (also called "minor planets") has sufficient mass to retain any atmosphere either; the comets and meteoroids have masses that are smaller still.

Of all the planets, Venus is believed to be most nearly like the Earth. It is 82% as massive as the Earth, and is thus capable of retaining an atmosphere composed mainly of carbon dioxide (CO_2). The surface temperature is about 425°C, far above the boiling point of water. At this temperature, organic molecules such as proteins and lipids could not survive, and life as we know it could therefore not exist. Yet Venus may resemble an earlier stage in the prebiotic history of the Earth. The atmosphere is not now strongly reducing, though it may once have been so. Complex organic substances may have been formed either on Venus' surface or in its upper atmosphere, but we will have to await further explorations before we can determine this with any certainty.

Mars has always attracted the most attention as a possible haven for life. With a mass roughly 1/7 that of Earth's, Mars is large enough to retain an atmosphere containing about 95% carbon dioxide, 3% nitrogen, 1% or more argon, and smaller amounts of other gasses. Water vapor is present in very small amounts, but these amounts vary with latitude, reaching a maximum over the Martian polar ice cap, and confirming that this ice cap contains frozen water. The seasonal melting and refreezing of parts of this ice cap raises the possibility that liquid water may somewhere have a transitory existence on Mars, but over the majority of the planet the atmospheric pressure is too low to permit water to persist without evaporation. A few Martian surface features have been attributed to the action of running water, and the isotopic ratios of argon-41 to argon-40 lend support to the hypothesis of an atmosphere that may once have been sufficiently dense to permit liquid water.

Jupiter, the largest planet, has no solid core, and is composed largely of hydrogen. Ammonia (NH_3) is the second most abundant material on Jupiter, and methane (CH_4) and formaldehyde ($HCHO$) have also been identified. The abundance of carbon-containing molecules is greater in Jupiter's "red spot," and some amino acids have even been detected there. Conditions on Jupiter may resemble the earlier stages of chemical evolution on Earth.

Saturn, Uranus, and Neptune resemble Jupiter in many ways. All are very large—sufficiently massive to retain hydrogen, the lightest gas, in their atmospheres. Both Saturn and Uranus have rings; Saturn's rings consist largely of frozen water. One of Saturn's many satellites, Titan, is the largest satellite of any planet in the Solar System, and the only one of sufficient mass to retain an atmosphere.

Pluto resembles the terrestrial planets (Mercury, Venus, Earth, and Mars) in size, and is believed to have a solid surface. Because of its distance from the Sun, Pluto is too cold to support life as we know it.

The only samples of extraterrestrial material that naturally occur on Earth are meteorites, the remnant material of meteors or "shooting stars." Meteorites may be classed as either iron meteorites (siderites), consisting largely of iron and nickel, or as stony meteorites (aerolites), containing silicate minerals, with 10% or more iron and nickel. Bacteria and other organisms have been detected in meteorites, but they belong to known earthly species and are probably terrestrial contaminants. The detection of amino acids and other organic molecules in meteorites suggests that the chemical reactions postulated to have taken place on the primitive Earth may also have taken place elsewhere; the mineralogy of these meteorites suggests that they were formed under strongly reducing conditions. The presence of both D- and L-amino acids argues strongly against terrestrial contamination, since the overwhelming majority of terrestrial amino acids are of the levorotatory or L form.

The Remote Detection of Life

The detection of life on remote planets raises many technical problems. Not the least of these problems is that we are not certain as to the kind of life we might expect to find or how best to detect it. Other difficulties arise from hypotheses that extraterrestrial life is likely to be microscopic in size and perhaps also so patchy in its distribution as to escape detection by instruments unfavorably situated.

The *Viking* mission to Mars was the first to seek evidence of microscopic life on a remote planet. Three ingenious experiments were designed for the *Viking* mission to detect chemical processes indicative of the activities of living organisms. The experiments, summarized in Item 25.1 on page 426, failed to find any unequivocal signs of life on Mars. Organic compounds that might represent decay products have also been sought in Martian soil (Biemann et al., 1976). The failure to detect these organic compounds, even in concentrations as low as a few parts per billion, suggests pessimism in the search for life on the red planet.

Is our own solar system unique, or do other stars have planets revolving around them, too? Estimates given by different astronomers differ widely: some suppose that nearly every star has a planetary system; others search in vain for a single proven case outside our own solar system. Barnard's star, one of the nearest stars to us in our own galaxy, shows small irregularities in its motion. Its motion seems to be perturbed by the revolution around it of at least one "unseen companion," either a large planet with a mass greater than Jupiter's, or else two such planets, "unseen" because they are not self-illuminating.

D. THE EVOLUTION OF PROCARYOTIC CELLS

The eucaryotic cells of animals and plants were described in Chapter 1, and their features were shown in Fig. 1.1 and Item 1.1. The cells of bacteria and blue-green algae are much simpler; we call them **procaryotic** cells (Fig. 25.5; see also Box 26–A). Procaryotic cells lack the nuclear envelope, endoplasmic reticulum, mitochondria, plastids, vacuoles, golgi complexes, centrioles, and microtubules of eucaryotic cells. The lack of a nuclear envelope leaves the boundary between nucleus and cytoplasm ill defined. Only a single chromosome, containing DNA but no protein, is present. True cilia or flagella, with the "9 + 2" internal structure, are absent, though a simpler type of flagellum is sometimes present.

The First Cells

The first living organism was probably a **bacterium**, meaning a procaryotic cellular organism lacking chlorophyll *a*. Under the assumption that the atmosphere was strongly reducing and lacked free oxygen, we should expect the first bacteria to have been **anaerobic** (oxygen-intolerant). True or obligate anaerobes are killed by free oxygen; they lack the very complex enzymes needed to detoxify oxygen-rich compounds. **Aerobic** (oxygen-tolerant) organisms include all eucaryotic animals and plants

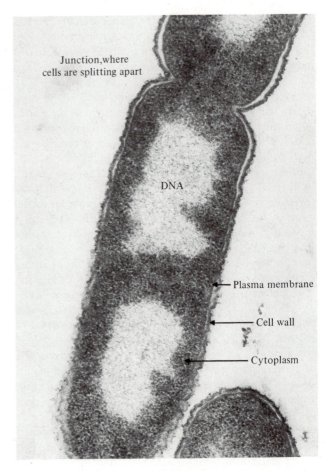

Fig. 25.5   The procaryotic cell of the bacterium *Bacillus subtilis*. Reprinted with permission from Jeffrey J. W. Baker and Garland E. Allen, *Study of Biology*, 3e © 1977. Reading, Massachusetts: Addison-Wesley (Fig. 5.23).

as well as many bacteria. Of these, obligate aerobes are absolutely dependent upon oxygen, while those that are facultative can exist either with or without free oxygen in their surroundings. In today's oxidizing atmosphere, it is not surprising that aerobic forms predominate, while obligate anaerobes are restricted to very special habitats. We may assume that anaerobic forms were more widespread in the reducing conditions of the remote past. But while some modern bacteria may always have been anaerobic, other modern anaerobes may be secondarily derived from aerobic species.

A majority of today's bacteria are both **chemotrophic** (deriving energy from energy-rich molecules obtained from the environment) and **heterotrophic** (dependent upon organic compounds as carbon sources). Most energy-rich organic compounds today are products of other organisms. This was once taken to mean that **autotrophs** (organisms that can derive their carbon compounds from CO_2 and carbonates) must have come first, since modern heterotrophs are dependent upon them. But we now realize that heterotrophs of the past probably subsisted on energy-rich organic compounds produced abiotically under reducing conditions. The diversity of nutritional types among modern bacteria (Box 25–C) presumably came later. Perhaps the following reactions were among those that provided energy to the earliest organisms.

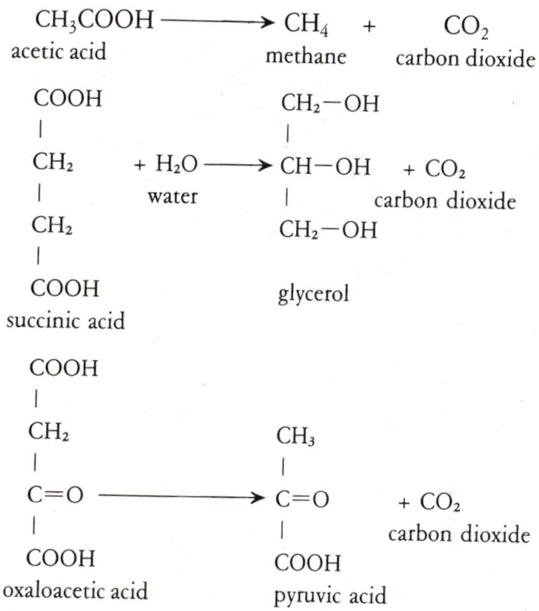

$$CH_3COOH \longrightarrow CH_4 + CO_2$$
acetic acid   methane   carbon dioxide

$$\begin{array}{c} COOH \\ | \\ CH_2 \\ | \\ CH_2 \\ | \\ COOH \end{array} + H_2O \longrightarrow \begin{array}{c} CH_2-OH \\ | \\ CH-OH \\ | \\ CH_2-OH \end{array} + CO_2$$
succinic acid   water   glycerol   carbon dioxide

$$\begin{array}{c} COOH \\ | \\ CH_2 \\ | \\ C=O \\ | \\ COOH \end{array} \longrightarrow \begin{array}{c} CH_3 \\ | \\ C=O \\ | \\ COOH \end{array} + CO_2$$
oxaloacetic acid   pyruvic acid   carbon dioxide

BOX 25–C   *Types of Nutrition among Bacteria and Other Organisms*

	Energy source	
	CHEMICAL ENERGY SOURCE (CHEMOTROPHIC)	SUNLIGHT AS ENERGY SOURCE (PHOTOTROPHIC)
Source of electrons:		
Inorganic reducing agents (Lithotrophic)	CO_2 or carbonates used as carbon sources: chemolithotrophic autotrophs Examples: some bacteria (Nitrobacteriaceae, Thiobacteriaceae, Siderocapsaceae). Organic carbon sources: **chemolithotrophic heterotrophs** Examples: some bacteria (Methylomonadaceae).	CO_2 or carbonates used as carbon sources: photolithotrophic autotrophs (photosynthesizers) Examples: true plants, blue-green algae, some bacteria (Chlorobiaceae, Chromatiaceae).
Organic reducing agents (Organotrophic)	Organic carbon sources: chemoorganotrophic heterotrophs Examples: animals, fungi, and most bacteria.	Organic carbon sources: photoorganotrophic heterotrophs Examples: some bacteria (Rhodospirillaceae).

We would suppose that the earliest bacteria must have been capable of nucleic acid replication and also of protein synthesis. A group of complex, multiple-ringed molecules that might have been beyond their synthetic capabilities are the **porphyrins** and their derivatives, including true chlorophylls, bacteriochlorophylls, cytochromes, heme, certain vitamins and cofactors, and a few important enzymes. Life without any of these porphyrin derivatives must have been restricted to anaerobic fermentation; the further oxidation of carbohydrates would have been difficult without atmospheric oxygen and impossible without cytochromes. Hydrogen peroxide (H_2O_2), a highly toxic bacterial poison, cannot be detoxified without such porphyrin-containing enzymes as catalase and peroxidases. Since peroxide is produced in small amounts whenever oxygen is present, life must have been strictly anaerobic. Photosynthesis would have been impossible without chlorophylls, and, without photosynthesis, sugars would not have been plentiful. Energy-yielding organic reactions that today use coenzymes derived from porphyrin compounds would have been very inefficient, using simple enzymes containing only protein, perhaps aided by the "capture" of abiotically synthesized porphyrins. Reactions that today use cytochromes to build up and store phosphate-bond energy in the form of ATP would not have occurred, since cytochromes are porphyrin derivatives too.

Considerable bacterial evolution may have taken place before the bacteria began to leave traces of themselves in the fossil record. The oldest recorded fossil is that of a bacterium, *Eobacterium isolatum* (Fig. 25.6), preserved in the Fig Tree Formation of South Africa, a formation now considered to be 3.5 billion years old. The discovery was made by Elso S. Barghoorn of Harvard University, who carefully examined many thin sections under the electron microscope.

Bacterial fossils are useful in confirming the great antiquity of bacteria as a group, but the morphological traits that are preserved are of little use in reconstructing a bacterial family tree. Most bacteriologists regard morphological traits as poor taxonomic characters, and fossils tell us little else. Bacterial fossils hardly reflect the overwhelming abundance and great physiological diversity of modern bacteria. The rarity of fossil bacteria may reflect in part their ability to decay rapidly and in part the fact that few searches are made for such fossils. Relatively few

Fig. 25.6 *Eobacterium isolatum,* the oldest known bacterium, from the Fig Tree formation of South Africa. Reprinted courtesy of E. S. Barghoorn and J. W. Schopf.

bacteriologists are primarily interested in phylogenies; those that are often rely more heavily on the comparative study of modern bacteria.

Modern Bacteria

All bacteria have procaryotic cells. Most are unicellular, but a few are found in simple, undifferentiated colonies, meaning that all the cells are basically alike, with no cellular specialization or division of labor among them.

Despite these underlying similarities, there is much diversity (Fig. 25.7). Bacteria exist in spherical cells (cocci), rod-shaped cells (bacilli), curved, sausage-shaped cells (vibrios), pear-shaped cells (coryneform), corkscrew-shaped cells (spirilla), and clusters of various sorts.

Most bacterial cells are surrounded by a cell wall, which may be either rigid or flexible depending upon its thickness. Cell walls that can be stained with a mixture of iodine plus crystal violet are called **Gram-positive**; they contain a thick layer of mucopeptides—complex heteropolymers of sugar and amino acid units. Typical **Gram-negative** cell walls, which do not retain this dye, have several layers containing proteins, polypeptides, and lipids. Many bacteria also possess either a protein or polysaccharide capsule surrounding the cell wall. Some bacteria contain flagella, but these invariably lack the complex "9 + 2" internal structure of eucaryotic flagella. Flagella that occur at the ends of rod-shaped cells are called **polar flagella**; those that occur all over the bacterial surface are called **peritrichous flagella.**

Bacteria exhibit a greater diversity of nutritional or metabolic types than either the plant or animal kingdoms (Box 25-C). A similar diversity, mentioned earlier, exists with regard to oxygen tolerance. Bacteria have been found in nearly all possible habitats on Earth, but they are most abundant in soils. Many bacteria inhabit other organisms, sometimes causing disease to their hosts.

The most widely followed classification of bacteria appears in *Bergey's Manual of Determinative Bacteriology*. In the latest (eighth) edition of this work, edited by Buchanan and Gibbons (1974), the bacteria are divided into 19 "parts," for which formal taxonomic names at the ordinal level may be substituted. Distinctions among these orders are based on characters of the cell wall, flagella, overall shape, and especially on various metabolic capabilities.

Fig. 25.7   Representative bacteria.

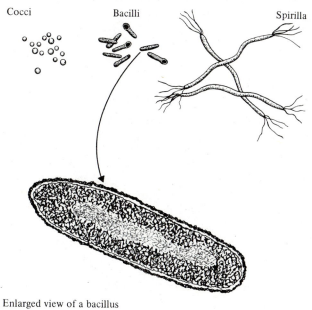

Cocci   Bacilli   Spirilla

Enlarged view of a bacillus

The methanogenic (methane-producing) bacteria, also called Archaebacteria, are physiologically simple, but they contain some unusual enzymes. Their unique form of metabolism produces methane gas according to the equation

$$8H + CO_2 \rightarrow CH_4 + 2\,H_2O.$$

hydrogens (from   carbon   methane   water
metabolism, via   dioxide
NADH)

Because they are strictly anaerobic, yet simple, it has long been suspected that they may represent the sole survivors from an age when most of the Earth still had a reducing atmosphere. The sequence of nucleotides in their ribosomal RNA supports this speculation, for it shows them to be quite distinct from all other procaryotic and eucaryotic organisms (Fox *et al.*, 1977).

The phototrophic bacteria (Rhodospiralles) display a spectrum of abilities. The most primitive group are photoorganotrophic, obtaining energy from sunlight, but depending on externally obtained organic compounds both as hydrogen donors and as carbon sources. Two photolithotrophic families are truly photosynthetic in that they can use carbon dioxide as their major carbon source, but they use hydrogen sulfide (H_2S) as a source of hydrogen and release elemental sulfur (S_2) in the process. The most efficient phototropism, common to blue-green algae and eucaryotic plants, uses water (H_2O) as a hydrogen donor and releases free oxygen. Chlorophyll *a* is required for this type of photosynthesis. The blue-green algae probably evolved from photosynthetic bacteria by the acquisition of this pigment.

Blue-Green Algae (Cyanophyta; Cyanobacteria)

The blue-green algae, or Cyanophyta (Fig. 25.8), are photosynthetic procaryotes using chlorophyll *a* as their sole photosynthetic pigment. Although their cell structure is procaryotic, they resemble other algae (and differ from bacteria) in using water as the hydrogen donor for photosynthesis, and in producing free oxygen (O_2) as a result. The free oxygen in the Earth's atmosphere is believed to have originated as a byproduct first of blue-green algal photosynthesis and later of the photosynthesis of eucaryotic plants.

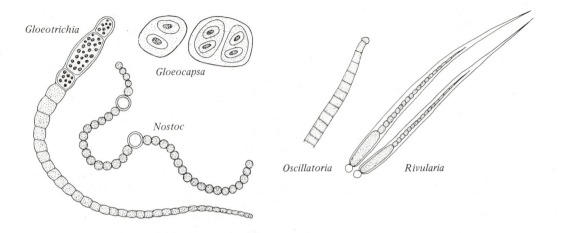

Fig. 25.8   Representative blue-green ''algae.''

Unlike the true plants in which photosynthetic and other pigments are restricted to special plastids, the pigments of blue-green algae occur on much simpler structures called chromatophores, which are scattered throughout the outer portion of the cytoplasm. Of these pigments, chlorophyll *a* and β-carotene are both universal among photosynthetic true plants, but absent in all bacteria. Several other pigments are unique to the Cyanophyta.

The cells of blue-green algae are procaryotic, resembling those of bacteria. Absent from these cells are the nuclear envelope, endoplasmic reticulum, mitochondria, plastids, or ''9 + 2'' structures characteristic of eucaryotic cells. The cells may occur singly, or in groups, colonies, or filaments.

Evolutionary trends among blue-green algae include advances in colonial organization, reproduction, and the invasion of new habitats. The most primitive forms are unencapsulated solitary cells reproducing by binary fission. The ancestral habitat is a matter of conjecture, but the majority of living blue-green algae are freshwater. The widespread ecological tolerance of blue-green algae has allowed them to invade a number of niches closed to most other organisms, including thermal springs, bare rock surfaces, and the surfaces of other organisms including animals as well as plants. This last mode of occurrence is called **epiphytic**, and organisms deriving support

from the surfaces of other organisms, without being parasitic upon them, are called **epiphytes**. Blue-green algae also enter into symbiotic associations (Chapter 9) with several other organisms, including a variety of green plants, fungi, and protozoans. Blue-green algae living symbiotically with green plants often contribute nitrates, which they obtain by the fixation of atmospheric N_2, an ability that eucaryotic plants lack. The associations with fungi are more intimate, forming lichens (Chapter 26).

The blue-green algae are divided into the two subclasses Coccogoneae and Hormogoneae. The most primitive Coccogoneae are planktonic unicells in which reproduction is restricted to the binary fission of cells and the fragmentation of colonies. Other Coccogoneae have a more advanced form of reproduction that includes the formation of true endospores.

The more advanced blue-green algae are placed in the subclass **Hormogoneae**. These are all truly filamentous, colonial forms in which the cells are arranged in uniserial rows known as trichomes. Trichomes are usually enclosed, either singly or in groups, in a gelatinous or mucilaginous sheath. The surrounding sheath, plus the trichome or group of trichomes contained within, is known as a true filament. Sometimes these filaments are branched, but more often they are straight. Sometimes the filaments are interwoven, forming a loose or dense sheetlike mat or a three-dimensional cluster or matrix.

Most characteristic of the subclass Hormogoneae is the method of reproduction, in which asexually reproducing fragments, known as hormogonia, detach from the tips of the trichomes.

The majority of Hormogoneae belong to the order Nostocales, which includes about two-thirds of all known blue-green algae. The only other order of Hormogoneae are the Stigonematales, often considered the most highly evolved group of blue-green algae because they are the only group in which true branching occurs.

A small number of filamentous blue-green algae form distinctive algal mats known as **stromatolites.** As a thin layer of calcium carbonate is deposited over a layer of photosynthesizing filaments, the light is prevented from reaching them, while a new layer of filaments grows over the calcium carbonate. The result is a very dense, thick mat made of alternating thin layers of calcium carbonate and dead algal filaments; only the uppermost layers contain living algal filaments. Fossil stromatolites occur in rocks of varying age, generally Paleozoic and Precambrian. The oldest known stromatolites, now considered to be at least 2.8 billion years old, are from the Bulawayan limestones of Zimbabwe.

The fossil record of blue-green algae has been reviewed by Schopf (1974). Nearly every family includes known fossil representatives. The majority of these date from the Precambrian, which Schopf calls the "Age of Blue-green Algae." The oldest alga-like fossils, which appear to belong to the Cyanophyta, are from the Fig Tree cherts of South Africa, approximately 3.2 billion years old. The Bulawayan stromatolites, 2.8 billion years old (early Precambrian), provide indirect evidence for the existence of filamentous forms presumably similar to the modern family Oscillatoriaceae. The Gunflint Iron Formation of Ontario, Canada, dated at about 1.9 billion years ago (middle Precambrian), contains a wide variety of fossils, most of them consisting of blue-green algae. These fossils show that the major groups of Cyanophyta had largely been established by this time.

By late Precambrian times, evolution of the blue-green algae had essentially been completed. The 1.0 billion year old Bitter Springs Formation of Australia contains representatives of virtually every living family, including at least ten hormogonean and three coccogonean forms comparable to modern genera. These and other fossils show the blue-green algae to have remained very conservative morphologically over the last billion years, and other studies have established independently the fact that the blue-green algae are biochemically conservative as well. One possible reason for the great conservatism of blue-green algae lies in their broad ecological tolerance, including tolerance for many environmental extremes inimical to the growth of other living organisms.

The decline in importance of the blue-green algae began in the late Precambrian with the evolution of sexually reproducing eucaryotes. During Phanerozoic time, the blue-green algae have been largely supplanted and replaced by other photosynthesizing organisms, namely the eucaryotic algae and higher plants. Except for a few specialized environments, such as thermal springs, the eucaryotic plants are now unquestionably the dominant forms of photosynthetic life.

E. VIRUSES

Viruses consist of minute infective particles without any evidence of cellular organization. Virus particles contain a single nucleic acid (DNA or RNA—never both) surrounded in most cases by protein. Viruses do not make their own carbohydrates or lipids. The lack of cellular organization (especially the absence of a true viral membrane and the presence of only one type of nucleic acid) serves to distinguish viruses from bacteria.

Viruses are known to infect animals, plants, fungi, and bacteria. The bacterial viruses are called **bacteriophages.** All viruses are obligate parasites. They reproduce by injecting their nucleic acid, generally without any protein, into the cells of the host, using the host's replicative machinery and the blueprint of viral nucleic acid to manufacture new virus particles (Fig. 25.9). Viral reproduction typically involves the death by rupturing (lysis) of host cells or tissues. A few bacterial viruses, known as temperate or lysogenic phages, combine with the host genetic material (chromosome) and are transmitted from one host cell to another and to succeeding bacterial generations. Then spontaneously, or with some external induction, the temperate phages become virulent, multiply rapidly, and rupture (lyse) the bacterial cell.

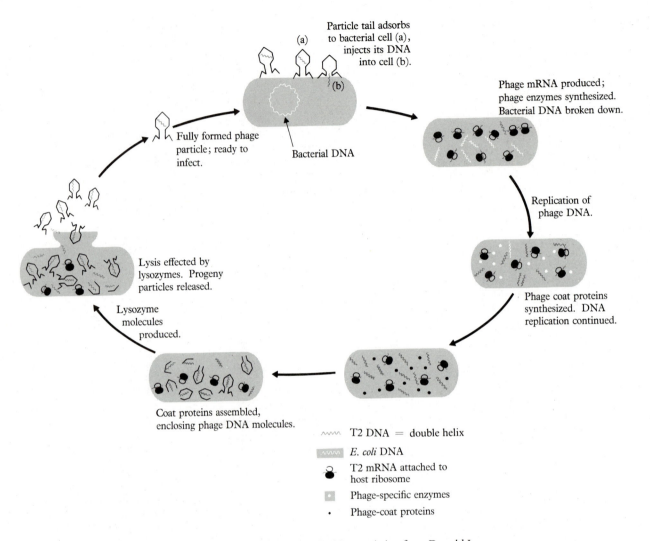

Fig. 25.9 The life cycle of a virus, bacteriophage T2. Reprinted with permission from Donald I. Patt and Gail R. Patt, *An Introduction to Modern Genetics,* © 1975. Reading, Massachusetts: Addison-Wesley (Fig. 7.6).

An important question regarding the evolution of viruses is whether they are more closely related to each other or to their hosts. Their degeneracy (loss of cellular organization, for instance) is perhaps explained by their parasitic way of life. Their apparent similarity to one another in parasitic habits and life cycle is to a great extent negated by the differences between DNA and RNA forms, between single- and double-stranded nucleic acids, and between helical, icosahedral (20-sided), and various other shapes. Perhaps the best evidence is provided by studies on the percentage of homologous sequences between viruses and their hosts, revealing in general a rather higher-than-random incidence of homology. Further evidence is provided by the fact that the so-called temperate phages referred to above can become incorporated into, and replicated as part of the bacterial chromosome. In this condition, the temperate phages are essentially similar to plasmids or episomes, small fragments of bacterial chromosomes capable of semi-autonomous existence within the bacterium.

Viruses are classified first on the nature of their nucleic acid (whether DNA or RNA), and secondly on the type of symmetry shown. The symmetry may be either helical (as in tobacco mosaic virus), cubical (as in poliovirus and other icosahedral forms), or "binal," with "head" and "tail" portions, as in most bacteriophages.

CHAPTER SUMMARY

Spontaneous generation—the origin of life from lifeless matter and without apparent cause—was rejected in the case of modern organisms as the result of Redi's experiments with flies and Pasteur's and Tyndall's experiments with microbes. Oparin's theory of primary abiogenesis teaches that life originated gradually from nonlife in the remote past, under reducing conditions, and that life itself has modified these conditions so greatly that living things today can no longer arise in this way—they arise only from previously living things. Oparin's theory passed a crucial test when S. L. Miller circulated a mixture of hydrogen, methane, ammonia, and water vapor over an electric spark for several days and recovered many organic compounds, including simple peptides. Nearly all important constituents of life have now been produced by experiments similar to Miller's.

Early chemical evolution is thought to have occurred in much the same way on Earth, with organic compounds produced abiotically until a "hot, dilute soup" was formed. Further chemical evolution, especially the origin of polymers, could occur only in those places where the dilute mixture could become concentrated, and several mechanisms have been proposed to account for this concentration. Oparin's own proposal is that small droplets were able to become separated from the larger body of fluid, and to enclose a somewhat more concentrated fluid within, a process known as coacervation. The membrane-like boundaries around these droplets may represent the first appearance of individuality. In the case of molecules that can exist either in right-handed or left-handed mirror images, life generally imposed a uniformity that resulted in the preponderance of one asymmetrical form over the other. A most crucial step, not yet fully understood, was the origin of a mechanism to permit the faithful replication of long-chain sequences in polymers such as proteins and nucleic acids.

Life may also have originated on other planets in much the same way as on the primitive Earth, but searches to date have discovered no unequivocal evidence of any extraterrestrial life. Most of the other planets in our Solar System have conditions hostile to life, though some of the early stages of chemical evolution did occur and could be studied to advantage. Exobiology, the detection and study of extraterrestrial life, has advanced with remote experiments on the surface of Mars, but the results of these experiments are inconclusive and unencouraging in the search for life. The planets of stars other than our Sun cannot yet be studied closely.

The first cells on Earth were procaryotic, and probably resembled bacteria. Though they may have lacked some of the more complex metabolic pathways of modern bacteria, such as the Krebs cycle, they certainly possessed membranes, nucleic acids, and mechanisms of protein synthesis necessary to ensure reproduction. Most modern bacteria also possess porphyrin-ring compounds such as catalase (an enzyme); more complex porphyrin-ring compounds include cytochromes, heme, and both bacterial and higher plant chlorophylls. Modern bacteria are diverse in shape, in flagellation, in cell wall construction, in oxygen tolerance, and in nutrition. Oxygen tolerance, autotrophic nutrition, use of light as an energy source, and use of inorganic reducing agents as hydrogen donors are all considered derivative traits. Blue-green algae are believed to have originated from photosynthetic bacteria that evolved a more efficient set of pigments, including chlorophyll *a*. Blue-green algae that reproduce by fragmented filaments (hormogonia) are considered more advanced than those that reproduce by fission only.

Viruses were formerly viewed as primitive forms of life or as degenerated parasites. Much evidence now points to their origin as fragments of the genetic material of the cellular organisms that they infect. Viruses may use either DNA or RNA as generic material, and they may be rod-shaped, cubical, icosahedral, or "binal," with a tail-piece, as in the bacterial viruses (bacteriophages).

FOR FURTHER KNOWLEDGE

ITEM 25.1 *The Search for Life on Mars: Three Exobiology Experiments*

In the search for life on Mars, the 1976 *Viking* mission conducted three experiments designed to test for various kinds of possible life. Each experiment was based on a different hypothesis: life on Mars might be photosynthesizing, or it might be heterotrophic. In either case, it might be expected to alter the composition of gasses in its immediate environment. In addition to these experiments, there was also visual monitoring of the *Viking* lander's surroundings for any relevant changes, and analysis of the soil for various organic chemicals. Neither the visual monitor nor the soil sample analyses revealed any signs of biological activity.

The three major experiments were designed to be performed automatically or by remote control. Even the analysis of experimental products had to be carried out by instruments that would transmit data back to Earth.

I. CARBON ASSIMILATION (PYROLYTIC RELEASE) EXPERIMENT

In this experiment, a sample of Martian soil was incubated in an atmosphere of radioactively labeled $^{14}CO_2$ and ^{14}CO, under simulated Martian sunlight provided by a xenon lamp. Any living matter that was present would be expected to incorporate radioactive carbon. After 120 hours of incubation, the radioactive gasses were vented from the test chamber, and the sample was pyrolyzed (heated to 625°C) to decompose any organic matter contained in it. The gas given off was analyzed for radioactivity. The experiment thus measured *the ability of Martian soil to incorporate radioactive CO_2 and CO into organic compounds*.

A certain rate of carbon assimilation was detected in this experiment, and the ability of the sample to assimilate carbon was greatly diminished when the sample was heated for three hours prior to testing. Nevertheless, the results have proven equivocal because at least one laboratory has succeeded in duplicating the experimental conditions and results on Earth, using a simulated Martian soil devoid of any life.

II. GAS EXCHANGE EXPERIMENT

In this experiment, a sample of Martian soil was placed in a chamber together with a nutrient medium under controlled conditions. The gas above the sample was analyzed for any changes in its composition that might reflect biological activity. Several changes were noticed, but these could all be attributed to inorganic reactions. In one of these changes, oxygen was released upon moistening of the soil sample. Several alternative explanations have since been advanced for this release of oxygen.

III. LABELED RELEASE EXPERIMENT

In this experiment, a nutrient medium was again added to a Martian soil sample. The nutrient medium in this case contained seven radioactively (^{14}C) labeled organic compounds (formate, glycolate, glycine, D-alanine, L-alanine, D-lactate, and L-lactate). The gasses in contact with the test chamber were analyzed for any evolved radioactivity that might indicate metabolic activity. An immediate increase in measured radioactivity occurred when the nutrient medium was first injected, and several other changes of lesser magnitude were also observed. None of these changes, however, could be attributed with any certainty to biological activity.

26

The Evolution of Eucaryotes

A. THE NATURE AND ORIGIN OF EUCARYOTIC CELLS

Perhaps the greatest unexplained transition in evolutionary history, subsequent to the origin of life itself, lies in the origin of eucaryotic cells. Eucaryotic cells (Fig. 1.1) differ profoundly from procaryotic cells (see Fig. 25.6) in many ways (Box 26–A): they are typically larger; they have an extensive series of internal membrane structures including endoplasmic reticulum, Golgi apparatus, and nuclear envelope; they possess microtubular organelles such as cilia and centrioles; they possess mitochondria; they often possess plastids; and their chromosomes, often multiple instead of single, contain proteins as well as DNA (see also Item 1.1 at the end of Chapter 1). The increase in size and the presence of internal membranes

BOX 26–A *Differences between Procaryotes and Eucaryotes*

	Procaryotic BACTERIA; VIRUSES, RICKETTSIAS; BLUE-GREEN ALGAE	**Eucaryotic** ALL PROTOZOA AND MULTICELLULAR ANIMALS; ALL TRUE FUNGI; ALL ALGAE (EXCEPT BLUE-GREEN); ALL HIGHER PLANTS
CELL SIZE	Small (1 to 10 micrometers)	Large (10 to 100 micrometers)
NUCLEUS	Not membrane-bound (no true nucleus)	Membrane-bound (with nuclear envelope)
GENETIC MATERIAL	Single "genophore" containing nucleic acids only; no histones; often no Feulgen reaction	Many true chromosomes containing proteins (histones, etc.) as well as nucleic acids; gives Feulgen reaction
CELL DIVISION	Direct, by binary fission; no spindle, centriole, or microtubules	Complex, by true mitosis using centrioles, spindle, or microtubules (25 nm diameter), etc.
SEXUAL REPRODUCTION	None, or else simple unidirectional transfer of material	Usually present; both sexes produce gametes by meiosis
MULTICELLULAR ORGANIZATION	Usually none; never from diploid zygotes; never by tissue differentiation	When (usually) present, from diploid zygotes by extensive tissue differentiation
OXYGEN USE	Varies from strict anaerobes (killed by O_2) to aerobes	All aerobic (need O_2); exceptions are clearly secondary modifications
METABOLISM	Great variations	All show Krebs cycle, Embden-Meyerhof reactions, etc.
OXIDATIVE ENZYMES	Located on cell membrane; mitochondria absent	"Packeted" into mitochondria
TRUE CILIA, FLAGELLA	None	Always, with "9 + 2" structure
PHOTOSYNTHESIS (WHEN PRESENT)	Enzymes on cell membrane (no chloroplasts); seldom produce O_2	Enzymes packeted into chloroplasts; always produce free O_2
DIGESTION	Uptake of small molecules from environment	Feed on whole organisms (predator and prey relationships)
CYTOPLASMIC MEMBRANE SYSTEM	None; no endoplasmic reticulum, no food vacuoles, no cyclosis	Endoplasmic reticulum; cytoplasmic streaming (cyclosis); formation of food vacuoles
DNA SYNTHESIS	Throughout life cycle of cell	At specific part of cell cycle only

Adapted from Margulis, 1970, pp. 13–14.

are interrelated features, for neither would be possible to any great extent without the other. The presence of a nuclear envelope between nucleus and cytoplasm has traditionally been considered the hallmark of the procaryote/eucaryote distinction.

Prevailing opinon before 1970 favored the view that procaryotic and eucaryotic organisms were closely related and that simpler, procaryotic algae had given rise to eucaryotic "higher algae." *Euglena* and its relatives were often singled out as possible ancestors of most (if not all) "higher" algae. These algae, in turn, were thought to have given rise to both "animals" (Protozoa) and fungi by the *loss* of their photosynthetic systems. The hypothetical ancestors of euglenoids ("uralgae") were never adequately characterized. The underlying assumption, rarely stated in full, was that eucaryotic algae had evolved from procaryotic algae by the *gradual differentiation* of such structures as mitochondria, plastids, microtubules, and folded membranes.

A radically different theory received few adherents until it was forcefully restated by Lynn Margulis (1970, 1974, 1981) with an impressive array of evidence. According to Margulis, the eucaryotic cell evolved as the result of endosymbiosis (internal symbiosis) among several genetically distinct procaryotes (Fig. 26.1). The mitochondrion, for example, is assumed by Margulis to have evolved from "aerobic, Gram-negative eubacteria containing the Krebs cycle enzymes and cytochrome system for total oxidation of carbohydrates to CO_2 and H_2O. . . ." Likewise, the plastid is assumed by Margulis to have arisen from a blue-green alga, and the microtubular structures from a motile procaryote identified as "perhaps" a spirochete. Margulis further describes the original host cell as "a microbe able to ferment glucose to pyruvate anaerobically via the Embden-Meyerhof pathway . . ."; elsewhere, she characterizes this original host as probably a mycoplasm.

Evidence for the theory of endosymbiosis is compelling in some respects, yet less convincing in others. The symbiotic origin of mitochondria and plastids is suggested by their possession of DNA independent of that in the nuclear chromosomes, often differing from nuclear DNA in the percentage of guanine and cytosine (GC percentage). Studies on the pairing ("hybridization") of nucleic acids *in vitro* show that mitochondrial or plastid

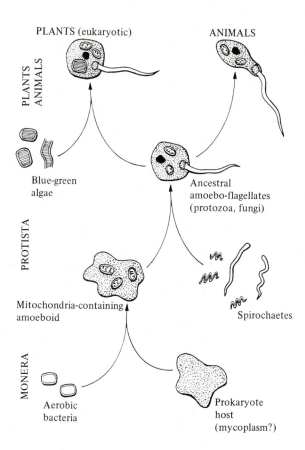

Fig. 26.1 The symbiotic origin of eucaryotic cells, according to Lynn Margulis's theory. Courtesy of Lynn Margulis, Boston University.

DNA is usually not homologous to the nuclear DNA, and shows other features (lack of histonelike proteins, for example) more suggestive of procaryotic than of eucaryotic cells. It is further curious that mitochondrial ribosomes sediment at 70S, the value typical of procaryotic ribosomes, rather than at 80S, the value more characteristic of eucaryotic ribosomes arising in the nucleus. There is evidence that mitochondria and plastids are self-replicating, i.e., they arise from pre-existing mitochondria and plastids, rather than being synthesized anew. Some mitochondrial or plastid proteins (including enzymes) are coded for by the organelles' own DNA, but many more are coded for by nuclear DNA, just as highly evolved symbionts frequently relinquish redundant biosynthetic pathways to the control of their hosts. The claim of a symbiotic origin for mitochondria and plastids has won the increasing support of evolutionists and of cellular and molecular biologists.

Evidence for the symbiotic origin of "9 + 2" organelles is much less compelling than it is for the symbiotic origin of mitochondria, and Margulis's specific identification of spirochetes as possible ancestors to "9 + 2" organelles has drawn few supporters indeed. There is no trace of "9 + 2" organization in any spirochete (or indeed in any bacterium), and the shape and method of contraction of spirochetes are unlike the shape and contraction of eucaryotic flagella. Bacterial "flagella" are stiff, rigid propellers that rotate, quite unlike the flexible eucaryotic flagella, which wave back and forth. The bacterial flagellar proteins (flagellins) are very different from the tubulin that makes up eucaryotic microtubules, and bacterial flagellins are never arranged in "9 + 2" arrays. Even the energy coupling reactions that induce movement are very different in procaryotes and eucaryotes.

Another anomaly of Margulis' theory is her choice of a mycoplasm as the original host. Mycoplasms are rather small (they are the smallest forms of life, other than viruses and rickettsias), while the host in any eucaryotic symbiosis would necessarily have been large (as large as a eucaryotic cell, which is several times larger than most bacteria).

We shall here adopt a modified symbiotic theory, in which mitochondria and plastids are assumed to have originated by endosymbiosis, as Margulis has suggested. The host organism is here assumed to have evolved its own contractile proteins and microtubules without benefit of endosymbiosis. The grouping of microtubules into a "9 + 2" arrangement is currently an enigma; no intermediate stages in this development are known.

How might an endosymbiont (proto-mitochondrion or proto-plastid) ever have entered a host cell? If it simply pierced the host's plasma membrane, as many intercellur parasites are inclined to do, it would be enclosed only by its own plasma membrane. Perhaps the host cell "ate" the symbiont, engulfing it in cellular extensions (pseudopods) and incorporating it into a vacuole, a process known as **phagocytosis**. This would explain why plastids and mitochondria are enclosed by *two* membranes: one derived from the host's plasma membrane, the other from the endosymbiont. Now, phagocytosis occurs in eucaryotic cells only, and it is dependent on the existence of contractile proteins like actin and myosin. The shape of the cell is meanwhile determined by individual microtubules, not organized into the more complex "9 + 2" arrangements; in many cases these microtubules are seen to radiate, like a mitotic spindle, from the centriole. Such microtubules, or the tubulin from which they are made, are unknown among procaryotes and may have evolved as a sort of skeletal system, giving structural support and shape to the otherwise amorphous cell. From these microtubules are derived the mitotic spindles of eucaryotes, and the various structures (cilia, flagella, and centrioles) in which microtubules are arranged in "9 + 2" configuration. Locomotion by means of eucaryotic flagella is thus dependent upon a complex arrangement of microtubules, while locomotion by pseudopods, which requires no such arrangement, probably evolved earlier.

Speculations concerning eucaryotic origins will have to be tested by further research into the nature of both eucaryotic organelles and their supposed procaryotic precursors. A speculative hypothesis, heuristically suggested as a falsifiable assumption to be tested, might run as follows:

1. The host organism, a bacterium, developed actinlike contractile proteins, enabling it to glide or creep more vigorously. The contractile proteins also permitted phagocytosis, i.e., the bulk incorporation of foreign particles or other organisms.

2. As motility and phagocytosis enabled the more efficient procurement of food, with less expenditure of energy than was previously possible, the host bacterium increased in size, along with the necessary expansion of the plasma membrane inward to form an internal membrane system (endoplasmic reticulum, nuclear envelope, etc.). This step crossed the traditionally defined boundary between procaryotes and eucaryotes.

3. The increase in mass demanded a more efficient system of motility. Contractile proteins such as actin and structural proteins such as tubulin became organized into multi-unit structures, which survived in proportion to their ability to increase locomotor efficiency. The ultimate achievement was the "9 + 2"

arrangement of the eucaryotic cilium or flagellum. (At a much later stage in evolution, the fully differentiated muscle tissues of multicellular animals arose in response to similar selection pressures.)

4. The energy-absorbing systems, particularly the "9 + 2" flagella, now demanded a more efficient production of energy in the form of ATP. This favored symbiotic associations with rapidly respiring, fully oxidizing inclusions ("proto-mitochondria"). These internal symbionts may first have been captured by phagocytosis, as if to be used as food, but greater advantage was found in maintaining them alive within the host cell. The vacuolar membrane (originally part of the host's plasma membrane) became the outer mitochondrial membrane, while the symbiont's plasma membrane became the inner mitochondrial membrane. (Perhaps it is this step that might serve as a more crucial, though equally arbitrary distinction between procaryotes and eucaryotes.)

5. The "9 + 2" flagellum specialized to function also as a centriole, through a series of stages carefully outlined by Margulis (1970). The spindle fibers of modern animal and plant cells are now extruded from centrioles derived from the basal bodies of flagella.

6. In the ancestors of plant cells only, a second symbiotic acquisition resulted in the incorporation of blue-green algae, which became plastids. Considerable evolution of pigments must also have taken place, including the loss of many pigments characteristic of blue-green algae, the retention of chlorophyll a, and the acquisition of several new pigments.

7. In plants and fungi, perhaps independently, a eucaryotic cell wall was evolved. The ancestral eucaryote is here assumed to have lacked a cell wall; the cell walls of fungi and plants differ markedly from those of bacteria, while animals lack cell walls altogether.

8. At a time and for a reason as yet undetermined, a radical restructuring of the genetic material took place during the transition from procaryotes to eucaryotes. Short segments of the genome became copied many times over, so that a large proportion of eucaryotic DNA consists of redundant or nearly redundant sequences. Also, the DNA became organized into separate and individual chromosomes, each containing proteins as well as the nucleic acid itself.

The above scenario is highly speculative, and differs from Margulis's theory in several important respects. For example, Margulis assumed mitochondria to have been acquired before "9 + 2" structures instead of after them, and she also assumed a symbiotic origin for the latter. The motility of eucaryote ancestors is here assumed to involve pseudopods before the acquisition of eucaryotic flagella; Margulis assumes the opposite sequence. This entire discussion is intended as much to provoke careful thought among students as it is to promote research.

B. PHYLOGENY AND CLASSIFICATION OF EUCARYOTES

The problem of devising a suitable phylogeny for the eucaryotes, or a phylogenetic classification, is related to the problem of eucaryotic origins. Alternative phylogenies of the eucaryotes differ principally in assuming different starting points.

Prior to 1969, many proposed phylogenies of the eucaryotes assumed that plants were necessarily more primitive than animals, especially since most classifications at the time included one-celled algae, fungi, and even bacteria among the "plants." Whittaker's (1969) classification clearly separated bacteria, blue-green algae, slime molds, and fungi away from the true plants. This, together with the theory that plastids originated through symbiosis, both indicate that eucaryotic photosynthesis evolved late. The lack of plastids is now usually assumed to be a primitive condition and their presence a secondary or derivative one.

The cell walls, mycelial growth habits, and absorptive, saprophytic nutrition of the fungi are likewise considered derivative characteristics. A free-living condition, independent (nonsymbiotic) nutrition, and the lack of cell walls are assumed to be primitive traits for eucaryotes. Procaryotes often do have cell walls, but they are of very different construction from those of eucaryotes. The procaryotic type of cell wall was presumably lost (or had never been present) in the ancestors of the eucaryotes.

Cilia of the eucaryotic type are here assumed to be a derived (apomorphous) characteristic, for reasons to be outlined later. In fact, ciliated protozoans possess many such derived characteristics.

For these various reasons, and others, we shall here assume that the primitive eucaryotes were free-living, nonphotosynthetic, unicellular organisms without either cell walls or plastids. According to Margulis's theory, the earliest eucaryotes would also have possessed flagella, but we have earlier outlined a somewhat different hypothesis under which the earliest eucaryotes were capable only of amoeboid locomotion and the flagellum was a later acquisition.

The phylogeny of eucaryotes assumed in this book is diagrammed in Fig. 26.2. This arrangement is but one of the many phylogenies that are possible: the fossil record is inadequate to decide among them. Presumably this is because many stages in the early evolution of eucaryotes have died out without a trace, leaving only their specialized descendants as survivors. Only the discovery of new morphological and biochemical characters will enable comparative morphologists to unravel the mystery of eucaryote phylogeny.

The classification of the kingdom Protista is currently unstable: hardly any two authors follow the same classification exactly, and the lines between the Protista and other kingdoms are not always drawn in the same way. The Protozoa, traditionally considered as one-celled "animals," form the basis of the kingdom Protista, for they are the only group that has always been included under this label. Whittaker (1969) also included the Euglenophyta, Chrysophyta, and Pyrrhophyta, traditionally considered as algae, and two phyla considered by many to represent unicellular fungi. Copeland (1956) had included not only these, but also all remaining fungi, slime molds, and all eucaryotic algae except for the green algae (Chlorophyta). Haeckel's original use of the term Protista had even included the bacteria and the sponges!

In the classification followed here, the Protista are restricted to the Protozoa only. All organisms with plastids are placed in the kingdom Plantae, and various non-photosynthetic forms, including slime molds, are placed in the kingdom Mycota or Fungi.

C. KINGDOM PROTISTA

The Protista, or Protozoa, are in simplest terms the animal-like unicells. What makes them appear "animal-like" is that they lack a cell wall and usually have at least

Fig. 26.2 Phylogeny of the eucaryotes as assumed in this book. Many other arrangements are also possible.

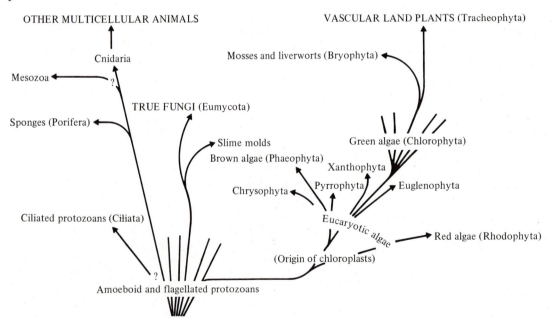

one motile stage in their life cylce. They are also, like animals, devoid of both plastids and chlorophyll pigments. Most are solitary, but a few form simple, undifferentiated colonies. Motility is important to all Protozoa except some parasitic ones, and the several mechanisms of motility serve to distinguish the various protozoan phyla. As here conceived, the Protista include the phyla Sarcodina, Mastigophora, Sporozoa, and Ciliata (Fig. 26.3). Other ''protists,'' besides Protozoa, are in this book placed in other kingdoms in an attempt to achieve a more ''vertical'' (Chapter 1) classification.

Phylum Sarcodina

Though the truly primitive eucaryotes are no longer living, the amoeboid protozoans of the Phylum Sarcodina possess many of the characteristics that a primitive eucaryote would be expected to have. An alternative theory considers the flagellated forms as more primitive. The amoeboid and flagellated forms are very closely related to one another. Intermediate forms exist between the two phyla, with alternately amoeboid and flagellated phases in their life cycles. The flagellates bear more resemblance to unicellular algae and fungi; the amoebalike forms bear more resemblance to the slime molds.

In the genus *Amoeba* and its relatives, no cell wall exists, and the cell as a whole has an irregular and highly varible shape that is constantly changing. As the shape undergoes these changes, cytoplasmic extensions known as **pseudopods** (or *pseudopodia*) enlarge in the general direction of movement, while the cytoplasm flows in this same general direction. The contraction of the plasma membrane and the flow of cytoplasm away from the trailing portion of the cell are facilitated by contractile fibers or fibrils. This creeping type of locomotion, using pseudopods, may be called amoeboid locomotion. It occurs primitively in the Sarcodina and persistently even among multicellular animals in their white blood cells and mesenchyme tissue.

Pseudopods serve their possessors for feeding as well as locomotion. The prey, usually a bacterium, is surrounded by pseudopods and captured whole. As the pseudopods come together, the prey becomes enclosed in a **food vacuole** (phagosome), which then fuses with a digestive vacuole. The enzymes that digest the prey are contained within this vacuole, and the products of enzymatic digestion are then absorbed into the cytoplasm across the vacuolar membrane. Undigested remains, if any, are left behind in the vacuole. The vacuole eventually migrates to the cell surface, fuses with the plasma membrane, and releases its undigested contents to the outside environment.

Major evolutionary events within the Sarcodina include the development in certain orders of a shell-like **test,** and changes in the shape of the pseudopods. Two orders have short, lobe-like pseudopods: one of these orders contains *Amoeba* and other naked forms without a test, while the other contains protozoans with a single-chambered test into which the entire body can be withdrawn. The remaining orders have greatly elongated pseudopods that are many times longer than their width. In the order Heliozoa, the pseudopods form stiff, needle-like rays (axopods) that radiate in all directions from the cell body. Solid tests are never formed, but sand grains, diatom tests, or other particulate matter may become trapped in the gelatinous capsule that surrounds the cell body. This shows what may have been a tendency among ancient Sarcodina to surround themselves with gritty or hard material that would decrease their value as food to a potential predator.

The order **Radiolaria** contains marine protozoans with a delicate, perforated test made usually of silica (SiO_2). The long, thin, radiating pseudopods emerge through the minute perforations and extend far beyond the test. Radiolarians are very abundant; their tests may accumulate in great numbers to form an organic-rich sediment called radiolarian ooze.

The order **Foraminifera** (or **Foraminiferida**) has the best fossil record of all the Protozoa. The long, thread-like pseudopods of these organisms communicate with one another by a series of anastomoses, a feature often used to define the group. The shell-like tests of foraminiferans are readily preserved as fossils, and this order enjoys one of the best fossil records of all groups of organisms.

Foraminiferal tests are most often composed of calcium carbonate ($CaCO_3$), but tests composed of other materials are also known. The most primitive foraminiferans, of the family Allogromiidae, are either naked or else possess a thin, delicate test made of chitin. Several families possess a test made of agglutinated foreign materials such as sand grains, diatom shells, and mineral

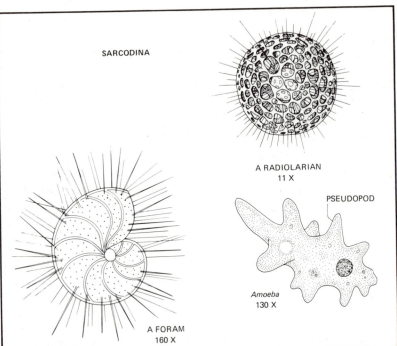

SARCODINA

A RADIOLARIAN
11 X

PSEUDOPOD

Amoeba
130 X

A FORAM
160 X

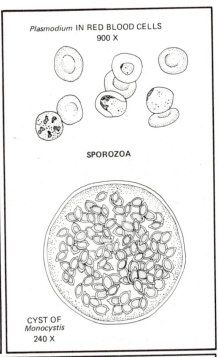

Plasmodium IN RED BLOOD CELLS
900 X

SPOROZOA

CYST OF
Monocystis
240 X

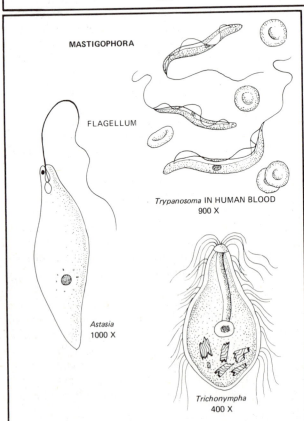

MASTIGOPHORA

FLAGELLUM

Trypanosoma IN HUMAN BLOOD
900 X

Astasia
1000 X

Trichonympha
400 X

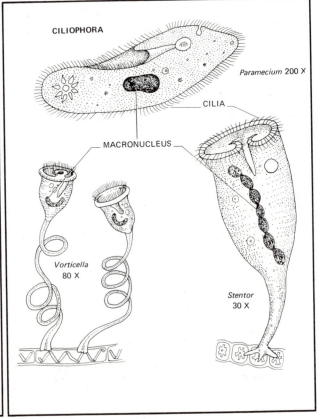

CILIOPHORA

Paramecium 200 X

CILIA

MACRONUCLEUS

Vorticella
80 X

Stentor
30 X

Fig. 26.3 (Facing page) Representative Protozoa. Reprinted with permission from John W. Kimball, *Biology*, 4e, © 1978. Reading, Massachusetts: Addison-Wesley (Fig. 35.3).

fragments. These agglutinated materials are held together by a cementing material and are underlain by a thin chitinous layer. Together, these materials diminish the value of foraminiferans as food to a potential predator. The cementing material, whether calcium carbonate, silica, or iron oxide, is primitively only a minor constituent, but in some advanced forms the cement becomes foremost in importance (perhaps as an adaptation for strengthening the test) and the agglutinated materials then become superfluous. Only in one family is a test built primarily of silica; the majority of Foraminifera have a test built largely of calcium carbonate over a thin chitin underlayer.

Phylum Mastigophora

Of the major types of Protozoa, the flagellates bear the closest resemblance to the plantlike unicells and the one-celled fungi. All these organisms share the possession of at least one whiplike flagellum that serves as the primitive organ of locomotion and has the typically eucaryotic "9 + 2" internal structure. Sexual reproduction always involves the fusion of haploid gametes, called **syngamy**.

The phylum **Mastigophora** or **Flagellata** includes all flagellated Protozoa. Most Mastigophora are solitary, but a few are colonial. Many are parasitic or otherwise symbiotic. None has any reliable fossil record.

Most flagellates have recognizable body shapes, but one group has a variable body shape that changes from moment to moment, perhaps indicating the closeness of the flagellates and the Sarcodina.

Phylum Sporozoa

The Sporozoa contain only certain parasitic Protozoa believed to be related to the Sarcodina. The mature forms are parasitic and nonmotile, reproducing by means of spores. The immature forms, however, are usually motile with some possessing flagella while others move by pseudopods. All sporozoans produce spores at some time in their life cycle. Several sporozoans are important as disease organisms; the malarial parasite, *Plasmodium falciparum* (Fig. 12.2) is an example.

Phylum Ciliata

The ciliated protozoans, phylum Ciliata or Ciliophora, are set apart from the other Protozoa by many specialized features. They are so distinctive that they are often placed in a taxon unto themselves, coordinate with all the remaining Protozoa combined.

The surfaces of ciliated protozoans are clothed in numerous hairlike **cilia** of short to moderate length. In cross-section, each cilium shows the same "9 + 2" internal structure as do eucaryotic flagella. Each ciliated cell has not one nucleus but two: a smaller "micronucleus" and a larger "macronucleus." Sexual recombination among the ciliates does not take place by fusion of gametes but rather by a unique type of conjugation involving reciprocal exchange. This complex process, together with the complexity of the ciliary apparatus, argues strongly for the evolutionary distinctness of the Ciliata.

D. KINGDOM MYCOTA (OR FUNGI)

Fungi have traditionally been treated as plants without chlorophyll, but many experts would now prefer to treat them as a separate kingdom, the Mycota. The fungi are characterized by the absence of chlorophyll or of plastids, by the absence of motile stages in all but the more primitive forms, by the presence of cell walls, and by the possession in all but the more primitive forms of a vegetative stage composed of undifferentiated branching filaments capable of absorptive nutrition, but incapable of phagocytosis.

Fungi are important as decay organisms in terrestrial and freshwater ecosystems. They tend to prefer moist conditions for optimal growth, though many have spores that are drought resistant. Most fungi live on dead or decaying organic matter; some are also parasitic since they invade living tissues. The vegetative state typically consists of a **mycelium**, a branching network of pervasive filaments through which nutrient molecules are absorbed from the surroundings. Occasionally the nutrients are not quite in usable form, in which case the fungus secretes to the outside an enzyme capable of breaking them down into a form that can readily be absorbed for use. The individual filaments of the mycelium, known as **hyphae** (singular, *hypha*), penetrate the food source, sometimes causing disease and destruction to the host tissues. The hypha is surrounded by a cell wall containing chitin or cellulose, or both. Reproductive structures and life cycles vary greatly, and form the basis for much of fungal classification, but in all cases there is a stage of the life cycle in which spores are formed. The cell membranes of fungi often break down at specified times, resulting in several different types of multinucleated aggregates known by such names as coenocytes, plasmodia, and heterokaryons. Familiar fungi are multicellular, and their cells (even their reproductive cells) lack flagella. Plant pigments, such as carotenes, xanthophylls, phycobilins and of course chlorophyll, are virtually absent, though carotenes do occur in the reproductive structures of some higher fungi. Exceptions do exist to most of the above characteristics among the more primitive fungi, many of which possess unicellular stages that are often flagellated. One family even has hyphae devoid of any cell wall.

The presence of a cell wall and the absence of flagella in the more typical fungi led to their being classified as plants that had lost (or never evolved) their chlorophyll. But the discovery of flagellated unicells among primitive fungi seems to support the alternative theory of a protozoan ancestry, an alternative strengthened by the lack of either plant pigments or of structural plastids. No fungal ancestor is known, but all mycologists (fungal specialists) do seem to agree that the familiar, multicellular, mycelial fungi evolved ultimately from flagellated unicells. The cell wall differs in detailed structure from the typical plant cell wall, and was probably an independent acquisition. In recent years, support has increased for the removal of the fungi from the plant kingdom, and their treatment as a separate kingdom.

Most specialists on fungi now consider them to represent a single major taxon, the Mycota, though Whittaker (1969) and Margulis (1974) have both proposed removal of certain primitive orders from this kingdom. The relations between slime molds and "true" fungi are likewise uncertain, but most mycologists now favor an arrangement similar to the one used here, with a kingdom Mycota subdivided into two major divisions, one for the slime molds and another for the true fungi.

Slime Molds

Related by some scientists to the Protozoa, and by others to the Fungi, are the so-called **slime molds.** The "true" slime molds, called either **Myxomycetes** or Mycetozoa, are now usually described as related to fungi. Their unicellular, vegetative stages consist of amoeboid cells called myxamoebae, or flagellated "swarm cells," or both. Once the populations reach sufficient density, these unicells aggregate into pairs and undergo sexual union (syngamy) to produce a diploid zygote. The zygote grows without cytoplasmic division, but its nucleus divides repeatedly to give rise to a multinucleated **plasmodium.** The plasmodium then creeps until it finds a suitable place, whereupon it forms fruiting structures that undergo meiosis to produce haploid spores. Each of these spores, upon germination, will form either a myxamoeba or a swarm cell, depending on the species, and in some cases depending also on environmental circumstances.

The cellular slime molds have a somewhat different life cycle (Fig. 26.4). The feeding stage consists of amoeboid unicells that live independently of one another. These behave rather like Protozoa, creeping on pseudopods and engulfing prey into food vacuoles by phagocytosis. When conditions no longer favor feeding, and if a sufficient number of amoeboid cells are present, aggregation begins: the various individual cells begin to creep toward one another, forming a very extensive colony or multicellular phase called a pseudoplasmodium. This colony then creeps over the surface, migrating toward a source of light and also toward an increase in temperature of even less than a thousandth of a degree! When migration ceases, the cells that were near the front end move downward and begin to form a cylindrical stalk, strengthened on the outside by cell walls composed of cellulose. Cells that had been near the rear of the migrating pseudoplasmodium now move upwards, atop the stalk, where they form a spherical fruiting structure called a sorocarp. Here the spores are formed, and each develops into a new amoeboid feeding stage.

The feeding stages of slime molds resemble various Protozoa, but the asexual reproduction by spores and the presence of cellulose in some stalks suggest affinity to the fungi. The resemblance between the life cycles of the true slime molds and the cellular slime molds is considered convergent by many workers.

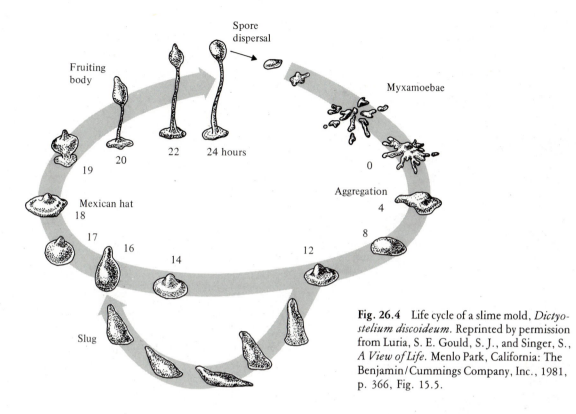

Fig. 26.4   Life cycle of a slime mold, *Dictyostelium discoideum*. Reprinted by permission from Luria, S. E. Gould, S. J., and Singer, S., *A View of Life*. Menlo Park, California: The Benjamin/Cummings Company, Inc., 1981, p. 366, Fig. 15.5.

True Fungi (Eumycota)

The true fungi, or Eumycota, include mushrooms, molds, yeasts, rusts, smuts, and their less familiar relatives (Fig. 26.5). Among the Eumycota, four classes still reproduce by means of flagellated cells (zoospores), betraying the aquatic ancestry of the fungi. The remaining fungi lack any motile stages.

The class Oömycetes, or water molds, reproduce by means of cells that bear two flagella. The anterior (leading) flagellum is of the "tinsel" type, meaning that it possesses hairlike projections (mastigonemes) all along its length. The posterior (trailing) flagellum is of the "whiplash" type, meaning that it consists of a single strand only, without any mastigonemes. Three further classes of primitive fungi also possess flagellated stages. One group has only the posterior whiplike flagellum, a second has only the anterior tinsel flagellum, and a third has two anterior flagella of the whiplash type.

The class Zygomycetes is a well-defined group of fungi whose reproductive cells are no longer flagellated, a feature that they share with the "higher" fungi.

The ascomycetes and basidiomycetes are sometimes together called the "higher fungi." They have several features in common, including the presence of specialized spore-bearing structures, the absence of flagellated stages, and the presence of cell walls, some of which form partial or complete septa dividing the hyphae into compartments. Perhaps the most remarkable similarity between these two major types of "higher" fungi is the formation, by various means, of dikaryotic cells—cells containing two nuclei.

The defining characteristic of the class Ascomycetes is the production of spores that are borne in sacs known as asci. Each ascus contains a definite number of spores, usually eight. Many familiar fungi are ascomycetes, including yeasts, black molds, blue molds, powdery mildews, cup fungi, morels, and truffles.

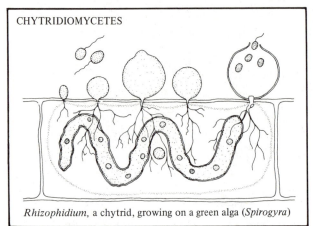

CHYTRIDIOMYCETES

Rhizophidium, a chytrid, growing on a green alga (*Spirogyra*)

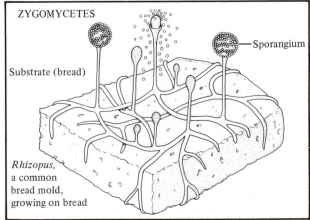

ZYGOMYCETES

Sporangium

Substrate (bread)

Rhizopus,
a common
bread mold,
growing on bread

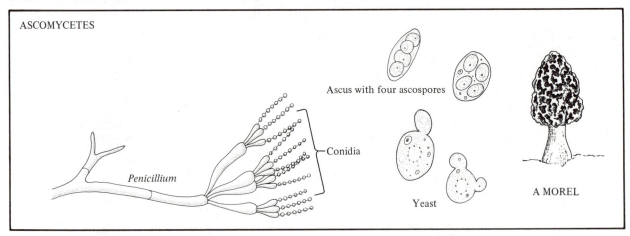

ASCOMYCETES

Penicillium

Conidia

Ascus with four ascospores

Yeast

A MOREL

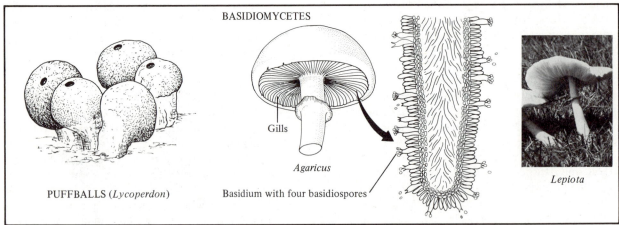

BASIDIOMYCETES

PUFFBALLS (*Lycoperdon*)

Gills

Agaricus

Basidium with four basidiospores

Lepiota

Fig. 26.5 (Facing page) Representative true fungi (Eumycota). The Chytridiomycetes are reprinted with permission from Preston Adams, Jeffrey J. W. Baker, and Garland E. Allen, *The Study of Botany,* © 1970. Reading, Massachusetts: Addison-Wesley (Fig. 21.15B). The rest of the figure is reprinted with permission from John W. Kimball, *Biology,* 4e, © 1978. Reading, Massachusetts: Addison-Wesley (Fig. 35.16).

The class Basidiomycetes contains the mushrooms, puffballs, stinkhorns, jelly fungi, rusts, and smuts. The defining characteristic of the class is the production of spores on the tips of club-shaped structures known as basidia. Each basidium typically bears four spores. Most mycologists believe that basidiomycetes evolved from ascomycetes, a conviction strengthened by the discovery of a fossil form intermediate between the two classes.

The ''fungi imperfecti,'' or deuteromycetes, are a small group of species. They reproduce by asexual spores called conidia. Sexual structures are unknown. Many deuteromycetes resemble the asexual stages of certain other fungi, and some of them may well be derived from either ascomycetes or basidiomycetes that have lost their sexual stages.

Lichens

A lichen is not an organism. It is a symbiotic association between an alga and a fungus—one of the firmest symbioses known. The fungus contributes moisture and certain minerals and trace nutrients to the association. The alga contributes the sugars produced through photosynthesis, and in some cases also the protein resulting from nitrogen fixation. Because of the alga, lichens prefer direct sunlight, which they tolerate very well. Lichens are well adapted to colonizing bare rock and other substrates on which no other organisms could live. They secrete peculiar lichen acids and other chemicals that enable them to break down the rock substance slowly and exploit its mineral supplies. Most lichen acids are produced by the intact lichen only, though some can be produced by the fungal partner alone.

Attempts to resynthesize lichens from their separated algal and fungal components have met with varied success. As long as conditions remain favorable for either or both species, the separated partners show not the slightest affinity for one another. But as soon as dryness or other stressful conditions are induced, the lichen association quickly reforms. Lichenizing fungi are mostly ascomycetes, but may also be basidiomycetes, deuteromycetes, or simpler fungi. Most lichen-forming algae are green algae, though a few blue-green algae may also form lichens. The green alga *Trebouxia* enters into many lichen associations, each with a different fungus.

E. EVOLUTION OF THE PLANT KINGDOM

Plants and Their Classification

The kingdom Plantae contains only those eucaryotic organisms possessing chlorophyll pigments, plastids, and also usually a cell wall in which cellulose is an important constituent. As so defined, the plant kingdom includes all the familiar land plants, the mosses and liverworts, and the many types of eucaryotic algae.

The ancestry of the plants is still an unsettled question, as is the matter of their relationships to other organisms. Direct development from procaryotic algae (blue-greens) was once a commonly held theory but now seems highly unlikely in view of the great gulf that separates procaryotes from eucaryotes in general. A close relationship between the unicellular, flagellated algae and the unicellular, flagellated Protozoa has been frequently suggested and still seems quite probable, raising the secondary question as to whether the algae descended from the Protozoa or the Protozoa from the algae. This last question may also be rephrased in terms of the presence or absence of plastids. Many biologists formerly assumed that the presence of plastids was primitive and their absence a derived condition among both animals and fungi. The question is still unsettled, but it now appears more likely that protozoans and fungi never possessed plastids. The acquisition of plastids among plants might therefore be considered the ''key character'' (Chapter 18) that gave rise to the plant kingdom. According to Margulis's (1970) theory, this acquisition took place by the capture of a procaryotic symbiont—a blue-green alga—which subsequently evolved into modern chloroplasts and other plastids.

The International Rules of Botanical Nomenclature specify that the plant kingdom is to be separated into "divisions" rather than "phyla." Eichler (1883) and his contemporaries divided the entire plant kingdom into four divisions, as follows:

Division Thallophyta. Plants without embryos.
  Class Algae. Thallophytes containing chlorophyll.
  Class Fungi. Thallophytes without chlorophyll.
Division Bryophyta. Plants with embryos but without vascular tissues: mosses and their relatives.
Division Pteridophyta. Vascular plants without seeds: ferns, horsetails, club mosses, and their relatives.
Division Spermatophyta. Seed plants.
  Class Gymnospermae. Seed plants without flowers, the seeds being naked rather than enclosed in an ovary.
  Class Angiospermae. Flowering plants, with the seeds protected in an ovary.

Eichler's classification was followed by botanists for the first half of the twentieth century. It has since been abandoned as a formal classification, but many of its terms have remained in current use: botanists still talk of algae, pteridophytes, and gymnosperms even though few would now recognize the Algae, Pteridophyta, or Gymnospermae as formal taxa.

Modern plant classifications have all recognized an increasing number of divisions. Tippo (1942) recognized 12, Cronquist (1968) recognized 16, and Bold (1973), an admitted splitter, recognized 28, many of them equivalent to the classes of his predecessors' classifications.

A major distinction recognized in this book, following Tippo (1942), divides the plant kingdom into two subkingdoms. The subkingdom **Thallophyta**, including all those plants commonly called "algae," are distinguished by the fact that their diploid stages develop either from unicellular zygotes or from multicellular structures in which all cells are fertile. The subkingdom **Embryophyta**, on the other hand, includes all those plants whose multicellular diploid phases develop from true **embryos**, defined as structures in which a reproductive cell (zygote) is surrounded by sterile, nonreproductive cells. The Embryophyta include the traditional divisions Bryophyta, Pteridophyta, and Spermatophyta. Klein and Cronquist (1967) have reexamined the classification of thallophytes, and Cronquist, Takhtajan, and Zimmerman (1966) have similarly reexamined the classification of the embryophytes.

The removal of the fungi from the plant kingdom leaves the Thallophyta essentially synonymous with the algae. Over half a dozen major groups of eucaryotic algae are now usually recognized; the classification used in this book recognizes seven, as follows:

Division Pyrrophyta: dinoflagellates and cryptomonads,
Division Chrysophyta: golden-brown algae and diatoms,
Division Xanthophyta: yellow-green algae,
Division Euglenophyta: *Euglena* and its relatives,
Division Rhodophyta: red algae,
Division Phaeophyta: brown algae,
Division Chlorophyta: green algae.

Tippo (1942) divided the Embryophyta into two major "phyla": Bryophyta, including the mosses and their relatives, and Tracheophyta, including all vascular plants, meaning those possessing vascular tissues (xylem, phloem). We shall here follow this distinction, but at the rank of superdivision. The divisions recognized among embryophytes correspond largely, but not exactly, to the divisions recognized by Bold (1973).

Many of the differences among major plant taxa involve changes in life cycles. Two major events mark every plant life cycle: the fusion of gametes (**syngamy** or **fertilization,**), which marks the start of the diploid phase, and meiosis (Chapter 2), which marks the beginning of the haploid phase. The haploid phase, from meiosis to fertilization, may or may not include a multicellular feeding stage, known in general as a **gametophyte.** The diploid phase, from fertilization to meiosis, may also include a multicellular feeding stage, known in general as a **sporophyte.** Some life cycles include a conspicuous sporophyte; others include a conspicuous gametophyte; still others include both. Many life cycles include also the asexual (vegetative) reproduction of the sporophyte, the gametophyte, or both. A general or overall trend in the evolution of the Embryophyta has been the gradual reduction of the gametophyte and the gradual increase in the importance of the sporophyte.

Algae

The term **algae** (from **alga**, the Latin word for seaweed) is used for plants that contain chlorophyll pigments but do not develop from multicellular embryos. In this book, all eucaryotic algae are included in the plant kingdom.

Four divisions of algae include only unicellular forms (Fig. 26.6). These four, the Pyrrophyta, Chrysophyta, Xanthophyta, and Euglenophyta, were included in the kingdom Protista by Whittaker (1969). While this arrangement has merit in emphasizing the close affinity between one-celled algae and one-celled protozoans, it leaves in the kingdom Plantae the three divisions of algae that contain multicellular organisms. Unfortunately, these three divisions (Rhodophyta, Phaeophyta, Chlorophyta) share few characters, and were probably derived independently, each from a different group of unicellular algae. Margulis (1974) has tried to overcome this difficulty by placing all eucaryotic algae within the kingdom Protista, leaving the plant kingdom synonymous with the subkingdom Embryophyta as here defined. Another possible solution, adopted in this book, retains all eucaryotic algae in the plant kingdom. Yet a third alternative would include in the plant kingdom only the green algae plus the Embryophyta, since most botanists now acknowledge that the higher plants evolved from green algae.

The division **Pyrrophyta** contains the dinoflagellates and the related cryptomonads. Common features include the possession of chloropylls *a* and *c,* of certain carotenoid pigments (dinoxanthin, etc.), of two laterally inserted flagella, and of starch as the main storage product. Some of these traits also occur among the brown algae. The dinoflagellates are characterized by two flagella arranged at right angles to one another. Most dinoflagellates are planktonic and marine. Detailed study of their cells reveals that they are a bit simpler than most eucaryotic cells. The chromosomes contain no histones, and remain condensed and clearly visible throughout both mitosis and interphase, as do also the nucleolus and nuclear envelope. Centrioles and spindle fibers are absent; the chromosomes seem to be pulled apart during mitosis by attachments (as yet undemonstrated) to the nuclear envelope.

The division **Chrysophyta** contains the golden-brown algae and diatoms. These algae have chlorophylls

a and *c,* fucoxanthin, and several unique xanthins. Storage products include fucosterol and chrysolaminarin. Certain storage products and pigments suggest affinities to the brown algae. The diatoms are abundant as plankton. In marine ecosystems, they are the most important primary producers. They often fossilize as large deposits of diatomaceous earth.

The yellow-green algae, or **Xanthophyta**, were formerly associated with the Chrysophyta, but are now considered distinct. They possess only chlorophyll *a,* and reportedly also a unique pigment called chlorophyll *e.*

The genus *Euglena* and its relatives, the so-called euglenoids, are now usually placed in a division of their own, closely allied to the Chlorophyta or green algae. They share with the Chlorophyta the possession of chlorophylls *a* and *b,* also such xanthins as astaxanthin and neoxanthin. But eugeloids are distinguished from green algae by their general absence of a cell wall (except for some proteinaceous strips), by their storage of paramylon instead of starch, and by their unique type of flagellum, which has but a single row of mastigonemes. *Euglena* is an actively swimming alga capable of either photosynthesis or ingestive nutrition; it was formerly classified by many protozoologists as a member of the Protozoa. Some euglenoids may lose their photosynthetic pigments and behave as animal-like protozoans; a few lack chlorophyll permanently.

Three algal divisions contain multicellular eucaryotes. Of these, the brown algae are dominant in cold marine waters, the red algae in tropical marine waters, and the green algae in freshwater floras at all latitudes. Each group also occurs outside its area of dominance.

The red algae, or **Rhodophyta** (Fig. 26.7) are perhaps the most distinctive of the eucaryotic algae. Though they are eucaryotic, they share a number of traits with the procaryotic blue-green algae: they have no "9 + 2" organelles such as flagella, and they possess certain pigments (phycocyanins, phycoerythrins) related to those of the blue-green algae. Many chemicals produced by red algae are very distinctive: chlorophyll *d,* for example, is found nowhere else, nor is "floridean starch," a form of amylopectin found in red algal cell walls. These unique substances, the lack of flagella, and the structure of red algal cell walls all suggest a red algal ancestry quite separate from the remainder of the eucaryotes.

All red algae have large, stationary female gametes and much smaller male gametes. Neither type of gamete

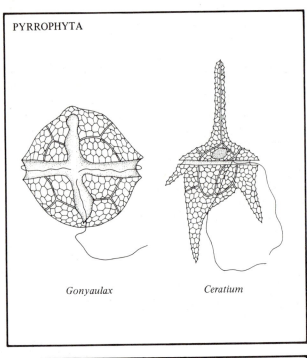

PYRROPHYTA

Gonyaulax

Ceratium

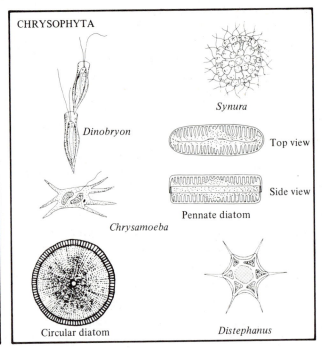

CHRYSOPHYTA

Dinobryon

Synura

Top view

Side view

Chrysamoeba

Pennate diatom

Circular diatom

Distephanus

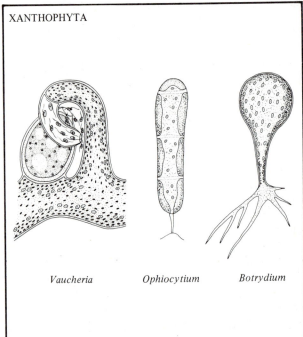

XANTHOPHYTA

Vaucheria

Ophiocytium

Botrydium

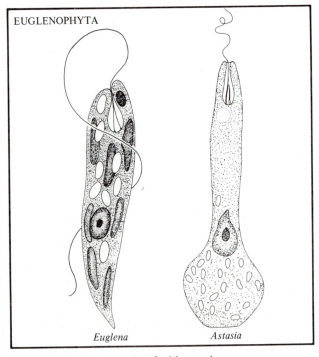

EUGLENOPHYTA

Euglena

Astasia

Fig. 26.6 Representative microscopic algae belonging to several groups. The *Euglena* is reprinted with permission from Donald D. Ritchie and Robert Carola, *Biology*, © 1979. Reading, Massachusetts: Addison-Wesley (Fig. 9.12). The rest of the figure is reprinted with permission from Preston Adams, Jeffrey J. W. Baker, and Garland E. Allen, *The Study of Botany*, © 1970. Reading, Massachusetts: Addison-Wesley (Fig. 20.14).

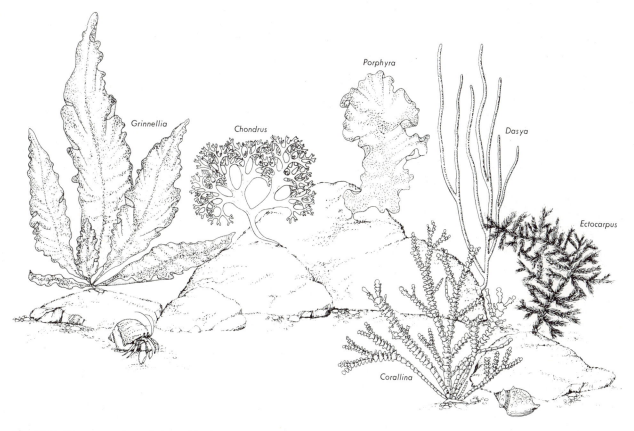

Fig. 26.7   Representative red algae (Rhodophyta). Reprinted with permission from Preston Adams, Jeffrey J. W. Baker, and Garland E. Allen, *The Study of Botany,* © 1970. Reading, Massachusetts: Addison-Wesley (Fig. 20.21).

is motile—motile cells are entirely absent among red algae; the male gametes are simply dispersed by currents and fertilize the female gametes only by chance.

The brown algae, or **Phaeophyta** (Fig. 26.8) are the dominant marine algae in temperate and colder waters, both in the intertidal zone and in mid-ocean waters. The Phaeophyta include the largest of algae—marine kelps that rival the tallest trees in their overall length. It is also among the brown algae that we meet with the greatest complexity of organization ever achieved by any thallophyte: division of the vegetative structure into as many as three separate regions, each differing from the others in cellular organization. A rather common morphology is division of the plant into three regions, superficially rootlike (the holdfast), stemlike

(the stipe), and leaflike (the blade); additional structures may also be present.

The brown algae are distinguished from other algae by their distinctive suite of pigments, including chlorophylls *a* and *c* and several unique xanthins, by their distinctive storage products, including laminarian starch, mannitol, fucosterol, and others, and by their biflagellated reproductive cells. These cells, the only flagellated cells among the brown algae, have two flagella inserted laterally; the forward or anterior flagellum is of the tinsel type, and the trailing or posterior flagellum is of the whiplash type. Alginic acid, or algin, is a fairly distinctive constituent of brown algal cell walls.

Of all the algae, the green algae or **Chlorophyta** are usually acknowledged by botanists to represent the type

Fig. 26.8 Representative brown algae (Phaeophyta). Reprinted with permission from Preston Adams, Jeffrey J. W. Baker, and Garland E. Allen, *The Study of Botany,* © 1970. Reading, Massachusetts: Addison-Wesley (Fig. 20.20).

of plant from which the various types of land plants evolved. Like all land plants, the typical green algae possess chlorophylls *a* and *b,* α- and β-carotenes, and a variety of xanthins. The main storage product is starch, while the main cell wall constituents are cellulose and pectin, with other constituents (mannan, xylan, chitin, calcium carbonate) variably present. The cells of at least one green alga divide with the formation of a floating cell plate (see Fig. 2.4), a character otherwise unique to higher plants.

Related to the typical green algae, or Chlorophyceae, are a small group of stoneworts, or Charophyceae, possessing γ-carotene instead of α-carotene, but otherwise quite similar to the typical green algae in all the characters listed above. They are morphologically distinct, however, and are presumably derived from Chlorophycean ancestors.

The Chlorophyta are the dominant group of algae in most freshwater ecosystems; a number of green algae are also marine. The most primitive green algae are the unicellular forms like *Chlamydomonas,* in which two whiplash flagella, equal in size, are together inserted apically. Reproduction in *Chlamydomonas* is **isogamous** (Fig. 26.9), meaning that male and female gametes are

Fig. 26.9 Relationships between the sizes of male and female gametes. Most algal groups are primitively isogamous, but trends toward anisogamy and oögamy are frequent in many groups. From *An Evolutionary Survey of the Plant Kingdom* by R. F. Scagel, R. J. Bandoni, G. E. Rouse, W. B. Schofield, J. R. Stein, and T. M. C. Taylor. © 1965 by Wadsworth Publishing Company, Inc. Belmont, California 94002. Reprinted by permission of the publisher.

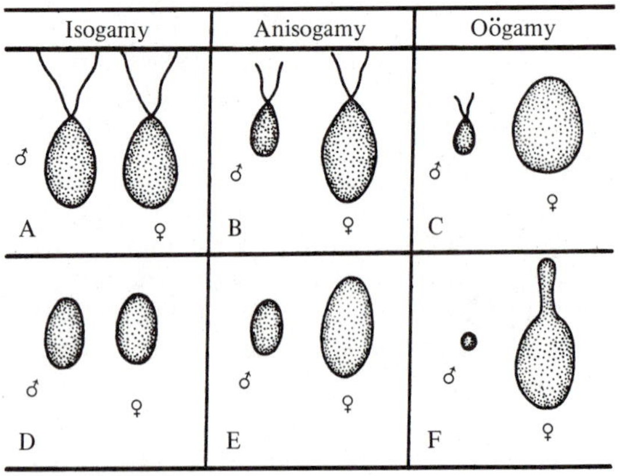

Isogamy	Anisogamy	Oögamy
A	B	C
D	E	F

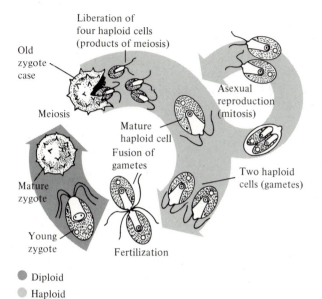

Liberation of
four haploid cells
(products of meiosis)

Old
zygote
case

Meiosis

Mature
haploid cell

Fusion of
gametes

Mature
zygote

Young
zygote

Fertilization

Asexual
reproduction
(mitosis)

Two haploid
cells (gametes)

● Diploid
● Haploid

Fig. 26.10 The life cycle of the one-celled green alga *Chlamydomonas*. Reprinted by permission from Luria, S. E., Gould, S. J., and Singer, S. *A View of Life,* Menlo Park, California: The Benjamin/Cummings Publishing Company, Inc., 1981, Fig. 8.11, p. 209.

morphologically indistinguishable. The life cycle of *Chlamydomonas* is shown in Fig. 26.10.

Among various divergent groups of Chlorophyta (Fig. 26.11), there has been a tendency for the female gametes to become larger than the male (**anisogamy**), and eventually very large and stationary (**oögamy**). (Parallel trends of this sort also occur among various brown algae.) There is also a tendency for an increase in morphological complexity in many lines, though regional differentiation never reaches the level that it does among the brown algae. Some groups have become marine, and a few have evolved other specializations: calcified cell walls, or an increase in the number of flagella. Some of the unicellular forms have also become colonial. The principle of mosaic evolution is very evident here: forms with advanced morphology may sometimes still be isogamous, marine forms may have simple morphology, and so forth (see Box 17–C). No group is simultaneously advanced in all these features.

Bryophytes

All the remaining plants to be considered in this chapter share a most important advance above the algal level: development from embryos. A plant **embryo** may be defined for our purposes as a reproductive cell (a zygote) surrounded by a protective layer of sterile, nonreproductive cells. Plants whose diploid phases develop in this manner are now usually recognized as a major taxon within the plant kingdom, the subkingdom **Embryophyta.** The multicellular diploid phase, or sporophyte, is further characterized by a tissue level of organization, in which groups of similarly specialized vegetative cells and their products are located together and function together harmoniously and with greater efficiency than is possible for undifferentiated, all-purpose cells. By way of contrast, the algae have no such tissue level of organization, and their multicellular diploid phases, when present, develop from zygotes or from other structures in which all cells are reproductively functional—there are no sterile or nonreproductive cells.

Mosses and related plants, known as **bryophytes** (Fig. 26.12), differ from algae by developing from embryos—they thus belong to the Embryophyta. They differ from all other embryophytes in lacking the well-developed vascular systems, with differentiation of xylem and phloem, that characterize all the more advanced Embryophyta. Lacking these vascular systems, the bryophytes are restricted both in size (rarely more than a few centimeters tall) and in habitat diversity. The lack of a well-developed root system, or of vascular tissues capable of conducting large quantities of fluid to peripheral parts, means that bryophytes must be restricted to rather moist situations, from moist shade to open water. Neither water nor nutrients can be transported from one part of the plant to another, except by slow diffusion from one cell to the next. Not only does this restrict bryophytes to a rather small size, it also means that all parts of the plant must carry on photosynthesis for themselves.

It is now generally assumed by most botanists that bryophytes evolved from algae. The green algae (Chlorophyta) are most often mentioned as possible ancestors because they have the same chlorophyll and carotenoid pigments, storage products, and cell wall constituents. A few botanists believe instead that bryophytes evolved degenerately from vascular land plants by the loss of their vascular systems, rather than from algae.

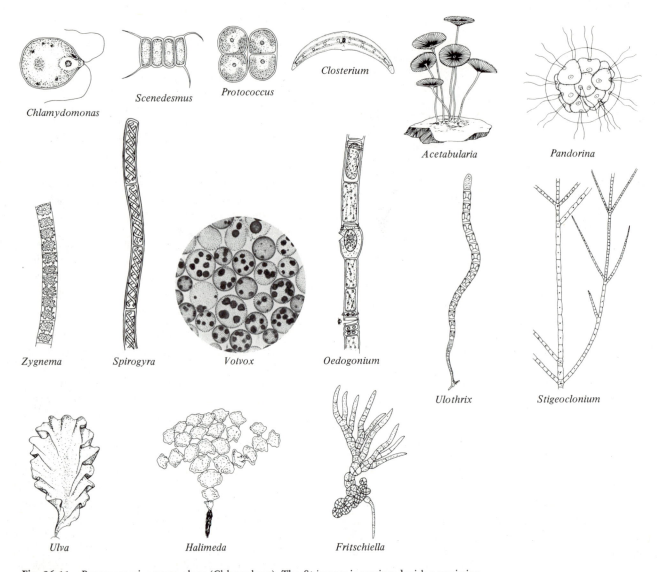

Fig. 26.11 Representative green algae (Chlorophyta). The *Spirogyra* is reprinted with permission from John W. Kimball, *Biology,* 4e, © 1978. Reading, Massachusetts: Addison-Wesley (Fig. 35.12). The rest of the figure is reprinted with permission from Preston Adams, Jeffrey J. W. Baker, and Garland E. Allen, *The Study of Botany,* © 1970. Reading, Massachusetts: Addison-Wesley (Fig. 20.25 and p. 88).

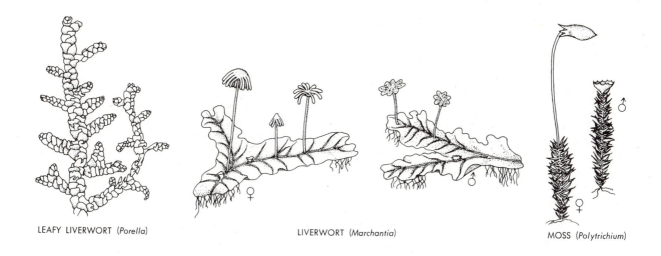

LEAFY LIVERWORT (*Porella*)　　　LIVERWORT (*Marchantia*)　　　MOSS (*Polytrichium*)

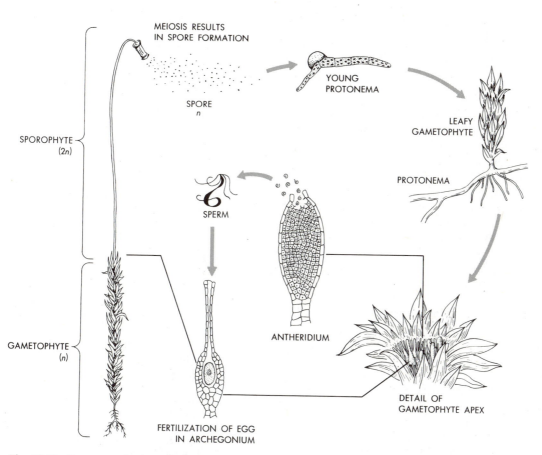

MEIOSIS RESULTS
IN SPORE FORMATION

SPORE
n

YOUNG
PROTONEMA

LEAFY
GAMETOPHYTE

PROTONEMA

SPOROPHYTE
(2*n*)

SPERM

ANTHERIDIUM

DETAIL OF
GAMETOPHYTE APEX

GAMETOPHYTE
(*n*)

FERTILIZATION OF EGG
IN ARCHEGONIUM

Fig. 26.12　Representative bryophytes. The lower half of the figure shows the life cycle of a moss.
Reprinted with permission from Preston Adams, Jeffrey J. W. Baker, and Garland E. Allen, *The
Study of Botany*, © 1970. Reading, Massachusetts: Addison-Wesley (Fig. 21.27, p. 89).

447

In all bryophytes there is a well-marked alternation of generations, with the haploid (gametophyte) generation dominant. The diploid (sporophyte) generation is always dependent as a virtual parasite upon the gametophyte. Reproduction is oögamous, with the female gamete (which becomes the zygote upon fertilization) always retained within the archegonium. Here the zygote develops into a small sporophyte, literally supported by the gametophyte and still held at its base within the archegonium. The plant that we see is thus usually a gametophyte, perhaps containing a small sporophyte growing upon it (Fig. 26.12). The sporophyte has no independent existence. It is incapable of life when separated from its gametophyte, though almost all sporophytes contain some chlorophyll and therefore may contribute somewhat to their own nutrition.

The two major groups of bryophytes are the mosses and the liverworts; most botanists now also recognize the hornworts as a third group.

Vascular Land Plants

Most of the familiar plants on this earth possess a much higher level of organization than any alga or bryophyte. A series of multicellular **roots** invade the substratum, from which they absorb moisture as well as certain mineral nutrients. A series of **leaves** conduct photosynthesis, resulting in the production of sugars. The roots and the leaves are connected by a **stem**, or series of branched stems, which contain two series of tubes: **xylem**, which conducts the water upwards from the roots, and **phloem**, which conducts the dissolved photosynthetic products downward from the leaves. Leaf, root, and stem differentiate separately as organs. Each is composed of a number of distinct tissues, the structure of which is the subject of the rather rich field of plant anatomy. Paramount among these tissue types are the vascular or conducting tissues, xylem and phloem, after which the plants we are discussing are called **vascular plants**. Armed with these vascular tissues, a plant can distribute to all its cells nutrients that it has absorbed by its roots or produced in its leaves. It can therefore hope to survive in situations unsuitable to either algae or bryophytes—most of the Earth's land surface is now covered with vascular plants. The vascular plants are furthermore capable of growing to a much larger size than was possible among their nonvascular ancestors, partly because of the conduc-

tion of fluids through the vascular tissue, and partly because the xylem is often developed into a stiff, rigid material that we all recognize as **wood**.

The similarities in the organization of vascular and other tissues among the land plants strongly suggest that they all evolved from a common source, a fact usually recognized by grouping them all into a taxon called **Tracheophyta**, the vascular plants. The Tracheophyta are often considered to be a division, but in this book we shall treat them as a superdivision, and the several major types of tracheophytes each as divisions.

Fig. 26.13   The hypothetical origin of vascular land plants from dichotomously branching green algae, according to Lignier. The "primitive land plant without leaves" is reprinted with permission from Preston Adams, Jeffrey J. W. Baker, and Garland E. Allen, *The Study of Botany,* © 1970. Reading, Massachusetts: Addison-Wesley (Fig. 21.1).

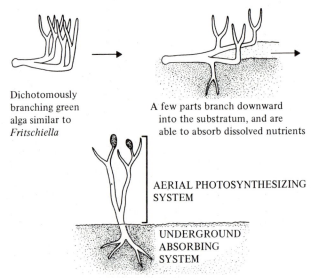

Dichotomously branching green alga similar to *Fritschiella*

A few parts branch downward into the substratum, and are able to absorb dissolved nutrients

AERIAL PHOTOSYNTHESIZING SYSTEM

UNDERGROUND ABSORBING SYSTEM

Primitive land plant without leaves. The upright portion is primarily photosynthesizing, and the underground portion primarily absorptive, but they are not yet differentiated into stems and roots. The reproductive structures are borne at the tips of certain upright telomes.

The origin of the tracheophytes has been the subject of much investigation. It is generally believed that they evolved from green algae (Chlorophyta), because only among these algae are all the pigments and other substances found that are characteristic of all higher plants. Many botanists favor a direct algal origin not involving bryophytes, while others point to the meristem tissue of hornworts as evidence that vascular plants evolved from algae via the bryophytes.

Considerable evidence favors the view that the early vascular plants were dichotomously branched, with two-fold branching exclusively. Some of the stages in the evolution of a vascular plant from a dichotomously branched ancestor are easily imagined (Fig. 26.13). Essential in these reconstructions is the differentiation of a surface-creeping, horizontal plant into above-ground aerial stems and a below-ground root system.

A number of Silurian and Devonian plant fossils document the early evolution of vascular plants. *Rhynia*, from the lower Devonian Rhynie flora of Scotland, was one of the simplest (Fig. 26.14). It consisted essentially of a creeping horizontal axis repeatedly giving rise to vertical shoots that branched dichotomously. The vascular tissue was simple in shape, and the reproductive structures (sporangia) were terminal in position. No roots were present; instead, small rhizoids arose from the horizontal stem, often in tufts, and invaded the substrate in the manner of bryophytes. *Rhynia* apparently lived in freshwater bogs, from which its erect stems protruded.

The simplest of living vascular plants are *Psilotum* and *Tmesipteris*. *Psilotum* bears many resemblances to the Rhynie plants: the stem is green and carries out photosynthesis; its whole surface is provided with breathing pores (stomates); true roots and leaves are lacking. Branching is on the dichotomous pattern but asymmetrical. Many botanists consider *Psilotum* and *Tmesipteris* sufficiently close to the Rhynie plants to unite them all in a common division **Psilophyta**; other botanists (Cronquist, Takhtajan, and Zimmerman, 1966) consider the relationships sufficiently remote to recognize two separate divisions called Psilophyta and Rhyniophyta. Still others have noted certain similarities between *Psilotum* and primitive fernlike plants.

From a dichotomously branched ancestor, the early land plants underwent an adaptive radiation that gave rise to at least four major groups, including some twenty

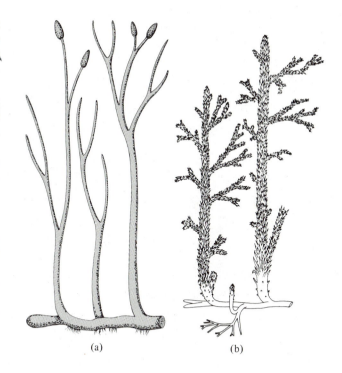

Fig. 26.14   Two early land plants: (a) *Rhynia;* (b) *Asteroxylon.* Part (a) is reprinted with permission from Donald D. Ritchie and Robert Carola, *Biology,* © 1979. Reading, Massachusetts: Addison-Wesley (Fig. 9.32). Part (b) is reprinted with permission from H. N. Andrews, *Studies in Paleobotany,* 1961. New York: John Wiley & Sons.

orders and numerous families and genera. By the end of Devonian times, the major groups of seedless vascular plants had already evolved. These plants, and their direct descendants, are here placed in four divisions: Psilophyta, Lepidophyta, Arthrophyta, and Pterophyta. The Pterophyta, or ferns, also became ancestral at a later time period to the seed plants, which collectively dominate the floras of the modern world.

The earliest vascular plants, as we have seen, had neither roots nor leaves—only simple, dichotomously branched, erect stems. These stems, which were offshoots of a creeping horizontal axis, functioned as both stem and leaf, since they were photosynthetic, and in several

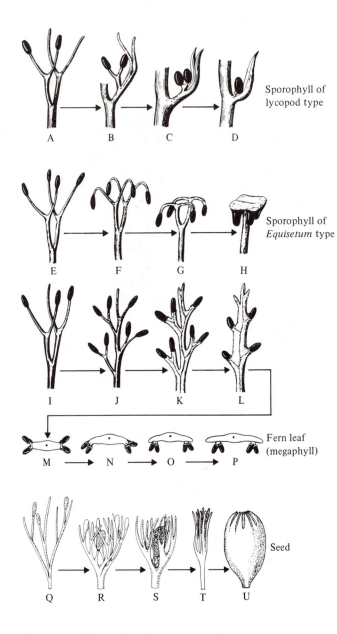

Sporophyll of lycopod type

Sporophyll of *Equisetum* type

Fern leaf (megaphyll)

Seed

Fig. 26.15 The origins of various types of plant structure, according to the telome theory. Adapted with permission from H. N. Andrews, *Studies in Paleobotany,* 1961. New York: John Wiley & Sons.

cases were provided with stomates. The spore-bearing structures, or **sporangia,** were borne at the tips of these erect stems, though some stems may also have been sterile (devoid of sporangia).

The German botanist Walter Zimmerman has termed each of these dichotomously branching, erect stems **telomes.** According to Zimmerman, the various types of leaves, branches, and reproductive structures, ultimately including flowers and seeds, were derived from such telomes (Fig. 26.15). This **telome theory** serves as a unifying concept for the study of seedless vascular plants and their evolution. In many ways, this telome

Fig. 26.16 Microphylls and megaphylls. (a) The differences between microphylls and megaphylls. (b) Evolution of microphylls, according to the enation theory. (c) Evolution of microphylls, according to the telome theory. Parts (b) and (c) are redrawn after W. N. Stewart.

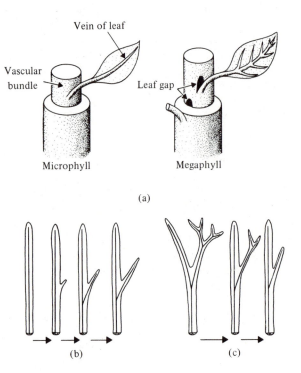

Vein of leaf

Vascular bundle

Leaf gap

Microphyll

Megaphyll

(a)

(b)

(c)

theory is reminiscent of the claim, made in the eighteenth century by the German poet-philosopher Goethe (Chapter 3), that all flower parts and other plant structures were modified leaves.

According to the telome theory, the broadening and flattening of a sterile telome would result in the evolution of a leaflike structure with a single vascular tube, or **vein**, running down its axis. Such leaves, termed **microphylls**, occur in the divisions Psilophyta, Lepidophyta, and Arthrophyta. Microphylls are typically small and simple, but some extinct lycopods had unusually large microphylls. Another type of leaf, the **megaphyll**, has many veins, and is derived, according to the telome theory, from the coalescence of a series of flattened telomes. Megaphylls are further recognized by the fact that they always occur together with a leaf gap—a break in the vascular bundle or stele (Fig. 26.16). Megaphylls of this type occur in ferns, where they bear the sporangia on their lower surfaces. The derivation of this fernlike leaf according to the telome theory is shown in Fig. 26.15.

A rival theory, the **enation** theory, accounts for the evolution of microphylls by the appearance of scalelike outgrowths from the surface of the primitive, erect stem. The subsequent enlargement of such an outgrowth and its invasion by an offshoot of the primitive vascular bundle would result in the production of a microphyll. The two alternative explanations for the evolution of microphylls are shown in Fig. 26.16. Many botanists currently accept the origin of microphylls by enation, while favoring the origin of megaphylls as described under the telome theory.

The division **Lepidophyta** contains the club mosses (Fig. 26.17) and their relatives. These plants, known as lycopods, have true roots and green leaves that are microphylls. Most microphylls are small, with sporangia borne within the angle between the leaf and stem. The living club mosses are small plants, but the Carboniferous, treelike lycopod *Lepidodendron* grew as high as 35 meters (110 feet).

The division **Arthrophyta** (Fig. 26.17d) is characterized by horizontal, underground rhizomes, aerial stems, reduced scalelike leaves, and the occurrence of these leaves in nodes or whorls at spaced intervals. The sole living genus, *Equisetum,* may well be considered a "living fossil" in that it is very similar to certain Carboniferous genera. Plants of the genus *Equisetum* are commonly called horsetails.

The modern ferns and their extinct relatives are together placed in the division **Pterophyta**. Members of this division are characterized by large, conspicuous leaves (megaphylls) supplied with many veins, and bearing sporangia usually on their lower surfaces, less often on their margins. The life cycle is in many ways similar to that of bryophytes, but with the sporophyte increasingly dominant and conspicuous, and the gametophyte correspondingly inconspicuous (Fig. 26.18).

Modern ferns typically have compound leaves (fronds), each divided into a stalklike petiole and a flattened blade. The reproductive structures (containing the sporangia) are borne on the ventral surface of each leaf. The life cycle of a typical fern is shown in Fig. 26.18.

Plants with Naked Seeds

The remaining divisions of plants possess seeds. A **seed** is a dispersive phase consisting of an embryo, with or without food reserves, enclosed in one or more protective coatings that have originated from coalesced telomes. Seeds have become the principal phase for dispersal of the higher plants—the phase in which they are disseminated, often over considerable distances. The seed plants share with their fernlike ancestors the possession of megaphylls with branching internal veination. (The reduced, needlelike leaves of pines and their relatives are considered megaphylls despite the lack of branching veination, partly because they have leaf gaps, and partly because they are believed to have evolved from megaphylls in which the veins did branch.)

Plants whose seeds are not enclosed in a fruit are called gymnosperms, a name that means "naked seeds." Botanists now usually arrange these plants (Fig. 26.19) in as many as five divisions: Pteridospermophyta, Cycadophyta, Ginkgophyta, Coniferophyta, and Gnetophyta. The male spores (microspores or pollen) of many gymnosperms are borne by the wind, and reach the female structures only by chance.

The seed ferns or pteridosperms (**Pteriodospermophyta**) are all extinct. They range in age from late Devonian to Jurassic. They were the dominant form of plant life during the Carboniferous, which was truly the age of

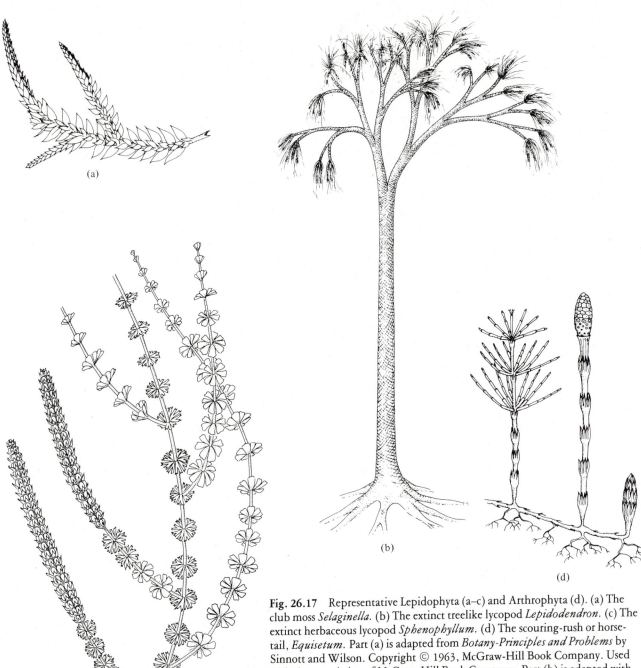

Fig. 26.17 Representative Lepidophyta (a–c) and Arthrophyta (d). (a) The club moss *Selaginella*. (b) The extinct treelike lycopod *Lepidodendron*. (c) The extinct herbaceous lycopod *Sphenophyllum*. (d) The scouring-rush or horse-tail, *Equisetum*. Part (a) is adapted from *Botany-Principles and Problems* by Sinnott and Wilson. Copyright © 1963, McGraw-Hill Book Company. Used with the permission of McGraw-Hill Book Company. Part (b) is adapted with permission from H. N. Andrews, *Studies in Paleobotany*, 1961. New York: John Wiley & Sons. Part (c) is adapted from *Cryptogamic Botany* by G. M. Smith. Copyright © 1938, McGraw-Hill Book Company. Used with permission of McGraw-Hill Book Company. Part (d) is reprinted with permission from Preston Adams, Jeffrey J. W. Baker, and Garland E. Allen, *The Study of Botany*, © 1970. Reading, Massachusetts: Addison-Wesley (p. 90).

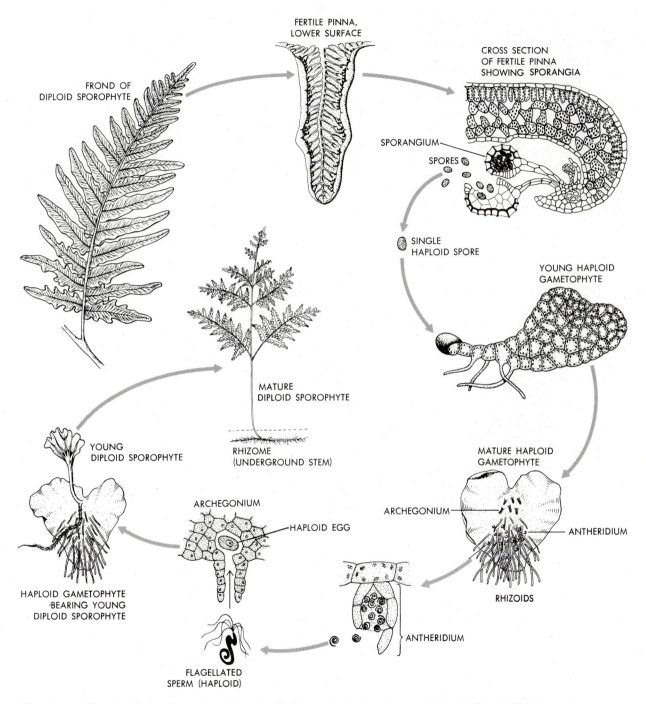

FERTILE PINNA, LOWER SURFACE

CROSS SECTION OF FERTILE PINNA SHOWING SPORANGIA

FROND OF DIPLOID SPOROPHYTE

SPORANGIUM

SPORES

SINGLE HAPLOID SPORE

YOUNG HAPLOID GAMETOPHYTE

MATURE DIPLOID SPOROPHYTE

RHIZOME (UNDERGROUND STEM)

YOUNG DIPLOID SPOROPHYTE

MATURE HAPLOID GAMETOPHYTE

ARCHEGONIUM

ARCHEGONIUM

ANTHERIDIUM

HAPLOID EGG

RHIZOIDS

HAPLOID GAMETOPHYTE BEARING YOUNG DIPLOID SPOROPHYTE

ANTHERIDIUM

FLAGELLATED SPERM (HAPLOID)

Fig. 26.18 Life cycle of a fern (Pterophyta). Reprinted with permission from Preston Adams, Jeffrey J. W. Baker, and Garland E. Allen, *The Study of Botany,* © 1970. Reading, Massachusetts: Addison-Wesley (Fig. 21.16).

(a)

Fig. 26.19 Plants with naked seeds: (a) a cycad, *Cycas* (Cycadophyta). (b) A *Ginkgo,* or maidenhair tree (Ginkgophyta). (c) A spruce, *Picea* (Coniferophyta). (d) A Pine, *Pinus* (Coniferophyta). Part (a) is reprinted with permission from John W. Kimball, *Biology,* 4e, © 1978. Reading, Massachusetts: Addison-Wesley (Fig. 36.9). Part (b) is a USDA Forest Service Photo. Parts (c) and (d) are reprinted with permission from Preston Adams, Jeffrey J. W. Baker, and Garland E. Allen, *The Study of Botany,* © 1970. Reading, Massachusetts: Addison-Wesley (Fig. 21.26).

(b)

(c)

(d)

seed ferns. The world's commercial supply of coal comes mostly from the fossilized remains of these plants. Many seed ferns had fernlike leaves, and were often misidentified as ferns until it was shown that they also possessed seeds.

The division **Cycadophyta,** or cycads and their relatives, includes the dominant plants of the Mesozoic and a small number of plants still living today. Typical cycads are vaguely palmlike in appearance, with an erect, unbranching trunk, topped by a crown of large fronds. The seeds are borne in a **strobilus** that closely resembles a large pine cone; the similarity in this case reflects homology, and is not merely superficial. The leaves are typically large and pinnately compound.

The ginkgos (**Ginkgophyta**) show similarities to both the cycads and conifers. The only living species, *Ginkgo biloba,* has fan-shaped leaves with dichotomously branched veins (Fig. 26.19).

The division **Coniferophyta** includes the most familiar and economically important gymnosperms: the pines, spruces, firs, hemlock, and yews. The division probably originated in the tropics, where a majority of genera still occur. A relative handful of common species also create vast expanses of coniferous forest (taiga) outside the tropics.

The **Gnetophyta** contains three very distinctive living genera, *Gnetum, Ephedra,* and *Welwitschia,* each usually placed in a separate order. *Gnetum* is a tropical plant that superficially resembles certain angiosperms, and *Ephedra* superficially resembles horsetails in many respects. *Welwitschia,* one of the strangest plants on Earth (see Fig. 20.2), has a giant taproot beneath the surface, and two large tapering leaves that shred into tattered strips in the plant's windblown near-desert habitat in Namibia (Southwest Africa). Many characteristics of the Gnetophyta approach those of the angiosperms, although several of these seem to be independently derived through parallel evolution.

Flowering Plants

The dominant plants in the modern world are unquestionably the flowering plants (division **Anthophyta**), also known as **angiosperms.** The angiosperms originated during Mesozoic times, and by the early Cretaceous a number of modern families had already become fairly well established. The ancestor of the angio-

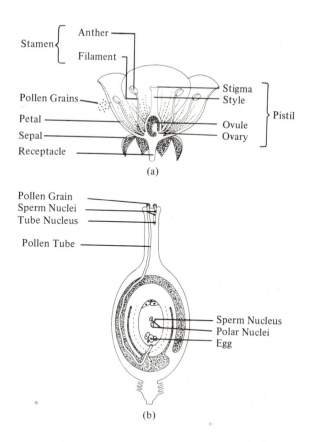

Fig. 26.20 (a) Structure of an angiosperm flower. (b) Fertilization in an angiosperm flower. Reprinted with permission from John W. Kimball, *Biology,* 4e, © 1978. Reading, Massachusetts: Addison-Wesley (Fig. 20.5).

sperms was undoubtedly a gymnosperm, probably a seed fern (pteridosperm).

The most notable of angiosperm characteristics is the presence of highly developed reproductive structures known as **flowers,** in which the seeds are first formed. When they have not become reduced through evolutionary specialization, flowers contain the following parts (Fig. 26.20):

A calyx, consisting of individual **sepals,** derived from leaves and often green in color. In the early stages of flower development, the sepals enclose and protect the unopened flower bud.

- **A corolla,** consisting of individual **petals.** The petals are the attractive parts of many flowers. Their beautiful colors and fragrances are legendary. These attractive features are especially pronounced in insect-pollinated flowers.

Stamens, which are the male pollen-producing structures. Typical stamens consist of a pollen-bearing **anther** at the top of a long **filament.**

One or more pistils, or female reproductive structures. Each pistil consists of a stigma, style, and ovary. The **stigma** provides a surface, often sticky, upon which the pollen falls; the **style** is a stalklike portion that

Fig. 26.21 Life cycle of an angiosperm (diagrammatic). Reprinted with permission from Jeffrey J. W. Baker and Garland E. Allen, *Study of Biology,* 3e, © 1977. Reading, Massachusetts: Addison-Wesley (Fig. 22.30).

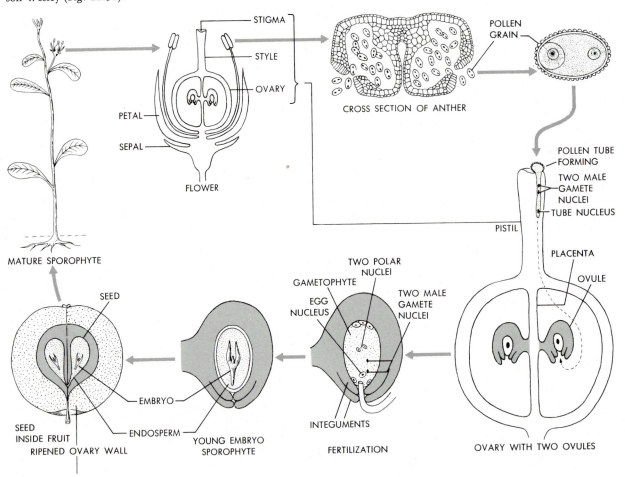

supports the stigma and down which the pollen tube must penetrate in order to achieve fertilization. The ovary contains one or more **carpels,** containing the female gametes or ovules; the carpels are believed to represent folded-up megasporophylls.

The life cycle of angiosperms shows other unique features (Fig. 26.21). The gametophyte generation has been reduced to insignificance. Meiosis in the male structures results in the production of pollen grains, which are the male spores. Pollen transport, or pollination, is achieved by one of several means (insects, wind, birds, etc.); the mechanism of pollination is usually constant for a given species. The pollen grains adhere to the stigmata, where they form small male gametophytes, containing only two or three nuclei. A **pollen tube** grows down from the pollen grain, penetrating the style; most of the cytoplasm and all the nuclei migrate down this tube, leaving behind the empty cell wall of the pollen grain. The tube nucleus leads the way at the growing tip, followed close behind by the generative nucleus, which is the true gamete. During the course of its descent, the generative nucleus divides into two nuclei, only one of which achieves fertilization by fusing with one of the several nuclei of the female gametophyte. The remaining generative nucleus has a more unique fate: it generally fuses with *two* of the nuclei of the female gametophyte to make a triploid (3N) tissue, the endosperm. This process of double fertilization is unique to angiosperms and argues strongly for the phyletic unity of the group.

Production of the seed follows fertilization. Seeds can be of various types, dispersed either by wind, by insects, by vertebrates, or by other agencies (Fig. 26.22).

Charles Darwin once referred to the origin of the angiosperms as an ''abominable mystery.'' It is now universally agreed that angiosperms originated from gymnosperms. Of the several groups of gymnosperms, the seed ferns or pteridosperms seem to be the best candidates for possible angiosperm ancestors. Many of the seed ferns were monoecious, with the sexes borne in separate strobili on a single plant. This, plus the proximity of the male and female parts to one another and to surrounding leaves, argues persuasively for the suitability of seed ferns as angiosperm precursors.

In the evolution of the angiosperm flower, the essential change, to which the name ''angiosperm''

refers, is that the seed is contained in a carpel and is thus protected. One widely supported theory for the evolution of such a protective structure is that the first angiosperms were pollinated by crawling insects, and the carpels evolved as a means of protecting the seed from the pollinating insect. Angiospermy (the enclosing of the seed in a carpel) and entomophily (pollination by insects) thus evolved together, and a hypothetical series of stages can be imagined, as follows:

1. Bearing of male and female sporophylls together in a common strobilus. Pollen primitively borne by wind (as in many gymnosperms); transfer of pollen inefficient (say, 2% success, though probably much less).

2. Pollen-eating insects destroy some pollen, but also accidentally transfer pollen from one flower to another, effecting cross-fertilization. As long as the pollen transfer is more efficient than before (say, 5% successfully transported), it would be selectively advantageous to the plant, even though much of the remaining 95% of the pollen were eaten.

3. Protection of seed against insect predation achieved by enclosing seed in a carpel. The carpel might also provide a food source for insects and increase the chances for pollination.

4. Evolution of the surrounding leaves into a calyx, composed of sepals, and a corolla, composed of petals. Radial pattern of organization selected by insects, which are preferentially attracted to such flowers and thus pollinate them more effectively. Attractiveness to insects also explains evolution of characteristic odors, or of colors other than green.

The great variety of insect life today is conditioned in large measure by the variety in their methods of exploiting angiosperms. But the very origin of angiospermy may well have been conditioned by a primitive association with insects. It is perhaps of interest that most primitive flowers are pollinated by insects belonging to the advanced, holometabolous orders, which probably did not come into existence until Mesozoic time.

Two quite distinct classes of angiosperms are recognized: the Dicotyledonae (dicots) and the Monocotyledonae (monocots). These names refer to that fact that the dicots have in their seeds two cotyledons, or leaflike

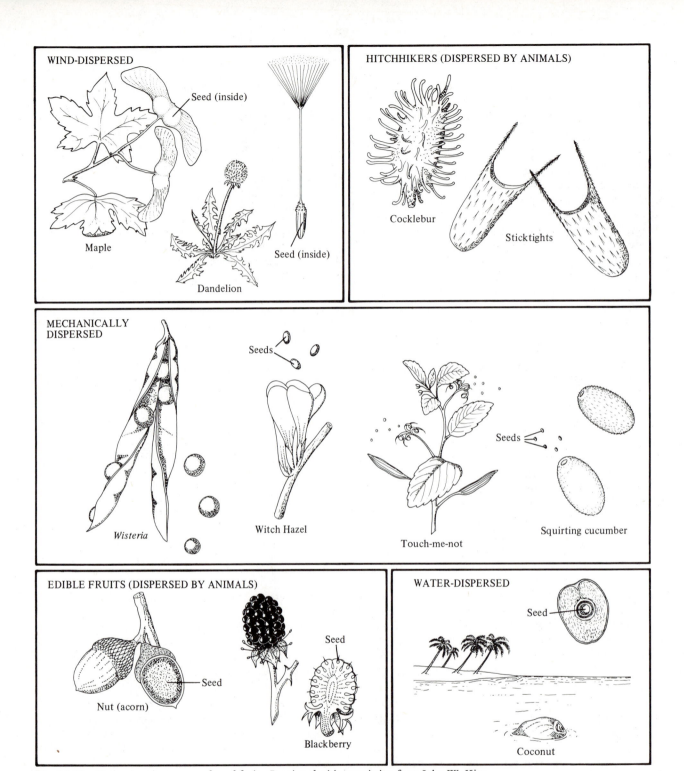

WIND-DISPERSED

Seed (inside)

Maple

Dandelion

Seed (inside)

HITCHHIKERS (DISPERSED BY ANIMALS)

Cocklebur

Sticktights

MECHANICALLY DISPERSED

Seeds

Wisteria

Witch Hazel

Seeds

Touch-me-not

Squirting cucumber

EDIBLE FRUITS (DISPERSED BY ANIMALS)

Seed

Nut (acorn)

Seed

Blackberry

WATER-DISPERSED

Seed

Coconut

Fig. 26.22 Various angiosperm seeds and fruits. Reprinted with permission from John W. Kimball, *Biology*, 4e, © 1978. Reading, Massachusetts: Addison-Wesley (Fig. 20.10).

structures ("seed-leaves") consisting of stored food, while monocots have only one. The majority of flowering plants are dicots, and have a netlike leaf veination, vascular bundles arranged in a ring, and flower parts typically in multiples of 2, 4, or 5. The monocots, on the other hand, have parallel veins in their leaves, scattered vascular bundles, and flower parts typically in multiples of 3 or 6. Monocots are mostly herbaceous (nonwoody), and are thought to have evolved from semi-aquatic dicots.

F. EVOLUTION OF THE ANIMAL KINGDOM

The animal kingdom traditionally included both the Protozoa and their multicellular descendants, the Metazoa. A currently more fashionable horizontal classification (Whittaker, 1969) separates the Protozoa at the kingdom level from the Metazoa. But regardless of where the boundary is drawn, it is generally agreed that multicellular animals evolved from the Protozoa. The family tree of animals, as accepted in this book, is shown in Fig. 26.23. The sponges and Mesozoa are considered to represent side branches that evolved from colonial protozoans; they are not part of metazoan ancestry. The true Metazoa, or Eumetazoa, consist of animals developing from two-layered embryos, and possessing a tissue level of organization as adults.

Sponges and Mesozoa

The simplest multicellular animals are the sponges (phylum **Porifera**) (Fig. 26.24a). Most zoologists consider them to represent a distinct side branch of the animal kingdom (subkingdom **Parazoa**), unrelated to the higher Metazoa.

Sponges live attached to the substratum. Water continuously enters the interior through a series of pores, and leaves through an excurrent opening or osculum. No specialized organs or tissues are formed, but several specialized types of cells are present. Each cell is in contact with the water at all times. There is very little division of labor; each cell is responsible for nearly all of its own life processes, including feeding and excretion.

Equally as advanced as the sponges, though divergent and presumably unrelated to them, is a small group known as the **Mesozoa** (Fig. 26.24b). They are all minute, microscopic animals of a few cells to a few dozen each. There is even less complexity among Mesozoa than among sponges, and there are fewer cell types.

Origin of the Metazoa

Multicellularity may have evolved only once, or as many as three times, among the animals. Two major theories have been advanced to account for the origin of the true Metazoa (or Eumetazoa). The older and more widely accepted **colonial theory** exists in many versions, in all of which the Metazoa evolved from colonial, flagellated protozoans. From a hollow ball, similar to that found in the green alga *Volvox* (see Fig. 26.11), a two-layered (diploblastic) form was derived by a process of invagination, as seen in the embryology of many metazoan animals. Haeckel called this diploblastic ancestor a *gastraea*. He applied the same names (ectoderm, entoderm*) to the two layers of cells in his hypothetical gastraea, in the freshwater coelenterate *Hydra*, and in higher metazoan embryos. Haeckel thought that bilaterally symmetrical animals evolved from the primitive gastraea by a process of elongation.

An alternative theory advocates a bilaterally symmetrical animal, perhaps an acoel flatworm, as the ancestral metazoan. According to this theory, bilateral symmetry is a primitive trait, and radial symmetry is secondarily derived.

Above the level of the sponges and Mesozoa, all the remaining multicellular animals are grouped together as Metazoa (or Eumetazoa). They all develop from multicellular embryos, and they all pass through a gastrula stage in their development. There is always a digestive cavity, the entrance to which is called a **mouth**. The cnidarians are the most primitive of the Eumetazoa, and the remaining Eumetazoa can all be imagined to be descendants of the cnidarians.

The Cnidarians or Coelenterates

The phylum **Cnidaria**, formerly called **Coelenterata**, contains simple eumetazoans possessing radial symmetry and only two layers of cells (ectoderm and endoderm). Special stinging structures (nematocysts) are contained within specialized stinging cells known as cnidariocytes. Two alternative body forms occur (Fig. 26.25). All members of the phylum are aquatic; the vast majority are

---

*Now usually spelled "endoderm."

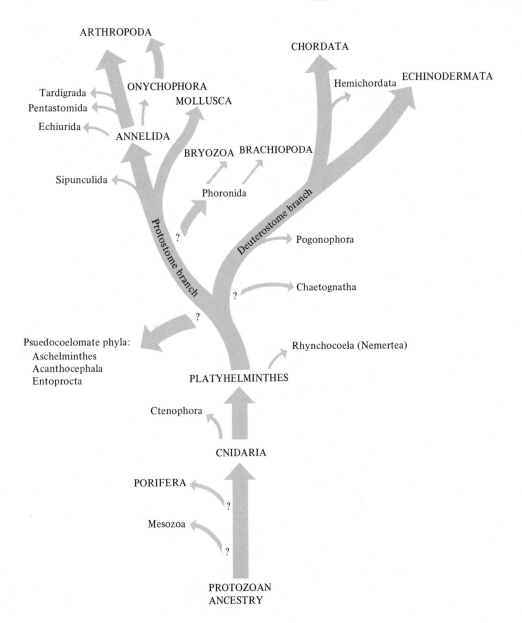

Fig. 26.23 A family tree of the animal kingdom, as assumed in this book.

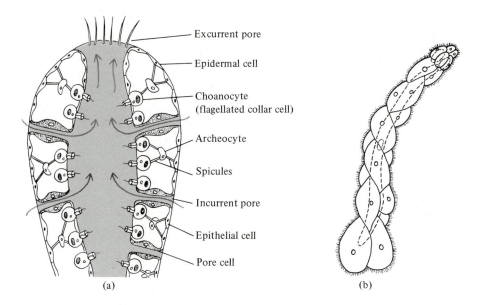

Fig. 26.24   (a) A simple type of sponge. (b) A mesozoan. Part (a) is reprinted with permission from Charles K. Levy, *Elements of Biology,* © 1978. Reading, Massachusetts: Addison-Wesley (Fig. 20.2).

Fig. 26.25   Comparison between a polyp and a medusa. The medusa is shown inverted from its true position for comparison. Reprinted with permission from John W. Kimball, *Biology,* 4e, © 1978. Reading, Massachusetts: Addison-Wesley (Fig. 37.4).

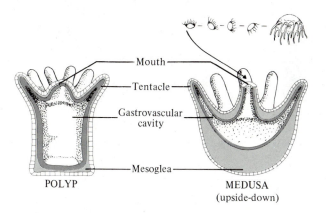

marine. Jellyfish, corals, and sea anemones belong to this phylum (Fig. 26.26).

Possibly related to the Cnidaria are the comb jellies or **Ctenophora.** Their symmetry is biradial, like that of a two-armed pinwheel, and their digestive cavity has eight T-shaped branches.

The Flatworms and Other Acoelomates

Above the level of the Cnidaria, **bilateral symmetry** becomes the rule; the exceptions to it are clearly secondary modifications. Bilateral symmetry is associated with the acquisition of a definite anterior (head) end, and with a differentiation between upper (dorsal) and lower (ventral) surfaces. Hyman (1940) attributes the origin of bilateral symmetry to the demands of locomotion across a surface such as the sea floor. An animal moving across such a surface would encounter no consistent difference between right and left; the same structures that would be

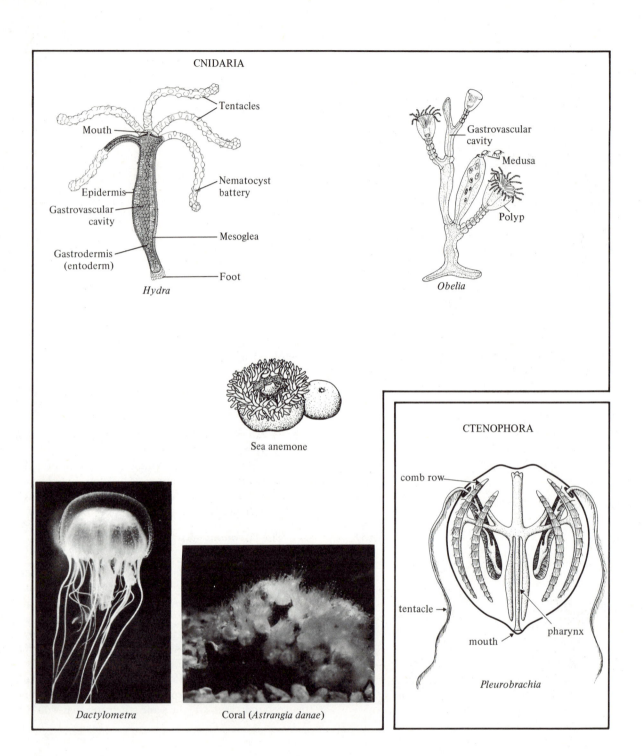

CNIDARIA

Tentacles

Mouth

Epidermis

Gastrovascular
cavity

Gastrodermis
(entoderm)

Nematocyst
battery

Mesoglea

Foot

Hydra

Gastrovascular
cavity

Medusa

Polyp

Obelia

Sea anemone

Dactylometra

Coral (*Astrangia danae*)

CTENOPHORA

comb row

tentacle

mouth

pharynx

Pleurobrachia

Fig. 26.26 (Facing page) Representative Cnidaria and Ctenophora. The *Hydra* is reprinted with permission from Donald D. Ritchie and Robert Carola, *Biology,* © 1979. Reading, Massachusetts: Addison-Wesley (Fig. 10.8). The *Obelia* and coral are reprinted with permission from John W. Kimball, *Biology,* 4e, © 1978. Reading, Massachusetts: Addison-Wesley (Fig. 37.2). The *Dactylometra* is used courtesy of William H. Amos.

adaptively advantageous on the right would also be advantageous in mirror image on the left. But the difference between top and bottom would be great, and so we see in flatworms and other similar groups a decided differentiation between upper and lower surfaces. The difference between front and rear would also be important; in particular, there would be selection for increasing sensory precision at the end that encountered novel stimuli first, meaning the front end. Selection would therefore tend to favor the concentration of sense organs (and subsequently of the nervous system as a whole) near the anterior (front) end, forming something recognizable as a **head**.

Among the bilaterally symmetrical metazoans, nearly all zoologists would now agree that those lacking a body cavity (or coelom, as defined later) are the most primitive. Beyond this point, there is disagreement: a number of zoologists place greater emphasis on the structure of the coelom and the manner of its formation, while others place greater emphasis on differences in early embryology.

The flatworms (**Platyhelminthes**) (Fig. 26.27) are bilaterally symmetrical. They have a simple central nervous system consisting of a double row of ganglia connected to one another in a network that resembles the rungs of a ladder. In their embryology, the flatworms are advanced beyond the Cnidaria in having a true middle layer, or mesoderm, making a three-layered (triploblastic) embryo. The mesoderm is solid and lacks any internal cavity or coelom.

In one important respect, the flatworms are still at the cnidarian level of efficiency: their digestive system still consists of a primitive gastrovascular cavity, with the mouth functioning as both entrance and exit. The food may leave just as it had entered, and wastes may re-enter just as they have left. There is no regional specialization of the digestive apparatus, for there is no way to ensure stepwise sequential action; all digestive enzymes must therefore be present simultaneously throughout the system.

Flatworms have no circulatory system, nor do they need any: their bodies are flat, never more than a few cell layers thick, so that no cell is ever very far from one of the body's surfaces.

In the course of creeping through the loose mud on the sea floor, some ancient groups of flatworms must have encountered other bottom-dwelling creatures. It is likely that some of these flatworms became occasional visitors, then more frequent visitors, upon or within these other organisms. This was the first step on the road to parasitism. Today, the majority of known flatworms are parasitic. Nearly all still maintain in their life cycle an aquatic stage, which then enters the body of a common bottom-dwelling species.

The phylum Rhynchocoela (Fig. 26.27) are a small group of worms whose greatest similarities are to the flatworms. But, unlike the flatworms, rhynchocoels have a digestive tract with separate mouth and anus. This means that digestion can now become an ''assembly line'' process, with enzymes acting sequentially instead of all at once. In the evolution of advanced digestive systems, with different regions specialized for different phases of digestion, the presence of a separate mouth and anus was undoubtedly the first step.

The Evolution of Body Cavities

Animals belonging to all the remaining phyla have most of their internal organs contained within a fluid-filled body cavity. In most cases, the body cavity is surrounded entirely by mesoderm; such a body cavity is called a **coelom**, from the Greek word *koilos,* meaning ''hollow.'' In a few phyla, however, the body cavity is not a true coelom, but rather a persistent blastocoel, lined on the inside with endoderm and on the outside with both mesoderm and ectoderm (Fig. 26.28). A body cavity of this sort is known as a pseudocoel. The pseudocoelomate phyla include the Aschelminthes, Entoprocta, and Acanthocephala (Fig. 26.29).

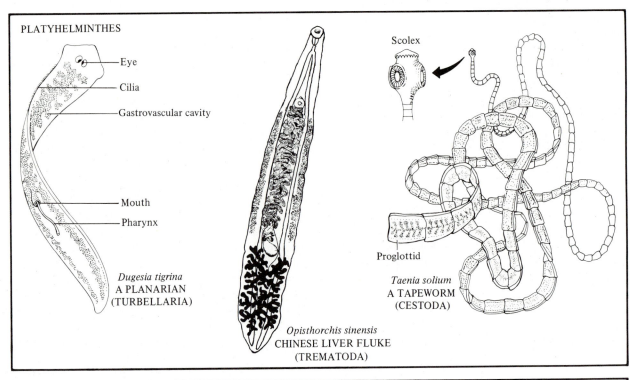

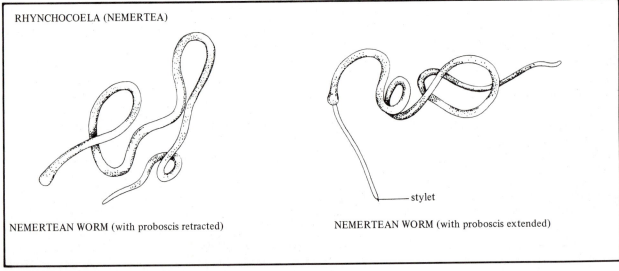

Fig. 26.27 Representative flatworms and rhynchocoels. The planarian and tapeworm are reprinted with permission from John W. Kimball, *Biology,* 4e, © 1978. Reading, Massachusetts: Addison-Wesley (Fig. 37.5). The Chinese liver fluke and nemertean worms are from *Invertebrate Zoology Laboratory Workbook,* 3rd edition, D. E. Beck and L. F. Braithwaite, Burgess Publishing Company, Minneapolis, Minn. Reprinted by permission of the publisher.

Parenchyma
tissue
(mesodermal)

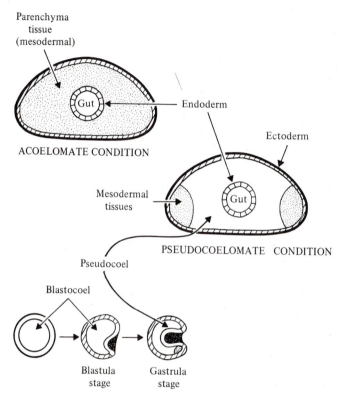

Endoderm

Ectoderm

ACOELOMATE CONDITION

Mesodermal
tissues

PSEUDOCOELOMATE CONDITION

Pseudocoel

Blastocoel

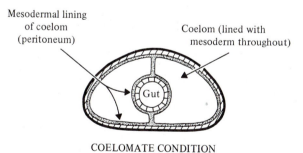

Blastula
stage

Gastrula
stage

Mesodermal lining
of coelom
(peritoneum)

Coelom (lined with
mesoderm throughout)

COELOMATE CONDITION

Fig. 26.28 Types of body cavities.

Various theories have been proposed for the origin of body cavities. According to one theory, the original coelom was formed from a persistent gonadal cavity, or gonocoel. According to another, the coelom originated as an outpocketing of the gut (an enterocoel). According to a third theory, the coelom originated, as it now does in the embryology of many higher metazoans, as a cavity or cleft within the mesoderm—a so-called schizocoel. There are difficulties with each of these theories, and each assumes a different group of metazoans as being more primitive (see Barnes 1980 or Hyman 1940 for more complete discussions).

According to the currently accepted theory, the body cavity originally formed a "**hydrostatic skeleton**" that aided in burrowing. Imagine a worm trying to burrow through the sediment on the ocean floor. Any mechanism that would give it hold or purchase upon its surroundings would be selected as an aid in locomotion. Now a body cavity would serve this purpose ideally, for by compressing this cavity from front to rear (by means of longitudinal muscles), and causing it to swell out in all other directions, the animal would temporarily anchor that portion of its body in the sediment around it. Under the above theory, one might suppose that any type of body cavity, whatever its embryological origin, would be equally suitable to serve as a hydrostatic skeleton. In fact, evolutionary opportunism might well be expected to lead to multiple solutions to the common functional requirement. This is in fact what we observe: some phyla have an enterocoel, arising embryologically from the gut, while others have a schizocoel that arises by splitting. Still other phyla have a body cavity that is not a true coelom, but rather a pseudocoel as described above. The truly coelomate condition probably evolved independently of the pseudocoelomate condition, perhaps more than once.

The Lophophorate Phyla

The three phyla Bryozoa, Brachiopoda, and Phoronida (Fig. 26.30) all share a unique apparatus for filter feeding known as a **lophophore**. A lophophore consists of a series of ciliated tentacles that create currents of water and also trap suspended food particles brought in by these currents. All lophophorate phyla have a simple digestive tract, complete with mouth and anus. Like all other animals treated from this point on, the lophophor-

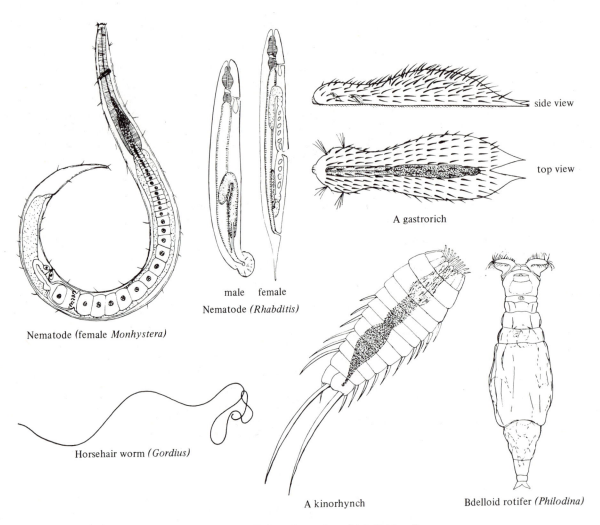

side view

top view

A gastrorich

Nematode (female *Monhystera*)

male   female

Nematode *(Rhabditis)*

Horsehair worm *(Gordius)*

A kinorhynch

Bdelloid rotifer *(Philodina)*

Fig. 26.29 Representative pseudocoelmates. The female nematode and bdelloid rotifer are reprinted with permission from R. W. Pennak, *Fresh-Water Invertebrates of the United States*, 2e, 1978. New York: John Wiley & Sons. All others are from *Invertebrate Zoology Laboratory Workbook*, 3rd Edition, D. E. Beck and L. F. Braithwaite, Burgess Publishing Company, Minneapolis, Minn. Reprinted by permission of the publisher.

Phoronida	Bryozoa	Brachiopoda
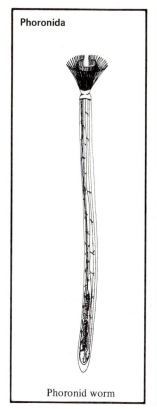	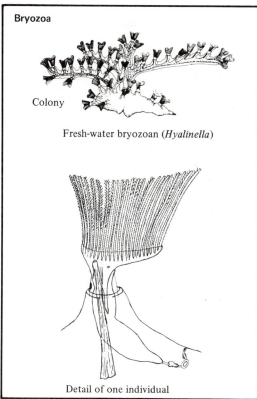	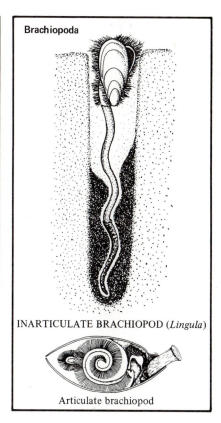
Phoronid worm	Colony Fresh-water bryozoan (*Hyalinella*) Detail of one individual	INARTICULATE BRACHIOPOD (*Lingula*) Articulate brachiopod

Fig. 26.30 Representative lophophorates. All drawings except *Lingula* are from *Invertebrate Zoology Laboratory Workbook,* 3rd Edition, D. E. Beck and L. E. Braithwaite, Burgess Publishing Company, Minneapolis, Minn. Reprinted by permission of the publisher.

ates have a true coelom, bounded on all sides by mesoderm.

The phylum **Phoronida** contains only two genera and about 15 species of wormlike animals that live in chitinous tubes. A newly metamorphosed phoronid worm burrows until it finds a suitable location; then it secretes its tube. The phoronids show us a possibly ancestral mode of life for lophophorates in general, perhaps even for other coelomate groups.

Animals of the phylum **Bryozoa** (''moss animals'') are sessile, colonial filter-feeders with retractable lopho-

phores. The fossil record of this group extends from Cambrian to Recent.

The phylum **Brachiopoda** contains animals sometimes known as lamp shells. Despite their superficial resemblance to clams, they can readily be distinguished by the fact that their shell has two unequal valves, with a plane of symmetry passing through the middle of each. The larger valve bears a stalk (the pedicle) by which the animal typically attaches to the substratum; the smaller (brachial) valve bears the spirally coiled lophophore.

There are two classes of brachiopods, both of them

ancient. The more primitive Inarticulata are today represented by a handful of genera in which the two valves are attached to each other only by a series of muscles. *Lingula,* the best-known genus, lives near the top of a long vertical burrow at the end of a long muscular pedicle (Fig. 26.30). Here it filter feeds by creating a current of water over its lophophore, but retracts quickly into its burrow if disturbed. The similarity to phoronid worms is evident; the major advance over the phoronids lies in the presence of the two valves, which probably originated as aids in burrowing. *Lingula* is certainly one of the oldest of animal genera; it is known from Ordovician deposits, and has hardly changed at all since.

The more diverse group of brachiopods are the Articulata, in which the two valves are hinged together. Most articulate brachiopods are oriented upside-down on their pedicles, for the pedicle valve, in uppermost position, corresponds to the ancestrally ventral surface, while the brachial valve, now in a lower position, represents the ancestrally dorsal surface.

The fossil record of the Brachiopoda extends from the Cambrian to the present, and the phylum is one of the major groups in Paleozoic faunas.

The Mollusca

The phylum **Mollusca** is the second largest phylum in the animal kingdom and contains some of the most familiar invertebrates, including snails, slugs, clams, oysters, scallops, mussels, squid, and octopus (Fig. 26.32).

Characteristic of most molluscs is the presence of a shell; those molluscs that lack a shell are clearly derived from shelled ancestors. The epidermis that secretes this shell is called the **mantle.** Beneath the posterior (hind) end of the shell, the mantle is retracted to form a **mantle cavity** whose presence is often taken as the defining characteristic of the entire phylum (Fig. 26.31). Within this mantle cavity are contained the gills, the anus, and the terminal opening of the urogenital system. A simple heart pumps blood through the gills, while a simple kidney (the nephridium) drains the coelomic cavity. The digestive system exhibits a number of specializations, including the presence of digestive glands and a coiled intestine, and the binding together of food clumps with mucus. The most remarkable digestive specialization is the presence of a protrusible scraping organ known as the **radula.** The radula contains a series of teeth embedded in

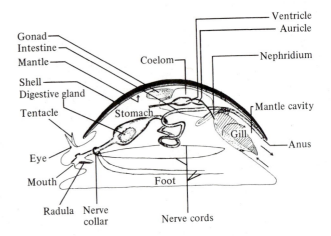

Fig. 26.31 Structure of a primitive mollusc, based largely on the anatomy of *Neopilina.* From *Invertebrate Zoology,* Fourth Edition by Robert D. Barnes. Copyright © by Saunders College/Holt, Rinehart and Winston. Copyright 1963, 1968, and 1974 by W. B. Saunders Company. Reprinted by permission of Holt, Rinehart, and Winston.

a straplike membrane. The membrane is moveable back and forth over a cartilage called the odontophore; this cartilage can be protruded from the mouth and applied to the surface on which the mollusc is feeding. The radular membrane with its teeth is drawn over the odontophore, producing a scraping motion that wears away at the substratum. The mollusc can thus feed on encrusted algae. Many living molluscs still retain this ancestral mode of eating. The radula and the mantle cavity with its gills were two of the "key characters" that enabled the evolutionary success of the molluscs. The gills served a respiratory function and thus allowed the molluscs to evolve body sizes much larger than any of the invertebrates considered thus far. No close homologues to either radula or mantle cavity are known in any other phylum.

Locomotion in the majority of molluscs is by creeping over the surface using a slime-covered muscular organ known as the **foot.** This form of locomotion was possibly inherited from flatworms or similar remote ancestors. The nervous system is rather simple and consists primitively of a ring that encircles the digestive tract just behind the mouth, along with several paired nerves emanating from this ring. Ganglia occur at various points along these nerves, the largest pair being the cerebral ganglia just above the digestive tract. Major sense organs often include a well-developed pair of eyes; nerves from these sense organs often lead directly to the cerebral ganglia.

Primitive molluscs have external fertilization and separate sexes. In several groups, copulatory structures have evolved from the lining of the mantle cavity, so that internal fertilization is common in many molluscs. Hermaphroditism has evolved in many orders, and snails are in fact the most famous hermaphrodites of the animal world.

Representative molluscs are shown in Fig. 26.32.

The class **Monoplacophora** includes the living genus *Neopilina* and its extinct relatives. *Neopilina* has a noncoiled conical shell with low profile. It also shows equivocal evidence of possible derivation from a segmentally organized ancestor such as an annelid. There are eight pairs of foot muscles, five pairs of gills, and ten runglike commissures between the major nerves.

The class **Gastropoda** includes the familiar snails and slugs. The fossil history of this class is very well documented and begins in the Cambrian. From a conical-shelled ancestor, the gastropods soon evolved a shell with a planispiral coil, meaning that the coils were all in a single plane along the midline (Fig. 26.33). The body was symmetrical. The coiled shell was carried over the head and tended to tip the animal forward. The key to the understanding of all subsequent gastropod evolution was a **torsion** of the body that brought the mantle cavity over the head and directed the coiled shell to the rear (Fig. 26.33). The shell also evolved a more compact shape, with a lowered center of gravity, by drawing out the spiral along a line perpendicular to the plane of coiling, producing an asymmetrical shape known as a trochoid. The apex of the spiral coil was drawn out to the right, so that the snail tended to tip to the right side, a problem that was easily corrected by tipping the entire shell to the left while twisting it slightly to the right. The apex is thus now deflected both upwards and to the rear, but is still directed to the right in most cases.

The asymetrical torsion of the body has had a profound effect on gastropod anatomy and its subsequent evolution. The mantle cavity is now located on the left side, and certain paired structures (gills, nephridia, etc.) have tended to disappear on the right side. The water current through the mantle cavity now runs from left to right across the animal's head. The nervous and digestive systems have likewise undergone torsion, and their anatomy is thus greatly modified. The position of the anus is particularly troublesome, for it is relatively close to the mouth, and fouling is thus an obvious problem. Several adaptations have evolved during the course of gastropod history that remedy this situation to varying degrees. At various times during gastropod evolution, the body has secondarily undergone a detorsion (untwisting). The shell has also been lost in a few cases and the body has once again become bilaterally symmetrical.

The tusk shells (class **Scaphopoda**) are bilaterally symmetrical molluscs with an exceedingly high-spired conical shell open at both ends. The mantle cavity is contained along the posterior margin, and the excurrent stream of water exits through the top, thus avoiding the problem of fouling that has plagued so many gastropods during their evolution. Scaphopods have a wedge-shaped foot with which they burrow into soft sand, where they feed on Foraminifera and other small organisms. The tip of their cone projects above the sea floor, and the excurrent stream of water is shot out there.

The class **Polyplacophora**, commonly known as chitons, are bilaterally symmetrical molluscs with poorly developed heads. Internally, they possess many characteristics associated with primitive molluscs, but externally they are unique in that their shell consists of eight partially overlapping plates. Chitons are slow-moving, and cling to rocky surfaces, which they scrape with their radulas, feeding upon the encrusted algae. The nervous system of chitons is of a type reminiscent of the Platyhelminthes, yet perfect as an ancestral condition from which the more complex nervous systems of other molluscs could readily have been derived.

A strange group of wormlike molluscs, the **Aplacophora,** are often considered to be allied to the chitons. They lack any shell, but they possess a radula and a posterior cavity that may represent a mantle cavity. There is also some suggestion of a muscular foot. The relation of these forms to the other molluscs is obscure. Particularly uncertain is the question as to whether the absence of a shell is a primitive or a degenerate feature.

The bivalves (class **Bivalvia** or **Pelecypoda**) are bilaterally symmetrical molluscs that were primitively adapted

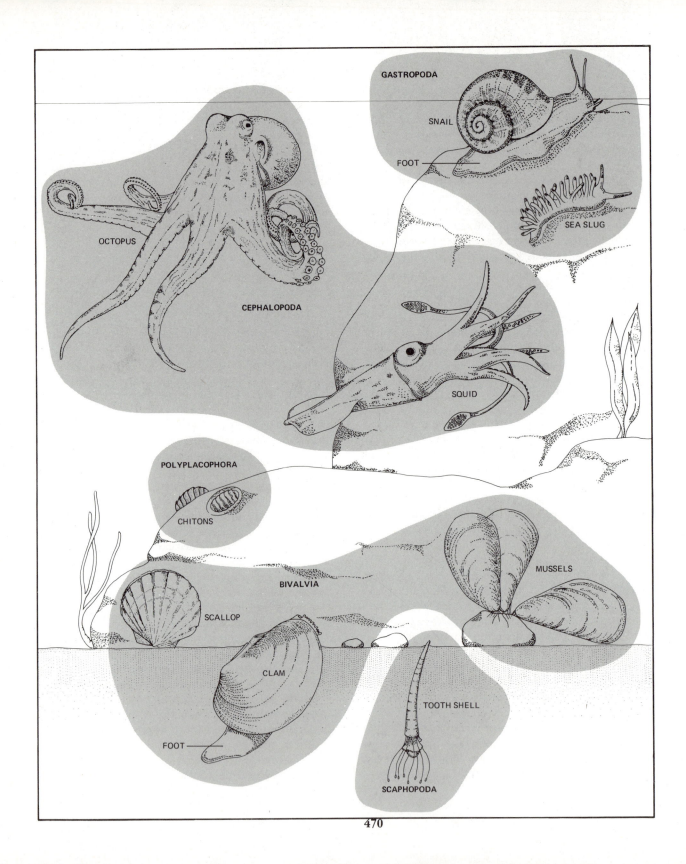

GASTROPODA

SNAIL

FOOT

SEA SLUG

OCTOPUS

CEPHALOPODA

SQUID

POLYPLACOPHORA

CHITONS

BIVALVIA

MUSSELS

SCALLOP

CLAM

FOOT

TOOTH SHELL

SCAPHOPODA

Fig. 26.32 (Facing page) Representative molluscs (phylum Mollusca). Reprinted with permission from John W. Kimball, *Biology*, 4e, © 1978. Reading, Massachusetts: Addison-Wesley (Fig. 37.12).

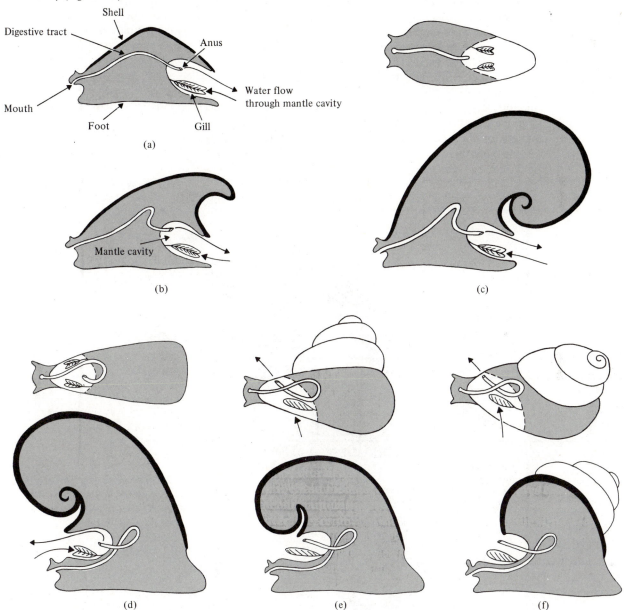

Fig. 26.33 Torsion of the gastropod shell. (a) Primitive conical shell; (b) coiling begins; (c) weight of coiled shell tilts animal toward rear; (d) torsion brings coiled shell and mantle cavity forward, but shell is top-heavy; (e) center of gravity is lowered by drawing apex of shell to the right; shell now tilts to right; (f) apex of shell is tilted upward and rearward; balance is achieved.

to a burrowing existence. Many members of the class still burrow, an activity in which the two-valved shell is a great asset. As an adaptation for burrowing, the body has become laterally compressed, the head greatly reduced, and the foot laterally compressed and thus hatchet shaped (the name Pelecypoda means ''hatchet foot''). The two valves of the shell are mirror images of one another, and each is asymmetrical; the plane of symmetry passes between the two valves.

The class **Cephalopoda** includes many actively swimming, generally predatory molluscs greatly modified from the ancestral molluscan type. Fossil cephalopods are among the most intensively studied of all fossil invertebrates. The living species, including the familiar octopus

and squid, as well as the chambered nautilus, are but a small fraction of the species known as fossils (Fig. 26.34).

Cephalopods have the most highly developed heads of all the marine invertebrates. Eyes are especially well developed. The mouth is provided with horny beaks in addition to the radula, and a series of tentacles surround its entrance. The animal is bent back upon itself, so that the mantle cavity exits forward, beneath the head, through a funnel known as the hyponome. This body shape evolved in shelled members of the class, but it persists even in those cephalopods in which the external shell has been lost. Cephalopods are capable of propelling themselves backwards by forcibly ejecting water from their mantle cavity through the hyponome.

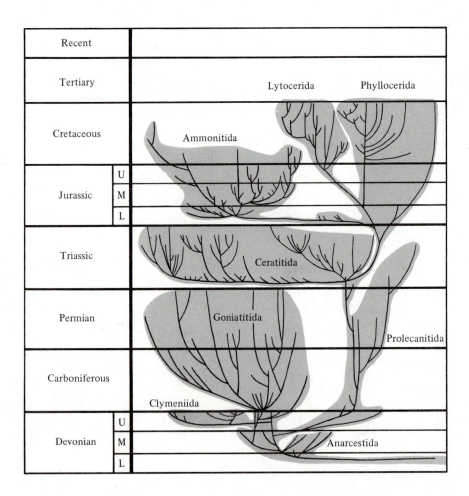

Fig. 26.34 Phylogeny of the extinct subclass Ammonoidea (class Cephalopoda).

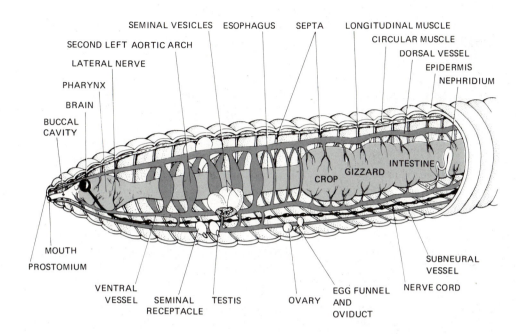

SEMINAL VESICLES   ESOPHAGUS   SEPTA   LONGITUDINAL MUSCLE
SECOND LEFT AORTIC ARCH   CIRCULAR MUSCLE
LATERAL NERVE   DORSAL VESSEL
PHARYNX   EPIDERMIS
BRAIN   NEPHRIDIUM
BUCCAL CAVITY
                                        INTESTINE
                               GIZZARD
                          CROP
MOUTH
PROSTOMIUM
                                                    SUBNEURAL VESSEL
VENTRAL VESSEL   NERVE CORD
SEMINAL RECEPTACLE   TESTIS   OVARY   EGG FUNNEL AND OVIDUCT

Fig. 26.35 Structure of an annelid worm. Reprinted with permission from Donald D. Ritchie and Robert Carola, *Biology*, © 1979. Reading, Massachusetts: Addison-Wesley (Fig. 10.21).

The familiar octopus and squid have no external shell, but the vast majority of cephalopods possessed shells. The shell was typically divided into compartments or chambers, of which only the largest and most recent was inhabited. When the animal grew too large for the chamber it occupied, it moved forward and secreted a new, larger chamber. The abandoned chambers were usually filled with gas and therefore buoyant; an air tube, or siphuncle, penetrated the various chambers from the last (occupied) chamber back to the first. The earliest cephalopods are believed to have had a straight or only slightly curved shell. The primitive orientation was vertical, the buoyant force of the gas-filled chambers insuring that the oldest chamber was at the top and the occupied chamber directed downward. Various mechanisms were evolved, some of them repeatedly, to achieve for the animal a more horizontal orientation. In many straight or slightly curved shells, the apical end was weighted down with mineral matter deposited either internally or externally. The shell was in some cases foreshortened or lost. In one group, the gas-filled chambers came to lie just above the animal's visceral mass (and there were also mineral deposits further back, serving as a counter-weight). In the most successful groups of cephalopods, the shell became curved and eventually coiled above the animal (see Figs. 17.3 and 18.5).

Annelida and Related Phyla

Some Metazoa, notably the phyla Annelida and Arthropoda, have their bodies divided into serially repetitive segments, a phenomenon known as **metamerism.** The Annelida are the most profoundly metameric of all animal phyla (Fig. 26.35), while the Arthropoda, which evolved from annelid ancestors, obey Williston's rule in having fewer segments with an increasing differentiation between them. Metamerism in annelids is highly adaptive because of its function in locomotion, but the pervasive segmentation of the body also affects the nerves, blood vessels, and most other body structures as well.

Locomotion in annelids is dependent upon segmentation of the coelom into separate fluid-filled compartments, one for each segment. Contraction of longitudinal muscles causes the fluid-filled coelomic compartment to swell sideways in all directions, anchoring that segment firmly against the substratum. Contraction of circular muscles causes the coelomic compartment to elongate along the body axis, pushing the anterior end of the body forward against the segments further posterior, many of which are at any given moment expanded against the substratum. By alternating the contraction of these two muscle groups rhythmically, the worm is able to move forcibly through sediment, pushing the sediment aside as it proceeds. The walking movements of arthropods depend upon muscles derived from these two sets in annelids. Even insect flight depends in many cases upon rhythmically alternating foreshortening and elongation of the coelomic compartments.

Annelids are metamerically organized invertebrates with a **prostomium** at the anterior end, a **pygidium** at the posterior end, and numerous identical (or nearly identical) segments in between. The prostomium, which includes the brain, is not considered a segment, neither is the pygidium. New segments originate just posterior to the prostomium, so that the first-formed segment is adjacent to the pygidium and the newest segment is the furthest anterior. In the course of evolution, the first few true segments may become coalesced with the prostomium, forming a head.

The digestive system contains both mouth and anus, and extends the entire length of the animal. The circulatory system is segmentally organized, with the blood circulating toward the head via the dorsal blood vessel and away from the head via the ventral blood vessel. The nervous system is segmentally organized, with paired ganglionic swellings in each segment. The nerve cord is ventral throughout the segmentally organized trunk, but an esophageal nerve ring connects to a dorsally situated cerebral ganglion (supraesophageal ganglion), also sometimes called a "brain."

The phylum **Onychophora** is transitional in many respects between the Annelida and Arthropoda. They are terrestrial creatures living in tropical or south temperate climates. Their superficial resemblance is to caterpilllars, but their numerous paired feet are not jointed and they lack the exoskeleton characteristic of arthropods. Their head is much better developed than in annelids, and includes a number of body segments coalesced with the prostomium. The paired mouthparts are segmentally organized and thus resemble the arthropods more than the annelids, and the presence of legs represents a definite advance beyond the annelid condition in the direction of arthropods. At least one arthropod specialist has suggested that onychophorans be placed in the phylum Arthropoda, but this would require a redefinition of that

phylum and present other difficulties explained in the next section. A fossil species from the Cambrian shows that the group is very ancient and has undergone little if any change since its origin.

The Phylum Arthropoda

The phylum **Arthropoda** is by far the largest of all animal phyla, including nearly 80% of the living members of the animal kingdom. They are derived from annelids, and have inherited a segmental or metameric plan of organization, usually with some regional specialization or modification of this underlying plan.

Arthropods are invariably enclosed in an external skeleton or **exoskeleton,** which is at least partly chitinous and may in some cases be calcified. The exoskeleton is an essential arthropod characteristic, and provides both protection and structural support. It does, however, create certain problems whose evolutionary solutions have brought about many other adaptations now characteristic of the Arthropoda and serving to distinguish them from the Annelida. Growth is one obvious problem for an animal with a hard outer skeleton. In arthropods, growth is made possible only by the periodic shedding of the exoskeleton, a process called molting or ecdysis. For a brief period of time during each molt, the individual is especially vulnerable to predation, and thus we see behavioral adaptations in many arthropods for concealment during molting. We also see reduction in the number of molts in certain cases, or the concentration of several successive molts into a more drastic metamorphosis.

Movability of the limbs, as in locomotion, is another problem for arthropods because most of the exoskeleton is generally rigid; movability of hardened mouthparts presents a very similar problem. These problems have been solved by leaving the generally rigid exoskeleton flexible at certain rather definite locations, thus forming joints. The paired appendages of arthropods are conspicuously jointed, and it is from this feature that the phylum derives its name (*arthros* = joint; *pod* = foot; i.e., joint-legged animals).

Many other arthropod characteristics are simply retentions from their annelid ancestors or slight modifications upon annelid patterns. Among these features are the generally metameric organization and the structure of the nervous system, with a ventral nerve cord bifurcating to run around the anterior end of the digestive tract and reuniting above to form a major ganglion that sometimes functions as a brain.

Some arthropod characteristics were made necessary by an initial increase in size over their annelid ancestors. The open circulatory system is one such characteristic; the various respiratory adaptations of the various arthropod groups were also necessitated by their initially larger size.

Arthropods were among the first animals to occur at the base of the Cambrian period. Indeed, stratigraphers customarily define the base of the Cambrian as the time of first appearance of trilobites, which were one of the more important groups of early arthropods.

The ''key character'' in arthropod evolution was undoubtedly the acquisition of a hardened exoskeleton. As already indicated, this was the feature that created the selection pressures that brought about the evolution of various other arthropod characteristics. The exoskeleton itself may have originated as a mechanism for structural rigidity, or as a means of protection, or it may have served both functions. There is even the possibility that a rigid exoskeleton aided in burrowing among ancestral bottom-living arthropods.

The actual identity of the ancestral arthropods is shrouded in mystery. Many paleontologists identify the trilobites as the ancestral arthropods, and certainly the derivation of certain marine arthropods (crustaceans, xiphosurans) and their terrestrial descendants (e.g., spiders) from trilobites seems quite plausible. Other scientists, especially entomologists (those who study insects), point to the terrestrial Onychophora as a possible transitional group between the Annelida and Arthropoda. It may just be that the Arthropoda are a polyphyletic taxon, with some of its members derived from trilobites and others from onychophorans.

In this book, the Arthropoda are divided into four subphyla: Trilobitomorpha, Crustacea, Chelicerata, and a fourth group containing insects and myriapods.

The extinct class **Trilobita**, and certain related forms, are here placed in a subphylum **Trilobitomorpha**. These early arthropods were exclusively marine bottom-dwellers who occasionally swam through the water above. They were actively moving creatures, and may have been scavengers or else detritus feeders. Their bodies were in all cases very much flattened, and their exoskeletons strengthened by the addition of calcium salts. Paired eyes were usually present dorsally, and a mouth was located on the underside. Numerous paired appendages were present and each was primitively Y-shaped (biramous), with a jointed leglike portion and a feathery fanlike portion.

The predominant marine arthropods of today are the **Crustacea** (Fig. 26.36), which include the lobsters, shrimp, crabs, barnacles, copepods, and their marine and freshwater relatives. The Crustacea are characterized by the presence of two pairs of antennae, three pairs of jaw-like post-oral appendages (the mandibles and two pairs of maxillae), and feathery gills that grow out from the legs. The crustacean exoskeleton is generally strengthened, as in trilobites, with calcium salts. The body may be divided into head, thorax, and abdomen, but more often the head and thorax are combined into a cephalothorax. Most crustaceans have paired eyes of the compound type, similar to those of insects, and also a median eye.

The subphylum **Chelicerata** (Fig. 26.37) includes a diverse group of arthropods lacking antennae, but having instead two pairs of modified appendages used in feeding: the chelicerae and the pedipalps. Four pairs of walking legs commonly follow the pedipalps, and the body is usually divided (except in mites) into a cephalothorax and abdomen. Many chelicerates are active predators; several have evolved poisons which immobilize their prey. Two of the three classes are marine, but the third and largest includes all those that have successfully invaded terrestrial habitats, such as scorpions, spiders, and mites.

The marine chelicerates include the class Pycnogonida (sea spiders) and the class Merostomata. Early merostomes, beginning in the Cambrian, were called eurypterids. They were active predators, and some grew very large. One Devonian genus was over three meters long, making it the largest arthropod ever. Jurassic and later merostomes are placed in the order Xiphosura, whose best-known representative is the living horseshoe crab, *Limulus*. *Limulus* is a scavenger of the ocean bottoms, and most other xiphosurans are believed to have had similar habits.

The largest and most successful group of chelicerates is the class **Arachnida**, which includes the scorpions, spiders, and their relatives. Most primitive among these are the scorpions, whose fossil record begins in the Silurian. Silurian scorpions were aquatic and possessed paired res-

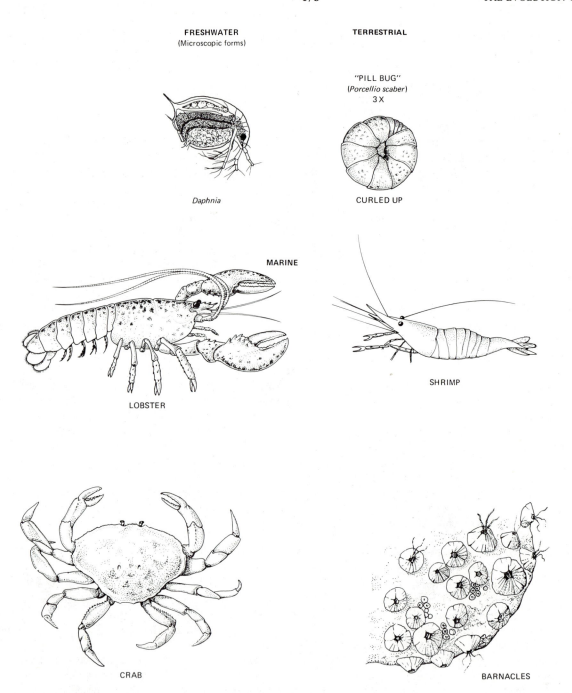

FRESHWATER
(Microscopic forms)

Daphnia

TERRESTRIAL

"PILL BUG"
(*Porcellio scaber*)
3 X

CURLED UP

MARINE

LOBSTER

SHRIMP

CRAB

BARNACLES

Fig. 26.36 Representative Crustacea. Reprinted with permission from John W. Kimball, *Biology*, 4e, © 1978. Reading, Massachusetts: Addison-Wesley (Fig. 37.19).

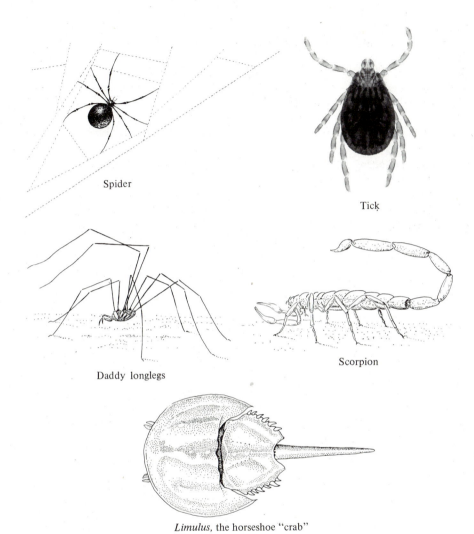

Spider

Tick

Daddy longlegs

Scorpion

Limulus, the horseshoe "crab"

Fig. 26.37 Representative Chelicerata. Reprinted with permission from John W. Kimball, *Biology,* 4e, © 1978. Reading, Massachusetts: Addison-Wesley (Fig. 37.17 and 37.18).

piratory structures known as **book gills.** By the Devonian period, the scorpions and a number of other arachnids had emerged onto the land, and had developed many of their present characteristics. Many arachnids produce a venom that immobilizes their prey and permits a peculiar type of predation and digestion: once the prey is immobilized, digestive fluids secreted from the midgut exude over its partially macerated tissues as these tissues are torn apart. The resulting liquids are sucked into the digestive tract, which needs no further adaptations for processing solid food. Among the adaptations that permitted the transition to land were the conversion of external book

gills into internal book lungs. One group (spiders) also paralleled the insects in the development of a tracheal system of air passages. Another key to arachnid success was the coalescence of several major ganglia into a more centralized brain. No sooner had the arachnids colonized the land (they were among the first animals in any phylum to do so) than they underwent a great adaptive radiation, or perhaps a series of such radiations; by late Carboniferous (Pennsylvanian) time, most of the 15 orders of arachnids had already made their appearance.

The most familiar arachnids are the spiders, which constitute the order Araneae. They are also most diverse in their adaptations, and they occupy the widest variety of habitats. One spider characteristic is the production of silk, which is secreted in liquid protein form and extruded through a series of spinnerets. The multi-stranded silk fiber solidifies upon stretching. Silk is put to a variety of uses. All spiders wrap their eggs in a protective layer of silk. All spiders also use their silk as a "drag-line" in the manner of mountain climbers, spinning a silken thread behind them as they proceed, and fastening it down every so often. If a spider looses its footing, it only needs to climb back up its dragline and proceed once more, or it can descend intentionally and still have a silken line by which it can ascend again. Many spiders cover a small piece of ground with a crisscross network of draglines in which prey may occasionally become entangled. From the occasional and originally accidental trapping of prey in draglines, spiders have evolved numerous specialized traps for their prey, including the highly advanced webs.

The order Acarina is the largest of the chelicerate orders and includes the mites and ticks. The vast majority are parasitic and extremely small.

Arthropods with elongated, wormlike bodies and many pairs of legs may be referred to as myriapods (Fig. 26.38). Included here are the centipedes (class Chilopoda), the millipedes (class Diplopoda), and two further classes, the Pauropoda and Symphyla. The last of these groups shows many similarities to insects.

Fig. 26.38 Two myriapods: a centipede and a millipede. Reprinted with permission from John W. Kimball, *Biology,* 4 e, © 1978. Reading, Massachusetts: Addison-Wesley (Fig. 37.20).

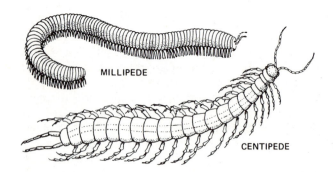

MILLIPEDE

CENTIPEDE

All myriapods possess mandibles and a single pair of antennae, characteristics that they share with insects. Many zoologists therefore believe the insects to be closely related to myriapods, and some have even suggested that the insects plus myriapods would constitute a natural or monophyletic taxon. Nearly all insects and myriapods are terrestrial and have air-breathing larval as well as adult stages, suggesting a terrestrial rather than a marine ancestry for the group as a whole. One feature shared by all insects and myriapods, and indicative of their terrestrial ancestry, is the presence of a tracheal system of air tubes (tracheae), opening to the outside by means of circular openings called spiracles. The occurrence of such tracheal systems in the Onychophora, combined with the generally similar body shape and terrestrial habits, indicate that onychophorans are likely ancestors to this group. If this hypothesis were confirmed by further discoveries, it would mean that the Arthropoda are polyphyletic, with crustaceans and chelicerates derived from trilobites while myriapods and insects were derived from onychophorans.

Evolution of the Insects

The class **Insecta** (Fig. 26.39) is by far the largest and most successful group of organisms on Earth, constituting about 75% of the animal kingdom. They flourish even in the most inhospitable environments on Earth, from the Arctic tundras to the hottest deserts; even the oceans have a few. On land, they are the dominant faunal element. The human attempt to subdue the Earth and its inhabitants has met with the firmest resistance from these hardy creatures. The same capacity for rapid evolution that has enabled them to survive since the Carboniferous and diversify to fill all the world's environments has also enabled the insects to adapt to such new environmental opportunities as the cultivation of crops and to develop resistance to pesticides.

The success of the insects has been made possible by a number of characteristics. Some of these are characteristics shared with other arthropods, such as the protective exoskeleton. The insect nervous system, built on the same plan as other arthropods, is more sophisticated, and their behavior correspondingly more complex. The insect respiratory system consists of numerous branched air tubules known as tracheae, opening to the outside through a series of nearly circular openings (spiracles) in either side of the abdomen. Such tracheal systems occur also

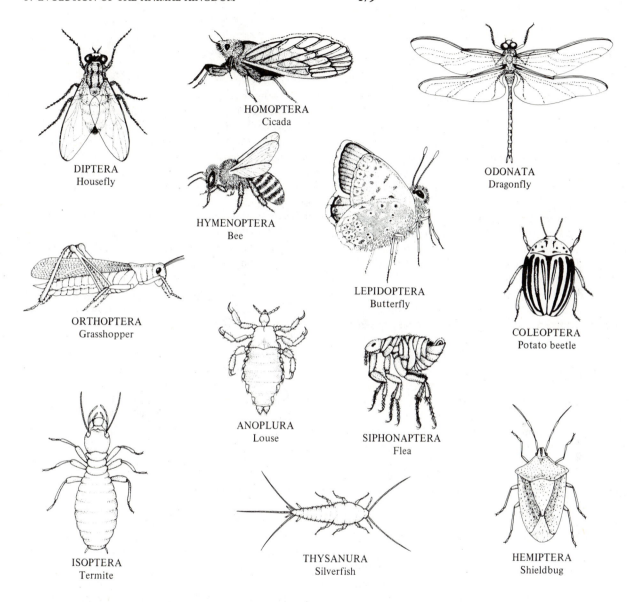

Fig. 26.39 The major orders of insects. Reprinted with permission from John W. Kimball, *Biology,* 4 e, © 1978. Reading, Massachusetts: Addison-Wesley (Fig. 37.21).

in myriapods and were independently evolved in arachnids. The antennae of insects were inherited from their
myriapod ancestors, and the primitive form of insect legs
is rather similar to the legs of certain myriapods. The tips
of these legs are provided with both hooks and adhesive
pads; the hooks are used on rough surfaces, the adhesive
pads on smooth ones. The well-developed sense organs of
insects include eyes of the compound type as well as very
sensitive chemoreceptors located at the tips of the legs
and in various other places.

The characters that most clearly distinguish insects
from other arthropods, especially from myriapods, include the division of the body into three major regions
(head, thorax, and abdomen). The head possesses antennae, compound eyes, and mouthparts that include the
mandibles and one pair of maxillae, the second pair having become fused into a lower lip (labium). The thorax
contains three segments, each of which bears a pair of
walking legs in adults. In most insects, the second and
third thoracic segments each carry a pair of wings. The
abdomen possesses as many as 11 segments in primitive
insects, but most modern insects have fewer segments.

The insects are here defined as mandibulate arthropods in which the body is divided into a head, a thorax of
three leg-bearing segments, and an abdomen of 11 or
fewer legless segments. This is the customary definition
and excludes the Symphyla and other myriapods, while
including the spring-tails and silverfish treated below.
Other definitions of the Insecta have been proposed;
some would include the Symphyla and perhaps other
myriapods as well, while one repeatedly proposed definition would restrict the class Insecta to the Pterygota
(winged forms).

Fossil insects are not so rare as is commonly believed.
Nearly every insect order is represented by fossils, and a
few are known by fossils only. One authority has claimed
that fossil insects outnumber the known fossils of all
other arthropods combined.

Coal swamp deposits of Carboniferous age are rich
in the variety of fossil insects represented. About six
orders are known from these deposits, including the
roaches (Blattaria). By Permian times, many additional
orders had made their appearance, including the true
bugs (Hemiptera) and the beetles (Coleoptera). The
order generally considered to be the highest, the Hymenoptera, made its appearance in the Jurassic, but a

few of the more delicate types of insects, including the
butterflies and moths (Lepidoptera), do not appear in
the fossil record until the Tertiary.

The historical record of insect origins has been lost,
but the comparative anatomy of the most primitive living
insects allows the origin and early evolution of the insects
to be reconstructed. Those insects that have neither wings
nor winged ancestors are collectively called apterygotes
(literally, "without wings"), though most entomologists
would agree that any taxon uniting these orders together
would be highly unnatural. The order Collembola
(spring-tails) are found in the Devonian of Scotland,
making them the oldest known insects. Other apterygote
orders include the minute Protura and Diplura, which
both inhabit forest floors and other humid places, and
Thysanura, commonly known as bristle-tails or silverfish.

The vast majority of insects have wings. Wings are
indeed the key to insect success, for those orders that
have always lacked wings have remained small and ecologically unimportant. The development of wings was
undoubtedly the greatest single event in insect history,
and opened to them new opportunities that they quickly
exploited. Despite predation upon them by spiders,
birds, bats, and other animals (even a few plants), ptery-
gote insects continue to flourish and are rarely surpassed
by any other competitors for the niches that they occupy.
Insects that have wings, or that are clearly descended
from winged ancestors, are called pterygotes, and are
usually considered a subclass Pterygota.

Important differences in embryological development, especially of the wings, may be used to subdivide
the Pterygota into two very large groups. The more
primitive of these groups, known either as Hemimetab-
ola or Exopterygota, are characterized by direct and gradual transformation of the larvae into miniature adults
(called nymphs) and finally into mature adults; this type
of development is sometimes called "incomplete metamorphosis." Larvae and adults are rather similar ecologically, usually subsisting on the same foods. No resting
stage, or pupal instar, intervenes between the larval and
adult stages. The name Exopterygota refers to the fact
that the wings develop externally and thus make their appearance rather gradually. The two largest groups of
hemimetabolous insects are the Orthoptera (grasshoppers, locusts, crickets, and their relatives) and the

Hemiptera (true bugs). The roaches, termites, and lice are also hemimetabolous.

The more advanced group of insects are called either Holometabola or Endopterygota. They are characterized by a "complete" metamorphosis in four stages: egg, larva, pupa, and adult. The resting stage, or pupal instar, is really the innovative part of this life cycle, and it is here that the wings develop internally, the feature to which the name Endopterygota refers. Upon emerging from the pupa, the newly hatched adult turns its wings out and inflates them; they harden soon after, and the winged adult flies off to begin a new life. The distinctness of wingless larval and winged adult stages, with a pupal instar between them, has allowed the evolution of separate larval and adult adaptations in the majority of holometabolous insects. The larva and adult are very different ecologically, feeding in most cases on different foods and otherwise occupying separate niches. Often the larva is the voracious eater, particularly in those species that do damage to foliage and crops, while the adult may feed much less destructively, except notably when it is providing food for the next generation of larvae.

There are nine orders of insects that exhibit the "complete" metamorphosis from a larva to a very different winged adult through an intervening pupal stage during which the wings develop internally. These include the Neuroptera (lacewings and ant lions), Mecoptera (scorpion flies), Trichoptera (caddis flies), Lepidoptera (butterflies and moths), Diptera (flies), Coleoptera (beetles), Siphonaptera (fleas), Strepsiptera, and Hymenoptera (bees, wasps, ants, etc.) The beetles (Coleoptera) are the largest insect order by far, making up nearly 40% of the class Insecta and over 25% of the entire animal kingdom.

The order **Hymenoptera** includes the ants, wasps, bees, and their relatives. The most primitive Hymenoptera are the sawflies and wood wasps, herbivorous forms with sawlike ovipositors. These ovipositors pierce the stems and leaves of plants and lay their eggs within. More advanced Hymenoptera, including gall wasps and ichneumon flies, have the first abdominal segment greatly constricted to form a flexible "waist" for which wasps are famous. They have generally long, pointed ovipositors that penetrate deep into plant tissues, often laying their eggs in a beetle or moth larva rather than in the plant tissue itself. The parasitic habit presumably

evolved by either an ovipositor or a larva fortuitously encountering the larva of another species, the selective advantage of this association making it an increasingly frequent occurrence.

The most advanced group of Hymenoptera have modified the ovipositor into a sting. The egg no longer runs through the entire length of the ovipositor, but exits instead at its base. These stinging Hymenoptera include the bees, wasps, and ants, characterized by a type of sex determination (haplodiploidy) in which diploid females develop from fertilized eggs while haploid males develop parthenogenetically from unfertilized eggs. Because of this, each female is more closely related to her mother's offspring (her sisters) than to her own offspring. This system favors female cooperation (see Chapter 24), which leads in turn to the evolution of complex societies.

Deuterostome Phyla

The bilaterally symmetrical animals considered up to this point have all been protostomes, in which the mouth always developed from an embryonic structure called the blastopore. The animals remaining to be considered are all deuterostomes, meaning that the blastopore comes to lie near the posterior end, while the mouth forms at the opposite (anterior) end (Box 26–B). The first three cleavages of deuterostome eggs are symmetrical (radial), and result in a two-tiered, eight-celled stage in which each cell in the upper tier rests directly above one in the lower tier. (If the upper four cells are facing north, east, south, and west, then so, too, are the lower four.) Deuterostome cleavage is usually indeterminate, meaning that one of these cells, carefully separated from the rest, is fully capable of growing into a complete (if smaller) embryo on its own. Protostomes, by way of contrast, have a spiral type of cleavage, resulting in a somewhat asymmetrical eight-celled stage in which the two tiers are staggered, each cell in the upper tier lying between two of the cells in the lower tier. (If the upper four cells are facing north, east, south, and west, then the lower four are facing northeast, southeast, southwest, and northwest.) Cleavage in protostome eggs is determinate in that either of the two cells formed from the first division is capable of forming only half an embryo rather than a whole one.

The method of coelom formation also differs: in the primitive members of all deuterostome phyla, the coe-

BOX 26–B   *Differences between Protostome and Deuterostome Phyla*

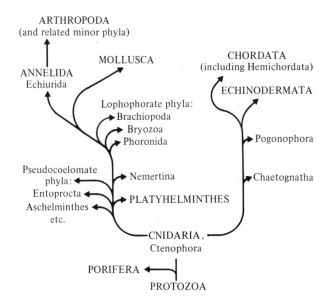

Protostomia	Deuterostomia
1. Early cleavages are spiral.	1. Early cleavages are radial.

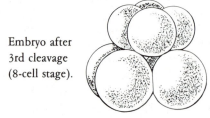

Embryo after
3rd cleavage
(8-cell stage).

Protostomia

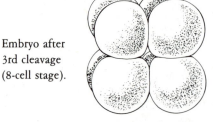

Embryo after
3rd cleavage
(8-cell stage).

Deuterostomia

2. These early cleavages are determinate, meaning that the resulting cells have already lost much embryonic potential. If the cells of a 4-celled embryo are separated, for example, each will produce only an abnormal one-fourth of an embryo.

2. These early cleavages are indeterminate, meaning that the resulting calls still have nearly all their embryonic potential. If the cells of a 4-celled embryo are separated, each will usually give rise to a small but well-formed, viable young animal. (This has been repeatedly confirmed in sea urchins; some recent evidence suggests it may not hold for all vertebrates.)

Protostomia	Deuterostomia
3. In the embryonic gastrula stage, the entrance from the outside to the embryo's interior, called the blastopore, represents the future mouth (anterior end) of the animal.	3. In the embryonic gastrula stage, the blastopore represents the future posterior end. The mouth, a secondary structure, develops at the other end.

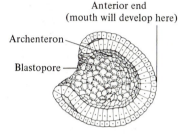

Anterior end
(mouth will develop here)

Archenteron—

Blastopore—

Gastrula stage of *Amphioxus*

Protostomia	Deuterostomia
4. A true body cavity (coelom) is not always present. When present, it usually develops by splitting (schizocoel). (The lophophorate phyla are an exception: their coelom usually arises from the gut.)	4. True body cavity (coelom) is always present. In primitive forms, it develops from the gut (enterocoel). (Vertebrates have a schizocoel and are therefore an exception to this rule.)
5. Adults are usually active, bilaterally symmetrical animals with well-developed heads. (There are many exceptions, however.)	5. Adults are in many cases sessile, radially symmetrical animals with poorly developed heads. (The larvae, however, are active and bilaterally symmetrical in all cases.)

lom arises as an outgrowth of the gut, while among those protostome phyla that possess a coelom its method of origin is usually by splitting from within. A biochemical distinction lies in the fact that most protostomes use phosphoarginine as a phosphate reserve, supplying needed phosphate groups for the reconstitution of ATP during muscle contraction, but deuterostomes use phosphocreatine in addition or instead for the same purpose.

Deuterostome phyla include the Echinodermata, Chordata, Hemichordata, and the minute arrowworms (Chaetognatha).

Phylum Echinodermata

The phylum **Echinodermata** (Fig. 26.40) is characterized by the metamorphosis of bilaterally symmetrical larvae first into asymmetrical and then usually into radially symmetrical organisms with a secondary five-fold (pentameral) symmetry. The familiar starfish is an example of pentameral symmetry. The body is frequently invested in rough dermal plates composed largely of calcium carbonate, a character to which the name of the phylum ("spiny skin") refers. There is never any well-developed head, nor is there ever an excretory system. A unique type of circulatory system, called the water-vascular system, is one of the hallmarks of the phylum. The general form of this water-vascular system is a series of water-filled tubes, including a circular tube around the mouth region, and a loop extending outward from this central ring into each of the five "arms" or rays. These loop-shaped branch tubes are called ambulacra; each ambulacrum frequently bears tiny tube feet that move under the influence of water pressure within them. The central ring is also connected to the seawater outside by a canal that penetrates the overlying dermal plate through

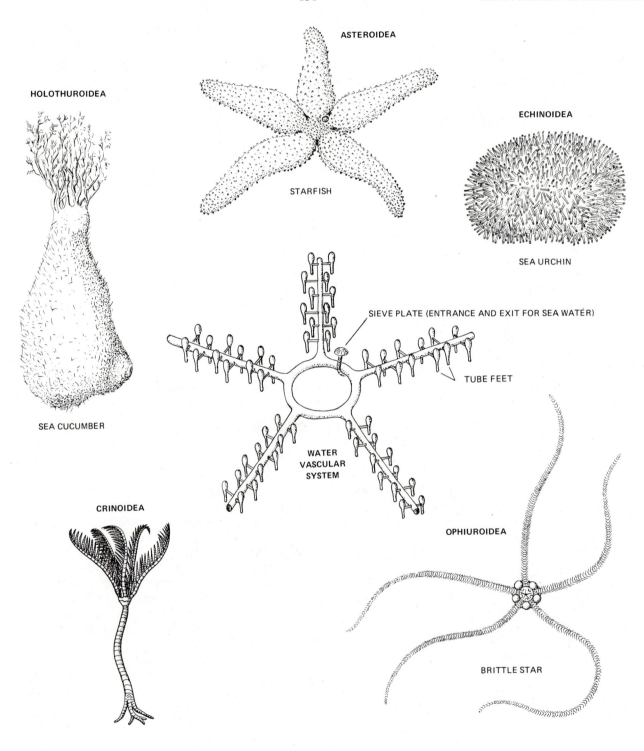

HOLOTHUROIDEA

SEA CUCUMBER

ASTEROIDEA

STARFISH

ECHINOIDEA

SEA URCHIN

SIEVE PLATE (ENTRANCE AND EXIT FOR SEA WATER)

TUBE FEET

WATER
VASCULAR
SYSTEM

CRINOIDEA

OPHIUROIDEA

BRITTLE STAR

Fig. 26.40   (Facing page) Representative echinoderms. All except crinoid are reprinted with permission from John W. Kimball, *Biology,* 4 e, © 1978. Reading, Massachusetts: Addison-Wesley (Fig. 37.23).

a pore called the madreporite. All echinoderms are marine.

Starfish may be the most familiar of echinoderms, but they are far from the most primitive. Unlike the predaceous starfish or the detritus-feeding scavengers such as the sea urchins, the more primitive echinoderms were stalked filter-feeders. These sessile invertebrates are today represented by only a few crinoids, compared to the many species of crinoids and several other classes that inhabited the Paleozoic seas.

Imagine, if you will, an upside-down starfish on a stalk, with its mouth turned upwards. Imagine also that the rays, or ''arms,'' each contain ciliated tentacles as extensions of the ambulacra, and that a ciliated groove, the ambulacral groove, runs along the midline of each ray toward the mouth. The various classes of sessile echinoderms are built on this general pattern. Their primitive way of life was undoubtedly as filter-feeders, their ciliated tentacles functioning in a manner similar to a lophophore. The ciliated tentacles trapped suspended food particles from the water, and the cilia carried these entrapped particles first to the ambulacral grooves, then along these grooves to the mouth.

A number of very primitive, sessile echinoderms of Paleozoic times had apparently not yet evolved pentameral symmetry. These asymmetrical echinoderms are now placed in a subphylum Homalozoa, and are sometimes referred to as ''carpoids.'' The four included classes have a combined range from Cambrian to Devonian time.

For a stalked, sessile filter-feeder, however, radial symmetry is a distinct advantage. From some asymmetrical group of ancestors, those that achieved the most nearly symmetrical body plan as adults had a selective advantage over those that did not, and the secondary radial symmetry evolved. Perhaps the number five was chosen in preference to even numbers as an adaptation for structural rigidity: with an even degree of symmetry (4 or 6, for instance), the weak joints between adjacent plates would lie opposite one another, 180° to the other side of the mouth, an odd degree of symmetry would avoid this problem. The radially symmetrical, sessile echinoderms are now set apart as the subphylum Crinozoa, including the classes Blastoidea, Crinoidea, Cystoidea, and five other classes. Except for the blastoids, which persisted into the Permian, and the Crinoids, which persist to the present day, the remaining Crinozoa range from the Cambrian to the Devonian.

The subphylum **Asterozoa** contains all those radially symmetrical, free-living echinoderms possessing a central disc from which five radiating ''arms'' (rarely a higher multiple of 5) extend. The growth of these ''arms'' or rays is radial, that is, outward from the central disc but in the same general plane. There is only one included class, the Stelleroidea, known from the Ordovician to the Recent. Included here are the familiar starfishes and the more delicate-rayed brittle stars.

The largest group of echinoderms are the **Echinozoa,** including the living classes Echinoidea (sea urchins and other echinoids) and Holothuroidea (sea cucumbers), as well as five extinct classes confined to the Paleozoic. In all Echinozoa, the rays develop nearly parallel to one another and perpendicular to the plane of the central disc. The Holothuroidea, or sea cucumbers, are scavengers of the sea floor. Their body consists of a large, cucumber-shaped sac tapering to a mouth, which is surrounded by a series of five rays resembling a tiny starfish. The largest class of echinoderms are the Echinoidea, including the sea urchins, sand dollars, and their relatives.

The Phylum Chordata

You and I are chordates—members of the phylum Chordata (Fig. 26.41). Chordates are deuterostomes; their closest relationships are to the other deuterostome phyla, with which they share the radial, indeterminate type of cleavage described previously. In addition to these deuterostome characteristics, the phylum Chordata is characterized by three features that may be taken together as the hallmark of the phylum:

Gill slits.   These are clefts or holes connecting the digestive tract to the outside. The slitlike nature of primitive chordate gills allowed them to be used as part of a straining apparatus in connection with filter feeding. The portion of the gut that contains these gill slits is always called the pharynx.

Notochord. This stiff but flexible rod of gelatinous connective tissue serves two functions. In the adults of certain lower chordates, the notochord allows the body to undulate from side to side, but prevents it from collapsing like an accordion when the muscles contract. Also, in the embryos of all chordates, the notochord induces the formation of the nervous system. The name ''Chordata'' refers to this notochord.

Dorsal, hollow nerve cord. The nerve cord of chordates runs along the back (dorsally) and is hollow, in contrast to the ventral, solid nerve cords of arthropods and many other invertebrates. The nerve cord always lies just above the notochord.

The above three features need not all be present in the adult stage, and they need not all be present at the same time. You still have your dorsal, hollow nerve cord (your brain is an enlarged part of it), but your gill slits and notochord were confined to your embryonic stages and have since disappeared or been greatly modified.

Primitive chordates were aquatic and swam by side-to-side undulations of the body. Their segmentally organized muscle blocks contracted first on this side, then on that. This mechanism requires a stiff but flexible rod to prevent accordionlike collapse, and the notochord provides just such a rod. The muscles must contract in a rhythmic series of alternately staggered waves in order to be effective—this requires coordination by a nervous system. The dorsal position of the nerve cord in chordates was selectively favored because the muscles to be coordinated were also dorsal.

The gill slits had a very separate history; their structure in primitive chordates shows that they were first used as part of a straining device for filter feeding. In at least one lower chordate, the body surface, not the gills, functions as the principal respiratory structure. Special organs of aquatic respiration (i.e., gills) evolved in various animal groups, but only in chordates did they evolve from a straining apparatus containing numerous slits.

Related to the Chordata—perhaps only distantly—are a group of animals called the **Hemichordata** (Fig. 26.41). Some zoologists consider them a subphylum of Chordata, but most treat them as an independent phylum. Like the undisputed chordates, they possess radial cleavage and other deuterostome characteristics, and they also have segmentally organized muscles and a phar-

ynx containing round gill slits. There is a brief nerve cord that is both dorsal and hollow. There appears to be no true notochord, though an anterior extension of the digestive tract has been rather tenuously suggested as a possible homologue. The ciliated larvae show strong resemblances to the echinoderms; the first larval hemichordate to be discovered was in fact mistakenly described as a larval sea urchin!

The best known hemichordates are called ''acorn worms'' (class Enteropneusta). These wormlike animals burrow with the aid of a muscular proboscis at their anterior end. They filter-feed by passing water in through their mouth and out through their gill slits. Food particles are trapped on the inner-facing, ciliated surfaces of these gill slits.

Another group of hemichordates, known as pterobranchs (class Pterobranchia) includes sessile filter-feeders that use lophophore-like tentacles as a means of collecting suspended food particles.

A third class of hemichordates are the extinct graptolites (class Graptolithina) found in Paleozoic rocks. Graptolites were minute, colonial filter-feeders, similar in general to the colonial pterobranchs. Their floating colonies were dispersed throughout Paleozoic seas, so that they make useful stratigraphic indicators. As was mentioned in Chapter 17, the number of biserial rows (''arms'' of the colony) decreased from several dozen to four and finally to two, and the orientation of these rows gradually changed throughout their evolution from downward-hanging to upward-reaching (see Fig. 17.4).

Aside from the questionably related hemichordates, two groups of chordates are unquestioned relatives of the vertebrates: the tunicates (Urochordata) and the sea lancets (Cephalochordata).

The **Urochordata**, or tunicates, illustrate an important transition in chordate evolution. Typical adults, often called ''sea squirts,'' live as sessile filter-feeders, passing large amounts of water into their pharynx and out through their numerous gill slits and living on the suspended organic matter trapped by the sheets of mucus that coat the inner surfaces of their gills. The bulk of their bodies is given over to an enormous pharynx or gill-basket, surrounded by an excurrent chamber known as the atrium.

The larval tunicate, on the other, hand, is a much more actively swimming creature with a well-formed tail. This tail contains a series of muscles, a stiffening noto-

HEMICHORDATA

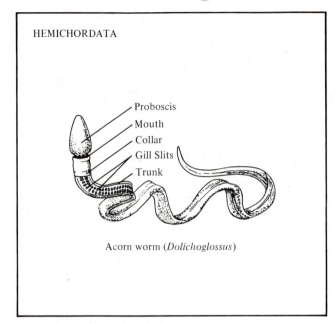

Proboscis
Mouth
Collar
Gill Slits
Trunk

Acorn worm (*Dolichoglossus*)

UROCHORDATA

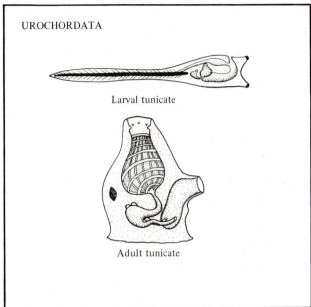

Larval tunicate

Adult tunicate

CEPHALOCHORDATA

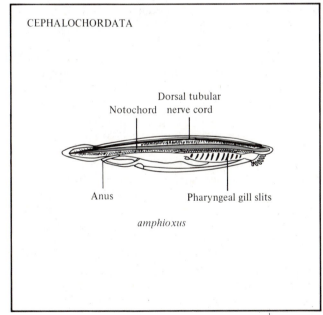

Notochord

Dorsal tubular
nerve cord

Anus

Pharyngeal gill slits

amphioxus

VERTEBRATA

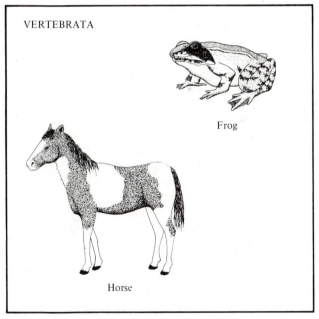

Frog

Horse

Fig. 26.41   Representative Chordata and Hemichordata. The acorn worm is reprinted with permission from Donald D. Ritchie and Robert Carola, *Biology,* © 1979. Reading, Massachusetts: Addison-Wesley (Fig. 10.41). The amphioxus is reprinted with permission from Jeffrey J. W. Baker and Garland E. Allen, *Study of Biology,* 3 e, © 1977. Reading, Massachusetts: Addison-Wesley (Fig. 23.12).

chord, and a dorsal, hollow nerve cord with nerve con-
nections to each group of muscles. There is also a head
region with such structures as a brain, a simple pharynx
(with a gill-basket foreshadowing that of the adult), and
an adhesive organ. During its larval existence, the young
tunicate filter-feeds, but occasionally searches the nearby
substratum for a suitable place to spend its adult life.
When it finds a suitable place, it attaches with its adhesive organ and later undergoes metamorphosis into an
adult. If it fails in its search, it swims away and looks elsewhere. During metamorphosis the entire tail is resorbed,
including the notochord, the muscles, and much of the
nervous system.

Most evolutionists now believe that the ancestral
tunicate was a sessile filter-feeder resembling the typical
adult of modern forms, and that the motile larva was at
first an adaptation for achieving dispersal of the adults to
more suitable locations. In subsequent chordate evolution, the larval stage was able to gain more self-sufficien-
cy, and the adult stage was gradually abandoned. We can
see certain stages in this sequence preserved within the
urochordate subphylum. In the majority of tunicates,
both larval and adult stages are present. The motile larval
stage is most important to those tunicates that spend
their adult lives on rocky surfaces and other hard-to-find
substrata, but the larval stage has been lost in a few genera that live as adults on sandy bottoms where the finding of a suitable site poses few problems. One small
group of tunicates retains their juvenile morphology
throughout their lives, their gonads undergoing maturation while motility persists. By dispensing with the sessile
adult stage, these neotenic tunicates show us the sort of
change that would have occurred in the ancestry of the
two remaining chordate subphyla, namely the Cephalochordata and the Vertebrata. This is not necessarily a case
of direct ancestry—parallelism is a more likely explanation.

The closest vertebrate relatives are the sea lancets or
Cephalochordata. The best known species, commonly
called amphioxus, often filter-feeds while half buried in
bottom sediments. Its segmentally organized muscle
blocks enable the animal to swim or to burrow with its
spear-shaped tail. A stiff notochord runs the entire
length of the body from head to tail. The bulk of the animal, only 6–8 cm long, is given over to a large pharynx

with numerous (about 100) gill slits. Ciliated tentacles
(buccal cirri) surround the mouth. Currents of water
draw suspended particles into the pharynx and through
the gill slits; food particles are trapped on the ciliated lining of the pharynx between the adjacent gill slits. Sheets
or rods of mucus, secreted along a groove known as the
endostyle, sweep over these cilia periodically and collect
the trapped food particles, which are then carried along
into the intestine for further digestion. The numerous
similarities to the simplest vertebrates include the presence of a circulatory system on the vertebrate pattern
(though without a true heart) and the presence of organs
homologous to the vertebrate liver, thyroid gland, and
pituitary. In many (but not all) respects, the cephalochordates are close in structure to the sort of ancestor that
the vertebrates would require.

Let us now review the currently favored hypothesis
of deuterostome evolution (Fig. 26.42). The deuterostome ancestor is believed to have been a tentacle-bear-
ing filter-feeder living attached to the substratum. Pterobranchs and crinoids are among the groups that have
remained close to this ancestral mode of existence in
which feeding currents are created by the beating of tentacles and by ciliary action. Those echinoderms that retained this ancestral mode of life usually became radially
symmetrical. In some form close to the ancestry of the
Chordata, a shift occurred from the use of tentacles to the
use of gill slits in filter-feeding. The resultant group of
organisms may have resembled the adult stages of modern tunicates, or they may have looked more like acorn
worms.

The actively swimming, motile stage probably
originated as a larval adaptation to the dispersal of the
sessile adult stage, as in modern tunicates. The chances of
finding a good spot would be increased if the larva had a
longer time in which to search, and this in turn gave a
selective advantage to those species with a larval stage of
longer duration. Prolongation of the larval stage may
have led to selection pressure for better feeding adaptations in the larva itself. The sessile adult stage ultimately
became dispensible and was finally bypassed entirely
through the acceleration of gonadal development into
the larval stage. The resultant chordate resembled either
a cephalochordate or a tunicate or both. This same trend
may have developed independently in more than one

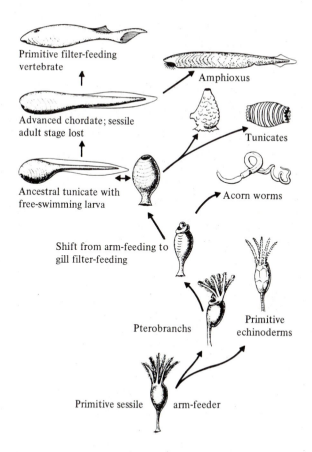

Primitive filter-feeding vertebrate

Amphioxus

Advanced chordate; sessile adult stage lost

Tunicates

Ancestral tunicate with free-swimming larva

Acorn worms

Shift from arm-feeding to gill filter-feeding

Pterobranchs

Primitive echinoderms

Primitive sessile arm-feeder

Fig. 26.42 Phylogeny of the Chordata and related phyla. Reprinted from *The Vertebrate Story* by Alfred S. Romer, by permission of The University of Chicago Press. Copyright © 1933, 1937, and 1944 by The University of Chicago Press.

group. The major advances of early vertebrates over cephalochordates were the increased development of a head (cephalization) and the protection of the nervous system by bone.

Jawless Vertebrates

The subphylum **Vertebrata** is characterized by the possession of three additional features above and beyond those that characterize chordates in general: a complete circulatory system including a heart, a more fully devel-

oped head region containing a brain and sense organs, and a backbone or **vertebral column.** The vertebral column consists of individual **vertebrae,** which develop in a position surrounding the notochord and usually also the nerve chord. Functionally, the vertebral column usually replaces the notochord, which then degenerates and is often lost entirely in adults.

The vetebrate head has an expanded brain and various sense organs including an inner ear for balance, organs of olfaction (smell), and usually also a well-developed pair of eyes. The expansion of the brain has precluded the notochord from reaching the tip of the head. Most of the delicate nervous system is protected by the skeleton: the brain is protected by a braincase, while extensions of the vertebrae usually protect the spinal cord.

The earliest vertebrates were jawless and belonged to a class called the **Agnatha.** Fossil agnathans were often heavily armored and are known as **ostracoderms.** One early group of ostracoderms includes the genus *Cephalaspis* and its relatives. They were flat-bottomed mud feeders with a vacuum-cleaner type of mouth. *Cephalaspis* is believed to have lived by swimming either just above or just below the surface of the mud, sucking in loose sediment, and straining out the organic particles on the inner surfaces of its many gills. Other groups of ostracoderms became actively swimming surface feeders. The earliest members of many agnathan groups were heavily armored, and in each group there was a trend for the gradual reduction of the bony armor. The two families of living agnathans, known as **cyclostomes,** contain naked scavengers, predators, and parasites. Perhaps they were derived from mud-burrowing forms that occasionally found their way into the carcass of a dead fish or other animal. During subsequent evolution, these forms abandoned the straining of mud for the burrowing into the bodies of other fish, and some eventually became parasitic.

Jawed Fishes

One of the most significant adaptive changes in the history of the vertebrates was the origin of jaws, an event which transformed the vertebrates from a minor group of filter-feeders and parasites to a successful group of actively swimming predators (Fig. 26.43). The gills of primitive vertebrates were supported by a series of carti-

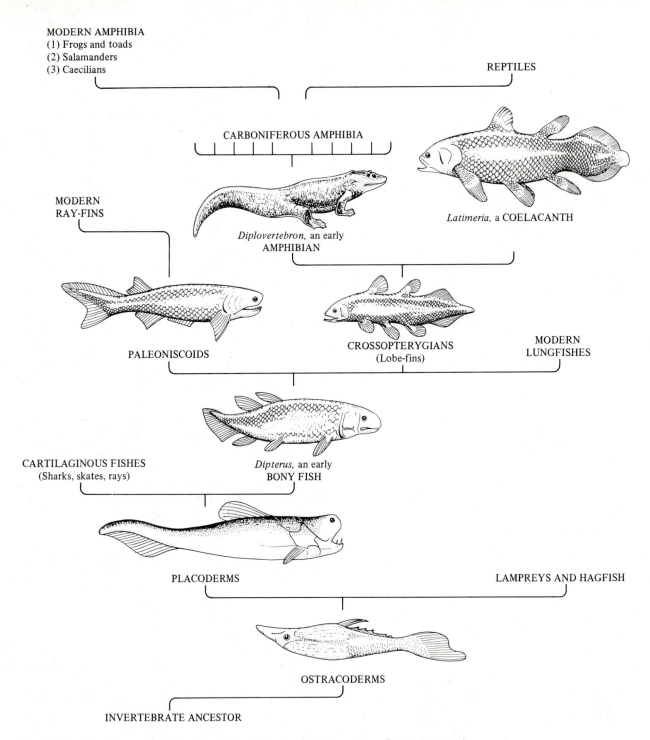

MODERN AMPHIBIA
(1) Frogs and toads
(2) Salamanders
(3) Caecilians

REPTILES

CARBONIFEROUS AMPHIBIA

MODERN
RAY-FINS

Diplovertebron, an early
AMPHIBIAN

Latimeria, a COELACANTH

PALEONISCOIDS

CROSSOPTERYGIANS
(Lobe-fins)

MODERN
LUNGFISHES

CARTILAGINOUS FISHES
(Sharks, skates, rays)

Dipterus, an early
BONY FISH

PLACODERMS

LAMPREYS AND HAGFISH

OSTRACODERMS

INVERTEBRATE ANCESTOR

Fig. 26.43 Family tree of fishes and amphibians. Reprinted with permission from John W. Kimball, *Biology*, 4 e, © 1978. Reading, Massachusetts: Addison-Wesley (Fig. 38.1).

laginous gill arches that became bony among some of their descendants. At first, all the gill arches were the same, and all the gill slits between them were similar. The transformation of one of these gill arches into a pair of jaws was accompanied by the loss of a few gill slits and also in most cases by the reduction of the gill slit just behind the jaws.

The extinct class **Placodermi** includes the armored, jawed fishes that dominated the Devonian seas. Most were predatory, and many were small, but a few genera reached enormous size: one had a massive skull that was nearly a meter long. Several placoderm orders independently evolved the same trend toward reduction of armor that characterized various jawless vertebrates as well.

Derived from the placoderms were the cartilage fishes (class **Chondrichthyes**), including the sharks and their relatives. These predators of the seas are actively swimming, lungless fishes with large, yolky eggs, and with special copulatory structures attached to the pelvic fins in males. The dermal armor of placoderms has become reduced in modern sharks to no more than a series of minute placoid scales.

The bony fishes, class **Osteichthyes,** are the largest and most successful group of fishes in both marine and freshwater environments. The original adaptations of this group were, however, to freshwater environments. The earliest bony fishes, from the Silurian, were older than any other jawed fishes. The few unifying characteristics of the class include the presence of scales (though of different construction in various groups), the presence of a gill cover (operculum), the presence of lungs or lung homologues in most members of the class, and various technical characters of the skull.

Within the class Osteichthyes, there are evolutionary trends for the transformation of lungs into swim bladders, reduction of the outer layers of the bony scales, and transformation of the primitively asymmetrical tail into a symmetrical shape. The highest and most successful group of bony fishes are the teleosts (Teleostei), which range from Triassic to Recent. The teleosts are now the most successful group of fishes in all environments.

Two groups of bony fishes, the lungfish and the Crossopterygii, have fins with fleshy bases. Lungs, a primitive feature in these groups, have been retained. Many crossopterygians had internal nostrils, too. These fortunate preadaptations allowed certain crossopterygians to evolve into amphibians by making the transition from water to land.

Amphibians

The amphibians (class Amphibia) are characterized by lung-breathing adults that use legs for locomotion in most cases. Eggs are typically laid in the water and fertilized externally; they then develop into gill-breathing larvae (tadpoles) that later metamorphose into lung-breathing adults.

The transition from fish to amphibian was largely a matter of transforming the fleshy-based crossopterygian fin into a walking leg. The earliest amphibians, the Ichthyostegalia, may have been persistently aquatic, using their legs only to disperse themselves from one pond or stream to the next.

The legs of most amphibians are sprawling; they push the animal's body along, but do not hold it up from the surface except very briefly. To illustrate this primitive tetrapod stance, place your chest across a table or desk and your palms flat against the surface with your fingers spread and your thumb directed forward. Your upper arms, from shoulder to elbow, should be in line with one another, and your forearms, from elbow to wrist, should be vertical and parallel to one another. Primitive amphibians had a similar stance in all four legs; modern salamanders and certain reptiles still do.

The conquest of the land included adaptations for breathing and walking on land, but not for terrestrial reproduction. Amphibian eggs still had to be laid in the water (or in other moist places) and fertilized externally. The larvae of amphibians still breathe with gills and thus cannot survive on dry land except in a few rare cases where other means of keeping them moist have evolved. Amphibians are therefore still tied to their ponds and streams by the aquatic phases of their life cycles. The final step in the conquest of the land was made by the reptiles, who developed a type of egg that was self-contained and could be laid on land.

Reptiles

One of the greatest advances in vertebrate evolution was made by the ancestral reptiles in their development of a so-called **cleidoic** or **amniotic egg** that could be laid on land. The structure of this type of egg is shown in Fig. 26.44 and includes a waterproof shell that is nevertheless

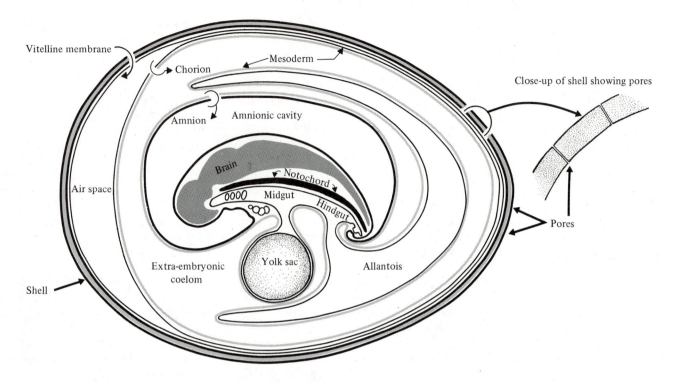

Fig. 26.44 Structure of the cleidoic or amniote egg. Shaded layer represents mesoderm.

permeable to air. Within this shell lie several embryonic membranes, of which one in particular, the amnion, surrounds the entire embryo and encloses it in its own watery medium. This self-contained, artificial "pond" also functions as a shock absorber, cushioning the delicate embryo against outside forces. Another membrane, the allantois, is highly vascularized and functions as a respiratory organ; it also serves as a repository for nitrogenous and other wastes that are simply left behind when the young animal hatches.

Development from an amniotic egg characterizes the three classes Reptilia, Aves, and Mammalia. It poses certain problems, of which reproduction is perhaps the most important. The egg shell, secreted by the mother, is impervious to fluids, including the fluid medium in which the sperm must swim. Fertilization must therefore take place before the shell is secreted, and this necessi-

tates internal fertilization. The development of copulatory organs was therefore favored, too, and other changes brought about a more efficient urinary system. The body covering became waterproof by the addition of a horny layer of keratin, the same protein found in your hair and fingernails.

Following their first appearance in Pennsylvanian times, the reptiles quickly diversified and filled many vacant niches (Fig. 26.45). Many new groups appeared in the Permian and by the Triassic nearly all the major

Fig. 26.45 (Facing page) Adaptive radiation of the reptiles. Reprinted with permission from John W. Kimball, *Biology,* 4 e, © 1978. Reading, Massachusetts: Addison-Wesley (Fig. 38.9).

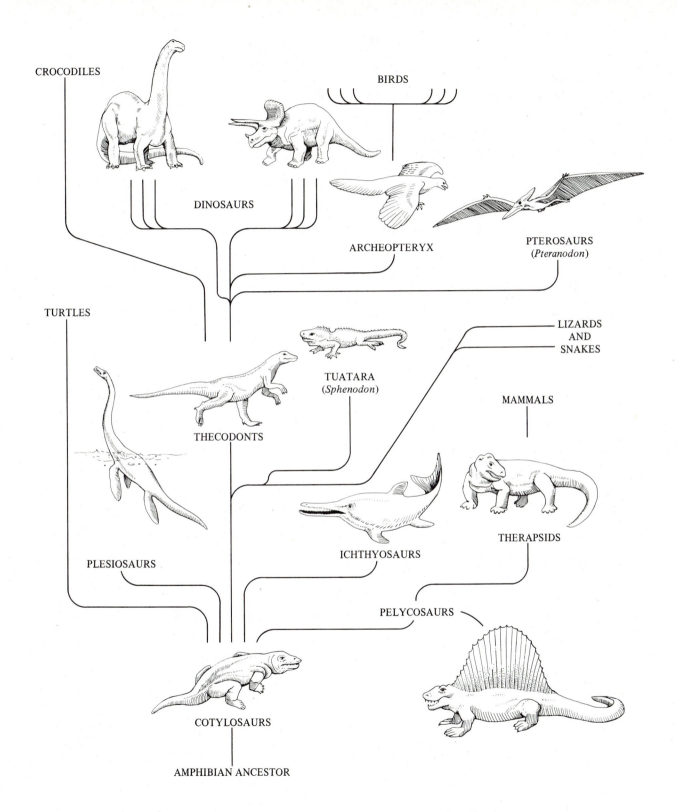

CROCODILES

BIRDS

DINOSAURS

ARCHEOPTERYX

PTEROSAURS
(*Pteranodon*)

TURTLES

LIZARDS
AND
SNAKES

TUATARA
(*Sphenodon*)

MAMMALS

THECODONTS

PLESIOSAURS

ICHTHYOSAURS

THERAPSIDS

PELYCOSAURS

COTYLOSAURS

AMPHIBIAN ANCESTOR

groups of reptiles had made their appearance. The Mesozoic Era became an ''Age of Reptiles,'' and the reptiles became the dominant vertebrates of the Mesozoic on land, in the seas, and even in the air. Several orders, including the ichthyosaurs and plesiosaurs, became predominantly aquatic. Lizards, snakes, and turtles diversified mostly on land. But the most successful reptiles of the Mesozoic were the five related orders of the subclass Archosauria. The archosaurs included a primitive group (the Thecodontia), the semi-aquatic Crocodilia, the flying reptiles (Pterosauria, including the pterodactyls),

and the two orders (Saurischia and Ornithischia) commonly known as dinosaurs (Fig. 26.46).

Birds

The most successful conquest of the air was made in late Jurassic time by the archosaur-derived birds (class Aves). The anatomy of birds shows so many similarities to their archosaurian progenitors that T. H. Huxley declared birds to be ''glorified reptiles.'' Most of the birds' claims to glory consist of a series of adaptations to flying, of which the possession of feathers is certainly the

Fig. 26.46   Evolution of the two orders of dinosaurs, with a diagram of the basic pelvic structure of each order.

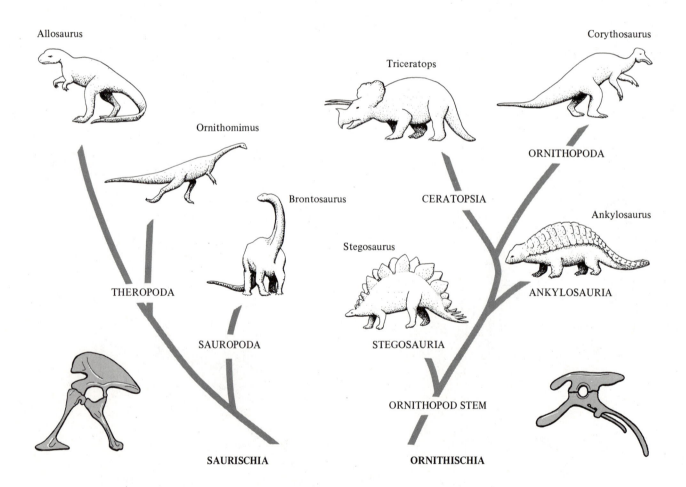

most characteristic feature. Flight by means of feathers was indeed the "key character" that set birds apart from their reptilian contemporaries, and from other flying creatures of both past and present. Feathers have a dual function, for they are both a flight surface and a means of insulation. As a flight surface they combine the virtues of lightness with those of strength and great air resistance; they also have the further advantage that a tear in the wing surface may destroy only a few feathers and thus do minimal damage. The efficacy of feathers as insulation is well attested to by our use of juvenile feathers (down) in quilts, sleeping bags, and the like.

The ancestral bird, *Archaeopteryx* (see Box 17–D), was first described by Richard Owen on the basis of some very well-preserved remains from an exceedingly fine-grained limestone of late Jurassic age. The skull was that of a reptile with a pair of long jaws bearing many sharp, simple teeth and with the same skull openings that typify dinosaurs and other archosaurian reptiles. The brain was somewhat expanded, but not so much as to exclude membership in the Reptilia. The hind limbs were much longer than the forelimbs, and a typically long, reptilian tail was present. The skeleton thus had some birdlike features, but was still basically reptilian in many other respects, as in the long tail and the presence of teeth. *Archaeopteryx* might well have been classified as a reptile, were it is not for the fact that its feathers were accurately preserved in the fine-grained limestone. The presence of feathers shows that *Archaeopteryx* had already achieved many important avian characteristics, including both flight (at least gliding flight) and warm-bloodedness, and is therefore to be counted as a primitive bird instead of a birdlike reptile.

The sustained flight of an animal as heavy as even a small vertebrate demands a certain minimum level of energy production and oxygen consumption that are both quite high. The feathers of *Archaeopteryx* show that it had achieved homeothermy, or "warm-bloodedness," in which the body temperature remains constantly high, hardly varying with changes in the level of activity. The maintenance of a high internal temperature depends upon the existence of an external insulation, and feathers provide this insulation very effectively. Homeothermy also requires a more efficient circulatory system than is known in any modern reptile; this was achieved in birds by the elimination of the left systemic arch and retention of only the right one.

Flying, like other forms of vertebrate locomotion, is a learned behavior pattern. As we saw in Chapter 24, this has as its price a juvenile period during which the learned behavior pattern had not yet been acquired. In this particular case, young birds cannot fly, and must therefore be protected by their parents. Parental care is thus important to most birds. The eggs also require a certain amount of warmth for their proper development, and this is usually provided by one or both parents' sitting on the eggs. Much of avian behavior and sociobiology reflects the adaptive needs for the formation of sexual partnerships that last until the eggs hatch, and the formation of parent-offspring bonds as well. The traits that we admire the most in birds, namely their bright plumage and colorful songs, are adaptations for locating and selecting mates.

Most modern birds have a series of additional adaptations that reduce their weight without sacrificing strength: loss of the teeth, reduction of the tail (but with frequent retention of long and decorative tail feathers), development of hollow bones, expansion of the lungs, etc. Extensions or outpocketings of the lungs can occur in several locations, including the hollow interiors of several bones. Modern birds orient primarily by vision rather than smell, and those parts of the brain that are concerned with vision and also with balance are accordingly increased. Most modern birds have a rigid, one-piece sternum, or breastbone, usually with a strong keel for the attachment of the major flight muscles. The collarbones (clavicles) of right and left sides generally fuse into a single bone called the furcula or "wishbone."

Mammals

The class Mammalia is set apart from the class Reptilia by a number of anatomical and physiological features, of which the maintenance of a high, stable internal body temperature (homeothermy) is of primary importance. Homeothermy was probably the "key character" in the evolution of mammals from reptiles—the feature whose evolution brought about all the subsequent changes. Homeothermy demands an outer insulation; in mammals this takes the form of hair or fur, morphologically different from reptilian scales but containing the same protein material. Skin glands associated with the maintenance of a constant internal temperature include sweat glands, which cool by the evaporation of their watery secretion, and sebaceous glands, whose oily secretions retard the evaporation of sweat and also provide waterproofing to the hairs. Homeothermy also demands a more efficient circulatory system, achieved in mammals by the elimination of the right systemic arch and retention of the left, just the opposite of the condition in birds. Homeothermy also demands a higher rate of food intake under most conditions and also a higher breathing rate in order to obtain adequate food and oxygen for the high-energy productivity that a constant temperature demands. Mammals cannot afford to stop breathing while they chew; the development of a secondary bony palate separates the nasal passages from the mouth cavity and thus permits breathing and chewing to take place at the same time. Certain mammal-like reptiles had already evolved the bony palate, and were probably well on the way to homeothermy.

The mammalian posture and gait were also achievements of the mammal-like reptiles. Instead of sprawling out from the body, the limbs of mammals are held beneath the body, with the knee pointing forward and the elbow back in most cases. The body is thus held high off the ground, even when standing still, an adaptation that enables mammals to run faster than reptiles. The speeding up of the body's metabolism also allows mammals to function more efficiently than reptiles at lower prevailing (ambient) temperatures.

Mammals are not only very active, but they are also highly intelligent creatures. Their brains are larger, but more significantly, much of their behavior is learned. The period of infant dependence is fairly long in mammals generally; young mammals depend upon their parents for both nutrition and protection. The hallmark of the class, and the basis of its name, is the fact that the young are nourished by a nutritive fluid, milk, secreted by the mother from paired mammary glands (Latin *mamma*, a breast).

The teeth of mammals have many cusps and are usually capable of shearing, especially in primitive mammals. Unlike the continually replaced teeth of reptiles and lower vertebrates, mammalian teeth are replaced only once, and molar teeth are not replaced at all. Teeth therefore have sufficient time to grow strong, permanent roots.

Reptiles, like other cold-blooded animals, keep getting bigger as they get older. The ends of most bones remain cartilaginous throughout life, which means that they keep on growing; there is no constant adult size.

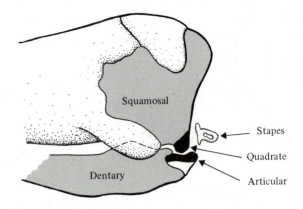

(a) Advanced mammal-like reptile

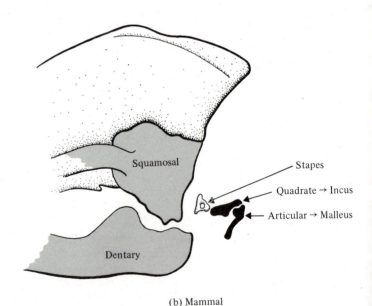

(b) Mammal

Fig. 26.47 Changes in the structure of the jaw region during the transition from reptiles to mammals.

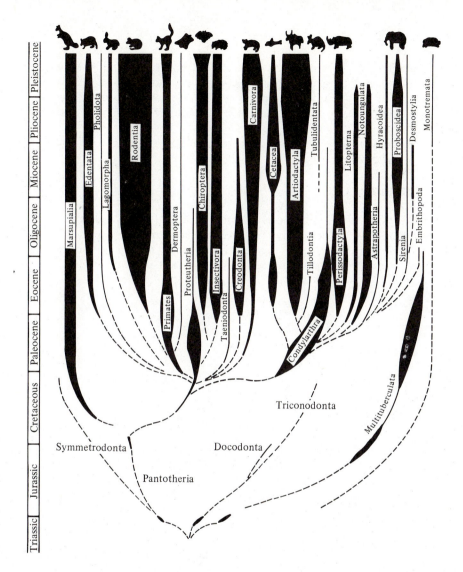

Fig. 26.48 Family tree of the mammals. Reprinted by permission from P. D. Gingerich, 1977, Patterns of evolution in mammalian fossil record. In A. Hallam, ed. *Patterns of Evolution as Illustrated by the Fossil Record.* Amsterdam: Elsevier.

Mammals have instead developed a manner of bone growth in which the ends of the major bones harden rather early, but a zone of rapidly growing cartilage persists behind each end. When this zone of rapid growth is finally replaced by bone tissue, adult size has been reached and growth ceases.

Mammalian jaws must be capable of much hard chewing. The dentary, one of the bones in the lower jaw, grew in importance during the evolution of mammal-like reptiles, and the remaining bones dwindled in importance as they were pushed further and further back (Fig. 26.47). Finally, only one other bone, the articular, remained in the lower jaw, and this bone was part of the joint itself. On the other side of the joint lay the quadrate bone, which abutted the squamosal bone and also the stapes. Vibrations of the lower jaw may have been transmitted to the movable quadrate and then to the stapes, which functioned as an auditory ossicle in the middle ear. The function of the quadrate as a transmitter of sound gradually came to overshadow its function as part of the jaw articulation, as the squamosal and dentary bones gradually assumed a greater share of this joint. The articular and quadrate bones, freed from the dual service of both chewing and hearing, became devoted to hearing exclusively, and are now known in mammals as the malleus and incus, respectively.

497

Perhaps the greatest evolutionary breakthrough within mammalian history was the evolution of the placenta, an adaptation that permits the growing fetus to be retained in the mother's uterus until it can be born alive. The placenta, through which the young is nourished, consists of maternal blood vessels and fetal blood vessels in close proximity, exchanging nutrients and dissolved gasses with one another by diffusion. The vast majority of mammals are today placentals, or eutherians; they are placed in several dozen orders (Fig. 26.48). The largest of these orders include the Insectivora (moles, shrews, etc.), Primates (lemurs, monkeys, apes, humans, etc.), Rodentia (squirrels, mice, and other mammals that gnaw), Chiroptera (bats), Carnivora (cats, dogs, bears, hyaenas, etc.), and Artiodactyla (even-toed hoofed mammals such as cattle, pigs, and camels). Also included are rabbits, anteaters, whales, elephants, horses, rhinoceroses, and many others, including several wholly extinct orders.

CHAPTER SUMMARY

For purposes of review, the classification at the end of the book serves as a convenient summary for the present chapter. The contrast of procaryotic and eucaryotic features in Box 26–A should also be examined.

27

Primates and Human Evolution

Chapter Outline

We have thus far reviewed evolutionary theories on the small scale as well as the large, and have discussed the ways in which various groups of animals and plants have evolved. We now come to the story of humankind's evolutionary history. To some people, this is really the central problem in evolutionary biology; it is often their major (or, alas, their sole) reason for being interested in the subject at all! The story of human evolution has been postponed to these last few chapters because of its direct and personal importance to all of us, and because it provides a very useful and familiar example to which all our earlier findings can be applied.

We begin with a description of the order Primates, the mammalian order to which we belong. We do so because many of the features that distinguish humans from other animals are simply exaggerated versions of traits common in greater or lesser degrees to all members of this order. The achievements of humankind will be greatly misunderstood unless we realize to what extent these achievements were based upon (or were extensions of) the earlier achievements of nonhuman primates. Human evolution, like many other stories, makes little sense out of context.

A. PRIMATE CHARACTERISTICS

The mammalian order Primates is neither the most primitive nor the most advanced of the mammalian orders. This statement may come as a shock to many people, who are accustomed to thinking of humans as Nature's "highest" creatures, and of our near relatives as therefore "second highest." But, as we saw in Chapters 17 and 18, evolution is far from a single, linear progression of increasingly "higher" forms. Remember that the *scala naturae* (Chapter 3) has been rejected by most modern evolutionists: evolution is more like a bush than a ladder.

Proximity to humans is no measure of "progress" except in a very limited and anthropocentric way. Also recall the discussion in Chapters 17–18 of primitive and advanced characters, mosaic evolution, and opportunism. *Some* of humankind's distinguishing features represent definite and far-reaching advances over primitive mammalian conditions. But in many other features, humans are distinctly primitive in the sense of being closer to the primitive mammalian condition, while such "lower" mammals as cows and elephants have in many ways diverged more greatly! Besides, humans are far from the most typical of primates (we are in many ways the *least* typical), and the degree of primitiveness or divergence of an order is hardly to be measured by the condition of one of its least typical and most specialized members.

We shall shortly come to those characteristics in which primates have advanced over the more primitive placental mammals. For the moment, however, let us mention what we are often more apt to forget, namely that primates have really stayed quite close to the primitive pattern of life of the early placental mammals. The primitive placentals had a forelimb with fairly good powers of mobility: the shoulder joint was supported by a bone known as the clavicle, which many mammals have lost, but which primates have retained (see Fig. 27.2). The two bones in the forearm (radius and ulna) were separate, allowing the hand to be turned "palms up" (supination) or "palms down" (pronation); this ability, again, has been lost among many mammals, but not among primates. In the hindlimbs, there was generally less mobility than in the forelimbs, but movement in several directions was still possible; many mammals are nearly incapable of moving their hindlimbs except in a fore-to-aft plane, but the primates have retained the original free mobility at the hip joint. In both forelimbs and hindlimbs, primates have retained, without loss or fusion, all of the primitive mammalian elements in both ankle and wrist; many other mammals have reduced the number of wrist or ankle bones. The primitive and unspecialized features of primates are not limited to the locomotor system: primates have retained, for example, a fairly complete, primitive set of mammalian facial muscles, while rodents, edentates, and several orders of hoofed mammals have independently lost several of these muscles. Primates have also retained a rather primitive configuration of carotid arteries in the head. Primates have also, with the exception of one suborder, retained a rather primitive form of placenta. Primate teeth, on the whole, have undergone fewer changes than have the teeth of many other mammalian groups, such as rodents, edentates, whales, elephants, horses, and advanced artiodactyls, to name just a few. Primates do not hibernate. Primates do not have tusks or horns. Primates do not have the partitioned stomachs of cud-chewing or ruminant artiodactyls. Primates are in many ways among the least specialized of the placental mammalian orders!

A large number of mammalian orders are characterized by a single adaptive feature or complex (a "key character," Chapter 18) whose origin marks also the taxonomic origin of the order. Available fossils do not permit the insectivore–primate boundary to be drawn with any precision, but we *can* specify the the evolutionary trends followed by the primate ancestors as they crossed this boundary. These trends can be described in three interrelated adaptive complexes (Box 27–A), all of them reflecting a general and early adaptation to life in the trees. To be sure, a number of primates have abandoned this original mode of existence, but they still bear many adaptations that once suited their ancestors to an arboreal existence.

Arboreal locomotion is here given primacy as the feature in terms of which all others may best be understood. Locomotion in the trees is practiced also by many other mammalian groups (primates are therefore not unique in this feature), but the method of purchase is remarkably different in the two cases: most arboreal mammals (squirrels, for instance) have sharp claws that dig into the bark, while the paws (we can hardly call them "hands") are applied against the surface of the tree, either flat or with the knuckles slightly raised. Primates, on the other hand, habitually wrap their fingers around all branches except the largest. This tendency to grasp with the hands and feet is possessed to a certain extent by primitive and not necessarily arboreal mammals (the opossum, for instance), but in the primates the grasping ability is further developed and enhanced by the ability to oppose the thumb (and, in nonhuman primates, the big toe as well) to the other digits. Grasp a make-believe branch with your hand (a baseball bat will do fine), or else grasp a make-believe fruit (a tennis ball or baseball is ideal); now look at your hand. You will probably find your thumb on one side, opposed to your

BOX 27–A *Primate Characteristics*

A. Arboreal Locomotion and Its Direct Consequences

1. Arboreal locomotion
2. Grasping hands and feet
3. Opposable thumb and/or big toe (usually both)
4. Individual mobility of all digits
5. Friction-skin on palm and sole
6. Claws reduced (usually) to fingernails
7. Clavicle retained and strengthened

B. Reliance on Vision and on Intelligence

8. Primarily visual orientation
9. Binocular, stereoscopic vision
10. Eyes rotated forward (orbital convergence)
11. Postorbital bar develops
12. Portions of brain concerned with vision expanded; brain develops calcarine fissure
13. Cerebral cortex expanded; higher intelligence
14. Manipulation-and-visual-inspection behavior; "curiosity"
15. Emphasis on learned behavior

C. Reproductive Consequences

16. Single births become typical; twins uncommon
17. Uterus simplex
18. Only one pair of nipples retained—in pectoral position
19. Extended, intensive period of parental care
20. Penis pendulous (hangs freely)

---

remaining four fingers on the other side. The fingerprints (palmar surfaces) of your thumb and your other fingers are probably facing each other, a position that nonprimates generally find impossible. The individual mobility of all the fingers (not just the thumb) is enhanced in all primates, enabling them to grasp objects of various sizes and shapes. Most primates possess hands *and* feet capable of grasping (Fig. 27.1; the human foot is very atypical). This is why many early naturalists referred to the primates as "quadrumana," literally meaning "four-handed." Because of the manner in which primates tend to grasp things (initially, a behavioral trait), skin surfaces that increase friction and prevent slippage have been favored by selection: genotypes that produced the most friction were favored at the expense of those that produced the least. This is why the palm and sole (and in a few primates the underside of the tail as well) generally have a special type of naked (hairless) "friction-skin," with a series of largely parallel ridges similar to human fingerprints. The claws, unnecessary with this type of grasping, tended to get in the way, and were modified (in most primates, but with a few exceptions) into flattened fingernails.

Consider now the limb movements in a specialized ground-runner like the horse: they are generally restricted to a single plane in which the limb moves fore-to-aft only. Primates, living as they do (or did) in the trees, find this unsuitable: the next foothold (branch) or the next food object (fruit) may be in nearly any position, and the limb must therefore be free to move in any plane. The freedom of rotation at the shoulder is made possible largely by the presence of the collarbone (*clavicle*) (Fig. 27.2). This bone holds the shoulder joints further apart, allowing the *glenoid cavity,* the socket into which the head of the armbone or humerus fits, to be pointed in a more lateral direction, away from the midline of the body. Look at the trunk of a mammal such as the dog, which has a greatly reduced clavicle, or the horse, which has none at all: the trunk is deep from top to bottom, and compressed from side to side, with the glenoid cavities directed ventrally and the arms generally held parallel to one another, capable of very little lateral spreading, or none at all. Look now at the trunk of a higher primate: it is flat from back to front, and wide from side to side, and the glenoid cavities are held far apart from one another and directed laterally. The

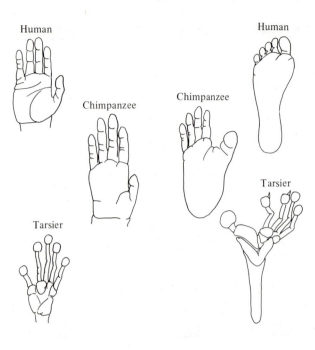

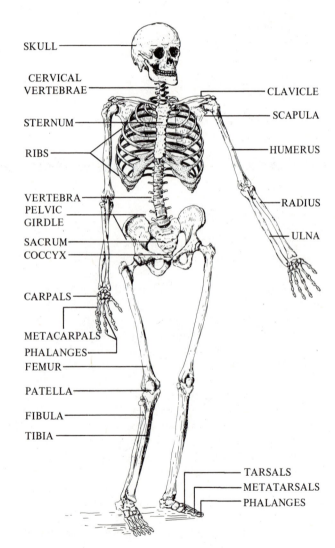

Fig. 27.1 Hands (left) and feet (right) of three primates. All the thumbs and most of the big toes are opposable. Reprinted with permission from Charles K. Levy, *Elements of Biology*, © 1978. Reading, Massachusetts: Addison-Wesley (Fig. 17.6).

Fig. 27.2 The human skeleton. Note the strong clavicle. Reprinted with permission from Jeffrey J. W. Baker and Garland E. Allen, *Study of Biology*, 3 e, © 1977. Reading, Massachusetts: Addison-Wesley (Fig. 11.10).

shoulder blades (scapulae) are no longer parallel, as they are in truly primitive mammals, and the arms are capable of a great degree of lateral spreading. All this is the consequence of possessing a strong clavicle.

Primitive mammals, and most of their ground-living descendants, orient largely by smell. When they encounter their surroundings, they encounter them with their noses first. Their brains reflect this: a great part of the forebrain is given over to those parts concerned with the sense of smell. Look at a dog or a horse face to face, and you are looking up its long snout. Primates, on the other hand, orient largely by sight. When they encounter their surroundings, it is first and foremost with their eyes. Those portions of the brain concerned with vision are enlarged and expanded; the parts concerned with smell are correspondingly reduced. Look at a primate face to face, and you are looking into its eyes. This is primarily because the eyes have become rotated forward (orbital convergence), with their visual fields greatly overlapping. The reasons for these differences are largely adaptive; on the ground, odors tend to linger, while vision is often obstructed by vegetation. But in the trees, the breeze carries many odors away, and many of the most important visual objects (the next foothold, for example) give off no perceptible odor at all. The next foothold, furthermore, must be judged for distance as well as direction: a miscalculation of distance may result in a fatal fall. Primate vision must therefore be *binocular* (with two eyes) and *stereoscopic* (with depth perception); orbital convergence helps bring these both about, and so does the partial instead of total crossing over of optic nerve fibers from either eye to the opposite side of the brain. In addition, olfaction works as well by night as by day, and the early terrestrial mammals may have been nocturnal. Vision, on the other hand, is much more effective by day, suggesting that ancestral primates, like the majority of living ones, may have been active by day (diurnal).

Visual orientation and locomotion are functionally related by an important primate behavior pattern. A primate, upon encountering a strange object, usually inspects it first visually. If the object is easily moved, it is plucked by the hands and brought to the front of the face where the eyes may inspect it more closely. The fingers may hold the object (delicately if necessary), rotate it around, stroke it, or poke at it, observing its reaction (if any) all the while. If the object is deemed suitable as food, the hands convey it to the mouth. The importance of this behavioral complex in primate evolution cannot be overestimated, for it conditioned the further evolution of the hand, the eye, the brain—both the sensory fields related to vision and tactile senses, and the motor fields related to finger movements—and a host of subsidiary traits more or less indirectly. Consider by way of contrast a more typical ground-living mammal (a dog will do): strange objects are inspected by smelling them, and food is ingested, after being smelled, by bringing the mouth to the food, instead of the food to the mouth. The forefeet, like the hindfeet, are in most mammals simply organs of locomotion; in primates, they are also (or primarily, even in locomotion) organs of prehension.

The visual orientation of primates is seen in many parts of the anatomy. The retina is more complex than in other mammals, and a depressed area of great visual acuity (the *area centralis* or fovea) develops within it. The eyes are rotated forward, and in most primate families it is enclosed from behind by a bony postorbital bar. The portions of the brain concerned with vision, tactile sensations, or finger manipulations are greatly increased. The interrelatedness of vision, manipulation, and brain development with the aforementioned behavioral trait of visual inspection has made primates more "curious" and also more intelligent. The complexities of arboreal locomotion placed an early selective premium on intelligence. The portions of the brain most significantly affected by the selective increase in intelligence are the cerebral hemispheres, especially their outer surface layer or cerebral cortex.

Primate behavior patterns are learned to a much greater extent than is true in primitive mammals. This includes locomotor behavior, which is characteristically a learned behavior pattern. The emphasis on learned behavior has had other consequences for primate evolution, not necessarily obvious to the casual observer. Learned behavior, as we mentioned in Chapter 24, is purchased at a price: there is a juvenile period during which this behavior has not yet been learned, or has only been learned imperfectly. This rather vulnerable juvenile period demands that primates depend upon their parents for care even longer than is generally the case among mammals. For this reason and others, most primates tend to form social units that include individuals of all ages and sexes mixed together. Primate youngsters are "curious," and learn largely through play with other group members. Primate mothers are often busy with their children, feeding them, grooming them, playing with them, carrying them. A primate infant, clinging to its mother's fur, is a hindrance to her free movements; two or more infants at the same time would be unbearable. These and similar requirements (having to climb trees and keep up with the group even during the late stages of pregnancy, for example) have placed a high selective value on single births, which are far more common than multiple births in all but one or two primate species. In terms of population biology (Chapter 9), primates are generally *K*-selected, and compensate for their single births by parental care that is both extensive and intensive. This care has had further consequences on the reproductive anatomy and on reproductive behavior. The uteri, which are paired in primitive mammals, fuse together as a single median uterus, a condition known as *uterus simplex*. The nipples, which are multiple in mammals that give birth to large litters, are reduced to a single pair. This pair is furthermore pectoral in position, near the shoulders rather than the hips, and the nursing position is in most primates front-to-front, with the heads of mother and infant near one another. The birth of one young at a time reinforces the mother primate's ability to care for each infant rather intensively, and for the infant to learn and play under mother's (often also under father's) protection. The father, incidentally, has a penis that hangs freely, rather than being slung against the ventral body wall as in most mammals; the adaptive significance of this last primate characteristic is obscure.

The discussion above follows what may be called the arboreal theory of primate origins, first proposed at the end of the last century by G. E. Smith and F. W. Jones, especially in the latter's book *Aboreal Man* (1906). This arboreal theory has been the cornerstone of most twentieth-century explanations of primate origins. Cartmill (1974), however, believes that the early adaptation of the ancestral primates was not to arboreal locomotion but to visual predation, predation by sight. The features of complex B in Box 27–A are, according to Cartmill, of greater importance and greater primacy than those in complex A. Cartmill therefore excludes from his definition of the order Primates three early families that he

judges to lack orbital convergence; Szalay and others have criticized this exclusion.

B. CLASSIFICATION OF THE PRIMATES

The task of classifying the primates is beset by many problems, of which the knottiest is certainly our own loss of perspective when it comes to discussing our near relatives. Divergence of opinion among taxonomists is great, and there is greater diversity in the degree of splitting and lumping practiced by those who would classify this order. Many taxonomic names have multiple synonyms.

Up to 1955, there was a nearly universal tendency to place (or force) fossil primates into living suborders and even into living families. Simpson (1955) and Romer (1966) first suggested, and Szalay (1973) later emphasized, the possibility of a wholly extinct suborder for the forms here called Plesiadapoidea. Up to 1966, it was nearly a universal practice to divide the majority of primates into two suborders; present classifications usually contain more than two. Terms such as "lower," "higher," and "intermediate" were often used, betraying an unspoken tendency to treat primates *as if* they could be arranged in a linear sequence, similar to the old *scala naturae* (Chapter 3). Even contemporary scientists often speak of the array of living primates as representing a linear series or a trend. Both Simons (1972) and Minkoff (1974) have argued against this habit.

The classification in Box 27–B represents a synthesis of recent opinions, none of them in exact agreement on all points. Among previously published classifications, it is closest to that of Romer (1966).

Tree Shrews

The tree shrews, or Tupaiidae (Fig. 27.3), are now generally acknowledged to be the most primitive living placental mammals of all. When first described, they were placed in the order Insectivora. The distinctiveness of the tupaiids from other insectivores and their similarities to the primates were emphasized earlier in this century by the anatomical studies of Albertina Carlsson and W. E. LeGros Clark. Van Valen (1965) has reviewed all the evidence and concluded that tree shrews are not primates at all, but are instead primitive insectivores. Romer

BOX 27–B *A Brief Classification of the Primates*

Order primates

Suborder *Plesiadapoidea:

*Phenacolemuridae, M.Pal.–L.Eoc., N.A.
*Carpolestidae, M.Pal.–L.Eoc., N.A.; U.Pal., Eur.
*Plesiadapidae, M.Pal.–L.Eoc., N.A.; U.Pal., Eur.

Suborder Lemuroidea:

*Adapidae, L.–M.Eoc., N.A.; ?L.,M.–U.Eoc., Eur.; U.Eoc., As.
Lorisidae, L.Mioc., E.Af.; Plioc., India; Recent: Africa south of the Sahara, S. Asia.
Lemuridae, Pleist.–Recent, Madagascar.
Daubentoniidae, Pleist.–Recent, Madagascar.

Suborder Tarsioidea:

*?Microsyopsidae, L.–U.Eoc., N.A.; M.Eoc., Eur.
*Anaptomorphidae, U.Pal., Eur.; L.–M.,?U.Eoc., N.A.
*Omomyidae, U.Pal.–L.Mioc., N.A.; M.Eoc., Eur.; M.–U.Eoc., As.
Tarsiidae, M.–U.Eoc., Eur.; Recent: E. Indies.

Suborder Platyrrhina:

Cebidae, ?U.Olig., Mioc.–Recent, S.A.; Pleist., W. Indies.
Callithricidae, U.Olig., Pleist.–Recent, S.A.

Suborder Catarrhina:

Superfam. *Parapithecoidea
*Parapithecidae, L.Olig., N.Af.

Superfam. Cercopithecoidea
Cercopithecidae, L.Mioc.–Recent, Af.; L.Plioc.–Pleist., Eur.; L.Plioc.–Recent, As.; Pleist.–Recent, E. Indies.

Superfam. Hominoidea
*Oreopithecidae, U.Mioc., E.Af.; L.Plioc., Eur.
Pongidae, ?U.Eoc., As.; L.Olig.–Recent, Af.; M.Mioc.–L.Plioc., Eur.; U.Mioc.–Recent, As.; Pleist.–Recent, E. Indies.
   Recent genera: *Hylobates* (gibbons), S.Asia. *Pongo* (orang-utan), Borneo. *Pan* (gorilla, chimp), equatorial Af.
Hominidae:
   *?*Ramapithecus,* U.Mioc.–L.Plioc., As.; L.Plioc., E.Af., Eur.
   **Australopithecus,* Pleist., Af., As., E.Indies.
   Homo, Pleist., Old World except Australia; Recent, World Wide.

ABBREVIATIONS:

L., Lower	Olig., Oligocene	A., America
M., Middle	Mioc., Miocene	Af., Africa
U., Upper	Plioc., Pliocene	As., Asia
Pal., Paleocene	Pleist., Pleistocene	Eur., Europe
Eoc., Eocene	N., S., E., W., compass directions	*extinct groups

Modified after Romer (1966).

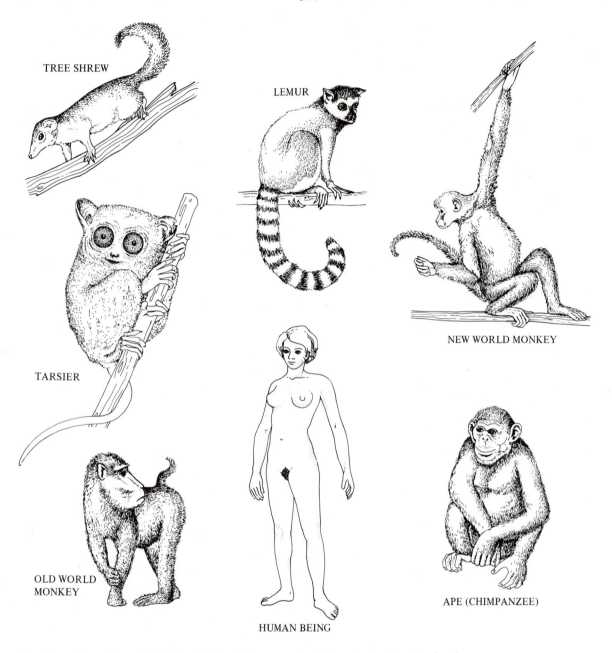

Fig. 27.3 A tree shrew (family Tupaiidae) and representative living primates. Reprinted with permission from Jeffrey J. W. Baker and Garland E. Allen, *Study of Biology,* 3e, © 1977. Reading, Massachusetts: Addison-Wesley (Fig. 23.20).

(1966) assigned the Tupaiidae to the insectivore suborder Proteutheria, of which they are the only living representatives. The fact that these primitive placental mammals should ever have been classified among primates not only shatters our feeling of dignity, it also emphasizes once again the fact that primates are among the more primitive placental mammals.

Plesiadapoidea

The Plesiadapoidea (or Paromomyiformes) were probably not ancestral to any of the modern families of primates (Fig. 27.4). Three families, all extinct, are generally included in this group (Szalay, 1972, 1973); some authorities include additional families as well. The best-known genera had enlarged incisor teeth that were pro-

cumbent in the lower jaw (inclined forward rather than vertical); the beginnings of a diastema, or tooth-free gap, separated the incisors from the cheek teeth. Unfortunately, these were the later and more specialized genera; the earlier genera had less dental specialization, but in some cases the teeth are incompletely known. The best-known genus, *Plesiadapis* (Fig. 27.5a), had a brain that was quite primitive compared to that of any living primate. Analysis of postcranial remains makes it rather clear that these animals were thoroughly arboreal. In addition to being arboreal, the Plesiadapoidea seem to have been largely though not exclusively frugivorous (fruit-eating). Like all primates, their diet probably included some insect matter, too: a mixed insectivorous–frugivorous diet is very likely.

Fig. 27.4 A family tree of the primates, after Minkoff (1974).

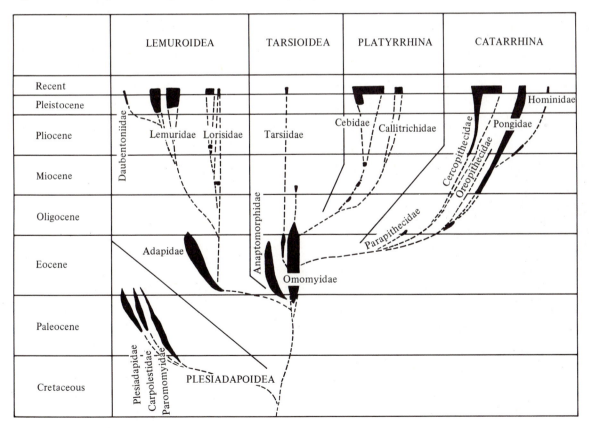

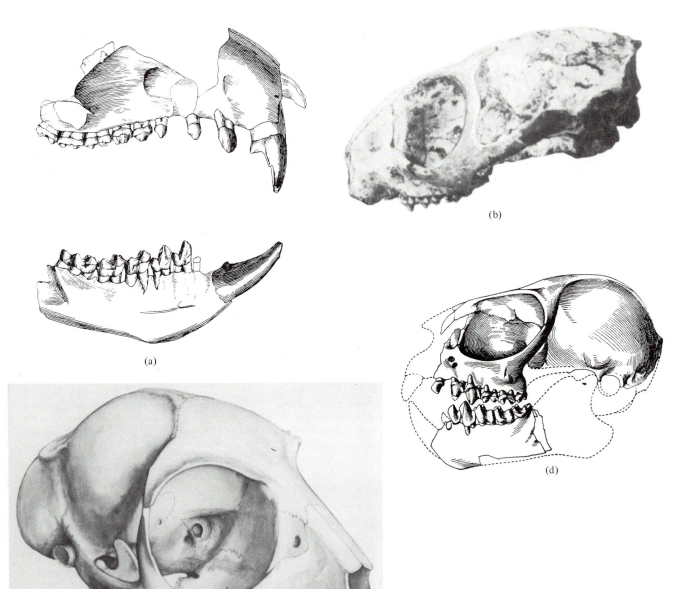

Fig. 27.5 Four Tertiary primates: (a) *Plesiadapis* (family Plesiadapidae); (b) *Pronycticebus* (family Adapidae); (c) *Necrolemur* (family Tarsiidae); (d) *Tetonius* (family Anaptomorphidae). Parts (b) and (c) are reprinted with permission of Macmillan Publishing Co. Inc., from *Primate Evolution* by E. L. Simons, Copyright © 1972 by Elwyn L. Simons.

Tarsioidea

The Tarsioidea (Fig. 27.5c and d) include the living *Tarsius* and its extinct relatives. *Tarsius* is a small, nocturnal, arboreal primate with elongated hindfeet. The elongation, for which the calcaneum and navicular bones are largely responsible, increases the leverage with which the tarsier may leap from perch to perch. "Vertical clinging and leaping," in which tarsiers are most expert, is considered by some to be a rather primitive form of locomotion, but others believe that it was independently evolved in the several groups which exhibit it. One Eocene genus (*Hemiacodon*) shows some parallel tendencies.

The Omomyidae, an extinct family of tarsioids, ranged from Paleocene to Miocene and are known from both Old and New Worlds. Except for the Plesiadapoidea, they are the oldest known primates. The lower Eocene omomyids of America and Europe were rather similar to one another, but subsequent American and European omomyids seem to have diverged from each other. Some American forms show possible resemblances to the Platyrrhina; the European forms may include the ancestors of the Catarrhina and perhaps the

Lemuroidea as well. The Omomyidae thus occupy an important place in primate history, for they are possibly ancestral to *all* primates other than the Plesiadapoidea. Omomyid teeth have been reduced to a maximum of two pairs of incisors and usually three pairs of premolars; these same maxima hold for non-plesiadapoid primates in general, with few exceptions.

Lemuroidea

The lemuroid primates (suborder Lemuroidea or Strepsirhini) are known as fossils from the Eocene of North America and Europe. Today they flourish largely on the island of Madagascar, though they are also found in continental Africa and in southern Asia.

Early Tertiary lemuroids are assigned to the family Adapidae (Fig. 27.6) the oldest of these show some resemblance to the Omomyidae, which may indicate a recent divergence from that family. Later adapids show very close resemblance in many respects to the modern lemurs of Madagascar. One genus, *Pronycticebus* (Fig. 27.5b), may be close to the ancestry of the Lorisidae.

The family Lorisidae includes slow, forest-dwelling, deliberate climbers such as the slow loris of Asia and the

Fig. 27.6 *Notharctus,* an Eocene lemuroid of the family Adapidae.

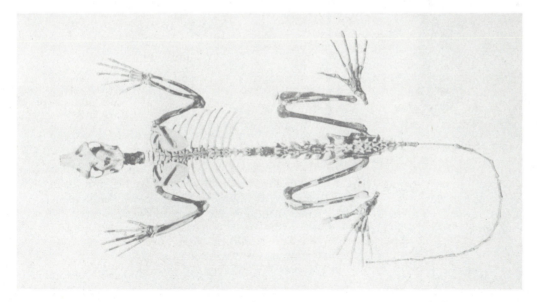

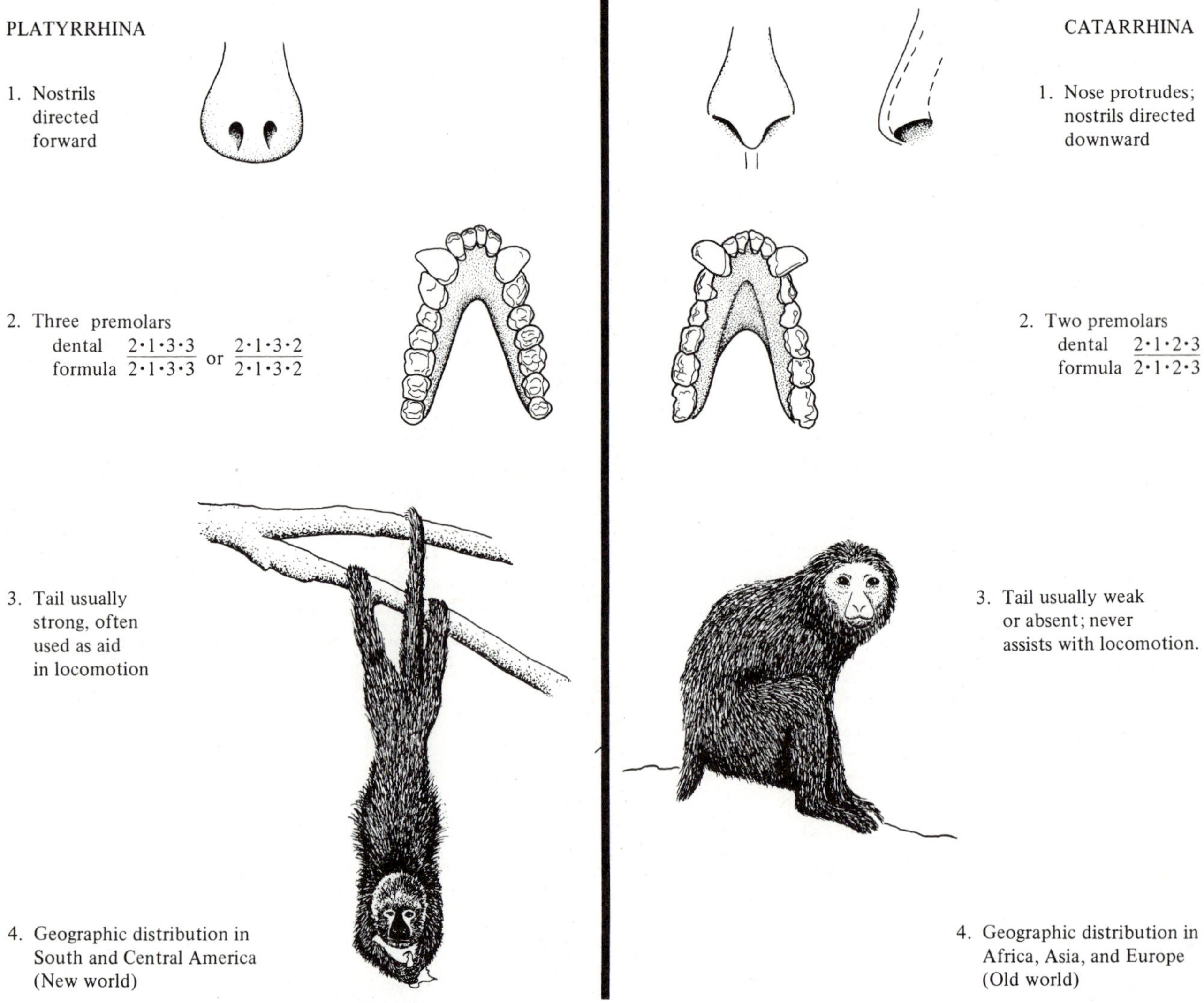

PLATYRRHINA   CATARRHINA

1. Nostrils   1. Nose protrudes;
   directed   nostrils directed
   forward   downward

2. Three premolars   2. Two premolars
   dental $\dfrac{2 \cdot 1 \cdot 3 \cdot 3}{2 \cdot 1 \cdot 3 \cdot 3}$ or $\dfrac{2 \cdot 1 \cdot 3 \cdot 2}{2 \cdot 1 \cdot 3 \cdot 2}$ dental $\dfrac{2 \cdot 1 \cdot 2 \cdot 3}{2 \cdot 1 \cdot 2 \cdot 3}$
   formula   formula

3. Tail usually   3. Tail usually weak
   strong, often   or absent; never
   used as aid   assists with locomotion.
   in locomotion

4. Geographic distribution in   4. Geographic distribution in
   South and Central America   Africa, Asia, and Europe
   (New world)   (Old world)

Fig. 27.7 Differences between Platyrrhina and Catarrhina.

potto of Africa, and also the swift, actively leaping gala-gos of African scrub forests and savannahs.

The remaining lemuroids are confined to the island of Madagascar. The variation among these species is con-siderable, and includes differences in chromosomes, ecology, and behavior, as well as morphology. The small mouse lemurs (*Cheirogaleus, Microcebus*) possess many features in common with the Lorisidae (Szalay and Katz, 1973; Minkoff, 1974). The aye-aye, *Daubentonia*, is so

unique that it is placed in a family of its own. Its incisors are enlarged and rodentlike, with a wide diastema sepa-rating them from the flattened cheek teeth.

New World Monkeys (Platyrrhina)

The remaining primates (suborders Platyrrhina and Catarrhina) were formerly grouped together, but most primatologists now realize that the two suborders evolved independently. Most of the resemblances between these two groups are now attributed to parallelism.

Platyrrhines differ from catarrhines in a number of features (Fig. 27.7). The names themselves refer to a difference in the shape of the nose: in the Platyrrhina it is flat, with the nostrils opening onto the surface; in the Catarrhina it protrudes from the face (like yours) and the nostrils open downward. Another difference is that the Platyrrhina always have three premolars in each jaw, while the Catarrhina have only two. A third difference lies in the importance of the tail, which is relatively stronger in platyrrhines where it is often used for balancing or as an aid in locomotion. In the Catarrhina, the tail is too weak to function in any important role except gesturing. Lastly, a geographical separation has confined the Platyrrhina to the tropical parts of the New World, while the Catarrhina have always been confined to the Old World (except, of course, for humans).

Platyrrhines (Fig. 27.8) are divided into two families, Cebidae and Callitrichidae. The Cebidae are the larger and presumably more ancestral group. They fit the popular image of a ''monkey'' very well. The marmosets (Callitrichidae) are smaller and appear much less monkeylike. They are thought to be derived from the Cebidae.

C. OLD WORLD MONKEYS AND APES (CATARRHINA)

The Catarrhina, or Old World monkeys, apes, and humans are distinguished from the Platyrrhina both by their noses and by their dental formulae. The noses of catarrhines always protrude from the face, with the nostrils opening downward. The premolars are reduced to two in each jaw. The tail is never prehensile, and its movements are in general weak.

Cercopithecoid monkeys

The largest group of living Catarrhina are the Old World monkeys, or Cercopithecoidea (Fig. 27.9). These

Fig. 27.8 Representative New World monkeys (Platyrrhina): (a) and (b) squirrel monkeys (*Saimiri sciureus*); (c) capuchin or ''organ grinder'' monkey (*Cebus capucinus*). Photographs (a) and (b) by Eli C. Minkoff.

(a)   (b)   (c)

(a)   (b)

(c)

Fig. 27.9   Representative Old World monkeys (Cercopithecidae): (a) guereza or colobus monkey (*Colobus polykomos*); (b) DeBrazza's guenon (*Cercopithecus neglectus*); (c) male and female mandrills (*Papio sphinx*). All photographs by Eli C. Minkoff.

are distinguished in part by the persistence of the tail, but even more strikingly by a "bilophodont" type of molar tooth, with a connection between the two anterior cusps and another between the two posterior cusps. Another dental specialization includes the enlargement of the upper canine and third lower premolar (C^1 and P_3) to form a knife-like shearing apparatus (Fig. 27.10). The geographic range of cercopithecoid monkeys includes much of Africa and tropical Asia except for the desert regions; this range extends north to Gibraltar in the west and Japan in the east.

Fig. 27.10 Two dental specializations of the Cercopithecidae: (a) bilophodont type of molar tooth (crown view); (b) shearing action of C^1 against P_3.

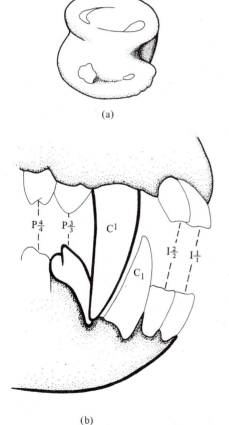

(a)

(b)

The Superfamily Hominoidea: Living Apes

The superfamily Hominoidea includes those catarrhines with five-cusped (nonbilophodont) molars and a fully reduced tail. To the apes (Pongidae) and humans (Hominidae), a third family is sometimes added for the extinct genus *Oreopithecus* and its presumed relatives.

The living apes, or Pongidae (Fig. 27.11), comprise three to five genera that differ strikingly in their locomotor adaptations.

The gibbons (subfamily Hylobatinae, "lesser apes") live on the Asian mainland in forested areas and spend nearly all their time in trees. They travel in small family groups, typically containing one adult male, one adult female, and their offspring. Gibbons are skillful acrobats, hurling themselves from branch to branch with great speed and agility. They are true brachiators in that only the arms, and not the legs, are used in locomotion until the final perch is reached. A light build and two exceedingly long arms are among the adaptations for this type of locomotion. Gibbon legs are not well suited for supporting weight, and when a gibbon is forced to walk along the ground it will do so rather clumsily, waving its long arms overhead as an aid to balancing.

The orang-utan (*Pongo*) and the African apes (*Pan*) together constitute the subfamily Ponginae, or Great apes. Orang-utans live on the island of Borneo and on the nearby island of Sumatra in the East Indies. Like the gibbons, they spend much of their time in trees, and their family groups are small. Their locomotion has been described as a "quadrumanual clambering" in which all four limbs are used about equally in the support of the body. Only one limb, or at most two, is moved at a time, while the others maintain their hold. (The gibbon, by way of contrast, often holds on with one arm only and frequently flings its body through the trees without any point of support at all). The body may be held at any angle. When orang-utans find it necessary to walk on the ground, they rest their weight on the outer margins of the feet rather than the plantar surface, while their curved toes (and opposable big toe) are curled in a grasping position.

Closest to humans in morphological, behavioral, and biochemical characteristics are the African apes of the genus *Pan,* including the gorilla (*Pan gorilla*) and the chimpanzee (*Pan troglodytes*). Both are very similar anatomically, differing principally in size. Both exhibit

Fig. 27.11 Representative apes (family Pongidae): (a) siamang gibbon (*Symphalangus syndactylus*); (b) and (c) lowland gorilla (*Pan gorilla*); (d) orang-utan (*Pongo pygmaeus*). Part (d) is reprinted with permission from William C. Schefler, *Biology: Principles and Issues,* © 1976. Reading, Massachusetts: Addison-Wesley (Fig. 13.33d). All other photographs by Eli C. Minkoff.

the same unique form of locomotion, **knuckle-walking**. In this type of locomotion, the body is held on an angle in a semi-erect but still largely quadrupedal position, with most of the weight borne by the legs (see Fig. 27.15). The long arms, in this position, barely reach the ground, and some of their weight is borne on the knuckles. The hand faces toward the rear, and is free to carry objects.

The gorilla (*Pan gorilla*) is the largest of living primates, attaining a weight of up to 270 kg (or 600 lb). Its popular image as a ferocious Goliath is totally false. George Schaller, who has studied gorillas in their natural habitat, has found them to be quite timid. They are exclusively vegetarians, eating a wide variety of leaves, fruits, stems, and other miscellaneous plant products. They have a rather fearsome-looking threat display, and this, together with their large size, probably accounts for their popular reputation. Their fury, however, is largely or entirely vented in the rather harmless uprooting of vegetation. Gorillas live in family groups containing several adult males, a larger number of adult females, and an assortment of juveniles and infants. The larger and older adult males, or "silverbacks" (so called because of the graying of the hair between the shoulders) are temporary associates of these groups, spending the greater amount of their time in solitary existence.

Chimpanzees (*Pan troglodytes*) are in nearly all respects (except size) very similar to gorillas. They approach closest to humans of all the living primates, most notably in intelligence. Their problem-solving ability exceeds that of any other nonhuman species, and "insight" learning plays a very important role in this problem-solving ability. Several captive chimps have now been taught to communicate with each other and with humans by means of a gesture language, AMSLAN, or American Sign Language for the Deaf. Wild chimpanzees have been observed by Jane van Lawick-Goodall to use tools that they fashion themselves, an otherwise human trait. One such tool, a stick or twig stripped of its leaves and moistened with saliva at the tip, is used as a termite-fishing tool: it is inserted into a termite nest in the hopes of finding a termite adhering to the saliva-covered tip when the stick is withdrawn.

Fossil Apes

Teeth of two questionable apes, *Amphipithecus* and *Pondaungia*, occur in upper Eocene deposits in Burma.

The earliest undoubted catarrhines are found in the lower Oligocene Fayûm deposits in northern Egypt. Six such genera are known, of which two show affinities to cercopithecoid monkeys, one to gibbons, and three to dryopithecine apes. The other mammals of the Fayûm fauna are of characteristically African groups such as the elephantlike Proboscidea, the hyraxes (Hyracoidea), sea cows (Sirenia), primitive whales (Cetacea), etc. Rodents, insectivores, and artiodactyls are each represented by indigenous African families. Lemuroid primates are conspicuous by their absence; perhaps they occurred in Africa at the time, but lived farther to the south. The mammals present in this fauna show few if any affinities to earlier or contemporaneous mammals of other continents, and their degree of divergence from other mammals suggests that Africa had been an island continent during the early Tertiary. Nearly every mammalian order or suborder of African origin is present here for the first time—the Catarrhina included.

Two of the Fayûm primates, *Apidium* and *Parapithecus,* seem to be allied to one another. Both are now considered to be related to cercopithecoid monkeys, or perhaps to the problematical *Oreopithecus.* Another Fayûm primate, *Aeolopithecus,* was a lightly built form, apparently related to the modern gibbons (Hylobatinae). The remaining three genera (*Oligopithecus, Propliopithecus, Aegyptopithecus*) seem definitely to be apelike. Simons (1972) refers them to the subfamily Dryopithecinae, to be discussed shortly. Postcranial fragments assigned to *Aegyptopithecus* are nearly identical to the corresponding bones of *Dryopithecus.*

The Miocene and Pliocene deposits of the Old World contain the remains of *Pliopithecus*, an early gibbon. Several fossil monkeys are also known from deposits of this age, as is the problematical genus *Oreopithecus.* This genus shows a curious mixture of monkeylike, apelike, and even hominidlike features. It is about the size of a monkey and appears to have practiced a form of "quadrumanual clambering" not unlike the modern orang-utan, *Pongo.*

Best known of the Miocene-Pliocene apes was the genus *Dryopithecus* (subfamily Dryopithecinae, Fig. 27.12). *Dryopithecus* was a very successful, widespread genus, whose remains have been found on three continents. One Asian species of *Dryopithecus* seems to be close to the ancestry of *Pongo* (the orang-utan), while an African species, *D. africanus,* seems close to the ancestry

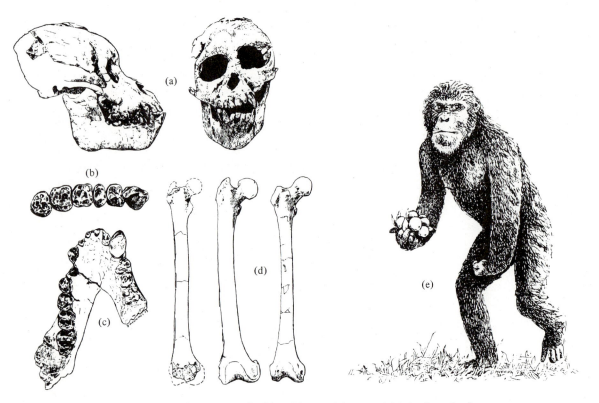

Fig. 27.12   The remains of *Dryopithecus:* (a) skull of *D. africanus;* (b) upper right cheek teeth of *D. africanus;* (c) lower jaw of *D. major;* (d) femur of *D. africanus* (left) compared with modern chimpanzee (center) and *Homo sapiens* (right); (e) reconstruction of *D. africanus.* Reprinted by permission from McKern, Sharon, and McKern, Thomas, *Living Prehistory, An Introduction to Physical Anthropology and Archaeology.* Menlo Park, Calif.: Cummings Publishing Company (now the Benjamin/Cummings Publishing Company), pp. 108–109, Fig. 2.

of *Pan* (gorillas and chimpanzees). Speculation as to the locomotor habits of *Dryopithecus* have varied greatly: brachiation, knuckle-walking, and terrestrial semi-erect bipedalism have all been suggested. It is quite possible that *Dryopithecus* was capable of doing all of these things (and others) for short periods of time, but lacked any of the anatomical specializations that accompany *habitual* brachiation or *habitual* knuckle-walking. Its wrist shows that *Dryopithecus* often did suspend at least part of its weight from above.

Dryopithecus was generally about the same size as a modern chimpanzee, or in some cases a bit larger. The limb bones, however, show that *Dryopithecus* was more lightly built. The teeth were more primitive than in any modern pongid. The molars were built on a pattern common to all later hominoids, but the last lower molar was distinctly larger than the first two. As we shall soon see, *Dryopithecus* was quite similar to *Ramapithecus*, which has often been considered a hominid.

See Fig. 27.13 for a comparison of various primate skulls.

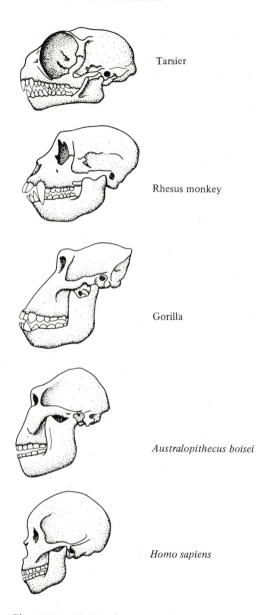

Tarsier

Rhesus monkey

Gorilla

Australopithecus boisei

Homo sapiens

Fig. 27.13 Skulls of various primates.

D. THE FAMILY HOMINIDAE

The superfamily Hominoidea contains, aside from the enigmatic fossil genus *Oreopithecus,* two major families: the Pongidae, or apes, and the Hominidae, or humans. The distinctions between these two families are explained in Box 27–C and Fig. 27.14. In most cases, but not all, the pongid condition is primitive and the hominid condition advanced. Upright locomotion (Fig. 27.15) can easily be interpreted as the "key character" (Chapter 18) in the sense that most of the other hominid advancements are best understood as consequences of upright locomotion. The key character was decidedly *not* an increase in intellegence or brain size, as was commonly believed some fifty years ago—the early hominids had brains differing but little from their pongid contemporaries.

The major pongid specialization is likewise a specialized form of locomotion, or rather several. Although only one modern ape (the gibbon) is a true brachiator, many pongid adaptations, (such as the long arms), make more sense if we assume that all apes had brachiating ancestors. Controversy surrounds this point, however, and many scientists do doubt that any of the great apes ever brachiated, while others insist that their ancestors must have brachiated at least occasionally.

The words "man" and "human" are much too vague and imprecise for a discussion of hominid fossils. To refer to creatures varying in their degree of humanity, a more precise terminology is used, as follows:

hominoid: any member of the superfamily Hominoidea, including both apes and humans.

hominid: any member of the family Hominidae, here understood to include only creatures that walked erect.

These taxonomic terms are usually preferred because of widespread agreement as to their meaning. There is much less agreement concerning the term "human." In this book, "human" is a virtual synonym of "hominid."

Most authorities now recognize three genera within the Hominidae. The earliest hominids, of the genus *Ramapithecus,* are barely distinguishable from the pongid genus *Dryopithecus.* Their membership in the Hominidae is not beyond question. In particular, the

BOX 27–C   *Differences between Hominids and Pongids*

Family Pongidae (apes)	Family Hominidae (humans)
A. LOCOMOTION AND POSTURE	
1. Brachiation and modifications thereof (knuckle-walking, quadrumanual clambering)	1. Upright, bipedal walking
2. Arms longer than legs (arm/leg ratio >110%)	2. Legs longer than arms (arm/leg ratio below 90%)
3. Foramen magnum toward rear of skull (''8 o'clock'' position)	3. Foramen magnum beneath skull; skull balanced on top (''6 o'clock'')
4. Arched vertebral column with anticlinal vertebra	4. S-shaped vertebral column with lumbar curve
5. Head supported by muscles and nuchal ligament; strong areas of attachment (nuchal or lambdoidal crest)	5. Head balanced atop vertebral column; no nuchal crest
6. Pelvis taller; no basin	6. Pelvis becomes basin-shaped
7. Gluteus maximus abducts leg	7. Gluteus maximus enlarged; extends leg
8. Femur straight; weight borne over inner condyle	8. Femur slightly tilted; weight borne over outer condyle
9. Calcaneum unmodified	9. Calcaneum modified to bear weight (deeper in vertical plane)
10. Big toe (hallux) opposable	10. Toes reduced; hallux no longer opposable
B. DENTAL FEATURES, etc.	
11. Jaw muscles strong	11. Jaw muscles typically weaker (except robust australopithecines)
12. Canine teeth large; protruding above other teeth	12. Canine teeth reduced and not protruding above other teeth
13. Dental arcade square; tooth rows parallel (Fig. 27.14)	13. Dental arcade parabolic to elliptical; tooth rows not parallel
14. Mandibular symphysis strengthened by simian shelf	14. Mandibular symphysis strengthened by chin
15. Molars erupt in unison	15. Molars erupt one at a time; delayed
C. OTHER CHARACTERISTICS	
16. Less habitual tool use	16. Habitual tool use
17. No language (in nature)	17. Language
18. Lower intelligence, smaller cranial capacity (400–900 cc)	18. Higher intelligence, larger cranial capacity (but not initially)
19. Estrus in female	19. No female estrus
20. Body hairier	20. Body less hairy

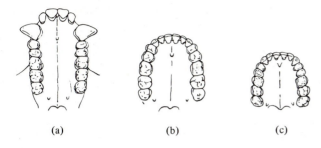

Fig. 27.14 Upper jaws of (a) chimpanzee; (b) *Australopithecus;* (c) *Homo sapiens.* Reprinted with permission from John Buettner-Janusch, *Origins of Man,* © 1966, New York: John Wiley & Sons, Inc.

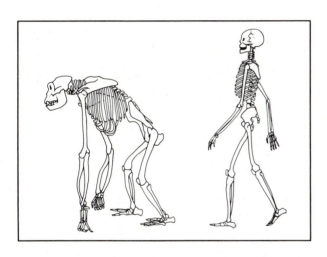

Fig. 27.15 Skeletons of a chimpanzee knuckle-walking and a human walking upright (not to scale). Reprinted by permission from Ayala, Francisco J., and Valentine, James W., *Evolving.* Menlo Park, Calif.: The Benjamin/Cummings Publishing Company, Inc., 1976, p. 387, Fig. 12.8.

claim that they walked erect is based on indirect inference.

Aside from *Ramapithecus,* the remaining hominids are here placed in the two genera *Australopithecus* and *Homo* (Box 27–D). *Australopithecus* includes Pliocene to early Pleistocene hominids with brains no larger than that of a chimpanzee or gorilla. All mid-Pleistocene and later hominids are placed in the genus *Homo,* but so are certain earlier ones. The brains of later species of *Homo* surpassed those of *Australopithecus,* but the extent of this difference was initially small. *Homo habilis,* a contemporary of *Australopithecus,* was once considered by some workers to be a species of *Australopithecus.* The mid-Pleistocene hominids with cranial capacities from about 1000–1200 cm^3 are placed in *Homo erectus. Homo sapiens* includes the late Pleistocene and Recent hominids with cranial capacities above 1200 cm^3.

Origins of the Hominidae: *Ramapithecus*

Among the many Miocene and Pliocene hominoids of Africa, Europe, and Asia, the vast majority are assigned to the genus *Dryopithecus.* Barely distinguishable from *Dryopithecus,* and frequently confused with that genus in the past, are a series of specimens now assigned to the genus *Ramapithecus* (Fig. 27.16). *Ramapithecus* was originally described as a genus by G. E. Lewis of Yale University, who considered it to possess certain humanlike features. In subsequent decades, many new specimens were found, but unfortunately several of these were mistakenly assigned to *Dryopithecus,* and the two genera became confused. Finally, Simons and Pilbeam (1965) studied all the known material of Mio-Pliocene hominoids, and were thereby able to correct the many misidentifications and confusions that had oc-

BOX 27-D   *A Classification of the Family Hominidae*

Family Hominidae

RAMAPITHECUS
R. punjabicus (Synonyms include *R. brevirostris.*) Southern Asia.
R. wickeri (Synonyms include "*Kenyapithecus.*") Africa.

AUSTRALOPITHECUS
A. afarensis. East Africa.

A. africanus (Synonyms include *Plesianthropus transvaalensis.*) South Africa, East Africa.
A. robustus (Synonyms include "*Paranthropus.*") South Africa.
A. boisei (Synonyms include "*Zinjanthropus.*") East Africa.

HOMO
H. habilis. Africa, perhaps also Asia.
H. erectus (Synonyms include "*Pithecanthropus,*" "*Sinanthropus.*") Europe, Africa, Asia.
H. sapiens (Synonyms include *H. neanderthalensis,* etc.) Europe, Africa, Asia: subsequently spread to all other continents.

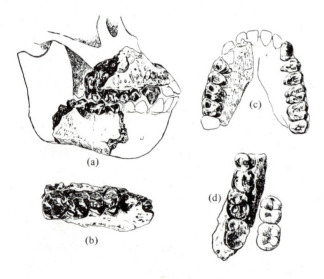

(a)

(b)

(c)

(d)

Fig. 27.16   Remains of *Ramapithecus:* (a) fragments of upper and lower jaws, with a reconstruction of the surrounding portions of the skull shown in outline; (b) right upper cheek teeth; (c) reconstruction of the palate and upper tooth rows; (d) last lower premolar and lower molar teeth of *R. punjabicus.* Reprinted by permission from McKern, Sharon, and McKern, Thomas, *Living Prehistory, An Introduction to Physical Anthropology and Archaeology.* Menlo Park, Calif.: Cummings Publishing Company (now the Benjamin/Cummings Publishing Company), p. 111, Fig. 3.

curred earlier. They showed *Ramapithecus* to be truly distinct from *Dryopithecus,* despite the earlier confusion. They also found *Ramapithecus* to show features indicative of later hominids, such as a shift in molar proportions.

Did *Ramapithecus* walk erect? Did it possess any of the other characteristic hominid traits? Direct evidence of erect posture or other distinctive hominid traits is lacking, but the shift in molar proportions and their delayed eruption do indicate some shift in diet, possibly from one of relatively more succulent plant matter (including many fruits as well as leaves) to one of drier and more fibrous vegetable material such as stems, roots, and grasses. If this was the case, it might indicate a shift from a forest to a more open grassland or savannah habitat. This ecological shift may well have preceded any morphological shift (such as those indicating customarily upright locomotion), and it is on these grounds that

Ramapithecus may be considered a hominid, even in the absence of any direct evidence of its upright posture.

The image of *Ramapithecus* that emerges is that of a terrestrial creature not very different from contemporary *Dryopithecus,* but differing in diet and perhaps also in ecological preference for a more open habitat. Faunal evidence seems to suggest that *Ramapithecus* may have lived on the forest margins, since the other animals fossilized in the same deposits include both forest and savannah species mixed together.

In the last decade, new doubts on the antiquity of the family Hominidae, and therefore on the status of *Ramapithecus* as a hominid, have been expressed by a number of biochemists. These biochemists point out that the proteins and DNA of *Homo* and *Pan* match each other almost exactly, and that our divergence from apes must therefore have been much more recent than *Ramapithecus.* Biochemical evidence seems to date the pongid-hominid divergence at about 5 million years ago, a figure that conflicts with the 20 million year age assigned to *Ramapithecus.* One authority has recently suggested that *Ramapithecus* is an ape close to the ancestry of the orang-utan.

Australopithecus

Hominids of the genus *Australopithecus* have been found in deposits ranging in age from upper Pliocene to second glacial (Mindel). The first finds were made in South Africa (Fig. 27.17), but subsequent discoveries were also made in East Africa and elsewhere in the Old World. Taxonomic splitters have referred these specimens to over half a dozen species and nearly as many genera.

The latest discovery consists of a new species, called *Australopithecus afarensis,* from the Afar region of Ethiopia. According to its discoverers, *Australopithecus afarensis* is a perfect intermediate between *Ramapithecus* and later species of *Australopithecus,* both in time and in morphology. *Australopithecus afarensis* is also suitable as an ancestor of *Homo habilis* and may represent the common ancestor of all later hominids. The Afar fossils are about 3.7 million years old. More fragmentary fossils of about the same age are known from Lothagam, Kanapoi, and Baringo, all in Kenya.

A later and better-known species, *Australopithecus africanus* (Figs. 27.18 and 27.19) contains erect creatures with essentially human dentition. The original skull was found in a cave at Taung, South Africa, by Dr. Raymond Dart in 1924. It had a rather human dentition, with reduced canines, and a very low foramen magnum, in a human rather than an apelike position. A heated controversy ensued as to whether Dart's fossils were "really human." Subsequent discovery of pelvic girdles (Fig. 27.19) and other postcranial remains showed quite conclusively that *Australopithecus* was an upright bipedal walker. Cultural remains associated with *Australopithecus* include simple stone tools along with a variety of bones, teeth, and horns. Certain leg bones and jawbones were intentionally fractured to produce sharp edges for cutting. The variety of species represented shows that *Australopithecus* was an able hunter. Baboons found as fossils in South African caves and at one Angola locality often have their heads smashed in, apparently with intention (Dart, 1949; Minkoff, 1972).

Fig. 27.17 African localities of early humans and other fossils. Reprinted by permission from Isaac, G. L. and McCown, E. R., *Human Origins.* Menlo Park, Calif.: The Benjamin/Cummings Publishing Company, Inc., 1976, p. 5, fig. 2.

Ternifine

Omo
Hadar
Lothagam, Kanapoi
East Rudolf
Kanam,
Rusinga
Kanjera
Olduvai
Fort
Garusi
Ternan
Peninj

Sterkfontein
Swartkrans
Makapansgat
Taung
Kromdraai

× Miocene fossil localities

Plio-Pleistocene human localities:

○ Plio-Pleistocene

● Middle Pleistocene

Other Plio-Pleistocene faunal localities:

△ Lower Pleistocene

▲ Middle Pleistocene

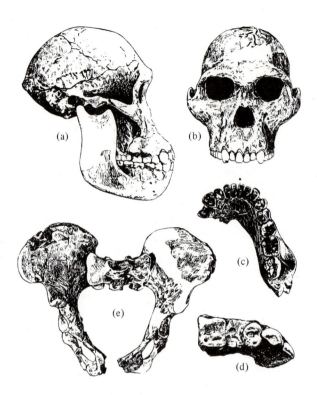

Fig. 27.18 *Australopithecus africanus:* (a)
female skull from Sterkfontein with reconstructed lower jaw; (b) frontal view of same
skull; (c) female mandible from Makapan;
(d) portion of right upper tooth row from
Sterkfontein, showing reduced canine; (e)
pelvic girdle from Sterkfontein. Reprinted by
permission from McKern, Sharon, and
McKern, Thomas, *Living Prehistory, An
Introduction to Physical Anthropology and
Archaeology,* Menlo Park, Calif.: Cummings
Publishing Company (now the Benja-
min/Cummings Publishing Company), p.
131, fig. 8.

Fig. 27.19 Pelvic girdles of *Pan, Australopithecus,* and *Homo.* Note that the pelvis of
Australopithecus agrees more closely with the
human type, especially in the region of the
sciatic notch. Reprinted with permission from
John Buettner-Janusch, *Origins of Man,* ©
1966, New York: John Wiley & Sons, Inc.

From the remains of other animals associated with
Australopithecus, it is evident that we are dealing no
longer with forest creatures, but with inhabitants of the
open savannahs. Though language and other aspects of
culture do not readily fossilize, we can surmise from the
successful hunting abilities of *Australopithecus* that some
degree of cooperative hunting was practiced. This in turn
implies at least enough symbolic language, whether spoken or by gestures, to communicate ideas such as ''I'll go
around this way, you go around that way, and we'll meet
on the other side.''

A quite different larger or ''robust'' australopithecine was originally described under the name *Paranthropus;* it is now usually called *Australopithecus robustus*
(Fig. 27.20). *A. robustus* was somewhat larger than *A.
africanus,* but the major differences between the two
species are in their molar teeth and their jaw muscles. In
A. robustus, the molars were much larger, even relative

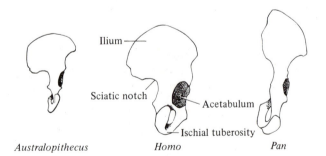

Ilium

Sciatic notch

Acetabulum

Ischial tuberosity

Australopithecus   *Homo*   *Pan*

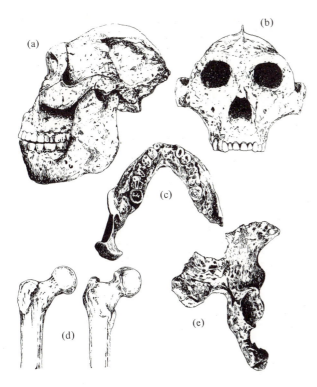

Fig. 27.20 *Australopithecus robustus* from Swartkrans, South Africa. (a) and (b) skull; (c) mandible; (d) top portion of femur (on left) compared to that of *Homo sapiens* (on right); (e) right half of pelvis. Reprinted by permission from McKern, Sharon, and McKern, Thomas, *Living Prehistory, An Introduction to Physical Anthropology and Archaeology.* Menlo Park, Calif.: Cummings Publishing Company (now the Benjamin/Cummings Publishing Company), p. 133, fig. 9.

to the overall size, and the grinding area of the molar surfaces was especially increased. The jaw muscles, and the places for their attachment, were also much stronger. A bony **sagittal crest** was often built up along the midline of the skull on top of the braincase.

In 1954, another robust australopithecine was discovered in Olduvai Gorge, Tanzania, by Mary and Louis Leakey, who named it *Zinjanthropus boisei*. It proved to be similar to *Australopithecus robustus*. The enlarged and deepened jaws, the stronger and more widely flared zygomatic arches, and the restructuring of the head to accommodate the enlarged temporalis muscle were all reminiscent of *A. robustus* and were indeed carried to further extremes (Fig. 27.21). Again, the enlargement of the grinding area of the molar teeth suggests a dietary shift toward grittier or tougher food. Some anthropologists have suggested that both robust australopithecines be placed in the same species, *Australopithecus robustus;* perhaps a larger number favor treating the robust forms as two species, *A. robustus* (South African) and *A. boisei* (East African). The fact that *A. boisei* was more extreme in its morphology does not necessarily mean that it was a descendent of *A. robustus,* although this very relationship has been postulated. Relative dating of the sites in South Africa with those of East Africa suggest that *A. robustus* may indeed have been earlier rather than contemporaneous with *A. boisei,* and evidence from other localities also suggests (but does not prove) that the distinguishing features of the robust australopithecines became progressively more exaggerated during the course of their evolution.

A variety of fragmentary hominid remains, probably those of *Australopithecus,* are scattered about East Africa in several additional localities. Hominids, it seems, oc-

Fig. 27.21 *Australopithecus boisei* from Olduvai Gorge, Tanzania.

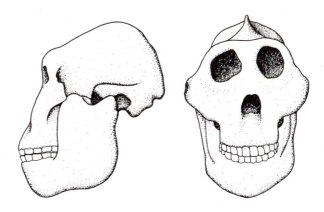

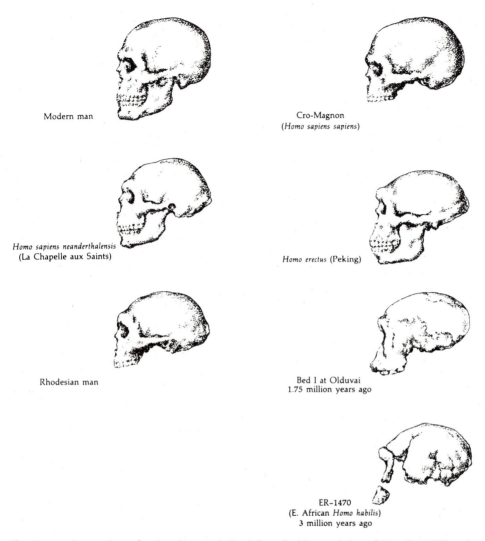

Modern man

Cro-Magnon
(*Homo sapiens sapiens*)

Homo sapiens neanderthalensis
(La Chapelle aux Saints)

Homo erectus (Peking)

Rhodesian man

Bed I at Olduvai
1.75 million years ago

ER-1470
(E. African *Homo habilis*)
3 million years ago

Fig. 27.22 Comparison of various human skulls. Adapted with permission of Howells, William, *Evolution of the Genus Homo*. Menlo Park, Calif.: The Benjamin/Cummings Company, Inc., 1973.

curred over much of East Africa during the last several million years.

Australopithecus, we now realize, coexisted with *Homo* for much of its history. The earliest representatives of *Homo* are older than the majority of *Australopithecus* remains. If we did descend from *Australopithecus,* it was

from the most primitive members of that genus, and not from the more advanced ones.

Homo habilis

Subsequent to the discovery of the large East African australopithecines, another, smaller hominid was discovered at Olduvai Gorge. The new hominid (Fig. 27.23) was quite small, and its brain, at 600 cm³, was within the size range of *Australopithecus,* yet well above the average (400 cm³). Louis Leakey named this new hominid *Homo habilis.* The discovery of *Homo* in deposits also containing *Australopithecus boisei* was of great importance in showing that two or more hominids could indeed live together contemporaneously. Previous generations of paleoanthropologists had doubted this, claiming that competition between hominid species would have been so intense that only one species could possibly survive if two had ever coexisted.

A most important discovery was made by Richard Leakey, the son of Louis and Mary Leakey. His specimen (no. 1470) was dated radiometrically at 3.2 million years, nearly a million years older than the *Australopithecus* deposits of South Africa. The specimen shows many structural features that we might expect in a common ancestor to *Australopithecus* and *Homo*, but is definitely closer to the latter genus. The discovery of this specimen confirmed Louis Leakey's earlier belief in the great antiquity of the genus *Homo.* According to Richard Leakey, specimen no. 1470 shows that *Homo* and *Australopithecus* were distinct lineages over the last 2 to 3 million years, and that neither genus descended from the other. Pilbeam and Gould (1974) have shown the estimated cranial capacity of this specimen to be exactly what a common structural ancestor to the two genera should have had: from a brain the size of this specimens', the phylogenetic increase in brain size took place allometrically in both *Australopithecus* and *Homo*, but with greater positive allometry (i.e., more rapidly) in the latter case.

Two specimens from the lower Pleistocene Trinil beds of Indonesia have been tentatively referred to *Homo habilis* by Tobias and von Koenigswald (1964). Their similarity to the teeth of *Homo habilis* from Africa is great, but no other cranial or postcranial remains are associated. If the Indonesian fossils do belong to *Homo habilis,* it shows that this species was already geographically widespread some 1.7 million years ago.

Homo habilis was a very small hominid, with a cranial capacity ranging from 400 cm³ in the earlier forms to around 700 cm³ in the later ones. This is far below the cranial capacity normally associated with other species of *Homo*, but Louis Leakey has pointed out that present-day microcephalic individuals of *Homo sapiens* have no less than normal intelligence with cranial capacities as low as 400 cm³. Louis Leakey claimed that the cranial capacity of *Homo habilis* was sufficient for the possession of culture, but as yet no cultural remains have been discovered associated with fossils of this species.

Homo erectus

Hominids from the middle Pleistocene are known from various localities on three continents (Asia, Africa, Europe), but the best-preserved remains are in Asia. The earliest discoveries were made in Java during the 1890s; these were designated as *Pithecanthropus erectus.* Present-day authorities would now place them in the genus *Homo* and call them *Homo erectus. Homo erectus* was a geographically widespread species whose known occurrences include sites in Germany, Hungary, Morocco, Tanzania, and perhaps even South Africa in addition to China and Indonesia.

Homo erectus (Fig. 27.22) was fully erect in posture, and its postcranial skeleton was virtually indistinguishable from modern *Homo sapiens*. The skull was somewhat flattened, at least in the Asian forms (from Java and Peking), with the occiput or back end drawn out posteriorly. The bones were often heavy, and the brow ridges

Fig. 27.23 *Homo habilis* from Olduvai Gorge, Tanzania: (a) lower jaw; (b) portion of left foot. Reprinted by permission from McKern, Sharon, and McKern, Thomas, *Living Prehistory, An Introduction to Physical Anthropology and Archaeology.* Menlo Park, Calif.: Cummings Publishing Company, (now the Benjamin/Cummings Publishing Company), p. 136, fig. 11.

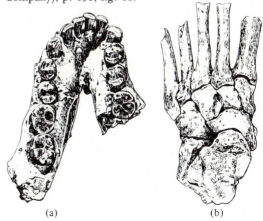

(a)   (b)

(supraorbital tori) were thick. Compared to *Australopithecus,* the stature was increased from under 140 to over 165 cm (from about 4½ to 5½ feet), and the brain was enlarged to a range from about 750 to just under 1200 cm^3, with an average close to 1000 cm^3. The increase in the brain capacity had, however, taken place from side to side and front to back only, not vertically, thus explaining the broad but flat shape of the braincase.

Cultural remains associated with *Homo erectus* include stone chopping tools of several different types. These differed somewhat from place to place, but were always more sophisticated than the Oldowan pebble-tools associated with *Australopithecus.* Stones found associated with *Homo erectus* sites often show evidence of having been heated (some of their fractures seem heat induced, for example), and this in turn suggests the use of fire. Charred animal bones have also been found at one site (Choukoutien), indicating that animal food may have been cooked before eating.

Homo sapiens

Toward the close of the second or Great Interglacial period, a new species, *Homo sapiens*, made its appearance. At Steinheim in Germany, Swanscombe in England, Tabun and Skhul in Israel, Shanidar in Iraq, Broken Hill in Zambia, Ngangdgong in Indonesia, and several other localities, early remains of *Homo sapiens* are preserved. The major difference separating all these forms from *Homo erectus* is an increase in the vertical height of the cranial vault, associated with an increase in cranial capacity beyond the rather arbitrary criterion of 1200 cm^3. In some of these forms, the heavy brow ridges also became reduced, but in certain others the brow ridges remained strong and may even have thickened a bit more.

The third (Riss/Würm) interglacial period of Europe, and the subsequent (Würm or fourth) glacial period, are marked by a peculiar group of humans (Fig. 27.23) known as Neanderthals, or more specifically as Classical Neanderthals. The brow ridges were prominent in these forms, and the bones were on the whole thicker and more heavy-set. It was previously thought that Neanderthals were only semi-erect, and walked stooped over, a myth that persists in most present-day cartoons of "cave men." We now realize that this early misconception was largely due to the pathological status of the original finds, one of which apparently included an arthritic individual.

That the Neanderthals were skillful hunters is shown by their association with the bones of very large game animals (reindeer, mammoth), and also with cave bears. That they were culturally advancd is shown by their tool culture (of the Mousterian type) and by their burial of the dead in preferred orientations. Quite unlike the brutish "cave-man" image left to us by past generations of anthropologists, Neanderthals were quite modern in both cultural and morphological terms. It has even been said that if a Neanderthal man were given a bath, a shave, and a suit of clothes, and paraded down a street in any of our major cities, he would scarcely attract any attention as being any different from the other pedestrians. To be sure, if one were to look carefully, one would probably notice the more heavy-set face, especially the brow ridges, but from the neck down even a careful medical check-up would probably reveal nothing very different from any modern human.

The second half of the fourth glacial period, in Europe at least, is dominated by a more modern form of *Homo sapiens,* the so-called Cro-Magnon populations (Fig. 27.23). With these fossils are associated a number of more advanced stone tools and various symbolic forms of art including both sculpture and cave painting. The great cave paintings at Lascaux and Altamira were undoubtedly painted by the same Cro-Magnon people that ushered in the late Paleolithic cultural period.

What Made Us Human?

What is it that has made us distinctively human? What sets us most clearly apart from the other animals? How did our peculiarly human traits evolve?

Among those features that set us apart from the other animals, we may mention upright locomotion, tool use, language, culture, the ability to reason, and the retention of certain juvenile traits throughout life. Most discussion through the years has emphasized brain development on the one hand, and upright locomotion and tool use on the other. The theory that brains are more important may have been initiated by Aristotle when he characterized man as a "rational animal." But the ability to reason is present in chimpanzees and other apes, if by "ability to reason" we mean the ability to figure out in advance a multi-step procedure that has never been performed before, and correctly predict its consequences (by "insight," Chapter 24). Culture does reach a pinnacle in humans, exalted several magnitudes in complexity over that of any nonhuman primate; yet traditionally transmitted behavior occurs even in rhesus monkeys.

Language is a distinctively human trait. Symbolic expression is the key to human language, and, although symbolic expression can be taught to apes, it is natural to humans alone and is exhibited by every human society. The importance of symbolism in the evolution of human culture has been repeatedly emphasized by many anthropologists, notably Leslie White.

Still, our cultural prejudice tells us that brains are of overriding importance. Grafton Elliot Smith, a turn-of-the-century neurologist, argued forcefully for our cerebral uniqueness, of which, he said, upright posture and the use of tools are mere outward manifestations. Smith's cultural prejudice led him astray when a skull with a modern brain and an apelike jaw was found near Piltdown in England. Smith took the Piltdown skull as proof that our brain had evolved first, but the skull was later found to have been a hoax. The same cultural prejudice led many anatomists to reject *Australopithecus* as a hominid in the first decades after its discovery, for its brain was the size of an ape's, and its other claims to humanity were therefore not taken seriously. But our knowledge of *Australopithecus,* and of early *Homo,* has improved greatly and shows beyond all doubt that small-brained creatures were already walking erect and using tools millions of years ago. Our cultural prejudice, in other words, has taught us the wrong thing.

In the turn-of-the-century heyday of Haeckel's theory of recapitulation, Louis Bolk proclaimed that we evolved by a process of "fetalization"—the retention of fetal and juvenile characteristics into our adult lives. We are, according to this theory, half-grown apes; the largeness of our brain and the flexure of our brain upon our spinal cord are but two fetal characters that we have extended into our adult lives. Gould (1977) has supported this theory anew, claiming that the enlargement of our brains was nothing more than the prolongation of fetal growth rates far beyond birth.

The theory that standing up is more important than brains is at least as old as Plato's characterization of humans as "bipeds without feathers." Benjamin Franklin was quoted by Boswell as characterizing us as "tool-making animals." Friedrich Engels devoted an entire chapter of his posthumous *Dialektik der Natur* to an essay entitled "The role played by labour in the evolution of man"; in this chapter he outlines the theory that upright posture freed our hands for the use of tools and was thus the impetus for our evolution. The fossil evidence has now vindicated this viewpoint by showing that small-brained ancestors already walked upright and used tools.

Animal remains and simple tools preserved at *Australopithecus* sites show that these hominids were already very effective hunters of antelope, baboon, and other game. Their language, culture, or reasoning ability did not fossilize, but all these are implied by the cooperation and foresight needed for successful hunting. Cooperative hunting was made possible by upright locomotion, and it was this locomotion that set us apart from the apes initially. Upright locomotion is characteristic of all hominids, and was the "key character" in the origin of the Hominidae. Its selective advantage arose from the need to run fast—in the escape as well as in the hunt. Since chimpanzees use a few simple tools, we may assume that hominid ancestors did so, too. But the further and habitual use of tools became possible only by the freeing of the hands from any use in locomotion. Likewise, the carrying of tools from place to place, or the fashioning of tools in anticipation of an indefinite future use, are both distinctively human traits that were part of the hunting ecology of early hominids. The requirements of the hunt dictated the selective advantages of tool use, foresight, rational behavior, cooperation, and planning. Cultural evolution (treated further in the next chapter) and the development of language facilitated cooperation even further. It was all of these things together that made us distinctively human.

E. HUMAN RACES

In addition to being geographically widespread, *Homo sapiens* is geographically variable. The major geographical subdivisions of *Homo sapiens* are called races. A race may be defined as a biologically distinct subdivision of a species, inhabiting a definable geographic region and possessing certain characteristic gene frequencies. A race in this sense is nearly the same as a geographic subspecies. More traditional definitions of "race," like more traditional definitions of "species," were rooted in essentialism and typology, and tended to emphasize obvious physical characteristics such as skin color and hair form.

Most misconceptions about races involve the scientific error or fallacy of typological thinking, under which a race is considered a collection of individuals conforming more or less to a Platonic "type" or *eidos*. The fallacy is dispelled when we realize that a race is not just a collec-

Fig. 27.24 A case of albinism, a recessive human trait occurring in all racial groups with very low frequency. Mrs. Olivia Stone is holding her twin children, one of whom is an albino. This case and others demonstrate that race cannot be based on skin color only, or on any other single feature. Photo no. 930097, courtesy of United Press International.

tion of individuals, but a group of interbreeding populations. Populations belonging to different races often differ in certain physical features. It is a mistake, though, to state that *all* members of a particular racial group have certain characteristics; such a statement betrays typological thought as well as a frequent element of prejudice (Fig. 27.24).

Racial Diversity

In describing the racial variation of the human species, it is convenient to have names by which different racial groups can be designated. This means, in effect, a racial classification, but (alas for the novice) no two anthropologists seem able to agree on exactly the same classification—a failure that itself constitutes an argument against the existence of racial "types."

Racial classifications have been published by competent anthropologists who recognize as few as two or three racial groups, and by others who recognize 34 or even more (Box 27–E). Moreover, these racial groups can be described only in terms of averages, for a broad overlap exists between any two racial groups in almost every feature. If a particular racial group has a high average value of some measured characteristic, there will always be many individuals within that race having low values that are neither pathological nor the results of interracial mixture.

The barest of racial classifications recognize the Caucasoids ("whites") of Europe and the Near East, the Negroids ("blacks") of Africa, and the Mongoloids of Asia as distinct racially. The aborigines of Australia are a racially distinct group whose closest resemblances seem to be to the Caucasoids of the Indian subcontinent. The so-called Bushmen of Africa are racially distinct from surrounding Negroid populations; their somewhat more reddish skin and "peppercorn" hair serve to highlight their separateness. The Mongoloids, who by nearly any biological criterion (numbers, adaptability) must be reckoned as the most successful of racial groups, include not only the rather yellow-skinned inhabitants of Asia north and east of the Himalayas, but also the Eskimos and the more ruddy-skinned "Indians" of the Americas. The Polynesians and other Pacific islanders are hard to place in any racial classification, and their points of resemblance to Mongoloids or to Australians are variously interpreted.

BOX 27-E *A Sampling of Different Racial Classifications*

Five Races (Boyd, 1950)

1. *European (Caucasoid)*—high frequencies of Rh cde and CDe, moderate frequencies of the other blood-group genes; M usually slightly above and N below 50 per cent

2. *African (Negroid)*—very high frequency of Rh cDe, moderate frequencies of the other blood-group genes

3. *Asiatic (Mongoloid)*—high frequency of B, few if any cde

4. *American Indian*—mostly homozygous O but sometimes high frequencies of A; absence of B, few if any cde, high M

5. *Australoid*—moderate to high A, few or no B or cde, high N

Nine Races (Garn, 1961)

1. *Amerindian*—the pre-Columbian populations of the Americas

2. *Polynesian*—islands of the eastern Pacific, from New Zealand to Hawaii and Easter Island

3. *Micronesian*—islands of the western Pacific, from Guam to Marshall and Gilbert Islands

4. *Melanesian-Papuan*—islands of the western Pacific, from New Guinea to New Caledonia and Fiji

5. *Australian*—Australian aboriginal populations

6. *Asiatic*—populations extending from Indonesia and Southeast Asia, to Tibet, China, Japan, Mongolia, and the native tribes of Siberia

7. *Indian*—populations of the subcontinent of India

8. *European*—populations of Europe, the Middle East, and Africa north of the Sahara; now world wide

9. *African*—populations of Africa south of the Sahara

Thirty-four Races (Coon, Garn, and Birdsell, 1950, as modified by Dobzhansky, 1962)

1. *Northwest European*—Scandinavia, northern Germany, northern France, the Low Countries, United Kingdom, and Ireland

2. *Northeast European*—Poland, Russia, most of the present population of Siberia

3. *Alpine*—from central France, south Germany, Switzerland, northern Italy, eastward to the shores of the Black Sea

4. *Mediterranean*—peoples on both sides of the Mediterranean, from Tangier to the Dardanelles, Arabia, Turkey, Iran, and Turkomania

5. *Hindu*—India, Pakistan

6. *Turkic*—Turkestan, western China

7. *Tibetan*—Tibet

8. *North Chinese*—northern and central China and Manchuria

9. *Classic Mongoloid*—Siberia, Mongolia, Korea, Japan

10. *Eskimo*—arctic America

11. *Southeast Asiatic*—South China to Thailand, Burma, Malaya, and Indonesia

12. *Ainu*—aboriginal population of northern Japan

13. *Lapp*—arctic Scandinavia and Finland

14. *North American Indian*—indigenous populations of Canada and the United States

15. *Central American Indian*—from southwestern United States, through Central America, to Bolivia

16. *South American Indian*—primarily the agricultural peoples of Peru, Bolivia, and Chile

17. *Fuegian*—nonagricultural inhabitants of southern South America

18. *East African*—East Africa, Ethiopia, a part of Sudan

19. *Sudanese*—most of the Sudan

20. *Forest Negro*—West Africa and much of the Congo

21. *Bantu*—South Africa and part of East Africa

22. *Bushman and Hottentot*—the aboriginal inhabitants of South Africa

23. *African Pygmy*—a small-statured population living in the rain forests of equatorial Africa

24. *Dravidian*—aboriginal populations of southern India and Ceylon

25. *Negrito*—small-statured and frizzly-haired populations scattered from the Philippines to the Andamans, Malaya, and New Guinea

26. *Melanesian-Papuan*—New Guinea to Fiji

27. *Murrayian*—aboriginal population of southeastern Australia

28. *Carpentarian*—aboriginal population of northern and central Australia

29. *Micronesian*—islands of the western Pacific

30. *Polynesian*—islands of the central and the eastern Pacific

31. *Neo-Hawaiian*—an emerging population of Hawaii

32. *Ladino*—an emerging population of Central and South America

33. *North American Colored*—the so-called Negro population of North America

34. *South African Colored*—the analogous population of South Africa

The majority of African populations south of the Sahara form a distinct racial group, variously called Negroid, Negro, Congoid, African, or "Black." In these populations, the level of melanin in the skin is usually high, and the skin color is therefore a very dark brown or nearly black. The hair is almost always black, with an extreme tendency to curl dictated in part by the narrow diameter of each fiber. There is a high frequency (usually near 60%) of the Rh blood group cDe, with lesser frequencies of cde, cDE, and CDe (in that order), while CDE, a genotype present with low frequency in nearly all other races, is virtually absent (Box 27–F). In the ABO blood group system, the frequency of A is relatively low (10–20%). Alleles M and N in the MN blood group system are about equally prevalent. Hemoglobin S is present in frequencies between 5 and 20% in most populations (in other races it is quite rare). Skulls are on the average, are narrow (dolichocephalic).

The Bushmen and Hottentot populations of southern Africa are sometimes considered distinct from the other inhabitants of the African continent. Hairs are black and narrow in diameter, but they tend to curl in a horizontal rather than a vertical plane, forming individual whorls or "peppercorn" hair rather than a continuous dense mat. Among Rh blood groups, cDe is most common (nearly 90%), while cde (normally the second most frequent genotype in other African populations) is almost unknown. Skulls are on the average broader than in other African populations, and the brownish skin has a slightly reddish tinge, two features that in the past have been mentioned as evidence for Mongoloid affinities. Hottentot women often show a remarkable growth of fat deposits on the buttocks, a condition called steatopygia.

The Caucasoid racial group, variously called Caucasian, "White," or European, comprises the native populations of Europe, northern Africa, southwestern Asia, and the Indian subcontinent. There is considerable variation within this widespread group of populations but the hair, often described as "wavy," rarely reaches the extremes of straightness and thickness seen in Mongoloids or those of curliness and thinness seen in Negroids. Hair color varies greatly and may be blond, red, brown, or black, with brown hair color most prevalent in Europe and black most prevalent among native Caucasoids elsewhere. Skin color varies also, but levels of melanin are generally low in most European populations.

This type of skin is most often described as "white," but its slight transparency allows the red color of blood to show through somewhat. Among Rh blood groups, CDe and cde are most common. In the ABO blood group system (Fig. 27.25), O is most frequent (about 55–70%), A moderately so (about 20–30% in most populations), and B variable (from near-zero among Basques to about 30% in northern India). In the MN blood group system, M slightly outnumbers N in most populations. Skulls are on the average narrow (dolichocephalic).

The Mongoloid racial group, also called Asian, Asiatic, or Oriental, includes most of the inhabitants of Asia north of the Himalayas and east of Assam. The Eskimos of Arctic America are also included in this group. Partially distinct from the Asian Mongoloids, but presumably related to them, are the native inhabitants of the Americas, the so-called American Indians or "Amerindians." Though they are distinct in other ways, Amerindians share with Asian Mongoloids the following racial features: high frequency of CDe and cDE, low frequency of cde, skulls broader on the average than in other racial groups, a thickness of the outer layer (stratum corneum) of the skin that prevents the blood vessels from showing through, a tendency to produce yellowish and reddish pigments (phaeomelanins), dark, coarse hair, and a generally flattened face with a nonprotruding nose. This last feature developed more in the Asian than in the American populations.

In Asian Mongoloids, the generally yellow appearance of the skin (it actually varies from a pale near-white to a deep brown) results from the combination of phaeomelanins with the failure of the blood to show through the thick stratum corneum. Also characteristic of Asian Mongoloids are the high frequency of the blood group B, and the presence of an **epicanthic fold** over the eyes. This fold, over the inner corner of each eye, seems to give to the eyes a "slanted" appearance. The noses of Asian Mongoloids are especially flat, with nostrils more constricted than in other racial groups. These features have been explained as adaptations to the cold, windy interior desert regions of central Asia, though only a small fraction of Asian Mongoloids live under these conditions today.

The American Indians differ from Asian Mongoloids in having generally a more reddish skin, less flattened face and nose, a higher proportion of blood group

BOX 27–F *Selected Blood Group Frequencies for Specified Racial Groups*

Table 1 *ABO* Blood Group System

Population	Number of persons tested	i	I^A	I^B
Americans (white)	20,000	0.67	0.26	0.07
Icelanders	800	0.75	0.19	0.06
Irish	399	0.74	0.19	0.07
Scots	2,610	0.72	0.21	0.07
English	4,032	0.71	0.24	0.06
Swedes	600	0.64	0.28	0.07
French	10,433	0.64	0.30	0.06
Basques	400	0.76	0.24	0.00
Swiss	275,644	0.65	0.29	0.06
Croats	2,060	0.59	0.28	0.13
Serbians	6,863	0.57	0.29	0.14
Hungarians	1,500	0.54	0.29	0.17
Russians (Moscow)	489	0.57	0.25	0.19
Hindus	2,357	0.55	0.18	0.26
Buriats (N. Irkutsk)	1,320	0.57	0.15	0.28
Chinese (Huang Ho)	2,127	0.59	0.22	0.20
Japanese	29,799	0.55	0.28	0.17
Eskimos	484	0.64	0.33	0.03
American Indians (Navajo)	359	0.87	0.13	0.00
American Indians (Blackfeet)	115	0.49	0.51	0.00
W. Australians (aborigines)	243	0.69	0.31	0.00
African Pygmies	1,032	0.55	0.23	0.22
Hottentots	506	0.59	0.20	0.19

SOURCE: Sinnott, Dunn, and Dobzhansky, Copyright © 1958. *Principles of Genetics,* 5th ed. New York: McGraw-Hill Book Company.

Table 2 *Rh* Blood Group System

Population	CDE	CDe	CdE	Cde	cDE	cdE	cDe	cde
EUROPE								
English	0.1	43.1	0	0.7	13.6	0.8	2.8	38.8
Danes	0.1	42.2	0	1.3	15.1	0.7	1.8	38.8
Germans	0.4	43.9	0	0.6	13.7	1.0	2.6	37.8
Italians	0.4	47.6	0.3	0.7	10.8	0.7	1.6	38.0
Spaniards	0.1	43.2	0	1.9	12.0	0	3.7	38.0
Basques	0	37.6	0	1.5	7.1	0.2	0.5	53.1

(continued)

BOX 27-F *(continued)*

Table 2 *Rh* Blood Group System

Population	CDE	CDe	CdE	Cde	cDE	cdE	cDe	cde
AFRICA								
Egyptians	0	49.5	0	0	9.0	0	17.3	24.3
Hutu	0	8.3	0	1.6	5.7	0	62.9	21.6
Kikuyu	0	7.3	0	1.4	9.9	1.4	59.5	20.4
Shona (Rhodesia)	0	6.9	0	0	6.4	0	62.7	23.9
Bantu (S. Africa)	0	4.7	0	5.8	8.5	0	59.6	21.4
Bushmen	0	9.0	0	0	2.0	0	89.0	0
ASIA								
Yemenite Jews	0.5	56.1	0	1.0	7.9	0	6.4	28.2
East Pakistan	1.6	63.3	0	6.5	7.6	0	3.9	17.1
South Chinese	0.5	75.9	0	0	19.5	0	4.1	0
Japanese	0.4	60.2	0	0	30.8	3.3	0	5.3
AUSTRALASIA								
Australian aborigines	2.1	56.4	0	12.9	20.1	0	8.5	0
Papuans	1.6	94.4	0	0	2.0	0	2.0	0
Javanese	1.2	84.0	0	0	8.3	0	6.5	0
Marshallese	0	95.1	0	0	4.4	0	0.5	0
AMERICA								
Eskimos (Greenland)	3.4	72.5	0	0	22.0	0	2.1	0
Ojibwa	2.0	33.7	0	0	53.0	3.2	0	8.0
Blood	4.1	47.8	0	0	34.8	3.4	0	9.9
Navajo	1.3	43.1	0	0	27.7	0	28.0	0

SOURCE: Based on data from Mourant (1954).

M and of blood group *O* (nearly 100% in many populations), the absence of blood group *B*, and the lack of the epicanthic fold. The frequency of blood group *A* is most often very low, but occasionally very high.

The Aborigines of Australia are dark-skinned, but in other respects they seem closer to the Caucasoids than to any of the other major racial groups. For instance, among the Rh blood groups, *CDe* is the most frequent type and *cDE* is also frequent, both traits more reminiscent of Caucasoids (or Mongoloids) than Negroids. The hair has a thickness and wavy texture more similar to that of Caucasoids than to other races. Hair color is often dark but varies greatly; the relatively high incidence of blond hair provides a further resemblance to Caucasoids. The head is narrow, on the average, like that of Caucasoids or Negroids rather than Mongoloids. There are, in addition,

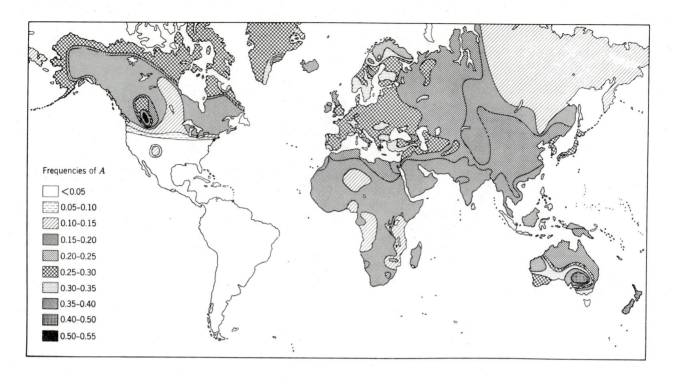

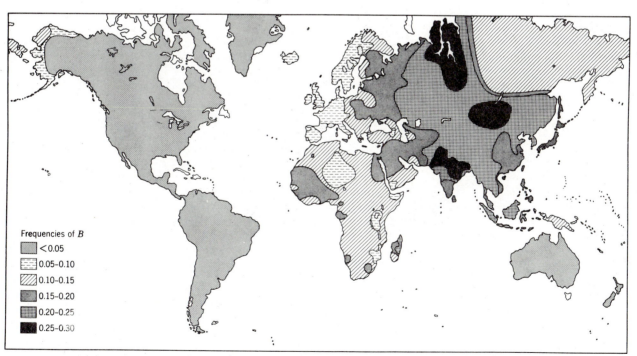

Fig. 27.25 Frequency maps of various blood groups in aboriginal populations of the world, up to about A.D. 1500. Reprinted with permission from John Buettner-Janusch, *Origins of Man,* © 1966, New York: John Wiley & Sons, Inc.

533

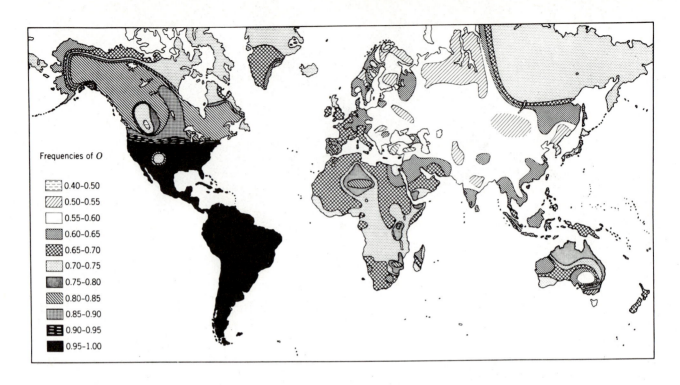

Frequencies of O

- 0.40–0.50
- 0.50–0.55
- 0.55–0.60
- 0.60–0.65
- 0.65–0.70
- 0.70–0.75
- 0.75–0.80
- 0.80–0.85
- 0.85–0.90
- 0.90–0.95
- 0.95–1.00

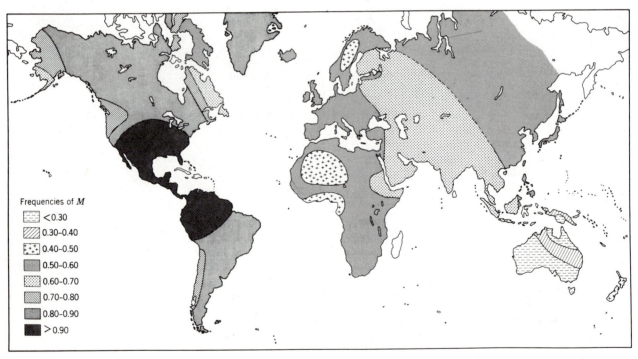

Frequencies of M

- < 0.30
- 0.30–0.40
- 0.40–0.50
- 0.50–0.60
- 0.60–0.70
- 0.70–0.80
- 0.80–0.90
- > 0.90

Fig. 27.25 (continued)

534

many unique features that set the Australians apart from any other racial group: total or near total absence of blood groups *B, cde,* and *cdE,* and high relative frequencies of *A* and *N.* The language and culture of the Australians also suggests that they have long been separated from contact with other racial groups, an isolation that may have allowed considerable genetic drift and deviation by selection from other racial groups.

The islands of the Pacific are peopled by a number of populations that are most difficult to place in any racial classification. These peoples show some resemblances to both Mongoloids and Australoids, including high frequencies (around 95%) of *CDe,* and they are further remarkable in a nearly total lack of several other types (*CdE, Cde, cdE, cde*), some of which are also absent in Asia and some in Australia. But the Pacific peoples are widely scattered, and their languages fall into three groups (Melanesian, Micronesian, Polynesian) that parallel to some extent their racial features. Racial origin of Pacific people from American Indians has been claimed, as has their origin as the result of "hybridization" of other races.

Are Racial Differences Adaptive?

The existence of a human species subdivided into races raises the question of how our species came to be so subdivided and so variable. Did the races evolve? We believe that they did. Are their differences adaptive? There is ample evidence to show that they are, at least to a large extent.

The existence of microgeographic variation, that is, variation *within* each major racial group, shows that racial variation follows a pattern typical of geographic variation in general. There is, say, among European Caucasoids, a considerable amount of variation within each population, but also a consistent variation from one population to the next in such features as stature (standing height), hair color, skin color, and blood group frequencies. Cavalli-Sforza found, for example, considerable variation among the blood-group frequencies of various populations from a single country, Italy (Fig. 27.26).

More remarkable, though, is the fact that this microgeographic variation takes a very definite form: it follows the widely known ecogeographic rules (Chapter 13) to a very great extent. Skin color, for example, is at its lowest average value in Scandinavia, and increases progressively as we move south through Central Europe to the shores of the Mediterranean, then East through the Middle East, Iran, and Pakistan, then south through India to Sri Lanka. This great and rather continuous gradation within members of a single major racial group (Caucasoid) is paralleled also among Asian Mongoloids (increasing in pigmentation, say, from Japan to Indonesia). These clines would seem to be following Gloger's rule on color, and it might be claimed that in general Caucasoids are adapted to cold, moist areas (therefore lightly pigmented), Mongoloids to dryness, and Negroids to warm, moist areas (therefore darkly pigmented). Conformity to Bergmann's and Allen's rules also occur: cold-adapted populations tend to have larger body sizes and shorter limbs than comparable warm-adapted populations; on the average, the limbs of Mongoloids are shorter while those of certain Negroid populations are much longer.

Variations in physiological traits also support the hypothesis that racial differences are adaptive. Most convincing is the evidence based on tolerances to environmental extremes. During the Korean War (1950–1954), for example, the U.S. Army noticed an interesting and previously undetected difference between Euro-American ("white") and Afro-American ("black" or "Negro") soldiers, all generally healthy males, ages 18–35: in the bitter cold of the Manchurian winter, Afro-American soldiers were much more susceptible to frostbite than were Euro-American soldiers. Classic physiological studies by German physiologists using European Caucasoid subjects had shown that, as the surrounding air temperature dropped, blood vessels to the extremities would generally constrict (vasoconstriction), but that this trend reversed itself when the temperature was lowered still further, the vasodilation helping to protect the hands and feet against frostbite. Apparently, in a large number of Afro-American subjects, the vasoconstriction never reversed itself, and frostbite was all too often the result. Subsequent experiments have confirmed this racial difference, and have also shown that Eskimos are among the best equipped physiologically to withstand cold-temperature stress: when the hands and arms are immersed in ice-cold water, the body temperature of Eskimos drops the least, on the average.

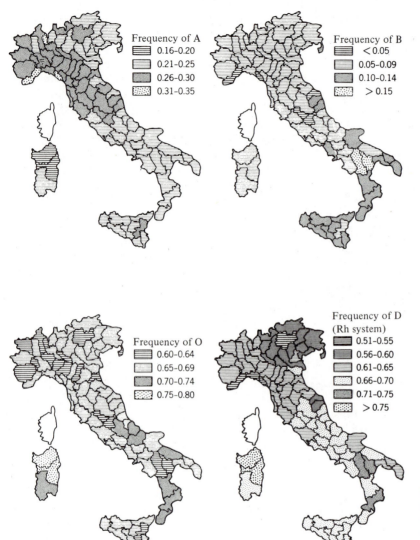

Fig. 27.26 Microgeographic variation of blood group frequencies in Italy. Reprinted with permission from John Buettner-Janusch, *Origins of Man,* © 1966, New York: John Wiley & Sons, Inc.

In warm, tropical or semi-tropical climates, on the other hand, Afro-Americans have proven physiologically more well-adapted than Euro-Americans, on the average, in terms of temperature-regulating ability and resistance to heat prostration. In controlled U.S. Army tests, Afro-American soldiers proved better able to withstand strenuous outdoor labor under the hot summer sun, without overheating. The Euro-American soldiers, on the other hand, lost weight faster (on the average) under these conditions, and more often collapsed unconscious.

That most conspicuous of all racial features, skin color, may itself be explained as the result of a physiological adaptation. Under tropical conditions, there is more than ample sunlight to achieve vitamin D synthesis beneath the skin, but there is a danger of skin cancer from overexposure to the sun's ultraviolet radiation. Melanin in the skin masks out much of the sun's ultraviolet rays, which accounts for the generally dark skins not only of Negroids, but also of Caucasoids (in India and Sri Lanka), Mongoloids (in Indonesia and the Philippines), and Pacific peoples (Melanesians in New Guinea and the western Pacific) inhabiting tropical climates. On the other hand, the less direct rays of the sun in more northern latitudes, combined with the frequent cloud cover in Europe, screen out much of the ultraviolet, and make vitamin D synthesis the more critical factor. The native peoples of Europe have therefore lightly pigmented skin that absorbs a maximal amount of sunlight for vitamin D synthesis, but even here a sunburning or suntanning mechanism prevents overexposure of the skin to ultraviolet radiation, either by producing temporary skin pigmentation (tanning) or by producing a painful sensation that causes most subjects to avoid any more exposure to the sun. The skin of Eskimos is considerably more opaque to sunlight, but Eskimos traditionally tend to eat large quantities of whole fresh fish, containing large amounts of fully synthesized vitamin D, and their need for sun exposure is thus greatly lessened.

Other racial traits that have been interpreted as climatic adaptations include the densely matted, curly hair of Negroids, which is said to serve as sort of a natural "pith helmet," protecting the head against the falling objects more often encountered in tropical climates. The epicanthic folds, flat faces, and restricted nostrils of Mongoloid peoples have all been attributed to adaptive protection against the harsh, dry desert winds and sandstorms of the Central Asian desert regions, an explanation that is satisfactory only if one assumes such desert regions to have been the ancestral home of the Mongoloid races. The "pygmy" races of tropical forest regions may also be mentioned in support of the climatic adaptation hypothesis: tropical forest peoples, of whatever racial origin, tend to be of rather short stature, dark skin, and densely matted hair, a combination which recurs among the (Negroid) Congo Pygmies of Zaire, the (probably Mongoloid) "Negrito" of the Philippines and Malaysia, the short-statured Caucasoids of Sri Lanka and southern India, the Amazonian Indians, and the Melanesians.

Probably most convincing from a scientific point of view is the evidence provided from studies of genetic polymorphism. These studies reveal consistent differences among racial groups in gene frequencies for polymorphic traits such as blood groups and, more important, for genes carrying immunity or partial immunity to certain diseases. The gene for sickle-cell anemia, for example, is more prevalent in those regions where malaria occurs, particularly in Africa. Malaria resistance in Mediterranean latitudes, in an arc extending from India to Italy, is brought about largely by a totally different gene, responsible for the disease known as thalassemia (or Cooley's anemia). The predominant malaria-resistant gene of Negroid populations is thus different from the predominant malaria-resistant gene of Caucasoids.

Other genes have racially restricted or racially prevalent distributions, suggesting climatic adaptations, but in most cases the adaptive advantages (probably restricted to heterozygotes) remain to be elucidated. This is certainly the case for the *ABO* blood group alleles (known on statistical grounds to favor greater or lesser incidence of a wide variety of diseases), Rh alleles, favism (deficiency of the enzyme G6PD, glucose-6-phosphate dehydrogenase, prevalent among Mediterranean Caucasoids), Tay-Sachs disease (infantile amaurotic idiocy, prevalent among Ashkenazic Jews), and others. Traits that are equally rare among all racial groups (including albinism, phenylketonuria, alkaptonuria, and others) may never or only rarely confer any adaptive advantage, under any climatic condition.

The adaptiveness of racial differences in relation to climate provides only a partial answer to the question of racial origins. The most favored condition in each climate consistently and repeatedly survived in greater numbers and became more and more prevalent. But a further inquiry into the origin of particular races can be answered only through a study of fossils.

Theories of Racial Origins

There are two general and mutually exclusive theories of racial origins, which may be termed the monophyletic (or "hat rack") school and the polyphyletic (or "candelabra") school. The monophyletic school was most popular during the first half of the present century. Its adherents taught that racial differences were of very recent origin. Adherents to this school debated among themselves as to the location of the "ancestral home" of *Homo sapiens,* presuming that there had to be one. At each grade or level of human evolution, it was held that only one population could be the "real" ancestor of later

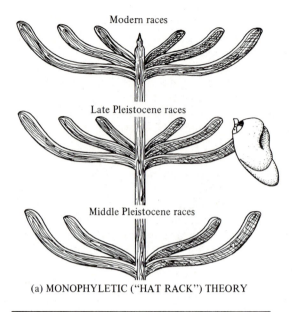

(a) MONOPHYLETIC ("HAT RACK") THEORY

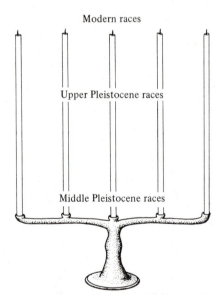

(b) POLYPHYLETIC ("CANDELABRA") THEORY

Fig. 27.27 Comparison between monophyletic (a) and polyphyletic (b) theories of racial origins.

stages, while the others were sterile side branches. The resultant phylogenies tended to resemble a certain kind of hat rack (Fig. 27.27a). According to the monophyletic school of thought, the "ancestral" race should be the oldest and most primitive, and other races should be derived from it. Although various suggestions were put forth, no substantial agreement was ever reached as to which race was supposed to be primitive or ancestral. Of course, where science leaves off, popular prejudices may well take over, and many scientists of past generations were able to find "scientific" evidence for a greater abundance of "primitive" or "advanced" features in this race or that, almost always confirming that their own racial group was the most advanced. Many of the fallacies in their arguments are explained by Gould (1977, 1981).

Although monophyletic theories had traditionally been favored, there were important dissenters, too, including Franz Weidenreich, Roy Chapman Andrews, and Carleton Coon. These authors instead favored polyphyletic theories, according to which the races of humankind are very old—at least as old as *Homo sapiens,* or perhaps older. Humans have always been geographically variable, and racial differences are part and parcel of this geographic variation. In its modern form, the polyphyletic theory of racial origins assumes that *Homo erectus* had been a geographically variable species, and that each race (or subspecies) of *Homo sapiens* could trace (a large proportion of) its ancestry from one of the races of *Homo erectus.* Phylogenetic trees drawn by adherents of this school resemble candelabras instead of hat racks (Fig. 27.27b), and they tend to designate each and every fossil population as ancestral to *some* modern race.

Coon (1962) places all mid-Pleistocene and later hominids in two species, *Homo erectus* (earlier) and *Homo sapiens* (later), using a cranial capacity of 1200 cm³ as the arbitrary boundary between them. Each of these species is divided by Coon into five racial groups (subspecies), with the five races of *Homo erectus* each ancestral to one of the five races of *Homo sapiens* (Fig. 27.28). Coon's five races are called Capoid (Bushmen and Hottentots), Congoid (other indigenous peoples of sub-Saharan Africa), Caucasoid, Mongoloid, and Australoid. Coon refers various fossils to these racial groups: Peking Man, for instance, is called a Mongoloid, while Java Man is called an Australoid.

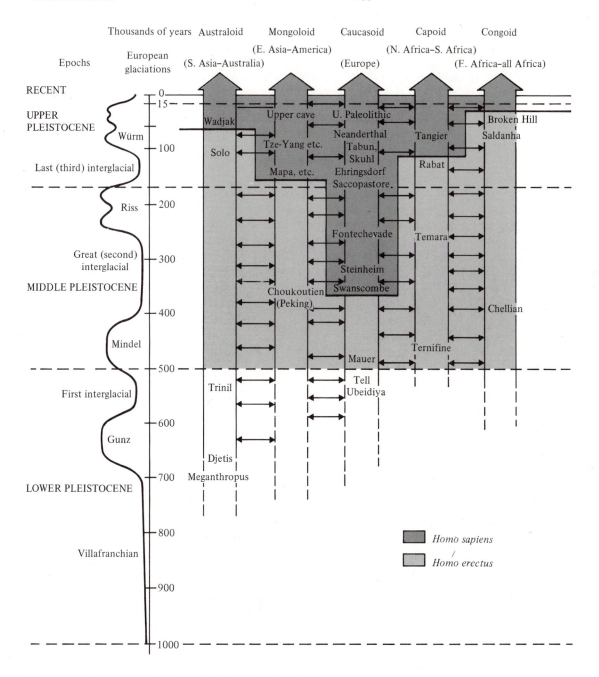

Fig. 27.28 A representation of Coon's theory of racial origins.

The identification of racial traits among fossils is fraught with difficulties, however: many racial features, including skin color, hair form, epicanthic folds, not to mention blood groups or gene frequencies, are simply not preserved in fossils. Even those features that are preserved (largely those of the skull) are found only in isolated specimens, never in whole populations. Preserved individuals may well be atypical, a great difficulty when dealing with cranial features that usually show great degrees of overlap between large samples taken from different racial groups.

Although Coon (1962) wrote some twenty years ago, one of the major issues he raised was never fully resolved. The problems of polytypic species evolving through time are not unique to the genus *Homo*. Coon's treatment of racial origins raises a problem that may be common to all evolving polytypic species, a problem referred to as Coon's Dilemma.

Imagine, if you will, an evolving polytypic species (*Homo erectus*) with five geographical subspecies. Each of these five subspecies maintains a low level of intermittent gene flow with adjacent subspecies whenever they come into contact. Yet each maintains its characteristic differences despite this gene flow, partly as the result of climatically controlled selection. Now imagine that this entire polytypic species is evolving through time (diachronically) as a single anagenetic species. New mutations that first appear in any of the five subspecies soon spread to the remaining subspecies, so that if any one subspecies spearheads the evolution by being the most frequent source of these mutations, none of the other subspecies can lag very far behind. This might be a reasonable description of anagenesis in any large, geographically variable polytypic species; even the number of subspecies may be changed without substantially affecting the model.

The dilemma arises when the evolving polytypic species crosses a species boundary, such as that between *Homo erectus* and *Homo sapiens*. Not all of the subspecies will cross the boundary at the same time, yet they will continue to interbreed with one another whenever they come into contact. During the transition, the more advanced subspecies belong to *Homo sapiens,* while the less advanced subspecies belong still to *Homo erectus,* yet they can interbreed, in violation of Mayr's definition of the species (Chapter 14). During the third (Riss) glacial

period, for example, Coon postulates that the Caucasoids had already proceeded well beyond the *sapiens* threshold, a threshold that Capoids, Congoids, and Australoids had not yet reached.

Coon's dilemma makes Mayr's biological definition of the species category inconsistent with our ability to define species taxa morphologically. In order to define a *sapiens* threshold at any specified cranial capacity (like 1200 cm^3), we must allow certain exceptions to the rule that different species do not interbreed under natural conditions. (Populations just above and just below the threshold would belong to different species, and yet would be capable of interbreeding in certain cases.) Alternatively, we could try to salvage Mayr's definition of the species by redefining a *sapiens* threshold in terms of a time boundary (like 150,000 years ago, at the close of the third glaciation), but we would no longer be able to assign a fossil to either species on the basis of its morphology unless we knew its age.

CHAPTER SUMMARY

The order Primates is characterized by no single character, but rather by a suite of 20 or so interrelated evolutionary tendencies reflecting early adaptations for life in the trees. Primate locomotion uses grasping hands and feet with friction-producing surfaces and opposable thumbs and/or big toes. Primates are visually oriented animals, with eyes rotated more forward (orbital convergence) and visual fields overlapping to permit binocular, stereoscopic vision. The cerebral cortex of the brain is enlarged; behavior is complex, and most of it is learned. One young is typically born at a time; the uteri fuse to form a uterus simplex. Extensive parental care compensates for the extended juvenile period and for single births.

The primates originated just prior to the beginning of the Tertiary. In this book, they are divided among five suborders. Suborder 1, the Plesiadapoidea, are extinct and largely Paleocene in occurrence. Suborder 2, the Tarsioidea, occur predominantly as Eocene fossils; one genus, *Tarsius,* also lives in the East Indies. Suborder 3, Lemuroidea, contains many Eocene fossils (family Adapidae), the lorises and galagos of Africa and tropical Asia (family Lorisidae), and a great variety of lemurs (two or three families) living on the island of Madagascar. Suborder 4, Platyrrhina, contains New World monkeys and marmosets, with three pairs of premolars and forward-pointing nostrils. Suborder 5, Catarrhina, contains Old World monkeys, apes, and humans, all with two pairs of premolars and downward-pointing nostrils.

Catarrhines first occur in the Oligocene Fayûm deposits of Egypt. They are divided into Cercopithecoidea (Old World monkeys) and Hominoidea (apes and humans). Living apes (Pongidae) include the brachiating gibbons, the "quadrumanual" orang-utan, and the knuckle-walking gorilla and chimpanzee.

Hominids, members of the family Hominidae, are distinguished from pongids (apes) by such characters as bipedal locomotion, reduction of the toes, evolution of a so-called lumbar curve (making the spinal column slightly S-shaped), and a forward shifting of the foramen magnum under the expanded braincase. Of these features, the achievement of upright bipedal locomotion is considered primal, and most of the others can be viewed as adjustments to this new form of locomotion. Pongid specializations away from the hominid condition include a "squaring up" of the tooth-rows and enlargement of the canine teeth.

The earliest hominids are assigned to the genus *Ramapithecus,* which is barely distinguishable from the geographically widespread pongid genus *Dryopithecus* of Miocene and Pliocene times. The major difference between *Dryopithecus* and *Ramapithecus* is a delay of molar tooth eruption associated with an increasing rate of tooth wear. This may indicate a shift in diet to tougher or grittier food, which may in turn indicate an ecological shift toward more open savannahs. Such a shift could have intensified selection pressures leading to the attainment of upright bipedalism, though postcranial remains of *Ramapithecus* have yet to confirm this supposition.

The genus *Australopithecus* is much better understood. Its postcranial skeleton shows that it walked fully upright: a lumbar curve was present, and the foramen magnum had shifted to the bottom of the skull. Its brain, on the other hand, was no larger than a chimpanzee's. *Australopithecus* is known from many South African and East African localities. A Pliocene species, *A. afarensis,* may have been ancestral to both later *Australopithecus* and *Homo.* Pleistocene *Australopithecus* represent at least two distinct lineages. One of these, *A. afri-*

canus, was a slender omnivore who hunted. Tools made from bones, horns, and teeth have been found associated with this form. The other lineage was that of the "robust" australopithecines, with enlarged jaw muscles, deepened jaws, and increased surface area of the molar teeth, probably indicating a still grittier diet that presumably included more vegetable matter. These features were already developed in *A. robustus,* but became even more exaggerated in the somewhat later *A. boisei.*

The genus *Homo* has progressively diverged from *Australopithecus* throughout its history of over 4 million years. *Homo habilis,* the oldest species, may have evolved from the Pliocene species *Australopithecus afarensis.* The major trend within the genus *Homo* was the great enlargement of the brain from a volume of around 450 cm^3 in *Homo habilis,* to around 1000 cm^3 in *Homo erectus,* to around 1400 cm^3 in modern *Homo sapiens.* At first, the increase in cranial capacity was in length and breadth, but not yet in height; *Homo erectus* therefore had a rather flattened skull that was broad from side to side and drawn out in back. *Homo erectus* was much taller than *Australopithecus* and was fully as erect as modern *Homo sapiens.* The major advance of *Homo sapiens* over *Homo erectus* was an increase in the vertical height of the braincase associated with an increase in cranial capacity. A cranial volume of 1200 cm^3 is arbitrarily taken as the dividing line between *Homo erectus* and *Homo sapiens.*

A race is a biologically distinct subdivision of a species inhabiting a definable geographic region and possessing certain characteristic gene frequencies. Modern definitions of race are based on populations rather than types. Great variation exists within any racial group, and considerable overlap occurs between any two racial groups. Races are subspecies within a species; they interbreed freely whenever they come into contact with one another.

There is much disagreement among experts as to the classification of human populations into racial groups. Even the number of these groups varies greatly. The most commonly recognized major racial groups include the Caucasoid ("white") racial group native to Europe and western Asia, the Mongoloid or Oriental racial group native to the rest of Asia, and the Negroid ("black") racial group native to Africa. American Indians are either considered Mongoloids or a distinct racial group. Smaller racial groups recognized in some classifications include the Bushman–Hottentot group, the Australoids, and the Melanesian, Micronesian, and Polynesian peoples of the Pacific.

The adaptiveness of racial differences is shown by the existence of microgeographic variation within each major racial group and by conformity to the ecogeographic rules of Bergmann, Allen, and Gloger related to body size, body shape, and color. The dark skins of tropical populations in all major racial groups probably evolved as an adaptation to reduce the risk of skin cancer from overexposure to the sun's ultraviolet rays, while the light skins of Caucasoids may be an adaptation to permit sunshine-induced vitamin D synthesis. Physiological traits also vary racially; these include resistance to such physiological stresses as temperature extremes. The adaptiveness of geographic variation is shown also by variations in the frequencies of genetically polymorphic traits such as blood groups.

Theories of racial origins fall into two groups: monophyletic theories, which assume one racial group to be ancestral to all the others, and polyphyletic theories, which assume all racial groups to be of great antiquity. Coon's polyphyletic theory assumes that the five racial groups of *Homo sapiens* (Congoid, Capoid, Caucasoid, Mongoloid, Australoid) were derived from five similar racial groups within *Homo erectus*. This theory raises important dilemmas for evolutionary taxonomy, for it makes the biological definition of the species category inconsistent with morphologically definable species taxa.

28

Cultural Evolution and the Future

A. CULTURAL AND BIOLOGICAL EVOLUTION

Culture may be defined as an integrated complex of learned beliefs, behavior patterns, and artifacts traditionally transmitted from one generation to the next. The adaptive integration of beliefs, artifacts, and various behavior patterns need not be emphasized here; it is treated extensively in most anthropology courses. Traditional transmission is often considered to be one of the hallmarks of culture: culture is not only learned by imitation, it is also consciously taught by each generation to the next. Comparisons between cultures, known as **cross-cultural comparisons**, have repeatedly revealed that most aspects of culture can vary greatly.

Language is one of the more important aspects of culture. Each society or group of societies has its own spoken language, and many languages of the world fall into groups indicative of their common cultural heritage. Features common to related languages usually reflect common derivation and are in that sense homologous. The functional similarities common to *all* languages may be analogies rather than homologies, reflecting the functional necessities of any system of verbal communication among humans. People can learn the languages of other cultures. Ruling people can force their language upon subjugates. Words can be adopted from other languages, along with the objects to which they refer; thus we have words like camel (Arabic), bamboo (Malay), boomerang (Australian), or tomato (Mexican). These several observations reinforce the view that language is culturally determined.

Complex cultural features such as language may undergo cultural evolution. Homologies, analogies, trends, parallelism, and adaptive radiation are among the evolutionary features common to cultural and biological evolution. Cultural innovation compares in many ways to mutation, while cultural diffusion is in many ways comparable to gene flow. A major difference lies in the cultural transmission (''inheritance'') of ''acquired characteristics.'' Another important difference lies in the absence of species boundaries (i.e., of reproductive isolation) in the case of cultural evolution—all societies are capable of acquiring cultural traits from other societies with which they come into contact.

Culture is more than just an outcome of biological evolution—it has become increasingly important as a modifier of biological evolution, and as such it represents a well-developed feedback system. It is culture, more than genetics, that determines among humans the choice of mates, the mode of life, the size of social groups, and the ability to migrate. These, in turn, determine the risks of life and death, and the number of offspring. This has always been so, and its importance is dramatically illustrated in industrial societies by advances in medicine and by culturally induced hazards. Any change in the risks of life and death may be viewed as a change in the force of selection. Medical and other cultural advances have greatly relaxed many forms of selection, while others have been intensified by the way in which we live (epidemics, traffic and industrial accidents). As the world's population increases, we should expect a further shift from r-selection to K-selection (see Chapter 9 for an explanation of these terms). The former prevailed in most places before 1900; the latter will certainly intensify as pollution and overpopulation increase.

Much of what follows is an overview, necessarily simplified, of human cultural evolution and its effects on biological evolution.

B. HUMAN SOCIOBIOLOGY

Human sociobiology, the study of human social organization in biological terms, is a relatively new field distinct from social anthropology (the study of human social organization in anthropological terms). Cooperation between these two fields can only benefit both.

Mating Preferences and Taboos

Humans do not mate at random. Culture, to a great extent, determines who is a desirable or a permissible mate. Of course, individual preferences will also determine the choice of mates, but only within culturally determined limits. Sexual selection, in other words, is now to a great extent culturally controlled. The very standards of human beauty and sexual attractiveness are culturally determined and thus vary widely from one civilization to another or even from one sub-culture to the next. The preference to mate with a person sharing the same cultural background is also reinforced culturally. Positive assortative mating for genetically determined traits (Box 28–A) can be demonstrated when people mate preferentially with others who share the same eye color, hair color, or stature. In many cases, negative assortment exists between near relatives.

One of the few cultural traits that approaches universality is the prohibition against matings between very close relatives, the so-called **incest taboo.** Various explanations for this taboo have been given, including those dealing with the social and/or psychological maladjustment of the marital partners. Another theory holds that marriage outside one's own group (exogamy) both encouraged and was encouraged by intergroup trade, commerce, and friendliness. Reduction of intergroup warfare would have served, under this theory, as a selective force favoring exogamy over mating within the group. Marriage fosters economic ties between families, but there is no such gain if the marriage is between family members.

A tempting hypothesis to explain the incest taboo is that matings between close relatives often cause a depression in fitness (known as **inbreeding depression**), and that all cultures are aware of this fact. This hypothesis is supported by the realization that the incest taboo is a sanction against intercourse even more than against marriage. Further evidence comes from the legends used to justify or explain the taboo in each society, for these legends often tell of sterile or physically malformed offspring resulting as ''punishments'' for the violation of the taboo. On the other hand, it is by no means evident that all cultures are aware of inbreeding depression or of its causal basis; some societies even seemed unaware that intercourse causes pregnancy and the ensuing birth. According to some anthropologists (e.g., Malinowski), certain primitive tribes even denied the male role in procreation. The few known exceptions to the incest taboo also argue against the inbreeding depression hypothesis, for these exceptions are all confined to the royalty, or

BOX 28-A *Assortative Mating in Humans*

Assortative mating for many phenotypic traits can be demonstrated in humans. In Sweden, for example, a study of 483 couples showed that husbands and wives tend to be alike in eye color (blue *versus* dark) more often than would be expected if matings were totally at random:

Wife's eye color	Husband's eye color	Observed frequency		Expected frequency	
dark	dark	92	(19.0%)	73	(15.1%)
dark	blue	117	(24.2%)	136	(28.2%)
blue	dark	77	(15.9%)	96	(19.9%)
blue	blue	197	(40.8%)	178	(36.9%)

For traits that can be measured, a correlation coefficient between husbands and wives would measure the extent of assortative mating. A coefficient of zero would indicate random mating with respect to that trait, while a high positive correlation (close to 1.0) would indicate a high preference of people for phenotypically similar mates. Studies in various English-speaking countries have shown that a number of traits show statistically significant correlations between mates:

	Correlation coefficient (r)
Age at marriage	0.76
Memory test score	0.57
IQ score	0.47
Length of ear	0.40
Length of ear lobe	0.40
Circumference at waist	0.38
Neurotic tendency	0.30
Standing height	0.28
Eye color	0.26
Circumference at hip	0.22
Weight	0.21
Circumference at neck	0.20

Many of these traits, of course, have strong environmental components in addition to their inherited (genetic) components. A few characters, such as fingerprint ridge counts, head length, or skull proportions failed to show any significant correlation. One might therefore conclude that, in choosing a mate, people pay more attention to their partner's age, intelligence, or ear lobes than they do to their partner's eye color or neck circumference, and that they pay almost no attention at all to fingerprints.

---

uppermost stratum, of highly stratified societies (ancient Egypt, ancient Peru, and aboriginal Hawaii). Cleopatra was the product of many generations of repeated marriages between close relatives.

Origins of Human Social Organization

Human social organization forms the basis for family life, society, and complex civilizations, and is therefore of such importance to our species that numerous theories have been proposed to account for its origin. Each of these theories can be considered either as internalistic or externalistic, depending upon whether the unique features of human social organization are attrib-

uted to factors intrinsic to our species or imposed upon us by our external environment.

Various internalistic theories have attributed human social organization to such factors as territoriality, sexual attraction, and the lack of hair. Robert Ardrey (1966), for example, has proposed that territoriality (defense of a territory) manifests itself in our own species as possessive ownership, the basis of property, wealth, and civilization. Indeed, the "possession" of a mate is seen by Ardrey as part of this "territorial instinct." But humans are not really territorial, for they do not defend territory against conspecific intruders indiscriminately: guests are welcomed frequently into the most private of domiciles, and hospitality, the antithesis of territoriality in many ways, is valued by a majority of societies. For each society in which land or property is measured off, surrounded with a fence, and defended against intruders, there are many other societies that practice none of these things. In particular, the possession of land is characteristic only of agricultural societies, not of the many nonagricultural societies that are also included in our species.

The lack of hair forms the crux of Desmond Morris's "Naked Ape" theory. According to Morris (1967), the sparseness of body hair is the most characteristic feature of our species, and owes its origin to the intrinsic human desire for bodily contact and visual stimulation. The latter often takes the form of hair around the genitals, which adds to their conspicuousness.

Marshall Sahlins (1959, 1960) has repeatedly supported a theory that might be called "sexist" in a very descriptive and not necessarily pejorative sense. According to this theory, and to similar theories proposed by other post-Freudian ethologists, the bonds that cement human society together are primarily sexual. Sahlins lays particular emphasis on changes in the female sexual cycles in the shaping of human social organization. Lower primates have estrous cycles, in which the time of ovulation is made conspicuous by genital swelling, secretion of odoriferous chemicals, and behavioral changes. Females are generally uninterested in sexual activity, except during these brief periods of estrus. The human female has a menstrual cycle in which the time of ovulation is not conspicuously marked, and she is continually receptive to sexual contacts throughout her cycle. Inconspicuous ovulation has been explained by Burley (1979) as a means to thwart the cultural tendency of many women toward birth control by abstinence during ovulation, but according to Sahlins the continual receptivity leads to more frequent intercourse, thus fostering more permanent social bonds, which become the foundation for human societies at all levels of complexity. But the strongest bonds within primate societies are between parents and their offspring, not between sexual partners, and Sahlins' theory does little to explain these bonds. Even among adults, the numerous bonds within a primate society are not exclusively sexual.

The latest theory to emphasize relations between the sexes in the origin of human social organization is presented by Hrdy (1981). Male primates, according to

Hrdy, attempt to maximize their contribution to future generations by mating with as many females as possible, and occasionally by fighting other males and brutalizing infants who are not their own. Females, on the other hand, subvert the males' belligerence by frequent, nonexclusive mating and by other subtle means that befuddle his ability to distinguish his own genetic offspring from those of other males. Genetically selfish behavior, then, is the cause behind mating systems, group structures, and other aspects of human social organization. Hrdy's theory is indeed a plausible explanation of human sexual behavior, but it is somewhat less successful in explaining those social bonds that are not primarily sexual.

Among the externalistic or environmental theories on the origins of human social organization we may mention Elaine Morgan's extraordinary suggestion (Morgan, 1972) that humans were originally coastal shellfish eaters. In particular, she attributes our general loss of hair to the reduction of skin friction in the water, an adaptation otherwise known among whales, porpoises, and fast-swimming fishes. Evidence to support these suggestions has not been forthcoming: heaps of discarded shells, which would fossilize readily, are found only rather late in human prehistory. The loss of hair is also a very late adaptation in marine mammals; most seals, for instance, are fur-covered. The loss of hair among whales was compensated by an increase in subcutaneous fat tissue (blubber), a trend that our own lineage did not share.

The most widely supported theory, consistent with the evidence from the most diverse approaches (archaeology, ethnology, paleontology, comparative ethology, etc.), is that human social organization originated in response to the ecological requirements of hunting. Other primates do partake of meat on occasion. More revealing, however, is the great fact that the seemingly most primitive of modern societies have much of their lives centered around the hunt, and that paleolithic tools seem to have been used in hunting, in preparing animal remains after the hunt, and in preparing other tools for use in the hunt. According to this theory, the initial attainment of upright posture may even have resulted from the pursuit of small game animals, though on this point there is much uncertainty. Once upright posture had been attained, subsequent biological and cultural evolution, up to the origin of agriculture, can be consistently viewed as a series of adaptations to the ecology of hunting.

Nonagricultural Societies Today

Nonagricultural societies occur in various parts of the world today, and occurred even more extensively in the past. Studies of contemporary peoples who do not practice agriculture are sometimes used as models for the understanding of past stages in cultural evolution, under the assumption that early humans had a mode of existence not unlike that of modern hunters and gatherers. Examples of such peoples include the Bushmen of Africa, the Australian Aborigines, and the somewhat atypical Eskimo.

Most nonagricultural peoples live in small groups as **hunters and gatherers.** The men generally do the hunting, while the women do the gathering. Otherwise, hunter–gatherer societies tend to be rather egalitarian, with no social class distinctions and no clearly defined "leaders." The social units, known as bands, tend to be small (roughly 20 to 60 individuals), and range widely over the area they inhabit. They have no permanent homes and make only temporary encampments. The women and children stay close to the encampment, gathering plant food such as fruits, leaves, and fleshy roots, preparing this food in some cases. The men roam more widely, following after game animals (principally herbivores), which they hunt. Some hunter–gatherer societies also fish, with the men usually doing most of the fishing. The food gathered by women represents 60–80% of the calories eaten by the group, but the meat or fish caught by the men is needed as a protein supplement, and is brought back to camp to be shared with the women and children. Food is generally consumed on a hand-to-mouth basis, for most food is perishable, and there is no provision for its storage. There is no other accumulation of wealth, either, because all personal belongings must always be carried from place to place. Without accumulated wealth, there are no status or class distinctions. There is very little to steal or to protect from stealing. Nobody has (or needs) the power to coerce others to do their bidding. There are few specialized crafts or occupations; division of labor rarely reaches beyond that based on sex and age.

C. HUMAN PREHISTORY

Box 28–B represents an attempt to summarize all of human cultural history in a single chart. Necessarily, it is an oversimplification, with many details and many exceptions omitted, but it does make clear the general trend. Stages I, II, and III (with *a* and *b* combined) correspond approximately to the three cultural "stages" not very charitably designated as "savagery," "barbarism," and "civilization" by Lewis Henry Morgan, and the transitions between these stages correspond to the Neolithic Revolution (beginning in the Old World c. 15,000 B.C.) and the Urban Revolution (c. 6,000 B.C.) first so designated by V. Gordon Childe. Few societies remain pre-agricultural today, though before the last glacial retreat of the Pleistocene they presumably all were. Among the many agricultural societies, the most successful ones in cultural terms are those that build cities, practice writing and commerce, and recognize stratification into social classes. Industrial societies are relatively recent, beginning with England in the late 1700s. Although there are extensive variations in detail, and certainly in timing, the general pattern of human cultural development is for each society, in its own time, to proceed through phases of development comparable to those shown in Box 28–B.

Paleolithic People and Tools

Before the last retreat of the Wisconsin (or Würm) glaciers, none of the world's people practiced agriculture, cultivated cereals, or kept any domestic animals except possibly for dogs used in the hunt. Cultural remains from this period consist principally of stone tools of so-called Paleolithic types, defining a cultural era known as the **Paleolithic** or **Old Stone Age.**

The early beginnings of human tool use will probably never be known. Simple wooden tools, such as digging sticks, were undoubtedly used by early human cultures, but the conditions for their preservation are so rare and unusual that no truly ancient wooden tools have ever been found. (Even in later prehistory, we find spearheads and arrowheads designed to be shafted onto sticks, yet we find no traces of the shafts themselves.)

Raymond Dart has promoted the idea that the next level of achievement was in the use of tools made from

BOX 28–B *Stages of Cultural Evolution and the Resulting Types of Human Social Organization*

Ecology and economy	Social groups and stratification	Typical family structure	Typical male pursuits	Typical female pursuits
Types of Human Social Organization				
I. PRE-AGRICULTURAL				
Hunting and gathering	Bands, often patrilocal; egalitarian, no stratification	Usually small; often polygamous	Hunting, also some magic and rituals	Gathering vegetable food (80% of calories); domestic work
II. EARLY AGRICULTURAL				
Simple agriculture with few cereals and no irrigation	"Tribe" with differential wealth but no classes, incipient stratification	Polygamous; often large (depending on wealth)	Hunting, warfare, tool-making, planning	Agricultural and domestic work (house and child care)
III-a. HIGH AGRICULTURE				
Complex agriculture including irrigation, based on cereal crops (wheat, rice, corn)	Large and highly stratified units (kingdoms, feudal estates); upper and lower classes (themselves stratified)	Tends toward monogamy in all but uppermost class, which varies; many children	Agriculture, stock-breeding, some crafts, warfare, government, priesthood	Domestic work only, or helping husbands (in fields, etc.)
III-b. INDUSTRIAL				
Mechanized agriculture, mechanized manufacture (textiles, other goods)	Stratified but with middle class predominant; at least some mobility	Typically monogamous; fewer children in middle class	Variety of occupations including leadership roles	Variety of occupations (often menial); domestic work

EXAMPLES OF EACH TYPE

I. Australian aborigines, Bushmen, Eskimo, and many others.

II. Many tribal societies of Africa, Pacific Islands, and America before 1492.

III-a. High civilizations of ancient world (Egypt, Babylon, Greece, etc.), also of ancient Mexico, Peru, and China; feudal Europe and feudal Japan.

III-b. Europe since the Industrial Revolution (and as exported to the rest of the world.)

bones, teeth, or horns (osteodontokeratic tools). An antelope leg bone (femur), for example, would make an excellent club, and if broken it might make a jaggedly pointed weapon or a simple spear. Antelope teeth could be used, still in the jaw, for scraping hides, and the jawbone itself could easily be made into a simple knife. Some bone tools show scratches seemingly made by a sharper implement such as a stone tool. It was perhaps in the manufacture of either osteodontokeratic or wooden tools that stone tools first came into habitual use.

The oldest stone tools were undoubtedly used much as they were found. They may have been used to throw at fleeing prey, or to bludgeon an animal to death once it had been seized. Baboon skulls have indeed been found with blunt depression fractures that could have been made by osteodontokeratic or stone tools. Occasionally, a stone tool used as a weapon might break in half. Pebble tools of this type, called Oldowan tools, have often been attributed to *Australopithecus*. They may have originated by accidental breakage; their purposeful manufacture came later.

For much of African and European prehistory, stone tools of the so-called Acheulean type were made, presumably by *Homo erectus*. An Acheulean hand axe (Fig. 28.1) was an all-purpose tool, for it could function to carve, to cut, to pound, or to scrape. Its manufacture was achieved by the removal of flakes, which were struck off by striking two stones together—the so-called percussion-flaking technique. In both Europe and Africa, hand axes of the Acheulean type persisted for a long time. During this time they became flatter and their edges sharper, which made them more effective as cutting tools. In Asia, a rather different tradition of ''chopper'' tools developed independently.

At about the time that *Homo sapiens* appeared in the fossil record, a new means of tool making came into use. Flakes were no longer struck by the percussion method. Instead, a new technique, called pressure-flaking, developed. In this technique, a platform is first prepared, then a flake is carefully removed by applying pressure rather than by striking. Like the Acheulean hand-axes, the earliest pressure-flaked tools were **core tools**, meaning that the small flakes were discarded and the core was kept as the functional tool. Soon afterwards, however, it occurred to some tool-makers that the flake itself might have a sharper edge and thus be of more use than the core (Fig. 28.2). The resultant type of tool is

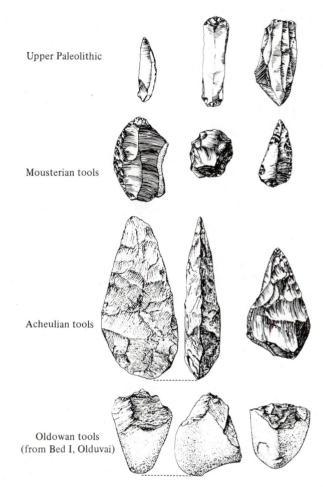

Upper Paleolithic

Mousterian tools

Acheulian tools

Oldowan tools
(from Bed I, Olduvai)

Fig. 28.1 Comparison of Oldowan, Acheulean, Mousterian, and Upper Paleolithic tool types. Reprinted with permission from *Genetics, Evolution, and Man*, by W. F. Bodmer and L. L. Cavalli-Sforza. W.H. Freeman and Company. Copyright © 1976.

called Mousterian, after the cave in France known as Le Moustier. Mousterian flake tools have often been found associated with Neanderthal fossils, and are considered characteristic of Neanderthal culture. A variety of special-purpose tools belong to this tradition, including scrapers, choppers, chisels, burins, and an implement that looked

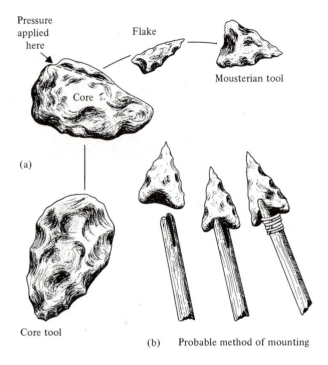

Pressure applied here

Flake

Mousterian tool

Core

(a)

Core tool

(b)   Probable method of mounting

Fig. 28.2   (a) the difference between core tools and early flake tools. (b) Probable method of mounting a Mousterian point on a stick or shaft.

like a simple awl or a drill. These hand-held tools were largely useful for the preparation of food after the kill had already been made. The weapons of the hunt undoubtedly included wooden spears, of which only one specimen, a spear-point from the village of Clacton in England, has been preserved to us. This Clactonian spear-point is the oldest wooden tool preserved in the archaeological record.

The replacement of Neanderthals by humans of modern aspect coincides in time with the replacement of Mousterian flake tools by still more advanced flake tools of Aurignacian, Solutrean, and Magdalenian types, collectively called upper Paleolithic (Fig. 28.3). The pace of cultural evolution became accelerated, with new tool traditions replacing one another even faster than before. (The Acheulean hand axe, by contrast, had endured for

about 700,000 years.) The cores from which flakes were struck no longer resembled the Acheulean hand axe, but rather were many-faceted prisms, allowing many flakes to be produced from a single core. Stone flakes were now sometimes inserted into wooden arrows and spear shafts. The throwing spear and the bow-and-arrow thus made their appearance in upper Paleolithic times, and hunting became a much more refined art. Cave paintings of this era depict game animals and hunting scenes involving both arrows and spears. Some drawings of mammoths show the heart clearly marked (Fig. 28.4), and an arrow may point to the heart. Here, perhaps, was a wish (or a prayer) that the arrow find its mark; perhaps it was a form of instruction in which the young hunters were being told where to aim. This simple directive of aiming for the heart shows that Paleolithic hunters were aware of animal physiology on at least a crude but effective level. The frequent drawing of pregnant females, both animal and human, shows that another important biological function was understood, and the drawing of reindeer mating rituals suggests that a causal connection was made between the act of mating and the subsequent birth many months later.

It was at approximately this level of cultural sophistication that the Americas were invaded across the Bering Strait. The Bering Land Bridge was available for migration from about 50,000 to 40,000 B.C., and again from about 27,000 to 8,000 B.C. Shortly after the land bridge became available in 27,000 B.C., we find the earliest paleo-Indian sites, such as that at Tlapacoya, Mexico (about 25,000 B.C.), including upper Paleolithic tool kits reminiscent of those found in sites of similar age in Siberia. The site at Tlapacoya is about 7,000 km (4,500 mi) from the Bering Strait, and about 2,000 years later in time (very approximately); this gives a possible migration rate of about 3.5 km (or 2.25 mi) per year, well within the range observed among modern hunter-gatherers.

Still later upper Paleolithic tool-kits of the Old World included a few even more specialized weapons, such as fishhooks, notched spear-points, and harpoons. These were often made from bone or ivory rather than stone, and often bore tiny hooks or barbs to prevent the weapon from shaking loose and falling out of the animal. Harpoons were independently invented in the New World, too.

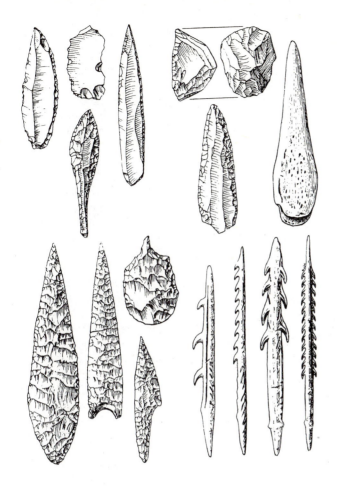

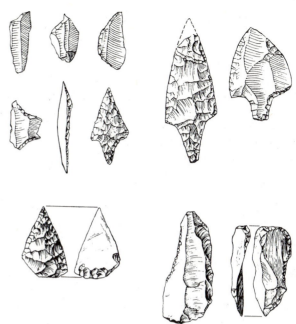

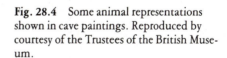

Fig. 28.3 An assortment of Upper Paleolithic tools, each used for a somewhat different purpose. Reprinted by permission from McKern, Sharon, and McKern, Thomas, *Living Prehistory, An Introduction to Physical Anthropology and Archaeology.* Menlo Park, Calif.: Cummings Publishing Company (now the Benjamin/Cummings Publishing Company), pp. 216–217, Fig. 7.

Fig. 28.4 Some animal representations shown in cave paintings. Reproduced by courtesy of the Trustees of the British Museum.

The Neolithic Revolution

In the history of human civilization, no changes have had so lasting and profound an effect on every aspect of human culture as the invention of agriculture. V. Gordon Childe has hypothesized that a Neolithic Revolution, beginning about 15,000 B.C., brought about the first domestication of plants and animals, while a much later Urban Revolution, beginning about 6,000 B.C., brought about the growth of "high civilizations" such as those of ancient Egypt and Mesopotamia. Other archaeologists have suggested that these terms designate early and late developments within a single process, the Agricultural Revolution, of which the early and late stages grade continuously into one another. Indeed, the parallel course of these developments in the Near East, in China, and in the New World illustrate that trends in cultural evolution, like trends in biological evolution, can take place in widely separated groups either at the same time or at different times.

The precondition for the origin of agriculture was the use of wild grasses, presumably as food. Simple agriculture may have originated when some seeds were spilled accidentally on fertile soil, and many new plants were later found growing from them. The "Neolithic Revolution" thus began with the cultivation of a few plants, notably cereals, and the keeping of animals such as dogs. It is now believed that wheat was first cultivated in the so-called Fertile Crescent of Iraq, where wild forms of wheat and goat grass (the ancestors of modern wheat) can still be found in the hills. The cultivation of wheat also began early in Egypt and in the Indus Valley of Pakistan, perhaps independently, or perhaps by cultural transmission of the technique (maybe even the seeds) from the Fertile Crescent. The cultivation of rice was first accomplished in China, perhaps twice independently. The cultivation of corn (maize) was first achieved in ancient Mexico and also independently in Peru and perhaps elsewhere in the New World. The domestication of animals, beginning in most cases with the dog, frequently accompanied or followed the domestication of plants.

These events all had a profound and lasting influence on human society, for in domesticating our animals and plants we had also domesticated ourselves. Group size increased, partly in response to the greater abundance of food, while group mobility declined. Shelters became more elaborate, since they were expected to last for a longer period of time. The more permanent tools necessary for agriculture were made and kept. In the Old World, but not in Mexico, beasts of burden came into use and belongings could therefore be carried from place to place. The grinding of cereal grains into flour brought with it a new technology. Stone tools were now ground to the desired sharpness; the term "Neolithic" was originally defined on the basis of such ground tools. Tool making, house building, and animal breeding became specialized endeavors, practiced by certain individuals who received for their services a portion of the harvest of other people's agricultural labor. Though hunting was still practiced, the cultivation of plants became increasingly important. Whoever sowed seed had either to remain or return there in order to gather the harvest. Control of the land, if not ownership, became important, and rules began to develop to govern the use and ownership of land.

The Neolithic Revolution seems to have begun in the Near East about 15,000 years ago. It spread gradually to Europe after the retreat of the Würm glaciation. This spread of culture was accompanied by gene migration (Menozzi *et al.*, 1978); it was not just a diffusion by cultural transmission. Much of life was still hand-to-mouth, because many of the foods deteriorated rapidly and could not be stored. The major exceptions were the cereal grains such as corn, wheat, or rice, which could easily be stored, and so became the original basis of accumulated wealth, social inequality, and ultimately social stratification into classes. Division of labor increased. The office of leader or chieftain emerged, often in the role of an economic redistributor of wealth, and secondarily as an arbiter of quarrels. Chieftains had little power at first, but their power grew rather steadily as more wealth was accumulated.

The Urban Revolution

The origin of what we call "civilization," but also of some social evils, occurred in the later phases of the Agricultural Revolution. Great intensification of cereal agriculture and the building of irrigation projects led to the origin of cities and the stratification of society into social classes. This phase of the Agricultural Revolution, often called the Urban Revolution, began about 6,000 B.C. in the Old World and about 1,500 B.C. in Mesoamerica.

Intensive cereal agriculture, which depends upon mass irrigation, was developed independently at least four times and perhaps seven times, in Mexico, Peru,

Egypt, Mesopotamia, the Indus Valley, and twice in China. Different cereal crops formed the basis of these several civilizations: corn (maize) in Mexico and Peru, rice in China, and wheat, accompanied by rye and other minor cereals, in Egypt, Mesopotamia, and the Indus Valley.

The Urban Revolution had a strong basis in agricultural technology. Cereal crops could be stored if kept dry, and wealth could thus be accumulated. Surpluses could be stored as a hedge against a future famine. Granaries (warehouses for storage of grain) were build as permanent, durable structures, and cites grew up around them. Many granaries also functioned as temples. Intensive cereal agriculture and the necessary irrigation to support it both required mass labor. Because of the need to maintain an irrigation system (canals, etc.) and to tend to the harvest, people settled down and built permanent settlements.

Perhaps the most profound effect on human civilization was in the development of social stratification of people into various classes. The ability to store and accumulate wealth gave rise to an upper class of land owners and managers. The need for mass labor gave rise to a lower class of workers or serfs, most of whom were rural peasants. The need for agricultural tools, weapons, and pottery gave rise to an initially small middle class of artisans and craft specialists: potters, carpenters, stone-cutters, bricklayers, metal craftsmen, and often soldiers and guards as well. These artisans generally clustered in the urban centers, along with the wealthy land owners and the managers of the granaries. Class prejudice arose initially as an urban prejudice against the rural peasants, but it was the labor of these agriculturalists that provided the surplus capable of maintaining the middle and upper classes in their nonagricultural activities. Land, water rights, and irrigation systems became objects demanding to be protected, as were the stores of grain in the granaries. Soldiers and guards arose as a new specialized occupation, and warfare became an important new means of acquiring new land. Conquerors often subjugated their conquered people and made slaves of them or recruited them into the lower-class peasantry.

With the increased attention given to sowing only good seed, and breeding only the best among domesticated animals, and also with the increased attention to class distinctions, more attention was now also paid to human reproduction and inheritance. The purity of seed,

of domestic stock, or (by extension) of human families were guarded with the utmost vigor. Sexual violation came to be viewed as a trespass tantamount to the sowing of inferior seed. The legitimacy of birth became a very important concern, for the inheritance of land and wealth depended on it. The status of women generally reached a low point in stratified agricultural societies: resented for the food they consumed, women were held (by men) to be necessary evils, good only for the production of offspring.

Racism probably sprang from such contexts, too. The subjugation of people for use in mass labor projects such as irrigation or the building of warehouses and temples engendered an attitude of looking down upon the lower classes, who were sometimes conquered people of different racial, ethnic, or linguistic background. Notions of the purity of racial stock were probably engendered by an equal concern for the purity of animal stock.

Writing first arose in connection with agricultural societies. The first written records were kept of grain transactions: the oldest samples of writing are balance sheets or tally sheets listing the amount of grain that each person had stored in the granary and the amount they were therefore entitled to withdraw. Of course, a certain fraction, called a tribute, had to be paid to the land owner, and middle-class artisans also demanded their payment in grain. Counting and mathematics arose from the keeping of such records. Attention to the harvest, and thus to the coming and going of seasons, became important. Calendars were developed (independently in the Middle East and in Mesoamerica) to keep track of the movements of heavenly bodies, and to allow planning for the harvest that was yet to come. Astronomy, the first real science, arose from a concern over celestial movements. Arithmetic developed from the tallying both of grain and of calendar days and months. Geometry was developed by the Egyptians, initially as a series of practical rules for re-surveying boundary lines between fields that had been flooded along the fertile banks of the Nile.

Religion, which had existed before in the form of magical beliefs and legends, now became institutionalized, often by the same political or managerial elite that controlled grain redistribution. A priestly class arose and temples were built, originally after the plan of grain storehouses. Prayers were now offered for a bountiful harvest or for favorable weather conditions. Awesome powers were attributed to the Sun, Moon, and other heavenly bodies, or to the deities that controlled their movements, for these movements were seen to control the coming and going of seasons and especially of harvests. The maintenance of the calendar became one of the most important tasks of the priestly class. Biblical and other sacred writings reflect these and other concerns over agricultural matters, over the changes of the skies and of the seasons, and over the purity of stock. The need for careful planning, for constant vigilance, for avoidance of waste, and for attention to seasonal changes and seasonal activities were repeatedly emphasized. Seasonal holidays and celebrations came into being, often as rituals associated with the planting of particular crops or with the harvest itself.

D. INDUSTRIAL SOCIETIES

Coming of the Industrial Revolution

From ancient times to the present, but especially from the Renaissance onward, improvements in agricultural technology increased the proportion of the population that could be channeled into nonagricultural pursuits. The middle class of artisans grew and expanded to include shopkeepers and merchants. Banking and commerce became increasingly important, and cities generally flourished in proportion to their role in trade. The increase in population that this permitted, and also the shift in the cultural center of Europe from the Mediterranean to the seafaring nations of the North, brought about the need for more clothing and for a more efficient textile industry. It was with the introduction of steam power and the mechanization of the textile industry in England that the Industrial Revolution really began. From England it spread to Germany and the rest of the Continent, to the United States and Canada, and later to Japan. It is now spreading to China and to parts of Latin America, regions that are still largely agricultural.

New technologies developed with the Industrial Revolution and spread throughout the industrialized countries; many of these technologies were also exported to the developing nations of the world. Transportation became fully mechanized with the development first of the train, and later of the automobile and the airplane. The telephone, the television, and the computer have revolutionized communications, and their effects have permeated nearly every aspect of our society.

Social organization has changed, too, as our society has become more and more complex. The increasing demand for manual labor pressed more and more people, including children, into nonagricultural employment. The status of women in society has begun to improve. Household appliances and convenience foods have allowed shortcuts to housekeeping that have enabled more and more women to find jobs in the world of commerce and industry. Prepared baby foods and especially baby bottles have enabled fathers to help mothers in the care of infants and children. As this trend continues, the distinctions between sexually defined occupational roles will undoubtedly diminish more and more.

Changes in Selective Forces

Human civilization has always modified the forces of selection, but the rate of modification increased markedly with the Urban Revolution and the development of stratified society. Soft, pasty, refined cereal products of increasing "purity" (low fiber content) gave rise to an increasing number of dietary ills, tooth decay, and overweight. Population densities increased to the point where epidemics became both more frequent and more devastating. The coming of the Industrial Revolution introduced many new man-made hazards in the form of industrial accidents.

Selection for many other traits meanwhile became increasingly relaxed. Individuals who would certainly have perished under pre-agricultural conditions (including blind, crippled, and feeble-minded individuals, who simply would have been abandoned) are now able to lead productive lives. The relaxing of these selection pressures means that more variant genes could persist. Our civilization selects more for genetic diversity than for genetic uniformity.

The Industrial Revolution brought about the further relaxation of certain selection pressures through advances in medical care. Communicable diseases, which once caused untold thousands of deaths, are now in many cases controlled. Infant mortality has diminished and life expectancy has risen. In general, selection by death before reproductive age has become far less important, while selective differences in fertility have become more important in determining fitness.

Some selective forces operating in four largely agricultural countries and in three industrialized countries are summarized in Box 28–C. Communicable diseases are far more important as selective forces in most of the world's nonindustrialized nations, while among industrialized nations, the degenerative diseases of the aged, such as heart disease and cancer, have become increasingly important. A more significant list, the causes of death among persons aged 15 to 44, shows that accidents, especially motor vehicle accidents, have become foremost as selective forces in the reproductively active age group of the population. Such accidents place a heavy selective premium on the enhancement of skills that have long been hallmarks of primate evolution: good coordination, good peripheral vision, general alertness, quick reaction time, and the ability to make intelligent decisions. A selective premium is also placed on whatever precludes people from driving. Industrial hazards and warfare also cause thousands of deaths each year. Pollution is increasing, and is becoming more and more important as a selective force in the world. Among the newest selective forces are those that result from the abuses of tobacco, marijuana, and various narcotic and hallucinogenic drugs, as well as the more traditional abuse of alcohol. All of these claim an increasing toll each year, thus selecting against whatever genetic background (if any) might predispose people to such self-abusive behavior. Selection by these agents or by pollutants will favor the evolution of increasing tolerance and insensitivity to such chemical substances—that is, if the necessary mutations arise. In general, the trend from r-selection to K-selection (Chapter 9) reaches its extreme in industrialized human societies.

Are We Still Evolving?

Much has been written on the question of continuing human evolution. Several authors, including H. J. Muller and Pierre Teilhard de Chardin, have emphasized the many selection pressures (famines, epidemics, predators) that civilization has greatly reduced and in some cases eliminated. Teilhard, an optimist, tells us that we have acquired control over our own evolution and that cultural evolution has taken over where biological evolution left off. Certainly, we could control our own cultural evolution, if only we could legislate human behavior sufficiently (an arguable proposition in itself). But what Teilhard seems to imply is that biological evolution has somehow ceased, or at least slowed down considerably. Few morphological changes have occurred in the last half million years of human evolution, and to this extent Teilhard is perfectly correct. There is no telling, of course, what changes (physiological and biochemical as well as morphological) may take place in the next million years or more.

H. J. Muller (1961, 1965) seems to agree with Teilhard that our biological evolution has largely come to a halt. But Muller, a pessimist on this issue, has repeatedly warned of the possible unfavorable consequences to our species. By relaxing the selection pressures that act on our species, Muller warns, we are undoing in a few generations what natural selection has done over millenia. We are breeding some of the **least** fit members of our species, including the lame, the blind, and those susceptible to communicable diseases, by allowing such individuals to survive and reproduce. The result, Muller warns us, can only lead to the genetic deterioration of our species, and an increase in our genetic load. We are simply becoming genetic cripples. According to Muller, our biological evolution has slowed down nearly to a halt; according to Teilhard it has simply been overtaken by a more rapid process of cultural evolution.

Perhaps a larger number of evolutionary biologists, notably Dobzhansky (1962 and elsewhere), have insisted that humans continue to evolve even today. Natural selection, the motive force of evolution, is still operating as long as we are subject to unequal rates of birth and death—as long as some of us have an increased chance, for whatever reasons, of leaving more offspring (Box 28–D). Despite numerous medical advances, children born with genetic defects still have greatly reduced chances of surviving to reproductive age or of leaving healthy offspring themselves. Dobzhansky's own figures are that achondroplastic (chondrodystrophic) dwarfs have only about one-fifth the chance of living to reproductive age and leaving viable offspring, compared to the general population. Despite the availability of insulin, diabetics also tend to have fewer offspring than the general population, and the same observation applies to others suffering from various genetic defects. On the other hand,

BOX 28–C *Major Causes of Death in Selected Countries*

Listed here are, in order of incidence for each country, those causes of death that account for 3% or more of total recorded deaths. In the nonindustrialized countries, note the high incidence of many infective diseases, including those that strike during childhood or during the peak reproductive years. These causes have a strong, direct selective influence on the gene pool of future generations. In the industrialized countries, note the higher incidence of cardiovascular diseases and other conditions that strike postreproductive individuals primarily. Such causes of death are of less selective significance because of the advanced age of many of their victims. Note also that, in the industrialized countries, accidents (especially motor vehicle accidents) are usually the major cause of death before or during the peak reproductive years and are therefore a major selective force.

Four Nonindustrialized Countries:

ANGOLA (1972)

1.	Ill-defined causes	16.7%
2.	Enteric infections	14.6
3.	Pneumonia	10.3
4.	Accidents	9.6
5.	Other noninfective	5.6
6.	Congenital anomalies	3.7
7.	Other perinatal causes	3.6
8.	Stroke	3.4
9.	Cancer	3.1

BOLIVIA (1966)

1.	Ill-defined causes	23.2%
2.	Other noninfective	13.4
3.	Pneumonia	11.0
4.	Other perinatal causes	9.5
5.	Whooping cough	5.8
6.	Tuberculosis	3.8
7.	Enteric infections	3.4
8.	Other infections	3.1

KENYA (1970)

1.	Whooping cough	13.2%
2.	Other noninfective	11.8
3.	Enteric infections	10.8
4.	Pneumonia	10.1
5.	Ill-defined causes	7.5
6.	Typhoid fever	6.1
7.	Measles	4.9
8.	Stroke	4.9
9.	Homicide, etc.	3.0

SRI LANKA (1968)

1.	Ill-defined causes	24.4%
2.	Other noninfective	19.1
3.	Other perinatal causes	8.7
4.	Enteric infections	5.3
5.	Pneumonia	5.0
6.	Accidents	4.1
7.	Cancer	3.9
8.	Other heart diseases	3.7
9.	Other infections	3.4

Three Industrialized Countries:

NETHERLANDS (1977)

1.	Cancer	25.9%
2.	Coronary heart disease	23.4
3.	Stroke	10.7
4.	Other noninfective	8.2
5.	Other heart diseases	6.9
6.	Ill-defined causes	5.4
7.	Accidents	4.7
8.	Bronchitis, etc.	3.4

POLAND (1977)

1.	Other noninfective	17.6%
2.	Cancer	17.1
3.	Other heart disease	11.8
4.	Coronary heart disease	9.5
5.	Ill-defined causes	8.2
6.	Stroke	6.8
7.	Accidents	6.0

Three Industrialized Countries: (continued)

UNITED STATES (1976)

1. Coronary heart disease	33.7%
2. Cancer	19.7
3. Stroke	9.8
4. Other noninfective	9.8
5. Accidents	5.2

Calculated from United Nations statistics published 1979 in the Demographic Yearbook for 1978, table 21.

BOX 28-D *Evidence of Ongoing Selection in Human Populations*

Several independent investigators working in different parts of the world and examining different traits have uncovered evidence of ongoing selection in human populations. Here are some examples.

ACHONDROPLASIA

Achondroplasia, also called chondrodystrophy, is a rare disease in which the cartilage tissue turns into bone at too young an age. The result is that the bones fail to grow to their normal length, producing a type of dwarfism in which the face has a near-normal appearance but the limbs and fingers are shortened. Achondroplasia occurs in various human populations around the world and in such other species as Pekingese dogs. It is inherited in simple Mendelian fashion as a dominant trait.

Mørch (1941) studied the incidence of achondroplasia in Denmark. He found that achondroplastic dwarfs enjoyed rather normal health as adults but produced children at only about one-fifth the rate of their nondwarf siblings. (The 108 dwarfs in this study had only 27 children, or about 0.25 offspring per dwarf parent. Their 451 nondwarf brothers and sisters, on the other hand, had 582 children, or about 1.27 offspring per nondwarf parent. The dwarfs thus reproduced at only 0.25/1.27, or 0.20 of the rate of their nondwarf siblings.) Possible reasons for this 80% reduction in fitness include the difficulties of childbirth among achondroplastic women, and the fact that achondroplastic dwarfs often do not marry or choose to remain childless. Cultural standards of beauty are one form of selection against achondroplastic dwarfs; culturally induced fear of producing achondroplastic offspring is another.

PORPHYRIA

The Afrikaners of South Africa are descended largely from Dutch colonists who arrived in the 17th century. About half of the nearly 2 million Afrikaners have the family names of only 40 original colonists. Among the Afrikaners, there is a relatively high incidence (1%) of a disease called porphyria, probably as the result of genetic drift among the founder population. The disease imparts a reddish color to both urine and feces. It also makes porphyrics acutely sensitive to many drugs, including barbiturates. Many porphyrics have died in South African hospitals as the result of drugs that were intended to help them—drugs that would have been beneficial to most "normal" patients. South Africa's highly renowned hospitals are thus responsible for a form of selection against a genetically determined trait that affects many thousands of South Africans.

ABO BLOOD GROUPS

Under most conditions, persons belonging to blood groups *A, B, AB,* or *O* enjoy equally good health. There are exceptions, however, in the form of diseases that are more prevalent among persons of a particular blood group or groups. Smallpox is one such disease, for it affects people of blood groups *A* and *AB* more than those of blood groups *B* and *O*. Vogel and Chakravartti (1971) studied the effects of smallpox epidemics that hit rural Bihar and West Bengal provinces (India) during the period of 1964–1966. Mortality in these largely unvaccinated populations was as high as 50% in some cases. The following were among the findings of this study:

1. Smallpox victims were more often of groups *A* and *AB,* as compared to their unaffected siblings or other near relatives.

2. Among the survivors of epidemics, scars from smallpox were larger and more striking among persons of blood groups *A* and *AB,* indicating a more severe level of the disease.

3. Most significantly, the proportions of blood groups *A* and *AB* fell, and the proportions of blood groups *B* and *O* rose, as the result of these epidemics.

(continued)

BOX 28–D.   *(continued)*

BIRTH WEIGHT

Around the world, even in cities offering the most advanced health care, the rate of infant mortality depends heavily on birth weight. Despite all the medical advances of the past 100 years, low-birthweight infants are still at a severe selective disadvantage. Infant survival rates are 98% or higher for the optimal birth weight interval of 3–4 kg (6–9 pounds). Survivorship is much lower for lower birth weights. Similar relationships have been documented in other studies in London, Singapore, and elsewhere around the world among various racial and ethnic groups.

There is evidence that infant survival is also related to altitude. At high altitudes, the lower oxygen tension results in somewhat lower infant survivorship under comparable conditions. There is thus a somewhat higher infant mortality in United States hospitals in Denver, Colorado, and other high-altitude Rocky Mountain cities, as compared to the rest of the United States. (On other criteria, these hospitals provide health care on a par with, or even better than, hospitals in lower-altitude states.) When Peru was ruled by Spain, a similar selection pressure is said to have been important in the decision to move the capital from Cuzco (the ancient Inca capital, in the highlands) to Lima (the modern capital, on the coastal plain).

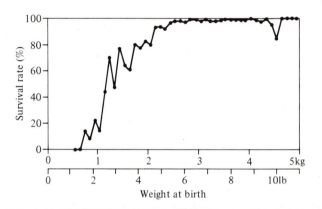

Rh-negative women seem to give birth to a greater-than-average number of offspring, thus perpetuating their genotype in the population despite the opposing selection, which in this case amounts to the increased risk of death (from hemolytic anemia) of *Rh*-positive infants born to such women. According to Dobzhansky and his supporters, we have neither violated nor vitiated the principle of natural selection. We have simply changed some (certainly a minority) of the selective forces operating on our own species.

selective forces has generally been carried out unwittingly, but it raises the question of more deliberate modification of these selective forces in the future. Can we control the future? To what extent, and by what means? Do we know which genotypes will be most fit in future environments? It is this author's firm belief that we can control the future, but only to a limited extent, and that our best hedge against future exigencies beyond our control is the preservation of the genetic variability of our populations.

"Futurology" and the Options Available

The study of the future, or of methods for predicting future events, has been given the somewhat barbarous name of "futurology." But as a science, futurology is still in its infancy. Historically, our ability to predict the course of future events has been notably poor.

E. PREPARING HUMANKIND FOR THE FUTURE

We hominids have been modifying the selective forces that act upon us ever since tools first came into use, if not before that. To be sure, this alteration of our species'

Cultural and technological innovation have had a profound influence on the selection pressures that beset the human species. Among the more important of the innovations were the development of agriculture, the rise of stratified societies, the industrial revolution, and the spread of modern medicine. None of these had been predicted in advance! Neither were penicillin, polio vaccine, air travel, airborne warfare, atom bombs, or smog predicted much in advance, yet each has had a profound effect on selection pressures. The exigencies of the future are hard to predict, except when they are merely extensions of the present. For example, the spread of modern medical practice to all parts of the globe can be safely predicted, but still its rate of spread cannot be foreseen.

Economists and population scientists (demographers) have built many a reputation on the ability to predict future events. The fortunes of stockbrokers and life insurance companies depend on their ability to predict future events, at least statistically. Yet few of these types of prediction can be extrapolated very far into the future. One reason is that extrapolations into the future are exceedingly sensitive to small changes in the values assumed, and a small inaccuracy in the assumed value can therefore lead to a large inaccuracy in the prediction. One very ambitious project was launched in 1974 by the Club of Rome: using the best data available to them at the time, along with the advice of several economists and demographers, this group set up an elaborate model of the world's population, world consumption patterns, and the world's economy in general. By careful adjustment of various parameters, the authors of this study were able to devise several alternative predictions for the state of the world in the year 2000. It is perhaps interesting that none of their possible scenarios predicts a stable future for the world, nor even one that oscillates within acceptable limits. If one reads these models as literally correct, then one is driven to the conclusion that, in one way or another, the world will be totally out of control by the year 2000. Of course, it is also possible that the model itself is mistaken, and that an entire team of well-trained and dedicated ''futurologists'' cannot predict even one generation into the future. Even the most ambitious of futurologists rarely extend their predictions two or more generations into the future, and two generations is but an instant in evolutionary time.

If the future cannot be predicted very far, then how do we know which genotypes will be most fit under future conditions? The true answer is that we simply do not know. Individuals born a mere hundred years ago with an inherited susceptibility to tuberculosis or smallpox had a greatly reduced chance of surviving to reproductive age, and the genes responsible for these traits would have been considered highly disadvantageous. Yet today, susceptibility to these diseases confers very little disadvantage within the industrialized nations, and even in the least developed parts of the world they confer far less disadvantage now than formerly. Who knows whether the same degree of improvement over the next hundred years will reduce the selective disadvantage of cancer, cystic fibrosis, or other presently incurable diseases? Who knows what technological improvements over the next hundred years may reduce the fatality rate of automobile accidents? Who knows what political events will make wars take either a far greater or a far smaller toll of human lives a hundred years from now? Who knows whether pollution will have worsened greatly or whether it will be brought under control?

In the absence of answers to these questions, perhaps the best hedge against future uncertainty is the preservation of population variability. Balanced polymorphism achieves this already, but the balance may of course shift if selective forces change. The eradication of malaria as a selective force in African populations will surely result in a reduction of sickle-cell anemia as well. Many rare genetic traits, such as albinism or phenylketonuria, exist in frequencies in the vicinity of one affected individual per every 40,000 to 60,000, corresponding to frequencies of between 1/200 and 1/250 for the responsible alleles. This is several thousands of times more frequent than would be expected of a recurrent mutation alone, showing that the gene frequency is maintained by some sort of polymorphism. Perhaps this is caused by an advantage (as yet undiscovered) to the heterozygotes, an advantage that would be lost to future generations if we were to eradicate the recessive allele somehow. A population with a greater degree of heterozygosity will in general be better able to cope with unexpected environmental changes than will a genetically uniform (homozygous) population.

If we were to try modifying the evolution of our species in a desired direction, what means would we have at our disposal? An attempt to categorize the various possible techniques appears in Box 28–E. Each has had its advocates; some are already being practiced. A necessary

BOX 28–E *Possibilities for Genetic Intervention*

I. DIAGNOSIS AND COUNSELING. These techniques permit informed decisions concerning the measures outlined in II and III.

1. Risk evaluation. Through a study of pedigrees and application of Mendelian principles, the probability of afflicted offspring can be assessed. For traits whose genetic basis is complex or not fully known, risks can be determined empirically from medical records.

2. Amniocentesis. By studing cells derived from the fetus within the amniotic fluid, many chromosomal abnormalities can be detected. Cultures made from these cells can also be tested for many single-gene defects; the number of defects detectible by these means is steadily increasing.

3. Mass screening of infants.

4. Mass screening of prospective parents. The next method listed is really an important special case of this.

5. Heterozygote detection. Development of techniques to enable heterozygotes to be identified, as is already possible with sickle-cell anemia and in a few other cases as well. Populations with significant frequencies of deleterious alleles could be screened, and heterozygotes discouraged from marrying one another and having children. As a bare minimum, heterozygous individuals should be educated to realize that the children of such marriages are medically at risk.

II. CHANGES IN THE GENE POOL (Eugenics in the broadest sense).

A. Alteration of individual genotypes. (Genetic engineering.)

1. Genetic surgery.

2. Recombination therapy, allowing replacement of a defective gene with its functional allele.

B. Alteration of the gene pool through selection. (Eugenics proper.)

1. Positive eugenics (eutelegenics), or increasing the contribution to the gene pool by desired genotypes. Semen from chosen males could, for example, be stored in sperm banks and used in artificial insemination. Egg banks are a more distant, future technology.

2. Negative eugenics, or decreasing the contribution to the gene pool by undesired genotypes. Among the methods available are abortion, sterilization, and various forms of murder including infanticide.

C. Asexual reproduction (cloning). The exact genetic copying of an individual would have a long-range effect similar to that of positive eugenics.

III. CHANGES IN GENE–ENVIRONMENT INTERACTIONS.

1. Medical engineering (euphenics). This type of medical intervention allows defective genes to be circumvented and a normal phenotype to be produced despite the genetic defect. Insulin for the diabetic is a familiar example of this; another example would be placing an individual who is homozygous for phenylketonuria (PKU) on a diet low in phenylalanine before any irreparable damage is done.

2. Environmental engineering (euthenics), or changing the environment to accommodate defective phenotypes. Wheelchairs for the lame and Braille for the blind fall into this category.

3. Social and behavioral engineering (eupsychics). This includes education and all other forms of behavioral modification in the broadest sense, which may be used to compensate for various deficiencies or to overcome them. Also included here are legislated or voluntary changes in social customs such as those concerned with mate selection, child rearing, and the like.

first step in each case of genetic intervention is always diagnosis and Mendelian analysis, followed in most cases by genetic counseling. Further measures are also possible; those that change the composition of the gene pool are rather permanent in their effects. On the other hand, changes in the interactions between gene and environment would generally be impermanent and would have to be perpetuated culturally and repeatedly practiced over each new generation.

Diagnosis and Counseling

The methods of diagnosis and counseling permit informed decisions regarding further intervention. Individuals at risk for particular traits, usually because of a family history in which the trait was recurrent, may present themselves to a genetic counseling clinic for advice when marriage or pregnancy is planned. Traits known to have a simple genetic basis can be analyzed through a study of pedigrees or family histories. By applying Mendel's principles, the probability of afflicted offspring can be calculated. For traits whose genetic basis is more complex, or not fully known, risks are determined empirically from studying the medical records of numerous similar cases. For instance, the offspring of diabetics have an empirically determined 84% chance of becoming diabetics themselves. Couples can therefore be advised on the probability that their next offspring will or will not have a particular defect in question.

Amniocentesis is a further diagnostic tool, in the case of pregnancies already in progress. A small amount of fluid is withdrawn from the amnionic cavity, and cells derived from the fetus are extracted from it. The fluid itself can be tested for the presence or absence of certain proteins, and the cells can be examined chromosomally or even grown in tissue culture. Chromosomal defects such as Down's syndrome (caused by any of several defects, including trisomy of chromosome no. 21) are already detected by this method, with the option of a therapeutic abortion available to the mother who wishes not to have the child as a result.

The above methods are in most cases applied only to those individuals at risk for a particular trait. They are not (as yet) used for the mass screening of large populations. Such mass screening methods could be practiced as routine public health measures, either by screening of infants or by screening of their parents. Routine screening

of infants in hospitals could easily be carried out. A blood sample, for instance, could be tested for the presence of various enzymes. If a particular enzyme were missing, euphenic measures could be instituted to supply the enzyme artificially, or to alter the diet in some suitable way. For example, patients lacking the enzyme to metabolize the amino acid phenylalanine (therefore in danger of developing phenylketonuria, PKU) could be placed on a phenylalanine-free diet, as is now done whenever the discovery is made in time.

Alternatively, mass genetic screening could be carried out on reproductive members of the population, say, all persons within a certain age group, or all persons applying for a marriage license. The most important form of screening in this case is heterozygote detection, for in this manner the risk of couples in which both husband and wife are heterozygous can be revealed. Traits of limited or delayed penetrance might also be detected at this stage, for instance a test that would enable early detection (and possibly thwarting) of Huntington's Chorea would be most welcome.

Genetic Engineering

We are now on the threshold of a new field, that of genetic engineering. If a defective gene can first be detected and identified, it might someday be possible to repair the defect, or to replace the defective gene with a properly functioning substitute. The first of these alternatives would be called genetic surgery, and would involve direct manipulation of the genotype. The second alternative may be called recombination therapy. A DNA segment containing the properly constituted gene is introduced, and the recombination of this gene with the patient's chromosome is brought about through a process similar to viral transduction. Neither technique has yet been perfected, and mass application of either technique would require far more knowledge of the human genome than we now possess.

The new, much-talked-about technology of DNA recombination (or "gene splicing") marks a new phase in the history of genetics, and in the history of our attempt to control our own evolution (and that of our domesticated species as well). A set of "restriction enzymes" is used to snip out a desired fragment of DNA from a much larger sequence, e.g., a single gene or gene complex from an entire chromosome. The fragment is then inserted into the DNA sequence of a host species—usually the bacterium *Escherichia coli.* (The host's DNA sequence must first be snipped by the same type of restriction enzyme, creating a gap into which the new fragment can be inserted.) If all goes well, the bacterium will be "fooled" into treating the inserted fragment as part of its own bacterial genome and making the gene product (first messenger RNA, then protein) for which the inserted gene codes.

As the decade of the 1980s opens, we have reached the stage in which DNA recombination has successfully been carried out in a few instances, producing in each case small quantities of the desired gene products. Soon we may be able to culture the recombinant bacteria, and permit them to make the product of the inserted gene on a commercial scale. This technique will enable us to manufacture many mammalian hormones (like insulin or somatostatin) in bacterial cultures—a great medical advance. Most insulin used today is obtained from pigs, but it is usually accompanied by other pig proteins to which many people are allergic. Some insulin is obtained from other animals, but humans may be allergic to them, too, and the hormone is still impure. Recombinant DNA technology will, in a few years, enable us to obtain nearly pure human insulin, of the type that nondiabetic humans normally produce, by culturing bacteria into which the human gene for insulin has been inserted. Another potential exists for somatostatin, also called pituitary growth hormone. People who cannot produce this hormone themselves are abnormally short pituitary dwarfs. The hormone can now be obtained in small quantities, but only at a very great expense (beyond the means of nearly all individuals), and still not in the quantities needed by even a single affected individual. Recombinant DNA technology may soon allow the commercial production of human growth hormone in bacterial cultures, at a price within means and in ample quantity.

A further step in recombinant DNA technology might someday allow a functional gene to be inserted into a human or domestic animal or plant whose own gene is defective. A functional insulin gene might, in other words, someday be implanted into the cells of a diabetic, allowing that individual to make insulin in the normal manner, and obviating the need for daily insulin injections from any source. Other human genetic defects, such as phenylketonuria (PKU), colorblindness, or sickle-cell anemia, could also eventually be cured by such genetic surgery. The same technique could also be used to insert desired genes into domestic animal and plant species. The most common and fast-growing cereal crops can only grow if the soil contains sufficient quantities of nitrogen in the form of nitrates. Usually these nitrates are supplied by petroleum-derived fertilizers. But a few plants (peas and other legumes) contain an enzyme that allows them to manufacture their own fertilizer by "fixing" atmospheric nitrogen and converting it into nitrates without using precious petroleum. If we could extract from a legume the gene that codes for this enzyme and insert it into the gene sequence of corn or wheat, we would be able to grow vast quantities of these cereal crops on land that is currently unsuitable for agriculture without the use of nitrate fertilizers derived from oil.

With all the potentialities that recombinant DNA technology holds, one might expect everyone to be in favor of the new techniques. But a number of prominent scientists and other citizens have instead sounded a cry of alarm over the dangers involved. What if an undesired gene were accidentally inserted into a bacterium? What if the bacterium were to escape from the laboratory into the surrounding atmosphere (an all too common occurrence)? The bacterial host usually used in these experiments is *Escherichia coli,* a normal inhabitant of the human intestinal tract! Such a bacterium could acquire a resistance to antibiotics very easily, and spread in epidemic proportions. These and other objections to recombinant DNA research will never be completely silenced, but their criticisms have already resulted in a number of important safeguards, including a series of guidelines put forth by the National Institutes of Health (NIH). One of the most important safeguards is the use of a special strain, K-12, of bacterial hosts; this strain is incapable of living outside the laboratory, and would therefore perish should it ever escape into the environment. The strain requires thymine in order to grow, and it is easily killed by ultraviolet light. Thymine can easily be supplied in the laboratory, but cannot usually be found in the outside environment.

A few mammalian genes have thus far been incorporated into laboratory cultures of the bacterium *Escherichia coli,* and the products of these genes have in some cases been synthesized by the bacteria. However, many further technological advances will be needed before defective DNA segments can be therapeutically repaired or replaced. Not only would the functional segment of

DNA have to be synthesized, but it would have to be incorporated first into a living system (perhaps a bacterium or a "tame" virus), and then controlled recombination between the properly constituted gene and the patient's defective gene would have to be effected.

In addition to the strictly scientific and technological questions, many ethical and legal questions have been raised by the new recombinant DNA technology. Among these questions are:

1. How can we ensure that accidental recombination of the "wrong" genes might not create a monster, or a new and dangerous disease organism whose growth cannot be controlled?

2. To whom is the scientist ultimately responsible? (To his profession only, or to his funding agency, or to society at large?)

3. Does it restrict scientific freedom to insist that scientists not perform experiments whose results may be hazardous or unpredictable? Does scientific freedom have its limits?

4. Who is the best judge of highly technical scientific issues affecting public policy? (This question, which has also arisen in connection with nuclear reactor safety rules, is basically political. But most political leaders lack the scientific expertise to decide such complex issues, and most scientists who are experts in the field would have a vested interest in the research. Political leaders often consider the advice of scientific "experts," but when the "experts" disagree among themselves, how is the nonscientist to decide which expert advice to follow?)

5. If a new type of organism is created through recombinant DNA technology, could it be patented by its originator? Patents have always been regarded as both a safeguard and a stimulus for inventors, but can anyone legally restrict someone else's use of a living species?

6. Who is at fault if a recombinant organism escapes from a laboratory and causes harm to persons outside? Who is responsible, and to what extent, to make sure that this never happens?

7. Is tampering with the human genotype morally acceptable, or are there ethical or religious objections? Should parents seek to cure their diabetic children by inserting a gene for insulin? Should parents be allowed to choose their children's hair color or eye color? Who decides which operations are permissible?

These questions, and others, are important issues that society will have to face in the coming few decades.

Cloning

A **clone** is a group of organisms or cells consisting of a single individual or cell and its asexually produced offspring. Barring mutation, all members of a clone are genetically identical. Identical twins are a clone of two, since they arose through cleavage of the same fertilized egg.

Human beings do not reproduce asexually. If, however, a technology was developed to allow such asexual reproduction (for example, by implanting a diploid nucleus from a somatic cell into an egg cell from which the original nucleus had been removed), a human clone could conceivably be produced. While various social and legal questions (inheritance, parental responsibilities, etc.) would need to be answered, the technique of cloning would allow the copying of desired genotypes. At this point, we must again ask, which are the desired genotypes, and who shall select them? Will society really benefit from multiple copies of any individual? As an added thought: great individuals (by any criterion) owe at least some of their greatness to favorable circumstances or favorable environment, neither of which could be guaranteed to their clonal offspring. If Darwin had been cloned, would his clonal offspring have had a Hooker or a Lyell with whom to associate, or a Wallace to prod them into publication? Would they have had the fortune to be sent on the *Beagle* voyage? Could the formative years of Darwin's early nineteenth-century upbringing be duplicated in another decade? How readily could the talents that produced the *Origin of Species* be used in a different generation to produce other, equally impressive results?

Eugenics

Eugenics, or changes in the gene pool through selection, is hardly a new idea. Plato advocated a form of positive eugenics in *The Republic* (book 5), by carefully selecting only the best and the healthiest individuals of each sex and breeding them together. This is very much what we do with our breeds of domestic cattle, sheep, and poultry, where the criteria of perfection are rather simple, since they conform to a Platonic *eidos:* dairy cattle, for instance, are selected for high milk yield, butterfat content, resistance to disease, and certain other clearly desirable traits.

More recently H. J. Muller (1961, 1965) has advocated storing the semen from selected male donors in sperm banks. The sperm thus stored could be used to artificially inseminate women who Muller thought would eagerly volunteer in the hopes that their children would be born with superior genotypes. A much more distant possibility would be the storage of eggs in egg banks, followed by their artificial insemination and their implantation into the uteri of willing human incubators. Many legal, social, and technological problems will have to be dealt with before any of these prospects are realized; among these are questions of inheritance and of maternal responsibility.

Positive eugenics, in any case, depends upon our determination of desirable genotypes. Here there is no set standard, no Platonic ideal, by which we could judge human beings. Moreover, persons who excel in one or a few areas might happen to possess traits that we might not care to perpetuate. Perhaps Einstein was heterozygous for a genetic defect of some sort; would we wish to breed this defect in the hopes of achieving superior intelligence? Suppose a person were to inherit Edgar Allen Poe's mental illness instead of his creative genius? Worse, what if the technology were to fall into the hands of a dictator who wanted, for instance, to make multiple copies of himself? Who is a superior individual? Who should have the power to make this choice?

Most discussions of eugenics have focused on its negative measures, those which decrease the genetic contributions of undesired genotypes. Historically, this was the focus of the modern eugencis movement. Founded by Francis Galton (1822–1911), a cousin of Charles Darwin, the eugenics movement concerned itself initially with metric variation such as in height or intelligence. The biometrical (i.e., statistical) approach has remained with British genetics ever since, and has resulted in a theory of population genetics of which Hardy, Fisher, and Haldane were the founders. Attempts to devise statistically reliable measures of intelligence brought about a cooperation with psychologists such as Spearmann who were developing what we now know as IQ tests. With such respected intellectual companions as these, eugenics has always enjoyed a favorable reputation in England.

In Germany, on the other hand, hastily thought-out eugenics measures were instituted without adequate safeguards, beginning with the compulsory sterilization of mental defectives. Such were the beginnings of Hitler's campaign against those who did not belong to his "master race." By 1945, millions of innocent people had been slaughtered in the name of racial purity and Aryan superiority.

In the United States, the eugenics movement began as an attempt to identify, segregate, and sterilize mental defectives, but soon it allied itself with racism and especially with efforts to curb the new waves of immigration that came to our shores between 1890 and 1920. In the 1890s, Connecticut passed a law that declared that "no man and woman either of whom is epileptic, or imbecile, or feeble-minded shall marry or have extra marital relations when the woman is under forty-five years of age," while in Kansas a doctor named F. H. Pilcher incurred the wrath of public opinion by sterilizing 44 boys and 14 girls at the Kansas State Home for the Feeble-Minded. An Indiana law passed in 1907 required the sterilization of "confirmed criminals, idiots, imbeciles, and rapists in state institutions when recommended by a board of experts"; fifteen other states passed similar laws, often for punitive as well as eugenic reasons. Between 1909 and 1929, 6,255 people were sterilized under such laws in California alone (Haller, 1963).

In 1910, a Eugenics Record Office was established on Long Island, N.Y., under the direction of C. B. Davenport. Davenport was influential in many eugenics societies, including the American Eugenics Society. It was during this period that the writings of American eugenicists became distinctly racist and anti-immigrationist in tone. The eugenicist P. Hall, wrote in 1910 that "The same arguments which induce us to segregate criminals and feebleminded and thus prevent breeding apply to excluding from our borders individuals whose multiplying here is likely to lower the average of our people."

The Nobel Prize-winning geneticist, H. J. Muller (1890–1967) was the most prominent American geneticist to ally himself with the cause of eugenics. On many occasions, notably in 1963, Muller expressed a note of warning to the American people: too much care for the genetically defective would thwart the very selection that naturally tends to eliminate them from every population. Thus, if we continue to alleviate genetic defects through medical intervention, we are unwittingly perpetuating these defects in the gene pool and increasing the genetic load of our population. The pessimistic picture painted by Muller included a population divided into two groups, one consisting of people so defective genetically that their very lives had to be maintained by extraordinary means, and the other consisting of people who were themselves normal, but whose entire lives had to be devoted to the care and sustenance of the first group.

Beginning in the 1920s and culminating in the succeeding three decades, a wave of antiracism and antihereditarianism overtook much of the American scientific community. Some eugenicists had made extravagant claims for the hereditary nature of such traits as alcoholism, criminality, and vice, and these claims had begun to backfire. The behaviorist school of psychology was distinctly antihereditarian. The absolute heredity of IQ scores was coming to be doubted and environmentalism was coming to be strongly influential in such fields as anthropology. Many practicing geneticists, except for Muller, denounced racism and with it the entire eugenics movement. The Catholic Church, long a foe of compulsory sterilization, began to oppose eugenics, and has continued to do so rather consistently. The rise of Hitler's Germany during the 1930s and its defeat during the 1940s made vivid to many Americans the dangers of a eugenics program without adequate safeguards. The American eugenics movement lost much support, which it has only begun slowly to regain. Geneticists such as Theodosius Dobzhansky (1893–1975) and S. E. Luria (1912–   ) began to oppose eugenics, arguing that there was nothing wrong with altering the environment so that formerly deleterious genes were no longer so harmful or debilitating. Dobzhansky, for example, made it plain that he favored measures that would permit the bearers of hereditary defects to overcome their handicaps, making them into phenotypic copies (phenocopies) of the normal, healthy members of the population. Once the phenotype could be controlled culturally, the relaxation of selection against the previously defective genotype should hardly be of any great concern.

The various measures suggested for negative eugenics are all rather simple: persons with undesired genotypes should undergo compulsory sterilization, or perhaps abortion should they become pregnant before sterilization can be effected. Infanticide, long practiced in China, as in many other cultures, would also be effective. The moral and legal problems are enormous if not insurmountable. Who is to decide that another person must undergo involuntary sterilization? And what safeguards should exist against the abuse of this power?

The dangers of a misguided eugenic program cannot be overemphasized. The decision as to which genotypes should be eliminated must be made by a sufficiently large and diverse group, including members of all minority groups. Otherwise, the possibility of legalized genocide becomes all too real, because a racial minority could be ordered sterilized, and thus wiped off the face of the Earth in a single generation, simply by declaring them to be mentally inferior or genetically unfit. Racists have long argued for the mental or genetic inferiority of other races, and Nazi Germany stands as a reminder that a misguided program of eugenics can indeed result in genocide. In fact, Nazi Germany is the only nation ever to have instituted a full-scale program of eugenics.

The constraints on any program of eugenics should again be made clear. Selection against a dominant trait with complete penetrance could be made totally effective in a single generation (Chapter 11). Selection against a recessive trait would at best proceed very slowly, and the rate would decline asymptotically as the trait became increasingly rare. Partial or inefficient selection, limited penetrance, epistasis, and environmental interactions would all tend to make the process even less efficient. A eugenic dictator who ordered the complete elimination of all albinos, for instance, would require about 200 generations (roughly 6,000 years) of constant vigilance merely in order to reduce the already low gene frequency in half. The process would take about twice as long if the eugenic authorities were successful in finding only half the albinos—for instance if the other half were able to hide or otherwise elude them. For a polygenic trait, selection would be slower still, and any environmental influence on the trait would slow down the selective process still further. Selection for height or IQ would fall into this last category, and would be so slow as to be barely perceptible. One geneticist calculated that it would take about 400 years of constant, unrelenting and totally efficient selection to raise IQs by about 4 points. The same result could also be achieved, in a group of disadvantaged children, by improving their education for as little as four years, a calculation based on data from the Philadelphia public schools.

The risks involved in negative eugenics are great. The benefits are of small magnitude, except in the case of completely penetrant dominant traits. A proponent of negative eugenics might be hard put to show that the benefits of such a program would outweigh the risks, given the availability of such alternative measures as are listed in Box 28–E.

Euphenics, Euthenics, and Eupsychics

Medical engineering can in some cases be used to circumvent defective genes and produce a normal phenotype (a phenocopy) in spite of them. This practice is called **euphenics,** and includes such measures as giving insulin to diabetics or installing pacemakers in defective hearts. Optimistically, early intervention may someday be able to supply the right enzyme (or enzyme product) at the right time, and this might in some cases make unnecessary the dependence upon continual administration of insulin, for example.

Euthenics consists of changing the environment to accommodate the defective phenotypes. Wheelchairs for the lame and Braille for the blind fall into this category. Special furniture or special cars can be obtained (but often at great expense) to accommodate a variety of special needs, and public buildings are increasingly being fitted with such accommodations (ramps for wheelchairs, etc.). Anything that ameliorates the hardship of those already suffering is desperately to be wished for.

Social and behavioral engineering, or **eupsychics,** may also serve to ameliorate conditions of hardship or to compensate for them. This includes the special education of afflicted individuals, and also the education and social conditioning of other members of society so as better to accommodate the needs of the afflicted.

Overpopulation and Pollution

Darwin, in 1859, noticed that every species tends to increase in numbers at a very high rate, a rate so prodigious that, if the overwhelming numbers of any species were not systematically destroyed, that species would soon cover the face of the Earth and literally crowd itself to death. Our species is certainly no exception to this rule, for in 1800 there were only about 800 million human inhabitants of our Earth, and by 1900 the number had nearly doubled. Between 1900 and 1968, the world population again more than doubled, from 1.6 billion to 3.5 billion, according to United Nations estimates. The rate of increase, moreover, is itself increasing: between 1750 and 1850, half a billion people were added to the Earth's population in a hundred years, yet between 1960 and 1968, half a billion people were added in only eight years! In the next thirty years, the world population will again double, and the Earth, which now supports some four billion inhabitants, will be called upon to support some eight billion!

The numbers of any species are kept in check by predation, disease, the shortage of food, the extreme of climate, and other such environmental hazards. When these checks are removed or reduced, as they have been for our species, an increase in numbers invariably follows. This can best be seen in those instances when an animal or plant species is accidentally or intentionally transported to a new continent, favorable in climate and other physical factors, where the natural enemies that kept its numbers down in its native land are absent. Rabbits, for example, were accidentally introduced into New Zealand not long ago, and already they have multiplied so greatly, in the absence of wolves and other common predators, as to become a great nuisance to sheep ranchers, and even to threaten many of the native animals with extinction. "Lighten any check, mitigate the destruction ever so little," declared Darwin, "and the number of the species will almost instantaneously increase to any amount."

The problem of pollution may seem to have little to do with the problem of overpopulation, but, in fact, they are quite intimately related. If one man defecates in a stream, the animals, plants, and especially the bacteria living in the stream can quite readily decompose the organic matter, recovering nutrients from it, and rendering the remainder harmless. But when a large city dumps its raw sewage into the same stream, the living organisms of that stream cannot cope with the drastic change in their environment, and so they die, and their rotting corpses make the situation even worse. Pollution is thus a problem of degree, and not of absolutes. Though it sometimes can be attributed to either ignorance or greed, it is also a certain consequence of overpopulation, and will never be controlled as long as the population of the world continues to increase. Of course, at any given level of population, it is possible to pollute the environment to a greater or lesser degree. *Some* degree of pollution is inevitable, since digestive wastes, dead bodies, and even the simplest of hand-made and manufactured products must somehow be disposed of. All of these can be injurious above a certain threshold quantity—pollution is largely a problem of quantities. It is this kind of social and economic concern, and not the strictly scientific concerns of Chapter 9, that many people have in mind when they speak of "ecology."

Our Uncertain Future

What are the evolutionary consequences of overpopulation, and of pollution? One obvious result is that we are now subject to different environmental stresses, and selection will consequently favor different genotypes. Under future conditions of K-selection, those genotypes that enable us better to withstand life in closer and more crowded quarters will be favored, as will those genotypes that permit us to feed more and more on artificially produced foods and to avoid the various types of accidental and violent deaths to which people living in overcrowded conditions are so prone. Under K-selection, longevity and tolerance of crowded, stressful, and polluted conditions will be favored more and more.

Of course, all this is irrelevant, unless humankind indeed has a future and unless people can indeed learn to live with each other. Human powers, in both pollution and warfare, have grown so much beyond our ability to control them that it is already to our credit that we have not blown ourselves off the face of the Earth. If only we could learn to control these destructive forces, then future generations could grow up free from the fears of world pollution or world annihilation. Yes, we can control our own evolution, but we must first establish adequate safeguards against our extinction. The future of our species is for the next few generations to determine.

CHAPTER SUMMARY

Culture, including the learned beliefs, behavior patterns, and artifacts traditionally transmitted from each generation to the next, undergoes a form of evolution analogous to but distinct from the evolution of genetic material. Cultural evolution differs from biological evolution in recognizing no species boundaries between sympatric populations and in allowing for the transmission of culturally acquired characteristics. Cultural and biological evolution often interact, as when culture dictates or restricts mating preferences. Human social organization reflects the ecology first of hunters and gatherers and later of agriculturalists.

Nonagricultural societies generally range widely and have no permanent settlements. The women usually gather vegetable food, providing the group with most of its calories, while the men range more widely and hunt for animal food to supplement the group's protein intake. There is little division of labor beyond this, nor are there any distinct social classes or comparable status groupings. Prior to the last glacial retreat some 15,000 years ago, all human societies consisted of hunters and gatherers living under paleolithic technology, meaning that stone tools were shaped by flaking rather than by grinding. Tools made of bones, horns, and teeth came even before stone tools, and the early stone tools, such as the Acheulean hand axe, were crudely shaped by removing flakes from them. Later, it was discovered that the flakes themselves could be used as tools and were in fact sharper than the older core tools. Technological advances led to a wide diversity of tools, each for a somewhat different purpose; some of these were made with special facets enabling them to be hafted onto wooden handles or shafts.

The Neolithic Revolution began with the cultivation of plants and the domestication of animals. The Urban Revolution brought about intensive, irrigation-dependent cereal argiculture, trade specialization, accumulation of wealth, stratification of society into classes, and the growth of cities.

The mechanization of textile manufacture, of transportation, and of communications were all consequences of the industrial revolution. The Urban and Industrial Revolutions brought about changes in the selection pressures that control biological evolution. Population size increased, and deaths due to communicable diseases, especially those of infancy and early childhood, have steadily decreased. Industrial hazards, traffic hazards, war, pollution, and drug abuse have all taken increasingly higher tolls, while differences in fertility have become more and more important as selective forces. But though the particular selective forces have changed, we are still subject to differential rates of birth and death, and thus to continuing selection.

Prediction of the future has always proven difficult. If we are to prepare our species for the future, we must retain a healthy degree of population variability, which offers the best protection against unforseen exigencies. The many means at our disposal for controlling our own evolution all begin with diagnosis and genetic counseling. Beyond this, changes in the gene pool may be brought about by genetic engineering (including genetic surgery), by intentional selection, or by cloning. Intentional selection includes both positive eugenics (increasing the number of offspring left by favored genotypes) or negative eugenics (decreasing the number of offspring left by undesired genotypes). Interactions between genotypes and the environment may be made to produce normal or near-normal phenotypes without permanently changing genotypes. Such changes may be either partial or total, and they may include medical provisions (prosthetic devices such as wheelchairs, dietary or biochemical therapy, etc.), restructuring of the surrounding environment, or modification of behavior (including education). Some environmental changes are occurring unwittingly; overpopulation and pollution are among these. We must first bring these under control; then we can control our own future evolution.

Appendix:
A Classification
of Organisms

KINGDOM MONERA (procaryotic organisms)

> Phylum Archaebacteria (methane-producing bacteria)
> Phylum Bacteria (typical bacteria)
> Phylum Cyanophyta or Cyanobacteria (blue-green "algae")

KINGDOM PROTISTA (here restricted to eucaryotic unicells without plastids or cell walls, often called "Protozoa")

> Phylum Ciliata (protists motile by means of cilia)
> Phylum Mastigophora or Flagellata (protists motile by means of flagella)
> Phylum Sarcodina (protists motile by means of pseudopods)
> Phylum Sporozoa (protists reproducing by spores)

KINGDOM MYCOTA (fungi with cell walls and absorptive nutrition, reproducing with the aid of spores)

> **Subkingdom Gymnomycota**

> Phylum Mycetozoa or Myxomycota (slime molds)

> **Subkingdom Eumycota**

> Phylum Chytridiomycetes
> Phylum Hyphochytridiomycetes
> Phylum Oömycetes
> Phylum Zygomycetes
> Phylum Ascomycetes (fungi with spores borne in sacs)
> Phylum Basidiomycetes (fungi with spores produced on basidia)

KINGDOM PLANTAE (eucaryotic organisms with plastids)

> **Subkingdom Thallophyta (plants without embryos; eucaryotic algae)**

> > Division Chrysophyta (golden-yellow algae and diatoms)

Division Pyrrophyta (dinoflagellates)
Division Xanthophyta (yellow-green algae)
Division Euglenophyta (euglenoids)
Division Rhodophyta (red algae)
Division Phaeophyta (brown algae)
Division Chlorophyta (green algae)

**Subkingdom Embryophyta
(plants with embryos)**

Superdivision Bryophyta
Division Bryophyta (mosses)
Division Hepatophyta (liverworts)
Superdivision Tracheophyta (vascular land plants)
Division Psilophyta (psilophytes)
Division Lepidophyta (lepidophytes)
Division Arthrophyta (horsetails and their relatives)
Division Pterophyta (ferns)
Division Pteridospermophyta (seed ferns)
Division Cycadophyta (cycads)
Division Ginkgophyta (ginkgos)
Division Coniferophyta (conifers)
Division Gnetophyta (*Gnetum, Ephedra,* and *Welwitschia*)
Division Anthophyta or Angiospermae (flowering plants)
Class Dicotyledonae (plants with two cotyledons in the seed)
Class Monocotyledonae (plants with one cotyledon in the seed)

KINGDOM ANIMALIA (animals)

Subkingdom Parazoa

Phylum Porifera (sponges)

Subkingdom Mesozoa

Phylum Mesozoa (mesozoans)

Subkingdom Eumetazoa

Phylum Cnidaria (coelenterates or cnidarians)
Phylum Ctenophora (comb jellies)
Phylum Platyhelminthes (flatworms)
Phylum Rhynchocoela or Nemertea

Phylum Aschelminthes
Class Nemathelminthes (roundworms)
Class Kinorhyncha (kinorhynchs)
Class Gastrotricha (gastrotrichs)
Class Gordiacea (horsehair worms)
Class Rotifera (rotifers)
Phylum Acanthocephala
Phylum Entoprocta
Phylum Phoronida
Phylum Bryozoa ("moss-animals")
Phylum Brachiopoda
Phylum Mollusca (molluscs)
Class Monoplacophora (*Neopilina* and its relatives)
Class Aplacophora (solenogasters)
Class Polyplacophora (chitons)
Class Gastropoda (snails, etc.)
Class Scaphopoda
Class Bivalvia or Pelecypoda (bivalves)
Class Cephalopoda (cephalopods)
Phylum Annelida (segmented worms)
Phylum Sipunculida
Phylum Echiurida
Phylum Pentastomida
Phylum Tardigrada
Phylum Onychophora
Phylum Arthropoda
Subphylum Trilobitomorpha
Class Trilobita
Subphylum Crustacea
Class Crustacea
Subphylum Chelicerata
Class Pycnogonida
Class Xiphosura
Class Arachnida
Subphylum Unirama
Class Chilopoda (centipedes)
Class Diplopoda (millipedes)
Class Pauropoda
Class Symphyla
Class Insecta (insects)
Phylum Chaetognatha (arrow-worms)
Phylum Echinodermata (echinoderms)
Phylum Hemichordata (hemichordates)
Phylum Chordata (chordates)

Subphylum Urochordata
  Class Tunicata
Subphylum Cephalochordata
  Class Cephalochordata
Subphylum Vertebrata (vertebrates)
  Class Agnatha (jawless fishes)
  Class Placodermi (placoderms)
  Class Chondrichthyes (sharks, etc.)
  Class Osteichthyes (bony fishes)
    Subclass Acanthodii
    Subclass Actinopterygii (ray-finned fishes)
    Subclass Sarcopterygii (lobe-finned fishes)
  Class Amphibia (amphibians)
  Class Reptilia (reptiles)

  Subclass Anapsida
  Subclass Euryapsida (plesiosaurs, etc.)
  Subclass Synapsida (mammal-like reptiles)
  Subclass Lepidosauria (snakes, lizards, rhynchosaurs, etc.)
  Subclass Ichthyopterygia (ichthyosaurs)
  Subclass Archosauria (thecodonts, dinosaurs, crocodiles, etc.)
Class Aves (birds)
Class Mammalia
  Subclass Prototheria
  Subclass Theria
    Infraclass Metatheria (marsupials)
    Infraclass Eutheria (placental mammals)

Glossary

Adaptation   Any structure or behavior pattern that makes its possessor more suited to its environment or to its mode of life.

Adaptive radiation   Diversification (both structural and ecological) among the descendants of a single species or group of species. An example of evolutionary opportunism.

Advanced   In any lineage that exhibits a trend, the later or derived stages or conditions. The opposite of primitive.

Allele   One of the several alternative states or conditions of a gene.

Allen's rule   An ecogeographic rule: In a geographically variable species, the populations in the warmer parts of the range tend to have longer and thinner protruding parts (legs, ears, tails, beaks, etc.), while in colder parts of the range, these same parts are shorter and thicker.

Allopatric   Occurring in different places, or having ranges that do not overlap; said of species or populations.

Allopolyploid   A polyploid that has two or more distinct genomes, usually as a result of earlier hybridization.

Anagenesis   Phyletic evolution within a single lineage. The opposite of cladogenesis.

Analogy   Resemblance or similarity due to similar adaptation.

Aneuploidy   A change in the chromosome number caused by the piecemeal loss or addition of parts of the genome, usually one chromosome at a time.

Artificial selection   Selection by human agency among domesticated species.

Assortative mating   Nonrandom mating between individuals that are phenotypically similar (positive assortment) or dissimilar (negative assortment). Deviation from panmixia by virtue of a phenotypic mating preference.

Autopolyploid   A polyploid that has more than two sets of the same genome.

Balanced polymorphism   A type of polymorphism that occurs when two or more distinct genotypes exist stably as the result of the selective superiority of heterozygotes. *See* Polymorphism.

Batesian mimicry   A type of mimicry in which a palatable species resembles an unpalatable one.

Behavioral (ethological) isolation   A premating isolating mechanism whereby two species may meet but will refuse to mate because of differences in courtship behavior.

Bergmann's rule   An ecogeographic rule: In a geographically variable species, the populations inhabiting colder parts of the range tend to have larger average body size, while those in the warmer parts of the range tend to have smaller average body size.

Biological species definition   Definition of the species category on the basis of reproductive isolation: "Species are groups of naturally interbreeding populations that are reproductively isolated from other such groups."

Bottleneck effect A kind of genetic drift that occurs when a large population becomes small and then large again. The gene pool of the enlarged population may reflect accidental changes that occurred when the population was small.

Catastrophism A geological theory, now discredited, according to which the face of the earth was shaped by violent events of great magnitude. Such ideas were common until 1830, when Lyell successfully established uniformitarianism as the cornerstone of modern geology.

Category (taxonomic) A level in the taxonomic hierarchy to which taxa may be assigned; for example, genus, phylum, family, species.

Centrifugal selection *See* Disruptive selection.

Centripetal selection *See* Stabilizing selection.

Character displacement A phenomenon in which the differences between two closely related species are increased or accentuated in the region of their sympatric overlap.

Chromosomal aberrations Changes in chromosome structure involving more than one gene at a time. Specific types include inversion, deletion, duplication, and translocation.

Chronospecies *See* Paleospecies.

Cistron The unit of genetic function; the hereditary material necessary to control the synthesis of one polypeptide chain. One cistron contains several dozen to several hundred base pairs. *See also* Gene.

Clade A taxon or other group consisting of a single species and its descendents. A taxon or other group that had a single origin. Also known as a monophyletic group.

Cladogenesis Branching evolution, or the multiplication of lineages, as opposed to anagenesis.

Cline A character gradient in a geographically variable species.

Codon The unit of genetic coding; the hereditary material that determines a single amino acid; equivalent to three base pairs.

Coevolution The evolution of a species that closely follows, or tracks, the evolution of another.

Commensalism Symbiosis in which one of the partners (the host) is neither harmed nor benefitted.

Competition The use or attempted use of a limited environmental resource by two or more populations or species. An interaction in which an increase in either species causes a decrease in the other.

Compositionism The belief that higher levels of organization are characterized by new properties or phenomena not predictable from the simpler levels, specifically, that the behavior of organisms cannot be predicted from a knowledge of their physics and chemistry alone. The opposite of reductionism.

Convergence The evolutionary acquisition of similar features (analogies) in unrelated organisms as the result of similar adaptations.

Cope's "Law of the Unspecialized" The theory that new major taxa originate from the more generalized, rather than the more specialized, members of an ancestral taxon. Oversimplified by the phrase "survival of the unspecialized."

Cope's rule The empirical generalization that body size tends more often to increase than to decrease in the course of phyletic evolution. Also called "Cope's law," but do not confuse with the "law of the unspecialized."

Darwinian evolution *See* Darwinism

Darwinian fitness *See* Fitness.

Darwinism An evolutionary theory of which the two essential components are the existence of heritable variation and natural selection.

Deletion A type of chromosomal aberration; the loss of part of a chromosome.

Deme A local population of a species.

Derived In any lineage that exhibits a trend, the later stages or conditions. The opposite of primitive.

Developmental homology Resemblance or similarity due to derivation from a common embryological source, as the male and female genitalia from sexually indifferent precursors.

Diploid Having two sets of genes, so that chromosomes exist in homologous pairs.

Directional selection Selection that causes a change in the mean value of a polygenic trait.

Disruptive (centrifugal) selection Selection that causes an increase in the variance of a polygenic trait.

DNA (deoxyribonucleic acid) The biopolymer that is the principal constituent of nearly all genes and chromosomes.

Dollo's law The "law of irreversibility," which states that a species can never return exactly to an ancestral state of affairs, or, more broadly, that historical processes never repeat themselves exactly, or that complex events can never recur exactly.

Dominant Achieving phenotypic expression even when heterozygous; said of alleles. The opposite of recessive.

Duplication A type of chromosomal aberration; the duplication or repetition of part of a chromosome in such a way that the same sequence occurs twice.

Ecogeographic rules Any of several rules that govern geographic variation within a species and that correlate adaptations with climate or other environmental conditions. Among these are Allen's rule, Bergmann's rule, and Gloger's rule.

Ecological isolation A premating isolating mechanism whereby potential mates never meet because their species inhabit different areas of the environment, or the same areas at different times.

Eidos A Platonic form; a heavenly or ideal "type." *See* Typology.

Electrophoresis A technique that separates proteins or other molecules according to their speed of migration through an electrical field.

Embryonic and fetal mortality A postmating isolating mechanism that results in death before birth or hatching.

Emergence *See* Compositionism.

Endemic Confined to a particular place or region.

Epigenotype The polypeptides that result from the immediate transcription and translation of the genotype.

Epistasis The interaction of several genes to produce a phenotype. Sometimes used more strictly to refer to the "masking" by one gene of the phenotypic effects of another.

Essentialism *See* Typology.

Ethological isolation *See* Behavioral isolation.

Eucaryotic Possessing a well-defined nucleus surrounded by a membrane and also such organelles as mitochondria, vacuoles, and an endoplasmic reticulum.

Euploid Having an integral number of entire genomes; includes monoploidy (haploidy), diploidy, and polyploidy.

Evolution
1. Originally, a synonym for ontogeny, or for Charles Bonnett's preformationist theory.
2. According to Lamarck and his contemporaries, the unfolding of (evolutionary) potentials as each species ascends the *Scala naturae*.
3. From 1809 on, the transformation of one species into another; phyletic evolution.
4. According to many geneticists (e.g., Dobzhansky), changes in the gene frequencies of populations.
5. Anagenesis plus cladogenesis. Phylogeny and the changes in gene frequencies that produce phylogenetic change.

Evolutionary homology Resemblance or similarity due to derivation from a common ancestral structure, as the forelimbs of various vertebrates (all derived from fins).

Exclusion principle, competitive exclusion principle The theory that no two species with identical ecological requirements can long coexist in the same geographical area. The competition between them will be too great and unrelenting; one species will inevitably exclude the other unless subdivision of the niche permits ecological requirements to become somewhat different. *See* Character displacement.

Experimental science A method of science in which the material under investigation is carefully controlled so that only one variable at a time may change. The opposite of naturalistic science.

Extinction The termination of a lineage without issue. *See also* Pseudoextinction; Taxonomic extinction.

Finalism The theory that evolution is directed (by some rational force) toward a final goal or end, which thus determines its direction. *See* Teleology (of which this is a special case); *see also* Orthogenesis.

First-arriver principle The theory that the first species to colonize a new environment or to adapt to a particular type of existence acquires thereby a competitive advantage over any later arrivals, merely because it got there first.

Fitness The relative contribution of a genotype to the gene pool of the next generation.

Fossil Remains or other evidence of past life.

Founder principle A type of genetic drift in which the gene pool of a new population may differ from that of the parent population merely because of accidents of sampling; that is, the founders may have been unrepresentative of the parent population as a whole.

Gametic mortality A postmating isolating mechanism; sperm death before fertilization.

Gene The functional unit of hereditary material (DNA). In former usage, a cistron, codon, muton, or recon. Nowadays, a cistron.

Gene flow The transfer or exchange of genetic material from one population to another.

Gene pool The sum total of all the genes contained in a given population.

Generalized Having the potential to evolve into a variety of alternative conditions. The opposite of specialized.

Genetic drift The deviation of small populations from the Hardy–Weinberg equilibrium due to chance ("sampling error") alone. *See also* Founder principle; Bottleneck effect.

Genetic load The relative difference between the actual mean fitness of a population and the fitness that would result if the optimal genotype were ubiquitous.

Genome One complete set of chromosomes. Possession of one genome produces a haploid; possession of two genomes produces a diploid.

Genotype The genetic or hereditary information borne by an individual, as distinct from the individual's visible features (phenotype). Specifically, the nature of the genes themselves.

Genus (plural, **genera**) A taxon that includes one or more related species. *See* Hierarchy.

Geoffroyism A theory, now discredited, that genetic or evolutionary changes are directly induced by the environment; the theory of direct environmental influence on the genotype.

Geographic isolation The separation of populations by extrinsic geographical barriers; the occurrence of populations in nonoverlapping mappable areas.

Gloger's rule An ecogeographic rule: In a geographically variable species, the populations inhabiting warmer, moister parts of the range tend to be darkly colored (black and near-black); those inhabiting warmer, dryer parts of the range (e.g., deserts) tend to be colored in medium to light shades of red, yellow, and tan; and those inhabiting cold (typically moist) parts of the range tend to be colored very pale or white.

Grade   A taxon or other group consisting of all those species that have attained a certain criterion of advancement in a particular feature or features, often as the result of parallelism. A taxon or other group that had multiple origins.

Gradualism   The theory that evolution always proceeds gradually rather than in abrupt steps.

Haploid   Having only one set of genes; that is, having unpaired chromosomes only. *See* Diploid; Genome.

Hardy–Weinberg equilibrium   A distribution of genotypes in the proportion of $p^2AA + 2pq\ Aa + q^2aa = 1$, where p and q are the gene frequencies of dominant and recessive alleles, respectively.

Hardy–Weinberg law   The theory that in a large, random-mating population of diploids, in the absence of natural selection, unbalanced mutation, or unbalanced migration, the proportions of the several genotypes will tend to remain the same. (The proportions of these genotypes are given by the Hardy–Weinberg equilibrium.)

Heterosis   Increased vigor, or selective superiority, of heterozygotes.

Heterozygote   A heterozygous individual.

Heterozygous   Having, at a particular locus, two different alleles of one gene.

Hexaploid   A polyploid that has six sets of chromosomes or genomes.

Hierarchy   A specified ordering of biological categories, as follows:

Kingdom
  Phylum
    Class
      Order
        Family
          Genus
            Species

Each genus includes one or more related species, each family includes one or more related genera, and so forth. Subsidiary categories are also used: For example, a *subclass* would be just below a class; an *infraclass* would be two steps below a class (i.e., just below a subclass); and a *superclass* would be just above a class.

Historical science   *See* Idiographic science.

Holism   Another name for Compositionism.

Holophyletic   Consisting of a species and all of its descendants without exception.

Homology   A resemblance or similarity between organisms or their parts due to derivation from a common source. *See*

Developmental homology; Evolutionary homology; Serial homology.

Homozygote   A homozygous individual.

Homozygous   Having, at a particular locus, two identical alleles.

Horizontal classification   A system of classification that tends to unite transitional forms with their ancestors; classification that favors grades over clades. The opposite of vertical classification.

Hybrid   An offspring resulting from hybridization.

Hybrid inviability   A postmating isolating mechanism that produces hybrids that do not live to reproductive age.

Hybrid sterility   A postmating isolating mechanism that produces hybrids that are viable but fail to reproduce.

Hybridization   The crossing of individuals belonging to two different species or populations with different adaptive norms. In older usage, the crossing of any two unlike individuals, or any two different genotypes.

Hypothetico-deductive reasoning   The testing of hypotheses by deducing consequences from them and attempting to test or falsify these consequences.

Idiographic (historical) science   Science that attempts to explain unique events or occurrences, usually by referring to their past history. The opposite of nomothetic science.

Inbreeding   Mating between near relatives.

Independent assortment   The phenomenon whereby genes on different chromosomes assort independently, no combination being favored over any other.

Industrial melanism   The occurrence of darkly colored forms of a species in areas polluted with soot or other industrial wastes.

Inversion   A type of chromosomal aberration; the turning of a chromosome segment through 180° such that it reads in reverse sequence.

Irreversibility   *See* Dollo's law.

Isolating mechanism   A biological property of individuals that prevents the interbreeding of populations. *See* Premating isolating mechanism; Postmating isolating mechanism.

K-selection   Selection that favors large, stable populations at maximum density. The opposite of r-selection.

Key character   Any feature or adaptation, the attainment of which opens up new opportunities, thus often ushering in a new major taxon. When the key character can be identified, its origin is usually taken to represent or define the taxonomic origin of the resulting group.

King-of-the-mountain principle   *See* First-arriver principle.

Larmarckism   An evolutionary theory, now discredited, based upon volition or willful striving of organisms to fulfill their own needs, together with the cumulative effects of use and disuse.

Lineage A linear succession of species arranged in a sequence from ancestor to descendant.

Linkage Occurrence of two or more genes on the same chromosome.

Linnaean hierarchy *See* Hierarchy.

Macroevolution Evolution above the species level.

Mechanical isolation A premating isolating mechanism that occurs in species, such as insects, with hardened genital structures. Mating may be attempted, but the "lock-and-key" arrangement of genitalia prevents successful sperm transfer.

Mechanism The belief that vital forces do not exist and that living things are made of nothing but chemical substances and are subject to investigation by chemical means. The opposite of vitalism.

Meiosis A type of cell division that reduces the chromosome number in half; typical of the production of gametes in higher organisms.

Microevolution Evolution at and below the species level; changes in the gene frequencies of natural populations.

Mimicry Protective resemblance (including camouflage) of one species to another. *See* Batesian mimicry; Mullerian mimicry.

Monophyletic Of a group or taxon, derived from a single ancestral species whose origin represents the origin of the group as a whole.

Monotypic Containing only one taxon of the next lower rank. For example, monotypic genera contain only one species; monotypic species contain only one subspecies.

Müllerian mimicry A type of mimicry in which two or more unpalatable species resemble each other.

Mutation Any spontaneous, heritable change in the genotype; especially a sudden change in a single gene.

Muton The genetic unit of mutation, that is, the smallest unit of hereditary material capable of mutation; equivalent to a single base pair. *See* Gene.

Mutualism Symbiosis beneficial to both partners.

Natural selection Differential reproduction; differential contribution of genotypes to the gene pool of the next generation under natural conditions.

Naturalistic science A method of science in which nature is manipulated as little as possible, one frequent aim being the study of what happens under natural conditions. The opposite of experimental science.

Neo-Darwinism *See* Weismannism.

Neontology The study of living species, as opposed to Paleontology.

Niche The way of life of a species; the sum total of its ecological requirements, needs, and tolerances.

Niche preclusion *See* First-arriver principle.

Nomenclature The naming of taxa in accordance with certain rules. The International Code of Zoological Nomenclature is followed for animals; similar codes exist for plants and for bacteria.

Nomothetic science Science that deals with the establishment of universal laws. The opposite of idiographic science.

Normalizing selection *See* Stabilizing selection.

Ontogeny The life history of an individual.

Opportunism The theory that (a) all potential modes of existence will sooner or later be tried by *some* group, and every potential niche will sooner or later be colonized; (b) "things happen as they may, and not as is theoretically best" (i.e., opportunity is limited); and (c) evolution proceeds according to opportunity rather than plan.

Organicism *See* Compositionism.

Orthogenesis The theory that evolution proceeds in predetermined pathways without branching and is not always subject to natural selection. Finalism is a special case of this.

Paleontology The study of fossils, as opposed to Neontology.

Paleospecies A species evolving through time; a portion of a lineage. Also called chronospecies.

Panmixia Random mating within a population.

Parallelism The independent occurrence of the same or a similar trend in two or more lineages. The lineages are usually, though not necessarily, related to one another.

Paraphyletic Consisting of a species and some but not all of its descendants.

Parasitism Symbiosis harmful to one partner and beneficial to the other (the parasite).

Phenotype The visible features of an organism, as contrasted with its genotype.

Phyletic evolution The direct evolution of one species from another without branching. *See* Anagenesis.

Phyletic speciation *See* Phyletic evolution.

Phylogenetic Having to do with phylogeny.

Phylogeny Those aspects of evolution that can be depicted in family trees. The course of evolution, as distinct from the mechanism.

Pleiotropism The phenomenon whereby a single gene has many phenotypic effects; that is, it influences many visible characters rather than only one.

Polygenes Genes of small, additive effect. *See* Polygeny.

Polygeny The control of a visible character by more than a single gene. More specifically, the control of a measurement (such as overall body size) by a number of genes, each contributing a small, additive effect.

Polymorphism The presence within a population of two or more distinct genotypes in frequencies too large to be accounted for by recurrent mutation alone. *See* Transient polymorphism; Balanced polymorphism.

Polyphyletic Originating from more than a single ancestor. A polyphyletic taxon includes no species ancestral to all the other members of that taxon.

Polyploidy An increase in the chromosome number by the addition of entire genomes; hence, a type of euploidy.

Polytopic Occurring in several geographically discontiguous areas; said of subspecies.

Polytypic Of a taxon; containing two or more taxa of the next lower rank. For example, a polytypic species is one that is divided into two or more subspecies. The opposite of monotypic.

Population All the members of a species inhabiting a definite geographic range and sharing a common gene pool. *See* Deme.

Position effect A certain gene, if it occurs on different parts of the same chromosome or on a different chromosome, will often have altered phenotypic effects because of its interaction with different neighboring genes in each case.

Postmating isolating mechanisms Isolating mechanisms that act following sperm transfer. *See* Gametic mortality, Zygotic mortality, Embryonic and fetal mortality, Hybrid inviability, Hybrid sterility.

Preadaptation
1. The occurrence of genetic traits (by spontaneous mutation) prior to, or irrespective of, any adaptive value that they may have.
2. The appearance of phenotypic traits (morphological, behavioral, etc.) prior to the origin of their present adaptive function.

Premating isolating mechanisms Isolating mechanisms that prevent sperm transfer. *See* Ecological isolation, Seasonal isolation, Behavioral isolation, Mechanical isolation.

Primitive In any lineage that exhibits a trend, the early or ancestral stages or conditions. The opposite of advanced or derived.

Procaryotic Lacking a well-defined nucleus or other organelles characteristic of eucaryotic cells.

Pseudoextinction Termination of a taxon within an evolving lineage without true extinction; nonsurvival of a taxon by virtue of its becoming changed into something else.

Punctuated equilibrium model *See* Punctuationalism.

Punctuationalism The view that long periods of stasis are interrupted by sudden events of speciation, and that nearly all morphological change is confined to these sudden events.

r-Selection Selection that favors the ability for rapid colonization and rapid population increase. The opposite of K-selection.

Race A biologically distinct subdivision of a species, inhabiting a definable geographic region and possessing certain characteristic gene frequencies. A subspecies. Also, when modified, a distinguishable population or group of populations (for example, ecological race, host race).

Recapitulation Haeckel's theory, now largely discredited, according to which "ontogeny recapitulates phylogeny, and phylogeny is the mechanical cause of ontogeny." Also, a resemblance between an immature stage and the adult form of its ancestor.

Recessive Achieving phenotypic expression only when homozygous; said of a gene or allele. The opposite of dominant.

Recon The unit of genetic recombination; the smallest unit of hereditary material capable of recombining with other units; equivalent to a single base pair. *See* Gene.

Reductionism A belief that higher levels of organization can be explained in terms of simpler levels; specifically, that life can ultimately be explained in physical and chemical terms. The opposite of compositionism.

Reproductive isolating mechanism *See* Isolating mechanism.

Reproductive isolation The inability of different species to interbreed freely, even when sympatric. *See* Isolating mechanism.

Saltation Sudden origin of a new species from a single individual as the result of one or more large mutations ("macromutations").

Scala naturae An 18th century arrangement of organisms along a single ascending scale of perfection and/or complexity. Also called "scale of being," "chain of being," or "échelle des êtres."

Science A method of investigation that uses hypothetico-deductive reasoning by first making hypotheses and then attempting to falsify them.

Seasonal isolation A type of ecological isolation in which each species becomes reproductively active in a different season.

Segregation In genetics, the separation of genes or chromosomes during meiosis, especially among heterozygotes. Also, the consequential appearance of recessive phenotypes among the offspring of dominant heterozygotes.

Selection Consistent differences in the contribution of various genotypes to the next generation.

Serial homology Homology between or among repeated parts, as between the legs and mouthparts of insects, or as among successive vertebrae.

Sexual selection Selection by one sex of variations within the opposite sex.

Special creation A theory, current before Darwin's time, according to which each species was the product of a special and separate act of divine creation.

Specialized
1. Having a narrow range of ecological tolerance.
2. Incapable of evolving into an alternative condition, or having a reduced potential for further evolutionary change. The opposite of generalized.

Speciation Properly, the splitting of one species into two or more distinct species; the subdivision of a gene pool into several isolated gene pools; cladogenesis. More loosely, the origin of new species by any means, including both true speciation and "phyletic speciation."

Species A group of interbreeding natural populations reproductively isolated from other such groups. The interbreeding of populations occurs freely within species but not between species. Also, a category to which species taxa are assigned; a category at the level where reproductive isolation can be used to delimit taxa.

Stabilizing (centripetal) selection Selection that causes a decrease in the variance of a polygenic trait.

Stasis Persistence of a species over time without change.

Statistical reasoning A method of reasoning or theorizing that takes into account the fact that populations, rather than individuals, are the units of evolution, that populations are variable rather than uniform, and that such statistical abstractions as the arithmetic mean are abstractions only and have no separate existence.

Subspecies One of several geographic or similar subdivisions of a species, between which interbreeding takes place at a reduced level (compared to the interbreeding within each). A taxonomically distinct subdivision of a species.

Symbiosis Any association between two species living together. *See* Parasitism; Commensalism; Mutualism.

Sympatric Of species or populations, occurring in the same geographical area; overlapping in range.

Taxon A collective group of species or of populations at any rank, usually recognized or delimited by morphological and similar features. Examples include Mammalia, *Drosophila*, Vertebrata, and *Homo sapiens*. *See also* Category.

Taxonomic Having to do with taxa or with the Linnaean hierarchy. *See* Taxon, Hierarchy.

Taxonomic extinction Nonsurvival of a taxon, either by extinction or by pseudoextinction.

Taxonomy The study of classification.

Teleology The explanation of phenomena by the purposes (ends, goals; Greek *telos*) they serve.

Tetraploid A polyploid that has four sets of chromosomes or genomes.

Transformation *See* Transformism.

Transformism Older name (18th–19th century) for what is now called evolution, or more especially phyletic evolution. The origin of one species from another.

Transient polymorphism A type of polymorphism that occurs temporarily while one allele is in the process of replacing another in the population.

Translocation A type of chromosomal aberration; the transfer of part of a chromosome either to a different chromosome (heterologous translocation) or to a different part of the same chromosome (homologous translocation).

Trend The continued evolutionary change of a character within a lineage. Also called chronocline.

Triploid A polyploid that has three sets of chromosomes or three genomes.

Typological species concept An older species concept based on Typology, according to which the characters of the "type" (in theory a Platonic *eidos* existing only in the world of ideas, but in practice usually a museum specimen) were considered of paramount importance, and any deviation from the "type" was ignored as inessential, or else treated as pathological or even morally inferior.

Typology The Platonic/Aristotelian theory of ideas applied to biology. Typology places emphasis on the individual rather than the population, ignores variation as inessential (if not illusory), and bases classification upon "types," which were thought to represent the Platonic *eidos*.

Uniformitarianism The theory that natural laws are invariant with time, and that "the present is the key to the past." The opposite of catastrophism.

Vertical classification A system of classification that groups transitional forms with their descendants rather than with their ancestors, and that recognizes taxa that correspond to clades rather than to grades. The opposite of horizontal classification.

Vitalism The belief that a living thing is more than a collection of chemicals, and that living things therefore contain a "vital force" or "vital principle" that cannot be described in physical and chemical terms alone.

Weismannism Weismann's theory of the "continuity of the germ plasm" and "the omnipotence of natural selection" (ignoring variation). In denying the inheritance of acquired characteristics, this theory was an advance, but in ignoring the importance of variation it was retrogressive.

Williston's rule In the evolution of any serially repeated parts, the number of those parts tends usually to decrease, while division of labor causes differential changes, and hence diversification, among those which remain.

Zygotic mortality A postmating isolating mechanism; death of the zygote following fertilization.

Bibliography

The scientific literature on evolution is already so vast, and growing at so rapid a rate, that only a small fraction of the total can be listed here. To compound this problem, there are many works that deal with the evolutionary aspects of such other subjects as genetics, ecology, anatomy, physiology, biochemistry, and paleontology, or with the study of some particular group of organisms. To include all these would require listing a sizable proportion of the entire biological literature. Any selective listing involves some personal judgment as to which works to include and which to omit; the present list is no exception. I have tried to include those references, especially recent works and works with extensive bibliographies, that would aid students or scholars seeking further information in each area. I have also included many works that were seminal or that are now considered classics. In the list for each chapter, the works (usually books) that would be most useful in locating further references are given first, followed by other references.

CHAPTER 1: LIVING SYSTEMS

General References

Ayala, F. J., and Th. Dobzhansky, eds. 1974. *Studies in the Philosophy of Biology: Reductionism and Related Problems.* Berkeley, Calif.: University of California Press.

Beardsley, M. C., and E. L. Beardsley. 1965. *Philosophical Thinking, An Introduction.* New York: Harcourt Brace & World.

Beckner, M. 1959. *The Biological Way of Thought.* New York: Columbia University Press.

Hull, D. L. 1974. *Philosophy of Biological Science.* Englewood Cliffs, N.J.: Prentice-Hall.

Kuhn, T. S. 1962, 1970. *The Structure of Scientific Revolutions,* 1st and 2nd eds. Chicago: University of Chicago Press.

Margulis, L. 1974. Five-kingdom classification and the origin and evolution of cells. *Evol. Biol.* 7:45–78.

Popper, K. R. 1959. *The Logic of Scientific Discovery.* New York: Basic Books.

Other References

Bertalanffy, L. von. 1962. *Modern Theories of Development, an Introduction to Theoretical Biology.* New York: Harper.

Copeland, H. F. 1938. The kingdoms of organisms. *Quart. Rev. Biol.,* 13:384–420.

———. 1956. *Classification of the Lower Organisms.* Palo Alto, Calif.: Pacific Books.

Dobzhansky, Th. 1973. Nothing in biology makes sense except in the light of evolution. *Amer. Biol. Teacher.* 35: 125–129.

Dodson, E. O. 1971. The kingdoms of organisms. *Systematic Zool.* 20:265–281.

Durkheim, E. 1895. *The Rules of Sociological Method,* 8th ed. Trans. by S. A. Solovay and J. H. Mueller. New York: Free Press.

Haeckel, E. H. 1866. *Generelle Morphologie.* Berlin: Georg Reimer.

Hein, H. 1972. The endurance of the mechanism-vitalism controversy. *J. Hist. Biol.* 5:159–188.

Linnaeus, C. 1758. *Systema Naturae, per Regna Tria Naturae,* 10th ed. Stockholm: Laurentii Salvii.

Loeb, J. 1912. *The Mechanistic Conception of Life.* Chicago: University of Chicago Press.

Medawar, P. B. 1967. *The Art of the Soluble.* London: Methuen.

Monod, J. 1971. *Chance and Necessity, an Essay on the Natural Philosophy of Modern Biology.* New York: Alfred A. Knopf.

Olby, R. 1971. Schrödinger's problem: what is life? *J. Hist. Biol.* 4:119–148.

Ruse, M. 1973. *The Philosophy of Biology.* London: Hutchinson.

Schrödinger, E. 1944. *What is Life? The Physical Aspect of the Living Cell.* Cambridge: Cambridge University Press.

———. 1958. *Mind and matter.* Cambridge: Cambridge University Press.

Simpson, G. G. 1963. Historical science. *In* L. Albritton, ed., *The Fabric of Geology.* San Francisco: Freeman Cooper, pp. 24–48.

Stanier, R. Y., E. A. Adelberg, and J. L. Ingraham. 1976. *The Microbial World,* 4th ed. Englewood Cliffs, N.J.: Prentice-Hall.

Stanier, R. Y., and C. N. Van Niel. 1942. The concept of a bacterium. *Arch. Mikrobiol.* 42:17–35.

Whittaker, R. H. 1959. On the broad classification of organisms. *Quart. Rev. Biol.* 34:210–226.

———. 1969. New concepts of kingdoms of organisms. *Science.* 193:150–159.

Windelband, W. 1895. *History of Ancient Philosophy.* Trans. by H. E. Cushman. Reprinted 1956, New York: Dover.

CHAPTER 2: BASIC PRINCIPLES OF GENETICS

General References

Ayala, F. J., and J. A. Kiger, Jr. 1980. *Modern Genetics.* Menlo Park, Calif.: Benjamin-Cummings.

Strickberger, M. W. 1968, 1976. *Genetics,* 1st and 2nd eds. New York: Macmillan.

Other References

Goodenough, U. 1978. *Genetics,* 2nd ed. New York: Holt, Rinehart & Winston.

Jenkins, J. B. 1979. *Genetics,* 2nd ed. Boston: Houghton Mifflin.

Lehninger, A. L. 1975. *Biochemistry,* 2nd ed. New York: Worth Publishers.

Watson, J. D. 1976. *Molecular Biology of the Gene,* 3rd ed. Menlo Park, Calif.: W. A. Benjamin.

CHAPTER 3: BIOLOGY BEFORE DARWIN

General References

Burkhart, R. 1974. *The Spirit of System. Lamarck and Evolutionary Biology.* Cambridge, Mass.: Harvard University Press.

Glass, B., O. Temkin, and W. L. Straus, Jr., eds. 1959. *Forerunners of Darwin, 1745–1859.* Baltimore: Johns Hopkins Press.

Lamarck, J. B. 1809. *Philosophie Zoologique, ou Exposition des Considérations Relatives à l'Histoire Naturelle des Animaux, etc.* Paris. Trans. with an intro. by H. Elliot, 1963. New York: Hafner.

Lovejoy, A. O. 1936. *The Great Chain of Being.* Cambridge, Mass.: Harvard University Press.

Magner, L. N. 1979. *A History of the Life Sciences.* New York and Basel: Marcel Dekker.

Nordenskiold, E. 1928. *The History of Biology, a Survey.* New York: Tudor Publishing Co.; reprinted, 1977, St. Clair Shores, Mich.: Scholarly Press.

Russell, B. 1945. *A History of Western Philosophy.* New York: Simon and Schuster.

Singer, C. 1954. *A History of Biology Before 1900.* New York: Abelard Schuman.

Other References

Greene, J. C. 1959. *The Death of Adam: Evolution and Its Impact on Western Thought.* Ames, Iowa: Iowa State University Press.

Lucretius, T. C. 55 B.C. *On the Nature of the Universe (De Rerum Naturae).* Trans. 1951 by R. E. Latham. Baltimore: Penguin.

Mayr, E. 1972. Lamarck revisited. *J. Hist. Biol.* 5:55–94.

Osborn, H. F. 1894. *From the Greeks to Darwin.* New York: Macmillan.

Paley, W. 1802. *Natural Theology, or Evidences of the Existence and Attributes of the Deity, collected from the Appearances of Nature.* London: Wilks & Taylor.

Windelband, W. 1895. *History of Ancient Philosophy.* New York: Dover. Trans. by H. E. Cushman. Reprinted 1956.

Zirkle, C. 1946. *The Early History of the Idea of the Inheritance of Acquired Characters and of Pangenesis.* Philadelphia: American Philosophical Society.

CHAPTER 4: GEOLOGY BEFORE DARWIN

General References

Bowler, P. J. 1976. *Fossils and Progress: Paleontology and the Idea of Progressive Evolution in the Nineteenth Century.* New York: Scientific History Publications.

Geikie, A. 1905. *The Founders of Geology,* 2nd ed. London and New York: Macmillan.

Lyell, C. 1830, 1831, 1833. *Principles of Geology, Being an Attempt to Explain the Former Changes of the Earth's Surface by Reference to Causes Now in Operation,* 3 vols. London: John Murray.

Millhauser, M. 1959. *Just Before Darwin: Robert Chambers and 'Vestiges.'* Middletown, Conn.: Wesleyan University Press.

Other References

Albritton, C. C., ed. 1963. *The Fabric of Geology.* San Francisco: Freeman Cooper.

[Chambers, R.] 1844. *Vestiges of the Natural History of Creation.* London: Robert Chambers.

Fenton, C. L., and M. A. Fenton. 1945. *The Story of the Great Geologists.* Garden City, N.Y.: Doubleday Doran.

Gould, S. J. 1965. Is uniformitarianism necessary? *Amer. J. Sci.* 263:223–228.

Playfair, J. 1802. *Illustrations of the Huttonian Theory of the Earth.* Edinburgh: Cadell and Davies.

Raven, C. E. 1942. *John Ray, Naturalist: His Life and Works.* Cambridge: University Press.

Rudwick, M. J. S. 1972. *The Meaning of Fossils. Episodes in the History of Paleontology.* London: Macdonald; New York: American Elsevier.

Wilson, L. 1972. *Charles Lyell, the Years to 1841.* New Haven: Yale University Press.

Zittel, K. A. von. 1901. *History of Geology and Paleontology to the End of the Nineteenth Century.* London: Walter Scott; reprinted 1962, New York: Hafner.

CHAPTER 5: DARWINISM

General References

Appleman, P., ed. 1970. *Darwin. Norton Critical Edition.* New York: W. W. Norton.

Darwin, C. R. 1859. *On the Origin of Species by Means of Natural Selection, or the Preservation of the Favoured Races in the Struggle for Life.* London: John Murray.

_____. 1871. *The Descent of Man, and Selection in Relation to Sex.* 2 vols. London: John Murray.

_____. 1872. *On the Origin of Species by Means of Natural Selection, or the Preservation of the Favoured Races in the Struggle for Life,* 6th ed., with additions and corrections. London: John Murray.

Eiseley, L. C. 1958. *Darwin's Century: Evolution and the Men Who Discovered It.* Garden City, N.Y.: Doubleday.

Ghiselin, M. T. 1969. *The Triumph of the Darwinian Method.* Berkeley, Calif: University of California Press.

Gould, S. J. 1977. *Ever Since Darwin.* New York: Norton.

Irvine, W. 1955. *Apes, Angels, and Victorians.* New York: McGraw-Hill.

Other References

New studies in the history of biology, but especially in the Darwinian era, appear frequently in the *Journal of the History of Biology.*

Bowler, P. J. 1977. Darwinism and the argument from design: suggestions for a reevaluation. *J. Hist. Biol.* 10:29–43.

Darwin, C. R. 1839. *Journal of Researches into the Geology and Natural History of the Various Countries Visited by H.M.S. Beagle, Under the Command of Captain FitzRoy from 1832 to 1836.* London: Colburn.

_____. 1842. *The Structure and Distribution of Coral Reefs, Being the First Part of the Geology of the Voyage of the Beagle.* London: Smith, Elder.

_____. 1958. *The Autobiography of Charles Darwin, 1809-1882, with Original Omissions Restored.* Edited with appendix and notes by Nora Barlow. London: Collins.

Darwin, F., ed. 1887. *The Life and Letters of Charles Darwin.* London: John Murray, 3 vols.

_____, ed. 1909. *The Foundations of the Origin of Species: Two Essays Written in 1842 and 1844.* Cambridge: Cambridge University Press.

Darwin, F., and A. C. Steward, eds. 1903. *More letters of Charles Darwin. A Record of His Work in a Series of Hitherto Unpublished Letters.* London: Murray.

De Beer, G. R. 1960. Darwin's notebooks on transmutation of species. Parts I–IV. *Bull. Brit. Mus. (Nat. Hist. Ser.).* 2:23–183.

———. 1964. *Charles Darwin.* Garden City, N.Y.: Double-day.

Eiseley, L. C. 1959. Charles Darwin, Edward Blyth, and the theory of natural selection. *Proc. Amer. Philos. Soc.* 103:94–158.

Farley, J. 1974. The initial reactions of French biologists to Darwin's *Origin of Species. J. Hist. Biol.* 7:275–300.

Freeman, R. B. 1977. *The Works of Charles Darwin: An Annotated Bibliographical Hand List,* 2nd ed. Folkestone, England: Dawson.

Glick, T. F., ed. 1974. *The Comparative Reception of Darwin.* Austin, Texas: University of Texas Press.

Herbert, S. 1971. Darwin, Malthus, and selection. *J. Hist. Biol.* 4:209–217.

———. 1974. The place of man in the development of Darwin's theory of transmutation. Part I, to July 1837. *J. Hist. Biol.* 7: 217–258.

Himmelfarb, Gertrude. 1959. *Darwin and the Darwinian Revolution.* London: Chatto and Windus.

Hofstadter, R. 1959. *Social Darwinism in American Thought,* rev. ed. New York: G. Braziller.

Hull, D. L. 1973. *Darwin and His Critics: The Reception of Darwin's Theory of Evolution by the Scientific Community.* Cambridge, Mass.: Harvard University Press.

Huxley, L. 1900. *Life and Letters of Thomas Henry Huxley.* New York: D. Appleton.

Malthus, T. R. 1799. *An Essay on the Principle of Population as it Affects the Future Improvement of Society, etc.* London: J. Johnson; reprinted 1926, London: Macmillan.

Mayr, E. 1977. Darwin and natural selection. *Amer. Scientist.* 65:321–327.

McKinney, H. L. 1972. *Wallace and Natural Selection.* New Haven: Yale University Press.

Ospovat, D. 1979. Darwin after Malthus. *J. Hist. Biol.* 12:211–230.

Ruse, M. 1975. Charles Darwin's theory of evolution: An analysis. *J. Hist. Biol.* 8:219–241.

Schwartz, J. S. 1974. Charles Darwin's debt to Malthus and Edward Blyth. *J. Hist. Biol.* 7:301–318.

Schweber, S. S. 1979. Essay review: The young Darwin. *J. Hist. Biol.* 12:175–192.

Vorzimmer, P. J. 1975. An early Darwin manuscript: The "outlines and draft of 1839." *J. Hist. Biol.* 8:191–218.

Wallace, A. R. 1855. On the law which has regulated the introduction of new species. *Ann. Mag. Nat. Hist.* ser. 2, 16:186–190.

———. 1858. On the tendency of species to form varieties, and on the perpetuation of varieties and species by natural means of selection. *J. Linnean Soc. (Zool.)* 3:45–62.

———. 1889. *Darwinism. An Exposition of the Theory of Natural Selection.* London: Macmillan.

CHAPTER 6: GENETICS AND THE MODERN SYNTHESIS

General References

Carlson, E. A. 1966. *The Gene: A Critical History.* Philadelphia: Saunders.

Dobzhansky, Th. 1937, 1942, 1951. *Genetics and the Origin of Species,* 1st, 2nd, and 3rd eds. New York: Columbia University Press.

———. 1970. *Genetics of the Evolutionary Process.* New York: Columbia University Press.

Fisher, R. A. 1930. *The Genetical Theory of Natural Selection.* Oxford: Clarendon Press; reprinted 1958, New York: Dover.

Mayr, E. 1942. *Systematics and the Origin of Species.* New York: Columbia University Press.

———. 1963. *Animal Species and Evolution.* Cambridge, Mass.: Belknap Press, Harvard University Press.

Peters, J. A., ed. 1959. *Classic Papers in Genetics.* Englewood Cliffs, N. J.: Prentice-Hall.

Simpson, G. G. 1949. *The Meaning of Evolution.* New Haven: Yale University Press.

———. 1953. *The Major Features of Evolution.* New York: Columbia University Press.

———. 1967. *The Meaning of Evolution,* revised ed. New Haven: Yale University Press.

Stebbins, G. L. 1950. *Variation and Evolution in Plants.* New York: Columbia University Press.

Stern, C., ed. 1950. The birth of genetics. *Genetics,* vol. 35, suppl. pp. 1–47. [Includes the reprinted papers of Mendel's rediscoverers and Mendel's letters to Naegeli.]

Sturtevant, A. H. 1965. *A History of Genetics.* New York: Harper & Row.

Other References

Adams, M. B. 1968. The founding of population genetics: Contributions of the Chetverikov school 1924–1935. *J. Hist. Biol.* 1:23–39.

———. 1970. Towards a synthesis: Population concepts in Russian evolutionary thought, 1925–1935. *J. Hist. Biol.* 3:107–129.

Allen, G. E. 1978. *Thomas Hunt Morgan, The Man and His Science.* Princeton, N.J.: Princeton University Press.

Chetverikov, S. S. 1926. On certain aspects of the evolutionary process from the standpoint of genetics. *Zhurnal Exp. Biol.* 1:3–54 (in Russian); English translation published 1959 in *Proc. Amer. Philos. Soc.* 105:167–195.

Clausen, J., D. D. Keck, and W. M. Hiesey. 1940. *Experimental Studies on the Nature of Species. I. Effects of Varied Environments on Western North American Plants.*

Washington, D.C.: Carnegie Institute of Washington, Publ. no. 520:1–452.

Darwin, C. R. 1868. *The Variation of Animals and Plants Under Domestication.* London: John Murray, 2 vols.

De Vries, H. 1901. *Die Mutationstheorie.* Leipzig: Veit.

Dice, L. R. 1932. Variation in the geographic race of the deer-mouse, *Peromyscus maniculatus bairdii. Occas. Papers Mus. Zool. Univ. Mich.,* no. 239.

———. 1940. Ecologic and genetic variability within species of *Peromyscus. Amer. Nat.* 74:212–221.

Dobzhansky, Th. 1933. Geographical variation in lady-beetles. *Amer. Nat.* 67:97–126.

Graham, L. R. 1972. *Science and Philosophy in the Soviet Union.* New York: Alfred A. Knopf.

Haldane, J. B. S. 1932. *The Causes of Evolution.* London and New York: Harper; reprinted 1966, Ithaca, N.Y.: Cornell University Press.

Huxley, J. S., ed. 1940. *The New Systematics.* Oxford: Clarendon Press.

———. 1942. *Evolution, The Modern Synthesis.* London: Allen & Unwin.

Iltis, H. 1932. *Life of Mendel.* Trans. by E. and C. Paul. Reprinted 1966, New York: Hafner.

Kettlewell, H. B. D. 1961. The phenomenon of industrial melanism in Lepidoptera. *Ann. Rev. Entomol.* 6:245–262.

Klippart, J. H. 1858. An essay on the origin, growth, diseases, varieties, etc., of the wheat plant. 12th Annual Report, Ohio State Board of Agriculture, for 1857.

Mayr, E. 1940. Speciation phenomena in birds. *Amer. Nat.* 74:249–278.

———. 1973. The recent historiography of genetics. *J. Hist. Biol.* 6:125–154.

———. 1976. *Evolution and the Diversity of Life.* Cambridge, Mass.: Belknap Press, Harvard University Press.

Mayr, E., and W. B. Provine, eds. 1980. *The Evolutionary Synthesis: Perspectives on the Unification of Biology.* Cambridge, Mass.: Harvard University Press.

Mendel, G. 1866. *Experiments in Plant Hybridization.* Cambridge, Mass.: Harvard University Press. Also reprinted in collections by Peters (1959) and Stern and Sherwood (1966).

Rensch, B. 1947, 1954. *Neuere Probleme der Abstammungslehre,* 1st and 2nd eds. Stuttgart: Enke.

———. 1960. *Evolution above the Species Level.* New York: Columbia University Press.

Simpson, G. G. 1944. *Tempo and Mode in Evolution.* New York: Columbia University Press.

Stern, C., and E. R. Sherwood, eds. 1966. *The Origin of Genetics, A Mendel Source Book.* San Francisco: W. H. Freeman.

Sutton, W. 1903. The chromosomes in heredity. *Biol. Bull.* 4:213–251.

Wright, S. 1931. Evolution in Mendelian populations. *Genetics.* 16:97–159.

CHAPTER 7: THE SOURCES OF VARIABILITY

General References

See also the general references for Chapters 2 and 6.

Britten, R. J., and E. H. Davidson. 1971. Repetitive and non-repetitive DNA sequences and a speculation on the origins of evolutionary novelty. *Quart. Rev. Biol.* 46:111–138.

Grant, V. 1963. *The Origin of Adaptations.* New York: Columbia University Press.

Stebbins, G. L. 1950. *Variation and Evolution in Plants.* New York: Columbia University Press.

Other References

Benzer, S. 1957. The elementary units of heredity. *In* W. D. McElroy and B. Glass, eds. *The Chemical Basis of Heredity.* Baltimore: Johns Hopkins Press.

Britten, R. J., and E. H. Davidson. 1969. Gene regulation for higher cells: A theory. *Science.* 165:349–357.

———. 1976. DNA sequence arrangement and preliminary evidence on its evolution. *Federation Proc.* 35:2151–2157.

Britten, R. J., and D. E. Kohne. 1968. Repeated sequences in DNA. *Science.* 161:529–540.

Crow, J. F., and M. Kimura. 1965. Evolution in sexual and asexual populations. *Amer. Nat.* 99:439–450.

Dobzhansky, Th. 1946. Genetics of natural populations. XIII. Recombination and variability in populations of *Drosophila pseudoobscura. Genetics.* 31:269–290.

Dobzhansky, Th., and A. M. Holz. 1943. A re-examination of the problem of manifold effects of genes in *Drosophila melanogaster. Genetics.* 28:295–303.

Dobzhansky, Th., H. Levene, B. Spassky, and N. Spassky. 1959. Release of genetic variability through recombination. III. *Drosophila prosaltans. Genetics.* 44:75–92.

Dobzhansky, Th., and A. H. Sturtevant. 1938. Inversions in the chromosomes of *Drosophila pseudoobscura. Genetics.* 23:28–64.

Farris, J. S. 1978. Inferring phylogenetic trees from chromosome inversion data. *Systematic Zool.* 27:275–284.

Galau, G., M. E. Chamberlin, B. R. Hough, R. J. Britten, and E. H. Davidson. 1976. Evolution of repetitive and non-repetitive DNA. *In* F. J. Ayala, ed. *Molecular Evolution.* Sunderland, Mass.: Sinauer Associates, pp. 200–224.

Ghiselin, M. T. 1974. *The Economy of Nature and the Evolution of Sex.* Berkeley, Calif.: University of California Press.

Graham, D. E., B. R. Neufeld, E. H. Davidson, and R. J. Britten. 1974. Interspersion of repetitive and non-repetitive DNA sequences in the sea urchin genome. *Cell.* 1: 127–137.

Grant, V. 1958. The regulation of recombination in plants. *Cold Spring Harbor Sympos. Quant. Biol.* 23:337–363.

Hinegardner, R. 1976. Evolution of genome size. *In* F. J. Ayala, ed. *Molecular evolution.* Sunderland, Mass.: Sinauer Associates, pp. 179–199.

Hood, L., J. Campbell, and S. Elgin. 1975. The organization, expression, and evolution of antibodies and other multi-gene families. *Ann. Rev. Genet.* 9:309–353.

Kohne, D. E. 1970. Evolution of higher-organism DNA. *Quart. Rev. Biophys.* 3:327–375.

Lederberg, J., and E. L. Lederberg. 1952. Replica plating and indirect selection of bacterial mutants. *J. Bacteriol.* 63: 399–406.

Leigh, E. G., Jr., E. L. Charnov, and R. C. Warner. 1976. Sex ratio, sex change, and natural selection. *Proc. Natl. Acad. Sci.* 73:3656–3660.

Lerner, I. M. 1954. *Genetic Homeostasis.* Edinburgh: Oliver and Boyd.

Lewis, E. B. 1950. The phenomenon of position effect. *Adv. Genet.* 3:73–115.

Lewis, K. R., and B. John. 1963. *Chromosome Marker.* London: Churchill.

Manning, J. E., C. W. Schmid, and N. Davidson. 1975. Interspersion of repetitive and non-repetitive DNA sequences in the *Drosophila melanogaster* genome. *Cell.* 4:141–156.

Mather, K. 1941. Variation and selection of polygenic characters. *J. Genet.* 41:159–193.

_____. 1943. Polygenic inheritance and natural selection. *Biol. Rev. Cambridge Philos. Soc.* 18:32–64.

Matthysse, S., K. Lange, and D. K. Wagener. 1979. Continuous variation caused by genes with graduated effects. *Proc. Natl. Acad. Sci.* 76:2862–2865.

Maynard-Smith, J. 1968. Evolution in sexual and asexual populations. *Amer. Nat.* 102:469–473.

_____. 1978. *The Evolution of Sex.* New York: Cambridge University Press.

Neel, J. V., and E. D. Rothman. 1978. Indirect estimates and mutation rates in tribal Amerindians. *Proc. Natl. Acad. Sci.* 75:5585–5588.

Ohno, S. 1970. *Evolution by Gene Duplication.* New York: Springer Verlag.

Ohta, T. 1974. Mutation pressure as the main cause of molecular evolution and polymorphism. *Nature.* 252:351–354.

Pardue, M. L. 1975. Repeated DNA sequences in the chromosomes of higher organisms. *Genetics.* 79(suppl.):159–170.

Pavlovsky, O., and Th. Dobzhansky. 1966. Genetics of natural populations. XXXVII. The coadapted system of chromosomal variants in a population of *Drosophila pseudoobscura. Genetics.* 53:843–854.

Smith, G. 1974. Unequal crossover and the evolution of multigene families. *Cold Spring Harbor Sympos. Quant. Biol.* 38:507–513.

Solbrig, O. T. 1976. On the relative advantages of cross- and self-fertilization. *Ann. Missouri Botan. Garden.* 63: 262–276.

Spiess, E. B. 1959. Release of genetic variability through recombination. II. *Drosophila persimilis. Genetics.* 44:43–58.

Spiess, E. B., and A. C. Allen. 1961. Release of genetic variability through recombination. VII. Second and third chromosomes of *Drosophila melanogaster. Genetics.* 46:1531–1553.

Stebbins, G. L. 1957. Self-fertilization and population variability in the higher plants. *Amer. Nat.* 91:337–354.

_____. 1966a. Chromosomal variation and evolution. *Science.* 152:1463–1469.

_____. 1966b. *Processes of Organic Evolution.* Englewood Cliffs, N.J.: Prentice-Hall.

_____. 1971. *Chromosomal Evolution in Higher Plants.* Reading, Mass.: Addison-Wesley.

Turner, J. R. G. 1967. Why does the genotype not congeal? *Evolution.* 21:645–656. [Selection for and against tightness of linkage; supergenes.]

White, M. J. D. 1973. *Animal Cytology and Evolution,* 3rd ed. Cambridge: Cambridge University Press.

Williams, G. C. 1966. *Adaptation and Natural Selection: A Critique of Some Current Evolutionary Thought.* Princeton, N.J.: Princeton University Press.

Wilson, A. C. 1976. Gene regulation in evolution. *In* F. J. Ayala, ed. *Molecular evolution.* Sunderland, Mass.: Sinauer Associates, pp. 225–234.

Wollman, E. L., F. Jacob, and W. Hayes. 1956. Conjugation and recombination in *Escherichia coli* K-12. *Cold Spring Harbor Sympos. Quant. Biol.* 21:141–162.

Wright, Sewall. 1955. Classification of the factors of evolution. *Cold Spring Harbor Sympos. Quant. Biol.* 20:16–24D.

_____. 1964. Pleiotropy in the evolution of structural reduction and of dominance. *Amer. Nat.* 98:65–69.

CHAPTER 8: GENES IN POPULATIONS

General References

Ayala, F. J., and J. A. Kiger, Jr. 1980. *Modern Genetics.* Menlo Park, Calif.: Benjamin-Cummings.

Crow, J. F., and M. Kimura. 1970. *An Introduction to Population Genetics Theory.* New York: Harper & Row.

Li, C. C. 1955. *Population Genetics.* Chicago: University of Chicago Press.

Spiess, E. B., ed. 1962. *Papers on Animal Population Genetics.* Boston: Little, Brown.

_____. 1977. *Genes in Populations.* New York: Wiley.

Strickberger, M. W. 1968, 1976. *Genetics,* 1st and 2nd eds. New York: Macmillan.

Wilson, E. O., and W. Bossert. 1971. *A Primer of Population Biology.* Stamford, Conn.: Sinauer Associates.

Other References

Cavalli-Sforza, L. L. 1969. ''Genetic drift'' in an Italian population. *Sci. Amer.* 221:30–37.

Cavalli-Sforza L. L., I. Barrai, and A. W. F. Edwards. 1964. Analysis of human evolution under random genetic drift. *Cold Spring Harbor Sympos. Quant. Biol.* 29:9–20.

Dobzhansky, Th. and O. Pavlovsky. 1953. Indeterminate outcome of certain experiments on *Drosophila* populations. *Evolution.* 7:198–210.

_____. 1957. An experimental study of interaction between genetic drift and natural selection. *Evolution.* 11:311–319.

Dobzhansky, Th., and N. P. Spassky. 1962. Genetic drift and natural selection in experimental populations of *Drosophila pseudoobscura. Proc. Natl. Acad. Sci.* 48:148–156.

Falconer, D. S. 1960. *Introduction to Quantitative Genetics.* New York: Ronald Press.

Glass, B., M. S. Sacks, E. F. Jahn, and C. Hess. 1952. Genetic drift in a religious isolate: an analysis of the causes of variation in blood group and other gene frequencies in a small population. *Amer. Nat.* 86:145–160.

Kimura, M., and T. Ohta. 1971. *Theoretical Aspects of Population Genetics.* Princeton, N.J.: Princeton University Press.

_____. 1975. Polyallelic mutational equilibria. *Genetics.* 79:681–691.

_____. 1979. Population genetics of multigene family with special reference to decrease of genetic correlation with distance between gene markers on chromosome. *Proc. Nat. Acad. Sci.* 76:4001–4005.

Neel, J. V., and E. A. Thompson. 1978. Founder effect and number of private polymorphisms observed in Amerindian tribes. *Proc. Natl. Acad. Sci.* 75:1904–1908.

Spencer, W. P. 1947. Genetic drift in a population of *Drosophila immigrans. Evolution.* 1:103–110.

Thompson, E. A., and J. V. Neel. 1978. Probability of founder effect in a tribal population. *Proc. Natl. Acad. Sci.* 75: 1442–1445.

Wallace, B. 1968. *Topics in Population Genetics.* New York: W.W. Norton.

Wright, S. 1948. On the roles of directed and random changes in gene frequency in the genetics of natural populations. *Evolution.* 2:279–294.

_____. 1951a. The genetical structure of populations. *Ann. Eugenics.* 15:323–354.

_____. 1951b. Fisher and Ford on ''the Sewall Wright effect.'' *Amer. Sci.* 39:452–479.

_____. 1966. Polyallelic random drift in relation to evolution. *Proc. Natl. Acad. Sci.* 55:1074–1081.

_____. 1968–1978. *Evolution and the Genetics of Populations.* 4 vols. Chicago: University of Chicago Press.

CHAPTER 9: THE ECOLOGY OF POPULATIONS

General References

Odum, E. C. 1959. *Fundamentals of Ecology,* 2nd ed. Philadelphia: W. B. Saunders.

Pianka, E. R. 1974. *Evolutionary Ecology.* New York: Harper & Row.

Ricklefs, R. 1976. *The Ecology of Nature.* Newton, Mass.: Chiron Press.

Slobodkin, L. B. 1962. *Growth and Regulation of Animal Populations.* New York: Holt, Rinehart & Winston.

Other References

Ahmadjian, V. 1967. *The Lichen Symbiosis.* Waltham, Mass.: Blaisdell.

Allen, P. M. 1976. Evolution, population dynamics, and stability. *Proc. Natl. Acad. Sci.* 73:665–668.

Atwood, K. C., L. K. Schneider, and F. J. Ryan. 1951. Selective mechanisms in bacteria. *Cold Spring Harbor Sympos. Quant. Biol.* 16:344–355.

Connell, J. H. 1970. A predator-prey system in the marine intertidal region. I. *Balanus glandula* and several predatory species of *Thais. Ecol. Monogr.* 40:49–78.

Connell, J. H., and E. Orians. 1964. The ecological regulation of species diversity. *Amer. Nat.* 98:399–414.

Ehrlich, P. R., and P. H. Raven. 1964. Butterflies and plants: a study in coevolution. *Evolution.* 18:586–608.

Emlen, J. M. 1973. *Ecology: An Evolutionary Approach.* Reading, Mass.: Addison-Wesley.

Gause, G. F. 1934. *The Struggle for Existence.* Baltimore: Williams and Wilkins.

Gilbert, L. E., and P. H. Raven, eds. 1975. *Coevolution of Animals and Plants.* Austin, Texas: University of Texas Press.

Gilpin, M. E. 1975. *Group Selection in Predator-Prey Communities.* Princeton, N.J.: Princeton University Press.

Gilpin, M. E., and F. J. Ayala. 1973. Global models of growth and competition. *Proc. Natl. Acad. Sci.* 70:3590–3593.

Hutchinson, G. E. 1959. Homage to Santa Rosalia, *or why are there so many kinds of animals? Amer. Nat.* 93:117–125.

Janzen, D. 1966. Coevolution of mutualism between ants and acacias in Central America. *Evolution.* 20:249–275.

_____. 1980. When is it coevolution? *Evolution.* 34:611–612.

Jones, R. 1979. Predator-prey relationships with particular reference to vertebrates. *Biol. Rev. Cambridge Philos. Soc.* 54:73–97.

Keith, L. B. 1963. *Wildlife's Ten-Year Cycle.* Madison, Wisc.: University of Wisconsin Press.

King, C. E. 1964. Relative abundance of species and MacArthur's model. *Ecology.* 45:716–727.

Lack, D. 1954a. *The Natural Regulation of Animal Numbers.* New York: Oxford University Press.

_____. 1954b. *Population Studies of Birds.* Oxford: Clarendon Press.

Levins, R. 1968. *Evolution in Changing Environments.* Princeton, N.J.: Princeton University Press.

Lotka, A. J. 1925. *Elements of Physical Biology.* Baltimore: Williams and Wilkins.

_____. 1956. *Elements of Mathematical Biology.* New York: Dover. [Reprint of Lotka (1925).]

MacArthur, R. H., and J. Connell. 1966. *The Biology of Populations.* New York: Wiley.

May, R. M. 1972. Limit cycles in predatory-prey communities. *Science.* 177:900–902.

_____, ed. 1976. *Theoretical Ecology: Principles and Applications.* Philadelphia: W. B. Saunders.

Pielou, E. C. 1969. *An Introduction to Mathematical Ecology.* New York: Wiley-Interscience.

Powell, R. A. 1980. Stability in a one-predator, three-prey community. *Amer. Nat.* 115:567–579.

Ricklefs, R. 1973. *Ecology.* Newton, Mass.: Chiron Press.

Rosenzweig, M. L. 1969. Why the prey curve has a hump. *Amer. Nat.* 103:81–87.

_____. 1973. Evolution of the predator isocline. *Evolution.* 27:84–94.

_____. 1977. Aspects of biological exploitation. *Quart. Rev. Biol.* 52:371–379.

Rosenzweig, M. L., and R. H. MacArthur. 1963. Graphical representation and stability conditions of predator-prey interactions. *Amer. Nat.* 97:209–223.

Roughgarden, J. 1979. *Theory of Population Genetics and Evolutionary Ecology: An Introduction.* New York: Macmillan.

Slatkin, M., and J. Maynard-Smith. 1979. Models of coevolution. *Quart. Rev. Biol.* 54:233–263.

Slobodkin, L. B. 1974. Prudent predation does not require group selection. *Amer. Nat.* 108:665–678.

Volterra, V. 1926. Fluctuations in the abundance of a species considered mathematically. *Nature.* 118:558–560.

_____. 1931. *Lessons sur la Théorie Mathématique de la Lutte pour la Vie.* Paris: Gauthier-Villars.

Weinstein, M. S. 1977. Hares, lynx, and trappers. *Amer. Nat.* 111:806–807.

Wilson, E. O., and W. H. Bossert. 1971. *A Primer of Population Biology.* Stamford, Conn.: Sinauer Associates.

Winterhalter, B. P. 1980. Canadian fur bearer cycles and Cree-Ojibwa hunting and trapping practices. *Amer. Nat.* 115:870–879.

CHAPTER 10: ADAPTATION

General References

Alexander, R. M. 1968. *Animal Mechanics.* Seattle, Wash: University of Washington Press.

Cott, H. B. 1957. *Adaptive Coloration in Animals.* London: Methuen.

Daubenmire, R. F. 1963. *Plants and Environment.* New York: Wiley.

Gans, C. 1975. *Biomechanics.* Philadelphia: Lippincott.

Hildebrand, M. 1974. *Analysis of Vertebrate Structure.* New York: Wiley.

Schmidt-Nielsen, K. 1960, *Animal Physiology.* Englewood Cliffs, N.J.: Prentice-Hall.

Thompson, D. W. 1942. *On Growth and Form,* 2nd ed. 2 vols. Cambridge: The University Press.

Wickler, W. 1968. *Mimicry.* New York: McGraw-Hill.

Wigglesworth, V. B. 1957. *The Life of Insects.* New York: Methuen.

Other References

Barrett, J. A. 1976. The maintenance of non-mimetic forms in a dimorphic Batesian mimic species. *Evolution.* 30:82–85.

Benson, W. W. 1971. Evidence for the evolution of unpalatability through kin selection in the Heliconiinae (Lepidoptera). *Amer. Nat.* 105:213–226.

_____. 1972. Natural selection for Müllerian mimicry in *Heliconius erato* in Costa Rica. *Science.* 176:936–939.

Boyden, T. C. 1976. Butterfly palatability and mimicry: experiments with *Ameiva* lizards. *Evolution.* 30:73–81.

Brower, J. V. Z. 1958a. Experimental studies of mimicry in some North American butterflies. Part I. The monarch, *Danaus plexippus,* and viceroy, *Limenitis archippus archippus. Evolution.* 12:32–47.

_____. 1958b. Experimental studies of mimicry in some North American butterflies. Part II. *Battus philenor* and *Papilio troilus, P. polyxenes,* and *P. glaucus. Evolution.* 12:123–136.

_____. 1958c. Experimental studies of mimicry in some North American butterflies. Part III. *Danaus gilippus bernice* and *Limenitis archippus floridensis. Evolution.* 12: 273–285.

———. 1960. Experimental studies of mimicry. IV. The reaction of starlings to different proportions of models and mimics. *Amer. Nat.* 94:271–282.

———. 1963. Experimental studies and new evidence on the evolution of mimicry in butterflies. *Proc. 16th Internat. Congr. Zool.* 4:156–161.

Clarke, B., and P. M. Sheppard. 1960. The evolution of mimicry in the butterfly *Papilio dardanus. Heredity.* 14: 163–173.

———. 1966. A local survey of the distribution of industrial melanic forms of the moth *Biston betularia* and estimates of the selective value of these in an industrial environment. *Proc. Roy. Soc. London,* ser. B. 165:424–439.

Clarke, B., P. M. Sheppard, and I. W. B. Thornton. 1968. The genetics of the mimetic butterfly *Papilio memnon* L. *Philos. Trans. Roy. Soc. London,* ser. B. 254:37–89.

Dobzhansky, Th. 1968. Adaptedness and fitness. *In* R. C. Lewontin, ed. *Population Biology and Evolution.* Syracuse, N.Y.: Syracuse University Press, pp. 109–121.

Eisner, T., K. Hicks, M. Eisner, and D. S. Robson. 1978. "Wolf-in-sheep's-clothing" strategy of a predaceous insect larva. *Science.* 199:790–794.

Folk, G. E., Jr. 1974. *Textbook of Environmental Physiology,* 2nd ed. Philadelphia: Lea & Febiger.

Hecht, M. K., and D. Marien. 1956. The coral snake mimic problem: a reinterpretation. *J. Morphol.* 98:335–364.

Hochachka, P., and G. Somero. 1973. *Strategies of Biochemical Adaptation.* Philadelphia: W. B. Saunders.

Howard, R. R., and E. D. Brodie, Jr. 1973. A Batesian mimetic complex in salamanders: response of avian predators. *Herpetologica.* 29:33–41.

Kettlewell, H. B. D. 1955. Selection experiments on industrial melanism in the Lepidoptera. *Heredity.* 9:323–342.

———. 1956. Further selection experiments on industrial melanism in the Lepidoptera. *Heredity.* 10:287–301.

———. 1961. The phenomenon of industrial melanism in Lepidoptera. *Ann. Rev. Entomol.* 6:245–262.

Lewontin, R. C. 1978. Adaptation. *Sci. Amer.* 239:212–230.

Limbaugh, C. 1961. Cleaning symbiosis. *Sci. Amer.* 205: 42–49.

Rettenmeyer, C. W. 1970. Insect mimicry. *Ann. Rev. Entomol.* 15:43–74.

Rothschild, M. 1967. Mimicry: the deceptive way of life. *Nat. Hist.* 76:44–51.

———. 1960. Predator behaviour and the perfection of incipient mimetic resemblances. *Behaviour.* 16:149–158.

Schmidt-Nielsen, B., and K. Schmidt-Nielsen. 1951. A complete account of the water metabolism in kangaroo rats and an experimental verification. *J. Cell. Comp. Physiol.* 38:165–182.

Steward, R. C. 1977. Industrial melanism in the moths, *Diurnea fagella* (Oecophoridae) and *Allophyes oxyacanthae* (Caradrinidae). *J. Zool.* 183:47–62.

Sumner, F. B. 1934. Does "protective coloration" protect? Results of some experiments with fishes and birds. *Proc. Natl. Acad. Sci.* 20:559–564.

Terhune, E. C. 1976. Wild birds detect quinine on artificial Batesian models. *Nature.* 259:561–563.

Valentine, J. W. 1976. Genetic strategies of adaptation. *In* F. J. Ayala, ed. *Molecular Evolution.* Sunderland, Mass.: Sinauer Associates, pp. 78–94.

Wallace, B. 1953. On coadaptation in *Drosophila. Amer. Nat.* 87:343–358.

Williams, G. C. 1966. *Adaptation and Natural Selection: A Critique of Some Current Evolutionary Thought.* Princeton, N.J.: Princeton University Press.

Wright, S. 1949. Adaptation and selection. *In* G. L. Jepsen, E. Mayr, and G. G. Simpson, eds. *Genetics, Paleontology, and Evolution.* Princeton, N.J.: Princeton University Press, pp. 365–389.

CHAPTER 11: NATURAL SELECTION

General References

See also the general references for Chapters 2 and 6.

Bajema, C. J. 1971. *Natural Selection in Human Populations.* New York: Wiley.

Williams, G. C. 1966. *Adaptation and Natural Selection: A Critique of Some Current Evolutionary Thought.* Princeton, N.J.: Princeton University Press.

Other References

Barber, H. N. 1965. Selection in natural populations. *Heredity.* 20:551–572.

Brady, R. H. 1979. Natural selection and the criteria by which a theory is judged. *Systematic Zool.* 28:600–621.

Bumpus, H. C. 1898. The elimination of the unfit as illustrated by the introduced sparrow, *Passer domesticus. Biol. Lectures, Marine Biol. Labs., Wood's Hole.* 6:209–226.

Burley, N. 1977. Parental investment, mate choice, and mate quality. *Proc. Natl. Acad. Sci.* 74:3476–3479.

Cain, A. J., and P. M. Sheppard. 1950. Selection in the polymorphic land snail *Cepaea nemoralis. Heredity.* 4: 275–294.

———. 1954. Natural selection in *Cepaea. Genetics.* 39: 89–116.

Cain, A. J., *et al.* 1968. Studies on *Cepaea. Philos. Trans. Roy. Soc. London,* ser. B. 253:383–595.

Campbell, B. G., ed. 1972. *Sexual Selection and the Descent of Man, 1871-1971.* Chicago: Aldine.

Cant, J. G. H. 1981. Hypothesis for the evolution of human breasts and buttocks. *Amer. Nat.* 117:199-204.

Clarke, B., and P. M. Sheppard. 1962. Disruptive selection and its effects on a metric character in the butterfly *Papilio dardanus. Evolution.* 16:214-226.

Clegg, M. T., R. W. Allard, and A. L. Kahler. 1972. Is the gene the unit of selection? Evidence from two experimental plant populations. *Proc. Natl. Acad. Sci.* 69:2477-2478.

Cody, M. L. 1966. A general theory of clutch size. *Evolution.* 20:174-184.

Demetrius, L. L. 1977. Adaptedness and fitness. *Amer. Nat.* 111:1163-1168.

Dice, L. R. 1947. Effects of selection by owls of deer-mice (*Peromyscus maniculatus*) which contrast in color with their background. *Contrib. Lab. Vert. Biol., Univ. Mich.* 34:1-20.

Dobzhansky, Th. 1947. Genetics of natural populations. XIV. A response of certain gene arrangements in the third chromosome of *Drosophila pseudoobscura* to natural selection. *Genetics.* 32:142-160.

Dobzhansky, Th., and H. Levene. 1948. Genetics of natural populations. XVII. Proof of operation of natural selection in wild populations of *Drosophila pseudoobscura. Genetics.* 33:537-547.

Dobzhansky, Th., H. Levene, and B. Spassky. 1971. Effects of selection and migration on geotactic and phototactic behavior in *Drosophila.* III. *Proc. Roy. Soc. London,* ser. B. 180:21-41.

Dobzhansky, Th., and B. Spassky. 1967a. An experiment on migration and simultaneous selection for several traits in *Drosophila pseudoobscura.* Genetics. 55:723-734.

———. 1967b. Effects of selection and migration on geotactic and phototactic behavior of *Drosophila.* I. *Proc. Roy. Soc. London,* ser. B. 168:27-47.

———. 1969. Artificial and natural selection for two behavioral traits in *Drosophila pseudoobscura. Proc. Natl. Acad. Sci.* 62:75-80.

Dobzhansky, Th., B. Spassky, and J. Sved. 1969. Effects of selection and migration on geotactic and phototactic behavior in *Drosophila.* II. *Proc. Roy. Soc. London,* ser. B. 173:191-207.

Dunn, E. R. 1942. Survival value of varietal characters in snakes. *Amer. Nat.* 76:104-109.

Endler, J. A. 1980. Natural selection on color patterns in *Poecilia reticulata. Evolution.* 34:76-91.

Fox, S. F. 1975. Natural selection on morphological phenotypes of the lizard *Uta stansburnia. Evolution.* 29:95-107.

Ghiselin. M. T. 1974. *The Economy of Nature and the Evolution of Sex.* Berkeley, Calif.: University of California Press.

Gibson, J., and J. M. Thoday. 1962. Effects of disruptive selection. VI. A second chromosome polymorphism. *Heredity.* 17:1-26.

Gould, S. J. 1977. *Ontogeny and phylogeny.* Cambridge, Mass.: Harvard University Press.

Grant, P. R. 1972. Centripetal selection and the house sparrow. *Systematic Zool.* 21:23-30.

Grant, P. R., B. R. Grant, J. N. M. Smith, I. J. Abbott, and L. K. Abbott. 1976. Darwin's finches: population variation and natural selection. *Proc. Natl. Acad. Sci.* 73:257-261.

Hamilton, W. D. 1972. Altruism and related phenomena, mainly in social insects. *Ann. Rev. Ecol. Systemat.* 3:193-232.

Hecht, M. K. 1952. Natural selection in the lizard genus *Aristelliger. Evolution.* 6:112-124.

Hirshfield, M. F., and D. W. Tinkle. 1975. Natural selection and the evolution of reproductive effort. *Proc. Natl. Acad. Sci.* 72:2227-2231.

Huheey, J. E. 1976. Studies in warning coloration and mimicry. VII. Evolutionary consequences of a Batesian–Müllerian spectrum: a model for Müllerian mimicry. *Evolution.* 30:86-93.

Kimura, M. 1978. Changes of gene frequencies by natural selection under population number regulation. *Proc. Natl. Acad. Sci.* 75:1934-1937.

Kimura, M., and J. F. Crow. 1978. Effect of overall phenotypic selection on genetic change at individual loci. *Proc. Natl. Acad. Sci.* 75:6168-6171.

Klauber, L. G. 1956. *Rattlesnakes: Their Habits, Life Histories, and Influences on Mankind.* Berkeley, Calif.: University of California Press.

Lack, D. 1954. *The Natural Regulation of Animal Numbers.* Oxford: Clarendon Press.

Lande, R. 1980. Sexual dimorphism, sexual selection, and adaptation in polygenic characters. *Evolution.* 34: 292-307.

Leigh, E. G., Jr. 1977. How does selection reconcile individual advantage with the good of the group? *Proc. Natl. Acad. Sci.* 74:4542-4546.

Lerner, I. M. 1959. The concept of natural selection: a centennial view. *Proc. Amer. Philos. Soc.* 103:173-182.

MacArthur, R. H. 1962. Some generalized theorems of natural selection. *Proc. Natl. Acad. Sci.* 48:1893-1897.

Mather, K. 1943. Polygenic inheritance and natural selection. *Biol. Rev. Cambridge Philos. Soc.* 18:32-64.

Mather, K., and B. J. Harrison. 1949. The manifold effect of selection. *Heredity.* 3:1-52, 131-162.

Maynard-Smith, J. 1964. Group selection and kin selection. *Nature.* 201:1145-1147.

Petit, C., and L. Ehrman. 1969. Sexual selection in *Drosophila*. *Evol. Biol.* 3:177–233.

Prout, T. 1962. The effects of stabilizing selection on the time of development in *Drosophila melanogaster*. *Genetical Res.* 3:364–382.

Slatkin, M., and M. J. Wade. 1978. Group selection on a quantitative character. *Proc. Natl. Acad. Sci.* 75:3531–3534.

Thoday, J. M., and J. B. Gibson. 1962. Isolation by disruptive selection. *Nature*. 193:1164–1166.

Trivers, R. L., and H. Hare. 1976. Haplodiploidy and the evolution of the social insects. *Science*. 191:249–263.

Van Valen, L. 1975. Group selection, sex, and fossils. *Evolution*. 29:87–94.

Wade, M. J. 1976. Group selection among laboratory populations of *Tribolium*. *Proc. Natl. Acad. Sci.* 73:4604–4607.

Williams, G. C. 1975. *Sex and Evolution*. Princeton, N.J.: Princeton University Press.

Woodworth, C. M., E. R. Leng, and R. W. Jugenheimer. 1952. Fifty generations of selection for protein and oil in corn. *Agron. J.* 44:60–66.

Wright, S. 1949. Adaptation and selection. *In* G. L. Jepsen, E. Mayr, and G. G. Simpson, eds. *Genetics, Paleontology, and Evolution*. Princeton, N.J.: Princeton University Press, pp. 365–389.

_____. 1956. Modes of selection. *Amer. Nat.* 90:5–24.

Wright, S., and Th. Dobzhansky. 1946. Genetics of natural populations. XII. Experimental reproduction of some of the changes caused by natural selection in certain populations of *Drosophila pseudoobscura*. *Genetics*. 27:373–394.

Wynne-Edwards, V. C. 1962. *Animal Dispersion in Relation to Social Behaviour*. Edinburgh: Oliver & Boyd.

CHAPTER 12: PERSISTENCE OF VARIABILITY WITHIN POPULATIONS

General References

Ayala, F. J., ed. 1976. *Molecular Evolution*. Sunderland, Mass.: Sinauer Associates.

Lewontin, R. C. 1974. *The Genetic Basis of Evolutionary Change*. New York: Columbia University Press.

Wallace, B. 1975. Gene control mechanisms and their possible bearing on the neutralist-selectionist controversy. *Evolution*. 29:193–202.

Other References

New studies in enzyme polymorphisms in particular species or particular enzyme systems are appearing almost weekly in such journals as *Evolution, Genetics,* and *Proceedings of the National Academy of Sciences*. Only a small sample of these are listed here. Most of the earlier studies are omitted from the list that follows because they can easily be located using the general references cited above.

Allison, A. C. 1955. Aspects of polymorphism in man. *Cold Spring Harbor Sympos. Quant. Biol.* 20:239–255.

_____. 1961. Genetic factors in resistance to malaria. *Ann. N.Y. Acad. Sci.* 91:710–729.

Anderson, W., Th. Dobzhansky, O. Pavlovsky, J. Powell, and D. Yardley. 1975. Genetics of natural populations. XLII. Three decades of genetic change in *Drosophila pseudoobscura*. *Evolution*. 29:24–36.

Ayala, F. J. 1965. Evolution of fitness in experimental populations of *Drosophila serrata*. *Science*. 150:903–905.

_____. 1968. Genotype, environment, and population numbers. *Science*. 162:1453–1459.

Ayala, F. J., and M. E. Gilpin. 1974. Gene frequency comparisons between taxa: support for the natural selection of protein polymorphisms. *Proc. Natl. Acad. Sci.* 71:4847–4849.

Ayala, F. J., M. L. Tracey, L. G. Barr, J. F. McDonald, and S. Pérez-Salas. 1974. Genic variation in natural populations of five *Drosophila* species and the hypothesis of the selective neutrality of protein polymorphisms. *Genetics*. 77:343–384.

Ayala, F. J., J. W. Valentine, D. Hedgecock, and L. G. Barr. 1975. Deep-sea asteroids: high genetic variability in a stable environment. *Evolution*. 29:203–212.

Battaglia, B. 1958. Balanced polymorphism in *Tisbe reticulata*, a marine copepod. *Evolution*. 12:358–364.

Bengstron, S. A., A. Nilsson, S. Nordström, and S. Rundgren. 1976. Polymorphism in relation to habitat in the snail *Cepaea hortensis* in Iceland. *J. Zool.* 178:173–188.

Berger, E. M. 1977. Are synonymous mutations adaptively neutral? *Amer. Nat.* 111:606–607.

Brues, A. M. 1954. Selection and polymorphism in the A-B-O blood groups. *Amer. J. Phys. Anthro.* 12:559–597.

_____. 1964. The cost of evolution versus the cost of not evolving. *Evolution*. 18:379–383.

_____. 1969. Genetic load and its varieties. *Science*. 164:1130–1136.

Cain, A. J. 1951a. So-called non-adaptive or neutral characters in evolution. *Nature*. 168:424.

_____. 1951b. Non-adaptive or neutral characters in evolution. *Nature*. 168:1049.

Cain, A. J., and P. M. Sheppard. 1954. The theory of adaptive polymorphism. *Amer. Nat.* 88:321–326.

Carson, H. L., and W. E. Johnson. 1975. Genetic variation in Hawaiian *Drosophila*. I. Chromosome and allozyme polymorphism in *D. setosimentum* and *D. ochrobasis* from the island of Hawaii. *Evolution*. 29:11–23.

Clarke, B. C. 1964. Frequency-dependent selection for the dominance of rare polymorphic genes. *Evolution*. 18:364–369.

———. 1972. Density-dependent selection. *Amer. Nat.* 108:1–13.

Coyne, J. A., W. F. Eanes, J. A. Rashman, and R. K. Koehn. 1979. Electrophoretic heterogeneity of α-glycerophosphate dehydrogenase among many species of *Drosophila. Systematic Zool.* 28:164–175.

Crow, J. F., and M. Kimura. 1963. The theory of genetic loads. *Proc. 12th Internat. Congr. Genetics.* 3:495–506.

———. 1970. *An Introduction to Population Genetics Theory.* New York: Harper & Row.

Darlington, P. J. 1977. The cost of evolution and the importance of adaptation. *Proc. Natl. Acad. Sci.* 74:1647–1651.

Dobzhansky, Th. 1964. How do the genetic loads affect the fitness of their carriers in *Drosophila* populations? *Amer. Nat.* 98:151–166.

———. 1970. *Genetics of the Evolutionary Process.* New York: Columbia University Press.

Dobzhansky, Th., and F. J. Ayala. 1973. Temporal frequency changes of enzyme and chromosomal polymorphisms in natural populations of *Drosophila. Proc. Natl. Acad. Sci.* 70:680–683, 2176.

Dobzhansky, Th., R. C. Lewontin, and O. Pavlovsky. 1964. The capacity for increase in chromosomally polymorphic and monomorphic populations of *Drosophila pseudoobscura. Heredity.* 19:597–614.

Dobzhansky, Th., and B. Spassky. 1963. Genetics of natural populations. XXXIV. Adaptive norm, genetic load and genetic elite in *Drosophila pseudoobscura. Genetics.* 48:1467–1485.

Dobzhansky, Th., B. Spassky, and T. Tidwell. 1963. Genetics of natural populations. XXXII. Inbreeding and the mutational and balanced genetic loads in natural populations of *Drosophila pseudoobscura. Genetics.* 48:361–373.

Donehower, L., C. Furlong, D. Gillespie, and D. Kurnit. 1979. DNA sequence of baboon highly repeated DNA: evidence for evolution by nonrandom unequal crossovers. *Proc. Natl. Acad. Sci.* 77:2129–2133.

Dubinin, N. P., and G. G. Tiniakov. 1945. Seasonal cycles in the concentrations of inversions in populations of *Drosophila funebris. Amer. Nat.* 79:570–572.

Dubinin, N. P., and Yu. P. Altukhov. 1979. Gene mutations (de novo) found in electrophoretic studies of blood proteins of infants with anomalous development. *Proc. Natl. Acad. Sci.* 76:5226–5229.

Ehrman, L., and C. Petit. 1968. Genotype frequency and mating success in the *willistoni* species group of *Drosophila. Evolution.* 22:649–658.

Ehrman, L., B. Spassky, O. Pavlovsky, and Th. Dobzhansky.

———. 1965. Sexual selection, geotaxis, and chromosomal polymorphism in experimental populations of *Drosophila pseudoobscura. Evolution.* 19:337–346.

Ferris, S. D., and G. S. Whitt. 1980. Genetic variability in species with extensive gene duplication: the tetraploid catostomid fishes. *Amer. Nat.* 115:650–666.

Fincham, J. 1972. Heterozygous advantage as a likely genetical basis for enzyme polymorphisms. *Heredity.* 28:387–391.

Fisher, R. A. 1930. *The Genetical Theory of Natural Selection.* Oxford: Clarendon Press; reprinted 1958, New York: Dover.

Flake, R. H., and V. Grant. 1974. An analysis of the cost-of-selection concept. *Proc. Natl. Acad. Sci.* 71:3716–3720.

Ford, E. B. 1945. Polymorphism. *Biol. Rev. Cambridge Philos. Soc.* 20:73–88.

———. 1965. *Genetic Polymorphism.* Cambridge, Mass.: M.I.T. Press.

———. 1971. *Ecological Genetics,* 3rd ed. London: Chapman & Hall.

Garten, C. T. 1976. Relationships between aggressive behavior and genic heterozygosity in the oldfield mouse, *Peromyscus polionotus. Evolution.* 30:59–72. [Heterozygous populations favored.]

Gooch, J. L., and T. J. M. Schopf. 1973. Genetic variability in the deep sea: Relation to environmental variability. *Evolution.* 26:545–552.

Grant, V., and R. H. Flake. 1974. Solution to the cost-of-selection dilemma. *Proc. Natl. Acad. Sci.* 71:3863–3865.

Haldane, J. B. S. 1957. The cost of natural selection. *J. Genetics.* 55:511–524.

Hubby, J. L., and R. C. Lewontin. 1966. A molecular approach to the study of genic heterozygosity in natural populations. I. The number of alleles at different loci in *Drosophila pseudoobscura. Genetics.* 54:577–594.

Ingram, V. M. 1963. *The Hemoglobins in Genetics and Evolution.* New York: Columbia University Press.

Johnson, F. M., and A. Powell. 1974. The alcohol dehydrogenases of *Drosophila melanogaster:* frequency changes associated with heat and cold shock. *Proc. Natl. Acad. Sci.* 71:1783–1784.

Johnson, G. B. 1976. Genetic polymorphism and enzyme function. *In* F. J. Ayala, ed. *Molecular Evolution.* Sunderland, Mass.: Sinauer Associates, pp. 46–59.

Jukes, T. H. 1966. *Molecules and Evolution.* New York: Columbia University Press.

Kan, Y. W., and A. M. Dozy. 1978. Polymorphism of DNA sequence adjacent to human β-globin structural gene: Relationship to sickle mutation. *Proc. Natl. Acad. Sci.* 75:5627–5630.

Kimura, M., and J. L. King. 1979. Fixation of a deleterious allele at one of two "duplicate" loci by mutation pressure and random drift. *Proc. Natl. Acad. Sci.* 76:2858–2861.

Kimura, M., and T. Ohta. 1971. *Theoretical Aspects of Population Genetics.* Princeton, N.J.: Princeton University Press.

———. 1974. On some principles governing molecular evolution. *Proc. Natl. Acad. Sci.* 71:2848–2852.

———. 1978. Stepwise mutation model and distribution of allelic frequencies in a finite population. *Proc. Natl. Acad. Sci.* 75:2868–2872.

King, J. L. 1966. The gene interaction component of the genetic load. *Genetics.* 53:403–413.

———. 1973. The probability of electrophoretic identity of proteins as a function of amino acid divergence. *J. Molec. Evol.* 2:317–322.

King, J. L., and T. H. Jukes. 1969. Non-Darwinian evolution: Random fixation for selectively neutral alleles. *Science.* 164:788–798.

Laurie-Ahlberg, C. C., G. Maroni, G. C. Bewley, J. C. Lucchesi, and B. S. Weir. 1980. Quantitative genetic variation of enzyme activities in natural populations of *Drosophila melanogaster. Proc. Natl. Acad. Sci.* 77:1073–1077.

Layzer, D. 1980. Genetic variation and progressive evolution. *Amer. Nat.* 115:809–826.

Levene, H., O. Pavlovsky, and Th. Dobzhansky. 1958. Dependence of the adaptive values of certain genotypes of *Drosophila pseudoobscura* on the composition of the gene pool. *Evolution.* 12:18–23.

Levins, R. 1968. *Evolution in Changing Environments.* Princeton, N.J.: Princeton University Press.

Levins, R., and R. MacArthur. 1966. The maintenance of genetic polymorphism in a spatially heterogeneous environment: Variations on a theme by Howard Levene. *Amer. Nat.* 100:585–589.

Lewontin, R. C. 1962. Interdeme selection controlling a polymorphism in the house mouse. *Amer. Nat.* 96:65–78.

———. 1967. An estimate of average heterozygosity in man. *Amer. J. Human Genet.* 19:681–685.

Lewontin, R. C., and J. L. Hubby. 1966. A molecular approach to the study of genic heterozygosity in natural populations. II. Amount of variation and degree of heterozygosity in natural populations of *Drosophila pseudoobscura. Genetics.* 54:595–609.

Livingstone, F. B. 1964. The distribution of the abnormal hemoglobin genes and their significance for human evolution. *Evolution.* 18:685–699.

———. 1967. *Abnormal Hemoglobins in Human Populations.* Chicago, Aldine.

Maruyama, T., and M. Kimura. 1978. Theoretical study of genetic variability, assuming stepwise production of neutral and very slightly deleterious mutations. *Proc. Natl. Acad. Sci.* 75:919–922.

Maynard-Smith, J. 1968. "Haldane's dilemma" and the rate of evolution. *Nature.* 219:1114–1116.

Merritt, R. 1972. Geographic variation and enzymatic properties of lactate dehydrogenase allozymes in the fathead minnow. *Amer. Nat.* 106:173–184.

Milkman, R. D. 1975. Allozyme variation in *E. coli* of diverse natural origins. *In* C. L. Markert, ed. *Isozymes, IV, Genetics and Evolution.* New York: Academic Press, pp. 273–285.

Morton, N. E. 1960. The mutational load due to detrimental genes in man. *Amer. J. Human Genet.* 11:237–251.

Morton, N. E., and C. S. Chung. 1959. Are the MN blood groups maintained by selection? *Amer. J. Human Genet.* 11:237–251.

Mourant, A. E. 1954. *The Distribution of Human Blood Groups.* Oxford: Blackwell.

Muller, H. J. 1950. Our load of mutations. *Amer. J. Human Genet.* 2:111–176.

Nei, M. 1975. *Molecular Population Genetics and Evolution.* Amsterdam: North-Holland.

Nevo, E., T. Perl, A. Beiles, D. Wool, and U. Zeller. 1980. Genetic structure as a potential monitor of marine pollution. *Journées Étud. Pollutions.* 5:61–68.

Nevo, E., T. Shimony, and M. Libni. 1978. Pollution selection of allozyme polymorphism in barnacles. *Experientia.* 34:1562–1564.

Ohta, T. 1974. Mutational pressure as the main cause of molecular evolution and polymorphism. *Nature.* 252:351–354.

Ohta, T., and M. Kimura. 1975. Theoretical analysis of electrophoretically detectable polymorphisms: models of very slightly deleterious mutations. *Amer. Nat.* 109:137–145.

Ornston, L. N., and W. K. Yeh. 1979. Origins of metabolic diversity: evolutionary divergence by sequence repetition. *Proc. Natl. Acad. Sci.* 76:3996–4000.

Pierce, B. A., and J. Mitton. 1979. A relationship of genetic variation within and among populations: an extension of the Kluge–Kerfoot phenomenon. *Systematic Zool.* 28:63–70.

Place, A. R., and D. A. Powers. 1979. Genetic variation and relative catalytic efficiencies: lactate dehydrogenase B allozymes of *Fundulus heteroclitus. Proc. Natl. Acad. Sci.* 76:2354–2358.

Prakash, S., and R. C. Lewontin. 1968. A molecular approach to the study of genic heterozygosity in natural populations. III. Direct evidence of coadaptation in gene arrangements of *Drosophila. Proc. Natl. Acad. Sci.* 59:398–405.

Schwartz, D., and W. Laughner. 1969. A molecular basis for heterosis. *Science.* 166:626.

Selander, R. K., W. K. Hunt, and S. Y. Yang. 1969. Protein polymorphisms and genic heterozygosity in two European subspecies of the house mouse. *Evolution.* 23:379–390.

Selander, R. K., and W. E. Johnson. 1973. Genetic variation among vertebrate species. *Ann. Rev. Ecol. Systemat.* 4:75–91.

Sheppard, P. M. 1951. Fluctuations in the selective value of certain phenotypes in the polymorphic land snail *Cepaea nemoralis* (L.). *Heredity.* 5:125–134.

Singh, R. S., J. L. Hubby, and R. C. Lewontin. 1974. Molecular heterosis for heat-sensitive enzyme alleles. *Proc. Natl. Acad. Sci.* 71:1808–1810.

Somero, G. N., and M. Soulé. 1974. Genetic variation in marine fishes as a test of the niche-variation hypothesis. *Nature.* 249:670–672.

Spiess, E. B. 1957. Relation between frequency and adaptive values of chromosomal arrangements of *Drosophila persimilis. Evolution.* 11:84–93.

———. 1968. Low frequency advantage in mating of *Drosophila pseudoobscura* karyotypes. *Amer. Nat.* 102:363–379.

Spiess, E. B., and L. D. Spiess. 1967. Mating propensity, chromosomal polymorphism, and dependent conditions in *Drosophila persimilis. Evolution.* 21:672–688.

Spiess, L. D., and E. B. Spiess. 1969. Minority advantage in interpopulational meetings of *Drosophila persimilis. Evolution.* 21:672–688.

Stalker, H. D. 1960. Chromosomal polymorphism in *Drosophila paramelanica* Patterson. *Genetics.* 49:669–682.

Sved, J. A. 1968. Possible rates of gene substitution in evolution. *Amer. Nat.* 102:283–293.

Sved, J. A., T. E. Reed, and W. F. Bodmer. 1967. The number of balanced polymorphisms that can be maintained in a natural population. *Genetics.* 55:469–481.

Turner, J. R. G., M. S. Johnson, and W. F. Eanes. 1979. Contrasted modes of evolution in the same genome: allozymes and adaptive change in *Heliconius. Proc. Natl. Acad. Sci.* 76:1924–1928.

Valentine, J. W., and F. J. Ayala. 1976. Genetic variability in krill. *Proc. Natl. Acad. Sci.* 73:658–660.

Van Valen, L. 1963. Haldane's dilemma, evolutionary rates, and heterosis. *Amer. Nat.* 97:185–190.

Vigue, C. L., and F. M. Johnson. 1973. Isozyme variability in species of the genus *Drosophila.* VI. Frequency–property–environment relationships of allelic alcohol dehydrogenase in *D. melanogaster. Biochem. Genet.* 9:213–227.

Wallace, B. 1970. *Genetic Load.* Englewood Cliffs, N.J.: Prentice-Hall.

Wallace, B., and Th. Dobzhansky. 1962. Experimental proof of balanced genetic loads in *Drosophila. Genetics.* 47:1027–1042.

Wills, C. 1966. The mutational load in two natural populations of *Drosophila pseudoobscura. Genetics.* 53:281–294.

Yoshimaru, H., and T. Mukai. 1979. Lack of experimental evidence for frequency-dependent selection at the alcohol dehydrogenase locus in *Drosophila melanogaster. Proc. Natl. Acad. Sci.* 76:876–878.

Zouros, E. 1975. Electrophoretic variation in allozymes related to function or structure? *Nature.* 254:446–448.

CHAPTER 13: GEOGRAPHIC VARIATION

General References

Endler, J. A. 1977. *Geographic variation, speciation, and clines.* Princeton, N.J.: Princeton University Press.

Soulé, M. 1976. Allozyme variation: its determinants in space and time. *In* F. J. Ayala, ed. *Molecular Evolution.* Sunderland, Mass.: Sinauer Associates, pp. 60–77.

Other References

Allen, W. R., and P. M. Sheppard. 1971. Copper tolerance in some California populations of the monkey flower, *Mimulus guttatus. Proc. Roy. Soc. London,* ser. B. 177:177–196.

Antonovics, J. 1971. The effects of a heterogeneous environment on the genetics of natural populations. *Amer. Scientist.* 59:593–599.

———. 1976. The input from population genetics: "the new ecological genetics." *Systematic Bot.* 1:233–245.

Antonovics, J., and A. D. Bradshaw. 1970. Evolution in closely adjacent populations. VIII. Clinal patterns at a mine boundary. *Heredity.* 25:349–362.

Avise, J. C., and M. H. Smith. 1974. Biochemical genetics of sunfish. I. Geographic variation and subspecific intergradation in the bluegill, *Lepomis macrochirus. Evolution.* 28:42–56.

Ayala, F. J., M. L. Tracey, L. G. Barr, and J. G. Ehrenfeld. 1974. Genetic and reproductive differentiation of the subspecies *Drosophila equinoxialis caribbensis. Evolution.* 28:24–41.

Bishop, J. A. 1972. An experimental study of the cline in industrial melanism in *Biston betularia* (L.) (Lepidoptera) between urban Liverpool and rural North Wales. *J. Animal Ecol.* 41:209–243.

Blair, W. F. 1947. Estimated frequencies of the buff and grey genes (G,g) in adjacent populations of deer-mice (*Peromyscus maniculatus*) living on soils of different colors. *Contrib. Lab. Vert. Biol. Univ. Mich.* 36:1–16.

Bodmer, W. F., and L. L. Cavalli-Sforza. 1976. *Genetics, Evolution, and Man*. San Francisco: W. H. Freeman.

Brues, A. M. 1972. Models of race and cline. *Amer. J. Phys. Anthro*. 37:389–400.

Cain, A. J., and J. D. Currey. 1963. Area effects in *Cepaea*. *Philos. Trans. Roy. Soc. London*, ser. B. 246:1–81.

Clarke, B. C. 1966. The evolution of morph-ratio clines. *Amer. Nat*. 100:389–402.

Clausen, J., and W. M. Hiesey. 1958. *Experimental Studies on the Nature of Species. IV. Genetic Structure of Ecological Races*. Washington D.C.: Carnegie Institute of Washington. Publ. no. 615:1–312.

Clausen, J., D. D. Keck, and W. M. Hiesey. 1940. *Experimental Studies on the Nature of Species. I. Effects of Varied Environments on Western North American Plants*. Washington D.C.: Carnegie Institute of Washington. Publ. no. 520: 1–452.

———. 1948. *Experimental Studies on the Nature of Species. III. Environmental Responses of Climatic Races of Achillea*. Washington D.C.: Carnegie Institute of Washington. Publ. no. 581:1–129.

Dice, L. R. 1940. Ecological and genetic variability within species of *Peromyscus*. *Amer. Nat*. 74:212–221.

Dillon, L. S. 1962. Historical subspeciation in the North American marten. *Systematic Zool*. 10:49–64.

Endler, J. A. 1950. Gene frequencies in a cline determined by selection. *Biometrics*. 6:353–361.

Gabriel, K. R., and R. R. Sokal. 1969. A new statistical approach to geographic variation analysis. *Systematic Zool*. 18:259–278.

Gorman, G. C., M. Soulé, S. Y. Yang, and E. Nevo. 1975. Evolutionary genetics of insular Adriatic lizards. *Evolution*. 29:52–71.

Gould, S. J., and R. F. Johnston. 1972. Geographic variation. *Ann. Rev. Ecol. Systemat*. 3:457–498.

Gould, S. J., D. S. Woodruff, and J. P. Martin. 1974. Genetics and morphometrics of *Cerion* at Pongo Carpet: A new systematic approach to this enigmatic land snail. *Systematic Zool*. 23:518–535.

Hagmeier, E. M. 1958. Inapplicability of the subspecies concept to the North American marten. *Systematic Zool*. 7:1–7.

Haldane, J. B. S. 1948. The theory of a cline. *J. Genet*. 48:277–284.

Hall, E. R., and K. R. Kelson. 1957. *The Mammals of North America*. New York: Ronald Press.

Highton, R., and T. P. Webster. 1976. Geographic protein variation and divergence in populations of the salamander *Plethodon cinereus*. *Evolution*. 30:33–45.

Johnson, G. B. 1976. Genetic polymorphism and enzyme function. *In* F. J. Ayala, ed. *Molecular Evolution*. Sunderland, Mass.: Sinauer Associates, pp. 46–59.

Johnson, W. E., and R. K. Selander. 1971. Protein variation and systematics in kangaroo rats (genus *Dipodomys*). *Systematic Zool*. 20:377–405.

Johnston, R. F., and R. K. Selander. 1964. House sparrows: Rapid evolution of races in North America. *Science*. 144:548–550.

Koehn, R. K., R. Milkman, and J. B. Mitton. 1976. Population genetics of marine pelecypods. IV. Selection, migration, and genotypic differentiation in the blue mussel, *Mytilus edulis*. *Evolution*. 30:2–32.

Lidicker, W. Z. 1962. The nature of subspecific boundaries in a desert rodent and its implications for subspecies taxonomy. *Systematic Zool*. 11:160–171.

Littlejohn, M. J., and R. S. Oldham. 1968. *Rana pipiens* complex: Mating call structure and taxonomy. *Science*. 162:1003–1005.

Mayr, E. 1963. *Animal Species and Evolution*. Cambridge, Mass.: Belknap Press, Harvard University Press.

McNeilly, T. 1968. Evolution in closely adjacent plant populations. *Agrostis tenuis* on a small copper mine. *Heredity*. 23:99–108.

Merritt, R. 1972. Geographic variation and enzymatic properties of lactate dehydrogenase allozymes in the fathead minnow. *Amer. Nat*. 106:173–184.

Moore, J. A. 1944. Geographic variation in *Rana pipiens* Schreber of eastern North America. *Bull. Amer. Mus. Nat. Hist*. 82:345–370.

———. 1949. Geographic variation of adaptive characters in *Rana pipiens* Schreber. *Evolution*. 3:1–24.

———. 1950. Further studies on *Rana pipiens* racial hybrids. *Amer. Nat*. 84:247–254.

Nei, M. 1972. Genetic distance between populations. *Amer. Nat*. 106:283–292.

Nevo, E., and C. R. Shaw. 1972. Genetic variation in a subterranean mammal, *Spalax ehrenbergi*. *Biochem. Genet*. 7:235–241.

Prakash, S. 1972. Origin of reproductive isolation in the absence of apparent genic differentiation in a geographic isolate of *Drosophila pseudoobscura*. *Genetics*. 72:143–155.

———. 1973. Patterns of gene variation in central and marginal populations of *Drosophila robusta*. *Genetics*. 75:347–369.

Rogers, J. S. 1972. Measures of genetic similarity and genetic distance. *University of Texas Publ*. 7213:145–153.

Schnell, G. D., T. L. Best, and M. L. Kennedy. 1978. Interspecific morphological variation in kangaroo rats (*Dipodo-*

mys): Degree of concordance with genic variation. *Systematic Zool.* 27:34–48.

Scholander, R. F. 1955. Evolution of climatic adaptations in homeotherms. *Evolution.* 9:15–26.

Schopf, T. J. M., and J. L. Gooch. 1971. Gene frequencies in a marine ectoproct: A cline in natural populations related to sea temperature. *Evolution.* 25:286–289.

Selander, R. K., and R. F. Johnston. 1967. Evolution in the house sparrow. I. Intrapopulation variation in North America. *Condor.* 69:217–258.

Soulé, M. 1971. The variation problem: The gene flow-variation hypothesis. *Taxon.* 20:37–50.

Soulé, M., and S. Y. Yang. 1974. Genetic variation in side-blotched lizards on islands on the Gulf of California. *Evolution.* 27:593–600.

Thompson, J. N., and R. C. Woodruff. 1980. Increased mutation in crosses between geographically separated strains of *Drosophila melanogaster. Proc. Natl. Acad. Sci.* 77:1059–1062.

Vasek, F. 1968. The relationship of two ecologically marginal, sympatric *Clarkia* populations. *Amer. Nat.* 102:25–40.

White, M. J. D. 1978. *Modes of Speciation.* San Francisco: W. H. Freeman.

White, M. J. D., R. C. Lewontin, and L. E. Andrew. 1963. Cytogenetics of the grasshopper *Moraba scurra.* VII. Geographic variation of adaptive properties of inversions. *Evolution.* 17:147–162.

Williams, G. C. 1975. *Sex and Evolution.* Princeton, N.J.: Princeton University Press.

Wilson, E. O., and W. L. Brown. 1953. The subspecies concept and its taxonomic application. *Systematic Zool.* 2:97–111.

Wright, S. 1943. Isolation by distance. *Genetics.* 16:97–159.

CHAPTER 14: THE ORIGIN OF NEW SPECIES

General References

See also the general references listed for Chapter 6, especially the now-classical works of Mayr (1942, 1963).

Avise, J. C. 1976. Genetic differentiation during speciation. *In* F. J. Ayala, ed. *Molecular Evolution.* Sunderland, Mass.: Sinauer Associates, pp. 106–122.

Ayala, F. J. 1975. Genetic differentiation during the speciation process. *Evol. Biol.* 8:1–78.

White, M. J. D. 1978a. *Modes of Speciation.* San Francisco: W. H. Freeman.

Other References

Ahearn, J. N., H. L. Carson, Th. Dobzhansky, and K. Y. Kaneshiro. 1974. Ethological isolation among three species of the *planitibia* subgroup of Hawaiian *Drosophila. Proc. Natl. Acad. Sci.* 71:708–712.

Anderson, W. W., and L. Ehrman. 1969. Mating choice in crosses between geographic populations of *Drosophila pseudoobscura. Amer. Midl. Nat.* 81:47–53.

Avise, J. C., and F. J. Ayala. 1975. Genetic differentiation in speciose versus depauperate phylads: Evidence from the California minnows. *Evolution.* 30:46–58.

Ayala, F. J., and M. L. Tracey. 1973. Enzyme variability in the *Drosophila willistoni* group. VIII. Genetic differentiation and reproductive isolation between two subspecies. *J. Hered.* 64:120–124.

Ayala, F. J., M. L. Tracey, D. Hedgecock, and R. C. Richmond. 1974. Genetic differentiation during the speciation process in *Drosophila. Evolution.* 28:576–592.

Ball, R. W., and D. L. Jameson. 1966. Premating isolating mechanisms in sympatric and allopatric *Hyla regilla* and *Hyla californiae. Evolution.* 20:533–551.

Brooks, J. L. 1950. Speciation in ancient lakes. *Quart. Rev. Biol.* 25:131–176.

Brown, R. G. B. 1965. Courtship behavior in the *Drosophila obscura* group. II. Comparative studies. *Behaviour.* 25:281–323.

Brown, W. L., and E. O. Wilson. 1956. Character displacement. *Systematic Zool.* 5:49–64.

Bush, G. 1975. Modes of animal speciation. *Ann. Rev. Ecol. Systemat.* 6:339–364.

Carmody, G. et al. 1962. Mating preferences and sexual isolation within and between the incipient species of *Drosophila paulistorum. Amer. Midl. Nat.* 68:67–82.

Carson, H. L. 1973. Reorganization of the gene pool during speciation. *In* N. E. Morton ed. *Genetic Structure of Populations.* Honolulu: University of Hawaii Press.

Carson, H. L., and P. J. Bryan. 1979. Change in a secondary sexual character as evidence of incipient speciation in *Drosophila sylvestris. Proc. Natl. Acad. Sci.* 76:1929–1932.

Dobzhansky, Th. 1951. Experiments on sexual isolation in *Drosophila.* X. Reproductive isolation between *Drosophila pseudoobscura* and *D. persimilis* under natural and under laboratory conditions. *Proc. Natl. Acad. Sci.* 37:792–796.

Dobzhansky, Th., L. Ehrman, and P. A. Kastritsis. 1968. Ethological isolation between sympatric and allopatric species of the *obscura* group of *Drosophila. Animal Behaviour.* 16:79–87.

Dobzhansky, Th., and O. Pavlovsky. 1967. Experiments on the incipient species of the *Drosophila paulistorum* complex. *Genetics.* 55:141–156.

_____. 1971. An experimentally created incipient species of *Drosophila*. *Nature*. 230:289–292.

Dobzhansky, Th., O. Pavlovsky, and J. R. Powell. 1976. Partially successful attempt to enhance reproductive isolation between semispecies of *Drosophila paulistorum*. *Evolution*. 30:201–212.

Dobzhansky, Th., and B. Spassky. 1959. *Drosophila paulistorum*, a cluster of species *in statu nascendi*. *Proc. Natl. Acad. Sci.* 45:419–428.

Dodson, M. M., and A. Hallam. 1977. Allopatric speciation and the fold catastrophe. *Amer. Nat.* 111:415–433.

Ehrlich, P. R. 1961. Has the biological species concept outlived its usefulness? *Systematic Zool.* 10:167–176.

Ehrman, L. 1962. Hybrid sterility as an isolating mechanism in the genus *Drosophila*. *Quart. Rev. Biol.* 37:279–302.

_____. Direct observation of sexual isolation between allopatric and between sympatric strains of the different *Drosophila paulistorum* races. *Evolution*. 19:459–464.

Endler, J. A. 1977. *Geographic Variation, Speciation, and Clines*. Princeton, N.J.: Princeton University Press.

Fisher, R. A. 1930. *The Genetical Theory of Natural Selection*. Oxford: Clarendon Press; reprinted 1958, New York: Dover.

Fryer, G., and T. D. Isles. 1972. *The Cichlid Fishes of the Great Lakes of Africa: Their Biology and Evolution*. Edinburgh: Oliver & Boyd.

Gordon, M., and D. E. Rosen. 1951. Genetics of species differences in the morphology of the male genitalia of xiphophoran fishes. *Bull. Amer. Mus. Nat. Hist.* 95:409–464.

Gottlieb, L. D. 1976. Biochemical consequences of speciation in plants. *In* F. J. Ayala, ed. *Molecular Evolution*. Sunderland, Mass.: Sinauer Associates, pp. 123–140.

Grant, B. R., and P. R. Grant. 1979. Darwin's finches: population variation and sympatric speciation. *Proc. Natl. Acad. Sci.* 76:2359–2363.

Grant, V. 1971. *Plant Speciation*. New York: Columbia University Press.

Hubby, J. L., and L. H. Throckmorton. 1968. Protein differences in *Drosophila*. IV. A study of sibling species. *Amer. Nat.* 102:193–205.

Kaneshiro, K. Y. 1980. Sexual isolation, speciation, and the direction of evolution. *Evolution*. 34:437–444.

Kessler, S. 1966. Selection for and against ethological isolation between *Drosophila pseudoobscura* and *Drosophila persimilis*. *Evolution*. 20:634–645.

Key, K. H. L. 1968. The concept of stasipatric speciation. *Systematic Zool.* 17:14–22.

Kozhov, M. 1963. *Lake Baikal and Its Life*. The Hague: W. Junk.

Levin, D. A., and H. W. Kerster. 1967. Natural selection for reproductive isolation in *Phlox*. *Evolution*. 21:679–687.

Lewis, H. 1966. Speciation in flowering plants. *Science*. 152:167–172.

_____. 1973. The origin of diploid neospecies in *Clarkia*. *Amer. Nat.* 107:161–170.

Lewis, H. and M. R. Roberts. 1956. The origin of *Clarkia lingulata*. *Evolution*. 10:126–138.

Littlejohn, M. J., and R. S. Oldham. 1968. *Rana pipiens* complex: mating call structure and taxonomy. *Science*. 162:1003–1005.

Marshall, A. J. 1954. *Bowerbirds, Their Displays and Breeding Cycles*. Oxford: Clarendon Press.

Maynard-Smith, J. 1966. Sympatric speciation. *Amer. Nat.* 100:637–650.

Mayr, E. 1957. Species concepts and definitions. *In* E. Mayr, ed. *The Species Problem*. Washington, D.C.: Amer. Assoc. Advancement Sci., pp. 1–22.

Moore, J. A. 1946. Incipient intraspecific isolating mechanisms in *Rana pipiens*. *Genetics*. 31:304–326.

Moore, R. E. 1965. Ethological isolation between *Peromyscus maniculatus* and *Peromyscus polionotus*. *Amer. Midl. Nat.* 74:341–349.

Nevo, E., Y. J. Kim, C. R. Shaw, and C. S. Thaeler. 1973. Genetic variation, selection, and speciation in *Thomomys talpoides* pocket gophers. *Evolution*. 28:1–23.

Parsons, P. A. 1975. The comparative evolutionary biology of the sibling species, *Drosophila melanogaster* and *D. simulans*. *Quart. Rev. Biol.* 50:151–169.

Plunka, S. M., and J. H. Potter. 1977. The effect of mixed culturing on reproductive isolation between *Drosophila pallidosa* and *Drosophila ananassae*. *Amer. Nat.* 111:598–603.

Ripley, S. D. 1959. Character displacement in Indian nuthatches (*Sitta*). *Postilla, Yale Peabody Museum*, 42:1–11.

Rizki, M. T. M. 1951. Morphological differences between two sibling species. *Drosophila pseudoobscura* and *Drosophila persimilis*. *Proc. Natl. Acad. Sci.* 37:156–159.

Robertson, A. 1970. A note on disruptive selection experiments in *Drosophila*. *Amer. Nat.* 104:561–569.

Schoener, T. W. 1965. The evolution of bill size differences among sympatric congeneric species of birds. *Evolution*. 19:189–213.

Sokal, R. R., and T. J. Crovello. 1970. The biological species concept: a critical evaluation. *Amer. Nat.* 104:127–153.

Thoday, J. M. 1972. Disruptive selection. *Proc. Roy. Soc. London*, ser. B. 182:109–143.

Thoday, J. M., and J. B. Gibson. 1962. Isolation by disruptive selection. *Nature*. 193:1164–1166.

_____. 1970. The probability of isolation by disruptive selection. *Amer. Nat.* 104:219–230.

Thompson, J. N., and R. C. Woodruff. 1980. Increased mutation in crosses between geographically separated strains of *Drosophila melanogaster. Proc. Natl. Acad. Sci.* 77: 1059–1062. [Hybrid dysgenesis as an isolating mechanism.]

Vaurie, C. 1951. Adaptive differences between two sympatric species of nuthatches (*Sitta*). *Proc. 10th Internat. Ornithol. Congr.* (Uppsala, 1950), pp. 163–166.

White, M. J. D. 1968. Models of speciation. *Science.* 159:1065–1070.

———. 1978b. Chain processes in chromosomal speciation. *Systematic Zool.* 27:285–298.

Wiley, E. O. 1978. The evolutionary species concept reconsidered. *Systematic Zool.* 27:17–26.

Wills, C. 1977. A mechanism for rapid allopatric speciation. *Amer. Nat.* 111:603–605.

Wood, D., and J. M. Ringo. 1980. Male mating discrimination in *Drosophila melanogaster, D. simulans,* and their hybrids. *Evolution.* 34:320–329.

CHAPTER 15: HYBRIDISM AND POLYPLOIDY

General References

Grant, V. 1971. *Plant Speciation.* New York: Columbia University Press.

Stebbins, G. L. 1950. *Variation and Evolution in Plants.* New York: Columbia University Press.

Other References

Adams, H., and E. Anderson. 1958. A conspectus of hybridization in the Orchidaceae. *Evolution.* 12:512–518.

Anderson, E. 1949. *Introgressive Hybridization.* New York: Wiley.

——— 1953. Introgressive hybridization. *Biol. Rev. Cambridge Philos. Soc.* 28:280–307.

Avise, J. C., and M. H. Smith. 1974b. Biochemical genetics of sunfish. II. Genic similarity between hybridizing species. *Amer. Nat.* 108:458–472.

Crenshaw, J. W. 1965. Serum protein variation in an interspecies hybrid swarm of turtles of the genus *Pseudemys. Evolution.* 19:1–15.

Darlington, C. D. 1953. Polyploidy in animals. *Nature.* 171:191–194.

Dobzhansky, Th., and O. Pavlovsky. 1958. Interracial hybridization and breakdown of coadapted gene complexes in *Drosophila paulistorum* and *Drosophila willistoni. Proc. Natl. Acad. Sci.* 44:622–629.

Fritts, T. H. 1969. The systematics of the parthenogenetic lizards of the *Cnemidophorus cozumela* complex. *Copeia.* 1969: 519–535.

Haga, T. 1956. Genome and polyploidy in the genus *Trillium.* VI. Hybridization and speciation by chromosome doubling in nature. *Heredity.* 10:85–98.

Hubbs, C. L. 1955. Hybridization between fish species in nature. *Systematic Zool.* 4:1–20.

Kihara, H. 1965. The origin of wheat in the light of comparative genetics. *Japanese J. Genetics.* 40:45–54.

Lerner, I. M., and B. Schaal. 1970. Reticulate evolution in *Phlox* as seen through protein electrophoresis. *Amer. J. Bot.* 57:977–987.

Levin, D. A. 1975. Interspecific hybridization, heterozygosity, and gene exchange in *Phlox. Evolution.* 29:37–51.

Lewis, H., and M. Lewis. 1955. The genus *Clarkia. Univ. Calif. Publ. Bot.* 20:241–392.

Lowe, C. H., J. W. Wright, C. J. Cole, and R. L. Bezy. 1970a. Natural hybridization between the teiid lizards *Cnemidophorus sonorae* (parthenogenetic) and *Cnemodophorus tigris* (bisexual). *Systematic Zool.* 19:114–127.

———. 1970b. Chromosomes and evolution of the species groups of *Cnemidophorus* (Reptilia: Teiidae). *Systematic Zool.* 19:128–141.

Sibley, C. G. 1954. Hybridization in the red-eyed towhees of Mexico. *Evolution.* 8:252–290.

Stebbins, G. L. 1958. The inviability, weakness, and sterility of interspecific hybrids. *Advanc. Genetics.* 9:147–215.

———. 1959. The role of hybridization in evolution. *Proc. Amer. Philos. Soc.* 103:231–251.

Wilson, A. C., and L. R. Maxson. 1974. Two types of molecular evolution. Evidence from studies of interspecific hybridization. *Proc. Nat. Acad. Sci.* 71:2843–2847.

CHAPTER 16: THE GEOLOGICAL RECORD

General References

Kummel, B. 1970. *History of the Earth, an Introduction to Historical Geology,* 2nd ed. San Francisco: W. H. Freeman.

Raup, D. M., and S. M. Stanley. 1978. *Principles of Paleontology,* 2nd ed. San Francisco: W. H. Freeman.

Zeuner, F. E. 1970. *Dating the Past, an Introduction to Geochronology,* 4th ed.; reprint of 1946 edition. New York: Hafner.

Other References

Bada, J. L., and R. Protsch. 1973. Racemization reaction of aspartic acid and its use in dating fossil bones. *Proc. Natl. Acad. Sci.* 70:1331–1334.

Eichler, D. L. 1976. *Geologic Time.* Englewood Cliffs, N.J.: Prentice-Hall.

Gould, S. J. 1980. The promise of paleobiology as a nomothetic, evolutionary discipline. *Paleobiology.* 6:96–118.

Harland, W. B., A. G. Smith, and B. Wilkock. 1964. *The Phanerozoic Time Scale.* London: Geol. Soc. London.

Harper, C. T., ed. 1973. *Geochronology: Radiometric Dating of Rocks and Minerals.* Stroudsburg, Pa.: Dowden, Hutchinson, and Ross.

Piaget, J. 1969. *The Child's Conception of Time.* Trans. by A. J. Pomerans. London: Routledge & Kegan Paul.

Rettie, J. C. 1950. The most amazing movie ever made. *In* R. Lord and K. Lord, eds. *Forever the Land.* New York: Harper.

Schopf, J. W. 1974. Paleobiology of the Precambrian: The age of blue-green algae. *Evol. Biol.* 7:1–43.

Schopf, T. J. M., ed. 1972. *Models in Paleobiology.* San Francisco: Freeman Cooper.

CHAPTER 17: EVOLUTIONARY LINEAGES AND TRENDS

General References

Simpson, G. G. 1953. *The Major Features of Evolution.* New York: Columbia University Press.

———. 1967. *The Meaning of Evolution,* revised ed. New Haven: Yale University Press.

Other References

Bessey, C. E. 1915. The phylogenetic taxonomy of flowering plants. *Ann. Missouri Botan. Garden.* 2:109–164.

Carson, H. L. 1974. Three flies and three islands: parallel evolution in *Drosophila. Proc. Natl. Acad. Sci.* 71: 3517–3521.

Cope, E. D. 1886. *The Origin of the Fittest.* New York: Appleton.

———. 1896. *Primary Factors of Organic Evolution.* Chicago: Open Court.

Cracraft, J., and N. Eldredge, eds. 1979. *Phylogenetic Analysis and Paleontology.* New York: Columbia University Press.

Darwin, C. R. 1859. *On the Origin of Species by Means of Natural Selection, or the Preservation of the Favoured Races in the Struggle For Life.* London: John Murray.

Edinger, T. 1948. Evolution of the horse brain. *Mem. Geol. Soc. Amer.* no. 25.

Eldredge, N., and S. J. Gould. 1972. Punctuated equilibria: an alternative to phyletic gradualism. *In* T. J. M. Schopf, ed. *Models in Paleobiology.* San Francisco: Freeman Cooper, pp. 82–115.

———. 1974. Morphological transformation, the fossil record, and the mechanisms of evolution: a debate. Part II. The reply. *Evolutionary Biology.* 7:303–308.

Forsten, A. 1975. The fossil horses from the Texas Gulf Coastal Plain. *J. Paleontol.* 49:395–399.

Gould, S. J. 1973. The misnamed, mistreated, and misunderstood Irish elk. *Nat. Hist.* 82(3):10–19.

———. 1974. The evolutionary significance of "bizarre" structures: antler size and skull size in the "Irish elk," *Megaloceros giganteus. Evolution.* 28:191–200.

Hennig, W. 1966. *Phylogentic Systematics.* Trans. by R. Zangerl. Urbana, Ill.: University of Illinois Press.

Hull, D. L. 1970. Contemporary systematic philosophies. *Ann. Rev. Ecol. Systemat.* 1:19–54.

Maglio, V. J. 1972. Evolution of mastication in the Elephantidae. *Evolution.* 26:638–658.

———. 1973. Origin and evolution of the Elephantidae. *Trans. Amer. Philos. Soc.,* n.s., 63(3):3–149.

Mayr, E. 1969. *Principles of Systematic Zoology.* New York: McGraw-Hill.

———. 1972. Lamarck revisited. *J. Hist. Biol.* 5:55–94.

Minkoff, E. C. 1974. New consideration of an old hypothesis concerning lower primate evolution. *Amer. Nat.* 103: 519–532.

Mivart, St. George. 1871. *On the Genesis of Species,* 2nd ed. London: Macmillan.

Moore, R. C., C. G. Lalicker, and A. G. Fischer. 1952. *Invertebrate Fossils.* New York: McGraw-Hill.

Packard, A. 1972. Cephalopods and fish: the limits of convergence. *Biol. Rev. Cambridge Philos. Soc.* 47:241–307.

Quinn, J. H. 1955. Miocene Equidae of the Texas Gulf Coastal Plain. *Univ. Texas Publ.,* no. 5516:1–102.

Rensch, B. 1947. *Neuere Probleme der Abstammungslehre.* Stuttgart: Enke.

Salvini-Plawen, L. v., and E. Mayr. 1977. On the evolution of photoreceptors and eyes. *Evol. Biol.* 10:207–263.

Simpson, G. G. 1951. *Horses.* New York: Oxford University Press.

———. 1961. *Principles of Animal Taxonomy.* New York: Columbia University Press.

Stirton, R. A. 1940. Phylogeny of North American Equidae. *Univ. Calif. Publ., Bull. Dept. Geol. Sci.* 25:165–198.

CHAPTER 18: DIRECTIONAL VERSUS OPPORTUNISTIC EVOLUTION

General References

Simpson, G. G. 1967. *The Meaning of Evolution,* revised ed. New Haven: Yale University Press.

Other References

Bock, W. J. 1963. The role of preadaptation and adaptation in the evolution of higher levels of organization. *Proc. 16th Internat. Congr. Zool.* 3:297–300.

_____. 1970. Microevolutionary sequences as a fundamental concept in macroevolutionary models. *Evolution.* 24: 704–722.

Boyden, A. 1947. Homology and analogy: a critical review of the meanings and implications of these concepts in biology. *Amer. Midl. Nat.* 37: 648–669.

Butler, P. M. 1946. The evolution of carnassial dentitions in the Mammalia. *Proc. Zool. Soc. London.* 116:198–220.

Carson, H. L., D. E. Hardy, H. T. Spieth, and W. S. Stone. 1970. The evolutionary biology of the Hawaiian Drosophilidae. *Evol. Biol.* 4, suppl.

Cuénot, C. 1965. *Teilhard de Chardin, an Autobiographical Study.* Baltimore: Helicon.

Cuénot, L., and A. Tétry. 1951. *L'évolution Biologique, les Faits, les Incertitudes.* Paris: Masson.

De Beer, G. R. 1954. *Archaeopteryx* and evolution. *Advanc. Sci.* 11:160–170.

Denison, R. H. 1941. The soft anatomy of *Bothriolepis.* *J. Paleontol.* 15:553–561.

Dobzhansky, Th. 1975. Darwinian or ''oriented'' evolution? *Evolution.* 29:375–378.

Du Noüy, P. L. 1947. *Human Destiny.* New York: Longmanns.

Grassé, P. P. 1977. *Evolution of Living Organisms. Evidence for a New Theory of Transformation.* New York: Academic Press.

Jepsen, G. L. 1949. Selection, ''orthogenesis,'' and the fossil record. *Proc. Amer. Philos. Soc.* 93:479–500.

Kohn, A. J., E. R. Myers, and V. R. Meenakshi. 1979. Interior remodeling of the shell by a gastropod mollusc. *Proc. Natl. Acad. Sci.* 76:3406–3410.

Lack, D. 1947. *Darwin's Finches.* Cambridge: Cambridge University Press.

Morse, D. H. 1975. Ecological aspects of adaptive radiation in birds. *Biol. Rev. Cambridge Philos. Soc.* 50:167–214.

Ostrom, J. H. 1974. *Archaeopteryx* and the origin of flight. *Quart. Rev. Biol.* 49:27–49.

Salvini-Plawen, L. v., and E. Mayr. 1977. On the evolution of photoreceptors and eyes. *Evol. Biol.* 10:207–263.

Schaeffer, B., and M. Hecht, eds. 1965. Symposium: The origin of higher levels of organization. *Systematic Zool.* 14:245–342.

Slobodkin, L. B., and A. Rapoport. 1974. An optimal strategy of evolution. *Quart. Rev. Biol.* 49:181–200.

Teilhard de Chardin, P. 1955. *The Phenomenon of Man.* Trans. 1959 by B. Wall. New York: Harper & Row.

Wiedmann, J. 1973. Evolution or revolution of ammonoids at Mesozoic system boundaries. *Biol. Rev. Cambridge Philos. Soc.* 48:159–194.

Yoon, J. S., K. Resch, M. R. Wheeler, and R. H. Richardson. 1975. Evolution in Hawaiian Drosophilidae: chromosomal phylogeny of the *Drosophila crassifemur* complex. *Evolution.* 29:249–256.

CHAPTER 19: RATES OF EVOLUTION

General References

Simpson, G. G., 1953. *Major Features of Evolution.* New York: Columbia University Press.

Stanley, S. M. 1979. *Macroevolution: Pattern and Process.* San Francisco: W. H. Freeman.

Other References

Alexander, R. R. 1979. Differentiation of generic extinction rates among Upper Ordovician–Devonian articulate brachiopods. *Paleobiology.* 5:133–143.

Avise, J. C., and F. J. Ayala. 1975a. Genetic differentiation in speciose versus depauperate phylads: Evidence from the California minnows. *Evolution.* 30:46–58.

_____. 1975b. Genetic change and rates of cladogenesis. *Genetics.* 81:757–773.

Avise, J. C., J. C. Patton, and C. F. Aquadro. 1980. Evolutionary genetics of birds: comparative molecular evolution in New World warblers and rodents. *J. Hered.* 71:302–310.

Dayhoff, M. O. 1972. *Atlas of Protein Sequence and Structure,* vol. 5. Silver Spring, Md.: National Biomedical Research Foundation.

Fitch, W. M. 1976. Molecular evolutionary clocks. *In* F. J. Ayala, ed. *Molecular Evolution.* Sunderland, Mass.: Sinauer Associates, pp. 160–178.

Fortey, R. A. 1980. Generic longevity in lower Ordovician trilobites: Relation to environment. *Paleobiology.* 6:24–31.

Gingerich, P. D. 1976. Paleontology and phylogeny: Patterns of evolution at the species level in early Tertiary mammals. *Amer. J. Sci.* 276:1–28.

_____. 1977. Patterns of evolution in the mammalian fossil record. *In* A. Hallam, ed. *Patterns of Evolution as Illus-*

trated by the Fossil Record. Amsterdam: Elsevier, pp. 469–500.

Goodman, M. 1976. Protein sequences in phylogeny. *In* F. J. Ayala, ed. *Molecular Evolution.* Sunderland, Mass.: Sinauer Associates, pp. 141–159.

Goodman, M., G. W. Moore, J. Barnabas, and G. Matsuda. 1974. The phylogeny of human globin genes investigated by the maximum parsimony method. *J. Molec. Evol.* 3:1–48.

Gorman, G. C., A. C. Wilson, and M. Nakanishi. 1971. A biochemical approach toward the study of reptilian phylogeny: Evolution of serum albumin and lactic dehydrogenase. *Systematic Zool.* 20:167–186.

Gould, S. J., D. M. Raup, J. J. Sepkoski, T. J. M. Schopf, and D. S. Simberloff. 1977. The shape of evolution: A comparison of real and random clades. *Paleobiology.* 3:23–40.

Jukes, T. H., and R. Holmquist. 1972. Evolutionary clock: nonconstancy of rate in different species. *Science.* 177: 530–532.

King, M. C., and A. C. Wilson. 1975. Evolution at two levels. Molecular similarities and biological differences between humans and chimpanzees. *Science.* 188:107–116.

Langley, C. H., and W. M. Fitch. 1974. An examination of the constancy of the rate of molecular evolution. *J. Molec. Evol.* 3:161–177.

Maeda, N., and W. M. Fitch. 1981. Amino acid sequence of a myoglobin from lace monitor lizard, *Varanus varius,* and its evolutionary implications. *J. Biol. Chem.* 256: 4301–4309.

Maxson, L. R., V. M. Sarich, and A. C. Wilson. Continental drift and the use of albumin as an evolutionary clock. *Nature.* 225:30–32.

Maxson, L. R., and A. C. Wilson. 1975. Albumin evolution and organismal evolution in tree frogs (Hylidae). *Systematic Zool.* 24:1–15.

Ohta, T., and M. Kimura. 1971. On the constancy of the evolutionary rate of cistrons. *J. Molec. Evol.* 1:18–25.

Patterson, B. 1949. Rates of evolution in taeniodonts. *In* G. L. Jepsen, E. Mayr, and G. G. Simpson, eds. *Genetics, Paleontology, and Evolution.* Princeton, N.J.: Princeton University Press, pp. 243–278.

Raup, D. M., S. J. Gould, T. J. M. Schopf, and D. S. Simberloff. 1973. Stochastic models of phylogeny and the evolution of diversity. *J. Geol.* 81:525–542.

Raup, D. M., and L. G. Marshall. 1980. Variation between groups in evolutionary rates: a statistical test of significance. *Paleobiology.* 6:9–23.

Sarich, V. M., and A. C. Wilson. 1967a. Rates of albumin evolution in primates. *Proc. Natl. Acad. Sci.* 58: 142–148.

———. 1967b. Immunological time scale for hominoid evolution. *Science.* 158:1200–1203.

Schopf, T. J. M., D. M. Raup, S. J. Gould, and D. S. Simberloff. 1975. Genomic versus morphologic rates of evolution: influence of morphologic complexity. *Paleobiology.* 1:63–70.

Selander, R. K., S. Y. Yang, R. C. Lewontin, and W. E. Johnson. 1970. Genetic variation in the horseshoe crab (*Limulus polyphemus*), a phylogenetic relic. *Evolution.* 24: 402–414.

Wallace, D. G., L. R. Maxson, and A. C. Wilson. 1971. Albumin evolution in frogs: a test of the evolutionary clock hypothesis. *Proc. Natl. Acad. Sci.* 68:3127–3129.

Wiedmann, J. 1973. Evolution or revolution of ammonoids at Mesozoic system boundaries. *Biol. Rev. Cambridge Philos. Soc.* 48:159–194.

Wilson, A. C., G. L. Bush, S. M. Case, and M. C. King. 1975. Social structuring of mammalian populations and rate of chromosomal evolution. *Proc. Natl. Acad. Sci.* 72: 5061–5065.

Wilson, A. C., S. S. Carlson, and T. J. White. 1977. Biochemical evolution. *Ann. Rev. Biochem.* 46:573–639.

Wilson, A. C., and V. M. Sarich. 1969. A molecular time scale for human evolution. *Proc. Natl. Acad. Sci.* 63: 1088–1093.

Wilson, A. C., V. M. Sarich, and L. R. Maxson. 1974. The importance of gene rearrangement in evolution. Evidence from studies on rates of chromosomal, protein, and anatomical evolution. *Proc. Natl. Acad. Sci.* 71:3028–3030.

CHAPTER 20: EXTINCTION AND ITS CAUSES.

General References

Simpson, G. G. 1953. *Major Features of Evolution.* New York: Columbia University Press.

Other References

Alvarez, L. W., W. Alvarez, F. A. Savo, and H. V. Michel. 1980. Extraterrestrial cause for the Cretaceous–Tertiary extinction. *Science.* 208:1095–1108.

Axelrod, D. I., and H. P. Bailey. 1968. Cretaceous dinosaur extinction. *Evolution.* 22:595–611.

Ayala, F. J., D. Hedgecock, G. S. Zumwalt, and J. W. Valentine. 1973. Genetic variation in *Tridacna maxima,* an ecological analog of some unsuccessful evolutionary lineages. *Evolution.* 27:177–191.

Bakker, R. T. 1971. Dinosaur physiology and the origin of mammals. *Evolution.* 25:636–658.

Berger, W. H., and H. R. Thierstein. 1979. On Phanerozoic mass extinctions. *Naturwissenschaften.* 66:46–47.

Flessa, K. W., K. V. Powers, and J. L. Cisne. 1975. Specialization and evolutionary longevity in the Arthropoda. *Paleobiology*. 1:71–81.

Gartner, S., and J. P. McGuirk. 1979. Terminal Cretaceous extinction, scenario for a catastrophe. *Science*. 206:1272–1276.

Gould, S. J. 1977a. Eternal metaphors of palaeontology. *In* A. Hallam, ed. *Patterns of Evolution as Illustrated by the Fossil Record*. Amsterdam: Elsevier, pp. 1–26.

_____. 1977b. *Ontogeny and Phylogeny*. Cambridge, Mass.: Harvard University Press.

Lipps, J. H. 1970. Plankton evolution. *Evolution*. 24:1–21.

Martin, P. S., and H. E. Wright, eds. 1967. *Pleistocene Extinctions: The Search for a Cause*. New Haven: Yale University Press.

McKenna, M. C. 1960. Fossil mammalia from the early Wasatchian Four Mile Fauna, Eocene of northwest Colorado. *Univ. Calif. Publ. Geol. Sci.* 37:1–130.

Minkoff, E. C. 1979. Mammalian cohorts. *J. Nat. Hist.* 13:589–597.

Newell, N. D. 1952. Periodicity in invertebrate evolution. *J. Paleont.* 26:371–385.

Raup, D. M. 1976. Taxonomic survivorship curves and Van Valen's law. *Paleobiology*. 1:82–96.

_____. 1978a. Cohort analysis of generic survivorship. *Paleobiology*. 4:1–15.

_____. 1978b. Approaches to the extinction problem. *J. Paleontol.* 52:517–523.

Russell, D. A. 1979. The enigma of the extinction of the dinosaurs. *Ann. Rev. Earth and Planet. Sci.* 7:163–182.

Schopf, T. J. M. 1974. Permo-Triassic extinctions: relation to sea-floor spreading. *J. Geol.* 82:129–143.

Stanley, S. M. 1979. *Macroevolution, Pattern and Process*. San Francisco: W. H. Freeman.

Stirton, R. A. 1940. Phylogeny of North American Equidae. *Univ. Calif. Publ., Bull. Dept. Geol. Sci.* 25:165–198.

Van Valen, L. 1973. A new evolutionary law. *Evol. Theory.* 1:1–30.

Whitten, R. C., J. Cuzzi, W. J. Borucki, and J. H. Wolfe. 1976. Effect of nearby supernova explosions on atmospheric ozone. *Nature*. 263:398–400.

CHAPTER 21: THE SEARCH FOR OVERALL PATTERNS AND REGULARITIES

General References

Stanley, S. M. 1979. *Macroevolution, Pattern and Process*. San Francisco: W. H. Freeman.

Other References

Anstey, R. L. 1978. Taxonomic survivorship and morphological complexity in Paleozoic bryozoan genera. *Paleobiology*. 4:407–418.

Avise, J. C. 1977. Is evolution gradual or rectangular? Evidence from living fishes. *Proc. Natl. Acad. Sci.* 74:5083–5087.

Bessey, C. E. 1915. The phylogenetic taxonomy of flowering plants. *Ann. Missouri Botan. Garden.* 2:109–164.

Dollo, L. 1893. Les lois de l'évolution. *Bull Soc. Belge Géol. Paléontol. Hydrol.* 7:164–166. Translation in Gould (1970).

_____. 1922. Les céphalopodes déroulés et l'irréversibilité de l'évolution. *Bijdr. tot de Dierkunde.* 10:215–227.

Eldredge, N. 1971. The allopatric model and phylogeny in Paleozoic invertebrates. *Evolution.* 25:156–167.

Eldredge, N., and S. J. Gould. 1972. Punctuated equilibria: an alternative to phyletic gradualism. *In* T. J. M. Schopf, ed. *Models in Paleobiology*. San Francisco: Freeman Cooper, pp. 82–115.

_____. 1974. Morphological transformation, the fossil record, and the mechanisms of evolution: a debate. Part II. The reply. *Evol. Biol.* 7:303–308.

Gingerich, P. D. 1976. Paleontology and phylogeny: patterns of evolution at the species level in early Tertiary mammals. *Amer. J. Sci.* 276:1–28.

_____. 1977. Patterns of evolution in the mammalian fossil record. In A. Hallam, ed. *Patterns of Evolution as Illustrated by the Fossil Record*. Amsterdam: Elsevier, pp. 469–500.

Gingerich, P. D., and M. Schoeninger. 1977. The fossil record and primate phylogeny. *J. Human Evol.* 6:483–505.

Gould, S. J. 1970. Dollo on Dollo's law: irreversibility and the status of evolutionary laws. *J. Hist. Biol.* 3:189–212.

Gould, S. J., and N. Eldredge. 1977. Punctuated equilibria: the tempo and mode of evolution reconsidered. *Paleobiology.* 3:115–151.

Hecht, M. K. 1974. Morphological transformation, the fossil record, and the mechanism of evolution: a debate. Part I. The statement and the critique. *Evol. Biol.* 7:295–303.

Holmes, E. B. 1977. Is specialization a dead end? *Amer. Nat.* 11:1021–1026.

Stanley, S. M. 1973. An explanation for Cope's rule. *Evolution.* 27:1–26.

Stebbins, G. L., and F. J. Ayala. 1981. Is a new evolutionary synthesis necessary? *Science.* 213:967–971.

West, R. M. 1979. Apparent prolonged evolutionary stasis in the middle Eocene hoofed mammal *Hyopsodus*. *Paleobiology.* 5:252–260.

Williamson, P. G. 1981. Palaeontological determination of speciation in Cenozoic molluscs from Turkana Basin. *Nature.* 293:437–443.

CHAPTER 22: MORPHOLOGY, PHYLOGENY, AND CLASSIFICATION

General References

Hennig, W. 1966. *Phylogenetic Systematics*. Trans. by R. Zangerl. Urbana, Ill.: University of Illinois Press.

Mayr, E. 1965. Numerical phenetics and taxonomic theory. *Systematic Zool.* 14:73–95.

_____. 1969. *Principles of Systematic Zoology*. New York: McGraw-Hill.

Sneath, P. H. A., and R. R. Sokal. 1973. *Numerical Taxonomy, the Principles and Practice of Numerical Classification*. San Francisco: W. H. Freeman.

Sokal, R. R., and P. H. A. Sneath. 1963. *Principles of Numerical Taxonomy*. San Francisco: W. H. Freeman.

Other References

Alberch, P., S. J. Gould, G. F. Oster, and D. B. Wake. 1979. Size and shape in ontogeny and phylogeny. *Paleobiology.* 5:296–317.

Ashlock, P. D. 1971. Monophyly and associated terms. *Systematic Zool.* 20:63–69.

_____. 1974. The uses of cladistics. *Ann. Rev. Ecol. Systemat.* 5:81–99.

_____. 1979. An evolutionary systematist's view of classification. *Systematic Zool.* 28:441–450.

Brothers, D. J. 1978. How pure must a cladistic study be? A response to Nelson on Michener. *Systematic Zool.* 27:118–122.

Camin, J. H., and R. R. Sokal. 1965. A method for deducing branching sequences in phylogeny. *Evolution.* 19: 311–326.

Colless, D. H. 1967. An examination of certain concepts in phenetic taxonomy. *Systematic Zool.* 16:6–27.

Cracraft, J., and N. Eldredge, eds. 1979. Phylogenetic analysis and paleontology. New York: Columbia University Press.

Crowson, R. A. 1970. *Classification and Biology*. New York: Atherton.

Farris, J. S. 1970. Methods for computing Wagner trees. *Systematic Zool.* 19:83–92.

Farris, J. S., A. J. Kluge, and M. J. Eckardt. 1970. A numerical approach to phylogenetic systematics. *Systematic Zool.* 19:172–189.

Felsenstein, J. 1978. Cases in which parsimony or compatibility methods will be positively misleading. *Systematic Zool.* 27:401–410.

_____. 1979. Alternate methods of phylogenetic inference and their interrelationship. *Systematic Zool.* 28:49–62.

Fitch, W. M. 1977. On the problem of discovering the most parsimonious tree. *Amer. Nat.* 111:223–257.

Ghiselin, M. T., and L. Jaffe. 1973. Phylogenetic classification in Darwin's monograph on the subclass Cirripedia. *Systematic Zool.* 22:132–140.

Gingerich, P. D. 1979. Stratophenetic approach to phylogeny reconstruction in vertebrate paleontology. *In* J. Cracraft and N. Eldredge, eds. *Phylogenetic Analysis and Paleontology.* New York: Columbia University Press, pp. 41–77.

Gould, S. J. 1977. *Ontogeny and Phylogeny.* Cambridge, Mass.: Harvard University Press.

Harper, C. W. 1979. A Bayesian probability view of phylogenetic systematics. *Systematic Zool.* 28:547–553.

Harper, C. W., and N. I. Platnick. 1978. Phylogenetic and cladistic hypotheses: a debate. *Systematic Zool.* 27:354–362.

Hull, D. L. 1970. Contemporary systematic philosophies. *Ann. Rev. Ecol. Systemat.* 1:19–54.

_____. 1979. The limits of cladism. *Systematic Zool.* 28:416–440.

Jardine, J. and R. Sibson. 1971. *Mathematical taxonomy.* London: Wiley.

Klotz, L. C., N. Komar, R. L. Blanken, and R. M. Mitchell. 1979. Calculation of evolutionary trees from sequence data. *Proc. Natl. Acad. Sci.* 76:4516–4520.

LeQuesne, W. J. 1972. Further studies based on the uniquely derived character concept. *Systematic Zool.* 21:281–288.

_____. 1979. Compatibility analysis and the uniquely derived character concept. *Systematic Zool.* 28:92–92.

Lóvtrup, S. 1978. On von Baerian and Haeckelian recapitulation. *Systematic Zool.* 27:348–352.

Lundberg, J. G. 1972. Wagner networks and ancestors. *Systematic Zool.* 21:398–413.

Mickevich, M. F. 1978. Taxonomic congruence. *Systematic Zool.* 27:143–158.

Moody, P. A., V. A. Cochran, and H. Drugg. 1949. Serological evidence on lagomorph relationships. *Evolution.* 3:25–33.

Nastansky, N., S. M. Selkow, and N. F. Stewart. 1974. An improved solution to the generalized Camin–Sokal model for numerical cladistics. *J. Theoret. Biol.* 48:413–424.

Nelson, G. 1971. "Cladism" as a philosophy of classification. *Systematic Zool.* 20:373–376.

Platnick, N. I. 1979. Philosophy and the transformation of cladistics. *Systematic Zool.* 28:537–546.

Ruse, M. 1979. Falsifiability, consilience, and systematics. *Systematic Zool.* 29:530–536.

Simpson, G. G. 1963. The meaning of taxonomic statements. *In* S. L. Washburn, ed., *Classification and Human Evolution.* Chicago: Aldine, pp. 1–31.

Wiley, E. O. 1979. Cladograms and phylogenetic trees. *Systematic Zool.* 28:88–92.

Wilson, E. O., 1965. A consistency test for phylogenies based on contemporaneous species. *Systematic Zool.* 14: 214–220.

CHAPTER 23: BIOGEOGRAPHY

General References

Darlington, P. J. 1957. *Zoogeography, the Geographic Distribution of Animals.* New York: John Wiley.

MacArthur, R. H., and E. O. Wilson. 1967. *The Theory of Island Biogeography.* Princeton, N.J.: Princeton University Press.

Other References

Colbert, E. H. 1971. Tetrapods and continents. *Quart. Rev. Biol.* 46:250–269.

Cracraft, J. 1974. Continental drift and vertebrate distribution. *Ann. Rev. Ecol. Systemat.* 5:215–261.

Croizat, L. 1958. *Panbiogeography.* 2 vols. Caracas, Venezuela: Published by the author.

Croizat, L., G. Nelson, and D. E. Rosen. 1974. Centers of origin and related concepts. *Systematic Zool.* 23:265–287.

Dammerman, K. W. 1948. The fauna of Krakatau, 1883–1933. *Verhandel. Koninkl. Ned. Akad. Wetenschap., Afdel. Natuurk.* 44:1–594.

Darlington, P. J. 1965. *Biogeography of the Southern End of the World.* Cambridge, Mass.: Harvard University Press.

Darwin, C. R. 1859. *On the Origin of Species by Means of Natural Selection, or the Preservation of the Favoured Races in the Struggle for Life.* London: John Murray.

Keast, A, 1971. Continental drift and the evolution of the biota on southern continents. *Quart. Rev. Biol.* 46:335–378.

MacArthur, R. H., and E. O. Wilson. 1963. An equilibrium theory of insular zoogeography. *Evolution.* 17:373–387.

Matthew, W. D. 1915. Climate and evolution. *Ann. N. Y. Acad. Sci.* 24:171–318.

Mayr, E., and J. Diamond. 1976. Birds on islands in the sky: origin of the montane avifauna of Northern Melanesia. *Proc. Natl. Acad. Sci.* 73:1765–1769.

Nelson, G., and D. E. Rosen. 1981. *Vicariance Biogeography, a Critique.* New York: Columbia University Press.

Rosen, D. E. 1978. Vicariant patterns and historical explanation in biogeography. *Systematic Zool.* 27:159–188.

Stebbins, G. L. 1974. *Flowering Plants: Evolution above the Species Level.* Cambridge, Mass.: Harvard University Press.

Valentine, J. W. 1973. *Evolutionary Paleoecology of the Marine Biosphere.* Englewood Cliffs, N.J.: Prentice-Hall.

Woods, C. A. 1972. Comparative myology of jaw, hyoid, and pectoral appendicular regions of New and Old World hystricomorph rodents. *Bull. Amer. Mus. Nat. Hist.* 147(3):119–198.

CHAPTER 24: BEHAVIOR AND SOCIOBIOLOGY

General References

Alcock, J. 1979. *Animal Behavior, an Evolutionary Approach,* 2nd ed. Sunderland, Mass.: Sinauer Associates.

Wilson, E. O. 1975. *Sociobiology, the New Synthesis.* Cambridge, Mass.: Belknap Press, Harvard University Press.

Other References

Alexander, R. D., and P. W. Sherman. 1977. Local mate competition and parental investment patterns in the social insects. *Science.* 196:494–450.

Axelrod, R., and W. D. Hamilton. 1981. The evolution of cooperation. *Science.* 211:1390–1396.

Darlington, P. J. 1978. Altruism: its characteristics and evolution. *Proc. Natl. Acad. Sci.* 75:385–389.

Dawkins, R. 1976. *The Selfish Gene.* New York: Oxford University Press.

Eberhard, M. J. W. 1975. The evolution of social behavior by kin selection. *Quart. Rev. Biol.* 50:1–34.

Eibl-Eibesfeldt, I. 1975. *Ethology, the Biology of Behavior,* 2nd ed. Trans. by E. Klinghammer. New York: Holt, Rinehart & Winston.

Gans, C. 1979. Momentarily excessive construction as the basis for protoadaptation. *Evolution.* 33:227–233.

Maynard-Smith, J. 1976. Group selection. *Quart. Rev. Biol.* 51:277–283.

Ruse, M. 1979. *Sociobiology: Sense or Nonsense?* Dordrecht, Holland: D. Reidel.

Sahlins, M. 1976. *The Use and Abuse of Biology: An Anthropological Critique of Sociobiology.* Ann Arbor, Mich.: University of Michigan Press.

Schulman, S. R. 1978. Commentary: Kin selection, reciprocal altruism, and the principle of maximization—a reply to Sahlins. *Quart. Rev. Biol.* 53:283–286.

Stanley, S. M. 1979. *Macroevolution, Pattern and Process.* San Francisco: W. H. Freeman.

Trivers, R. L. 1971. The evolution of reciprocal altruism. *Quart. Rev. Biol.* 46:35–37.

Trivers, R. L., and H. Hare. 1976. Haplodiploidy and the evolution of the social insects. *Science.* 191:249–263.

Wade, M. J. 1978. A critical review of the models of group selection. *Quart. Rev. Biol.* 53:101–114.

Wilson, D. S. 1975. A theory of group selection. *Proc. Natl. Acad. Sci.* 72:143–146.

Wynne-Edwards, V. C. 1962. *Animal Dispersion in Relation to Social Behavior.* Edinburgh: Oliver and Boyd.

Yokoyama, S., and J. Felsenstein. 1978. A model of kin selection for an altruistic trait considered as a quantitative character. *Proc. Natl. Acad. Sci.* 75:420–422.

CHAPTER 25: THE ORIGIN AND EARLY EVOLUTION OF LIFE

General References

Fox, S. W., and K. Dose. 1972. *Molecular Evolution and the Origin of Life.* San Francisco: W. H. Freeman.

Oparin, A. I. 1938. *The Origin of Life.* Trans. by S. Morgulis. New York: Macmillan; reprinted 1953, New York: Dover.

Orgel, L. E., 1973. *The Origins of Life: Molecules and Natural Selection.* New York: John Wiley.

Other References

Barghoorn, E. S. 1971. The oldest fossils. *Sci. Amer.* 224:30–53.

Biemann, K., *et al.* 1976. Search for organic and volatile inorganic compounds in two surface samples from the Chryse Planitia region of Mars. *Science.* 194:72–76.

Buchanan, R. E., and N. E. Gibbons. 1974. *Bergey's Manual of Determinative Bacteriology,* 8th ed. Baltimore: Williams and Wilkins.

Calvin, M. 1969. *Chemical Evolution.* London: Oxford University Press.

Crick, F. H. C. 1968. The origin of the genetic code. *J. Molec. Biol.* 38:367–379.

Crick, F. H. C., and L. E. Orgel. 1973. Directed panspermia. *Icarus.* 49:341–348.

Dubos, R. J. 1950. *Louis Pasteur, Free Lance of Science.* Boston: Little, Brown.

Engels, F. 1940. *Dialectics of Nature.* Trans. by C. Dutt. New York: International Publishers; London: Lawrence & Wishart. [With preface and notes by J. B. S. Haldane.]

Folsome, C. E. 1976. Synthetic organic microstructures and the origins of cellular life. *Naturwissenschaften.* 63:303–306.

Fox, G. E., L. J. Magrum, W. E. Balch, R. S. Wolfe, and C. R. Woese. 1977. Classification of methanogenic bacteria by 16S ribosomal RNA characterization. *Proc. Natl. Acad. Sci.* 74:4537–4541, 5778.

Fox, S. W., ed. 1965. *The Origins of Prebiological Systems and of Their Molecular Matrices.* New York: Academic Press.

_____. 1973. Origin of the cell: experiments and premises. *Naturwissenschaften.* 60:359–369.

Gabriel, M. L., and S. Fogel, eds. 1955. *Great Experiments in Biology.* Englewood Cliffs, N.J.: Prentice-Hall.

Graham, L. R. 1972. *Science and Philosophy in the Soviet Union.* New York: Alfred A. Knopf.

Heinz, B., W. Reid, and H. D. Pflug. 1980. The lunisphere, a new model of pre-biotic evolution. *Naturwissenschaften.* 67:178–181.

Hora, H., and S. Osawa. 1979. Evolutionary change in 5S RNA secondary structure and a phylogenetic tree of 54 5S RNA species. *Proc. Natl. Acad. Sci.* 76:381–385, 4157.

Kuhn, H. 1976. Model consideration for the origin of life.

Naturwissenschaften. 63:68–80.

Klein, H. P., *et al.* 1976. The Viking biological investigation: Preliminary results. *Science.* 194:99–105.

Lacey, J. C., and K. M. Pruitt. 1969. Origin of the genetic code. *Nature.* 223:799–804.

Lehninger, A. L. 1975. *Biochemistry,* 2nd ed. New York: Worth Publishers.

Lohrmann, R., and L. E. Orgel. 1973. Prebiotic activation processes. *Nature.* 223:799–804.

Margulis, L. H. 1970. *The Origin of Eukaryotic Cells.* New Haven: Yale University Press.

_____. 1981. *Symbiosis in Cell Evolution.* San Francisco: W. H. Freeman.

Michaelis, W., and P. Albrecht. 1979. Molecular fossils of archaebacteria in Kerogen. *Naturwissenschaften.* 66:420–422.

Miller, S. L. 1953. A production of amino acids under possible primitive Earth conditions. *Science.* 117:528–529.

_____. 1955. Production of some organic compounds under possible primitive Earth conditions. *J. Amer. Chem. Soc.* 77:2351–2361.

Miller, S. L., and L. E. Orgel. 1973. *The Origins of Life.* Englewood Cliffs, N.J.: Prentice-Hall.

Oparin, A. I. 1965. The origin of life and the origin of enzymes. *Advanc. Enzymol.* 27:347–380.

_____. 1968. *Genesis and Evolutionary Development of Life.* New York: Academic Press.

Ponnamperuma, C., ed. 1972. *Exobiology.* Amsterdam: North-Holland.

Ragan, M. A., and D. J. Chapman. 1978. *A Biochemical Phylogeny of the Protists.* New York: Academic Press.

Rohlfing, D. L., and A. I. Oparin, eds. 1972. *Molecular evolution, Prebiological and Biological.* New York: Plenum.

Shklovskii, I. S., and C. Sagan. 1966. *Intelligent Life in the Universe.* San Francisco: Holden-Day.

Schopf, J. W. 1974. Paleobiology of the Precambrian: the age of blue-green algae. *Evol. Biol.* 7:1–43.

Vallery-Radot, R. 1923. *The Life of Pasteur.* Trans. by R. L. Devonshire. Garden City, N.Y.: Doubleday Page.

Woese, C. R., and G. E. Fox. 1977. Phylogenetic structure of the prokaryotic domain: the primary kingdoms. *Proc. Natl. Acad. Sci.* 74:5088–5090.

CHAPTER 26: THE EVOLUTION OF EUCARYOTES

General References

Barnes, R. D. 1980. *The Invertebrates,* 4th ed. Philadelphia: W. B. Saunders.

Bold, H. C. 1973. *Morphology of Plants,* 3rd ed. New York: Harper & Row.

Buchsbaum, R. 1975. *Life Without Backbones,* 2nd ed. Chicago: University of Chicago Press.

Colbert, E. H. 1980. *Evolution of the Vertebrates,* 3rd ed. New York: John Wiley.

Cronquist, A. 1968. *The Evolution and Classification of Flowering Plants.* Boston: Houghton Mifflin.

——. 1971. *Introductory Botany,* 2nd ed. New York: Harper & Row.

Hyman, L. H. 1940–1959. *The Invertebrates,* 6 vols, others forthcoming. New York: McGraw-Hill.

Moore, R. C., ed. 1953–1978. *Treatise on Invertebrate Paleontology.* 19 parts in 30 vols. New York: Geological Society of America.

Moore, R. C., C. G. Lalicker, and A. G. Fischer. 1952. *Invertebrate Fossils.* New York: McGraw-Hill.

Romer, A. S. 1966. *Vertebrate Paleontology,* 3rd ed. Chicago: University of Chicago Press.

Romer, A. S., and T. S. Parsons. 1977. *The Vertebrate Body,* 5th ed. Philadelphia: W. B. Saunders.

Scagel, R. F., R. J. Bandoni, G. E. Rouse, W. B. Schofield, J. R. Stein, and T. M. C. Taylor. 1965. *An Evolutionary Survey of the Plant Kingdom.* Belmont, Calif.: Wadsworth.

Schopf, J. W. 1969. Precambrian micro-organisms and evolutionary events prior to the origin of vascular plants. *Biol. Rev. Cambridge Philos. Soc.* 45:319–352.

Smith, G., ed. 1951. *Manual of Phycology, an Introduction to the Algae and Their Biology.* New York: Ronald Press; Waltham, Mass.: Chronica Botanica.

——. 1955. *Cryptogamic Botany,* 2nd ed. 2 vols. New York: McGraw-Hill.

Stebbins, G. L. 1974. *Flowering Plants: Evolution above the Species Level.* Cambridge, Mass.: Belknap Press, Harvard University Press.

Villee, C. A., W. F. Walker, Jr., and R. D. Barnes. 1973. *General Zoology,* 4th ed. Philadelphia: W. B. Saunders.

Young, J. Z. 1962. *The Life of Vertebrates,* 2nd ed. New York: Oxford University Press.

Other References

Axelrod, D. I. 1961. How old are the angiosperms? *Amer. J. Sci.* 259:447–459.

Copeland, H. F. 1956. *Classification of the Lower Organisms.* Palo Alto, Calif.: Pacific Books.

Cronquist, A., A. Takhtajan, and W. Zimmermann. 1966. On the higher taxa of Embryobionta. *Taxon.* 15:129–134.

Eichler, A. W. 1833. *Syllabus der Verlesungen über specielle und medicinisch-pharmaceutische Botanik,* 3rd. ed. Berlin: Borntraeger.

Esau, K. 1953. *Plant anatomy.* New York: John Wiley.

Klein, R. M., and A. Cronquist. 1967. A consideration of the evolutionary and taxonomic significance of some biochemical, micromorphological, and physiological characters in the thallophytes. *Quart. Rev. Biol.* 42:105–296.

Manton, S. M. 1977. *The Arthropoda: Habits, Functional Morphology, and Evolution.* Oxford and New York: Clarendon Press, Oxford University Press.

Margulis, L. H. 1970. *Origin of Eukaryotic Cells.* New Haven: Yale University Press.

——. 1974. Five-kingdom classification and the origin and evolution of cells. *Evol. Biol.* 7:45–78.

——. 1981. *Symbiosis in Cell Evolution.* San Francisco: W. H. Freeman.

Takhtajan, A. 1969. *Flowering Plants: Origin and Dispersal.* Trans. by C. Jeffrey. Edinburgh: Oliver & Boyd.

Tippo, O. 1942. A modern classification of the plant kingdom. *Chronica Bot.* 7:203–206.

Whittaker, R. H. 1969. New concepts of kingdoms of organisms. *Science.* 193:150–159.

Zimmerman, W. 1952. Main results of the telome theory. *Paleobotanist.* 1:456–470.

CHAPTER 27: PRIMATES AND HUMAN EVOLUTION

General References

Buettner-Janusch, J. 1966. *Origins of Man, Physical Anthropology.* New York: John Wiley.

Clark, W. E. LeG. 1959. *The Antecedents of Man.* Edinburgh: Edinburgh University Press; New York: Harper Torchbooks.

Dobzhansky, Th. 1962. *Mankind Evolving.* New Haven: Yale University Press.

Simons, E. 1972. *Primate Evolution.* New York: Macmillan.

Other References

Andrews, P. 1982. Hominoid evolution. *Nature.* 295:185–186.

Andrews, P., and E. N. Evans. 1979. The environment of *Ramapithecus* in Africa. *Paleobiology.* 5:22–30.

Boyd, W. 1950. *Genetics and the Races of Man.* Boston: Little, Brown.

Carlsson, A. 1922. Über die Tupaiidae und ihre Beziehungen zu den Insectivora und den Prosimiae. *Acta Zool.* 3: 227–270.

Cartmill, M. 1974. Rethinking primate origins. *Science.* 184: 436–443.

Clutton-Brock, T. H., and P. H. Harvey. 1979. Primates, brains, and ecology. *J. Zool.* 190:309–323.

Coon, C. S. 1962. *The Origin of Races.* New York: Alfred A. Knopf.

Coon, C. S., S. M. Garn, and J. B. Birdsell. 1950. *Races.* Springfield, Ill.: C. C. Thomas.

Dart, R. 1949. The predatory implementary technique of *Australopithecus. Amer. J. Phys. Anthro.,* n.s., 7:1–38.

Ferris, S. D., A. C. Wilson, and W. M. Brown. 1981. Evolutionary tree for apes and humans based on cleavage maps of mitochondrial DNA. *Proc. Natl. Acad. Sci.* 78: 2432–2436.

Goldsby, R. 1971. *Race and Races.* New York: Macmillan.

Goodman, M. 1963. Man's place in the phylogeny of the primates as reflected in serum proteins. *In* S. L. Washburn, ed. *Classification and Human Evolution.* Chicago: Aldine, pp. 204–234.

Goodman, M., and R. E. Tashian, eds. 1976. *Molecular Anthropology: Genes and Proteins in the Evolutionary Ascent of the Primates.* New York: Plenum Press.

Gould, S. J. 1977. *Ever Since Darwin.* New York: W. W. Norton.

———. 1980. *The Panda's Thumb.* New York: W. W. Norton.

———. 1981. *The Mismeasure of Man.* New York: W. W. Norton.

Howell, F. C. 1978. Hominidae. *In* V. J. Maglio and H. B. S. Cooke, eds. *Evolution of African Mammals.* Cambridge, Mass.: Harvard University Press, pp. 154–248.

Johanson, D. C., and T. D. White. 1979. A systematic assessment of early African hominids. *Science.* 203:321–332.

Johanson, D. C., and M. A. Edey. 1980. *Lucy, the Beginnings of Mankind.* New York: Simon & Schuster.

Jolly, A. 1972. *Evolution of Primate Behavior.* New York: Macmillan.

King, M. C., and A. C. Wilson. 1975. Evolution at two levels: molecular similarities and biological differences between humans and chimpanzees. *Science.* 188:107–116.

Leakey, M. D., and R. L. Hay. 1979. Pliocene footprints in the Laetolil beds at Laetoli, northern Tanzania. *Nature.* 278:317–323.

Minkoff, E. C. 1972. A fossil baboon from Angola, with a note on *Australopithecus. J. Paleont.* 46:836–844.

———. 1974. The direction of lower primate evolution: an old hypothesis revived. *Amer. Nat.* 108:519–532.

Pilbeam, D. 1972. *The Ascent of Man.* New York: Macmillan.

Pilbeam, D., and S. J. Gould. 1974. Size and scaling in human evolution. *Science.* 186:892–901.

Romer, A. S. 1966. *Vertebrate Paleontology,* 3rd ed. Chicago: University of Chicago Press.

Simons, E. L., and R. F. Kay. 1980. "Dawn ape" provides clue to social life. *Geotimes.* 25:18.

Simons, E. L., and D. R. Pilbeam. 1965. Preliminary revision of the Dryopithecinae (Pongidae, Anthropoidea). *Folia Primatol.* 3:81–152.

Simpson, G. G. 1955. The Phenacolemuridae, a new family of early Tertiary primates. *Bull. Amer. Mus. Nat. Hist.* 105:411–442.

Szalay, F. S. 1972. Cranial morphology of the early Tertiary *Phenacolemur* and its bearing on primate phylogeny. *Amer. J. Phys. Anthro.* 36:59–76.

———. 1973. New Paleocene primates and a diagnosis of the new suborder Paromomyiformes. *Folia Primatol.* 19: 73–87.

Szalay, F. S., and E. Delson. 1979. *Evolutionary History of the Primates.* New York: Academic Press.

Szalay, F. S., and C. C. Katz. 1973. Phylogeny of lemurs, galagos, and lorises. *Folia Primatol.* 19:88–103.

Tobias, P. V., and G. H. R. von Koenigswald. 1964. A comparison between the Olduvai hominines and those of Java, and some implications for hominid phylogeny. *Nature.* 204:515–518.

Uzzell, T., and D. Pilbeam. 1971. Phyletic divergence dates of hominoid primates: a comparison of fossil and molecular data. *Evolution.* 25:615–635.

Van Valen, L. 1965. Tree shrews, primates, and fossils. *Evolution.* 19:137–151.

CHAPTER 28: CULTURAL EVOLUTION AND THE FUTURE

General References

Cavalli-Sforza, L. L., and W. F. Bodmer. 1971. *The Genetics of Human Populations.* San Francisco: W. H. Freeman.

Fagan, B. M. 1974. *Men of the Earth, an Introduction to World Prehistory.* Boston: Little, Brown.

Harris, M. 1977. *Cannibals and Kings, the Origins of Cultures.* New York: Random House.

Howard, T., and J. Rifkin. 1977. *Who Should Play God?* New York: Dell.

Karp, L. E. 1976. *Genetic Engineering, Threat or Promise?* Chicago: Nelson-Hall.

Mertens, T. R. 1975. *Human Genetics: Readings on the Implications of Genetic Engineering.* New York: Wiley.

Setlow, J. K., and A. Hollaender, eds. 1979. *Genetic Engineering, Principles and Methods,* vol. 1. New York: Plenum Press.

Wade, N. 1977. *The Ultimate Experiment: Man-Made Evolution.* New York: Walker.

Other References

Ardrey, R. 1966. *The Territorial Imperative.* New York: Atheneum.

_____. 1976. *The Hunting Hypothesis*. New York: Atheneum.

Bajema, C. J. 1971. *Natural Selection in Human Populations: The Measurement of Ongoing Genetic Evolution in Contemporary Societies*. New York: Wiley.

Blum, H. F. 1978. Uncertainty in interplay of biological and cultural evolution: man's view of himself. *Quart. Rev. Biol.* 53:29–40.

Bodmer, W. F., and L. L. Cavalli-Sforza. 1976. *Genetics, Evolution, and Man*. San Francisco: W. H. Freeman.

Burley, N. 1979. The evolution of concealed ovulation. *Amer. Nat.* 114:835–858.

Campbell, D. T. 1975. On the conflicts between biological and social evolution and between psychological and moral tradition. *Amer. Psychologist* 30:1103–1126.

Dobzhansky, Th. 1962. *Mankind Evolving*. New Haven: Yale University Press.

Fox, R. 1980. *The Red Lamp of Incest*. New York: Dutton.

Haller, M. H. 1963. *Eugenics: Hereditarian Attitudes in American Thought*. New Brunswick, N.J.: Rutgers University Press.

Hrdy, S. B. 1981. *The Woman that Never Evolved*. Cambridge, Mass.: Harvard University Press.

Klingmüller, W. 1979. Genetic engineering for practical application. *Naturwissenschaften.* 66:182–189.

Malinowski, B. 1922. *Argonauts of the Western Pacific*. New York: E. P. Dutton.

_____. 1929. *The Sexual Life of Savages*. New York: Harcourt, Brace, and World.

Meadows, D. H., D. L. Meadows, J. Randers, and W. W. Behrens III. 1972. *The Limits to Growth*. New York: Signet, New American Library.

Menozzi, P., A. Piazza, and L. Cavalli-Sforza. 1978. Synthetic maps of human gene frequencies in Europeans. *Science.* 200:786–792.

Mörch, E. T. 1941. Chondrodystrophic dwarfs in Denmark. *Opera ex Domo Biol. Hered. Hum. Univ. Hafniensis.* 3: 1–200.

Morris, D. 1967. *The Naked Ape*. New York: Dell.

Muller, H. J. 1961. Human evolution by voluntary choice of germ plasm. *Science.* 134:643–649.

_____. 1965. Means and aims in human genetic development. *In* T. M. Sonneborn, ed. *The Control of Human Heredity and Evolution*. New York: Macmillan, pp. 100–122.

Richardson, W. N., and T. H. Stubbs. 1976. *Evolution, Human Ecology, and Society*. New York: Macmillan.

Sahlins, M. D. 1959. The social life of monkeys, apes, and primitive man. *In* J. N. Spuhler, ed. *The Evolution of Man's Capacity for Culture*. Detroit: Wayne State University Press, pp. 54–73.

_____. 1960. The origin of society. *Sci. Amer.* 203:76–87.

Service, E. R. 1962. *Primitive Social Organization, an Evolutionary Perspective*. New York: Random House.

_____. 1975. *Origins of the State and Civilization; the Process of Cultural Evolution*. New York: W. W. Norton.

Sonneborn, T. M., ed. 1965. *The Control of Human Heredity and Evolution*. New York: Macmillan.

Symons, D. 1979. *The Evolution of Human Sexuality*. New York: Oxford University Press.

Trivers, R. L. 1972. Parental investment and sexual selection. *In* B. Campbell, ed. *Sexual Selection and the Descent of Man, 1871–1971*. Chicago: Aldine, pp. 136–179.

Turner, J. R. G., and B. Glass. 1976. Commentary. Genetics, intelligence, and society. *Quart. Rev. Biol.* 51:85–88.

Vogel, F., and M. Chakravartti. 1971. ABO blood groups and smallpox in a rural population of West Bengal and Bihar (India). *In* C. J. Bajema, ed., *Natural Selection in Human Populations*. New York: Wiley, pp. 147–165.

Uyenoyama, M., M. W. Feldman, and L. L. Cavalli-Sforza. 1979. Evolutionary effects of contagious and familial transmission. *Proc. Natl. Acad. Sci.* 76:420–424.

Wilson, E. O. 1978. *On Human Nature*. Cambridge, Mass.: Harvard University Press.

Index